Advances in Semiconductor Nanostructures

Advances in Semiconductor Nanostructures

Growth, Characterization, Properties and Applications

Edited by

Alexander V. Latyshev
Rzhanov Institute of Semiconductor Physics of the Siberian Branch of the Russian Academy of Sciences, Novosibirsk, Russia
Novosibirsk State University, Novosibirsk, Russia

Anatoliy V. Dvurechenskii
Rzhanov Institute of Semiconductor Physics of the Siberian Branch of the Russian Academy of Sciences, Novosibirsk, Russia
Novosibirsk State University, Novosibirsk, Russia

Alexander L. Aseev
Rzhanov Institute of Semiconductor Physics of the Siberian Branch of the Russian Academy of Sciences, Novosibirsk, Russia

ELSEVIER

AMSTERDAM • BOSTON • HEIDELBERG • LONDON
NEW YORK • OXFORD • PARIS • SAN DIEGO
SAN FRANCISCO • SINGAPORE • SYDNEY • TOKYO

Elsevier
Radarweg 29, PO Box 211, 1000 AE Amsterdam, Netherlands
The Boulevard, Langford Lane, Kidlington, Oxford OX5 1GB, United Kingdom
50 Hampshire Street, 5th Floor, Cambridge, MA 02139, United States

British Library Cataloguing-in-Publication Data
A catalogue record for this book is available from the British Library

Library of Congress Cataloging-in-Publication Data
A catalog record for this book is available from the Library of Congress

ISBN: 978-0-12-810512-2

Publisher: John Fedor
Acquisition Editor: Anita A. Koch
Editorial Project Manager: Amy M. Clark
Production Project Manager: Maria Bernard
Designer: Victoria Pearson

Typeset by MPS Limited, Chennai, India

Contents

Part III Radiation Effects on Semiconductor Structures

List of Contributors

I.O. Akhundov Rzhanov Institute of Semiconductor Physics of the Siberian Branch of Russian Academy of Sciences, Novosibirsk, Russia; Novosibirsk State University, Novosibirsk, Russia

V.Sh. Aliev Rzhanov Institute of Semiconductor Physics of the Siberian Branch of the Russian Academy of Sciences, Novosibirsk, Russia

V.L. Alperovich Rzhanov Institute of Semiconductor Physics of the Siberian Branch of Russian Academy of Sciences, Novosibirsk, Russia; Novosibirsk State University, Novosibirsk, Russia

A.P. Anciferov Rzhanov Institute of Semiconductor Physics of the Siberian Branch of the Russian Academy of Sciences, Novosibirsk, Russia

A.L. Aseev Rzhanov Institute of Semiconductor Physics of the Siberian Branch of the Russian Academy of Sciences, Novosibirsk, Russia

A.K. Bakarov Rzhanov Institute of Semiconductor Physics of the Siberian Branch of the Russian Academy of Sciences, Novosibirsk, Russia

I.I. Beterov Rzhanov Institute of Semiconductor Physics of the Siberian Branch of the Russian Academy of Sciences, Novosibirsk, Russia

M.V. Budantsev Rzhanov Institute of Semiconductor Physics of the Siberian Branch of the Russian Academy of Sciences, Novosibirsk, Russia

A.A. Bykov Rzhanov Institute of Semiconductor Physics of the Siberian Branch of the Russian Academy of Sciences, Novosibirsk, Russia

A.V. Chaplik Rzhanov Institute of Semiconductor Physics of the Siberian Branch of the Russian Academy of Sciences, Novosibirsk, Russia

A. Chin National Chiao Tung University, Hsinchu, Taiwan

I.A. Derebezov Rzhanov Institute of Semiconductor Physics of the Siberian Branch of the Russian Academy of Sciences, Novosibirsk, Russia

D.V. Dmitriev Rzhanov Institute of Semiconductor Physics of the Siberian Branch of the Russian Academy of Sciences, Novosibirsk, Russia

S.A. Dvoretsky Rzhanov Institute of Semiconductor Physics of the Siberian Branch of the Russian Academy of Sciences, Novosibirsk, Russia

A.V. Dvurechenskii Rhzanov Institute of Semiconductor Physics of the Siberian Branch of the Russian Academy of Sciences, Novosibirsk, Russia; Novosibirsk State University, Novosibirsk, Russia

M.V. Entin Rzhanov Institute of Semiconductor Physics of the Siberian Branch of the Russian Academy of Sciences, Novosibirsk, Russia

V.M. Entin Rzhanov Institute of Semiconductor Physics of the Siberian Branch of the Russian Academy of Sciences, Novosibirsk, Russia

L.I. Fedina Rzhanov Institute of Semiconductor Physics of the Siberian Branch of the Russian Academy of Sciences, Novosibirsk, Russia

T.A. Gavrilova Rzhanov Institute of Semiconductor Physics of the Siberian Branch of the Russian Academy of Sciences, Novosibirsk, Russia

K.V. Grachev Rzhanov Institute of Semiconductor Physics of the Siberian Branch of the Russian Academy of Sciences, Novosibirsk, Russia

V.A. Gritsenko Rzhanov Institute of Semiconductor Physics of the Siberian Branch of the Russian Academy of Sciences, Novosibirsk, Russia; Novosibirsk State University, Novosibirsk, Russia

A.K. Gutakovskii Rzhanov Institute of Semiconductor Physics of the Siberian Branch of Russian Academy of Sciences, Novosibirsk, Russia

A.V. Haisler Rzhanov Institute of Semiconductor Physics of the Siberian Branch of the Russian Academy of Sciences, Novosibirsk, Russia

V.A. Haisler Rzhanov Institute of Semiconductor Physics of the Siberian Branch of the Russian Academy of Sciences, Novosibirsk, Russia

D.G. Ikusov Rzhanov Institute of Semiconductor Physics of the Siberian Branch of the Russian Academy of Sciences, Novosibirsk, Russia

D.R. Islamov Rzhanov Institute of Semiconductor Physics of the Siberian Branch of the Russian Academy of Sciences, Novosibirsk, Russia; Novosibirsk State University, Novosibirsk, Russia

E.V. Ivanova Ioffe Physical-Technical Institute, Russian Academy of Sciences, St. Petersburg, Russia

M.M. Kachanova Rzhanov Institute of Semiconductor Physics of the Siberian Branch of the Russian Academy of Sciences, Novosibirsk, Russia

V.V. Kaichev Boreskov Institute of Catalysis, Siberian Branch of the Russian Academy of Sciences, Novosibirsk, Russia

A.K. Kalagin Rzhanov Institute of Semiconductor Physics of the Siberian Branch of the Russian Academy of Sciences, Novosibirsk, Russia

A.N. Karpov Rzhanov Institute of Semiconductor Physics of the Siberian Branch of Russian Academy of Sciences, Novosibirsk, Russia

D.M. Kazantsev Rzhanov Institute of Semiconductor Physics of the Siberian Branch of Russian Academy of Sciences, Novosibirsk, Russia; Novosibirsk State University, Novosibirsk, Russia

S.S. Kosolobov Rzhanov Institute of Semiconductor Physics of the Siberian Branch of the Russian Academy of Sciences, Novosibirsk, Russia

A.S. Kozhukhov Rzhanov Institute of Semiconductor Physics of the Siberian Branch of the Russian Academy of Sciences, Novosibirsk, Russia

D.A. Kozlov Rzhanov Institute of Semiconductor Physics of the Siberian Branch of the Russian Academy of Sciences, Novosibirsk, Russia

V.D. Kuzmin Rzhanov Institute of Semiconductor Physics of the Siberian Branch of the Russian Academy of Sciences, Novosibirsk, Russia

Z.D. Kvon Rzhanov Institute of Semiconductor Physics of the Siberian Branch of the Russian Academy of Sciences, Novosibirsk, Russia; Novosibirsk State University, Novosibirsk, Russia

A.V. Latyshev Rzhanov Institute of Semiconductor Physics of the Siberian Branch of the Russian Academy of Sciences, Novosibirsk, Russia; Novosibirsk State University, Novosibirsk, Russia

A.S. Mardezov Rzhanov Institute of Semiconductor Physics of the Siberian Branch of the Russian Academy of Sciences, Novosibirsk, Russia

A.S. Medvedev Rzhanov Institute of Semiconductor Physics of the Siberian Branch of the Russian Academy of Sciences, Novosibirsk, Russia

N.N. Mikhailov Rzhanov Institute of Semiconductor Physics of the Siberian Branch of the Russian Academy of Sciences, Novosibirsk, Russia

A.G. Milekhin Rzhanov Institute of Semiconductor Physics of the Siberian Branch of the Russian Academy of Sciences, Novosibirsk, Russia; Novosibirsk State University, Novosibirsk, Russia

D.A. Nasimov Rzhanov Institute of Semiconductor Physics of the Siberian Branch of the Russian Academy of Sciences, Novosibirsk, Russia

I.G. Neizvestny Rzhanov Institute of Semiconductor Physics of the Siberian Branch of the Russian Academy of Sciences, Novosibirsk, Russia; Novosibirsk State Technical University, Novosibirsk, Russia

L.A. Nenasheva Rzhanov Institute of Semiconductor Physics of the Siberian Branch of the Russian Academy of Sciences, Novosibirsk, Russia

E.B. Olshanetsky Rzhanov Institute of Semiconductor Physics of the Siberian Branch of the Russian Academy of Sciences, Novosibirsk, Russia

T.V. Perevalov Rzhanov Institute of Semiconductor Physics of the Siberian Branch of the Russian Academy of Sciences, Novosibirsk, Russia; Novosibirsk State University, Novosibirsk, Russia

A.G. Pogosov Rzhanov Institute of Semiconductor Physics of the Siberian Branch of the Russian Academy of Sciences, Novosibirsk, Russia; Novosibirsk State University, Novosibirsk, Russia

D.A. Pokhabov Rzhanov Institute of Semiconductor Physics of the Siberian Branch of the Russian Academy of Sciences, Novosibirsk, Russia; Novosibirsk State University, Novosibirsk, Russia

V.P. Popov Rzhanov Institute of Semiconductor Physics of the Siberian Branch of the Russian Academy of Sciences, Novosibirsk, Russia

V.Ya. Prinz Rzhanov Institute of Semiconductor Physics of the Siberian Branch of the Russian Academy of Science, Novosibirsk, Russia

V.G. Remesnik Rzhanov Institute of Semiconductor Physics of the Siberian Branch of the Russian Academy of Sciences, Novosibirsk, Russia

S.V. Rihlicky Rzhanov Institute of Semiconductor Physics of the Siberian Branch of the Russian Academy of Sciences, Novosibirsk, Russia

E.E. Rodyakina Rzhanov Institute of Semiconductor Physics of the Siberian Branch of the Russian Academy of Sciences, Novosibirsk, Russia; Novosibirsk State University, Novosibirsk, Russia

D.I. Rogilo Rzhanov Institute of Semiconductor Physics of the Siberian Branch of the Russian Academy of Sciences, Novosibirsk, Russia

K.N. Romanyuk Rzhanov Institute of Semiconductor Physics of the Siberian Branch of Russian Academy of Sciences, Novosibirsk, Russia

N.S. Rudaya Rzhanov Institute of Semiconductor Physics of the Siberian Branch of Russian Academy of Sciences, Novosibirsk, Russia

I.I. Ryabtsev Rzhanov Institute of Semiconductor Physics of the Siberian Branch of the Russian Academy of Sciences, Novosibirsk, Russia

I.V. Sabinina Rzhanov Institute of Semiconductor Physics of the Siberian Branch of the Russian Academy of Sciences, Novosibirsk, Russia

V.K. Sandyrev Rzhanov Institute of Semiconductor Physics of the Siberian Branch of the Russian Academy of Sciences, Novosibirsk, Russia

A.A. Saraev Boreskov Institute of Catalysis, Siberian Branch of the Russian Academy of Sciences, Novosibirsk, Russia

O.I. Semenova Rzhanov Institute of Semiconductor Physics of the Siberian Branch of the Russian Academy of Sciences, Novosibirsk, Russia

D.V. Sheglov Rzhanov Institute of Semiconductor Physics of the Siberian Branch of the Russian Academy of Sciences, Novosibirsk, Russia

A.A. Shevyrin Rzhanov Institute of Semiconductor Physics of the Siberian Branch of the Russian Academy of Sciences, Novosibirsk, Russia; Novosibirsk State University, Novosibirsk, Russia

A.A. Shklyaev Rzhanov Institute of Semiconductor Physics of the Siberian Branch of Russian Academy of Sciences, Novosibirsk, Russia; Novosibirsk State University, Novosibirsk, Russia

V.A. Shvets Rzhanov Institute of Semiconductor Physics of the Siberian Branch of the Russian Academy of Sciences, Novosibirsk, Russia

N.L. Shwartz Rzhanov Institute of Semiconductor Physics of the Siberian Branch of the Russian Academy of Sciences, Novosibirsk, Russia; Novosibirsk State Technical University, Novosibirsk, Russia

G.Yu. Sidorov Rzhanov Institute of Semiconductor Physics of the Siberian Branch of the Russian Academy of Sciences, Novosibirsk, Russia

Yu.G. Sidorov Rzhanov Institute of Semiconductor Physics of the Siberian Branch of the Russian Academy of Sciences, Novosibirsk, Russia

S.V. Sitnikov Rzhanov Institute of Semiconductor Physics of the Siberian Branch of the Russian Academy of Sciences, Novosibirsk, Russia

E.V. Spesivcev Rzhanov Institute of Semiconductor Physics of the Siberian Branch of the Russian Academy of Sciences, Novosibirsk, Russia

A.S. Terekhov Rzhanov Institute of Semiconductor Physics of the Siberian Branch of the Russian Academy of Sciences, Novosibirsk, Russia

S.A. Teys Rzhanov Institute of Semiconductior Physics of the Siberian Branch of Russian Academy of Sciences, Novosibirsk, Russia

O.A. Tkachenko Rzhanov Institute of Semiconductor Physics of the Siberian Branch of the Russian Academy of Sciences, Novosibirsk, Russia

V.A. Tkachenko Rzhanov Institute of Semiconductor Physics of the Siberian Branch of the Russian Academy of Sciences, Novosibirsk, Russia

A.I. Toropov Rzhanov Institute of Semiconductor Physics of the Siberian Branch of the Russian Academy of Sciences, Novosibirsk, Russia

D.B. Tretyakov Rzhanov Institute of Semiconductor Physics of the Siberian Branch of the Russian Academy of Sciences, Novosibirsk, Russia

A.V. Tsarev Rzhanov Institute of Semiconductor Physics of the Siberian Branch of the Russian Academy of Sciences, Novosibirsk, Russia; Novosibirsk State University, Novosibirsk, Russia

I.E. Tyschenko Rzhanov Institute of Semiconductor Physics of the Siberian Branch of the Russian Academy of Sciences, Novosibirsk, Russia

I.N. Uzhakov Rzhanov Institute of Semiconductor Physics of the Siberian Branch of the Russian Academy of Sciences, Novosibirsk, Russia

V.S. Varavin Rzhanov Institute of Semiconductor Physics of the Siberian Branch of the Russian Academy Of Sciences, Novosibirsk, Russia

S.A. Vitkalov City College of the City University of New York, New York, NY, United States

A.I. Yakimov Rhzanov Institute of Semiconductor Physics of the Siberian Branch of the Russian Academy of Sciences, Novosibirsk, Russia; Novosibirsk State University, Novosibirsk, Russia

M.V. Yakushev Rzhanov Institute of Semiconductor Physics of the Siberian Branch of the Russian Academy of Sciences, Novosibirsk, Russia

A.S. Yaroshevich Rzhanov Institute of Semiconductor Physics of the Siberian Branch of the Russian Academy of Sciences, Novosibirsk, Russia

D.R.T. Zahn Technical University of Chemnitz, Chemnitz, Germany

M.V. Zamoryanskaya Ioffe Physical-Technical Institute, Russian Academy of Sciences, St. Petersburg, Russia

E.Yu. Zhdanov Rzhanov Institute of Semiconductor Physics of the Siberian Branch of the Russian Academy of Sciences, Novosibirsk, Russia; Novosibirsk State University, Novosibirsk, Russia

Yu.A. Zhivodkov Rzhanov Institute of Semiconductor Physics of the Siberian Branch of the Russian Academy of Sciences, Novosibirsk, Russia

Preface

At present the main challenges facing fundamental research and technological development are studies of a variety of semiconductor-based structures such as nanostructured and nanocomposite materials, composite systems, thin films, and heteroepitaxy on different substrates. Their performance is still driven by defect engineering procedures, novel growth techniques, and advanced diagnostics tools. The research and development intends to highlight the newest breakthroughs in terms of materials and their integration for electronic and photonic purposes, integrated device design and architecture, as well as advanced and innovative applications. Many new approaches are highly interdisciplinary and are settled at the interface between physics, chemistry, biology, material science, and electronics.

This book covers selected reviews written by scientists of Rzhanov Institute of Semiconductor Physics Siberian Branch of Russian Academy of Science in physics and technologies of semiconductor nanostructures, which promises to provide the next generation of nanoscale electronic, nanophotonic, nanosensor, and other functional devices. The book is devoted to new approaches in growth and processing of semiconductor nanostructures by molecular beam epitaxy, ion-beam implantation/synthesis, pulsed laser action, and three-dimensional nanostructure fabrication on all types of III–V, IV, and II–VI semiconductors, nanofabrication by bottom-up and top-down approaches, precise control by UHV-REM, HREM, STM, SEM, and AFM of structural characterization of quantum wells, nanowires, quantum dot heterostructures, and electrical, optical, magnetic, and spin phenomena studies.

There are outlines of modern trends in the field, covering a wide range of topical areas: zero-, one-, and two-dimensional electronic systems; semiconductor nanoheterostructures; atomic structures of surface and interface; molecular beam epitaxy; single photon and single electron devices; computer modeling; laser annealing; three-dimensional nanostructuring; and ion-beam synthesis of silicon-on-isolator.

The book contains four parts: (1) Low-Dimensional Systems: Theory and Experiment; (2) Surface, Interface, Epitaxy; (3) Radiation Effects on Semiconductor Structures; and (4) Electronic Advanced Materials.

Part I covers different aspects of advanced theoretical considerations and the experimental realization of modern solid state nanostructures. Theoretical investigations include surface scattering of electrons in two-dimensional systems, photogalvanic effects, and electron states in curved low-dimensional systems. The collective effects in one- and two-dimensional single- and multicomponent systems are analyzed, including plasma waves and Wigner crystallization. The screening of electrostatic fields by dipole excitons (including their Bose condensate), and modeling of quantum transport and single-electron charging in

GaAs/AlGaAs-nanostructures are also under consideration. Experimental results are surveyed for two-dimensional semi-metals in HgTe-based quantum wells, nonlinear electron conductivity in GaAs quantum wells in Corbino geometry at high filling factors, silicon-based heterostructures with quantum dots, and spectroscopy of vibrational states in low-dimensional and nanoelectromechanical systems.

The second part focuses on the basic concepts of crystal growth: atomic processes on a silicon surface, thermal smoothing of a surface, kinetic transformations of GaAs step-terraced surfaces at annealing in equilibrium conditions, atomic mechanisms of molecular beam epitaxy, atomic structure and defects of semiconductor low-dimensional heterosystems, strained Ge layers on silicon and silicon oxidized substrates, thin film growth of $Cd_xHg_{1-x}Te$, and Monte Carlo simulations of semiconductor nanostructure growth.

Part III presents the results on laser annealing of semiconductors, including data illuminating one of the long-standing problems in silicon related to the prevailing formation of extended defects in {113} planes upon various process exposures, and a survey of ion-beam synthesis and hydrogen transfer to produce silicon-on-insulator (SOI) structures.

The fourth part describes modern concepts of various semiconductor devices for application in electronics and photonics. It contains the review results on the operating principles of subminiature semiconductor emitters and the research results of the performance of those emitters, data on three-dimensional systems and nanostructures, the nature of defects responsible for transport in a hafnia-based resistive random access memory element, and an optical multiplexor based on waveguides in SOI systems.

One of the important points is that practically all the chapters are really reviews of the results obtained in collaboration with research groups from Russia, Germany, France, England, Japan, Holland, United States, Belgium, China, Israel, Brazil, and former Soviet Union countries (Ukraine, Belarus, Kazakhstan, etc.). There are common publications with researchers from these groups, and the chapters include appropriate references. The chapters presented really will cover recent developments in the field from all over the world.

We hope that this book will be useful for researchers, postgraduates, and students involved in studies in material science, surface physics, quantum phenomena, electrical transport, and the optical and magnetic physics of semiconductor nanoheterostructures.

We wish to acknowledge the help and support received from our colleagues and friends. We are indebted to Dr. A. Kalameitsev for taking care of the contact with the publishing company.

A.V. Latyshev, A.V. Dvurechenskii, A.L. Aseev

PART

I

Low-Dimensional Systems: Theory and Experiment

1

The Theory of Two-Dimensional Electronic Systems*

A.V. Chaplik, M.V. Entin

RZHANOV INSTITUTE OF SEMICONDUCTOR PHYSICS OF THE SIBERIAN BRANCH OF THE RUSSIAN ACADEMY OF SCIENCES, NOVOSIBIRSK, RUSSIA

Activities in physics of thin (including quantum) films, which later gave rise to physics of low-dimensional systems, had started at the Rzhanov Institute of Semiconductor Physics of the Siberian Branch of the Russian Academy of Sciences (ISP SB RAS) of the USSR in the 1960s as a natural evolution of studying the semiconductor surface. In theoretical aspects, there already had been publications on the electron properties of thin films in the presence of dimensional quantization (see review [1]) and inversion layers [2]. Specific features of electron scattering in quantum films were still a blank spot in the theory.[1]

In particular, there was no consistent quantum mechanics theory of surface scattering. At that time, a purely classical phenomenological boundary condition proposed by K. Fuchs was used. The first contribution of theoretical researchers of ISP SB RAS to the new field was the development of a microscopic theory of scattering of charge carriers on a rough surface of the quantum well [3].

1.1 Electron States and Conductivity in a Quantum Well With a Nonideal Boundary

The wave function of the electron obeys the condition $\psi(\rho, z = u(\rho)) = 0$ on the boundary $z = u(\rho)$, $\rho = (x, y)$. We proposed a method for considering a rough surface, which is based on a mixed boundary condition of the type $(\psi + u(\rho)\partial_z\psi)|_{z=0} = 0$ obtained by means of expanding the exact boundary condition with respect to the deviation of the boundary position from its ideal location.

Using the Green functions, we reformulated the problem of finding the wave function inside the quantum well and reduced it to solving an integral equation on a fictitious surface "charge density" $\mu(\boldsymbol{\rho}) = \int \frac{d^2p}{4\pi^2} e^{i\mathbf{p}\boldsymbol{\rho}} \mu(\mathbf{p})$:

$$(tg(d\sqrt{k^2 - p^2}))/\sqrt{k^2 - p^2}\mu(\mathbf{p}, \mathbf{p}_1) + \int \frac{d^2p'}{4\pi^2} u(\mathbf{p} - \mathbf{p}')\mu(\mathbf{p}', \mathbf{p}_1) = \delta(\mathbf{p} - \mathbf{p}_1)$$

* Investigations in the field of the electron theory of two-dimensional systems at the Rzhanov Institute of Semiconductor Physics of the Siberian Branch of the Russian Academy of Sciences.

[1] A more illustrative term "quantum film" was later replaced by "quantum well" under coercion of English-speaking researchers.

Advances in Semiconductor Nanostructures. DOI: http://dx.doi.org/10.1016/B978-0-12-810512-2.00001-9

Here, $k^2/2m = E$ is the energy of the electron and m is the effective mass. After that, the surface was assumed to have a random Gauss-correlated shape. The Green function was averaged on this surface for finding the density of states and the paired product of the Green functions, which were used to obtain the conductivity of the quantum well with surface scatterers. For this purpose, we constructed a diagram technique for the mean Green functions of this integral equation.

Surface fluctuations were assumed to be weak. Standard small parameters for kinetic calculations were used, namely, the small inverse time of relaxation as compared to the electron energy. As a result, we found widening of singularities of the density of states, transport time of relaxation, and conductivity. In particular, in the thin film limit, it was found that the mobility of electrons increases as $\propto d^{-6}$ with decreasing film thickness d.

The approach to surface scattering was further applied by many researchers. In particular, such a method was used to study the decay of acoustic waves on the moon [4], which is rather far from the field of our activities.

It should be noted that the approach based on the mixed boundary condition is an alternative to the pseudo-potential description of a rough surface [5]. The latter approach can be reformulated as follows [6]. Let the jump of the potential U_0 on the quantum well boundary be large, but finite. Then the wave function in a system with an ideally smooth external boundary can be expressed as $\psi \propto \exp(-\kappa z)$, where $\kappa^2 = \sqrt{2mU_0/\hbar}$. As the logarithmic derivatives inside and outside should be equal to each other, we have $\partial_z\psi/\psi|_{z=0} = -\kappa$. The initial potential can be presented as $U = U_0\theta(z - u(\rho))$. Expanding it with respect to the perturbation, we obtain:

$$U = U_0\theta(z - u(\rho)) = U_0\theta(z) - U_0 u(\rho)\delta(z).$$

Let us consider some matrix element of the perturbation $U_0 u(\rho)\delta(z)$ with respect to transverse wave functions of the quantum well $\psi_n(z)$:

$$\psi_n(z)U_0 u(\rho)\delta(z)\psi_m(z) \equiv ((\hbar^2/4m)\partial_z\psi_n)u(\rho)\partial_z\psi_n\delta(z).$$

The last formula does not contain U_0, and it yields the final result as $U_0 \to \infty$. Therefore, interaction with a rough surface can be also described by a Hamiltonian of the form:

$$H_s = (\hbar^2/4m)\overleftarrow{\partial}_z u(\rho)\delta(z)\overrightarrow{\partial}_z.$$

The arrow shows the direction of the derivative. This Hamiltonian was used in some investigations to describe scattering on the surface, e.g., for the photogalvanic effect (PhGE) in two-dimensional systems (see Section 1.4).

In the text that follows, the range of problems dealing with the electron theory of two-dimensional systems is significantly expanded. The main results obtained in this field by theoreticians of the Rzhanov ISP SB RAS are summarized in Section 1.2.

1.2 Charged Impurities in a Quantum Well

In Ref. [7], we studied the electron states in quantum layers with charged impurities and found stationary and virtual states of electrons on the impurity in the limit where the effective Debye radius is greater than the well thickness. There were two important facts noted in that paper.

First, the bound state of the electron in a two-dimensional system turned out to be four times deeper than that in a three-dimensional system. Later, the same behavior was predicted for the excitons in the quantum wells [8], which means that the exciton state exists in the GaAs quantum well at room temperature, with corresponding consequences for semiconductor optoelectronics, in particular, lasers and light-emitting diodes. Second, the impurity potential in wells placed into a vacuum (later, such systems were called free-suspended systems) does not contain the material permittivity ε at large distances from the impurity, which increases this potential by a factor of ε as compared to a three-dimensional system. This implies 10-fold enhancement of the Coulomb interaction in free-suspended systems, resulting in conversion them from systems with weak electron–electron interaction to the class of systems with strong interactions. It was only recently that such wells were grown in practice, and ISP SB RAS researchers started to study their electron properties [9,10].

1.3 Two-Dimensional Systems in a Strong Magnetic Field

Baskin et al. [11] studied a two-dimensional electron gas in a strong magnetic field with a random impurity potential. They considered the limits of the classical and quantizing fields. In the classical system, they used a limit of a smooth potential with a scale greater than the cyclotron radius r_c on short-range impurities with a size smaller than the cyclotron radius. The smooth potential model was later on called the "adiabatic transport" model.

It was shown in Ref. [11] that the electron motion is the sum of its rapid rotation over a cyclotron orbit and slow drift of the orbit center over the potential level lines. As all level lines except for one in a two-dimensional isotropic system should be closed, it was concluded that the drift conductivity of the system should vanish, except for the situation where the Fermi level exactly coincides with the energy of this level ("zero energy," which more exactly coincides with the median of the potential distribution function). Later on, this approach was extended by other researchers to the case where the magnetic field quantizes the electron motion. In particular, this method provided a possible explanation for the Hall quantum effect. In the presence of a constant sweeping electric field, the level line corresponding to the zero energy becomes unclosed and infinite in the direction normal to the electric field, while the Hall conductivity becomes equal to its value in the absence of the random potential.

It was demonstrated in the short-range impurity model that electrons either move over a circular orbit, or perform rosette-like motions around single impurities as long as the cyclotron radius is sufficiently small $Nr_c^2 < 0.68$, where $N$is the impurity concentration; in this case, however, electrons never travel over the entire sample. This means that there is no diagonal component of conductivity in a sufficiently strong magnetic field. Later on, suppression of conductivity in such systems was called the memory effect: the electron memorizes the starting point and returns to it many times until other processes become involved.

Another model considered in Ref. [11] was a system with point delta-functional impurities in a quantizing magnetic field. It was demonstrated that some of the states at the Landau level remain degenerate if the impurity concentration is sufficiently low. The corresponding criterion is determined by comparisons of the impurity concentration with the degeneration multiplicity of the Landau level $N < g = 1/2\pi a^2$, where $a = \sqrt{c\hbar/eB}$ is the magnetic length. This criterion determines the threshold at which the phase transition with vanishing of degenerate states occurs. Moreover, this model demonstrated the absence of diagonal elements of the conductivity tensor in the case with the Fermi level being located between the Landau levels.

1.4 Photogalvanic Effect and Quantum Pumps

The expansion of the current with respect to the static electric field **E** may contain even contributions in terms of the field:

$$j_i = \sigma_{ij}E_j + \alpha_{ijk}E_jE_k + \alpha_{ijkl}E_jE_kE_l + \ldots$$

If we assume that the field is harmonically varying in time and average this expression over time, then the linear terms disappear, while the quadratic and next even terms yield nonzero contributions [12–14]. The coefficient at the quadratic term α_{ijk} is a third-rank tensor. Thus, we can conclude that the stationary current under the action of an alternating field with a zero mean is possible in all media whose symmetry allows the existence of a third-rank tensor. Such three-dimensional media include crystals without the inversion center. In this case, the crystal is not necessarily polar and possesses a dipole moment. For example, all A_3B_5 semiconductors possess a third-rank tensor [12–14].

A definition of the PhGE was given in Ref. [12]: the emergence of a stationary photocurrent in a homogeneous medium under homogeneous stationary illumination without dragging stationary field, which is not related to momentum transfer from photons to electrons.

Let us consider this definition in more detail. The steady state is necessary to separate the PhGE from relaxation currents of previously generated polarization, e.g., capacitor discharge current or electret photodepolarization. The last definition distinguishes the PhGE from the photon drag effect (in other words, photon pressure effect), in the corpuscular language, associated with momentum transfer from photons to electrons. The inhomogeneity of the light field results in an inhomogeneous distribution of photoexcited carriers, their diffusion, and, as a consequence of differences in diffusion coefficients of different groups of charge carriers, emergence of a compensating electromotive force (EMF; Dember effect). The inhomogeneity in the Dember effect arises due to light absorption in the subsurface layer. This is the reason for the requirement for illumination homogeneity. Examples of inhomogeneous systems are the p−n transition or the Schottky barrier, which give rise to an EMF associated with separation of charge carriers under the action of a static field or a chemical potential gradient.

For the stationary acceleration of electrons in the PhGE the momentum transfer from a "third body" (impurities, phonons, and other scatterers) to electrons is responsible. In most cases, scatterers decelerate electrons. In the PhGE, however, they serve both as a decelerator and an "accelerator." (By the way, it is the force of friction of the wheels on the road that accelerates a car.)

Important properties of the PhGE are its sensitivity to light polarization, and its independence from the wave vector direction. In the major part of frequencies that are involved in physics of photoelectric phenomena, the photon momentum is small compared to a typical electron momentum. This means that phenomena induced by the wave vector are comparatively weak. The PhGE weakness has another origin: it is related to the smallness of the medium asymmetry, and also to the fact that asymmetry is manifested in the course of photon absorption only if interaction with a third body is taken into account.

The term "photogalvanic" effect is sometimes replaced in the literature by the terms "photovoltaic" or "ratchet." The "photogalvanic" variant (which we prefer) underlines the fact that the effect works as a current generator. The PhGE current exists without external drag field in the conducting media. The "photovoltaic" term presumes the voltage appearance in the medium, which is not an obligatory feature. At the same time, the voltage appearing in insulators as a result of optical rectification by the nonlinear polarizability of the second order bore no relation to PhGE. The "ratchet" effect term traces back to R. Feinman's reasoning about reversibility in statistics.

A quasi-stationary field can be considered as a real field. In such a field, all components change simultaneously. The photogalvanic tensor in a slowly changing field is also real. This PhGE type was called a linear PhGE. However, there is one more variant of the PhGE, which is a circular PhGE [15–19]. A simple everyday analog of this PhGE is a car whose translational motion is ensured by the rotation of its wheels. Another example is a spinning ping-pong ball reflected from a surface in accordance with the spinning direction (see Fig. 1.1).

In these models, rotation is converted to translational motion. In formal language, a pseudo-vector (angular momentum) or a force moment generates a velocity vector. In other words, a pseudo-vector generalized force initiates a flux vector response. In the linear

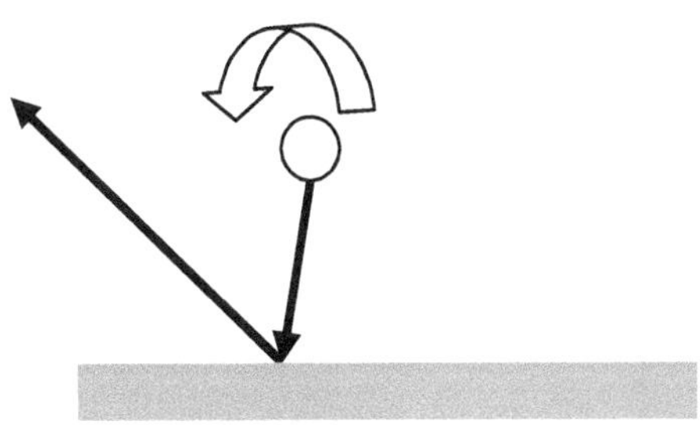

FIGURE 1.1 Simple model of a circular photogalvanic effect.

response theory, the coefficient of the relationship between the current density and the amplitude of the electric field of the electromagnetic wave should be a pseudo-tensor of the second rank:

$$j_i = \beta_{ij}[\mathbf{E} \times \mathbf{E}^*]_j.$$

In quantum language, the circular PhGE is associated with transfer of the spin from photons to electrons, and with the transformation of this nonequilibrium spin into translational motion. It is important for the circular PhGE that the electron should "remember" the field phase within the field period: if the field changes more slowly than the phase relaxes, the circular PhGE vanishes. In those cases where the circular PhGE arises due to creation of a nonequilibrium (in terms of the spin) state of electrons (this is not the only way of imparting the memory about the circular polarization to electrons, see, e.g., Refs. [6,20–23]), the circular PhGE has been recently called the spin-galvanic effect.

In microscopic language, the spin circular PhGE arises due to the correlation between the spin and the electron momentum induced by spin-orbital (SO) interaction. Owing to this interaction, transitions with spin flipping and a certain direction of the momentum are more advantageous than transitions with the opposite orientation of the momentum. It is this phenomenon that generates the electric current. In the two-dimensional case, the spin-galvanic effect is determined by SO interaction related to Rashba's Hamiltonian [24].

After studying the PhGE in bulk materials, it was proposed to use confined samples [25]. A qualitative model of such an effect (in the circular variant) is provided by Fig. 1.1. Owing to the nonequivalence of the layer surfaces combined with scattering on them, a current directed along the surface can arise (in contrast to the Dember EMF directed normal to the surface). For such a current, the crystal may possess an inversion center; at the same time, there should exist a vector parallel to the surface in the crystal + surface + field system. The PhGE in the classical films induced by the nonequivalence of the crystal surfaces was first studied theoretically [25]. This current may be caused either by the asymmetric cut of the crystal, or by a tilted position of the polarization vector with respect to the normal.

Experimentally, the most illustrative effect in a typical semiconductor system (bulk GaAs) turned out to be the surface PhGE, i.e., surface current generated owing to a finite depth of absorption in a semi-infinite sample [26,27].

The case where a usual film turned into a quantum well was considered in [6,22,23]. In such a system, the photocurrent arises as a response to an alternating electric field with a tilted polarization with respect to the normal. An inversion channel on a semiconductor surface was studied experimentally in Ref. [22] and theoretically in Ref. [23]. It was demonstrated that the current has a resonance character if the light frequency is close to the frequency of the transition between the quantum subbands. The resonance is symmetric for linear polarization and anti-symmetric for circular polarization. Some recent publications of Entin and Magarill should be noted [28–30]. They demonstrated, in Ref. [28], how the resonant photocurrent arises in the classical model of electron gas in a parabolic potential well with inhomogeneous friction across the well. If there is no transverse inhomogeneity and fluid friction, the alternating electric field with the normal and tangential components initiates circular motion or linear vibrations of electrons, depending on the field polarization (circular or linear), and on the offset from the resonance. More exactly, circular motion occurs under resonance conditions in the case of linear polarization, and under off-resonance conditions in the case of circular polarization. The presence of inhomogeneous friction results in different braking of circular motion on the top and bottom of the well, resulting in the drift of particles along the well (see Fig. 1.2). This behavior can also be seen in quantum problems for transitions between subbands of a double quantum well [29], and for the electron gas above the liquid helium surface [30] (see Fig. 1.3).

Various aspects of physics of bulk and surface PhGEs were studied both theoretically and experimentally in subsequent years. They were discussed in a number of reviews and books [31–34]. During recent years, this field of research has become one of the commonly recognized areas of solid state physics.

Despite the 30 years of history of the experimental detection of PhGE and explaining its nature, interest in this and related effects has not faded away. Among recent papers we should mention [35,36], where the emergence of a stationary current under the action of an electromagnetic field in an artificial two-dimensional network of asymmetric scatterers, which are semi-circular antidots or oriented segments with two different (specular and diffuse) sides, was theoretically analyzed. The study was performed by using the classical kinetic equation. It was shown that the last variant has an analytical solution. The analytical results were compared with each other and with the results of computer simulations. It was found that the polarization dependence of the photocurrent is not universal. In particular, it turned out that the photocurrent in the model of half-disks, in contrast to the model of segments, vanishes after averaging over polarizations. It was demonstrated that the differences in the analytical results and simulations emerge due to different formulations of the problem and drawbacks of molecular dynamics methods.

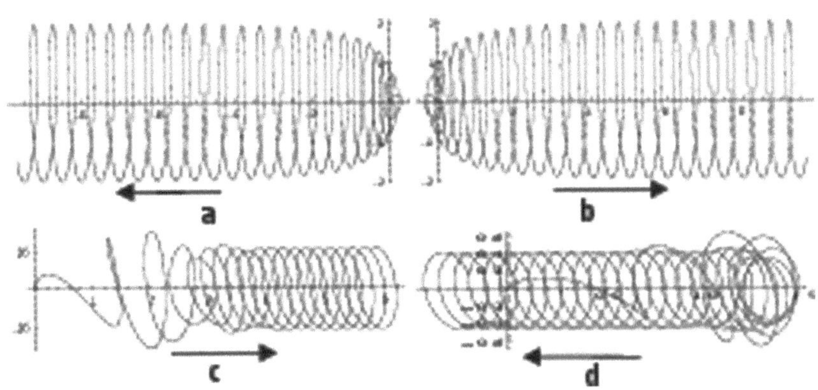

FIGURE 1.2 Mechanism of current emergence in a parabolic classical quantum well in the presence of a high-frequency field with vertical E_z and horizontal E_x components. The figure shows the electron trajectory in an alternating field. Weak friction is inhomogeneous across the well (in the z direction); (a) and (b) linear polarization in the exact resonance of the field and well frequencies with different signs of E_z; (c) and (d) circular polarization with different signs of the detuning from the resonance [28].

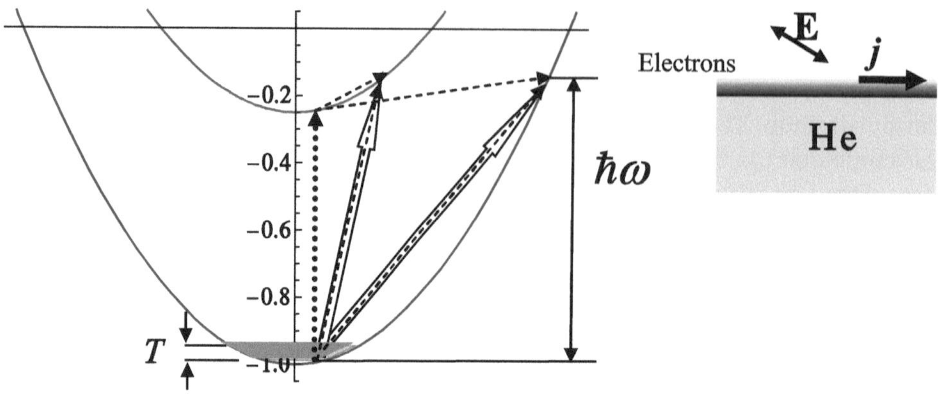

FIGURE 1.3 Transitions and hypothetical experiment aimed at detecting the surface photocurrent in a two-dimensional electron gas above the liquid helium surface. At the beginning, the electrons are located at the bottom of the lower subband. After absorption of a photon with a frequency ω and simultaneous change in the longitudinal momentum due to impurity scattering, the electron goes to the neighborhood of the bottom of the second subband, or (in the excited state) to the first subband. Both the vertical and horizontal components of the electric field **E** are involved into the transition. The change in the longitudinal momentum allows the electron to "remember" the polarization direction, leading to surface current generation [30].

The shape of the low-dimensional system can play a special role. In a one-dimensional system, the longitudinal motion of electrons is affected only by the longitudinal component of the electric field of the wave. Owing to its shape, the external homogeneous electromagnetic field is effectively converted to an inhomogeneous field with the inhomogeneity size comparable with the Fermi wavelength of the electron. The problem is to obtain a running potential wave and local electromagnetic fields in a low-dimensional system. The curvature induces spatial inhomogeneity of the acting electric field by increasing its wave vector with respect to the wave vector in the free space. The helical quantum wire transforms the wave with circular polarization into a wave running along the wire in the direction determined by the wave polarization sign and the wire helicity [35,36]. Transfer of the momentum from the wave to electrons leads to electron acceleration, and to current generation under the action of the light field. This phenomenon has an analog in vacuum electronics: a traveling wave tube. In such a situation, comparatively weak effects of electron drag by photons are drastically enhanced.

The emergence of an EMF in curved quantum wires under the action of a weak electromagnetic field was studied in Ref. [36]. Various types of curvature of planar one-dimensional systems (stiff corner, rounded corner, and Ω-shaped curve) were considered. The considerations were performed both in the classical approximation, on the basis of the kinetic equation, and by using the quantum approach where the electron energy is commensurable with the photon energy.

The electromagnetic wave with circular polarization in systems with periodic helical symmetry generates an induced effective running wave with the wave length determined by the helix period [36,37]. The photon drag effect produced by a strong running potential wave in a helical quantum wire was studied. The classical approximation in a weak external field and the strong wave mode where the wave generates the Bloch states of electrons were considered. The problem was solved by using the kinetic equation for the states renormalized by the wave with due allowance for scattering of electrons on impurities. It was demonstrated that the stationary drag current oscillates as a function of the Fermi level. If the Fermi level falls within the forbidden band gap, electrons are completely entrained by the wave, while the current is quantized by the product of the doubled electron charge and the number of filled zones divided by the period. In the weak external field, the current oscillates between the quantized values in band gaps and the classical values corresponding to the weak photon drag effect where the Fermi level is located in allowed bands. The deviation from the quantized values in the band gaps was found [37].

The influence of interelectron scattering on some kinetic coefficients in two-dimensional systems (conductivity, magnetic transport, and PhGE) was analyzed. In contrast to the commonly accepted viewpoint that the momentum transfer between electrons is not accompanied by changes in these kinetic coefficients, it was shown that taking into account interparticle scattering in many-sorted systems leads to additional friction, and to transformations of higher angular harmonics of the distribution function. As a result, the kinetic coefficients become temperature-dependent. As examples, the temperature dependence of the PhGE in periodic lattices of two-dimensional asymmetric oriented antidots, together with the conductivity and cyclotron resonance of a two-dimensional semi-metal, were investigated [38,39].

The photon drag effect in single-layer graphene was studied. A specific feature of graphene is a close-to-isotropic energy spectrum in individual valleys. The photon drag effect occurs due to transmission of the photon in-plane momentum to electrons. It was demonstrated that there is a resonance mechanism associated with selection of interband transitions by the Fermi cutoff surface in a degenerate electron gas in addition to the usual effect due to acceleration of electrons by light [40].

The PhGE is absent in single-layer graphene possessing an inversion center. Nevertheless, this fact does not prevent the emergence of a partial photogalvanic current in individual valleys. The mechanism of generation of valley current related to interelectron scattering was considered. The valley current is determined by trigonal asymmetry of the valley spectrum and even anisotropy of electron excitation. It was shown that the valley current is excited in the course of interelectron thermalization owing to azimuth-dependent selection of transitions by energy spectrum anisotropy [41].

1.5 Theory of a Nonadiabatic Quantum Pump

The emergence of a stationary current in a microscopic system under the action of a variable signal was termed the quantum ratchet. Such systems include microscopic systems subjected to the action of periodic forces that induce stationary pumping of electrons. The theory of perturbations in the external field is not the only aspect in studying quantum ratchets. The mere nonlinearity may lead to spontaneous violation of spatial symmetry, resulting in the emergence of a stationary current even if there is no preferential direction.

Similar effects in bounded quantum systems were called a quantum electron pump. A quantum pump is a quantum dot with field electrodes to which a periodically varying voltage is applied. An example of a hypothetical quantum pump is shown in Fig. 1.4.

If the system is spatially asymmetric or there is appropriate phasing of potentials, a finite charge flows between the source and drain in the system during the period of variation of parameters, i.e., a stationary current is generated. In this sense, the quantum pump is an analog of the PhGE. Under certain conditions, the charge flowing within the period of variation of parameters turns out to be quantized. Various methods of implementation of quantum pumps were proposed. For example, a system with an effective potential "step + local well-barrier oscillating in time" can be realized by using a quantum wire with an inhomogeneous thickness placed into an orthogonal alternating electric field.

Depending on parameters, there is a large variety of operation modes of a one-dimensional quantum pump based on a structure with two potential delta-functional barriers/wells whose height/depth oscillates in time. It was demonstrated that the current experiences oscillations with a period corresponding to the multiplicity of the wavelength of incident or excited electrons to the distance between the delta functions. The resonance on the quasi-stationary states between the barriers, on the zero energy, and on steady states (in the case of quantum wells) was considered. A symmetric pump with a supply of a phased alternating voltage was specially examined. Dependences of the current on the Fermi momentum, frequency, and phase difference between the alternating voltages on the field electrodes determining the barrier height (well depth) were obtained.

Nonlinear pump operation modes under the action of an intense variable perturbation were studied by numerical methods. The influence of the signal amplitude was obtained, which illustrates the behavior in the range of multiphoton oscillations [42], as well as resonance widening due to the finite magnitude of the field. The last case was studied both numerically and analytically [43,44].

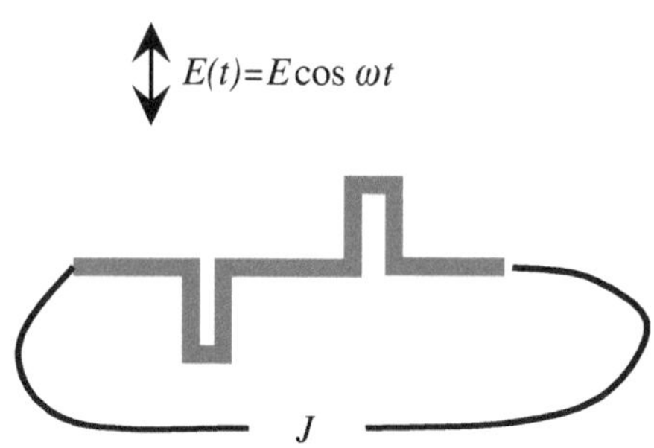

FIGURE 1.4 Example of a quantum pump: curved quantum wire (gray contour) placed into an external alternating homogeneous electric field $E(t)$; the electrons are subjected to an effective potential $V(x,t)$; the pump generates a direct current J.

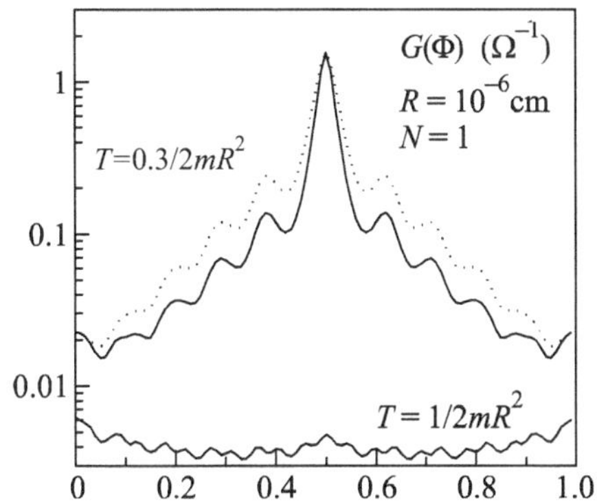

FIGURE 1.5 Conductance (response to a circular electric field) of a GaAs quantum ring with one electron at different temperatures *T* versus the stationary magnetic flux through the ring. The dotted curve shows the approximation of the relaxation time.

A two-frequency signal corresponding to a coherent PhGE was examined. It was shown that application of a two-frequency signal leads to generation of a stationary photocurrent even in a spatially symmetric system [45].

A new phenomenon was discovered in studying nonlinear regimes [46]: it turned out that the current exhibits a stepwise behavior at certain amplitudes of the signal (the derivative of the photogalvanic current vanishes). As such a system, the potential of one or two delta-functions with harmonically oscillating amplitudes was considered. It was found that a similar phenomenon also occurs for the conductance of the system with a single oscillating delta-function, namely, the conductance vanishes at certain values of parameters. This phenomenon, which was called the resonance high-frequency blockade, is not related to electron–electron interaction, in contrast to the Coulomb and spin blockade. We found that the reason for this phenomenon is the emergence of true localized quasi-energy states in open systems under the action of a strong periodic perturbation.

This topic also includes the issue of the induction current generation in a quantum ring [47]. The induction current is a correction in terms of $dB(t)/dt$ in the full circular current J to a nondecaying current over a ring J_0 in a magnetic field $B(t)$: $J = J_0(\Phi) - G(\Phi)E$, $\Phi(t) = -\pi R^2 dB(t)$, $E = -\pi R^2\, dB(t)/dt/c$. The point is that the slowly changing magnetic flux Φ through the ring in a magnetic field linearly depending on time can be treated as a periodic change in the boundary conditions on a conventionally "cut" ring, while the stationary current generated as a response can be interpreted as a current pumped by the quantum pump due to these changes. We studied the induction current in the case with electron–phonon interaction (see Fig. 1.5).

1.6 Electron States in Graphene

The electron states in graphene with a one-dimensional potential were studied. An approximate solution was obtained for a small angle of incidence with respect to the potential gradient. Problems of the electron transition through a barrier of the smoothed step type $U(x) = U\tanh(x/a)$ and a potential with a singularity $U(x) = -U/(|x| + d)$ were solved analytically. The energy and wave functions of bound states in the potential $U(x) = -U/(|x| + d)$ were obtained. Based on the solution of the previous problem, the electron states on the Coulomb impurity in a carbon nanotube were found. The coefficients and transmission/reflection phases for various potential barriers were determined. It was shown that the particle in the state with a zero projection of the momentum onto the nanotube axis is not reflected in an arbitrary potential [48].

Because of the absence of back scattering in these states, the conductance of a finite nanotube is equal to the conductance quantum when the Fermi level falls within the domain of states with a zero projection of the momentum. At the same time, the specific conductivity of an infinite nanotube turns to infinity. Thus, there is no localization of electrons in the nanotube. This stimulated us to consider an interferometer consisting of two long parallel nanotubes. The phonon-induced phase failure that does not lead to back scattering can violate the interference.

It was demonstrated that the basic mechanism is interaction with soft (unstable) phonons with frequencies of the order of the phonon time of flight through the nanotube [49].

1.7 Excitons in Graphene

In contrast to usual slotted semiconductors, the existence of excitons in graphene, which has no slots, seems to be impossible at first glance, because all states of the electron–hole pair are located in the allowed energy band.

However, a pair with similar velocities can form a bound state [50,51]. It turned out that the conical approximation is insufficient to describe electron–hole coupling. In this case, the kinetic energy of the pair $\varepsilon_c((\mathbf{p}+\mathbf{k})/2) - \varepsilon_v((\mathbf{p}-\mathbf{k})/2)$ with the electron momentum $(\mathbf{p}+\mathbf{k})/2$ and the hole momentum $(\mathbf{p}-\mathbf{k})/2$ should have a minimum as a function of the relative momentum $\mathbf{k}$. Situations where the pair is formed from one valley ($\mathbf{p} \ll 2\mathbf{K}$, where $\mathbf{K}$ is the vector to the valley center), or from opposite valleys $\mathbf{p} = 2\mathbf{K} + \tilde{\mathbf{p}}$ are potentially possible. It is shown that no exciton is formed for the case of a pair from the same valley. Nevertheless, electron–hole coupling is possible for an indirect exciton formed by a pair from different valleys with allowance for trigonal corrections to the conical spectrum. The pair coupling energy was found. It was shown that coupling is possible only in trigonal sectors of the space of the total momentum of the exciton $\mathbf{p}$, where $\cos 3\varphi_{\mathbf{p}} > 0$ ($\cos \varphi_{\mathbf{p}} = \mathbf{pK}/\mathrm{pK}$). The spectrum of moving excitons is illustrated in Fig. 1.6.

1.8 Electrons in Curved Low-Dimensional Systems

The interest in theoretical investigations of curved low-dimensional systems was stimulated by the progress of technologies which made it possible to create various low-dimensional systems with complicated geometrical shapes [52,53]. Curved systems were studied in Refs.

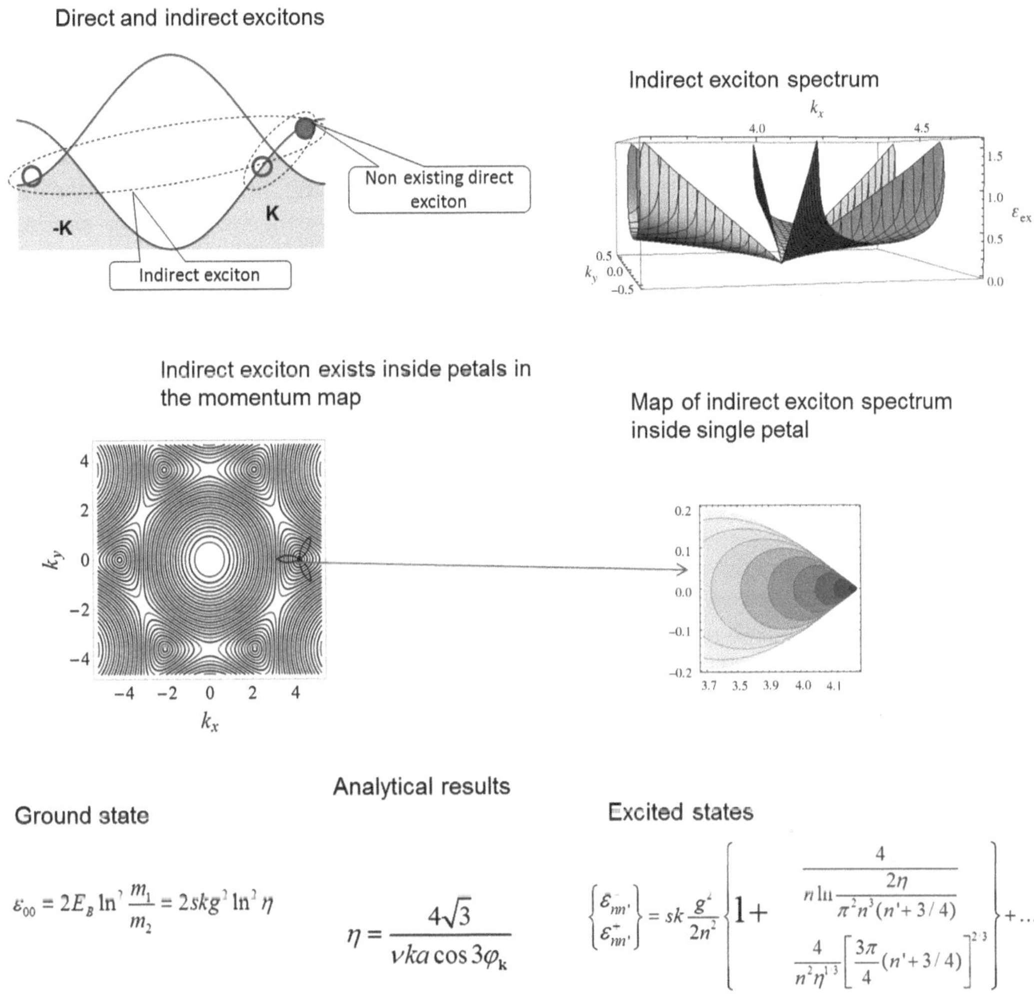

FIGURE 1.6 Illustration of the mechanism of the emergence of exciton states in graphene; ε_{nm} are the energies of the ground and excited states. The plus and minus signs are used to indicate the even parity of the states.

[54–65]. Constraints on electron motion over one or several coordinates lead to transverse quantization of the electron states. If the system is inhomogeneous, then the transverse levels become dependent on the longitudinal coordinate. Such a presentation is adequate if the characteristic size of the inhomogeneity is large compared to the layer or quantum wire thickness. Such adiabatic quantization can be applied to curved low-dimensional systems, which are locally planar (in the two-dimensional case) or straight-line (in the one-dimensional case). If the electron motion occurred rigorously over the surface or a curved line, it could have been described by an appropriate transformation of a two-dimensional or one-dimensional Schroedinger equation for a free particle to curvilinear coordinates. If it is necessary to take into account the finite character of the surface layer or quantum wire thickness, the effective Hamiltonian becomes much more complicated.

First of all, quantization of transverse states leads to large contributions to energy, which are proportional to $1/d^2$. If the thickness d tends to zero, these contributions increase, whereas the kinetic energy of longitudinal motion may remain finite. However, if the transverse characteristics of the transverse potential and the states along the surface are constant, these contributions are also constant and can be eliminated by means of energy shifting. The presence of curvature leads to additional contributions to energy. Remaining small compared to the quantization energy, these contributions are comparable with the longitudinal kinetic energy. The best-known contribution is the geometrical potential, which depends neither on the transverse potential characteristic, nor on the transverse state structure. As this is a local variable, it is determined only by the local geometrical characteristics of the curved system. In terms of its dimension, the geometrical potential should be proportional to the second power of the system curvature.

If the transverse kinetic energy of the electron is small as compared to transverse quantization intervals, then the geometrical potential is the main consequence of the system curvature. However, if the kinetic energy of the electron is comparable to or greater than these intervals, the effect of another factor can be manifested: centrifugal shifting of the electron to the external boundary of the quantum well makes a contribution to the energy, which is proportional to the curvature, kinetic energy, and layer thickness. Curved one-dimensional systems differ from two-dimensional systems because it is necessary to take into account their transverse geometry (cross-sectional shape). As the cross-section may experience rotations along the wire, the local one-dimensional Hamiltonian of the system should be determined both by the local curvature of the system and by twisting of the wire along its axis. Twisting and curvature also affect the spin degree of freedom, leading to additional contributions to SO interaction in curved low-dimensional systems. These issues are discussed in the review papers [57–64] and [65].

Mathematically, the challenge is to reduce a three-dimensional problem to a curved two-dimensional or one-dimensional problem. We derived Hamiltonians for the general case of a curved two-dimensional system and for a curved and twisted one-dimensional system. It was demonstrated that, in addition to the geometrical potential $V_g(q_1, q_2) = -\frac{\hbar^2}{8m}(\kappa_1 - \kappa_2)^2$ expressed via the principal curvatures of the surface, the Hamiltonian also contains an additional component $\hat{H}_c$ induced by shifting of the centroid of the transverse wave function to the external boundary of the quantum well by the centrifugal force. If the potential well is asymmetric, the potential energy is changed in proportion to the centrifugal force. In the simplest cases of a spherical surface and a circular cylinder, we obtain:

$$\hat{H}_c = -\frac{(q_3)_{nn}}{mR^3} l^2 \quad \text{(sphere)},$$

$$\hat{H}_c = -\frac{(q_3)_{nn}}{mR^3} \hat{l}_z^2 \quad \text{(cylinder)}$$

where $l, \hat{l}_z$ are the operators of the angular momentum and its projection onto the cylinder axis, and $(q_3)_{nn}$ is the mean value of the transverse coordinate over the wave function of transverse motion.

Thus, $\hat{H}_c$ leads to renormalization of the effective mass of the relative order δ/R, where δ is the mean value of the transverse coordinate (effective thickness of the layer); for a cylinder, naturally, only the azimuthal component of the mass is renormalized.

A similar Hamiltonian for a curved and twisted wire has the form:

$$\left[\frac{\hat{p}_3^2}{2m} - \overbrace{\frac{\kappa^2(q)}{8m}}^{a)} + \overbrace{\frac{1}{2m}(\hat{l}_z^2)_{nn}\zeta^2(q)}^{b)} + \overbrace{\frac{1}{m}\{\kappa l, \hat{p}_3^2\}(q_\perp)_{nn}}^{c)}\right]$$

Here $\zeta = \tau + \omega$, τ characterizes the geometrical twisting of the curve, and ω characterizes the twisting of the wire cross-section.

The Hamiltonian of SO interaction on a curvilinear surface was also derived.

Applications of curved systems were considered. In particular, it was demonstrated that excitons in a curved quantum wire tend to gather in regions with the greatest curvature, which may even result in their Bose condensation at reasonable parameters of the system.

1.9 Collective Effects in Low-Dimensional Systems (Plasma Waves, Wigner Crystallization, Screening)

It became obvious already during the first years of the development of the physics of low-dimensional systems that they play an important role in understanding phenomena of general physical interest. This primarily refers to phenomena where interparticle interaction plays an important or, sometimes, crucial role. As Sandomirskii aptly mentioned at that time, two-dimensional systems are good testing grounds for studying many-particle effects. The reason is extensive variability of almost all characteristic parameters of such structures; moreover, the researcher can change the main parameter (electron concentration) by several orders of magnitude.

1.9.1 Electrons in Inversion Layers

Historically, the first implementation of such a two-dimensional system with readily controlled properties was the inversion layer on the silicon surface in a MIS structure. The theory of the electron plasma in the inversion channel, which takes into account the effects of electron–electron interaction, was developed in Ref. [66]. For reachable bending of the bands and temperatures of the order of 1–10K, quantization of particle motion normal to the surface becomes essential; moreover, only one subband of transverse quantization may turn out to be populated. In this case, the bare vortex part of the electron–electron interaction, which takes into account electrostatic image forces, is a matrix $\Gamma_{nm,n'm'}(k)$, where n, m are quantum numbers of the subbands, and k is the two-dimensional vector of momentum transfer at scattering. Simple estimates show that this matrix can be also considered as a diagonal matrix:

$$\Gamma = \frac{2\pi\tilde{e}^2}{k}\delta_{nn'}\delta_{mm'}, \quad \tilde{e}^2 = \frac{2e^2}{\varepsilon_1 + \varepsilon_2}, \tag{1.1}$$

if the inversion layer thickness a is much smaller than the effective Bohr radius $a^* = \hbar^2/m\tilde{e}^2$. Here ε_1 and ε_2 are the permittivities of the basic crystal and the dielectric of the MIS structure, and m is the effective mass of the electron in the interface plane. If the condition $a \ll a^*$ is satisfied, the renormalized electron–electron and electron–ion vortex parts can be calculated by using the usual temperature diagram technique. For the zero frequency (e.g., screening of the field of the Coulomb impurity), the result has the form:

$$\tilde{\Gamma} = 2\pi\tilde{e}^2\left(k + \frac{2}{a^*}\sum_m f(\varepsilon_m)\right)^{-1}, \tag{1.2}$$

where f are the Fermi occupation numbers and ε_m are the energies of the transverse quantization levels. At $T=0$, the screening constant in the denominator of Eq. (1.2) is equal to $2N(\mu)/a^*$, where $N(\mu)$ is the number of ε_m levels below the Fermi level. Thus, the screening radius experiences jumps when filling of a new subband begins. For a nondegenerate gas occupying only the first subband, the expression for the Debye radius in a two-dimensional system is $r_D = T/2\pi\tilde{e}^2 N_s$ (N_s is the surface density of electrons). Let us emphasize that this expression, and also Eq. (1.2), are essentially different from their three-dimensional analogs. The renormalized e–e vortex at a finite frequency has the form:

$$\Gamma_{ee}(\omega, k) = \frac{2\pi\tilde{e}^2}{k + 2\tilde{e}^2\Pi(\omega, k)} \quad \Pi = \frac{m}{\pi}\left(1 - |\omega|/\sqrt{\omega^2 - k^2 v_0^2}\right), \quad \omega > kv_0 \tag{1.3}$$

and determines the law of dispersion of two-dimensional plasma waves:

$$\omega(k)^2 = \frac{2\pi\tilde{e}^2 N_s k}{m}\frac{(1 + ka^*/2)^2}{1 + ka^*/4}, \quad k \ll p_0. \tag{1.4}$$

Here $p_0 = mv_0$, v_0 is the Fermi velocity. At $ka^* \ll 1$, we have the known square-root dependence $\omega \sim \sqrt{k}$; at $ka^* \gg 1$, the plasmon dispersion reaches a zero-sound straight line $\omega = kv_0$. All this theory is applicable for a weakly nonideal plasma with a sufficiently large density: $N_s a^{*2} \gg 1$. Therefore, $\omega(k)$ reaches the zero-sound dispersion at $ka^* \gg 1$ already in the domain of applicability of Eq. (1.4): $k \ll p_0$.

Let us now calculate the corrections to the electron spectrum due to electron–electron interaction. The self-energy part $\sum$ calculated without renormalization of the Coulomb interaction has a logarithmic singularity at $p = p_0$ in the two-dimensional case. For instance, we have $\sum = -2\tilde{e}^2 p_0 E(p/p_0)/\pi$ in the domain $p < p_0$, where E is the elliptical integral of the second kind. At $|p - p_0| \le 1/a^*$, screening should be taken into account because a small transfer of momentum in the vortex part Γ_{ee} becomes fairly significant. Rather cumbersome transformations yield the following results: the quasi-particle energy counted from the chemical potential shifted by interaction is:

$$\varepsilon(p) = \xi(p)\left(1 + \frac{(\ln(2p_0a^*) - 2)}{\pi p_0 a^*}\right); \tag{1.5}$$

the decay rate of quasi-particles is described by the expression:

$$\gamma(p) = \frac{\xi|\xi|m}{4\pi p_0^2}\left(\ln\left(\frac{2p_0}{a^*|\xi|}\right) - \frac{1}{2}\right), \tag{1.6}$$

where $\xi = (p^2 - p_0^2)/2m$; Eqs. (1.5) and (1.6) refer to the domain $|\xi| \ll v_0/a^*$. It follows from here that the temperature dependence of the of electron–electron relaxation rate is described by the formula $T^2|\ln T|$. The most essential aspect here is the logarithmic factor in the expression for the decay of quasi-particles. The presence of this factor distinguishes the two-dimensional case from the three-dimensional case, and from the known result based on Landau's conclusion: in a strongly degenerate gas, the frequency of e–e collisions is found as a bimolecular reaction rate constant; therefore, it is proportional to the squared concentration of "active" particles, i.e., $(\nu(E_F))^2$, where ν is the density of states at the Fermi level. Formula (1.6) was applied many years later to interpret the experiments of Ref. [67], where relaxation of electron excitaton in a two-dimensional gas of the GaAs-AlGaAs heterostructure was measured. It was noted that good agreement with the experiments is obtained without any fitting parameters.

1.9.2 Plasma Oscillations of Multicomponent Two-Dimensional Systems

In the previous paragraph, we considered plasma oscillations of a two-dimensional electron gas in an ultra-quantum mode, where all electrons were located at the lowest level of transverse quantization. It is also of interest to consider systems that are "basically" two-dimensional, but exhibit to some extent the transverse degree of freedom [68]. As applied to the quantum well or inversion layer, this means that several transverse levels are populated, and transitions between these levels should be taken into account. In the case of a layered structure (multilayer superlattice), tunneling between the layers may be meant. Thus, there arises a problem of a two-dimensional electron plasma containing several components. This term is used to denote groups of electrons that differ either by the quantum number of transverse motion, or by the number of the layer in the superlattice. The spectrum of plasma waves in such systems has a number of interesting specific features. New branches of oscillations appear; some of them resemble ion-sound branches in the gas plasma (acoustic plasmons); others are similar to exciton-type excitation (intersubband plasmons). The Landau damping in such systems and the spectrum of their optical absorption also have certain specific features.

We consider a system with a single-particle spectrum of the quantum well type:

$$W_n(p) = \varepsilon_n + p^2/2m \tag{1.7}$$

and the wave functions of transverse motion $\varphi_n(z)$. Within the framework of the theory of a linear response to a perturbation $U^0(z,k)e^{i(k\rho-\omega t)}$, we have to find variation of the charge density induced by the perturbation. If the delay effects are neglected, the equation of a self-consistent field is actually the Poisson equation for the induced part of the potential. Taking into account that the density perturbation contains the total potential $U(z,k)$ and writing the formal solution of the Poisson equation via the Green function, we take the matrix elements over the functions $\varphi_n(z)$ and obtain the system of equations:

$$U_{ij} + \frac{2\pi\tilde{e}^2}{k\varepsilon}\sum_{nm} I_{ij,nm}(k)\prod_{nm} U_{nm} = U^0_{ij}, \tag{1.8}$$

where

$$\begin{aligned} I_{ij,nm} &= \int_{-\infty}^{+\infty} \varphi_i(z)\varphi_j(z)e^{-k|z-z'|}\varphi_n(z')\varphi_m(z')dzdz'; \\ \Pi_{nm} &= -\frac{1}{S}\sum_q \frac{f_n(q)-f_m(k+q)}{W_n(q)-W_m(k+q)+\omega+i\delta}, \end{aligned} \tag{1.9}$$

and S is the area of the system. The spectrum of the plasma waves is obtained from the requirement for vanishing the determinant of system (1.8) whose rank is equal to the squared number of transverse subbands taken into account. Let us discuss some results.

1.9.2.1 Single-Component Plasma in a Finite-Thickness Layer

The plasma frequency is expressed by the same formula as one used for an infinitely thin layer with $\tilde{e}^2$ being replaced by $\tilde{e}^2 I(k)$, where the form factor $I(k)$ is determined by the density distribution over the transverse coordinate $I(k) = \int \rho(z)\rho(z')e^{-k|z-z'|}dzdz'$. In the long-wavelength limit, we have $kL \ll 1$ (L is the characteristic thickness of the layer), $\omega^2 = \omega_{2D}^2(1\text{-const}\cdot kL)$, $\text{const} \sim 1$, ω_{2D} is the plasmon frequency in the infinitely thin layer.

1.9.2.2 Two Spatially Separated Layers of the Two-Dimensional Plasma

The distance between the layers is Δ, and the tunneling effects are neglected. There are two plasmon branches: optical and acoustic. In the limit $ka^*,\ k\Delta \ll 1$, we obtain

$$\omega_{opt}^2 = 2\pi\tilde{e}^2 k\left(\frac{N_1}{m_1}+\frac{N_2}{m_2}\right), \quad \omega_{ac}^2 = 4\pi\tilde{e}^2\Delta\frac{N_1N_2k^2}{N_1m_2+N_2m_1} \tag{1.10}$$

Here $m_{1,2}$ and $N_{1,2}$ are the electron masses and concentrations in the layers. If the Fermi velocities in the layers are different, then the acoustic branch experiences Landau damping even at a zero temperature: plasmon in the layer with the lower Fermi velocity decays on electrons of the other layer because the frequencies of the branches in the zero-sound domain tend to kv_{01}, kv_{02}, and we have $\omega_2 < kv_{01}$ at $v_{01} > v_{02}$. The optical branch at $T = 0$ remains nondecaying for all values of k, while the Landau damping in the acoustic branch is equal to zero at values of

k smaller than a certain critical value k_c. Such a critical value exists if the following condition is satisfied:

$$\Delta > \frac{1}{4}\frac{a_2(v_{01}^2 - v_{02}^2)^{1/2}}{v_{01} - (v_{01}^2 - v_{02}^2)^{1/2}} \tag{1.11}$$

If $v_{01} = v_{02}$, then the damping is equal to zero in both branches for all values of k.

1.9.2.3 Three Identical Equidistant Layers of the Two-Dimensional Plasma

There is one optical and two acoustic branches. At $k\Delta \ll 1$, we have:

$$\omega_{opt}^2 = 3\omega_0^2, \quad \omega_{ac1}^2 = 2k\Delta\omega_0^2, \quad \omega_{ac2}^2 = \frac{2}{3}k\Delta\omega_0^2; \quad \omega_0^2 \equiv \frac{2\pi\tilde{e}^2 N_s k}{m}.$$

1.9.2.4 Intersubband Plasmons in the Quantum Well

If the quantum well is symmetric with respect to the mid-plane and only two lower subbands are taken into account, the form factors I_{ijnm} from Eqs. (1.9) vanish if some of the subscripts, which take the value of 1 or 2, are encountered only once. The frequencies of optical and acoustic plasmons in the limit of small values of k are described by Eq. (1.10), and $1/2(dD/dk)_{k\to 0}$, where $D(k) = I_{1111}I_{2222} - I_{11,22}^2$, plays the role of the parameter Δ. The same replacement defines the domain of momentum where the Landau damping is absent in the acoustic branch. Another solution of the characteristic equation (determinant 4×4) corresponds to the intersubband plasmon:

$$\omega^2 = (E_2 - E_1)^2 + \frac{4\pi\tilde{e}^2 J(k)(E_2 - E_1)(N_1 - N_2)}{k}, \tag{1.12}$$

where $J(k) = I_{1212}$, $N_{1,2}$ are the subband populations. The second term in Eq. (1.12) tends as $k \to 0$ to a constant value of δ_p^2; δ_p is the depolarization shift. Thus, the resonance frequency of the intersubband transition differs from the distance between the levels of the single-particle spectrum (owing to dynamic screening effects) by the depolarization shift value. The optical and acoustic plasmons are also manifest as absorption peaks of the modulated RF electric field. The modulation period l determines the plasmon momentum $k = 2\pi/l$. The width of the acoustic resonance in the quantum well is greater than the width of the optical resonance owing to contributions of intersubband transitions to electron scattering. Plasma oscillations in double quantum wells and in multilayer superlattices were intensely studied in more recent experimental investigations and found many applications owing to the development of terahertz engineering. All the basic postulates of the above-described theory were validated.

1.9.3 Electron Crystal

Back in the 1930s, Eugene Wigner studied the cohesion energy of metals and demonstrated that the potential energy of interaction of the electron gas in the limit of the low electron density is significantly greater than its kinetic (Fermi) energy. In this case, in the gel model, where the positive compensating charge is uniformly spread over the space, which is commonly used in the theory of metals, the minimum of the total energy of the system is reached if the electrons form a regular crystal lattice (Wigner crystal) [69]. In real solids, however, the positive charge is concentrated on ions; if the distance between the ions becomes noticeably greater than the Bohr radius, then the ground state corresponds to a set of neutral atoms characterized by a dielectric-type electron spectrum. In the case of the inversion channel of the MIS structure, the neutralizing charge is located on the metallic electrode and is separated from two-dimensional electrons by a dielectric layer. Therefore, a two-dimensional Wigner crystal is expected to form in the limit $N_s a^{*2} \ll 1$ [70].

The Fourier image of the electron potential in the inversion layer whose thickness is much smaller than the thickness of the dielectric layer of the MIS structure has the form:

$$\varphi(q) = \frac{4\pi e}{q(\varepsilon_1 + \varepsilon_2 \coth(q\Delta))} \tag{1.13}$$

Therefore, the potential energy of the lattice whose nodes are defined by a set of planar vectors a_n is:

$$W = \frac{1}{2}\sum_{i \neq j} \int \frac{dq}{4\pi^2} e\varphi(q) e^{iq(a_i - a_j)}. \tag{1.14}$$

Writing the force matrix $T_{ij} = \partial^2 W / \partial u_i \partial u_j$ and solving the characteristic equation of the vibrational problem, we can find the dispersion of the phonons of the electron crystal. In the long-wavelength limit $k \ll \sqrt{N_s}$, we obtain the expressions for the longitudinal branch of oscillations:

$$\omega_{\|}^2 = \frac{4\pi e^2 N_s k}{m(\varepsilon_1 + \varepsilon_2 \coth(k\Delta))} \tag{1.15}$$

and for the transverse branch:

$$\omega_{\perp}^2 = s_{ij}^2 k_i k_j \tag{1.16}$$

in the order of magnitude, $s^2 = \tilde{e}^2 \sqrt{N_s}/m$, and a particular value of the sound velocity s depends on the lattice type. It is seen that the longitudinal branch in the continuum approximation corresponds to two-dimensional plasma oscillations whose dispersion law differs from the square-root dependence (cf Eq. 1.4) owing to the influence of image forces in the metallic electrode: at $k\Delta \ll 1$, we have $\omega_{\|} \sim k$. In a magnetic field H normal to the interface plane, the electron oscillation frequencies $\tilde{\omega}$ can be found from the classical equations of motion. The results are expressed via the oscillation frequencies at a zero magnetic field as:

$$\tilde{\omega}_{1,2}^2 = \frac{1}{2}(\omega_{\|}^2 + \omega_{\perp}^2 + \Omega^2) \pm \frac{1}{2}\sqrt{(\omega_{\|}^2 + \omega_{\perp}^2 + \Omega^2)^2 - 4\omega_{\|}^2 \omega_{\perp}^2}, \tag{1.17}$$

where $\Omega = eH/mc$. It is seen that the magnetic field eliminates degeneration of the vibrational spectrum at the point $k = 0$. At $k \ll \sqrt{N_s}$, we obtain:

$$\tilde{\omega}_1(k) = \Omega + \frac{\alpha k + s^2 k^2}{2\Omega}, \quad \tilde{\omega}_2(k) = \frac{\alpha^{1/2} s k^{3/2}}{\Omega}, \quad \alpha \equiv \frac{2\pi \tilde{e}^2 N_s}{m} \tag{1.18}$$

The magnetization of the Wigner crystal at $T \ll \Omega$ is related to the "crystal" branch $\tilde{\omega}_2$, whereas the first (plasma) branch makes an exponentially small contribution:

$$M(T) = -\frac{\Gamma(7/3)\xi(7/3)}{3\pi}T^{7/3}\left(\frac{e}{mc}\right)^{4/3}H^{1/3}(\alpha s^2)^{-2/3}. \tag{1.19}$$

The Wigner crystallization of two-dimensional electrons was experimentally observed by Kukushkin and his colleagues in the GaAs-AlGaAs structure in 1990 [71]. This state of the electron system plays an important role in the fractional quantum Hall effect at small filling factors.

1.10 Screening in Nanostructures

An important example of many-particle effects is screening of electron–electron and electron–impurity interaction in solids. A particular form of the screened potential is determined by the effective dimension of the system and by the energy spectrum of the electron gas. Therefore, it is clear that screening in nanostructures has many specific features. These issues were considered in research activities performed at our laboratory in 2007–10 [72–74]. The main results can be summarized as follows.

1.11 Quasi-One-Dimensional Systems

In the case of a nanotube, the screened potential can be expanded into a series in terms of cylindrical harmonics. The behavior of the zero and all other harmonics is qualitatively different. The zero harmonic, i.e., the axially symmetric part of the potential, experiences logarithmically weak screening:

$$V_0(z) \sim e^2/z\,\ln^2[z/a] \tag{1.20}$$

(a is the nanotube radius). All higher harmonics are screened in accordance with the dielectric type, and the effective permittivity depends on the harmonic number as:

$$\varepsilon_n = \left(1 + \frac{\pi\kappa_n}{ma^*n}\right)^2, \tag{1.21}$$

where κ_n is the polarization operator of electrons at the zero frequency and zero longitudinal momentum corresponding to virtual transition with a change of the azimuthal quantum number by n.

1.12 Double Quantum Well

The matrix elements of the bare and screened potentials are related by a matrix dielectric function. Its form follows from the system (1.8) at zero frequency. In the case of a two-component plasma (two parallel plasma layers), the potential of the Coulomb centrum located in one of the layers decreases at large distances as $1/\rho^3$, i.e., in the same manner as it happens in a single layer of two-dimensional electrons. However, the coefficient at ρ^{-3} now depends on the population of the layers, and it is no longer a universal constant $\tilde{e}^2a^{*2}$.

1.13 Multilayer Superlattice

In the sense of the coordinate dependence of the screened potential of the point charge, a three-dimensional situation is observed, i.e., the Yukawa law is valid:

$$U(\rho, n) = \frac{\tilde{e}^2}{2m} e^{-\kappa r_n}, \tag{1.22}$$

where $\kappa^2 = 2/a^* \Delta$, $r_n^2 = \rho^2 + (n\Delta)^2$, n is the layer number, and the charge is located in the layer $n = 0$. However, the "memory" about two-dimensionality of screening charge carriers is manifested as follows: the screening radius is independent of the concentration of charge carriers. Moreover, in directions close to the angles equal to 0 and $\pi/2$ with respect to the superlattice axis, there appear preexponential factors (different for these two directions) in Eq. (1.22), which can be interpreted as anisotropic renormalization of the charge. Thus, it may be argued that screening in a multilayer superlattice possesses certain features of two-dimensional and three-dimensional features simultaneously.

1.14 Screening by Dipole Excitons

Here we encounter an absolutely new situation: screening of a charged or short-range impurity is provided by neutral Bose particles, i.e., indirect excitons in a double quantum well. Moreover, the gas consisting of such excitons can form the Bose-Einstein condensate at low temperatures. The consideration is performed within the framework of both the self-consistent field approximation and the Gross–Pitaevskii equation. At temperatures above the condensation point, which are, nevertheless, sufficiently low so that the Boltzmann statistics cannot be used, screening at distances much greater than the dipole arm d has a dielectric character with the effective permittivity:

$$\varepsilon_{eff} = 1 + \frac{2d}{a^*}\left(e^{2\pi N/mT} - 1\right), \tag{1.23}$$

where N is the density of excitons. The most essential feature of this formula (in contrast to the Fermi systems) is an extremely strong (exponential) dependence of the screening parameter on the concentration of particles. It provides at least a qualitative explanation for the experimentally observed drastic increase in the mobility of indirect excitons with an increase in the excitation level.

If there is a condensate, the character of screening becomes qualitatively different and similar to screening in metals. The Fourier images of the bare and screened potentials are related by the momentum-dependent effective dielectric function:

$$\varepsilon_{eff} = 1 + \frac{4mN_c W_0(k)}{k^2}, \tag{1.24}$$

where $W_0(k)$ is interexciton interaction. If exciton–exciton interaction is assumed (as it is usually for the sake of simplicity) to be contact interaction, i.e., $W_0(k) = \text{const}$, then the screened potential decreases at large distances as ρ^{-5}.

Thus, screening in a system of dipole excitons is essentially different from its electron analog, and demonstrates new physical mechanisms of many-particle effects.

References

[1] B.A. Tavger, V. Ya, Demikhovskiĭ, Quantum size effects in semiconducting and semimetallic films, Sov. Phys. Uspekhi 11 (1969) 644–658.

[2] F. Stern, W.E. Howard, Properties of semiconductor surface inversion layers in the electric quantum limit, Phys. Rev. 163 (1967) 816–835.

[3] A.V. Chaplik, M.V. Entin, Energy spectrum and electron mobility in a thin film with non-ideal boundary, Sov. Phys. JETP 28 (1969) 514–517.

[4] R.E. Prange, T.W. Nee, Quantum spectroscopy of the low-field oscillations in the surface impedance, Phys. Rev. 168 (1968) 779–786.

[5] E.I. Urazakov, L.A. Fal'kovskiĭ, Propagation of a Rayleigh wave along a rough surface, Sov. Phys. JETP 36 (1973) 1214–1216.

[6] L.I. Magarill, M.V. Entin, Photogalvanic effect in size-quantized films, Poverhnost'. Fizika, Himija, Mehanika 1 (1982) 74–78 (in Russian).

[7] A.V. Chaplik, M.V. Entin, Charged impurities in very thin layers, Sov. Phys. JETP 34 (1972) 1335–1339.

[8] L.V. Keldysh, Coulomb interaction in thin semiconductor and semimetal films, JETP Lett. 29 (1979) 658–661.

[9] A.G. Pogosov, M.V. Budantsev, R.A. Lavrov, A.E. Plotnikov, A.K. Bakarov, A.I. Toropov, et al., Coulomb blockade and the thermopower of a suspended quantum dot, JETP Lett. 83 (2006) 122–126.

[10] A.G. Pogosov, M.V. Budantsev, A.A. Shevyrin, A.E. Plotnikov, A.K. Bakarov, A.I. Toropov, Resonance breakdown of a coulomb blockade due to the mechanical vibrations of a quantum dot, JETP Lett. 90 (2009) 574–577.

[11] E.M. Baskin, L.I. Magarill, M.V. Entin, Two-dimensional electron-impurity system in a strong magnetic field, Sov. Phys. JETP 48 (1978) 365–370.

[12] E.M. Baskin, M.D. Blokh, M.V. Entin, L.I. Magarill, Quadratic-in-field current and photogalvanic effect in crystals without the inversion center, Phys. Status Solidi B 83 (1977) K97–K100.

[13] E.M. Baskin, L.I. Magarill, M.V. Entin, Photogalvanic effect in crystals without inversion center, Sov. Phys. Solid State 20 (1978) 1403–1408.

[14] M.D. Bloh, L.I. Magarill, M.V. Entin, Theory of transport phenomena in non-centrosymmetric crystals subjected to strong electric field, Sov. Phys. -Semiconductors 12 (1978) 143–147.

[15] V.I. Belinicher, Space-oscillating photocurrent in crystals without symmetry center, Phys. Lett. A 66 (1978) 213–214.

[16] V.M. Asnin, A.A. Bakun, A.M. Danishevskii, E.L. Ivchenko, G.E. Pikus, A.A. Rogachev, Observation of a photo-emf that depends on the sign of the circular polarization of the light, JETP Lett. 28 (1978) 74–77.

[17] V.V. Lemanov, S.Kh Esayan, A.Yu Maksimov, V.T. Gabrielyan, Circular photovoltaic effect in the ferroelectric Pb5Ge3O11, JETP Lett. 34 (1981) 423–425.

[18] E.L. Ivchenko, G.E. Pikus, New photogalvanic effect in gyrotropic crystals, JETP Lett. 27 (1978) 604–608.

[19] V.M. Asnin, JIA. Bakun, A.M. Danishevsky, E.L. Ivchenko, G.E. Pikus, A.A. Rogachev, Circular photogalvanic effect in optically active crystals, Solid State Commun. 30 (1979) 565–570.

[20] L.I. Magarill, M.V. Entin, Photogalvanic effect on free electrons in classical films, Poverhnost', Fizika, Himija, Mehanika 4 (1984) 51–53 (in Russian).

[21] L.I. Magarill, M.V. Entin, Surface photogalvanic effect in metals, Sov. Phys. JETP 54 (1981) 531–535.

[22] G.M. Gusev, Z.D. Kvon, L.I. Magarill, A.M. Palkin, V.I. Sozinov, O.A. Shegai, et al., Resonant photovoltaic effect in an inversion layer at the surface of a semiconductor, JETP Lett. 46 (1987) 33–36.

[23] L.I. Magarill, M.V. Entin, Photogalvanic effect in inversion channel on the vicinal face, Sov. Phys. Solid State 31 (1989) 1299–1304.

[24] Yu.A. Bychkov, E.I. Rashba, Properties of a 2D electron gas with lifted spectral degeneracy, JETP Lett. 391 (1984) 78.E.I. Rashba, V.I. Sheka, in: G. Landwehr, E.I. Rashba (Eds.), Landau Level Spectroscopy, Elsevier, Netherlands, 1991, p. 178.F.T. Vas'ko, Spin splitting in the spectrum of two-dimensional electrons due to the surface potential, JETP Lett. 30 (1979) 541–544.

[25] L.I. Magarill, M.V. Entin, Photogalvanic effect in films, Sov. Phys.-Solid State 21 (1979) 743–747.

[26] V.L. Al'perovich, V.I. Belinicher, V.N. Novikov, A.S. Terekhov, Surface photogalvanic effect in gallium arsenide, JETP Lett. 31 (1980) 546–549.

[27] V.L. Al'perovich, V.I. Belinicher, V.N. Novikov, A.S. Terekhov, Surface photovoltaic effect in solids. Theory and experiment for interband transitions in gallium arsenide, Sov. Phys. JETP 53 (1981) 1201–1209.

[28] M.V. Entin, L.I. Magarill, Photogalvanic current in a parabolic well, JETP Lett. 97 (2013) 639–643.

[29] M.V. Entin, L.I. Magarill, Photogalvanic current in a double quantum well, JETP Lett. 98 (2013) 43–47.

[30] M.V. Entin, L.I. Magarill, Photogalvanic current in electron gas over a liquid helium surface, JETP Lett. 98 (2014) 816–822.

[31] V.I. Belinicher, B.I. Sturman, The photogalvanic effect in media lacking a center of symmetry, Sov. Phys. Uspekhi 23 (1980) 199–223.

[32] E.L. Ivchenko, G.E. Pikus, Sbornik statej k 100-letiyu so dnya rozhdeniya A.F. Ioffe, Problemy sovremennoj fiziki, Nauka, Leningrad, 1980.

[33] E.L. Ivchenko, G.E. Pikus, in: V.M. Tushkevich, V.Ya Frenkel (Eds.), Semiconductor Physics, Cons. Bureau, New York, NY, 1986, p. 427.

[34] B.I. Sturman, V.M. Fridkin, The Photovoltaic and Photorefractive Effects in Noncentrosymmetric Materials (Ferroelectricity and Related Phenomena), Gordon and Breach, Philadelphia, Reading, Paris, Montreux, Tokyo, Melbourne, 1992, 238 pp.

[35] M.V. Entin, M.M. Mahmoodian, L.I. Magarill, Physical effects in low-dimensional systems subjected to local high-frequency field, Phys. E 42 (2010) 1220–1223.

[36] M.V. Entin, L.I. Magarill, Photovoltage in curved one-dimensional systems, Phys. Rev. B 79 (2009) 075434-1-6.

[37] M.V. Entin, L.I. Magarill, Stationary drag photocurrent caused by strong effective running wave in quantum wires: quantization of current, Phys. Rev. B 81 (2010) 073305-1-4.

[38] A.D. Chepelianskii, M.V. Entin, L.I. Magarill, D.L. Shepelyansky, Ratchet transport of interacting particles, Phys. Rev. E 78 (2008) 041127(1-8).

[39] A.D. Chepelianskii, M.V. Entin, L.I. Magarill, D.L. Shepelyansky, Theory of photogalvanic effect in asymmetric nanostructure arrays, Phys. E 40 (2008) 1264–1266.

[40] M.V. Entin, L.I. Magarill, D.L. Shepelyansky, Theory of resonant photon drag in monolayer graphene, Phys. Rev. B 81 (2010) 165441-1-5.

[41] L.E. Golub, S.A. Tarasenko, L.I. Magarill, M.V. Entin, Valley separation in graphene by polarized light, Phys. Rev. B 84 (2011) 195408-1-4.

[42] M.M. Mahmoodian, L.S. Braginsky, M.V. Entin, One-dimensional two-barrier quantum pump with harmonically oscillating barriers: perturbative, strong-signal, and nonadiabatic regimes, Phys. Rev. B 74 (2006) 125317-1-9.

[43] M.M. Makhmudian, M.V. Entin, L.S. Braginskii, High-frequency resonant blockade in one-dimensional quantum pump with oscillating potential wells, JETP 105 (2007) 495–501.

[44] L.S. Braginskii, M.M. Makhmudian, M.V. Entin, Theory of one-dimensional quantum pump based on a two-barrier structure, JETP 100 (2005) 920–928.

[45] M.M. Mahmoodian, M.V. Entin, Theory of one-dimensional double-barrier quantum pump in two-frequency signal regime, Europhys. Lett. 77 (2007) 67002–67006.

[46] M.M. Mahmoodian, M.V. Entin, L.S. Braginsky, High-frequency blockade and related phenomena, Phys. E 40 (2008) 1205–1208.

[47] L.I. Magarill, M.V. Entin, L.I. Magarill, M.V. Entin, Inductive current in a quantum ring, JETP Lett. 80 (2004) 421–425.

[48] D.S. Miserev, M.V. Entin, Quantum mechanics of graphene with a one-dimensional potential, JETP 115 (2012) 694–705.

[49] D.S. Miserev, M.V. Entin, Dephasing in gapless carbon nanotubes and nanostrips and the suppression of interference in a quantum interferometer based on them, JETP Lett. 99 (2014) 410–414.

[50] M.M. Mahmoodian, M.V. Entin, Moving gapless indirect excitons in monolayer graphene, Nanoscale Res. Lett. 7 (2012) 599.

[51] M.M. Mahmoodian, M.V. Entin, Moving zero-gap Wannier-Mott excitons in graphene, Europhys. Lett. 102 (2013) 37012–37018.

[52] V. Ya Prinz, V.A. Seleznev, A.K. Gutakovsky, A.V. Chehovskiy, V.V. Preobrazhenskii, M.A. Putyato, Free-standing and overgrown InGaAs/GaAs nanotubes, nanohelices and their arrays, Phys. E 6 (2000) 828–831.V.Ya Prinz, et al., A new technique for fabricating three-dimensional micro- and nanostructures of various shapes, Nanotechnology 12 (2001) 399–402.

[53] O.G. Schmidt, K. Eberl, Nanotechnology. Thin solid films roll up into nanotubes, Nature 410 (2001) 168.

[54] H. Jensen, H. Koppe, Quantum mechanics with constraints, Ann. Phys. 63 (1971) 586–591.

[55] R.C.T. da Costa, Quantum mechanics of a constrained particle, Phys. Rev. A 23 (1981) 1982–1987.

[56] N. Ogawa, K. Fujii, A. Kobushukin, Quantum mechanics in Riemannian manifold, Progr. Theor. Phys 83 (1990) 894–905.

[57] M.V. Entin, L.I. Magarill, Electrons in a twisted quantum wire, Phys. Rev. B 66 (2002) 205308-1-5.

[58] L.I. Magarill, D.A. Romanov, A.V. Chaplik, Electron energy spectrum and the persistent current in an elliptical quantum ring, JETP 83 (1996) 361–367.

[59] L.I. Magarill, D.A. Romanov, A.V. Chaplik, Ballistic transport and spin-orbit interaction of two-dimensional electrons on a cylindrical surface, JETP 86 (1998) 771–779.

[60] L.I. Magarill, A.V. Chaplik, Influence of spin-orbit interaction of two-dimensional electrons on the magnetization of nanotubes, JETP 88 (1999) 815–818.

[61] M.V. Entin, L.I. Magarill, Spin-orbit interaction of electrons on a curved surface, Phys. Rev. B 64 (2001) 085330.

[62] L.I. Magarill, M.V. Éntin, Optical and photoelectric properties of helical quantum wires, JETP Lett. 78 (2003) 213–217.

[63] A.V. Chaplik, R.H. Blick, On geometric potentials in quantum-electromechanical circuits, New J. Phys. 6 (2004) 33-1-6.

[64] A.V. Chaplik, Electronic spectrum and ballistic transport in bent nanotubes, JETP Lett. 80 (2) (2004) 130.

[65] L.I. Magarill, A.V. Chaplik, M.V. Entin, Quantum size effects in semiconducting and semimetallic films, Phys. Usp. 48 (2005) 953–958.

[66] A.V. Chaplik, Energy spectrum and electron scattering processes in inversion layers, Sov. Phys. JETP 33 (1971) 997.

[67] O.T. Dial, R.C. Ashori, L.N. Pfeiffer, K.W. West, High-resolution spectroscopy of two-dimensional electron systems, Nature 448 (2007) 176–179.

[68] R.Z. Vitlina, A.V. Chaplik, Plasma oscillations of multicomponent two dimensional systems, Sov. Phys. JETP 54 (1981) 536–545.

[69] E.P. Wigner, On the interaction of electrons in metals, Phys. Rev. 46 (1934) 1002–1011.

[70] A.V. Chaplik, Possible crystallization of charge carriers in low-density inversion layers, Sov. Phys. JETP 35 (1972) 395–398.

[71] H. Buhmann, W. Joss, K. von Klitzing, I.V. Kukushkin, G. Martinez, A. Plaut, et al., Spectroscopic observation of Wigner crystallization of 2D electrons in a strong transverse magnetic field, JETP Lett. 52 (1990) 306–310.

[72] R.Z. Vitlina, L.I. Magarill, A.V. Chaplik, Screening of electron-electron interaction in semiconductor nanotubes, JETP Lett. 86 (2007) 123–125.

[73] V.M. Kovalev, A.V. Chaplik, Screening effects and Friedel oscillations in quantum-well nanostructures, JETP 107 (2008) 839–845.

[74] V.M. Kovalev, A.V. Chaplik, Screening of static perturbation in a system of dipole excitons, JETP Lett. 92 (2010) 185.

[75] M.V. Entin, L.I. Magarill, Photocurrent in nanostructures with asymmetric antidots: exactly solvable model, Phys. Rev. B 73 (2006) 205206,1-8.

[76] M.V. Entin, L.I. Magarill, Suppression of spin-orbit effects in 1D system, Europhys. Lett. 68 (2004) 853–859.

2

Two-Dimensional Semimetal in HgTe-Based Quantum Wells

Z.D. Kvon[1,2], E.B. Olshanetsky[1], D.A. Kozlov[1], N.N. Mikhailov[1], S.A. Dvoretsky[1]

[1]RZHANOV INSTITUTE OF SEMICONDUCTOR PHYSICS OF THE SIBERIAN BRANCH OF THE RUSSIAN ACADEMY OF SCIENCES, NOVOSIBIRSK, RUSSIA
[2]NOVOSIBIRSK STATE UNIVERSITY, NOVOSIBIRSK, RUSSIA

In this chapter results are reported from a study of a new two-dimensional electron system, a two-dimensional semimetal, that is observed in wide quantum wells based on mercury telluride, which have an inverted band spectrum. Magnetotransport experiments confirm the existence of a semimetal state in quantum wells with (013) and (112) orientations, and thicknesses of 18–21 nm. These experiments show that the band overlap is $\Delta = 3\text{–}5$ meV. A comparison of the experimentally determined Δ with a theoretical calculation of the energy spectrum reveals the fundamental role of strain effects in the formation of the semimetal state. Scattering processes in the two-dimensional semimetal are studied, and it is found that the jump in the electron mobility during electronic metal-two-dimensional semimetal transitions is caused by screening of electron scattering on impurities by holes. The substantial anomalous rise in the resistivity of the two-dimensional semimetal with an increasing temperature is caused by electron–hole scattering. This is the first observation of the direct effect of interparticle scattering (Landau mechanism) on the resistivity of metals. The properties of two-dimensional semimetals in the quantum Hall effect regime are examined. Primary attention is paid to the observed suppression of strong localization under the conditions of the quantum Hall effect. It is shown that, in a strong magnetic field, the two-component electron–hole plasma has fundamentally different topological properties from those of ordinary single-component (electron or hole) plasma. It is suggested that they lead to the appearance of an infinite set of conducting current states and to the suppression of localization.

2.1 Introduction

Over the more than 40-year history of two-dimensional electron systems, several dozens of different semiconductor structures have been found in which these systems can exist. The appearance of new structures with two-dimensional electron systems are no longer a significant event. But the creation of two-dimensional electron systems with high electron, as well as hole, mobility in quantum wells based on binary $Cd_{1-x}Hg_xTe/HgTe/Cd_{1-x}Hg_xTe$

Advances in Semiconductor Nanostructures. DOI: http://dx.doi.org/10.1016/B978-0-12-810512-2.00002-0

heterojunctions went far beyond the ordinary and even now are an advanced topic in physics of low-dimensional electron systems. This is primarily because of the unique (owing to the gapless nature of HgTe) energy spectrum of two-dimensional electron systems in HgTe quantum wells, which is determined to a great extent by relativistic effects and, therefore, by spin-orbital interactions. In particular, it has recently been possible to create two-dimensional semimetals in wells of this kind [1,2], and an entire series of interesting aspects of their behavior have already been discovered. This behavior is discussed and analyzed here.

2.2 Quantum Wells Based on HgTe Technology and Structure

The process of growing HgTe quantum wells by molecular-beam epitaxy has been reviewed by us elsewhere [3]. Thus, here, we go into those details of the wells which have a direct relation to the theme of the present review. The quantum wells structure is shown in Fig. 2.1 Wells with thicknesses of 18–21 nm and (013) and (112) surface orientations were studied. In all cases the substrate for the start of molecular-beam epitaxy was seminsulating GaAs

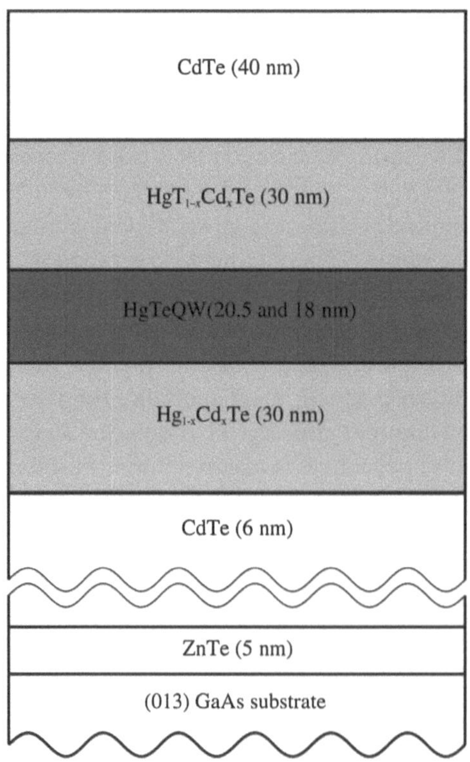

FIGURE 2.1 Schematic cross section of a structure with an undoped HgTe quantum well grown by molecular-beam epitaxy.

with the appropriate surface orientation. This was followed by buffer layers, the most important of which is a CdTe layer with a thickness of a few micrometers. In the final stage, a binary $Cd_{0.7}Hg_{0.3}Te/HgTe/Cd_{0.7}Hg_{0.3}Te$/heterojunction, which also contains the HgTe quantum well between the $Cd_{0.7}Hg_{0.3}Te$ barriers, is grown. The distinctive feature of all these wells is that they essentially have no doping.

2.3 Samples and Experimental Technique

For the magnetotransport measurements, Hall samples, 50 μm wide with a separation between the potentiometer contacts of 100 and 250 μm were fabricated from these double heterojunctions by standard photolithography. Then ohmic contacts were prepared by indium brazing. Beside these samples, extensive studies were made in field-gate transistor structures based on the samples (see Fig. 2.2). They were produced by plasma chemical deposition of a low temperature (growth temperature about 100 °C) bilayer dielectric onto the surface of a Hall sample with ohmic contacts. The bilayer dielectric consists of a 100 nm-thick layer of SiO_2 and a 200 μm-thick layer of Si_3N_4 with a TiAu metallic gate deposited on it. This structure made it possible to take measurements with different positions of the Fermi level, which was varied by applying a gate voltage ranging from 3 V to −3 V. The leakage currents through the dielectric were less than 10^{-12} A, or many orders of magnitude below

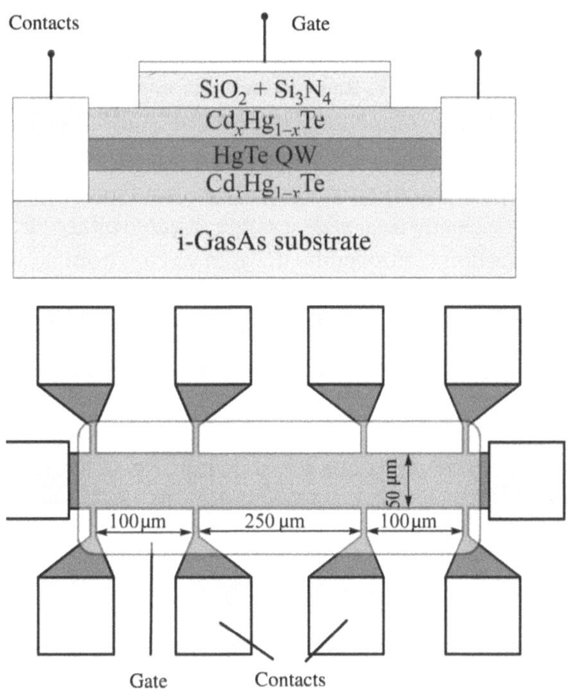

FIGURE 2.2 A transistor Hall structure with a field gate. The top figure is a schematic vertical cross section and the bottom, a view from above.

the currents flowing through the conducting channel (here, the quantum well) of the structure. Measurements were made at temperatures of 4.2–0.1K, in magnetic fields up to 10 T using a standard phase sensitive detector at frequencies of 3–12 Hz, and for measurement currents of 1–10 nA flowing through the sample to exclude heating effects.

2.4 The Semimetal State in Wide HgTe Quantum Wells With an Inverted Band Structure: Their Discovery and Nature

The clearest and most convincing experimental evidence for the existence of a two-dimensional semimetal in these wells is the observation of the temperature independent sign-changing Hall effect, which occurs precisely because of the simultaneous contributions of electrons and holes to the Hall signal [1,2]. Besides the sign-changing Hall effect, there is also a large positive magnetoresistance. All of these effects were discussed previously [3] for samples without a field gate. It is obvious that the transistor structure shown in Fig. 2.2 can be used to study the structure of two-dimensional electron systems during metal–semimetal transitions controlled by simple variation of the gate voltage, as well as for various relationships between the concentrations of two-dimensional electrons and holes. The dependences of the Hall component of resistivity ρ_{xy} on the magnetic field obtained with this device for different gate voltages V_g are shown in Fig. 2.3A. They depend fundamentally on the magnitude and sign of the gate voltage applied to the sample. Fig. 2.3A shows the plots of $\rho_{xy}(B)$ for gate voltages of -2, 0.19, and 2 V. They clearly fundamentally differ from each other. For $V_g = 2$ V, there is, initially, a linear rise and then a quantum Hall plateau, which is typical of single-component, high mobility electron gas. For $V_g = 0.19$ V, the sign-changing Hall effect appears and it indicates that the Fermi level falls into a semimetallic state. Finally, for $V_g = -2$ V, the $\rho_{xy}(B)$ curve again becomes linear, but with the opposite sign. This behavior of $\rho_{xy}(B)$ shows clearly that, with a range of gate voltages from $+2$ to -2 V, in this quantum well, there is a transition controlled by the gate voltage from a two-dimensional electron system with electron conductivity to a two-dimensional electron system with hole conductivity through an intermediate semimetal state, as illustrated in the inset to Fig. 2.3. The concentrations of two-dimensional electrons and holes as functions of the gate voltage, $N_S(V_g)$ and $P_S(V_g)$, obtained by fitting the experimental plots of the resistivity as a function of magnetic field, $\rho_{xx}(B)$, and the curve of Hall effect, $\rho_{xy}(B)$, to the standard magnetotransport formulas in the case of two groups of charge carriers with different signs are shown in Fig. 2.4 As expected, $N_S(V_g)$ is linear with the slope of $8.7 \cdot 10^{14}\ \mathrm{m}^{-2}\,\mathrm{V}^{-1}$, which corresponds to the dielectric capacitance. An entirely different pattern is observed for $V_g = 0.5$ V, when the Fermi level enters the region where the bands overlap (the two-dimensional semimetal). It is clear that, as the negative bias on the gate is increased, the hole concentration increases and that of the electrons decreases linearly with the slopes of $8.1 \cdot 10^{14}$ and $0.7 \cdot 10^{14}\ \mathrm{m}^{-2}\,\mathrm{V}^{-1}$, respectively. Note that the sum of the absolute values of these slopes roughly equal to the absolute magnitude of the slope of the $N_S(V_g)$ curve for $V_g \geq +1$ V, as is to be expected, since for $V_g \geq +1$ V the electrons are the only type of charge-carrier. In addition, the ratio $P_s(V_g)/N_s(V_g) \approx 11.6$ of the slopes for $V_g < 0$ should correspond to

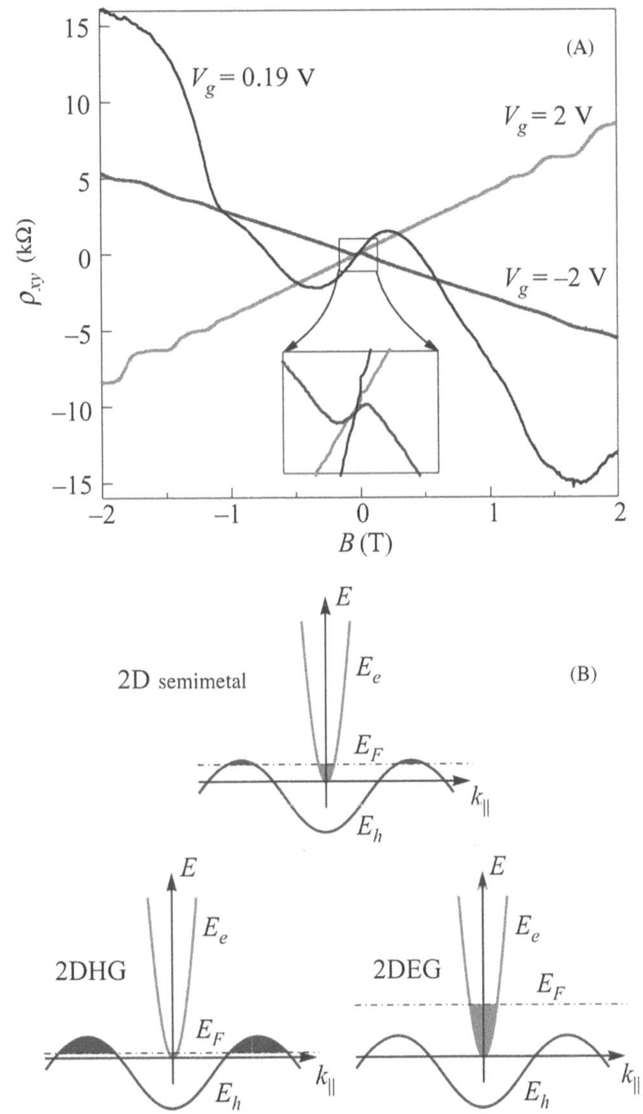

FIGURE 2.3 The Hall component of the resistance, $\rho_{xy}(B)$ for a (013) HgTe quantum well with a thickness of 20 nm for different values of the gate voltage (A); energy spectra of two-dimensional electron systems in an HgTe well: 2D semimetal, 2D electron gas (2DEG), and 2D hole gas (2DHG) (B).

the ratio of the state density of holes and electrons. Then, if the holes fill two valleys and the electrons [1,2], then for an electron mass $m_e \approx 0.025m_0$ [4], the hole mass is $m_h \approx 0.15m_0$, i.e., roughly, six times greater, which is consistent with other data [5]. We note also that, at the point where the densities of two-dimensional electrons and holes are equal, $P_S = N_S \approx 5 \cdot 10^{10}$ cm^{-2}. Thus, the data described above confirm the picture of the energy spectrum of the two-dimensional semimetal proposed in Ref. [1]. The origin of the overlap, however, remains unclear. Numerical

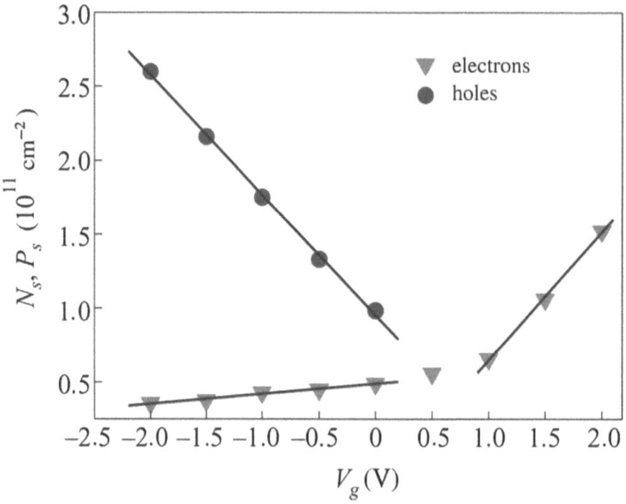

FIGURE 2.4 The concentrations of electrons N_S and holes P_S as functions of gate voltage for a 20-nm-thick (013) HgTe quantum well.

calculations of the spectrum [5,6] for wells with the (100) orientation and essentially the same thickness as those studied in this review, show that there is no overlapping of the bands, but, rather, there is an indirect gap of magnitude on the order of 10 meV. A study of quantum wells with the (112) orientation and thickness of 20 μm has made it possible to explain the situation. Field-gate transistor structures similar to those shown in Fig. 2.2 have also been fabricated using them. Fig. 2.5A is a plot of $\rho(V_g)$, the resistance of this structure as the gate voltage function, and Fig. 2.5B illustrates the behavior of $\rho_{xy}(B)$ for different gate voltages. These are qualitatively similar to those shown in Fig. 2.3 and demonstrate a transition from an electron to a hole two-dimensional system through a two-dimensional semimetal state controlled by the gate voltage. Fig. 2.5B shows how the concentrations of the two-dimensional electrons and holes change as V_g is varied. At the point where the concentrations are equal, $P_S - N_S = (3–6) \cdot 10^{10}$ cm^{-2}; i.e., they differ little from those for quantum wells with the (013) orientation. Accordingly, the overlap is also close to 5 meV. Thus, these results show clearly that the semimetal state develops in wide undoped quantum wells with an inverted spectrum independently of the well's surface orientation. In order to understand the reason for this behavior, we have calculated the energy spectrum for the (112) HgTe quantum well using effective mass approximation and employing an eight-band *kp* Hamiltonian.

The results of the calculation neglecting the quantum well deformation are shown in Fig. 2.6A. It can be seen that the spectrum contains an indirect gap similar to that introduced in Refs. [5,6]; so, there is no band overlap. Fig. 2.6B shows the results when the tensile stress owing to the mismatch between the lattice constants of HgTe and CdTe is included. At first sight, it seems that the lattice constant for the $Cd_{0.7}Hg_{0.3}Te$ barrier layers, immediately adjacent to the quantum well, has to be used. However, they are considerably thinner than the stress relaxation length, which usually exceeds 100 nm. Thus, in these calculations, we have

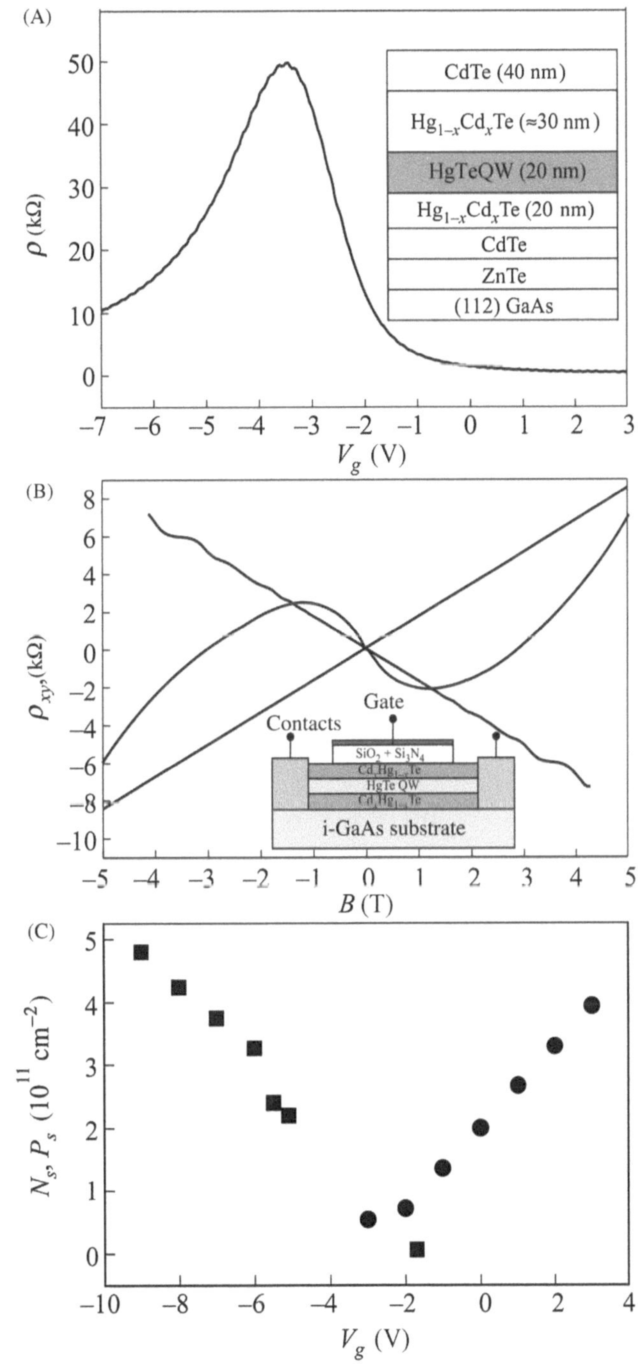

FIGURE 2.5 $\rho(V_g)$ for T = 10 K (A), the Hall component of the resistivity $\rho_{xx}(B)$ for a 20-nm-thick (112) HgTe quantum well for different gate voltages (B), and the electron N_S and hole P_S concentrations as functions of gate voltage for the same wells (C).

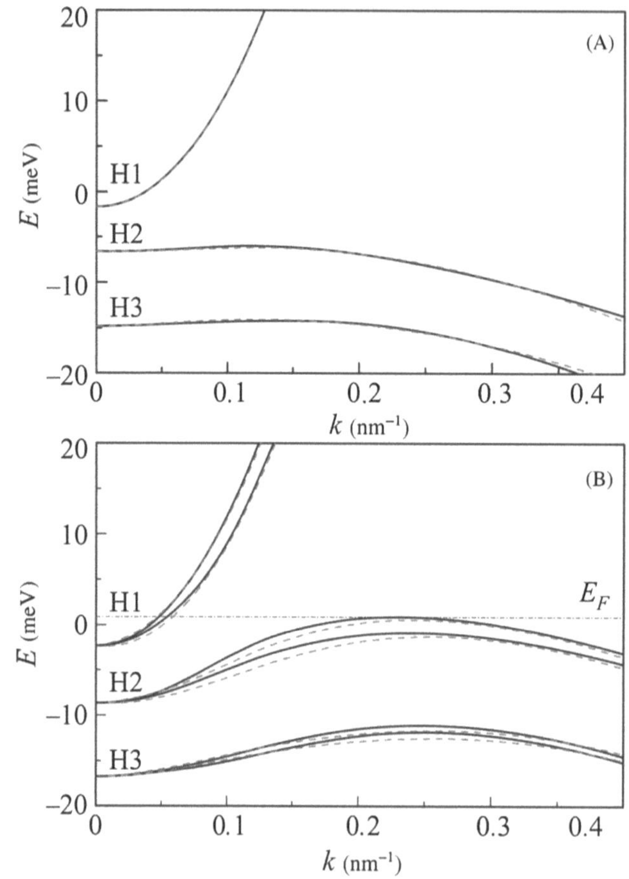

FIGURE 2.6 Calculated dispersion relation E(*k*) for a 20-nm-thick (112) HgTe quantum well, neglecting the mismatch of the lattice constants of HgTe and CdTe (A). For lattice constants of HgTe equal to 6.462 Å and CdTe, 6.482 Å (B), the smooth curves $(k_x,k_y) = (1,0)$ and the dashed curves $(k_x,k_y) = (0,1)$.

used the lattice constant of pure CdTe. It is quite clear that including the stress leads to the band overlap of 3.2 meV, close to the experimentally observed amount. The Fermi level corresponding to the condition $P_S = N_S \approx 4.4 \cdot 10^{10}\ \mathrm{cm}^{-2}$, which is also in fair agreement with the experiment, is also shown in Fig. 2.6B. Hence, these data support the important conclusion that the band overlap and, therefore, the semimetal state in wide (18−21 nm) HgTe quantum wells with an inverted band structure develop as a result of the tensile quantum wells stress caused by the mismatch between the lattice constant of HgTe and the CdTe, which serves as a thick buffer layer during the binary $Cd_{1-x}Hg_xTe/HgTe/Cd_{1-x}Hg_xTe$ quantum well growth.

2.5 Scattering Processes in a Two-Dimensional Semimetal

We now examine scattering processes in a two-dimensional semimetal. Fig. 2.7 shows the electron and hole mobilities as the gate voltage functions. The curves drawn through the

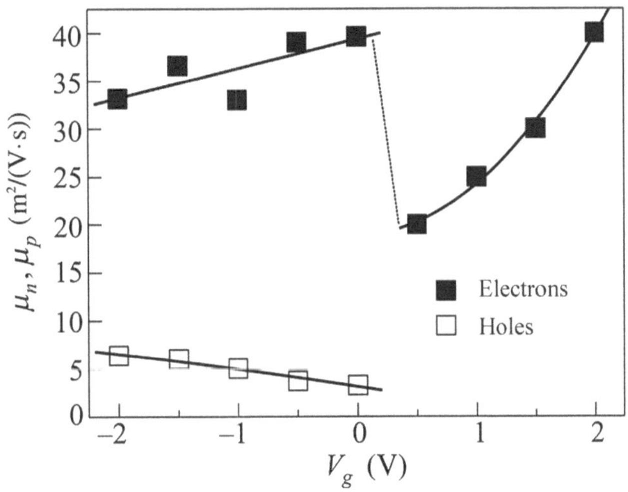

FIGURE 2.7 The mobilities of two-dimensional electrons $\mu_n(V_g)$ and holes $\mu_p(V_g)$ as functions of gate voltage for a two-dimensional semimetal in a 20-nm-thick (013) HgTe quantum well at $T = 0.2$ K.

experimental points are for clarity. For $+1 \leq V_g \leq +2$ V, the drop in the electron concentration is accompanied by a significant reduction in their mobility, in proportion to $N^{3/2}$. This kind of behavior is typical of other two-dimensional structures where the carrier concentration is controlled by a gate voltage and originates in the fact that the transport time for scattering on impurities depends on the concentration at $\tau_{tr} \sim N_S{}^{\alpha}$, where $\alpha = 1{-}2$. For gate voltages $0 \leq V_g \leq +0.5$ V, which correspond to approximate equality of the electron and hole concentrations, there is a sharp jump in the electron mobility, indicated here by a *dotted line*.

Further increases in the negative bias at the gate lead to a slight reduction in the electron mobility and a slight linear increase in the hole mobility. The jump in the electron mobility for $0 \leq V_g \leq +0.5$ V is of greatest interest. It can be seen from Fig. 2.7 that this jump coincides with the start of the range of gate voltages, where the observed electron and hole concentrations are roughly equal. This can be related to the screening of electron scattering on impurities by holes, and suggests the following explanation. For gate voltages $V_g \geq 0.5$ V, when $P_S \ll N_S$ or $P_S = 0$, the electron scattering over impurities is observed, as noted above. In a two-dimensional system based on HgTe, the important feature of this scattering is that the dielectric constant is large, $\varepsilon_{HgTe} > 100$ [7] (for quantum wells $\varepsilon_{QW} = (\varepsilon_{HgTe} + \varepsilon_{CdTe})/2$). Because of this and the small effective electron mass, screening wave vector q_{TF} will be much smaller than Fermi momentum k_F of the electrons when $V_g < 0.5$ V (when only the electron conduction band is filled). Then the electrons scatter over impurities that are weakly screened by the electrons proper. On crossing the Fermi level of the ceiling of the valence band, which begins at $V_g < 0.5$ V, and, therefore, when holes appear, there is a sharp (by more than an order of magnitude) increase in the density of states and, also, an equal increase in q_{TF}. In this case, screening of impurity scattering becomes substantial and leads to a jump in the mobility. The estimated effect is in satisfactory agreement with the experiment only if we assume that there are really anomalously large (hundreds) values of the dielectric permittivity in HgTe-based quantum wells. In a system containing two types of charge carriers, momentum relaxation can be caused by, among other factors, their mutual scattering (the Landau–Barber mechanism) [8]. Since only particles in the range of kT near the Fermi level participate in momentum relaxation from both sides in this mechanism, we should expect that the corresponding relaxation time will vary with temperature roughly as $\sim T^{-2}$. The resistance of the structure as a function of gate voltage in zero magnetic field for temperatures of 0.195–4.1K is plotted in Fig. 2.8A. All the $\rho(V_g)$ curves have distinct maxima. At lowest temperature $T = 0.195$K, this peak coincides roughly in the position with the gate voltage at which the electron and hole concentrations are equal. Another interesting feature of the curves in Fig. 2.8A is their distinctly asymmetric temperature dependence relative to the gate voltage. For gate voltages $V_g \geq 0.5$ V, the resistance is essentially independent from the temperature within this range of temperatures. At the same time, for $V_g \leq 0.5$ V, there is a distinct (by a factor of 1.5–3) increase in the resistance with the rising temperature up to 5K, which is greatest in the range of $-1 \leq V_g \leq -0.5$ V, and decreases with a rising negative bias on the

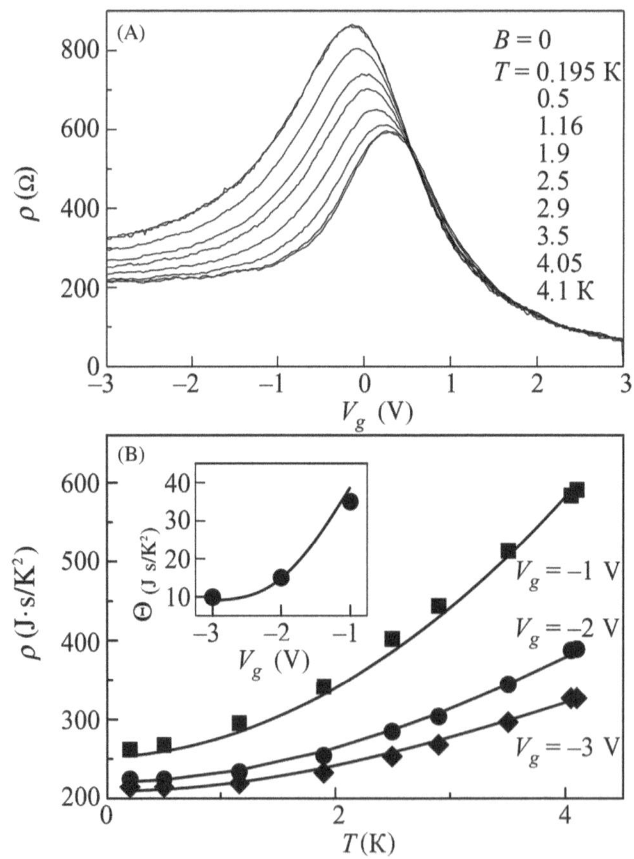

FIGURE 2.8 $\rho(V_g)$ curves for $B = 0$ and different temperatures in the range $T = 0.2$–4.1 K (the temperature increases from the bottom to the top curve) (a); $\rho(T)$ for $V_g = -1$, -2, and -3 V (b). The curves are calculated using Eq. (2.1); the solid symbols are experimental data. The inset shows the fitting parameter Θ as a function of gate voltage: the curves are calculated using Eqs. (2.2) and (2.3) and the points are experimental data.

gate. It is also clear that increasing the temperature shifts the maximum of the $\rho(V_g)$ curve by 0.5 V toward negative gate voltages. We believe that this behavior can be explained by the above-mentioned mechanism of momentum relaxation in a system with two types of carriers. Here there is no temperature dependence of the resistance for $V_g \geq 0.5$ V because, at these gate voltages, there is only one type of charge-carrier, the electrons, in the system (see Fig. 2.8A). To describe the behavior of the system for $V_g \leq 0.5$ V, we use an equation for the temperature dependence of the resistance [2] in a system with two types of carriers such that the momentum relaxation is caused by their mutual scattering:

$$\rho(T) = \rho_0 \frac{1 + (\eta/e)(N_s\mu_p + P_s\mu_n)}{1 + (\eta/e)(N_s - P_s)^2\mu_n\mu_p/(N_s\mu_n + P_s\mu_p)} \tag{2.1}$$

where ρ_0, N_s, P_s, μ_n, μ_p are the resistance of the system, the electron and hole concentrations, and their mobilities at $T = 0$; e is the electronic charge; and, η is the mutual drag coefficient of the electrons and holes. Mutual drag coefficient η is determined by electron scattering over holes. Beyond the dependence on the scattering mechanism for fixed concentrations of the electrons and holes, the probability of this process falls as the square root of the temperature, i.e., $\eta = \Theta T^2$, where Θ is a constant. In Fig. 2.8B the solid data points represent $\rho(T)$ derived from the experimental curves in Fig. 2.8A for $V_g = -1$, -2, and -3 V. The temperature dependence of the resistance at $T \leq 0.5$K saturates for all three gate voltages. This justifies using the mobilities and concentrations of the electrons and holes at these temperatures in Eq. (2.1) for fitting the curves in Fig. 2.8B. Thus, the entire procedure for fitting each of these curves will depend only on parameter Θ and the expression for η. The fits are shown in Fig. 2.8B as smooth curves. The inset to Fig. 2.8B shows the dependence of fitting parameter Θ on the gate voltage (solid points). Assuming that the interparticle scattering is neutral, and that the electron spectrum is quadratic and isotropic, for η we can write:

$$\eta = (k_B T)^2 \frac{D}{N_s P_s} F\left(\sqrt{\frac{N_s g_h}{P_s}}\right) \tag{2.2}$$

where

$$F(x) = \int r dr J_0^2(xr)(J_0^2(r) - J_1^2(r)) \tag{2.3}$$

$g_h = 2$ is the number of hole valleys, $J_n(r)$ is the Bessel function, and D is a constant determined by the features of the interaction between the electrons and holes, and is independent from their concentration. The curve in the inset to Fig. 2.8B was obtained using Eqs. (2.2) and (2.3), which imply that $\Theta = k_B^2 DF(\sqrt{N_S g_h/P_S})/N_S P_S$. Here we used the experimentally determined values of the electron and hole concentrations, and $k_B{}^2 D \approx 4.5 \cdot 10^3$. The good agreement between the experiment and theory confirms the validity of our assumption that the observed temperature dependence of the resistance originates in mutual scattering of electrons and holes. It should be noted that this is the first time that a direct influence of interparticle scattering on the transport time in metals has been observed. In ordinary situations, such as in bismuth or antimony, the occurrence of this mechanism is not unambiguous because of the competing influence of scattering on phonons [9].

2.6 Quantum Hall Effect

Here we present some results from a study of two-dimensional semimetals under quantum Hall effect conditions. Fig. 2.9A contains plots of the longitudinal ρ_{xx} and Hall ρ_{xy} resistances as functions of gate voltage at a fixed magnetic field. Distinct plateaux with $\rho_{xy} = -h/\nu e^2$ can be seen at $\nu = -2, -1, 1$, and 2, accompanied by deep minima in ρ_{xx} on the hole and electron sides of the curve. As the gate voltage is changed, on crossing the point where the electron and hole concentrations are equal (or, in other words, the charge neutrality points), the longitudinal resistance has a distinct maximum, while ρ_{xy} passes through zero from a plateau with h/e^2 on the electron side to a plateau with $-h/e^2$ on the hole side. Thus, as can be seen in the figure, ρ_{xx} is an even and ρ_{xy} an odd function of B. Fig. 2.9B shows components ρ_{xx} and ρ_{xy} as functions of the gate voltage obtained from the experimentally determined components of the resistivity tensor. The customary quantum Hall effect plateau $\sigma_{xy} = e^2\nu/h$ is

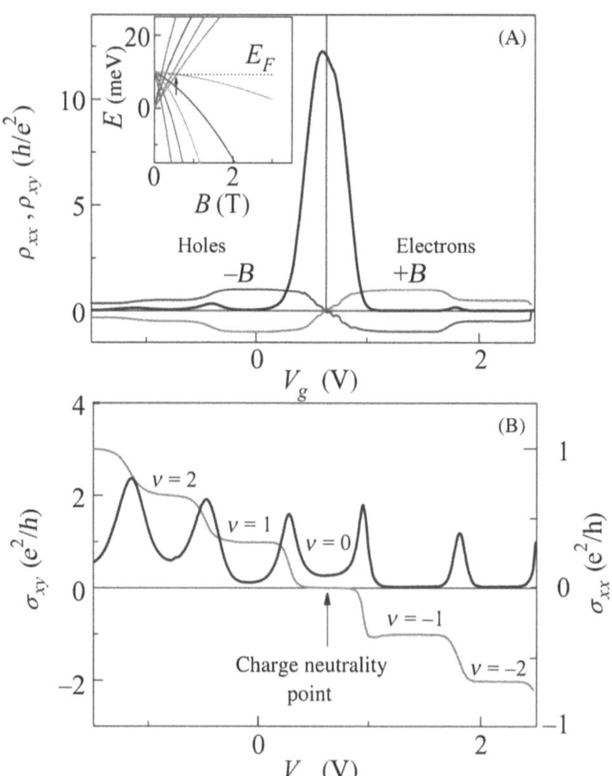

FIGURE 2.9 Diagonal ρ_{xx} and Hall ρ_{xy} components of the resistivity tensor as a function of gate voltage for a fixed magnetic field of 2.8 T. The Hall component of the resistivity tensor is indicated for two signs of the magnetic field. Inset: a Landau level diagram for the electron and hole subbands. The last Landau levels from the electron and hole subbands intersect at $B = Bc$. E_F is the Fermi level at the charge neutrality point (A). The diagonal σ_{xx} and Hall σ_{xy} conductivities as functions of gate voltage for the fixed magnetic field and $T = 90$ mK. The arrow indicates the position of the charge neutrality point, where $N_S = P_S$ (B).

seen, along with the corresponding minima in σ_{xy}. However, it should be noted that the plateaux in σ_{xy} are not entirely flat on the hole side and the minima in σ_{xx} are not very deep, which is explained by the higher effective hole mass, and the low hole mobility. The height of the peaks in σ_{xx} is extremely close to e^2/h for all the Landau levels, as would be expected in the ordinary quantum Hall effect. The state at the charge neutrality point at $\nu = 0$ is the most interesting. Fig. 2.9B shows that σ_{xy} has a flat plateau in the neighborhood of the charge neutrality point, while σ_{xx} has a deep minimum. Both the plateau and this minimum at the charge neutrality point strongly depend on the applied magnetic field and the temperature. Fig. 2.10 shows the changes, observed in σ_{xy} and σ_{xx} as the temperature and magnetic field are varied.

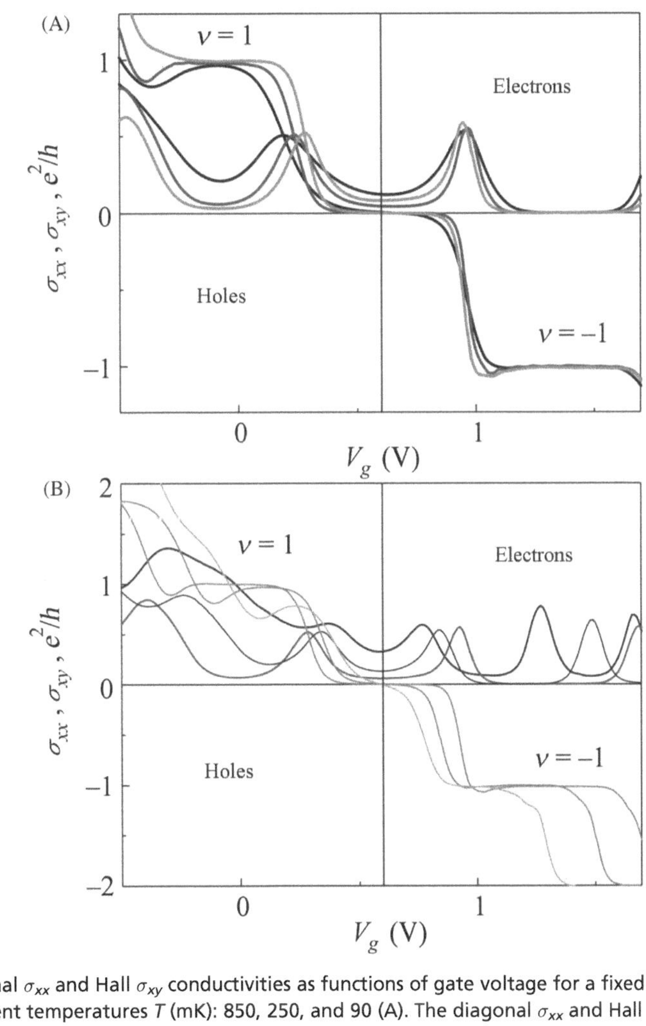

FIGURE 2.10 Diagonal σ_{xx} and Hall σ_{xy} conductivities as functions of gate voltage for a fixed magnetic field B = 2.8 T and different temperatures *T* (mK): 850, 250, and 90 (A). The diagonal σ_{xx} and Hall σ_{xy} conductivities as functions of gate voltage for different magnetic fields *B*(T):1.5, 2, and 2.5 and *T* = 50 mK. The vertical line indicates the position of the charge neutrality point (B)

All the features of the quantum Hall effect show up more distinctly as the temperature is reduced and the magnetic field is increased, especially the plateau and the minima on the hole side, i.e., to the left of the charge neutrality point. The positions of all the minima, except for that of $\nu = 0$, shift with the magnetic field owing to increased Landau level degeneracy.

Here we concentrate on the magnetoresistance in the charge neutrality point. Fig. 2.11A shows that, at the charge neutrality point, there is a sharp increase in ρ_{xx} with the magnetic field at different temperatures. It should be noted that, in all the plots of the resistance as a function of magnetic field, there is a singularity $\rho_{xx} \sim h/4e^2$ at $B_c \approx 1.4$ T. At magnetic fields below 1.4 T the resistance is essentially independent from temperature, while, for $B > 1.4$ T, the magnetic field dependence of the resistance is stronger for lower temperatures. This kind of temperature dependence may indicate the activation behavior associated with the appearance of a cyclotron gap in the spectrum, as shown in the insert to Fig. 2.9. Nevertheless, an analysis shows that the temperature dependence of the resistance is not consistent with an activation dependence of the form $\rho(T) \sim \exp(\Delta/2kT)$, where Δ is the activation gap. The inset to Fig. 2.11A shows that $\rho(T)$ at low temperatures strongly deviates from an activation dependence, and approaches saturation at temperatures below 0.5. Thus, we must conclude that the observed temperature dependence cannot be explained by the appearance of a gap in the spectrum at the charge neutrality point; therefore, other mechanisms must be found. In addition, if a cyclotron gap opens up in the spectrum, then an exponential growth must be observed in the resistance with the increasing magnetic field. In fact, at the charge neutrality point, the proposed gap in the spectrum is given by $\Delta = \hbar\omega_c^e/2 + \hbar\omega_c^p/2 - \Delta_0$, where $\hbar\omega_c^{e,p}$ is the cyclotron energy of the electrons or holes; $\omega_c^{e,p} = eB/(m_{e,p}c)$ is the cyclotron frequency; $m_{e,p}$ is the effective mass; and $\Delta_0 \approx 5$ meV is the overlap of the conduction and valence bands. Given that $m_e = 0.025m_0$ and $m_p = 0.15m_0$, we find that, in the charge neutrality point, $\Delta > 7.4$ meV for $B > 3$ T. The magnetoresistance in the charge neutrality point is compared to $\rho(T) \sim \exp(\Delta/2kT)$, including the linear magnetic field dependence of the gap size Δ in Fig. 2.11B. The increase in ρ_{xx} with the rising magnetic field is qualitatively consistent with the appearance of gaps in the spectrum, but the experimentally determined value of Δ is roughly a factor of 30 smaller than the above-mentioned value. Furthermore, it follows from the experiment that the gap decreases as the temperature is lowered, which is even more anomalous. According to the scheme illustrated in the inset to Fig. 2.9, when $B > B_c$, the gap that opens up in the spectrum should lead to repopulation of the Landau levels and, thereby, to the transformation of the system into an insulator, with $N_s = P_s \rightarrow 0$. In accordance with the above-presented scheme, this should happen for $\hbar\omega_c^e/2 + \hbar\omega_c^p/2 = \Delta_0$. For $\Delta_0 \approx 5$ meV, we obtain $B_c \approx 1.4$ T, in good agreement with the position of the singularity in the $\rho_{xx}(B)$ curves in Fig. 2.11. Thus, on the one hand, an analysis of the $\rho_{xx}(B)$ curves in the charge neutrality point is inconsistent with the existence of a gap at $\nu = 0$ and, on the other hand, it indicates the opposite.

In order to understand the behavior of the system at the charge neutrality point, it is necessary to consider the fluctuations in the filling factor around $\nu = 0$ owing to the random potential of charged impurities. Local fluctuations in filling factor ν lead to the formation of percolation paths along the contours of $\nu = 0$. Here it is important that a conducting grid of states, formed by similar circuits, may be formed with two neighboring percolation clusters

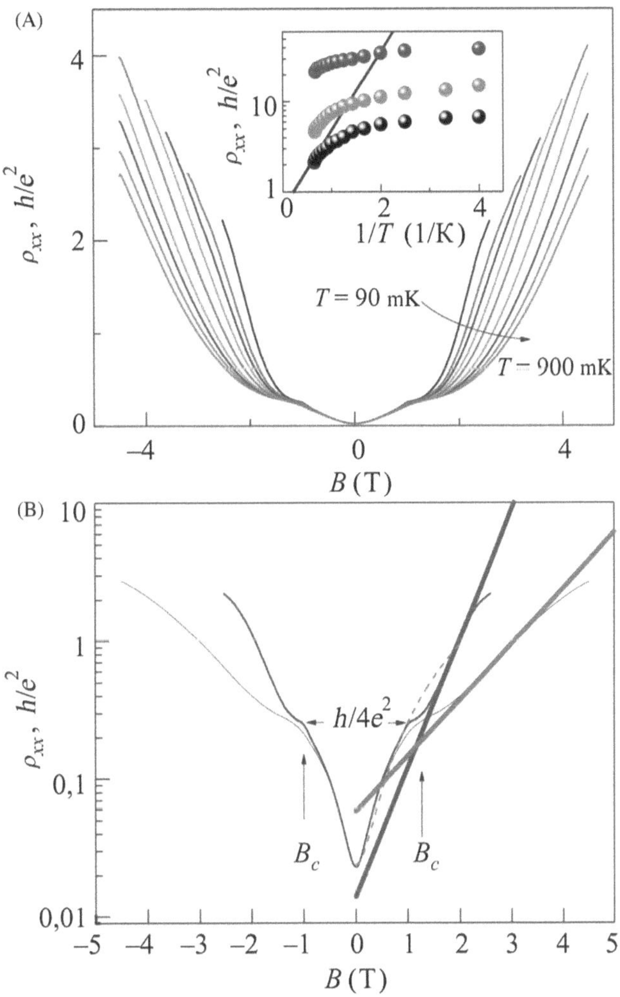

FIGURE 2.11 Magnetoresistance at the charge neutrality point for different temperatures. Inset: resistance as a function of $1/T$ for fixed values of the magnetic field B(T), from bottom to top: 4, 6, and 7. The straight line in the inset is an Arrhenius dependence with $\Delta = 0.3$ meV (A). Magnetoresistance at the charge neutrality point for temperatures of 90 and 900 mK. The smooth curves are fits to $\rho(T) \sim \exp(-\Delta/2kT)$. The dashed curve is a fit obtained using percolation theory for spiral states in the case of a random quasiclassical magnetic field [13] (B).

being connected through saddle points, as shown in Fig. 2.12. The conductivity of the system is by electrons and holes moving along the $\nu = 0$ curves; in the quasiclassical case, this can be represented as spiral trajectories [10] similar to those shown in Fig. 2.12. Here we must emphasize the similarity in the descriptions of the conductivity of a two-dimensional electron–hole system near a charge neutrality point, and the transport of two-dimensional electrons in a random magnetic field with an average value of zero [11–13]. The features of transport along spiral trajectories on $\nu = 0$ contours and the scattering of these trajectories at saddle points can explain the large magnetoresistance seen at the charge neutrality point in

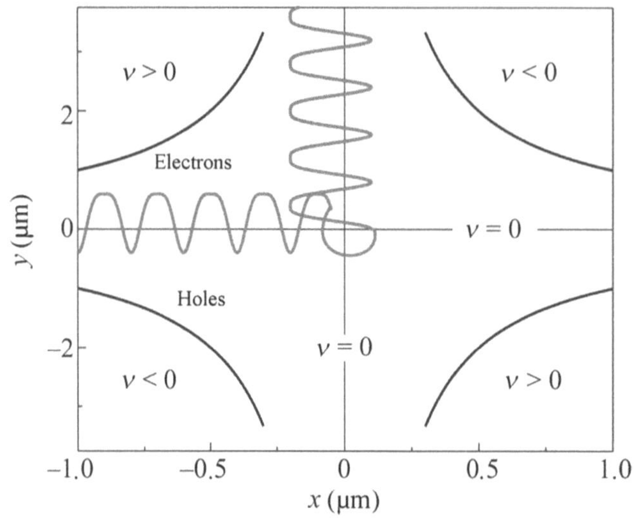

FIGURE 2.12 Schematic illustration of electron-hole spiral states propagating along $\nu = 0$ contours at the charge neutrality point in a strong magnetic field and of the geometry of a saddle point between neighboring percolation clusters.

Fig. 2.11. An analytical solution of the model in Ref. [12] was obtained for two different regimes, corresponding to a small and large amplitude of the random magnetic field. In the limit of a small amplitude, the random magnetic field with $\alpha \ll 1$, where $\alpha = d/R_c$, $R_c = \hbar k_F/m_{e,p}\omega_c$ is the Larmor radius, and k_F is the Fermi wave vector, $\sigma_{xx} = e^2/h(k_F d/4\alpha^2)$. The limit of weak disorder $\alpha \ll 1$, in our case, corresponds to a low-amplitude random magnetic field. The smooth curves in Fig. 2.3 show the result of comparing our values to the percolation theory along the spiral states. Good agreement is obtained for $d = 60$ nm, which seems a quite reasonable value. The model in Ref. [12] also predicts the transition from a weak to a strong random magnetic field at $\sigma_{xx} \sim h/4e^2$, which also corresponds to the position of the above-mentioned features at B_c, which are indicated by the arrows in Fig. 2.11B. For strong disorder, $\alpha \gg 1$, the theory in Ref. [12] predicts that $\sigma_{xx} \sim \alpha^{-1/2}(\ln \alpha)^{-1/4}$, while the model in Ref. [13] provides a different result. Although the experiment does not allow us to establish which of these approximations is more realistic, they do both predict a rapid rise in the magnetoresistance with the magnetic field, in accord with our observations. Yet another important comment should be made regarding the approximation of a strong magnetic field. We expect that domains with a nonuniform charge density within the sample should vanish once a critical field is attained where the magnetic length becomes smaller than the characteristic size of these domains and the spiral states should be suppressed. Under these conditions, the conductivity with a variable jump length between islands with different filling factors should predominate. Another possible scenario in a high magnetic field is the simultaneous occurrence in the transport of jump conductivity with a variable jump length and conductivity along spiral states and oppositely directed trajectories. But the final choice between these models requires a more detailed theory.

We now consider the similarities and differences between the behavior of our system in the quantum Hall effect and the quantum Hall effect in graphene. Here, there is some similarity in terms of the transport properties at the charge neutrality point [14,15]. But there is an important difference in that electrons and holes are present in graphene at the Dirac point only in the magnetic field, while an HgTe quantum well is always filled with electrons and holes. This means that we can study the properties of this system when $B < B_c$, and compare them to the theory of spiral states for a low-amplitude random magnetic field. At fields above B_c the situation becomes similar to graphene, although the nature of the gap at $\nu = 0$ is very different.

2.6.1 Cyclotron Resonance in a Two-Dimensional Semimetal

In this section the experimental results of microwave cyclotron resonance (CR) study of electrons and holes in 20.5 nm HgTe quantum wells are presented. The resonance has been studied by measuring microwave photoresistance in the frequency range of 35–170 GHz. The effective cyclotron masses of electrons and holes have been determined. The studied experimental samples were the same as for magnetotransport measurements. The principal feature of the samples is the presence of microwave semitransparent gates fabricated by depositing thin 10 nm Ti/Au films. Cyclotron photoresistance appears in the system due to the CR absorption of microwave radiation, and the corresponding resonance heating of the system.

The magnetic field dependence $R_{ph}(B)$ of the photoresistance signal is a curve with one (in the case of metal) or two (in the case of semimetal) main peaks, whose positions depend on the frequency of the incident radiation and the type and density of carriers. As is clearly seen in Fig. 2.13A the field dependence of the CR photoresistance for two-dimensional electrons has the form of a broadened resonance whose position ($B_{peak} = 0.06–0.17$ T) depends not only on the microwave frequency, but also on the gate voltage. It can also be seen that

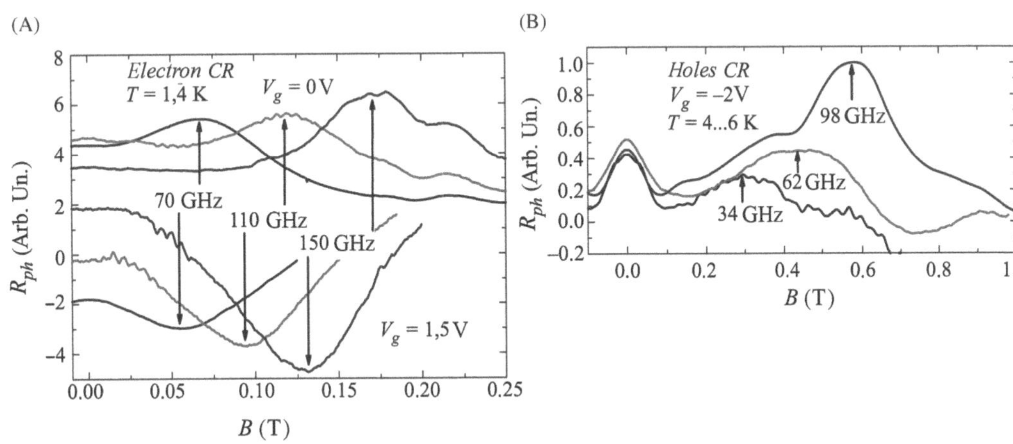

FIGURE 2.13 (A) The cyclotron resonance photoresistance signal $R_{ph}(B)$ at incident frequency $f = 70$, 110, and 150 GHz for gate voltage $Vg = 0$ (semimetal state, upper panel) and 1.5 V (metallic state, lower panel). (B) The cyclotron resonance photoresistance signal $R_{ph}(B)$ at fixed gate voltage $V_g = -2$ V and incident radiation frequency $f = 34$, 62, and 98 GHz.

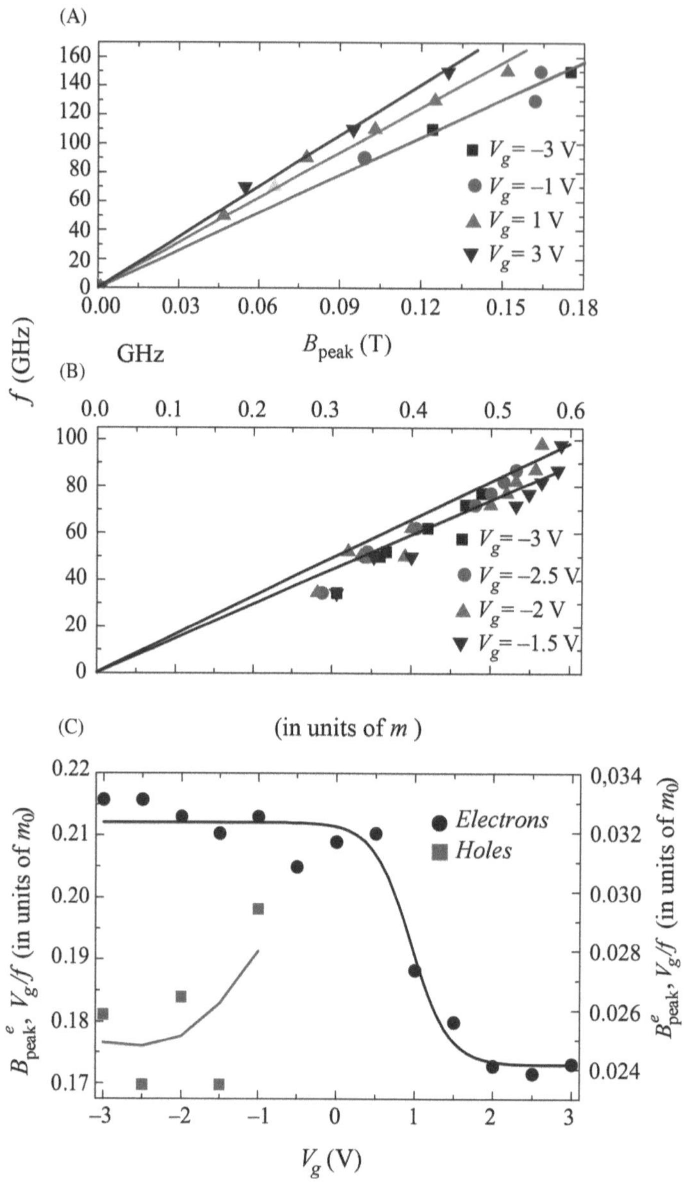

FIGURE 2.14 Frequency dependence of the position of the cyclotron resonance photoresistance peak at different gate voltages for (A) electrons and (B) holes. The symbols and solid lines are experimental points and linear fittings, respectively. (C) Slope of the linear dependences in panels (A) and (B) in units of the bare electron mass (multiplied by $e/2\pi m_0$) versus the gate voltage ($B_{peak}(V_g)/f$) for the (right vertical axis) electron and (left vertical axis) hole cyclotron resonance.

the sign of the CR photoresistance is given by the sign of dR/dT, certainly indicating the warming up mechanism of the CR photoresistance in the system under investigation. As expected, the peaks of the hole CR appear in much higher magnetic fields ($B_{peak} = 0.3-0.6$ T) (Fig. 2.13B). The experimental dependence of the peak position of the CR photoresistance on the incident radiation frequency is shown in Fig. 2.14A and B.

As is shown, the dependence for electrons is well approximated by a straight line passing through zero. This line corresponds to the coincidence of microwave frequency f and the cyclotron frequency given by formula $\omega_c = eB/m^*$, where m^* is the effective mass of electron or hole. In contrast to electrons, some points for holes corresponding to low-frequency measurements lie below the fitting line passing through zero and high-frequency points. This deviation is most probably associated both with a large nonlinearity of the temperature coefficient of resistance in the ranges of gate voltages $V_g = -3-0$ V and magnetic fields $B = 0.2-0.4$ T, which leads to the distortion of the shape of the hole CR photoresistance peak, with stronger broadening of the peak at low frequencies.

The dependence of slopes B_{peak}/f of straight lines in Fig. 2.14A and B on gate voltage V_g is shown in Fig. 2.14C. Let us first analyze the behavior of B_{peak}/f for electrons. When the position of the Fermi level corresponds to the free electron metal, the slope weakly depends on the gate voltage and corresponds to the effective cyclotron mass $m_e = 0.024m_0$, which well agrees with the previously measured value in doped HgTe of the same thickness [16] at $N_s < 3 \times 10^{11}$ cm^{-2}.

However, as soon as the Fermi level falls into the band crossing region (the semimetallic state), B_{peak}/f initially exhibits a noticeable increase, and then it is saturated. Such an increase could be attributed to the electron–hole interaction. However, the electron–hole interaction in the semimetal is known to lead only to the broadening of the cyclotron line [17]. Thus, the above-mentioned shift of the CR peak cannot be explained by the electron hole interaction. The second possible reason can be associated with plasma effects caused by the interaction between the electron and hole plasma modes.

Let us now discuss the hole CR. The effective cyclotron masses of the two-dimensional holes found from the values of B_{peak}/f range within $0.17-0.2m_0$ with a mean value of $0.18m_0$. This coincides with the estimate of the effective mass from the measurements of the density of states [2]. No specific conclusions about the dependence of B_{peak}/f on the gate voltage for two-dimensional holes can be drawn from Fig. 2.14 in view of large experimental errors.

References

[1] Z.D. Kvon, E.B. Olshanetsky, D.A. Kozlov, N.N. Mikhailov, S.A. Dvoretsky, Two-dimensional electron–hole system in a HgTe-based quantum well, JETP Lett. 87 (2008) 502.

[2] E.B. Olshanetsky, Z.D. Kvon, M.V. Entin, L.I. Magrill, N.N. Mikhailov, S.A. Dvoretsky, Scattering processes in a two-dimensional semimetal, JETP Lett. 89 (2009) 290.

[3] Z.D. Kvon, E.B. Olshanetsky, N.N. Mikhailov, D.A. Kozlov, Two-dimensional electron systems in HgTe quantum wells, Low Temp. Phys. 35 (2009) 6.

[4] Z.-D. Kvon, S.N. Danilov, N.N. Mikhailov, S.A. Dvoretsky, W. Prettl, S.D. Ganichev, Cyclotron resonance photoconductivity of a two-dimensional electron gas in HgTe quantum wells, Phys. E 40 (2008) 1885.

[5] K. Ortner, X.C. Zhang, A. Pfeuffer-Jeschke, C.R. Becker, G. Landwehr, L.W. Molenkamp, Valence band structure of HgTe/$Hg_{1-x}Cd_xTe$ single quantum wells, Phys. Rev. B 66 (2002) 075322.

[6] X.C. Zhang, A. Pfeuffer-Jeschke, K. Ortner, V. Hock, H. Buhmann, C.R. Becker, et al., Rashba splitting in n-type modulation-doped HgTe quantum wells with an inverted band structure, Phys. Rev. B 63 (2001) 245305.

[7] N.N. Berchenko, M.V. Pashkovskii, Mercury telluride—a zero-gap semiconductor, Sov. Phys. Usp. 19 (1976) 462.

[8] V.F. Gantmakher, Y.B. Levinson, Carrier Scattering in Metals and Semiconductors. Modern Problems in Condensed Matter Sciences, vol. 19, North-Holland, Amsterdam, Oxford, New York, Tokio, 1987, p. 459. Seiten. Dfl. 280.00. ISBN 0−444−87025−3.

[9] V.F. Gantmakher, V.T. Dolgopolov, Temperature dependence of the electron and hole mean free paths in antimony, JETP 33 (1971) 1215.

[10] J.E. Muller, Effect of a nonuniform magnetic field on a two-dimensional electron gas in the ballistic regime, Phys. Rev. Lett. 68 (1992) 385.

[11] D.K.K. Lee, J.T. Chalker, D.Y.K. Ko, Localization in a random magnetic field: the semiclassical limit, Phys. Rev. B 50 (1994) 5272.D.K.K. Lee, J.T. Chalker, Unified model for two localization problems: electron states in spin-degenerate Landau levels and in a random magnetic field, Phys. Rev. Lett. 72 (1994) 1510; S.C. Zhang, D.P. Arovas, Effective field theory of electron motion in the presence of random magnetic flux, Phys. Rev. Lett. 72 (1994) 1886.

[12] F. Evers, A.D. Mirlin, D.G. Polyakov, P. Wolfle, Semiclassical theory of transport in a random magnetic field, Phys. Rev. B 60 (1999) 8951.

[13] D.V. Khveshchenko, Magnetoresistance of two-dimensional fermions in a random magnetic field, Phys. Rev. Lett. 77 (1996) 1817.

[14] D.A. Abanin, K.S. Novoselov, U. Zeitler, P.A. Lee, A.K. Geim, L.S. Levitov, Dissipative quantum Hall effect in graphene near the Dirac point, Phys. Rev. Lett. 98 (2007) 196806.

[15] J.G. Checkelsky, L. Li, N.P. Ong, Zero-energy state in graphene in a high magnetic field, Phys. Rev. Lett. 100 (206801) (2008); J.G. Checkelsky, L. Li, N.P. Ong, Divergent resistance at the Dirac point in graphene: evidence for a transition in a high magnetic field, Phys. Rev. B 79 (115434) (2009).

[16] E. Vasiliadou, G. Muller, D. Heitmann, et al., Collective response in the microwave photoconductivity of Hall bar structures, Phys. Rev. B 48 (1993) 1714.

[17] S. Sassine, Yu Krupko, E.B. Olshanetsky, et al., Microwave radiation induced collective response in Si/SiGe heterostructures with a 2D electron gas, Solid State Commun. 142 (11) (2007) 631.

[18] V.L. Bonch-Bruevich, S.G. Kalashnikov, Physics of Semiconductors, VEB, Berlin, 1982.

3

Nonlinear Two-Dimensional Electron Conductivity at High Filling Factors

A.A. Bykov[1], S.A. Vitkalov[2]

[1]RZHANOV INSTITUTE OF SEMICONDUCTOR PHYSICS OF THE SIBERIAN BRANCH OF THE RUSSIAN ACADEMY OF SCIENCES, NOVOSIBIRSK, RUSSIA [2]CITY COLLEGE OF THE CITY UNIVERSITY OF NEW YORK, NEW YORK, NY, UNITED STATES

3.1 Introduction

The recent interest in a comprehensive study of nonlinear magnetotransport in two-dimensional (2D) electron systems was stimulated by an observation of the Zener tunneling of highly mobile 2D electrons between Landau levels, which is induced by the Hall electric field in GaAs/AlGaAs heterojunctions [1]. The effect was originally found in Hall bar geometry and appeared at high filling factors as oscillations of differential resistance $r_{xx}(B)$ induced by dc electric current I_{dc}. The positions of the oscillations in the magnetic field B obeyed the following relation: $\gamma R_c e E_H = l\hbar\omega_c$, where $\gamma \approx 2$, l is an integer, ω_c is the cyclotron frequency, R_c is the cyclotron radius, and E_H is Hall electric field. Later, the Zener oscillations of magnetoresistance r_{xx} were found in highly-doped GaAs quantum wells [2], in double quantum wells [3], and in hole gas [4].

Another intriguing nonlinear phenomenon, observed in 2D electron systems placed in crossed electric and quantizing magnetic fields, is the electronic state with zero-differential resistance $r_{xx} = 0$ (ZDR state) [5–7]. The experimental data have demonstrated that, in Hall bar geometry, the initial decrease of longitudinal differential resistance r_{xx} with applied dc current I_{dc} terminates at $I_{dc} = I_{th}$, corresponding to $r_{xx} = 0$. At $I_{dc} > I_{th}$ the differential resistance stays at zero value in a wide range of electric currents I_{dc}, significantly exceeding the threshold value I_{th}. The initial drop of the resistance is associated with a quantal heating induced by the spectral diffusion of 2D electrons in crossed electric and magnetic fields [8–11]. The transition into the ZDR state is attributed to the local instability of the electric current at $I_{dc} > I_{th}$ [12].

Two distinct sample geometries are widely used in studying the magnetotransport in 2D electron systems: Hall bar geometry and Corbino geometry. In Hall bar geometry, components of the resistivity tensor ρ are measured. Components of the conductivity tensor σ are then computed using the relations $\sigma_{xx} = \rho_{xx}/(\rho_{xy}{}^2 + \rho_{xx}{}^2)$ and $\sigma_{xy} = \rho_{xy}/(\rho_{xy}{}^2 + \rho_{xx}{}^2)$. The Corbino geometry is less universal, since only the diagonal component σ_{xx} of the conductivity tensor can be measured. In the linear regime, the measurements of the conductivity of

Advances in Semiconductor Nanostructures. DOI: http://dx.doi.org/10.1016/B978-0-12-810512-2.00003-2

uniform and isotropic conductors are equivalent in both geometries. Measurements of the nonlinear response can yield significantly different results because of the distinctly different electric field distribution in the two geometries.

3.2 Nonlinear Properties of 2D Electron Corbino Disks

This chapter presents the results of experimental investigations of nonlinear properties of 2D electron Corbino disks. The goals of the investigation are the observation of Zener tunneling between Landau levels, and the electronic state with zero-differential conductance (ZDC) in the regime when the Hall electric field is absent. The experiments show that the conductance oscillations, which are induced by Zener tunneling, occur only in "narrow" 2D rings. In "wide" rings, the oscillations are strongly damped due to an averaging of the oscillation amplitude induced by the electric field variations in the radial direction. In contrast to the resistance oscillations in the Hall bar geometry, which are periodic in the inverse magnetic fields, the Zener oscillations in 2D rings are periodic in $1/B^2$. This is related to the fact that the Hall electric field increases linearly with B, whereas the electric field in Corbino disks does not depend on the magnetic field. Additionally, experimental data shows that, at low temperature $T \sim 1.6$ K, the differential conductance of the Corbino disks reaches a zero value in a wide range of applied dc voltages.

The heterojunctions studied are the single GaAs quantum wells which are sandwiched between AlAs/GaAs superlattices [13,14]. The width of the GaAs quantum wells is 13 nm. The structures are grown by molecular beam epitaxy on (100) GaAs substrates. Fig. 3.1 shows a schematic picture of the Corbino disks. Electric contacts are formed by the deposition, optical lithography, and annealing of AuGe eutectic. The radius of the inner (outer) contact is $r_1(r_2)$. Two types of the rings are studied: "narrow" and "wide," shown in Fig. 3.1. Rings of the same type demonstrate the same results. This chapter presents the results for two samples: (1) "narrow" ring, and (2) "wide" ring.

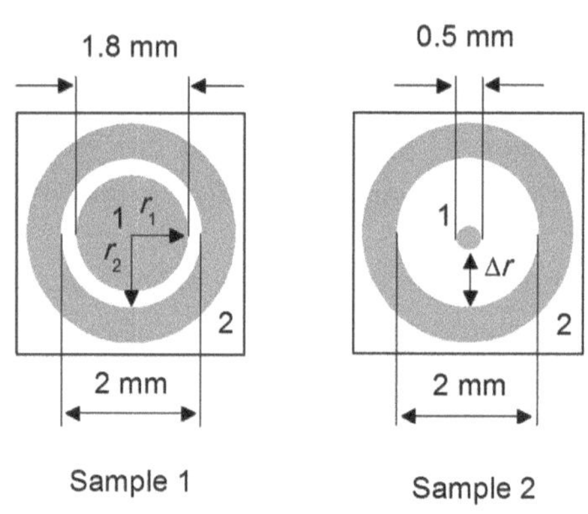

FIGURE 3.1 Schematic picture of "narrow" (sample 1) and "wide" (sample 2) Corbino disks. *White color* indicates 2D electron gas regions. Electric contacts are *gray*. Numbers 1 and 2 label the inner and outer electric contacts.

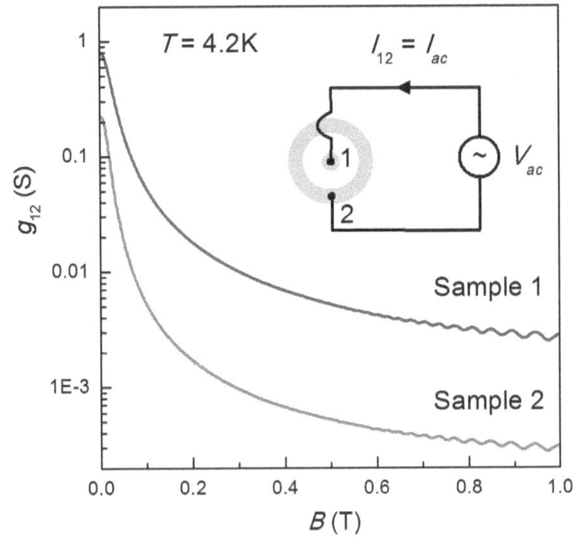

FIGURE 3.2 Dependencies $g_{12}(B)$ for "narrow" (Sample 1), and "wide" (Sample 2) Corbino rings at $T = 4.2$K. The scheme of conductance measurements is shown in the inset.

The conductance of Corbino disks $g_{12} = I_{ac}/V_{ac}$ (Fig. 3.2) is determined by measurements of alternating current I_{ac} at a frequency from 10 Hz to 1 KHz using a lockin amplifier. The amplitude of ac voltage V_{ac}, applied between inner and outer contacts, was kept constant. The voltage amplitude, providing the linear response of the 2D electron systems, was less than 1 mV. Differential conductance g_{12} was measured the same way by applying a dc voltage V_{dc} in addition to voltage V_{ac} (Fig. 3.3). The measurements were taken at temperatures 1.6K and 4.2K in magnetic fields $B < 2$ T. The density and the mobility of 2D electrons are $n_e \approx 8.0 \times 10^{15}$ m^{-2}, $\mu \approx 150$ m^2/Vs at $T = 4.2$K.

Fig. 3.2 presents magnetic field dependences of the conductance of "narrow" and "wide" 2D electron Corbino disks. At $T = 4.2$K and $B > 0.5$ T, Shubnikov de Haas (SdH) conductivity oscillations are observed. At smaller magnetic fields, the longitudinal conductivity decreases significantly with field B. Fig. 3.3 shows that, at small magnetic fields, an application of the dc voltage induces conductance oscillations, which are considerably stronger in the "narrow" Corbino disks. An analysis of positions of the oscillation maxima reveals $1/B^2$ periodicity. This is shown in the inset to Fig. 3.3B. The appearance of the oscillating differential conductance of Corbino disks in response to V_{dc} suggests that the dc bias-induced Zener transitions are a possible origin of the observed oscillations.

At strong magnetic fields, the electric field applied to high-mobility 2D electrons in the Hall bar geometry is mostly Hall electric field $E_H \propto J_{dc}B$, and is directed perpendicular to electric current density J_{dc}. The following relation determines the position of the Zener oscillations [1]: $B_l = \gamma(2\pi/n_e)^{1/2}(m_e^*/e^2)(J_{dc}/l)$, where m_e^* is the electron effective mass. In Corbino disks, the electric field is determined by the voltage applied between the inner and outer contacts $V_{12} = V_{dc}$ and the distance between them. The electric field $E_{dc}(r)$ is not uniform and

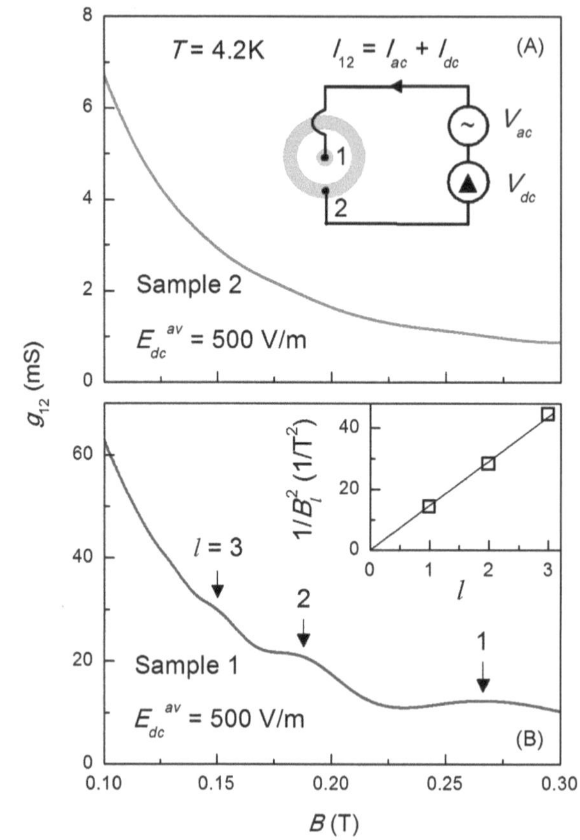

FIGURE 3.3 (A) Dependence $g_{12}(B)$ for a "wide" (Sample 2) Corbino disk at $E_{dc}{}^{av} = 500$ V/m; $T = 4.2$K. The measurement scheme of the differential conductance is demonstrated in the inset. (B) Dependence $g_{12}(B)$ for a "narrow" (Sample 1) Corbino disk at $E_{dc}{}^{av} = 500$ V/m; $T = 4.2$K. The positions of maxima for $l = 1$, 2, and 3 are indicated with arrows. The dependence of $1/B_l^2$ on l. is shown in the inset.

decreases with the distance from the inner contact. The mean value of the electric field is $E_{dc}{}^{av} = V_{dc}/(r_2 - r_1)$. In terms of electric field E_{dc}, the conditions for Zener transitions occur at $B_l^2 = \gamma(2\pi n_e)^{1/2}(m^*/e)(E_{dc}/l)$. This behavior is observed in "narrow" disks in which the electric field E_{dc} weakly depends on radius r and is equal to $E_{dc}{}^{av}$. In "wide" Corbino rings, electric field $E_{dc}(r)$ varies considerably with radius r and the Zener oscillations are averaged out.

Taken at a fixed magnetic field $B = 0.1$ T, dependencies of the conductance on the mean electric field demonstrate qualitatively similar behavior (Fig. 3.4). Fig. 3.5A presents dependencies $g_{12}(E_{dc}{}^{av})$ at B differing from 0.2 to 0.3 T. The figure shows that the increasing magnetic field shifts value $E_{dc}{}^{av}$, corresponding to Zener maximum $l = 1$, to a higher electric field. Fig. 3.5B shows that value $E_{dc}{}^{av}$ of maximum $l = 1$ is proportional to B^2. A numerical fit of the shown dependence indicates the slope $\gamma \approx 1.8$, which is close to 2, as expected for the Zener transitions between Landau levels at high filling factors [1,2].

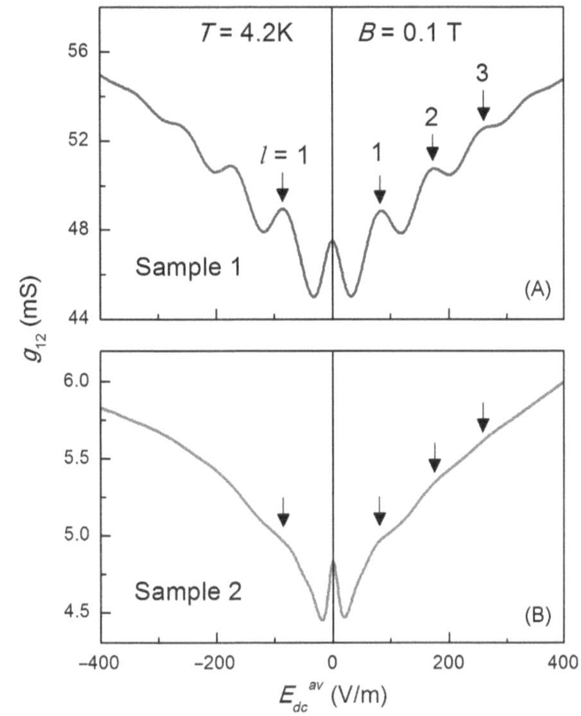

FIGURE 3.4 Dependencies $g_{12}(E_{dc}^{av})$ for "narrow" (A) and "wide" (B) Corbino disks at $T = 4.2K$ and $B = 0.1$ T. The positions of maxima for $l = 1$, 2, and 3 are indicated with arrows.

3.2.1 Nonlinear conductance in "narrow" Corbino disks

Figs. 3.6 and 3.7 present dependences of the differential conductance $g_{12}(B)$ of 2D electrons in the "narrow" Corbino disks on the magnetic field B taken at $T = 1.6K$ at different E_{dc}, as labeled. For the studied samples, the width of the conducting ring Δr was much less than the averaged radius of the ring: $\Delta r = (r_2 - r_1) \ll (r_2 + r_1)/2$. Due to this property, E_{dc} between contacts was nearly independent from t radius r and equal to $E_{dc} = V_{12}/\Delta r$. At $E_{dc} = 0$, the conductance $g_{12}(B)$ demonstrates SdH oscillations in $B > 0.3$ T, as shown in Fig. 3.6. An application of electric field $E_{dc} = 250$ V/m decreases the amplitude of the quantum oscillations significantly and, at strong magnetic fields, the conductance of the structure approaches values that are very close to zero. As shown in Fig. 3.7, a further increase of the dc electric field produces additional peaks in the dependence $g_{12}(B)$, which are labeled by arrows. These maxima result from Zener tunneling between Landau levels, which are induced by applied electric field E_{dc}. The positions of the maxima obey the relation $\gamma R_c e E_{dc} = l\hbar\omega_c$, shown in the inset of Fig. 3.7.

Fig. 3.8 presents the dependences of $g_{12}(E_{dc})$ for different magnetic fields, as labeled. At magnetic field $B = 0.261$ T, the initial drop of the differential conductance with the E_{dc} is due

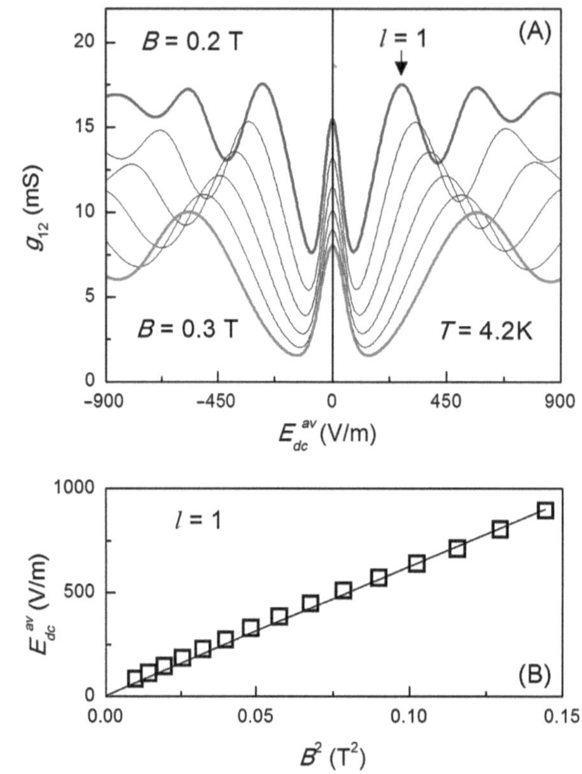

FIGURE 3.5 (A) Dependencies $g_{12}(E_{dc}{}^{av})$ for the "narrow" (Sample 1) Corbino disk at different magnetic fields from 0.2 to 0.3 T with step 0.02 T; $T = 4.2$K. The position of maximum for $l = 1$ is indicated with the arrow. (B) Dependence of the position of the maximum for $l = 1$ on B^2. Straight line corresponds to relation $\gamma R_c e E_{dc} = \hbar\omega_c$ with $\gamma = 1.8$.

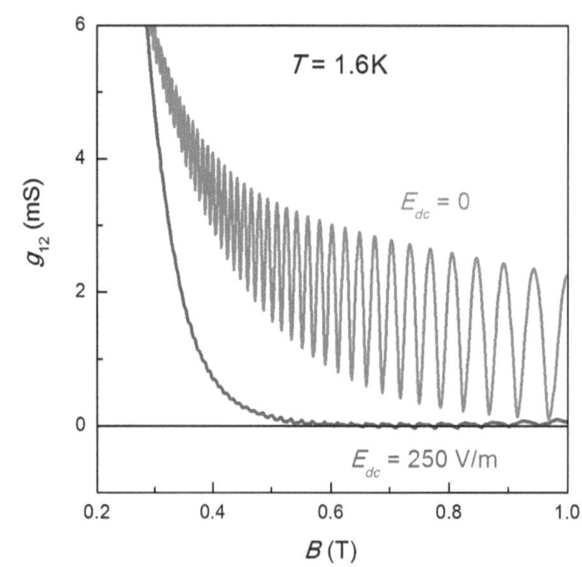

FIGURE 3.6 Dependence of conductance g_{12} of a 2D electron Corbino disk on a magnetic field at temperature $T = 1.6$K at different dc electric fields, as labeled.

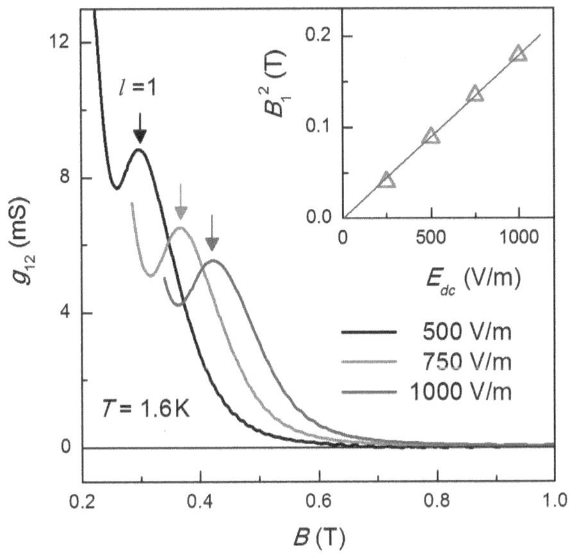

FIGURE 3.7 Dependence of conductance g_{12} of a 2D electron Corbino disk on a magnetic field at temperature $T = 1.6$K at different dc electric fields, as labeled. The positions of maximum B_l at $l = 1$ in different electric fields are indicated with arrows. The inset presents the dependence of $B_1{}^2$ on dc electric field E_{dc}. The solid line corresponds to the relation $\gamma R_c e E_{dc} = \hbar\omega_c$.

to the intralevel quantal heating [8,11]. The increase of the differential conductance at higher electric field is related to interlevel electron transitions [1,15]. The maximum marked by the arrow in Fig. 3.8A corresponds to Zener tunneling between Landau levels at $l = 1$. At a higher magnetic field $B = 0.847$ T, the differential conductance demonstrates similar behavior at small electric fields, but, at higher dc biases, the conductance retains values near zero, $g_{12} \approx 0$, in a wide range of electric field E_{dc}. This is the ZDC state. As shown in the figure, the state with $g_{12} = 0$ does not occur at $T = 4.2$K.

Fig. 3.9 presents $I-V$ dependences of the "narrow" Corbino disk for two different magnetic fields, as labeled. The figure shows that, when the 2D electron systems enters the state with ZDC, electric current I_{dc} saturates and becomes independent from electric field E_{dc}. A comparison between dependences $g_{12}(E_{dc})$ and $I_{dc}(V_{dc})$ taken at $T = 1.6$K and magnetic field $B = 0.847$ T indicate that electric current I_{dc} reaches a saturation value I_s at electric field $E_{dc} > E_{th}$.

We consider that, in the studied Corbino disks, the $g_{12} = 0$ state occurs due to a local instability of electric field E_{dc} [12]. The dominant nonlinear mechanism, leading to the instability, is a peculiar Joule heating (quantal heating), which occurs in systems with a discrete spectrum [8,11]. The instability develops at the conditions of a negative differential conductivity corresponding to the negative slope of the N-shaped $I-V$ dependence shown in the inset of Fig. 3.9. In the case of the N-shaped $I-V$ dependence, a spatially uniform electric field distribution is not stable and, typically, should evolve into a structure containing domains of a weak E_1 and a strong E_2 [16]. At these conditions, both moving and static

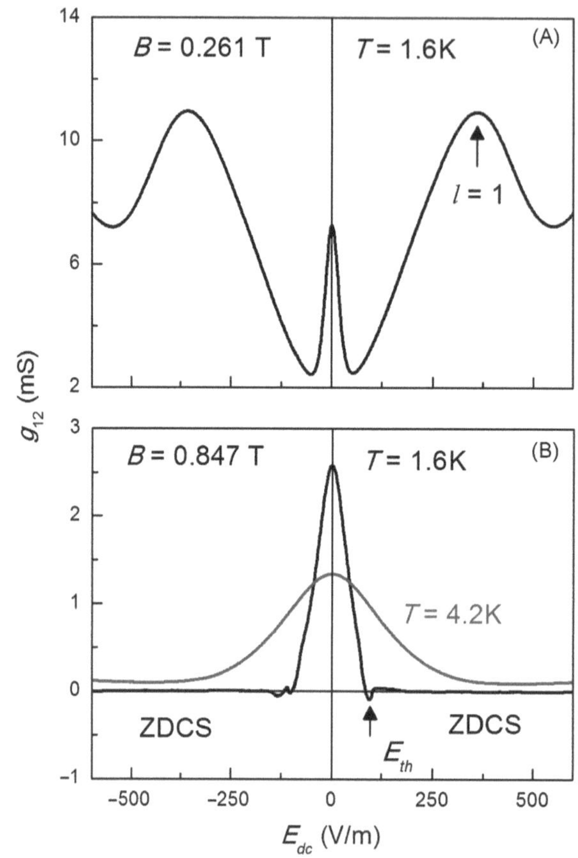

FIGURE 3.8 (A) Dependence of differential conductance g_{12} on dc electric field E_{dc} in magnetic field $B = 0.261$ T at $T = 1.6$K. The maximum corresponding to Zener transition at $l = 1$. (B) is indicated by the arrow. Dependence of differential conductance g_{12} on electric field E_{dc} in magnetic field $B = 0.847$ T at different temperatures is shown, as labeled.

domains may occur. In the first case, in a conductor with a fixed voltage applied, there are electric current oscillations. This is known as the Gunn effect [17]. In the case of static domains, the constant electric current saturates with the applied voltage. There is a similarity between nonlinear transport in Gunn diodes and in the 2D electron systems presented in this chapter. We note, however, that despite the similarity, the nonlinear mechanisms leading to the local instability of electric field E_{dc} are different in these two systems.

In summary, this chapter presents experimental investigations of the effect of dc electric fields on the conductance of 2D electron systems placed in a strong perpendicular magnetic field. The experiments show that the application of the electric field to 2D Corbino disks, fabricated from selectively doped heterostructures GaAs/AlAs, induces oscillations of the conductance which are due to Zener tunneling between Landau orbits, and the electronic state with zero-differential conductance occurs due to a local electric field instability [18,19].

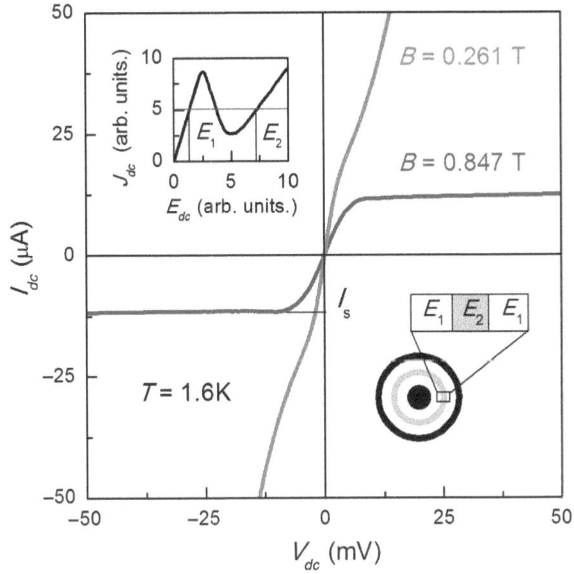

FIGURE 3.9 Dependence of electric current I_{dc} on dc voltage V_{dc} at temperature T = 1.6K in different magnetic fields, as labeled. The inset in the upper left corner shows the suggested N-shaped dependence $J_{dc}(E_{dc})$, indicating two electric fields E_1 and E_2 corresponding to the same value J_{dc}. The possible distribution of the electric field corresponding to the electron state with zero-differential conductance in a 2D Corbino disk is shown in the inset in the lower right corner.

References

[1] C.L. Yang, J. Zhang, R.R. Du, J.A. Simmons, J.L. Reno, Zener tunneling between Landau orbits in a high-mobility two-dimensional electron gas, Phys. Rev. Lett. 89 (2002) 076801.

[2] A.A. Bykov, J.-Q. Zhang, S. Vitkalov, A.K. Kalagin, A.K. Bakarov, Effect of dc and ac excitations on the longitudinal resistance of a two-dimensional electron gas in highly doped GaAs quantum wells, Phys. Rev. B 72 (2005) 245307.

[3] A.A. Bykov, Zener tunneling between Landau levels in a double quantum well at high filling factors, JETP Lett. 88 (2008) 394.

[4] Y. Dai, Z.Q. Yuan, C.L. Yang, R.R. Du, M.J. Manfra, L.N. Pfeiffer, et al., Magnetotransport in Zener tunneling regime in a high-mobility two-dimensional hole gas, Phys. Rev. B 80 (2009) 041310.

[5] A.A. Bykov, J.-Q. Zhang, S. Vitkalov, A.K. Kalagin, A.K. Bakarov, Zero-differential resistance state of two-dimensional electron systems in strong magnetic fields, Phys. Rev. Lett. 99 (2007) 116801.

[6] A.T. Hatke, H.-S. Chiang, M.A. Zudov, L.N. Pfeiffer, K.W. West, Zero differential resistance in two-dimensional electron systems at large filling factors, Phys. Rev. B 82 (2010) 041304(R).

[7] A.A. Bykov, E.G. Mozulev, S.A. Vitkalov, Zero differential resistance in a double quantum well at high filling factors, JETP Lett. 92 (2010) 475.

[8] I.A. Dmitriev, M.G. Vavilov, I.L. Aleiner, A.D. Mirlin, D.G. Polyakov, Theory of microwave-induced oscillations in the magnetoconductivity of a two-dimensional electron gas, Phys. Rev. B 71 (2005) 115316.

[9] J.-Q. Zhang, S. Vitkalov, A.A. Bykov, A.K. Kalagin, A.K. Bakarov, Effect of a dc electric field on the longitudinal resistance of two-dimensional electrons in a magnetic field, Phys. Rev. B 75 (2007) 081305(R).

[10] N.R. Kalmanovitz, A.A. Bykov, S. Vitkalov, A.I. Toropov, Warming in systems with a discrete spectrum: spectral diffusion of two-dimensional electrons in a magnetic field, Phys. Rev. B 78 (2008) 085306.

[11] J.-Q. Zhang, S. Vitkalov, A.A. Bykov, Nonlinear resistance of two-dimensional electrons in crossed electric and magnetic fields, Phys. Rev. B 80 (2009) 045310.

[12] A.V. Andreev, I.L. Aleiner, A.J. Millis, Dynamical symmetry breaking as the origin of the zero-dc-resistance state in an ac-driven system, Phys. Rev. Lett. 91 (2003) 056803.

[13] K.J. Friedland, R. Hey, H. Kostial, R. Klann, K. Ploog, New concept for the reduction of impurity scattering in remotely doped GaAs quantum wells, Phys. Rev. Lett. 77 (1996) 4616.

[14] D.V. Dmitriev, I.S. Strygin, A.A. Bykov, S. Dietrich, S.A. Vitkalov, Transport relaxation time and quantum lifetime in selectively doped GaAs/AlAs heterostructures, JETP Lett. 95 (2012) 420.

[15] M.G. Vavilov, I.L. Aleiner, L.I. Glazman, Nonlinear resistivity of a two-dimensional electron gas in a magnetic field, Phys. Rev. B 76 (2007) 115331.

[16] B.K. Ridley, Specific negative resistance in solids, Proc. Phys. Soc. 82 (1963) 954.

[17] J.B. Gunn, Microwave oscillations of current in III–V semiconductors, Solid State Commun. 1 (1963) 88.

[18] A.A. Bykov, D.V. Dmitriev, I.V. Marchishin, S. Byrnes, S.A. Vitkalov, Zener tunneling between Landau orbits in two-dimensional electron Corbino rings, Appl. Phys. Lett. 100 (2012) 251602.

[19] A.A. Bykov, S. Byrnes, S. Dietrich, S. Vitkalov, I.V. Marchishin, D.V. Dmitriev, Zero-differential conductance of two-dimensional electrons in crossed electric and magnetic fields, Phys. Rev. B 87 (2013) 081409(R).

4

Silicon-Based Nanoheterostructures With Quantum Dots

A.V. Dvurechenskii[1,2], A.I. Yakimov[1,2]

[1]RHZANOV INSTITUTE OF SEMICONDUCTOR PHYSICS OF THE SIBERIAN BRANCH OF THE RUSSIAN ACADEMY OF SCIENCES, NOVOSIBIRSK, RUSSIA [2]NOVOSIBIRSK STATE UNIVERSITY, NOVOSIBIRSK, RUSSIA

THIS CHAPTER REVIEWS THE MAIN RESULTS ON GeSi QUANTUM DOT HETEROSTRUCTURE IN THE FOLLOWING DIRECTIONS:

1. Homogeneity and density of the arrays of quantum dots tuned with low-energy ion-beam actions at epitaxy and postgrowth pulsed laser annealing. Approaches to grow spatially ordered groups of quantum dots and quantum dot molecules.
2. Electronic structure of Si-based quantum dots. Hole, electron, and excitonic states in single and multiple quantum dots.
3. Hole transport in dense arrays of quantum dots.
4. Electron spin states and spin relaxation in ensembles of tunnel-coupled quantum dots.
5. Quantum dot photodetectors for near and mid-infrared operation.

4.1 Introduction

Modern semiconductor microelectronics is mainly based on silicon structures. The advances in silicon technology have been the main driver for today's electronics and photovoltaics industries. After the incorporation of many new materials, the core technological platform for classical computation remains based on silicon. In photonics, silicon dominates solar cell applications. At the same time, it is becoming increasingly evident that integrated photonics is an upcoming area, expected to become almost as important as electronic chip technology. The most application areas have become possible as a result of the continuous striving for miniaturization where nanofabrication becomes a natural avenue. To expand the range of required material properties, the coupling of silicon with other materials is an established way to use advanced silicon technology.

The most attractive approach is based on heteroepitaxy of different materials, which has already provided a large class of heterostructures with unique functional characteristics [1,2]. Concerning heteroepitaxy on silicon, there are many materials currently used for research, beginning with monoatomic semiconductors (Ge and C), through semiconductor compounds (GaAs, InAs, AlAs, GaN, AlN, CdTe, HgTe, PbTe, and SnTe), and ending with solid solutions

Advances in Semiconductor Nanostructures. DOI: http://dx.doi.org/10.1016/B978-0-12-810512-2.00004-4

based on semiconductor materials. A mismatch of lattice parameters of two materials leads to the emergence of nonhomogeneous elastic strain fields, resulting in several effects: (1) formation of defects during relaxation of elastic strains (dislocations); (2) strains are the driving force for the transition from a continuous film growth to surface islanding and their self-organization maintaining size and shape during heteroepitaxy; (3) strains strongly modify the electron properties of the heterostructures. One of the most important problems of heteroepitaxy is the formation of defect-free (dislocation-free) heterostructures. It is often impossible to avoid defect generation completely; in this case, the challenge is to reduce the density of defects in the working region of the heterostructure as much as possible, in order to minimize deterioration of electrophysical and optical properties of the material.

Heterostructures with similar values to the lattice constant (e.g., AlGaAs/GaAs [1]) have practically no defects. Quantum dot (QD) nanoheterostructures free from dislocations can be formed owing to the small size of nanocrystals coherently embedded into the crystal lattice of another material [2,3]. The idea of using morphological evolution of the surface during the growth of lattice mismatched heteroepitaxial systems for producing an array of nanoscale crystallites with a transition from two-dimensional to three-dimensional growth based on the Stranski–Krastanov mode turned out to be very fruitful. This idea was implemented for the first time in 1992 for the Ge/Si system [4]. Based on those results, it was concluded that a new class of nanostructures had been found, which possess the electron properties inherent in zero-dimensional systems [4,5]. The technology of silicon-based dislocation-free Ge/Si QD heterostructures was developed for further successive investigations [6].

Another class of structures includes Si-based quantum well nanoheterostructures with Si and Ge quantum wells. Sufficiently perfect structures of this kind were formed by molecular beam epitaxy on relaxed SiGe buffer layers with a gradient of the composition over the thickness, which ensured reduction of the dislocation density, and high values of carrier mobilities in quantum wells [7].

4.2 Homogeneity and Density of the Arrays of QDs

The challenges for the worldwide scientific community in the field of growth of QD nanostructures are: (1) to increase the homogeneity of the QDs array in terms of size, while retaining unified shapes and elementary composition; (2) to control the spatial sites of QDs (spatially ordered arrays) and the density of QDs for obtaining extremely dense or low-density arrays; (3) to reduce the density of extended (dislocations) and point defects in QD nanostructures. Of particular interest are small-sized QDs ($\leq$10 nm) as artificial atoms with a small number of quantum states. The large energy gap between discrete levels in such QDs ensures device operation at close-to-room temperatures. To increase the efficiency of lasers, light-emitting diodes, and photodetectors, it is necessary to provide the maximum concentration of electrons in the system. In view of the requirement of a small QD size, this condition meets the formation of arrays with the layer density about 10^{12} cm^{-2} at the characteristic QD size of the order of 10 nm.

At the present time, there are several approaches to tune QD density and QD size distribution. Most of these approaches are based on controlling the processes of nucleation and subsequent growth of QDs from molecular beams. Obviously, the QD density depends on the temperature and density of molecular beams. However, the dependence of the homogeneity of the QD ensemble on the flux of the material supplied onto the surface is less obvious. For the Ge/Si system, it was found that an increase in the growth rate by two orders of magnitude leads to a fivefold increase (up to $(3 \div 6) \times 10^{11}\ \mathrm{cm}^{-2}$) in the density of pyramidal shape Ge nanocrystals (hut-clusters with a base of 15 nm and a height of 1.5 nm), and to reduction of their size dispersion [8]. The greatest dispersion in terms of size (15%) was observed for structures grown at comparatively low temperatures ($\leq$500°C) and high deposition rates.

A high degree of homogeneity (within $\pm$3%) was observed in the case of high-temperature (600°C) growing of an ensemble of large QDs (shaped as dome-clusters with a base of 70 nm and a height of 15 nm) whose density is usually not very high ($3 \times 10^{9}\ \mathrm{cm}^{-2}$) [9]. Such QD densities are often insufficient for obtaining appropriate performance from various devices.

In addition to variation in the molecular beam parameters and substrate temperature, other approaches are also developed to control QD nucleation and growth. They include: (1) preliminary deposition of sub-monolayer coatings of doping elements (e.g., antimony or oxygen [10,11] for the Ge/Si system and Al for the InAs/GaAs system [9]; (2) irradiation by low-energy intrinsic ions during heteroepitaxy [12,13]; (3) preliminary growth of solid solution layers [14]; (4) formation of nucleation sites in the case of tilting of the growth surface from singular planes or etching of local regions after precision lithography operations [15].

4.2.1 Effects of Low-Energy Ion Beam Action During Epitaxy

Physical processes and phenomena inherent in interaction of low-energy ($\sim$100 eV) ions with solid surfaces form the fundamentals for the development of many directions in the physics of condensed matter. One of these directions is deposition of thin films with simultaneous irradiation by low-energy ions, or deposition directly from a low-energy ion beam. In this approach, the energy of particles in the low-energy ion beam is much more than the thermal energy of particles in the molecular beam. The use of "superthermal" particles for film deposition leads to essential changes in the film growth kinetics and resultant physical properties of films of various materials, including elementary semiconductors, solid solutions, semiconductor compounds, and metals [16–20]. These phenomena involve certain physical effects, such as reduction of the epitaxial growth temperature, reduction of the surface relief depth, a move in the critical film thickness at which a transition from two-dimensional layered to a three-dimensional mechanism of film growth occurs, and growth of ordered metastable phases during heteroepitaxy at a large lattice mismatch of the materials. Both intrinsic ion beams of the deposited material and ion beams of inert gases used for irradiation during the deposition were studied in experiments. As for Ge/Si structures, nucleation and growth of $Ge_{0.5}Si_{0.5}$ solid solutions on Si[100] from molecular beams with simultaneous irradiation by a continuous beam of low-energy (300 eV) ions of inert gases were studied [16]. It was found that irradiation by low-energy ions during film deposition provided a decrease in the epitaxy temperature, allowed smoother surfaces to be obtained, and allowed significant improvement in the film crystalline quality.

Significant progress in understanding microscopic mechanisms on the solid surface during epitaxy from ion-molecular beams was ensured by studying pulsed actions of low-energy ions. This approach was proposed and implemented in studying epitaxy on silicon from ion-molecular beams [19,20]. The basic underlying idea was to apply a short-time ion action in order to bring additional energy to surface atoms, and to study morphological and structural evolutions during subsequent film growth. It was shown that the pulsed low-energy ion action initiates surface reconstruction: metastable surface phases formed during the growth transform to more stable states. The investigations were performed on the Si[111] surface whose typical feature is coexistence of (7×7) and (5×5) superstructures during growth at temperatures above 500°C. It was shown that a pulsed ion beam action leads to the transition $(5 \times 5) \Rightarrow (7 \times 7)$ [18]. This approach allows the study of the kinetics of morphological evolution due to pulsed ion beam action, and discovery of the mechanism of these surface modifications.

Nucleation of QDs due to a pulsed low-energy (~ 100 eV) beam action of intrinsic ions at certain time instants in terms of the degree of filling of a monolayer during Ge/Si heteroepitaxy from molecular beams resulted in the formation of a dense QD array (up to 10^{12} cm^{-2}); moreover, the array was more homogeneous (the inhomogeneity decreased to 10%) (Fig. 4.1). The idea of the approach is to provide a pulsed action of intrinsic ions onto the substrate at the beginning of growth of each monolayer, and the pulse duration is chosen to be much shorter than the time needed for one monolayer to grow. It was this approach that ensured the high homogeneity of the QD array. Continuous irradiation by ions during growth resulted in the formation of a less homogeneous QD array, even compared to its growth without ion irradiation.

A model of ion-beam-induced growth of Ge nanocrystals on the Si surface was proposed in Ref. [13]. This model includes deposition from a molecular beam, surface diffusion of adatoms, and generation of interstitial atoms and vacancies by an ion beam in the subsurface of the growing layer. The model takes into account both the strains induced by the lattice mismatch of the Ge and Si, and the strain induced by the formation of vacancies and interstitial complexes in the subsurface layer due to ion irradiation. It was assumed that the strain in the surface layer alters the energy of the adatom bond on the growing layer surface. Calculations based on this model showed that centers of predominant nucleation of three-dimensional islands on the surface are local tensile regions arising above the complexes of interstitial atoms or vacancies induced by the ion action. In fact, this means that nucleation of Ge islands occurs in regions of the best matching of the Ge and Si lattice constants. The homogeneity is actually provided by forced instantaneous (the ion pulse duration is much shorter than the monolayer growth time) nucleation of islands with their subsequent simultaneous enlargement.

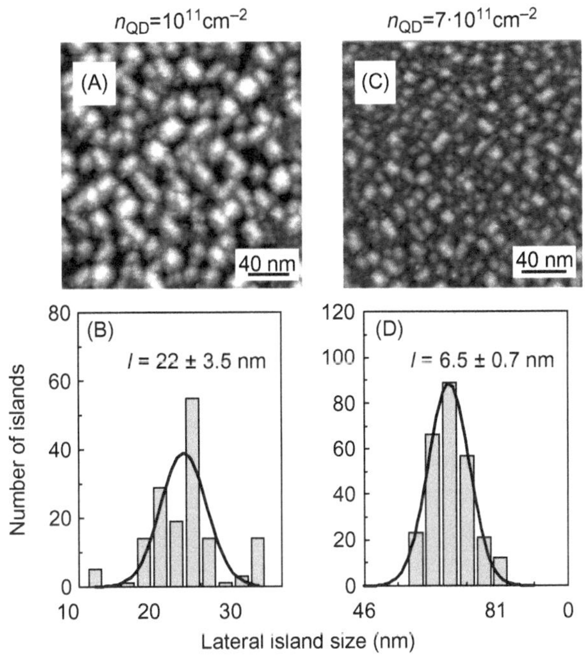

FIGURE 4.1 (A and C) Scanning tunneling microscopy (STM) images of the surface with three-dimensional islands formed after deposition of five germanium monolayers on the Si[100] substrate, and (B and D) the corresponding size distributions of islands for two types of experiments: (A and B) conventional molecular beam epitaxy of germanium on the Si[100] substrate, and (C and D) molecular beam epitaxy with pulsed irradiation by Ge^{+} ions. Conditions: the molecular germanium flux density is $7 \times 10^{13}\ cm^{-2}\ s^{-1}$, the temperature of the substrate is 620K, the ion beam energy is 140 eV, the pulse duration is 0.5 s, and the ion flux density is $3 \times 10^{11}\ cm^{-2}\ s^{-1}$ [13].

4.2.2 Laser Annealing of Nanostructures

All the above-considered approaches were based on controlling the QD nucleation and growth during epitaxy. A quite different approach implies a possibility of correcting QD inhomogeneity in an already grown structure. This approach is based on using pulsed laser annealing of Ge/Si layers with QDs, aimed at improving the homogeneity of their structural and electronic properties [21]. The method of pulsed laser modification of Ge nanocrystals in Si included the following aspects: in situ studies by optical methods and pyrometry of the dynamics of the nanosecond laser action on Ge/Si heterostructures with QDs, determining the threshold of nonlocal melting of Ge/Si structures as a function of pulsed laser beam density, finding the threshold of local melting of Ge QDs and the thickness of the layers of the Ge–Si solid solution formed by means of mixing under a pulsed laser action on the basis of simulations, studying the elemental composition and mechanical stresses in Ge/Si heterostructures with embedded QD layers before and after the pulsed laser action, and finding the energy spectrum of charge carriers in layers of Ge QDs modified by laser irradiation. Pulsed laser treatment was provided by a ruby laser with a wavelength $\lambda = 694$ nm.

The energy density of the pulsed beam was close to 1 J/cm^2, which corresponded to the threshold of melting of the subsurface region of the Si monocrystal (~5 nm), and the pulse duration was 80 ns.

Raman spectroscopy was used to obtain data on the elemental composition and mechanical stresses in Ge/Si QDs heterostructures before and after the pulsed laser action. The experimental data show that a pulsed laser action leads to reduction of the fraction of Ge in nanocrystals and partial relaxation of mechanical stresses in the system. The degree of homogeneity of the structural characteristics of QDs after nanosecond laser irradiation was found to improve, which was not observed after thermal annealing during a longer time. Thus, for instance, after treatment of a sample with a mean QD size equal to 8 nm by 10 laser pulses, the width of the peak of the Raman light scattering of Ge–Ge bonds decreased by a factor of 1.5 [21].

The energy spectrum of holes was studied by admittance spectroscopy. A comparison of the results of self-consistent calculations of the band structure of the samples with experimental data allows two important conclusions to be drawn. First, the laser action leads to an increase in the hole-binding energy in QDs. A deeper level of the hole means an increase in the depth of the potential well in the Ge valence band. This fact can be attributed to an increase in the mean QD size associated with mixing of Ge and Si on the Ge/Si interface. Second, the dispersion of the energy levels in the QD array in laser-annealed samples is smaller, which testifies to the enhancement of the homogeneity of nanocrystal sizes under the laser action. Thus, after treatment of the samples with 10 pulses of a ruby laser, a twofold decrease in the energy dispersion of hole energy levels in the ground state of Ge QDs was observed (Fig 4.2).

A model of melting and crystallization of Ge nanocrystals in a crystalline matrix of silicon under transient heating at a pulsed laser action was developed in Ref. [22]. The melting temperature of bulk Ge (958.5°C) is lower than the melting temperature of crystalline Si (1410°C). So, pulsed laser action can provide melting of Ge QDs, leaving the ambient silicon

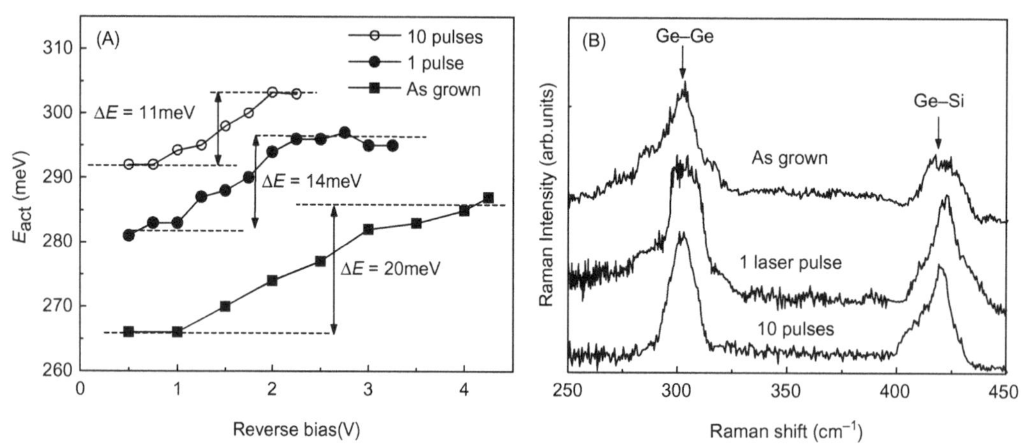

FIGURE 4.2 (A) Bias-dependent activation energies of hole emission rate before and after pulsed laser processing. (B) Raman spectra of Ge QD samples before and after pulsed laser annealing [21].

matrix in the solid state. The time of subsequent solidification of the molten material is determined by heat diffusion to the Si substrate from the heated Si layer with embedded Ge QDs. When the thickness of the heated layer is much smaller than the substrate thickness, the cooling time is comparable with the duration of the pulsed laser action and is determined by the substrate parameters (light absorption coefficient, thermal conductivity) [23]. The proposed model describes the kinetics of melting and crystallization of Ge nanocrystals, taking into account the nanocrystal's sizes, and shapes, as well as the elastic strains induced by the Ge and Si lattice mismatch. It was found that the melting temperature of nanocrystals increases as their size decreases. If the contribution of elastic strains is taken into account, the melting point of Ge nanocrystals can be either higher (for nanocrystals smaller than 15 nm) or lower (for nanocrystals greater than 15 nm) than the melting point of bulk Ge (Fig. 4.3). For pulsed laser action parameters close to experimental values (pulse duration 80 ns and energy density 0.85 J/cm^2), nanocrystals whose size exceeds 15 nm were found to tend to melt, while nanocrystals of smaller sizes remained in the crystalline state. Melting of "large" nanocrystals is accompanied by a decrease in their size owing to partial dissolution in the Si matrix during the time of melt crystallization after the laser action. As a result, the as-grown dispersion of nanocrystal sizes defined by the growth parameters is expected to become smaller. However, their layer density should not change significantly because of the small probability of complete dissolution of large clusters.

It should be noted that simulations predict that selective on size melting of nanocrystals occurs in a rather narrow range of energy densities: from 0.8 to 0.9 J/cm^2. Higher energy densities of laser action correspond to the regime of complete melting of nanocrystals, regardless of their sizes. As the ratio of the surface area of the nanocrystal to its volume increases as its size decreases, it is easier for small clusters to dissolve in the ambient Si matrix at a fixed time of melt existence. This can lead to complete dissociation of nanocrystals smaller than a certain critical size determined by the length of Si diffusion in the Ge melt. Thus, for both regimes of laser treatment considered above, the dispersion of nanocrystal

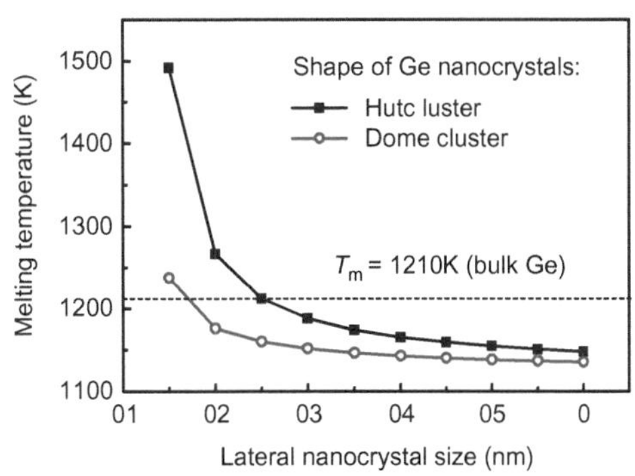

FIGURE 4.3 The dependence of the melting point of a Ge nanocrystal embedded in a Si matrix on its base size, calculated for hut and dome clusters, taking into account elastic deformations caused by the mismatch of the lattice constants of the Ge nanocrystal and Si matrix [22]. The horizontal dotted straight line corresponds to a melting point of bulk Ge of 1210 K.

sizes can be expected to decrease. In the latter case, however, the layer density of nanocrystals should decrease substantially owing to their complete dissolution.

The kinetics of phase transformations induced by pulsed laser treatment [22] demonstrates the possibility of decreasing the size dispersion in heterostructures with QDs whose melting temperature in the bulk material is lower than the melting point of the ambient matrix.

4.2.3 Spatially Ordered Groups of QDs

The idea of the growth of QD ordered groups (molecules) was based on the following assumptions. An array of SiGe nanodisks embedded into the Si layer is initially formed by molecular beam epitaxy. That is the first stage of substrate preparation, when germanium is deposited onto the Si[100] substrate at a temperature of 700°C, which results in nucleation and growth of three-dimensional Ge islands shaped as dome-clusters. At the second stage, thin Si layers are deposited onto these three-dimensional Ge islands at the same substrate temperature. Silicon deposition stimulates diffusion of Ge from the apex of the three-dimensional islands onto the surface of the wetting layer. As a result, nanodisks are formed. The size of these nanodisks and their elemental composition can be controlled by changing the temperature and deposition rate. Thus-formed nanostructures with embedded strain sources in the form of SiGe nanodisks are used as a basis for nucleation and growth of an array of QDs. The thickness of the cap Si layer over GeSi nanodisks was varied to find the regime when the strain generated by the layer of SiGe nanodisks stimulates nucleation and spatial alignment of QDs on the Si cap layer. The Ge nanocrystal nucleation on the Si cap layer occurs at definite sites, above the disks, thus creating ordered groups of QDs.

Growth regimes were found to ensure the formation of symmetric QD molecules, typically consisting of four elongated QDs ordered along the [010] and [100] directions (Fig. 4.4) [24]. It was demonstrated that this spatial configuration is caused by the anisotropic character of the elastic energy density distribution over the nanodisk surface. The morphological transition from fourfold QD molecules to continuous fortress-like quantum rings with an increasing amount of deposited Ge was revealed. Experimental data are well described by a simple thermodynamic model based on the accurate evaluation of the strain-dependent part of the surface chemical potential. The spatial arrangement of QDs inside molecules is attributed to the effect of elastic property anisotropy.

4.3 Electronic Structure of Ge/Si QDs

In contrast to quantum-well (QW) and quantum-wire heterostructures, the properties of electrons and holes in QDs cannot be described as a gas of quasiparticles. This case is a fruitful concept of localized states. The electron or hole localization radius in a nanocluster is comparable with the cluster size, and it frequently exceeds the Bohr radius of single-impurity atoms with shallow levels in homogeneous bulk semiconductors. However, the energy level in a QD may be deep, and this is one more feature of QDs as deep-level

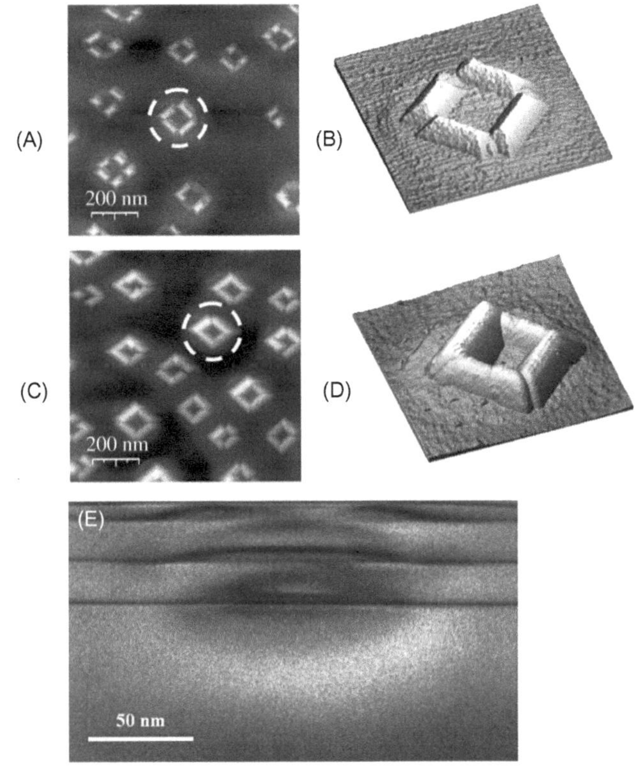

FIGURE 4.4 STM images of surface ($1 \times 1\ \mu m^2$) with the groups of closely spaced SiGe islands grown at 600°C on a three-layered QD superlattice template with different amounts of deposited Ge: (A) four monolayers, (C) five monolayers. Image sides are oriented along the [100] directions. Panels (B) and (D) correspond to three-dimensional representations of the fourfold symmetric QD molecules. (E) Cross-sectional transmission electron microscopy micrograph of virgin substrate with a threefold stack of buried SiGe QDs [24].

impurity centers. The study of QD heterostructures now constitutes a separate branch of condensed-matter physics. Ge QDs in a Si matrix form a type-II band lineup. The large ($\sim$0.7 eV) valence-band offset in this heterojunction leads to an effective localization of holes in Ge regions, whereas electrons are in states of the continuous spectrum of the Δ valleys of the Si conduction band.

4.3.1 Hole States

Usually, the behavior of holes in two-dimensional systems is more complicated than that of electrons because of the complex valence band structure in III–V and IV semiconductors. The hole state in Ge QD is built mainly from valence-band states, namely, heavy-hole states $|3/2, \pm 3/2\rangle$, the light-hole states $|3/2, \pm 1/2\rangle$, and the split-off hole states $|1/2, \pm 1/2\rangle$. Here, $|J, J_z\rangle$ are the eigenstates of the effective angular momentum J and its projection J_z. The strong mixing between the light- and heavy-hole states is a result of the uncertainty in all components of quasi-impulse owing to hole spatial confinement both in vertical and lateral directions. The elastic strain due to the lattice mismatch between QDs and the surrounding matrix has essential effects on the electronic structure of QDs. In uniaxially strained semiconductors, strain lifts the degeneracy of the valence band, making the heavy-hole states the highest valence band [25]. For this reason, the contribution of the heavy-hole states in the ground state of QDs must be predominant. The hole energy spectra in self-assembled pyramid-shaped Ge nanoislands in Si have been numerically investigated using the effective mass approximation [26,27] and the tight-binding approach [28,29]. In the first case, a six-band $\mathbf{k}\cdot\mathbf{p}$ formalism was used to study single-particle hole states. The finite element calculations of three-dimensional spatial distribution of strain components were performed with the approach described in Ref. [30]. In a tight-binding method, the strain distribution was found in terms of atomic positions, using the valence-force-field model with a Keating interatomic potential. As demonstrated by different approaches, in Ge nanoclusters with a lateral size of $\sim$10 nm and height of $\sim$1 nm the contribution of the states with $J_z = \pm 3/2$ amounts to about 85–90%.

Calculated energies of the ground state and the next nine excited states of the hole spectrum are presented in Fig. 4.5 as a function of Ge pyramid base length [28]. The wave functions found are characterized by the absence of nodal surfaces perpendicular to the growth direction. From the wave function space configuration one may conclude that the ground state has an *s*-like wave function, and the first two excited states have *p*-like wave functions oriented along [110] and [$\bar{1}$01]. The absence in wave function density of nodal surfaces perpendicular to the growth direction is clear evidence that the difference between all the found states is determined only by quantization in the plane of the pyramid base. Therefore, optical transitions between these states can be active under normal incidence infrared (IR) radiation, which agrees with experimental results [31].

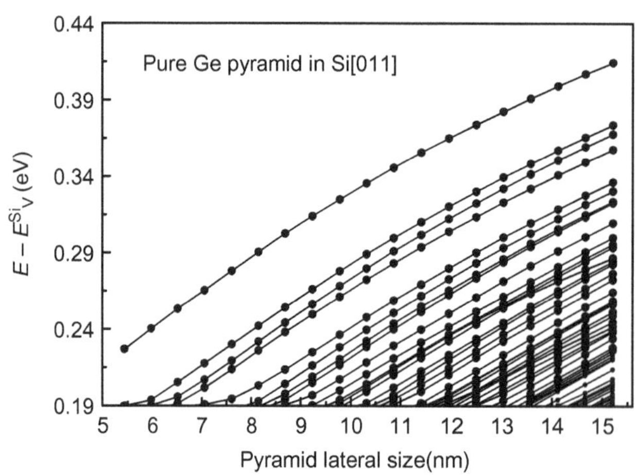

FIGURE 4.5 Energy spectrum of holes in the pyramidal Ge/Si QD as a function of pyramid base length *l* with aspect ratio ($h/l = 0.1$) being constant [28]. The energy is counted with respect to the valence band edge in bulk Si. The ground state is the upper curve.

Coupled QDs represent another example of a zero-dimensional system in which the interdot coupling extends the analogy between QDs and natural atoms to artificial and natural molecules. Recently, considerable interest has been devoted to double QDs occupied by two electronic spins. Quantum computation requires coherent coupling between the dots, and the coherence to be preserved over sufficiently long time-scales. The spin degrees of freedom (rather than charge) are of special interest for information processing, since they can take advantage of the comparatively long spin-coherence time and, hence, can serve as the quantum bits (qubits). The formation of double Ge QDs in a Si matrix [32] is exemplified in Fig. 4.6, which also shows the image of the Si surface on which the array of Ge nanoclusters with lateral sizes of 10–20 nm are formed.

Theoretical investigations of the energy spectrum and electronic structure of two vertically stacked Ge/Si QDs loaded with a single hole or with a pair of interacting holes along with the dependence of basic physical characteristics on dot size and interdot separation were performed in Refs. [26,27,33]. The model structure is shown in Fig. 4.7. The study focused on the splitting of the lowest spin singlet and triplet states, and spatial correlations between holes. Here the term "spin" is rather an index for two Kramers degenerate states. The two holes were treated by six-band $\mathbf{k}\cdot\mathbf{p}$ calculations combined with the configuration–interaction model taking into account the realistic situation when both Si matrix and Ge nanoclusters are inhomogeneously strained. This approach permits an accurate treatment of the direct Coulomb and exchange interaction, as well as the correlation effects, by allowing

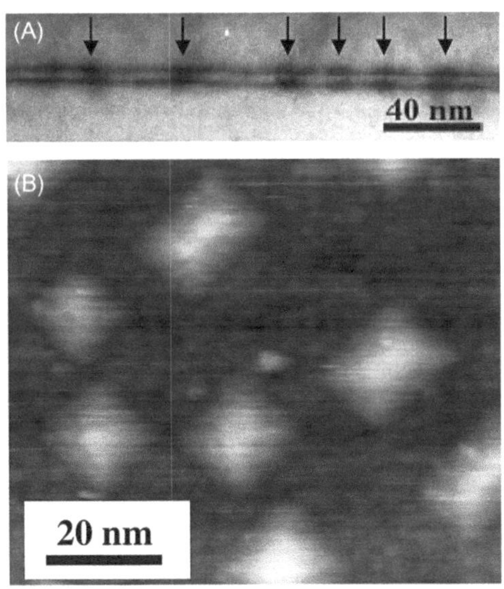

FIGURE 4.6 (A) Transmission electron microscope image of the cross-section of a real structure with pairs of vertically coupled Ge quantum dots (marked by arrows) in Si. (B) Scanning tunnel microscope image of the Si[001] surface with (light regions) an array of self-organized Ge nanoclusters. *Reproduced from A.I. Yakimov, G.Yu. Mikhalyov, A.V. Dvurechenskii, A.I. Nikiforov, Hole states in Ge/Si quantum-dot molecules produced by strain-driven self-assembly, J. Appl. Phys. 102 (2007) 093714, with the permission of AIP Publishing.*

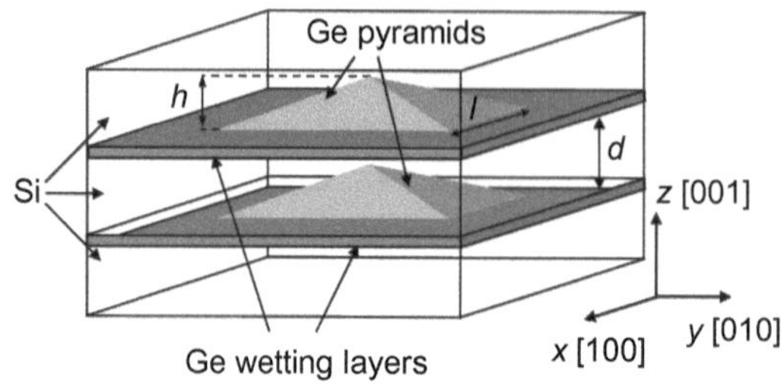

FIGURE 4.7 Schematic picture of a Ge/Si double QD used for simulation of hole states.

interconfiguration coupling. It was shown that asymmetry of strain distribution and spin–orbit coupling of the valence band introduce characteristic features in both single- and many-particle hole states, which are not captured by the usual one-band heavy-hole approximation. According to quantum mechanics, an electron level in each of two identical atoms approaching each other is split into two levels. In this case, the ground state of a diatomic molecule corresponds to a bonding molecular orbital σ_g, whereas an excited state corresponds to an antibonding molecular orbital σ_u. A similar situation was observed for double QDs containing electrons [34], which are artificial analogs of molecules. Our studies revealed that the picture for hole molecular orbitals in coupled strained Ge/Si QDs is significantly different. Fig. 4.8 shows a single-particle spectrum of holes for paired Ge/Si QDs. The most interesting property of the spectrum is the crossing of levels with different symmetry which occurs with changing dot separation d. As a result, at large d the antibonding molecular ground state appears.

In Fig. 4.9, we show the calculated values of singlet–triplet splitting for double QDs of different size as a function of dot separation. As the space between dots increases, the singlet–triplet splitting drops to zero and then restores again, implying level anticrossing with a zero anticrossing energy gap. We demonstrated that the many-particle level anticrossing occurs at the interdot distance at which the single-particle ground state changes its symmetry from bonding to antibonding. The results were interpreted in terms of asymmetry-induced localization of single-particle orbitals caused by the interplay of effects of strain and hole subband mixing. The symmetry breaking for many-particle states as a joint action of correlations and strain is observed with an increasing dot separation. We hope these results may inspire further investigation exploring the singlet–triplet entanglement in two-hole systems for possible device applications.

The experimental evidence of the existence of the antibonding ground state of holes in Ge/Si double QDs was obtained in Ref. [35]. The idea of this experiment was to use the strong dependence of the matrix element of interlevel transitions induced by linearly polarized light on the parity of the initial and final states. To implement these transitions, we proposed the separation of layers of Ge nanoclusters by a thin layer of SiGe film with a low Ge content, rather than by Si. This layer serves as a two-dimensional potential well in which the energy spectrum of holes is quantized along the growth direction (z axis). The ground state is formed by the heavy-hole subband and the corresponding wave function ψ_f is even

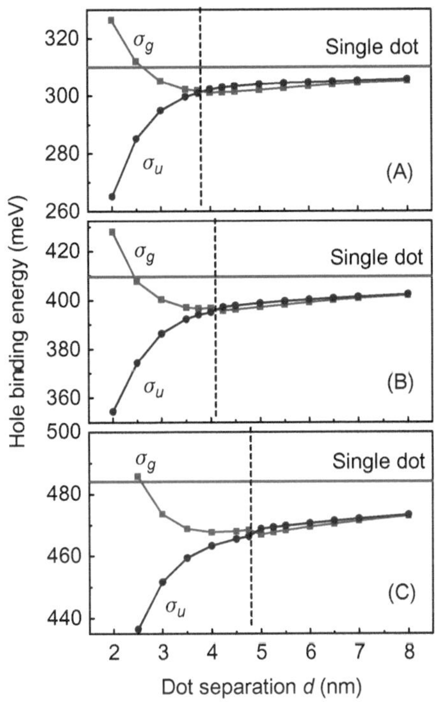

FIGURE 4.8 Evolution of hole-binding energy in bonding σ_g and antibonding σ_u states as a function of the distance between QDs, calculated for lateral pyramid size: (A) $l = 10$ nm, (B) 15 nm, and (C) 20 nm [27]. As a reference, the hole ground-state energy of the isolated single dot is also shown by the *horizontal lines*. All energies are counted with respect to the valence band edge in bulk Si. The *vertical dot lines* are used to indicate the distance at which the ground-state symmetry changes.

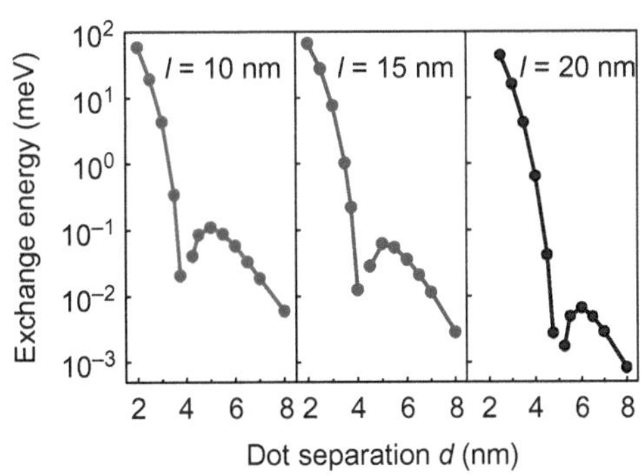

FIGURE 4.9 The singlet–triplet splitting plotted in semilogarithmic scale as a function of dot separation d, and dot base length l [34].

(symmetric) along the z direction. In the long-wavelength dipole approximation, only transitions between the states with the opposite parities are allowed, i.e., $\sigma_u \rightarrow \psi_f$. Fig. 4.10 shows the transmission of the z-polarized radiation for the samples, in which the QDs in each of the double layers are separated by the $Si_{0.85}Ge_{0.15}$ QW with the thicknesses $d = 2.8$, 3.8, and 4.8 nm. The absorption bands near 200 meV correspond to the transitions of the $\sigma_g \rightarrow \psi_f$ and $\sigma_u \rightarrow \psi_f$ type. The absorption amplitude at $d = 2.8$ nm is small (0.3%), and is hardly visible against the noise background. We attribute this line to the $\sigma_g \rightarrow \psi_f$ transition, which is forbidden in the dipole approximation. The probability is that this transition is nonzero because the wave function of a hole in the bonding state is not completely even due to the anisotropy of the shape of Ge nanoclusters and inhomogeneous distribution of elastic deformations in the system. With an increase in d to 3.8 nm, the minimum transmission is shifted to lower energies, and the absorption probability increases by a factor of 5. As the QDs move away from each other, the absorption of z-polarized radiation decreases again. We believe that the strong increase in absorption and, above all, the nonmonotonic behavior of the absorption amplitude with d indicate the formation of the antibonding ground state of the hole in $Ge/Si_{0.85}Ge_{0.15}/Ge$ double QDs. As a result, the $\sigma_u \rightarrow \psi_f$ transition becomes active.

4.3.2 Electron States

The simple consideration of the conduction band in Ge/Si heterostructures disregards possible modification of the band structure due to inhomogeneous strain in the dots and the surrounding matrix. As discussed earlier, in Ge/Si QDs, the strain appears due to the 4% lattice mismatch between Si and Ge. Tensile strain in the nearby Si causes splitting of the sixfold-degenerate Δ valleys into the fourfold-degenerate in-plane Δ_{xy} valleys and the twofold-degenerate Δ_z valleys along the [001] growth direction. The lowest conduction band

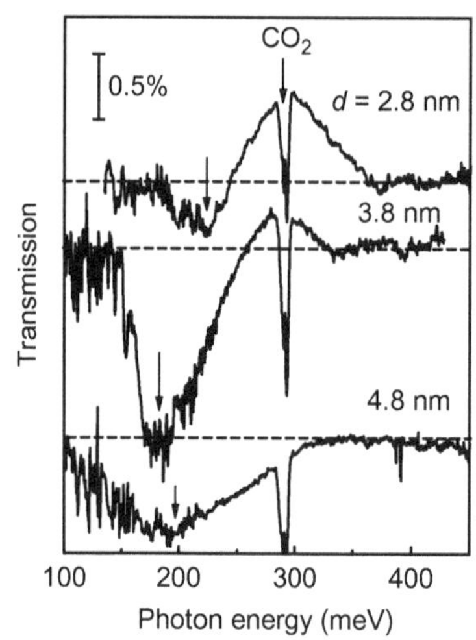

FIGURE 4.10 Transmission of the z-polarized radiation for double Ge QDs separated by a $Si_{0.85}Ge_{0.15}$ quantum well with well thicknesses $d = 2.8$, 3.8, and 4.8 nm. Arrows show the intraband transitions of holes from the ground state in quantum dots to the ground state of the $Si_{0.85}Ge_{0.15}$ layer [36].

edge just above and below the Ge island is formed by the Δ_z valleys yielding the triangle potential well for electrons in Si near the Si/Ge boundary. Thus, one can expect three-dimensional localization of electrons in the Si region near the dot apexes [36–41] and close to the dot base edges [39,42]. The distribution of strain in Si and the energy of localized electronic states are determined by the shape, size, and composition of Ge/Si QDs. The electron binding energy in a strain-induced potential well in an isolated small-sized Ge/Si pyramid was predicted to be very small (<10 meV) [37]. This value is expected to enlarge vastly in multilayer Ge/Si structures with vertical stacking of Ge islands due to accumulation of strain energy from different dot layers in a stack, and an increase in the potential well depth. The experimental realization of such a structure is demonstrated in Fig. 4.11. Fig. 4.12 shows simulation results of the electronic structure for a multilayer QD stack. We considered six identical vertically-aligned Ge nanoclusters with four {105}-oriented facets and base orientation along [100] (x) and [010] (y) directions [41]. The island base length is $l = 20$ nm, the height is 2 nm. An in-plane tension and compression along the growth direction z in the Si above and below the Ge islands was observed. In the lateral direction the strain in Si relaxes from the center at a scale comparable to the diameter of underlying Ge QD, and then changes its sign, demonstrating that the Si is laterally compressed around the edges of the Ge islands. One can see that the potential well for electrons is more shallow in the layer plane, and just its depth determines the electron binding energy E_b. The electron ground state is formed by states of two lower Δ_z valleys localized in the central Si layer close to the QD apex

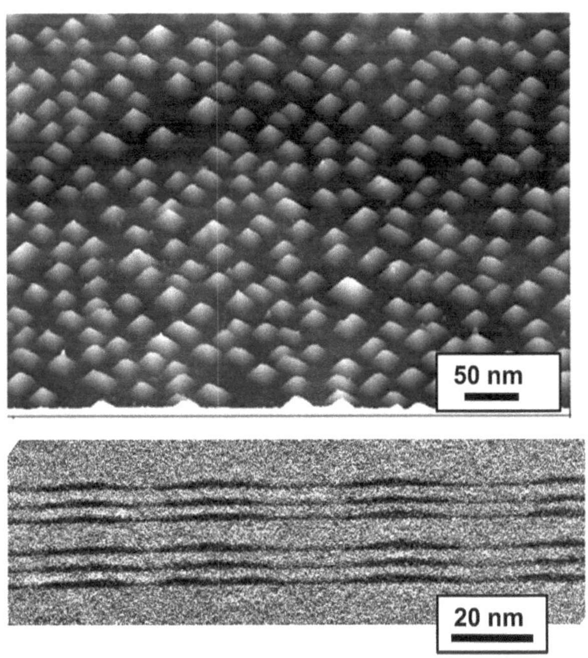

FIGURE 4.11 Three-dimensional scanning tunneling microscopy (top) and cross-sectional transmission electron microscopy (bottom) images of vertically-aligned Ge QDs [41]. Each Ge island in the upper layers grows on top of the islands in the lower layers, indicating a nearly perfect vertical correlation of the dots.

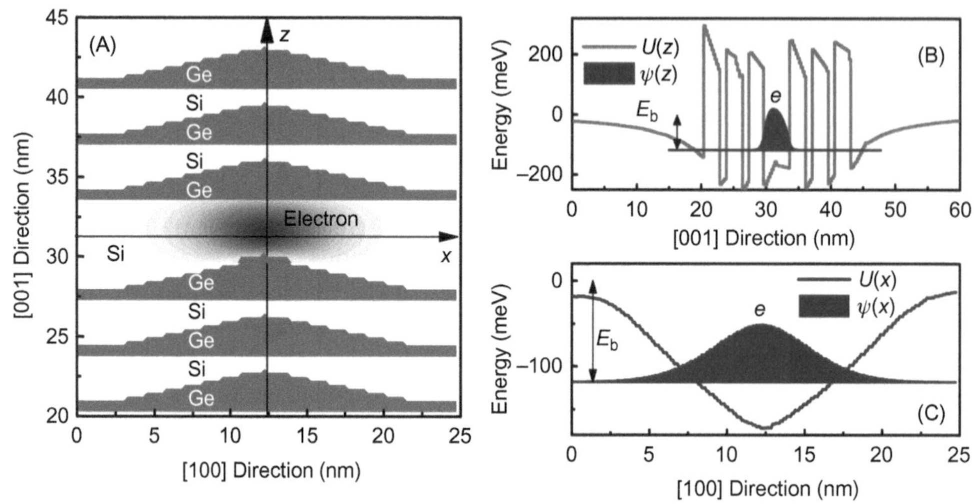

FIGURE 4.12 (A) Two-dimensional view of the model structure with six vertically-aligned Ge QDs separated by Si. Calculated wave function of the ground electron state across the x–z plane of symmetry, cutting through the pyramids and the wetting layers, is shown in *red* (gray in print versions). (B) and (C) The strain-induced confinement potential U for electrons with respect to the conduction band edge of unstrained Si along the vertical z axis and the x axis. In a latter case the potential profile is plotted for the central Si separation layer above the tip of the Ge island. The electron wave functions ψ are shown by the *red areas* (gray in print versions) [41].

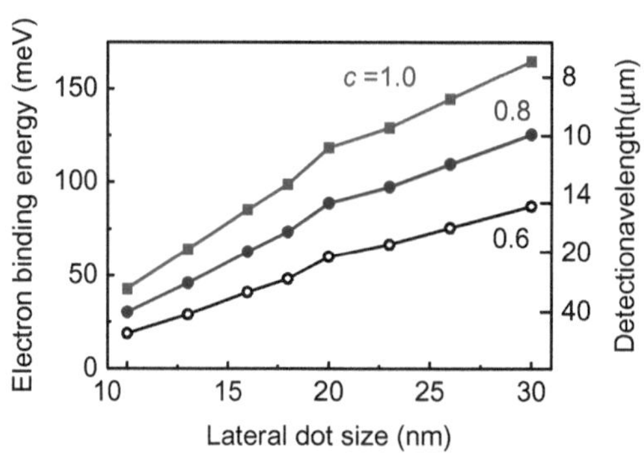

FIGURE 4.13 The ground-state electron-binding energy E_b and respective detection wavelength λ_c in the sixfold stack of Ge_cSi_{1-c}/Si QDs as a function of pyramid base length and Ge content (c), with aspect ratio being constant (0.1). The thickness of Si layers is the same as in Fig. 4.11. The energy is counted with respect to the conduction band edge in unstrained bulk Si. Here we assume that λ_c relates to the binding energy as λ_c (μm) = 1.24/E_b(eV) [41].

(red clouds (gray in print versions) in Fig. 4.12). Depending on the QD size, the excited electron states were found to be located either in Si spacers or in the region close to the base edges of QDs, in agreement with the results of Refs. [39,42]. For $l = 15–25$ nm, the first excited state is a p-like state and is localized also in the central Si spacer. The calculated values of the electron binding energy for the ground state E_b are shown in Fig. 4.13 for different Ge content in the dots. From the photoresponse characteristic point of view, the structures with E_b from 90 to 155 meV act as detectors with the wavelength threshold from 14 to

8 μm. Clearly, it can be seen that there is a range of reasonable dot sizes and compositions for which the photoresponse is expected in a long-wave atmospheric window.

The experimental evidence for three-dimensional electron confinement in strained Ge/Si QDs was obtained in Refs. [41,43,44]. In our studies, we exploited the fact that two-dimensional QWs with electrons are known to be insensitive to radiation, whose polarization vector lies in the plane of the quantum well. This is because of the orthogonality of the envelopes of electron wave functions in the size quantization sub-bands. With respect to two-dimensional systems, the additional in-plane confinement in a strain-induced potential well of the Si region adjacent to the dots would make the Ge/Si QDs sensitive to the normal incident radiation. Our experiments with *n*-type layers of vertically-ordered Ge/Si revealed the normal-incidence photoresponse in the mid- and far-IR spectral range, thus giving evidence for the zero-dimensional nature of the confined electron states [41,43,44]. The experimental evidence of the deformation mechanism of the formation of localized electronic states was obtained by establishing a direct relation between strains in Si and the binding energy of the electron in layers of Ge/Si QDs [44]. The idea of the experiment was based on the control of elastic strains in the multilayer Ge/Si heterostructure by the precision change in the thickness of silicon spacers *d*. The dependence of the strain on the distance between QDs is due to the superposition of strain fields from neighboring layers of QDs. It is manifest in the additional extension of Si in the plane of the heterostructures at small *d* values, when sources of deformation are located closely one above another. Fig. 4.14 shows the spectra of the lateral photocurrent in *n*-type samples with different thickness of the

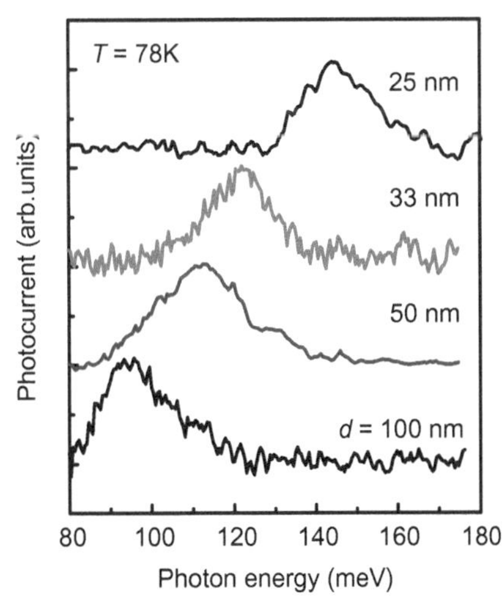

FIGURE 4.14 Normal-incidence photocurrent spectra in samples with different thicknesses of the silicon spacer *d*. The samples were grown by molecular beam epitaxy on silicon-on-insulator substrates and consisted of 10 layers of vertically-aligned Ge quantum dots separated by Si spacers. In order to obtain *n*-type doping, Si layers were doped with antimony to a bulk concentration of 10^{17} cm^{-3}. The Ge islands have a form of domes with a base 80 nm in diameter, a height of 20 nm, and a layer density of 3×10^9 cm^{-2}. It is very important that quantum dots are located one above another. Only in such vertical structures at certain distances between layers of quantum dots are strain fields summed, and an increase in the depth of the potential well for electrons is expected [44].

silicon spacer *d*. The heterostructures consisted of 10 layers of dome-shaped GeSi islands with a base 80 nm in diameter, a height of 20 nm, and content of germanium $c = 0.5$. The photoresponse observed in the range of 90–150 meV was attributed to transitions of electrons from bound states to states of the continuous spectrum. Here, the main result is the observation of the shift of the maximum of the photocurrent toward shorter wavelengths with a decrease in *d*; this means an increase in the binding energy of the electron at the Si/Ge heterointerface owing to tension in the Si layers.

4.3.3 Exciton States in Coupled QDs

The search for the physical mechanisms that increase the efficiency of radiative recombination, and the processes of the interband absorption of light in indirect-band semiconductors such as silicon and germanium, has been one of the topical problems of the physics of semiconductors for a long time. A significant increase in the oscillator strength for interband transitions will allow for the integration of opto- and microelectronic devices on a single silicon substrate. Certain hopes for solving this problem were set on Ge/Si heterostructures with QDs in which the weakening of the selection rules in the quasimomentum was expected owing to the size limitation of the motion of charge carriers in all of the three directions. However, Ge/Si[001] QDs are second-type heterostructures, where the charge carriers of different signs are separated by the heterointerface. Thus, the interband transitions in such a system are indirect not only in the space of wave vectors, but also in real space. In order to obtain a high efficiency of optical interband transitions (large oscillator strength) in Ge/Si structures with QDs, it is necessary to solve the following problems. First, localized electronic states should be formed near a QD. The wave vector **k** is not a “good” quantum number for a localized state: the state localized in space can be represented as a superposition of delocalized states with different wave vectors **k**. As a result, interband transitions in such a system become direct in the space of wave vectors, because the mean momenta of both an electron in the bound state in the conduction band of Si and a hole remaining in Ge are zero. Second, it is necessary to ensure maximum overlapping of the wave functions of localized electrons and holes and, thus, the possibility of direct optical transitions in real space. This condition means that a hole should penetrate from Ge to Si, and an electron, from Si to Ge. We demonstrated that both conditions can be fulfilled in vertically coupled Ge/Si QDs [45,46].

The double QD was simulated by a model presented in Fig. 4.7. In the dipole approximation, the oscillator strength for the transition of the electron from the valence band to the conduction band is proportional to $R = |\langle f|\mathbf{e}\cdot\mathbf{p}|i\rangle|^2$, where $|f\rangle$ and $|i\rangle$ are the final and initial states, respectively; $\mathbf{e}$ is the light polarization vector; and $\mathbf{p}$ is the momentum operator. The electron wave function at the Δ_1 point of the conduction band has the form $|f\rangle = \psi_e(\mathbf{r}_e)u_c(\mathbf{r}_e)$, where $\psi_e(\mathbf{r}_e)$ is the envelope wave function, and $u_c(\mathbf{r}_e)$ is the Bloch function at the bottom of the Δ_1 band. In the valence band, $|i\rangle = \psi_h(\mathbf{r}_h)u_v(\mathbf{r}_h)$, where $u_v(\mathbf{r}_h)$ is the Bloch function at the top of the $\Gamma_{25'}$ band. Since the periodic parts u_c and u_v vary rapidly over the characteristic length of the variation of the envelope functions ψ_e and ψ_h, the oscillator strength is reduced

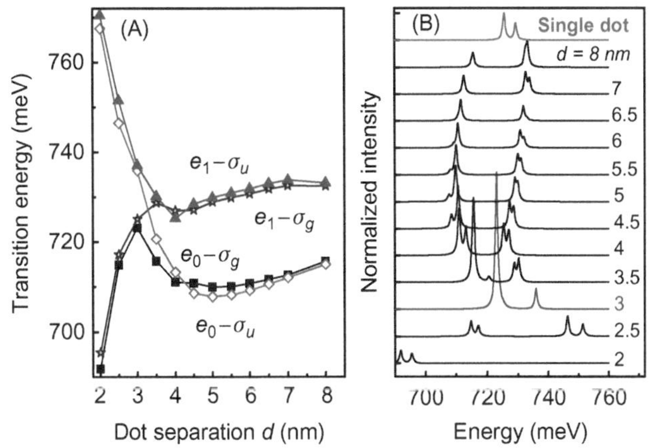

FIGURE 4.15 (A) The energy levels of the first four exciton states in double QD, and (B) corresponding optical resonances versus the quantum dot separation *d*. For comparison, the electronic states containing the e_0-h_0 and e_1-h_0 transitions for single quantum dots are also shown. The intensities of the resonances are normalized to the e_0-h_0 transition intensity in a single quantum dot with the same size [46].

to evaluating the electron–hole wave function overlap integral squared $\langle\psi_e|\psi_h\rangle^2$ [25], and $R\sim\|p\|^2\langle\psi_e|\psi_h\rangle^2$, where $\|p\|$ is the numerical version of the momentum matrix elements between Bloch functions. We also took into account that hole states represent the six-component complex functions and are built from the heave-hole states, light-hole states, and the split-off states.

The calculations were performed for the first two states of the holes (σ_g and σ_u) and electrons (e_0 and e_1). The electron–hole Coulomb interaction was taken into account. The four exciton transitions with the lowest energies ($e_0-\sigma_g$, $e_1-\sigma_g$, $e_0-\sigma_u$, $e_1-\sigma_u$) were analyzed. Here, e_0 and e_1 are the ground and excited electronic states, respectively, and σ_g and σ_u correspond to the bonding) and antibonding molecular orbitals of the holes, respectively. Fig. 4.15 shows the energies of the exciton transitions and the relative intensities of these transitions for single and double QDs with the lateral size 15 nm and various separations d of paired QDs. The most interesting result shown in Fig. 4.15 is an increase in the intensity of the $e_0-\sigma_g$ transition for the separation $d=3$ nm. In particular, the probability of the exciton transition in the double QD with $l=15$ nm and $d=3$ nm is five times larger than the probability of this transition in the single QD. The physics of such behavior can be understood by considering the envelop electron and hole functions. Detailed analysis showed that in this case, a significant fraction of the electron and hole wave functions is in the same spatial region [46], thus ensuring the possibility of direct transitions in a real space.

4.4 Hole Transport in Dense Arrays of QDs

The charge-transfer mechanisms and electron correlations in arrays of Ge/Si QDs with areal density $(1-4)\times10^{11}$ cm^{-2} were discussed in Refs. [47–53]. A motivation for choice of QDs for investigation of correlation effects is clear from the following arguments: (1) it is possible to fix the dot density and vary only carrier wave functions by successively loading holes into different quantum states of the dots, being deep in the insulator side of the metal-insulator phase transition; (2) self-assembled QDs are all located at the same position in the growth direction, i.e., there is no vertical disorder; (3) as distinct from the majority of impurities in semiconductors, ensembles of QDs can serve as a system of multicharged localization centers, in which the role of Coulomb potential is especially important. Measurements of hopping transport of holes in modulation-doped Si field-effect structures with Ge nanometer-scale dots embedded in proximity with the p-type conductive channel have been reported in Ref. [47]. It was found that the activation energy of hopping conductivity in the boron impurity band of the doped Si layer changes with increasing QD density, passing through a minimum, due to trapping of holes by the QDs. At low temperatures ($T \approx 6$K), a small ($\sim 10-20\%$) contribution to conductance which intriguingly oscillated with gate voltage was detected, and was ascribed to direct hopping of holes between dots.

In Refs. [48–53] we described a set of experiments in which we have studied hopping transport in doped or gated Si structures containing a two-dimensional array of Ge self-assembled QDs as a conductive channel. The dots are separated from each other by weakly doped silicon, and the only conduction mechanism at low temperatures is tunneling of holes between them. Pseudomorphic Ge islands grown epitaxially on a Si[001] surface exhibit a large band discontinuity in the valence band, and can be viewed as doping "artificial atoms." As we demonstrated by our experiments, QDs provide a system in which the number of confined holes, the structure of energy levels, the shape of wave functions, and the strength of the interaction can be controlled.

Conductance oscillations with gate voltage resulting from successive loading of holes into the dots in field-effect structures were observed in Refs. [48–50]. With scanning the gate voltage the Fermi level moves across the zero-dimensional density of states. The maximum current occurs when the given hole level in the dots is half-filled, as this maximizes the product of possible initial and final states for tunneling processes, and avoids increasing the energy of the system due to the appearance of extra charge in a final dot. Based on measurements of the temperature dependence of the conductance maxima, the charge-transfer mechanism in the channels was identified as being due to Efros–Shklovskii variable-range hopping between the dots, with the typical hopping energy determined by interdot Coulomb interaction. In this regime, the hopping distance increases as temperature is lowered, and the temperature dependence of conductivity is given by $G(T) = G_0(T)\exp[-(T_0/T)^x]$ with $x = 1/2$ [54], where $kT_0 = Ce^2/\kappa\xi$ is the characteristic interaction energy scale, C is a numerical constant, k is the Boltzmann constant, κ is the relative permittivity of the host lattice, and ξ is the carrier localization length. The characteristic spatial dimension of the hole wave functions, as well as the charging energies of the dots, were determined from the conductance data. It was found that the effective localization radius of charge carriers in QDs varies nonmonotonically upon filling Ge QDs with holes [53], which was explained by the successive filling of electron (hole) shells.

When the number of QDs in the active channel is reduced, the drain current peaks become more pronounced and show Coulomb blockade oscillations up to 200K (Fig. 4.16).

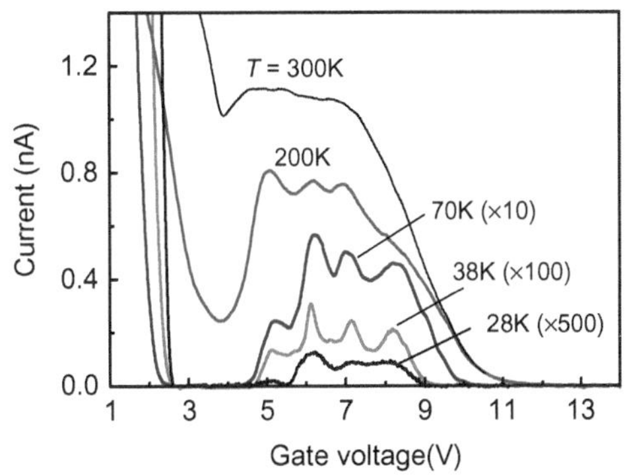

FIGURE 4.16 Gate voltage dependence of drain current at various temperatures in Ge/Si QD metal-oxide-semiconductor field-effect transistor (MOSFET). The device contains an array of 10^3 10 nm diameter Ge QDs embedded into the active MOSFET channel. *Reproduced from A.I. Yakimov, A.V. Dvurechenskii, V.V. Kirienko, A.I. Nikiforov, Ge/Si quantum-dot metal-oxide-semiconductor field-effect transistor, Appl. Phys. Lett. 80 (2002) 4783–4785, with the permission of AIP Publishing.*

The device contains an array of only one thousand 10 nm diameter Ge QDs embedded into the active metal-oxide-semiconductor field effect transistor (MOSFET) channel. The observed four maxima right above the transport continuum were attributed to hole transport through the excited state of Ge/Si QDs [50]. A very similar fine-structure consistent with four maxima, separated by the Coulomb blockade energy, was observed in InAs/GaAs QD field-effect heterostructures [55]. The charge-transfer mechanism in the Ge/Si transistor channel was identified as being due to nearest-neighbor hopping of holes between the dots, with the typical hopping energy determined by disorder in the QD ensemble. A dot charging energy of 43 meV and disorder energy of 21 meV were determined from the oscillation period and the temperature dependence study of current maxima, respectively.

The large spatial scale of intersite correlations allows one to examine the role of Coulomb interaction experimentally by making use of intentially introduced screening [56]. If one places a metal plane parallel to the conductive channel at some distance, the interaction may be described by including images of the real charges in the screening electrode. If the separation of initial and final states in a hop is small compared with the distance between a charge and its image, the screening electrode makes little difference, and the interaction remains monopolar. At large distances, however, a charge and its image behave as a dipole and interactions fall off more rapidly with distance. The effect of proximity of a bulk conductor on hopping transport was studied in Ref. [50]. We found that putting a metal plane close to the dot layer causes a crossover from Efros–Shklovskii variable-range hopping conductance ($x = 1/2$) to two-dimensional Mott behavior ($x = 1/3$) as the temperature is reduced. At the crossover temperature the hopping activation energy is observed to fall off. The metal plane was shown not to affect the conductance of samples which initially show Mott hopping.

Fig. 4.17 shows the conductance in units of e^2/h, the quantum of conductance, of unscreened samples (no metal plane) with different dot occupation plotted versus $T^{-1/2}$.

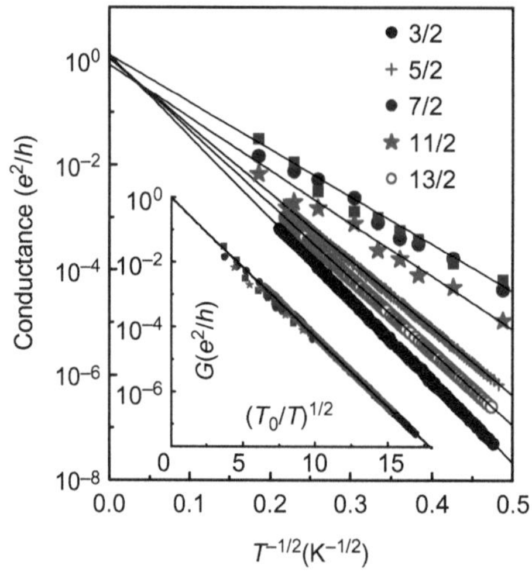

FIGURE 4.17 The conductance $G(T)$ versus $T^{-1/2}$ for different dot occupation with holes. The inset shows the same data plotted versus $(T_0/T)^{1/2}$ [49]. The average filling factor was controlled either by a remote boron δ-doping Si layer, or by the gate potential in MOSFET structures.

An impressive feature is that all curves extrapolate to the same prefactor $G_0 \approx e^2/h$. This is more evident when $G(T)$ is plotted against the dimensionless parameter $(T_0/T)^{1/2}$ (see inset of Fig. 4.17). In this plot, the data collapse onto a single universal curve with intersect $G_0 = (1.05 \pm 0.05)\ e^2/h$. These features are inconsistent with a model of single-particle hopping and provide experimental evidence for interaction-driven many-electron excitations in the transport properties of two-dimensional arrays of Ge/Si coupled QDs.

4.5 Spin Phenomena in an Array of Ge/Si QDs

Spin phenomena in various materials are nowadays one of the promising branches of solid state physics. Practical use of the spin degree of freedom includes the following operations: (1) generating nonequilibrium spin polarization (recording information); (2) manipulating spins with the aid of external fields; and (3) detection of the resulting spin state (reading the result). The development and production of functional devices that perform these operations is the basis of so called spintronics. One of the key characteristics of such devices is the decoherence time of a spin state.

A large number of papers have been dedicated to studying the peculiarities of spin-dependent phenomena in low-dimensional structures using different experimental techniques, e.g., photoluminescence, photoconductivity in a magnetic field, and magnetoresistance [57–61]. Electron spin resonance (ESR) is a direct method for studying spin relaxation that allows us to determine the value and nature of spin–orbit interaction and spin relaxation. Deviation of the g factor from its free electron value yields information as to the value of spin–orbit interaction in a studied object. The spin relaxation time is found from the

ESR signal linewidth, which can depend on the decoherence time T_2 of the spin system (the transverse spin relaxation time), and on spin-lattice relaxation time T_1 (the longitudinal relaxation time). For systems with a high spin density, where spin–spin interaction plays a major role, T_2 influences the linewidth. In some cases, however, the latter is proportional to $1/T_1$.

The creation of low-dimensional structures has encouraged studies of spin dynamics in semiconductors over the last two decades. Theoretical estimations showed that the spin relaxation time increases when the dimensionality of a system is lowered. This suggestion was proved experimentally in semiconductor compounds A3B5.

Special attention has been given to heterostructures with QDs, since control of the spin state is easier in such systems due to the restriction of charge carrier motions in all three dimensions. For the same reason, the most effective ways of spin relaxation are suppressed in QDs, leading to long periods of spin information storage [62].

QDs heterostructures are now regarded as promising objects in creating logical elements for quantum computation (QD qubit) and channels of spin information transmission. Operations of this kind can be performed only if there is coupling between QDs. We therefore focus on systems of tunnel-coupled QDs.

QD qubits have been fabricated in both II–V Group and Group IV heterostructures. QDs formed in GaAs/AlAs quantum wells are most advanced because of two advantages compared to silicon: a smaller electronic effective mass (which means that length scale of the devices can be larger, easing lithography experiments), and essentially perfect lattice matching between GaAs and AlGaAs (minimization of misfit dislocation). The main complication of GaAs is the presence of nuclear spins, which cannot be eliminated, because neither Ga nor As has a spin-zero isotope.

A major advantage of Si and Ge for use in qubits is that they each have an abundant spin-zero nuclear isotope, so that isotopic purification would enable the elimination of nuclear spins as a decoherence mechanism of spin relaxation. Even in natural-abundance Si, which contains some nonzero nuclear spin isotopes, electron coupling to nuclear spin is orders of magnitude weaker than in GaAs. Si also has much weaker spin–orbit coupling. Together, these two differences lead to longer spin relaxation and coherence time [57].

4.5.1 Spin Relaxation in Single QD

A flip-flop of an electron spin inside an individual QD can occur due to the absorption or emission of phonons, or due to phonon scattering. At sufficiently low temperatures, we can limit ourselves to acoustical phonons only. At temperatures of $T \geq 4K$, two-phonon (Raman) processes must be considered in calculations of spin relaxation time. In this case, relaxation proceeds via an intermediate excited state. Spin relaxation in QD depends on the intensity of the external magnetic field, which is capable of measuring wave functions of localized charge carriers [63]. Fluctuations in the electric field, arising due to the movement of charge carriers through the ensemble of QDs, also contribute to the relaxation process.

4.5.2 Spin Relaxation During Resonance Tunneling Between QDs

It has been shown in theory for a Ge/Si system containing germanium QDs that tunneling in the ground state occurs mainly by preserving the spin orientation: only one spin flip-flop is observed in 100 tunneling events. The probability of a spin flip-flop increases for excited states: on average, five to ten tunneling events include one flip-flop. Calculations show that the probability of a flip-flop after tunneling depends on the form of the QD: as the ratio of the height to the base of a pyramidal Ge nanocrystal grows, the probability of a flip-flop is predicted to rise [63].

4.5.3 Spin Relaxation in Two-Dimensional Structures With Asymmetrical QDs and Asymmetrical Tunnel-Coupled QDs

In an asymmetrical quantum well (asymmetry results from differences in the properties of upper and lower well boundaries, or from the application of an external electric field) we observe spin–orbit interaction, the energy of which is proportional to the electrons' velocity and spin value (the Bychkov–Rashba interaction) [64]. Such spin–orbit interaction is formally described in terms of the effective magnetic field (the Bychkov–Rashba field), which depends on the value and orientation of an electron's quasi-impulse $\vec{k}$: $H_{BR} = \alpha\left[k \times \vec{e}_z\right]/g\mu_B$, where $\vec{e}_z$ is the unit vector showing the electric field's orientation; α is the Bychkov–Rashba coefficient describing the strength of the spin–orbit interaction; g is the g factor; and μ_B is a Bohr magneton. The Bychkov–Rashba field lies in the two-dimensional gas plane and is determined by the orientation of the electrons' quasi-impulse in a two-dimensional plane.

ESR experiments on two-dimensional conductivity electrons have proved the existence of the Bychkov–Rashba field in an asymmetrical SiGe/Si/SiGe quantum well. Asymmetry was achieved via the addition of a dopant on one side of a quantum well, or by applied electric field [65]. The authors discovered the anisotropy of the linewidth and g factor of two-dimensional conductivity electrons in the SiGe/Si/SiGe structure, which was attributed to the emergence of the Bychkov–Rashba field. With an external magnetic field perpendicular to the quantum well's plane, the broadening of the ESR line induced by the Bychkov–Rashba field disappears, and the most intensive and narrow line was observed. Extremely narrow lines (≈ 0.03 G) are observed in structures with higher mobility and arise due to dynamical line narrowing. It was shown that the anisotropy of the ESR linewidth for structures with high mobility depends on the suppression of spin relaxation by cyclotron motion.

Based on our investigations, we found the Bychkov–Rashba coefficient $\alpha = 0.55 \times 10^{-12}$ eV cm, which was three orders of magnitude lower than in structures with quantum wells made from A3B5 semiconductors. This agrees with the observation that in Si, spin–orbit interaction is much weaker than in A3B5 semiconductor compounds. The anisotropy of the g factor rises as the concentration of carriers increases, and is explained by the growth of electrons' quasi-impulse. Extrapolation toward low carrier density yields the value of the g factor for two-dimensional electrons on the edge of the energy subband in Si: $g_{00} = 2.00073 \pm 0.00010$. Estimating the Bychkov–Rashba field in the SiGe/Si/SiGe quantum wells yielded a value of $H_{BR} \approx 10-20$ G. This is lower than the value in A3B5 materials, due to the weak spin–orbit interaction in Si. The spin relaxation rate in the SiGe/Si/SiGe system with quantum wells is therefore much lower than in analogous A3B5 heterostructures.

ESR studies of two-dimensional conductivity electrons in SiGe/Si/SiGe quantum wells with flowing current (voltage applied along the layer of two-dimensional electrons) show that the g factor changes with increasing current density. Moreover, the direction of ESR line shift in a magnetic field (toward either a higher or a lower field) depends on the current direction in the structure. This is explained by the change in the Bychkov–Rashba field's contribution to the Zeeman splitting that result from a change in the direction of movement and the average velocity of electrons.

An attempt to determine the mechanism of carriers' spin relaxation from spin echo measurements of free two-dimensional electrons in SiGe/Si/SiGe was made in Ref. [66] with spin echoes measured by a pulsed ESR spectrometer. Time T_2 of the two-dimensional electron gas confined in a SiGe/Si/SiGe quantum well was about 3 μs at liquid helium temperature. The decoherence of the spin state was attributed to the scattering of the electron quasi-impulse, which changes the direction of the effective Bychkov–Rashba field. This suggestion is based on the high ESR linewidth anisotropy, and on the following relation between relaxation times: $T_2 > T_1$ in a magnetic field perpendicular to the two-dimensional electron layer. A random change in the effective magnetic field lying in the structure's growth plane leads to a variation in the direction of the spin precession and to loss of spin coherence. This spin echo experiment is equivalent to single-qubit operations on an array of two-dimensional electrons.

An attempt to localize electrons at QDs and perform ESR studies of electrons, localized in a regular array of Ge/Si QDs grown on a specially prepared Si substrate with sites of predominant nucleation, was made in Ref. [67]. The investigated structures were regular arrays of Ge/Si QDs with a density of $\sim 10^{10}$ cm^{-2}, grown on Si[100] substrates. Nucleation sites of Ge nanocrystallites were formed by electron or holographic lithography. Arrays of QDs grown on virgin substrates without lithography were studied as well. In both cases, an ESR line with an isotropic g factor close to that of the conduction electrons in Si ($g = 1.998$) was observed when the samples were irradiated by light with photon energies less than the energy gap of Si. Spin relaxation times T_1 and T_2 were similar to those found in quantum well SiGe structures with two-dimensional electron gas (~ 1 μs). The authors attribute the observed signal to electrons localized in Si near the Ge QDs. This conclusion seems to be unconvincing, because the isotropic g factor is typical of free electrons, and the expected relaxation times of a localized state in Si must be several orders of magnitude longer. The authors note that as a result of the large sizes of the QDs (100 nm), the energy quantization could be insufficient and could thus lead to electron delocalization.

An ESR signal of electrons localized at Ge QDs in the system Ge/Si was found for the first time in the samples prepared with Stranski–Krastanov growth mode. Germanium nanocrystals had a pyramidal shape (hut clusters), with square (rectangular) bases and an average base size of 20 nm. The density of QDs was $\approx 10^{11}$ cm^{-2}. The observed ESR signal had the anisotropic g factor typical for a localized state. A strong dependence of linewidth on the orientation of the external magnetic field was revealed. In a magnetic field perpendicular to the QD growth plane the minimum linewidth was observed [68].

It was found that despite the up to 10% dispersion of QD sizes and the corresponding dispersion of the energy levels of electron states, the ESR lines are narrow (up to 0.8 G). This result is important for practical use in the spintronics of heterostructures with QDs, based on present day growth technologies.

A system of tunnel-coupled Ge/Si QDs is similar to an object with asymmetric quantum wells. The main difference is that electrons spend most of their time in the localized state. However, hopping transport through the ensemble of pyramidal (asymmetric) QDs also leads to the appearance of the Bychkov–Rashba interaction: $H_{BR} = \alpha_{QD}\Delta k[\vec{n} \times \vec{e_z}]/g\mu_B$, where α_{QD} is the Bychkov–Rashba coefficient for a system with QDs, $\vec{n}$ is the unit vector showing the electric field orientation, and Δk is the uncertainty in the electrons' impulse [69].

Spin relaxation takes place due to the interaction of an electron's own magnetic moment with the effective magnetic field, the vector of which lies in the QDs ensemble plane. The magnetic field appears during tunnel transitions of electrons between QDs, which are asymmetric with respect to their laying plane. This field in two-dimensional structures with broken structural symmetry depends on the direction of the hopping transfer of electrons in the system.

During tunneling, an electron's spin is turned by a small angle, resulting in spin flip-flop after a sufficient number of hops between QDs. However, since the hopping between QDs are random in nature (the direction of a hopping changes occasionally), the direction of the effective magnetic field changes after each hop. This means that the phase does not change on average, but its root mean square does change over time. The transverse relaxation time T_2 determines the ESR linewidth in the case of homogeneous broadening. Since, in our case, the ESR line is broadened inhomogeneously, we can estimate only the lower limit for the transverse relaxation time. The narrowest ESR line has $\Delta H \approx 0.8$ G, and $T_2 \sim 10^{-7}$ s. Estimating the effective magnetic field value yields $H_{BR} \approx 30$ G, which is close to the corresponding value in Si/Ge structures with quantum wells. It proves that the Bychkov–Rashba constants for structures with quantum wells and with QDs are similar [70].

Direct measurements of longitudinal T_1 and transverse T_2 relaxation times in Ge/Si structures with QDs were performed by the spin echo method (pulsed ESR) [71]. There were two characteristic values of T_1 and T_2 relaxation times in the ensemble of Ge/Si QDs: (1) $T_1 \approx 10$ μs and $T_2 \approx 20$ μs; (2) $T_1 \approx 0.1$ μs and $T_2 \approx 1$ μs. The unusual relation $T_2 > T_1$ was observed in both cases. This relation proves the proposed model of spin relaxation due to spin precession in the effective magnetic fields that arise during tunneling of charge carriers between QDs.

The existence of two characteristic relaxation times is connected with the disordering of QDs in the laying plane. There are pairs of QDs with strong tunnel coupling, in which the spin relaxation is rather short. The remaining QDs have a slower rate of relaxation.

Spin relaxations in QDs of two types with different shape were investigated: (1) hut clusters with a height-to-base ratio of $h/l = 0.1$; and (2) dome clusters ($h/l = 0.2$). Spin states in an array of Ge/Si QDs with different shapes of nanocrystals were studied. It was found that the ESR signal (the *g* factor and linewidth) depend on the shape of the nanocrystals: in hut clusters the transverse spin relaxation time (T_2) was as high as 20 μs, while in dome clusters the T_2 value was only 3 μs. The same ratio was also found for T_1 times: 10 μs in the hut clusters, which was one order of magnitude longer than in the dome clusters.

4.5.4 Electron Localization in Ge/Si Heterostructures With Double QDs

The organization of quantum calculations requires tunneling coupling between electron states. As was found, the tunneling coupling in plane laying asymmetrical QDs leads to strong reduction of spin relaxation time due to hopping transitions of electrons and the appearance of Bychkov–Rashba interaction. The natural way to create QDs with sufficient tunneling coupling and suppressed Bychkov–Rashba interaction is by the growth of stacking QDs with vertical alignment and small width of separating layers. In these structures, the strain distribution has a great influence on the energy spectrum of electrons localized near the QDs. For quantum calculations it needs selective access to individual qubits for implementation of one-qubit and two-qubit operations (e.g., the controlled-NOT gate or combinations of single-spin rotations of two coupled electrons and the SWAP gate). From this point of view, it would be better to distinguish these electrons in the *g*-factor value to perform separate single-spin rotations.

Strain in the SiGe QD system can be used to obtain effective localization of electrons with different *g* factors. Inhomogeneous strain distribution in Si surroundings of Ge QDs can lead to the formation of potential wells for electrons, not only near the QD apexes but also in other Si regions surrounding Ge QDs. It was shown that the strain in stacked QD structures with double QD layers induces the formation of a deep potential well near the base edge of QDs where the localization of electrons in Δ^{100} and $\Delta^{\bar{1}00}$ valleys is possible [42]. At special parameters of the structure this potential well can be deeper than the potential well near the apex of QD. These results open a possible way of electron localization in Δ^{100} and $\Delta^{\bar{1}00}$ valleys through the tuning of the strain distribution in stacked QD structures. The localization of electrons in these Δ valleys can provide another *g*-factor value of electrons, and makes single- and two-qubit operation implementation possible in the future.

The possibility of localization of electrons with different *g* factors in the Ge/Si QD system was demonstrated in Ref. [42]. ESR measurements of Ge/Si heterostructures with double QDs show that the place of electron localization can be controlled by changing the spacer thickness between QD layers. Different spatial localization of electrons leads to different *g* factors. These results can be used in the future for the building blocks of a quantum computer. Fig. 4.18 shows the double QD space configuration to achieve electron localization near the apex or near the edge of a pyramidal shaped nanocrystal.

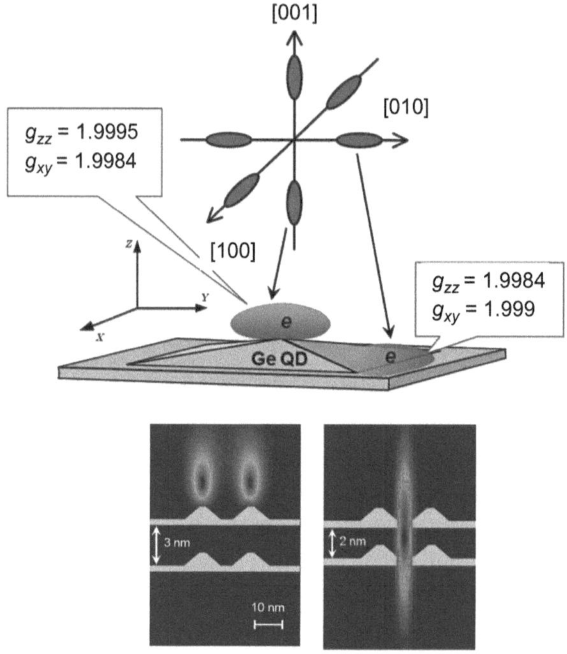

FIGURE 4.18 Scheme of possible electron localization places in the vicinity of Ge QDs [42]. The bottom panels show wave functions of the ground electron state calculated for structures with dot separation $d = 2$ and 3 nm.

4.6 Ge/Si QDs for Near- and Mid-Infrared Photodetection

4.6.1 Near-Infrared Photodetectors Operating at 1.3–1.55 μm

The design of fiber optic communication lines is one of the most important directions in the development of highly-efficient methods for transmitting information in television and telephone networks, the Internet, and optical computers. By replacing electrical wires with optical fibers, data rates can be enhanced from 10 Mb/s up to the order of 10 Gb/s, with much lower power budgets. The optical fibers used in optical telecommunications networks are predominantly produced from silica. High-purity silica is characterized by an absorption spectrum with three transparency windows at wavelengths around 0.85, 1.30, and 1.55 μm. The near-atmospheric transparency window lies in the same wavelength range. Silicon has a high absorption in the range of ultraviolet and near infrared, but from a wavelength of 1.1 μm Si becomes transparent. In recent years, several optoelectronic components have been integrated on silicon platforms, opening prospects in silicon photonics [72]. Among them, Ge QDs epitaxially grown on Si have reached responsivities comparable with those of commercial III–V devices. By properly tailoring growth conditions, it turns out to be possible to apply Ge/Si QDs for 1.3–1.55 μm optical communication applications [73–75].

Near-IR Ge/Si quantum dot photodetectors (QDIPs) utilize the interband transitions of electrons from the valence band ground state level of the Ge QDs to the conduction band.

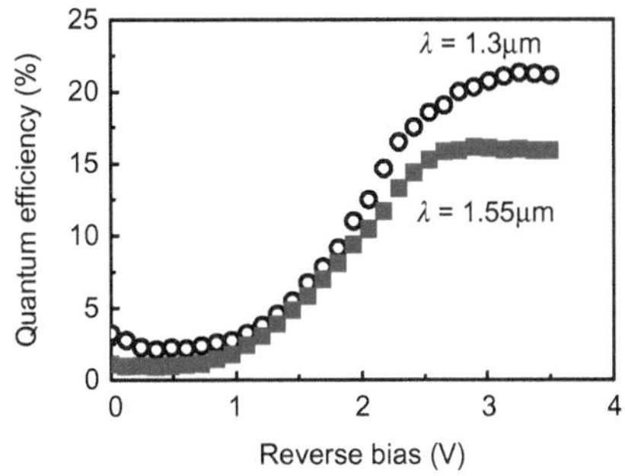

FIGURE 4.19 Quantum efficiency of a 4 mm waveguide near-IR Ge/Si photodetector as a function of reverse bias [75]. QDIP was designed as a vertical *p–i–n* diode on a SOI substrate in combination with waveguide lateral geometry and contain 36 Ge quantum-dot layers. The areal density of QDs is 1×10^{12} cm^{-2}; the dot size in the plane of growth is 8 nm.

Electromodulated reflectance study of self-assembled 10 nm sized Ge/Si QDs demonstrated that the energy of the spatially direct transitions between the electrons confined in Δ_{xy} valleys of the Ge conduction band and the localized hole states at the Γ point is about 0.8 eV [76], resulting in the respective detection wavelength of ≈ 1.55 μm. The generated electron–hole pairs then contribute to photocurrent via thermionic emission. In our samples, the sheet density of QDs varied from 0.3×10^{12} to 1.0×10^{12} cm^{-2}, and their lateral size was about 8–10 nm. Photodetectors were fabricated in the form of *p–i–n* photodiodes or *n–p–n* bipolar phototransistors with a base containing an embedded Ge/Si multilayer heterostructure with Ge self-assembled nanoislands [73]. QDIPs were produced either in the form of conventional vertical devices with illumination from the side of the *p–n* junction (or the Ge/Si heterojunction), or in the form of edge-illuminated lateral waveguides on silicon-on-insulator (SOI) substrates [74,75]. For a reverse bias of 1 V, the dark current density reaches 2×10^{-5} A/cm^2. This value was the lowest in the data on dark current densities available in the literature for Ge/Si photodetectors at room temperature [77]. The quantum efficiency of photodiodes and phototransistors subjected to illumination from the side of the plane of the *p–n* junctions is found to be 3% at a wavelength of 1.3 μm [73]. To increase the interaction length between the light and the QD layers, and to provide the intrachip interconnections, a vertical stacking of 36 layers of coherent Ge nanoislands was inserted into a waveguide obtained with a SOI structure [75]. The sample was processed into ridge waveguide. Devices with lengths from $L = 0.1$ to 5 mm were fabricated. The room temperature quantum efficiency of a 4 mm device versus reverse bias U_b is shown in Fig. 4.19. The light is coupled through the edge of the detector. The maximum external quantum efficiency achieved is 16% for $\lambda = 1.55$ μm, and 21% for $\lambda = 1.3$ μm at $L > 3$ mm and $U_b > 3$ V.

4.6.2 Mid-Infrared Photoresponse of Ge/Si QDs

There is an increasing need for sources and detectors for the mid-infrared spectral region due to the broad range of the medical and industrial applications, such as measurement of skin temperature, detection of cancer or infection, air pollution monitoring, meteorological research, and remote temperature sensing. Quantum well infrared photodetectors (QWIPs) utilizing intersubband transitions have been successful in these applications [78]. The intersubband transition energy in the quantum well is easily tunable by varying the QW width and barrier height. Also, there is a potential for the fabrication of uniform detector arrays with large areas. However, QWIPs have drawbacks such as intrinsic insensitivity to normal incidence radiation and a relatively large dark current.

It is generally assumed that, similar to QWIPs, the QD infrared photodetector is a photodetector with a majority carrier conduction in which the signal detection is not only determined by the absorption quantum efficiency, but also by the photoconductive gain. With respect to QWs, the additional in-plane confinement of carriers and the peaked density of states in QDs lead to attractive properties in the mid-wave (3–5 μm) and long-wave (8–12 μm) IR regions where the Earth's atmosphere has its major transmission windows. The potential advantages of QDIPs as compared with two-dimensional systems are the following [79,80]: (1) increased sensitivity to normally incident radiation as a result of breaking of the polarization selection rules, so eliminating the need for reflectors, gratings, or optocouplers; (2) expected large photoconductive gain associated with a reduced capture probability of photoexcited carriers due to suppression of electron–phonon scattering; and (3) small thermal generation rate, resulting from the zero-dimensional character of the electronic spectrum, that renders a much improved signal-to-noise ratio and raises operating temperature. Most of the demonstrations of QDIPs were achieved with *n*-type III–V self-assembled heterostructures. The 640×512 pixel IR QDIP imaging focal plane array (FPA) with noise equivalent temperature difference of 40 mK has already been developed for the InAs/GaAs material system [81]. Only limited studies of *p*-type QDIPs have been reported [82]. The attractive features of *p*-QDIPs include a well-preserved spectral profile, compared to a conventional *n*-type response strongly dependent on applied bias, increased density of states, and lower dark current due to the higher hole effective mass.

SiGe-based QDIPs represent another attractive type of device due to its compatibility with the standard Si-readout circuitry [83,84]. The most important advantage of SiGe-based devices is that the SiGe epitaxial growth process is compatible with both front- and back-end silicon complementary metal-oxide-semiconductor fabrication technologies. Another important advantage of photodetectors based on Ge/Si QDs, compared with II–VI and III–V materials of hybrid matrices is the match of the coefficient of thermal expansion with the silicon readout circuit. The "classical" Ge/Si QDIP with vertical photocurrent consists of a p^+–p–p^+ silicon structure with planar arrays of Ge QDs embedded in the undoped or lightly doped with boron Si region (Fig. 4.20A). Usually, Si barriers contain a boron delta-doping layer located near the QD plane to provide holes to the dots. The layers of QDs play the role of the QDIP active element. Heavily doped p^+-Si regions serve as the emitter and collector of the detector. The photoresponse of Ge/Si heterostructures with QDs in the mid-wave atmospheric window was observed by several groups [85–88] and attributed to the transitions from the hole states bound in Ge QDs to continuum states of the Si matrix, and subsequent transportation by an internal or built-in electric field. The photocurrent (PC) excited by

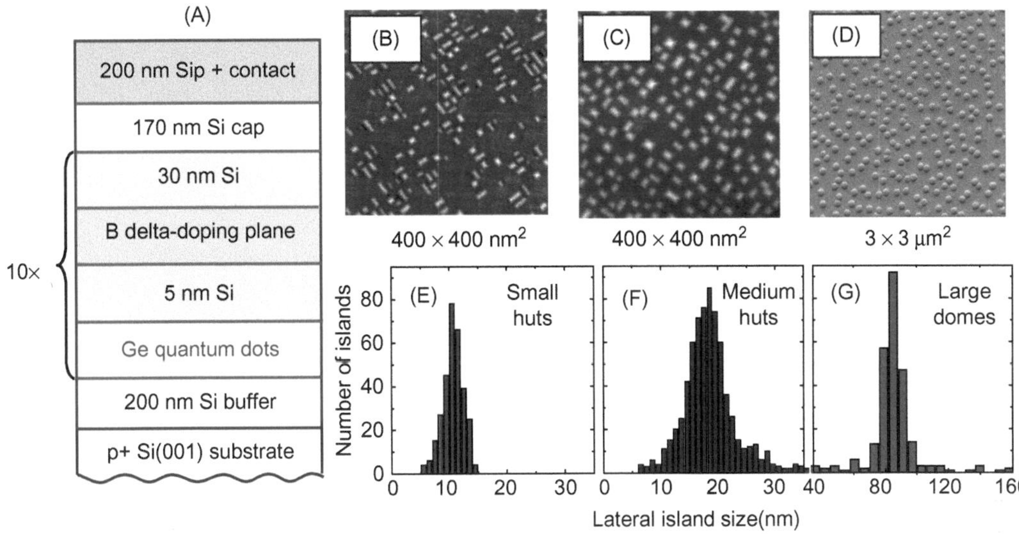

FIGURE 4.20 (A) Layer sequence of the Ge/Si photodetectors. STM images (B and C), AFM image (D), and size distribution histograms (E, F, and G) from topmost uncapped Ge layer of small-dot (B and E), medium-dot (C and F), and large-dot (D and G) samples. The image size is 400×400 nm^2 in (B) and (C), and 3×3 μm^2 in (D). Small and medium QDs have the form of huts with the areal density $n_{QD} \approx 10^{11}$ cm^{-2}, while the large dots are domes with $n_{QD} \approx 3 \times 10^9$ cm^{-2}. The areal doping density (10^{12} cm^{-2}) in boron delta-doping layers was the same for all samples. Although some of the small and medium islands did not have a square base, we used their geometrical mean, $l = \sqrt{a \times b}$ (*a* and *b* are the island base lengths), as a convenient measure of their size.

long-wave IR radiation was demonstrated by our group for a Ge/Si QDIP grown on a relaxed SiGe pseudosubstrate [89]. The detector exhibits tunable photoresponse in both 3–5 μm, and 8–12 μm spectral regions with responsivity values up to about 1 mA/W at a bias of −3 V, and operates under normal incidence radiation with background limited performance (BLIP) at 100K. A broad spectral PC response ranging from 2 to 12 μm was found in Ge QDs placed on a strained SiGe QW, which, in turn, is incorporated in a Si matrix [90].

As mentioned earlier, an expected remarkable property of QDs which originated from their discrete energy spectrum is a suppression of carrier relaxation rates due to the phonon bottleneck effect [91,92]. If the energy separation between discrete quantized levels in the dot is designed such that it exceeds the energy of optical phonons, then the electron–phonon scattering becomes strongly quenched, leading to a long photoexcited carrier lifetime. A reduced capture probability of optically-generated carriers will result in enhancement of the photoconductive gain and detector response. Clearly, this effect can be particularly important in heterostructures with small QDs having markedly separated bound states.

The phonon bottleneck has been initially proposed for electrons and observed experimentally in *p*-type Ge/Si QDIPs in Ref. [93]. We studied the effect of QD size on the mid-infrared photo- and dark current, photoconductive gain, and hole capture probability in 10-period Ge/Si QDIPs. The dot dimensions were varied by changing the Ge coverage and the growth temperature during molecular beam epitaxy of Ge/Si[001] system in the

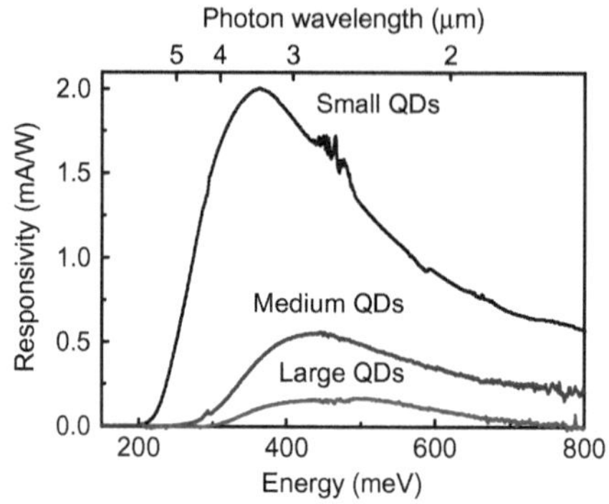

FIGURE 4.21 Zero bias photoresponse measured at $T = 80K$ for samples with QDs of different sizes. From theoretical analysis, we found that the ground hole state is located about 303, 272, and 202 meV from the Si barrier edge for the large-dot, medium-dot, and small-dot samples, respectively. All these values agree well with the onset of the photoresponse. Thus, the observed photocurrent is associated with the dominant transitions from the ground state within the Ge dots to continuum states at the Si valence band edge. Possible optical excitations from the other bound states have much smaller oscillator strength due to the presence of the nodal planes in wave function density, and make a small contribution to PC. *Reproduced from A.I. Yakimov, V.V. Kirienko, V.A. Armbrister, A.A. Bloshkin, A.V. Dvurechenskii, Phonon bottleneck in p-type Ge/Si quantum dots, Appl. Phys. Lett. 107 (2015) 213502, with the permission of AIP Publishing.*

Stranski–Krastanov growth mode. The large QDs have the shape of a lens or a dome with a circular base (Fig. 4.20D). The typical dot height is 18 nm and their base diameter is around 84 nm with the standard deviation $\pm 5\%$. The medium-sized hut-shaped Ge dots have a typical base length $l = 18 \pm 3$ nm, and about 2 nm height. For small dots, base length is $l = 11 \pm 2$ nm (Fig. 4.20B), with ≈ 1 nm height. Due to Si diffusion, the dots obtained are not pure Ge. The average Ge content of $c = 0.43$ in large QDs, 0.61 in QDs with intermediate sizes, and 0.65 in small QDs, was determined from Raman scattering experiments. Fig. 4.21 depicts the PC spectra measured at a zero bias voltage. All QDIPs show pronounced photovoltaic behavior caused by a built-in electric field of charged delta-doping planes [94]. The broad nature of the photoresponse suggests that the photocurrent is associated with a bound-to-continuum transition [83]. As the dot size decreases, the peak position shifts towards less photon energy by ~ 100 meV. This property follows directly from the general properties of QDs: decreasing the dot dimensions will reduce the hole-binding energies, bringing them closer to the continuum states. We found that the PC for small dots integrated over the mid-infrared atmospheric window from 3 to 5 μm exceeds the signals from medium and large QDs by factors of about 8 and 32, respectively. An improvement of peak responsivity was detected over the wide bias range [93].

The gain G and hole capture probability are shown in Fig. 4.22. For the sample with small QDs, the gain is much higher than unity, and the probability that a hole is captured by a QD after its optical generation is small, displaying a suppression factor of 3–30 depending on applied voltage and dot size. At a positive bias of +1 V, $G \approx 100$. This value is much higher than that achieved in GaAlAs/GaAs and InGaAs/GaAs QW photodetectors, in which G varies from 0.1 to 10 [78]. Reduction of the hole capture probability, enhancement of photocurrent,

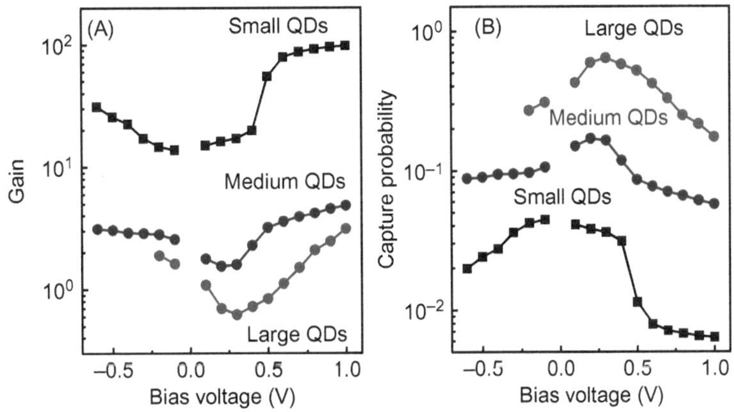

FIGURE 4.22 Photoconductive gain (A), and hole capture probability (B), as a function of applied bias [94]. The gain is defined by the ratio of the hole lifetime and the transit time across the active QD region to the contact electrode. *Reproduced from A.I. Yakimov, V.V. Kirienko, V.A. Armbrister, A.A. Bloshkin, A.V. Dvurechenskii, Phonon bottleneck in p-type Ge/Si quantum dots, Appl. Phys. Lett. 107 (2015) 213502, with the permission of AIP Publishing.*

and gain with the decrease of the dot dimensions has been explained by a phonon bottleneck mechanism for holes [93]. A carrier optically excited from the ground state to the continuum is first captured to the higher-laying excited states of the QD, and then reduces in energy towards the ground state. The calculated energy gaps between the ground and the first excited state are $\Delta E = 37$ and 60 meV for the samples containing medium and small QDs, respectively. In a device with the largest dots, $\Delta E = 0.4$ meV. The energy level spacing for the small Ge dots exceeds all LO phonon energies corresponding to Si–Si (54 meV), Si–Ge (52 meV), and Ge–Ge (37 meV) vibrations, and therefore holes can display an effective bottleneck in the relaxation towards the ground state due to the lack of phonons needed to satisfy the energy conservation rule. An opposite energy ratio is observed for QDs with intermediate dimensions where the Si–Si and Si–Ge phonon energies are larger than ΔE, and the Ge–Ge energy is very close to ΔE. In this case, the single-phonon mediated scattering process is allowed due to interaction with a Ge–Ge vibration mode, in agreement with the experimental observations. In a large-dot device, the hole spectrum has a QW-like character with a quasi-two-dimensional density of states, and the relaxation between the discrete eigenstates can be fast due to the presence of a great number of suitable final states. These results indicate that a proper choice of QD size can serve as a promising way to optimize the Ge/Si QDIP performance.

Photovoltaic sensors operating without external bias voltage are preferable for application in FPAs with low energy consumption and power dissipation, as they have the advantage of reduced noise equivalent temperature difference [95]. At zero or low bias, the detector noise is minimal and the detectivity is limited only by the thermal fluctuations. Under such conditions, the presence of the built-in electric fields near QDs, which are caused by the asymmetry of the shape of QDs, the inhomogeneous distribution of elastic deformations in the

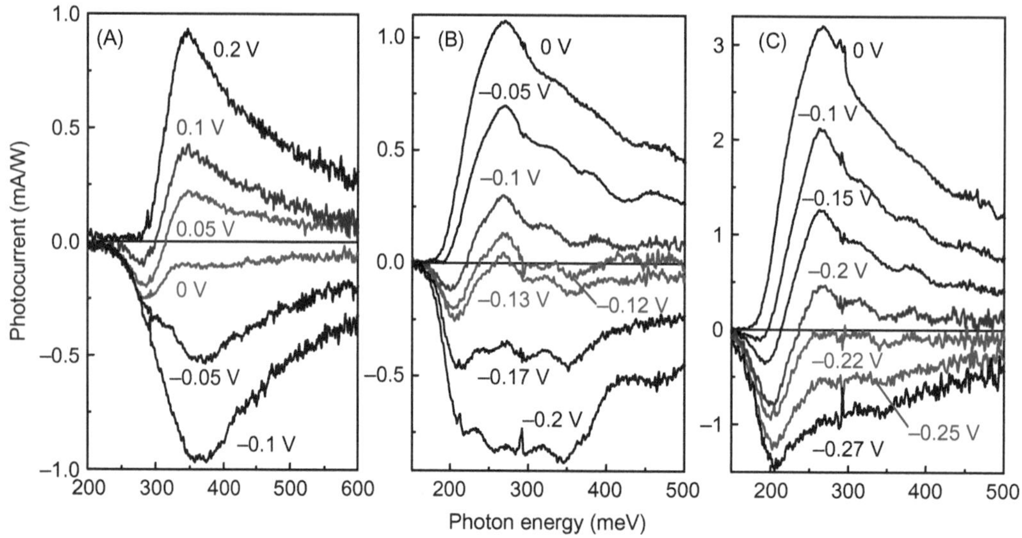

FIGURE 4.23 Mid-infrared photocurrent spectra at different voltages in the samples with different impurity concentrations of boron in the δ-doped layers: (A) $p_B = 2 \times 10^{11}$, (B) 4×10^{11}, and (C) 8×10^{11} cm^{-2}. The sample temperature was 80K [97].

medium, and the space charge of the ionized impurity atoms, is a significant physical factor. In this respect, fundamental problems arise concerning the establishment of the mechanisms and characteristics of formation of the photocurrent at low applied bias voltages. In Ref. [96], the PC spectra in boron delta-doped silicon structures with Ge QDs were studied in weak external electric fields. The weak fields are those that do not exceed the internal built-in electric fields in the sample. It was found that the direction of the photocurrent of holes in the photovoltaic mode is reversed with the increase in the impurity concentration in the structures. The most interesting result is the observation of the dependence of the photocurrent sign on the wavelength of the radiation exciting the intraband transitions (Fig. 4.23).

The effects of growth conditions and delta-doping on the performance of 10-period mid-infrared Ge/Si QDIPs were investigated in Refs. [94,97,98]. Ge QDs fabricated by molecular beam epitaxy at 500°C were overgrown with Si at different temperatures T_{cap} ranging from 300°C to 750°C for different samples, with temperature ramps before and after QD growth [97]. With increasing T_{cap} the photocurrent peak wavelength shifts from 2.3 to 3.9 μm, reflecting enhanced Si–Ge intermixing at higher overgrowth temperatures. The best characteristics were demonstrated by the sample with $T_{cap} = 600$°C. The BLIP temperature was about 90K, with 300K background and a 53 degrees field of view. For the device with a remote delta-doping density $p_B = 0.8 \times 10^{12}$ cm^{-2}, high photovoltaic detectivity of 0.62×10^{11} cm $\cdot$ Hz$^{1/2}$/W at a wavelength of 3 μm was achieved at $T = 90$K. This value is larger than that reported previously for InAs/InGaAs/GaAs QD FPAs [81]. The possibility of the Ge/Si device to operate in a photovoltaic mode makes it compatible with existing Si-readout circuits in FPA applications.

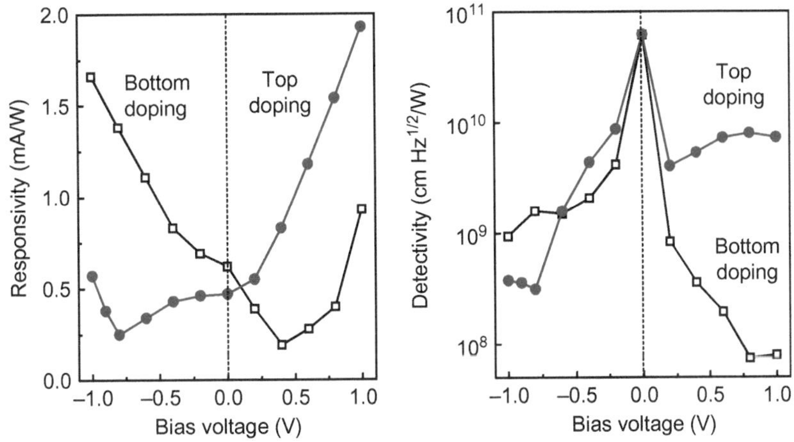

FIGURE 4.24 Measured peak responsivity and detectivity as a function of applied bias for Ge/Si QDIPs with bottom ($\lambda = 3.0$ μm) and top ($\lambda = 3.4$ μm) doping configuration. *Reproduced from A.I. Yakimov, V.A. Timofeev, A.A. Bloshkin, V.V. Kirienko, A.I. Nikiforov, A.V. Dvurechenskii, Influence of delta-doping on the performance of Ge/Si quantum-dot mid-infrared photodetectors, J. Appl. Phys. 112 (2012) 034511, with the permission of AIP Publishing.*

Within the sample set, we examined devices with different positions of the delta-doping layer with respect to the QD plane (t), different distances between the delta-doping layer and the QD plane, and different doping densities. All detectors show pronounced photovoltaic behavior, implying the presence of an internal inversion asymmetry. The measured responsivity and detectivity of the two QDIPs as a function of applied voltage are depicted in Fig. 4.24. In one of the detectors, the delta-doping plane is below the dot layer, and in the second, it is above this layer. In both samples, the distance between the QD layer and the doping plane is $t = 5$ nm, the sheet hole density $p_B = 0.8 \times 10^{12}$ cm^{-2}. The important observation is a reversal of the voltage dependence of responsivity with respect to zero bias when the δ-doping plane is moved from the bottom to the top of the dot layer. This result undoubtedly points out that the main reason for the asymmetric photoresponse is the existence of a built-in electric field due to the charged δ-doping plane [95], and is not caused by the asymmetry of the strain distribution, dot shape, and wetting layer.

Devices with a lower doping density ($p_B = 4 \times 10^{11}$ cm^{-2}) or with a shorter distance between the doping layer and QDs ($t = 2$ nm) are found to operate better in a photoconductive mode (at finite bias) with the highest peak detectivity of about 0.6×10^{11} cm · Hz$^{1/2}$/W at $T = 90$K and $U_b = 0.2$ V bias. The best performance is achieved for the device $p_B = 10^{12}$ cm^{-2} and $t = 5$ nm in a photovoltaic regime ($U_b = 0$ V). The QD layers were grown at 250°C and contain about 3×10^{11} cm^{-2} Ge huts in each layer with the island lateral size of $l \approx 6$ nm. At a sample temperature of 90K and no applied bias, a responsivity of 2 mA/W and detectivity of 1.1×10^{11} cm · Hz$^{1/2}$/W at a wavelength of 3.3 μm were measured under normal incidence infrared radiation.

Fig. 4.25 shows typical dark current–voltage (I–V) characteristics for the Ge/Si QDIPs device surrounded by a cold shield and measured at different temperatures along with the

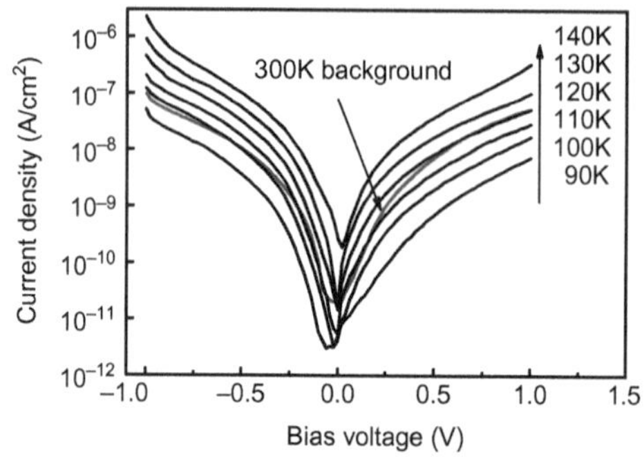

FIGURE 4.25 The bias dependence of dark current density measured at temperatures from 90K to 140K. The *dashed line* represents the response to a 300K background radiation (field of view = 180 degrees). The device has a top doping configuration with $t = 10$ nm and $p_B = 8 \times 10^{11}$ cm^{-2}. *Reproduced from A.I. Yakimov, V.A. Timofeev, A. A. Bloshkin, V.V. Kirienko, A.I. Nikiforov, A.V. Dvurechenskii, Influence of delta-doping on the performance of Ge/Si quantum-dot mid-infrared photodetectors, J. Appl. Phys. 112 (2012) 034511, with the permission of AIP Publishing.*

300K-background window current measured at 2π FOV at 90K. The asymmetry of $I-V$ characteristics is attributed to the difference in the effective barrier height experienced by the hole under positive and negative bias conditions. It can be seen that the background-limited-performance temperature is about 110K. The dark currents in Ge/Si QDIPs were found to be 3–4 orders of magnitude less than those in InAs/GaAs based photodetectors [99,100], due to the larger values of the hole effective mass and the deeper hole-binding energies in Ge/Si QDs. The high dark differential resistance at zero bias voltage ($r_0 \approx 10^{10}$ Ohm over the device area $A = 1 \times 1$ mm^2 at 90K) was responsible for the extremely low noise level ($\sim 10^{-15}$ A/Hz$^{1/2}$). The values $r_0 A \approx 10^8$ Ohm cm^2 obtained in our structures are significantly superior to the results reported for MCT systems.

Acknowledgments

The authors are much obliged to many collaborators. Among them are A.I. Nikiforov, A.V. Neneashev, A.F. Zinovieva, V.A. Zinoviev, V.V. Kirienko, N.P. Stepina, P.L. Novikov, Zh.V. Smagina, A.A. Bloshkin, V.A. Armbrister, V.A. Kudryavtsev, A.S. Deryabin, V.A. Volodin, L.V. Kulik, S.A. Teys, A.A. Shklyaev, A.K. Gutakovskii, V.A. Timofeev, R. Groetzschel, K.-H. Heinig, G.D. Ivlev, E.I. Gatskevich, P.A. Kuchinskaya, V.A. Markov, R. Rubinger, N.A. Sobolev, J.P. Leitão, M.C. Carmo, N.I. Morozova. The work was funded by Russian Scientific Foundation (grant 14-12-00931).

References

[1] Zh.I. Alferov, The history and future of semiconductor heterostructures, Semiconductors 32 (1998) 1–14.

[2] A.L. Aseev (Ed.), Nanotechnologies in Semiconductor Electronics, Siberian Branch of the Russian Academy of Science, Novosibirsk, 2004 (in Russian).

[3] N.N. Ledentsov, V.M. Ustinov, V.A. Shchukin, P.S. Kop'ev, Zh.I. Alferov, Quantum dot heterostructures: fabrication, properties, lasers (Review), Semiconductors 32 (1998) 343–365.

[4] A.I. Yakimov, V.A. Markov, A.V. Dvurechenskii, O.P. Pchelyakov, Coulomb staircase in a Si/Ge structure, Philos. Mag. B 65 (1992) 701–705.

[5] A.I. Yakimov, V.A. Markov, A.V. Dvurechenskii, O.P. Pchelyakov, Conductance oscillations in Ge/Si heterostructures containing quantum dots, J. Phys. Condens. Matter. 6 (1994) 2573–2582.

[6] O.P. Pchelyakov, Yu.B. Bolkhovityanov, A.V. Dvurechenskii, A.I. Nikiforov, A.I. Yakimov, B. Voigtländer, Molecular beam epitaxy of silicon-germanium nanostructures, Thin Solid Films 367 (2000) 75–84.

[7] Yu.B. Bolkhovityanov, A.S. Deryabin, A.K. Gutakovskii, L.V. Sokolov, A.P. Vasilenko, Dislocation interaction of layers in the Ge/Ge-seed/Ge_xSi_{1-x}/Si(0 0 1) (x ~ 0.3–0.5) system: trapping of misfit dislocations on the Ge-seed/GeSi interface, Acta Mater. 61 (2013) 5400–5405.

[8] A.I. Yakimov, A.I. Nikiforov, A.V. Dvurechenskii, V.V. Ulyanov, V.A. Volodin, R. Groetzschel, Effect of growth rate on the morphology and structural properties of hut-shaped Ge islands in Si(001), Nanotechnology. 17 (2006) 4743–4747.

[9] K.L. Wang, D. Cha, J. Liu, C. Chen, Ge/Si self-assembled quantum dots and their optoelectronic device applications, Proc. IEEE 95 (2007) 1866–1883.

[10] C.S. Peng, Q. Huang, W.Q. Cheng, J.M. Zhou, Y.H. Zhang, T.T. Sheng, et al., Improvement of Ge self-organized quantum dots by use of Sb surfactant, Appl. Phys. Lett. 72 (1998) 2541–2543.

[11] A.I. Nikiforov, V.V. Ul'yanov, O.P. Pchelyakov, S.A. Teys, A.K. Gutakovskii, Growth and structure of Ge nanoislands on an atomically clean silicon oxide surface, Phys. Solid State 46 (2004) 77–79.

[12] A.V. Dvurechenskii, V.A. Zinoviev, Zh.V. Smagina, Self-organization of an ensemble of Ge nanoclusters upon pulsed irradiation with low-energy ions during heteroepitaxy on Si, JETP Lett. 74 (2001) 267–269.

[13] J.V. Smagina, V.A. Zinovyev, A.V. Nenashev, A.V. Dvurechenskii, V.A. Armbrister, S.A. Teys, Self-assembly of germanium islands under pulsed irradiation by a low-energy ion beam during heteroepitaxy of Ge/Si(100) structures, JETP 106 (2008) 517–527.

[14] N.V. Vostokov, Yu.N. Drozdov, Z.F. Krasil'nik, D.N. Lobanov, A.V. Novikov, A.N. Yablonskii, et al., Influence of a predeposited $Si_{1-x}Ge_x$ layer on the growth of self-assembled SiGe/Si(001) islands, Phys. Solid State 47 (2005) 26–29.

[15] J. Stangl, V. Holý, G. Bauer, Structural properties of self-organized semiconductor nanostructures, Rev. Mod. Phys. 76 (2004) 725–783.

[16] S.W. Park, J.Y. Shim, H.K. Baik, Growth mode of epitaxial $Si_{0.5}Ge_{0.5}$ layer grown on Si(100) by ion-beam-assisted deposition, J. Appl. Phys. 78 (1995) 5993–5999.

[17] D. Marton, K.J. Boyd, A.H. Al-Bayati, S.S. Todorov, J.W. Rabalais, Carbon nitride deposited using energetic species: a two-phase system, Phys. Rev. Lett. 73 (1994) 118–121.

[18] G. Rosenfeld, R. Servaty, C. Teichert, B. Poelsema, G. Comsa, Layer-by-layer growth of Ag on Ag(111) induced by enhanced nucleation: a model study for surfactant-mediated growth, Phys. Rev. Lett. 71 (1993) 895–898.

[19] A.V. Dvurechenskii, V.A. Zinov'ev, V.A. Markov, R. Crötzchel, K.-X. Heinig, Effects of pulsed irradiation by low-energy ions during homoepitaxy of silicon from a molecular beam, JETP Lett. 64 (1996) 742–747.

[20] A.V. Dvurechenskii, V.A. Zinov'ev, V.A. Markov, Mechanism of structural changes of Si(111) surfaces subjected to low-energy ion pulses during molecular-beam epitaxy, JETP 87 (1998) 1116–1121.

[21] A.I. Yakimov, A.V. Dvurechenskii, V.A. Volodin, M.D. Efremov, A.I. Nikiforov, G.Yu Mikhalyov, Effect of pulsed laser action on hole-energy spectrum of Ge/Si self-assembled quantum dots, Phys. Rev. B 72 (2005) 115318.

[22] V.A. Zinoviev, A.V. Dvurechenskii, Zh.V. Smagina, G.D. Ivlev, E.I. Gatskevich, V.L. Malevich, Melting of nanocrystals embedded in a crystal matrix heated by nanosecond laser pulses, JETP 115 (2012) 436–444.

[23] A.V. Dvurechenskii, G.A. Kachurin, E.V. Nidaev, L.S. Smirnov, Pulse Annealing of Semiconductor Materials, Nauka, Moscow, 1982 (in Russian).

[24] V.A. Zinovyev, A.V. Dvurechenskii, P.A. Kuchinskaya, V.A. Armbrister, Strain-induced formation of fourfold symmetric SiGe quantum dot molecules, Phys. Rev. Lett. 111 (2013) 265501.

[25] C.G. Van de Walle, Band lineups and deformation potentials in the model-solid theory, Phys. Rev. B 39 (1989) 1871–1883.

[26] A.I. Yakimov, A.A. Bloshkin, A.V. Dvurechenskii, Asymmetry of single-particle hole states in a strained Ge/Si double quantum dots, Phys. Rev. B 78 (2008) 165310.

[27] A.I. Yakimov, A.A. Bloshkin, A.V. Dvurechenskii, Calculating the energy spectrum and electronic structure of two holes in a pair of strained Ge/Si quantum dots, Phys. Rev. B 81 (2010) 115434.

[28] A.V. Dvurechenskii, A.V. Nenashev, A.I. Yakimov, Electronic structure of Ge/Si quantum dots, Nanotechnology 13 (2002) 75–80.

[29] A.I. Yakimov, G.Yu Mikhalyov, A.V. Dvurechenskii, Molecular ground hole state of vertically coupled GeSi/Si self-assembled quantum dots, Nanotechnology 19 (2008) 055202.

[30] S. Christiansen, M. Albrecht, H.P. Strunk, H.J. Maier, Strained state of Ge(Si) islands on Si: finite element calculations and comparison to convergent beam electron-diffraction measurements, Appl. Phys. Lett. 64 (1994) 3617–3619.

[31] A.I. Yakimov, A.V. Dvurechenskii, A.I. Nikiforov, Yu.Yu Proskuryakov, Interlevel Ge/Si quantum dot infrared photodetector, J. Appl. Phys. 89 (2001) 5676–5681.

[32] A.I. Yakimov, G.Yu Mikhalyov, A.V. Dvurechenskii, A.I. Nikiforov, Hole states in Ge/Si quantum-dot molecules produced by strain-driven self-assembly, J. Appl. Phys. 102 (2007) 093714.

[33] A.I. Yakimov, A.A. Bloshkin, A.V. Dvurechenskii, Double-occupancy probability and entanglement of two holes in double Ge/Si quantum dot, JETP Lett. 92 (2010) 36–39.

[34] L. He, G. Bester, A. Zunger, Singlet-triplet splitting, correlation, and entanglement of two electrons in quantum dot molecules, Phys. Rev. B 72 (2005) 195307.

[35] A.I. Yakimov, V.A. Timofeev, A.I. Nikiforov, A.V. Dvurechenskii, Antibonding ground state of holes in double vertically coupled Ge/Si quantum dots, JETP Lett. 94 (2011) 744–747.

[36] O.G. Schmidt, U. Denker, K. Eberl, O. Kienzle, F. Ernst, Effect of overgrowth temperature on the photoluminescence of Ge/Si islands, Appl. Phys. Lett. 77 (2000) 2509–2511.

[37] A.I. Yakimov, N.P. Stepina, A.V. Dvurechenskii, A.I. Nikiforov, A.V. Nenashev, Excitons in charged Ge/Si type-II quantum dots, Semicond. Sci. Technol. 15 (2000) 1125–1130.

[38] A.I. Yakimov, A.V. Dvurechenskii, A.I. Nikiforov, A.A. Bloshkin, A.V. Nenashev, V.A. Volodin, Electronic states in Ge/Si quantum dots with type-II band alignment initiated by space-charge spectroscopy, Phys. Rev. B 73 (2006) 115333.

[39] D. Grützmacher, T. Fromherz, C. Dais, J. Stangl, E. Müller, Y. Ekinci, et al., Three-dimensional Si/Ge quantum dot crystals, Nano Lett. 7 (2007) 3150–3156.

[40] R.O. Rezaev, S. Kiravittaya, V.M. Fomin, A. Rastelli, O.G. Schmidt, Engineering self-assembled SiGe islands for robust electron confinement in Si, Phys. Rev. B 82 (2010) 153306.

[41] A.I. Yakimov, V.V. Kirienko, V.A. Armbrister, A.A. Bloshkin, A.V. Dvurechenskii, Electronic states in vertically ordered Ge/Si quantum dots detected by photocurrent spectroscopy, Phys. Rev. B 90 (2014) 035430.

[42] A.F. Zinovieva, A.I. Nikiforov, V.A. Timofeev, A.V. Nenashev, A.V. Dvurechensii, L.V. Kulik, Electron localization in Ge/Si heterostructures with double quantum dots detected by an electron spin resonance method, Phys. Rev. B 88 (2013) 235308.

[43] A.I. Yakimov, V.V. Kirienko, A.A. Bloshkin, V.A. Armbrister, P.A. Kuchinskaya, A.V. Dvurechenskii, Localization of electrons in dome-shaped GeSi/Si islands, Appl. Phys. Lett. 106 (2015) 032104.

[44] A.I. Yakimov, V.V. Kirienko, A.A. Bloshkin, V.A. Armbrister, A.V. Dvurechenskii, Strain-induced localization of electrons in layers of the second-type Ge/Si quantum dots, JETP Lett. 101 (2015) 750–753.

[45] A.I. Yakimov, A.A. Bloshkin, A.V. Dvurechenskii, Enhanced oscillator strength of interband transitions in coupled Ge/Si quantum dots, Appl. Phys. Lett. 93 (2008) 132105.

[46] A.I. Yakimov, A.A. Bloshkin, A.V. Dvurechenskii, Excitons in Ge/Si double quantum dots, JETP Lett. 90 (2009) 569–573.

[47] A.I. Yakimov, C.J. Adkins, R. Boucher, A.V. Dvurechensii, A.I. Nikiforov, O.P. Pchelyakov, et al., Hopping conduction and field effect in Si modulation-doped structures with embedded Ge quantum dots, Phys. Rev. B 59 (1999) 12598–12603.

[48] A.I. Yakimov, A.V. Dvurechenskii, V.V. Kirienko, A.I. Nikiforov, C.J. Adkins, Oscillations of hopping conductance in an array of charge-tunable self-assembled quantum dots, J. Phys. Condens. Matter. 11 (1999) 9715–9722.

[49] A.I. Yakimov, A.V. Dvurechenskii, V.V. Kirienko, Yu.I. Yakovlev, A.I. Nikiforov, C.J. Adkins, Long-range Coulomb interaction in arrays of self-assembled quantum dots, Phys. Rev. B 61 (2000) 10868–10876.

[50] A.I. Yakimov, A.V. Dvurechenskii, V.V. Kirienko, A.I. Nikiforov, Ge/Si quantum-dot metal-oxide-semiconductor field-effect transistor, Appl. Phys. Lett. 80 (2002) 4783–4785.

[51] A.I. Yakimov, A.V. Dvurechenskii, A.V. Nenashev, A.I. Nikiforov, Evidence for two-dimensional correlated hopping in arrays of Ge/Si quantum dots, Phys. Rev. B 68 (2003) 205310.

[52] A.I. Yakimov, A.V. Dvurechenskii, A.V. Nenashev, A.I. Nikiforov, A.A. Bloshkin, M.N. Timonova, Two-dimensional phononless VRH conduction in arrays of Ge/Si quantum dots, Phys. Status Solidi C 1 (2004) 51–54.

[53] A.I. Yakimov, A.V. Dvurechenskii, A.I. Nikiforov, A.A. Bloshkin, Phononless hopping conduction in two-dimensional layers of quantum dots, JETP Lett. 77 (2003) 376–380.

[54] B.I. Shklovskii, A.L. Efros, Electronic Properties of Doped Semiconductors, Springer-Verlag, Heidelberg, 1984.

[55] K.H. Schmidt, M. Versen, U. Kunze, D. Reuter, A.D. Wieck, Electron transport through a single InAs quantum dot, Phys. Rev. B 62 (2000) 15879–15887.

[56] F.W. Van Keuls, X.L. Hu, H.W. Jiang, A.J. Dahm, Screening of the Coulomb interaction in two-dimensional variable-range hopping, Phys. Rev. B 56 (1997) 1161–1169.

[57] M.A. Eriksson, S.N. Coppersmith, M.G. Lagally, Semiconductor quantum dot qubit, MRS Bull. 38 (2013) 794–801.

[58] L. Gordon, J.R. Weber, J.B. Varley, A. Janotti, D.D. Awschalom, C.G. Van de Walle, Quantum computing with defects, MRS Bull. 38 (2013) 802–807.

[59] S. Frolov, S.R. Plissard, S. Nadj-Perge, L.P. Kouwenhoven, E.P.A.M. Bakkers, Quantum computing based on semiconductor nanowires, MRS Bull. 38 (2013) 809–815.

[60] W.D. Oliver, P.B. Welaner, Materials in superconducting quantum bits, MRS Bull. 38 (2013) 816–825.

[61] D.A. Hite, Y. Colombe, A.C. Wilson, D.T.C. Allcock, D. Leibfried, D.J. Wineland, et al., Surface science for improved ion traps, MRS Bull. 38 (2013) 826–833.

[62] R. Hanson, L.P. Kouwenhoven, J.R. Petra, S. Tarucha, L.M.K. Vandersypen, Spins in few-electron quantum dots, Rev. Mod. Phys. 79 (2007) 1217–1265.

[63] A.F. Zinovieva, A.V. Nenashev, A.V. Dvurechenskii, Hole spin relaxation during the tunneling between coupled quantum dots, Phys. Rev. B 71 (2005) 033310.

[64] Yu.A. Bychkov, E.I. Rashba, Properties of a 2D electron gas with lifted spectral degeneracy, JETP Lett. 39 (1984) 78–81.

[65] Z. Wilamowski, W. Jantsch, H. Malissa, U. Rössler, Evidence and evaluation of the Bychkov-Rashba effect in SiGe/Si/SiGe quantum wells, Phys. Rev. B 66 (2002) 195315.

[66] A.M. Tyryshkin, S.A. Lyon, W. Jantsch, F. Schäffler, Spin manipulation of free two-dimensional electrons in Si/SiGe quantum wells, Phys. Rev. Lett. 94 (2005) 126802.

[67] H. Malissa, W. Jantsch, G. Chen, D. Gruber, H. Lichtenberger, F. Schäffler, et al., Investigation of the spin properties of electrons in zero-dimensional SiGe structures by electron paramagnetic resonance, Mater. Sci. Eng. B 126 (2006) 172–175.

[68] A.F. Zinovieva, A.V. Dvurechenskii, N.P. Stepina, A.S. Deryabin, A.I. Nikiforov, R.M. Rubinger, et al., Spin resonance of electrons localized on Ge/Si quantum dots, Phys. Rev. B 77 (2008) 115319.

[69] B.I. Shklovskii, Dyakonov-Perel spin relaxation near the metal-insulator transition and in hopping transport, Phys. Rev. B 73 (2006) 193201.

[70] A.V. Dvurechenskii, A.F. Zinovieva, A.V. Nenashev, Electron spins in an array of tunnel-coupled quantum dots, Bull. Russian Acad. Sci. Phys. 76 (2012) 189–193.

[71] A.F. Zinovieva, A.V. Dvurechenskii, N.P. Stepina, A.I. Nikiforov, A.S. Lyubin, Direct measurements of spin relaxation times of electrons in tunnel-coupled Ge/Si quantum dot arrays, Phys. Rev. B. 81 (2010) 113303.

[72] F. Priolo, T. Gregorkiewicz, M. Galli, T.F. Krauss, Silicon nanostructures for photonics and photovoltaics, Nat. Photon 9 (2014) 19–32.

[73] A.I. Yakimov, A.V. Dvurechenskii, A.I. Nikiforov, S.V. Chaikovskii, S.A. Tiis, Ge/Si photodiodes with embedded arrays of Ge quantum dots for the near infrared (1.3–1.5 μm) region, Semiconductors 37 (2003) 1345–1349.

[74] A.I. Yakimov, A.V. Dvurechenskii, V.V. Kirienko, N.P. Stepina, A.I. Nikiforov, V.V. Ulyanov, et al., Ge/Si waveguide photodiodes with built-in layers of Ge quantum dots for fiber-optic communication lines, Semiconductors 38 (2004) 1225–1229.

[75] A.I. Yakimov, A.V. Dvurechenskii, V.V. Kirienko, A.I. Nikiforov, Ge/Si photodiodes and phototransistors with embedded arrays of germanium quantum dots for fiber-optic communication lines, Phys. Solid State 47 (2005) 37–40.

[76] A. Yakimov, A. Nikiforov, A. Bloshkin, A. Dvurechenskii, Electromodulated reflectance study of self-assembled Ge/Si quantum dots, Nanoscale Res. Lett. 6 (2011) 208.

[77] J. Osmond, G. Isella, D. Chrastina, R. Kaufmann, M. Acciarri, H. von Känel, Ultralow dark current Ge/Si(100) photodiodes with low thermal budget, Appl. Phys. Lett. 94 (2009) 201106.

[78] B.F. Levine, Quantum well infrared photodetectors, J. Appl. Phys. 74 (1993) R1–R81.

[79] V. Ryzhii, Physical model and analysis of quantum dot infrared photodetectors with blocking layer, J. Appl. Phys. 89 (2001) 5117–5124.

[80] J. Phillips, Evaluation of the fundamental properties of quantum dot infrared detectors, J. Appl. Phys. 91 (2002) 4590–4594.

[81] S.D. Gunapala, S.V. Bandara, C.J. Hill, D.Z. Ting, J.K. Liu, S.B. Rafol, et al., Demonstration of 640 × 512 pixels long-wavelength infrared (LWIR) quantum dot infrared photodetector (QDIP) imaging focal plane array, Infrared Phys. Technol. 50 (2007) 149–155.

[82] Y.F. Lao, S. Wolde, A.G. Unil Perera, Y.H. Zhang, T.M. Wang, J.O. Kim, et al., Study of valence-band intersublevel transitions in InAs/GaAs quantum dot-in-well infrared photodetectors, Appl. Phys. Lett. 104 (2014) 171113.

[83] A.I. Yakimov, A.V. Dvurechenskii, Germanium self-assembled quantum dots in silicon for mid-infrared photodetectors, in: V. Ryzhii (Ed.), Intersubband Infrared Photodetectors, World Scientific Publishing Co., Singapore, 2003, pp. 281–298.

[84] A.I. Yakimov, Ge/Si heterostructures with Ge quantum dots for mid-infrared photodetectors, Optoelectron. Instrum. Data Process. 49 (2013) 467–475.

[85] A.I. Yakimov, A.V. Dvurechenskii, Yu.Yu Proskuryakov, A.I. Nikiforov, O.P. Pchelyakov, S.A. Teys, et al., Normal-incidence infrared photoconductivity in Si p-i-n diode with embedded Ge self-assembled quantum dots, Appl. Phys. Lett. 75 (1999) 1413–1415.

[86] C. Miesner, O. Röthig, K. Brunner, G. Abstreiter, Intra-valence band photocurrent spectroscopy of self-assembled Ge dots in Si, Appl. Phys. Lett. 76 (2000) 1027–1029.

[87] D. Bougeard, K. Brunner, G. Abstreiter, Intraband photoresponse of SiGe quantum dot/quantum well multilayers, Phys. E 16 (2003) 609–613.

[88] E. Finkman, N. Shuall, A. Vardi, V.L. Thanh, S.E. Schacham, Interlevel transitions and two-photon processes in Ge/Si quantum dot photocurrent, J. Appl. Phys. 103 (2008) 093114.

[89] A. Yakimov, V. Kirienko, V. Armbrister, A. Dvurechenskii, Broadband Ge/SiGe quantum dot photodetector on pseudosubstrate, Nanoscale Res. Lett. 8 (2013) 217.

[90] A.I. Yakimov, A.I. Nikiforov, V.A. Timofeev, A.A. Bloshkin, V.V. Kirienko, A.V. Dvurechenskii, Midinfrared photoresponse of Ge quantum dots on a strained $Si_{0.65}Ge_{0.35}$ layer, Semicond. Sci. Technol. 26 (2011) 085018.

[91] U. Bockelmann, G. Bastard, Phonon scattering and energy relaxation in two-, one-, and zero dimensional electron gases, Phys. Rev. B 42 (1990) 8947–8951.

[92] H. Benisty, C.M. Sotomayor-Torrés, C. Weisbuch, Intrinsic mechanism for the poor luminescence properties of quantum-box systems, Phys. Rev. B 44 (1991) 10945–10948.

[93] A.I. Yakimov, V.V. Kirienko, V.A. Armbrister, A.A. Bloshkin, A.V. Dvurechenskii, Phonon bottleneck in p-type Ge/Si quantum dots, Appl. Phys. Lett. 107 (2015) 213502.

[94] A.I. Yakimov, V.A. Timofeev, A.A. Bloshkin, V.V. Kirienko, A.I. Nikiforov, A.V. Dvurechenskii, Influence of delta-doping on the performance of Ge/Si quantum-dot mid-infrared photodetectors, J. Appl. Phys. 112 (2012) 034511.

[95] H. Schneider, M. Walter, C. Schönbein, R. Rehm, J. Fleissner, W. Pletschen, et al., QWIP FPAs for high-performance thermal imaging, Phys. E 7 (2000) 101–107.

[96] A.I. Yakimov, V.V. Kirienko, V.A. Timofeev, A.V. Dvurechenskii, Bidirectional photocurrent of holes in layers of Ge/Si quantum dots, JETP Lett. 100 (2014) 91–94.

[97] A.I. Yakimov, A.A. Bloshkin, V.A. Timofeev, A.I. Nikiforov, A.V. Dvurechenskii, Effect of overgrowth temperature on the mid-infrared response of Ge/Si(001) quantum dots, Appl. Phys. Lett. 100 (2012) 053507.

[98] A. Yakimov, V. Kirienko, V. Timofeev, A. Bloshkin, A. Dvurechenskii, Influence of delta-doping on the hole capture probability in Ge/Si quantum dot mid-infrared photodetectors, Nanoscale Res. Lett. 9 (2014) 504.

[99] Z. Ye, J.C. Campbell, Z. Chen, E. Kim, A. Madhukar, Voltage-controllable multiwavelength InAs quantum-dot infrared photodetectors for mid- and far-infrared detection, J. Appl. Phys. 92 (2002) 4141–4143.

[100] S. Chakrabarti, S. Adhikary, N. Halder, Y. Aytac, A.G.U. Perera, High-performance, long-wave ($\sim 10.2\ \mu m$) InGaAs/GaAs quantum dot infrared photodetector with quaternary $In_{0.21}Al_{0.21}Ga_{0.58}As$ capping, Appl. Phys. Lett. 99 (2011) 181102.

5

Electron Transport: From Nanostructures to Nanoelectromechanical Systems

A.G. Pogosov[1,2], M.V. Budantsev[1], A.A. Shevyrin[1,2], E.Yu. Zhdanov[1,2], D.A. Pokhabov[1,2]

[1]*RZHANOV INSTITUTE OF SEMICONDUCTOR PHYSICS OF THE SIBERIAN BRANCH OF THE RUSSIAN ACADEMY OF SCIENCES, NOVOSIBIRSK, RUSSIA* [2]*NOVOSIBIRSK STATE UNIVERSITY, NOVOSIBIRSK, RUSSIA*

5.1 Introduction

The prospects for semiconductor electronics aimed at miniaturization with a high processing rate have stimulated progress in the technology that makes it possible to fabricate nanoscale semiconductor structures. The great interest in such structures results from a variety of their electron-transport features, including quantum, ballistic, single-electron, and correlation effects. The scope of the transport phenomena studied has been considerably expanded by the development of recently available nanostructures detached from the bulk. It has been shown, e.g., that separation of a nanostructure from the bulk with a high dielectric constant significantly enhances electron–electron interaction. A distinguishing feature of such systems is the presence of mechanical degrees of freedom, such as bending and vibration. Electrical and mechanical degrees of freedom are often coupled to each other. Such a mutual influence of mechanical and electrical properties of suspended nanoconductors gives rise to novel effects with their possible applications, which are reflected in the name: nanoelectromechanical systems (NEMS). The study of NEMS has begun only recently, but this research field is developing rapidly, with the participation of the authors of this chapter.

This chapter reviews work reflecting the progress in the study of electron transport in nanostructures, covering a range from ordinary structures embedded in the bulk to suspended NEMS.

5.2 Electron Transport in Antidot Lattices

In the majority of ordinary conductors, e.g., in normal metals, low-temperature transport properties are determined by the electron momentum relaxation rate, which is in turn, determined by the scattering on lattice imperfections, or on randomly distributed impurities.

Advances in Semiconductor Nanostructures. DOI: http://dx.doi.org/10.1016/B978-0-12-810512-2.00005-6

Modern molecular beam epitaxy makes it possible to obtain two-dimensional electron gas (2DEG) with a momentum relaxation length far exceeding 10 μm. On the other hand, modern electron lithography, accompanied by anisotropic plasma etching, is capable of introducing artificial scatterers with sizes of less than 100 nm into the 2DEG obtained. Electron transport in such systems is determined by these introduced artificial scatterers, rather than by random impurities. The classical dynamics of an electron colliding with the boundaries of artificially introduced potential resembles a ball racing over a billiard table. With this analogy in mind, the systems described have been named electron billiards.

Examples of the most intensively investigated electron billiards are periodic lattices of scattering disks, antidots, which can be thought of as artificial crystal models. Such lattices were first investigated in Ref. [1]. The interest in studying them was initially motivated by the idea that their transport properties would reflect the quantization of electronic wave functions in such artificial crystals, leading to an energy spectrum in the form of the Hofstadter butterfly [2]. However, experimental investigations [3,4] have demonstrated that what mainly determines the conductivity of the crystals is classical chaotic electron dynamics, and the systems themselves are, in fact, examples of so-called Sinai billiards [5].

The results presented below were obtained on square antidot lattices. The experimental samples were fabricated from the GaAs/AlGaAs heterojunction containing 2DEG with an electron density of $n = (3-5) \times 10^{11}$ cm^{-2} and a mobility of $\mu = (5-8) \times 10^5$ cm^2/V s (see Fig. 5.1). The sample's mesa in the form of Hall bars with dimensions of $L \times W = 100 \times 50$ μm was defined by optical lithography. Using electron lithography and subsequent anisotropic plasma etching, the Hall bars were covered with square lattices of antidots with a diameter of $2a = (0.1-0.3)$. The real size of the antidots was somewhat larger, due to the depletion layers with a width of 0.1 μm surrounding the antidots. The magnetoresistance of the antidot lattices was measured using the four-probe technique in the magnetic field perpendicular to the 2DEG plane.

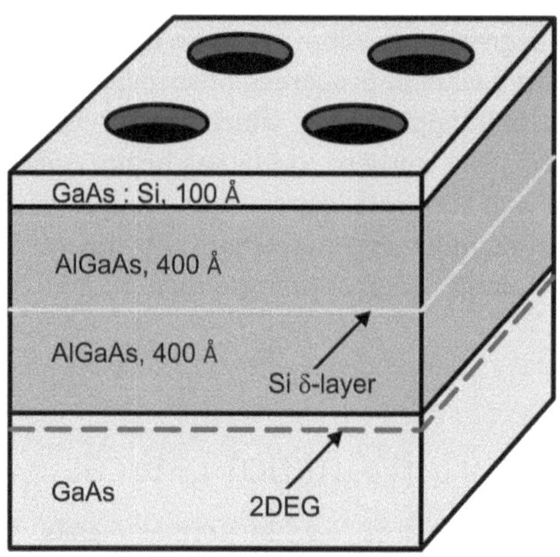

FIGURE 5.1 The AlGaAs/GaAs heterostructure containing two-dimensional electron gas. The etched antidots are represented on top of the heterostructure.

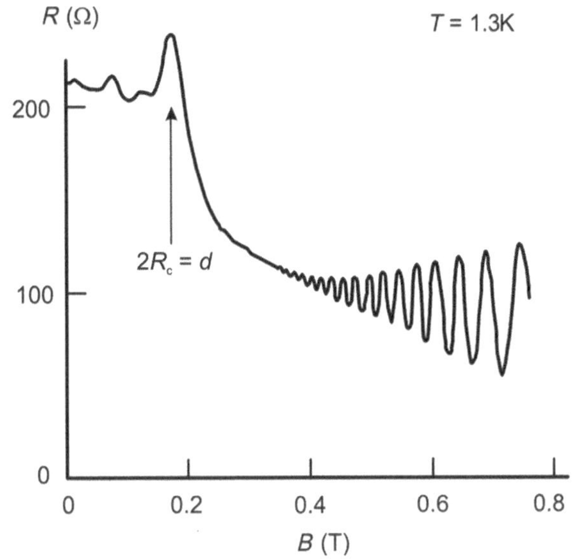

FIGURE 5.2 Magnetoresistance of the square antidot lattice, $d = 1.3$ μm.

The magnetoresistance of the square lattice of antidots (lattice period $d = 1.3$ μm) measured at a temperature of $T = 1.3$K is shown in Fig. 5.2. In a low magnetic field, one can see oscillations where the peak, at $B \approx 0.2$ T, has the maximum amplitude. This magnetic field corresponds to the commensurability between the cyclotron orbit diameter and the lattice period $2R_c = d$. The observed low-field anomalies of magnetoresistance are called commensurability oscillations. The study of commensurability oscillations over a wide temperature range has shown that they survive up to temperatures of about 50–70K, and this indicates their classical origin. In magnetic fields higher than 0.4 T, one can see Shubnikov–de Haas oscillations caused by Landau quantization, which are suppressed at a liquid helium temperature of 4.2K.

A numerical simulation of electron motion in the lattice of antidots was performed in order to point out classical trajectories responsible for the observed commensurability oscillations (Fig. 5.3). The analysis of electron trajectories enables us to conclude that, besides chaotic disordered trajectories, a set of dynamically stable regular trajectories occupying a finite volume in the phase space exists, which is the signature of systems with dynamic chaos.

The physical model proposed in Ref. [3] explains the main commensurability peak by a skipping trajectory running away along the antidot arrays that appears at $2R_c \approx d$. The Poincare surface of such a trajectory is presented in Fig. 5.4, where the correspondence between the initial conditions (φ,ψ) and the number of successive jumps to neighboring antidots, until a trajectory breaks out of an array, is shown in color. The large white area corresponds to the trajectories skipping without a break along the array at distances exceeding 100 lattice periods. Among other things, the calculated figure reflects the fact that the stability area of the skipping trajectories has a fractal structure. Moving away from the stability area in the phase space leads to the gradual destruction of regular motion. Taking into

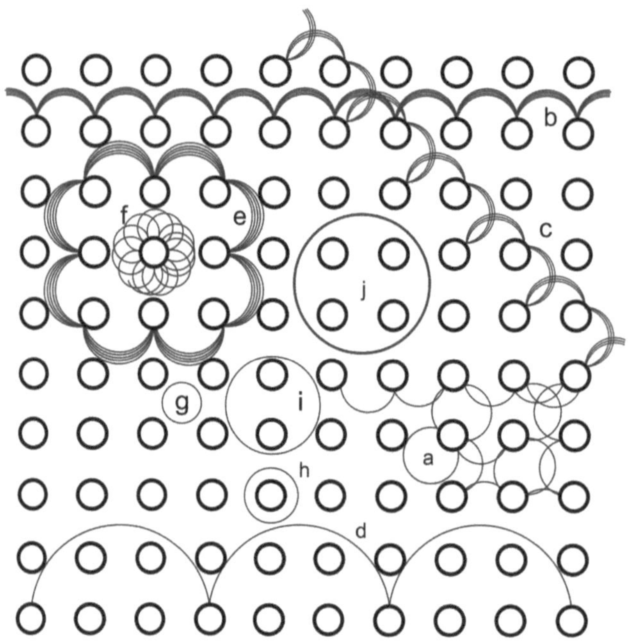

FIGURE 5.3 Some types of possible electron trajectories in a square lattice of antidots: (a) ordinary diffusion trajectory; (b–d) runaway trajectories; (e) skipping localized trajectory; (f) rosette trajectory; (g–j) pinned orbits.

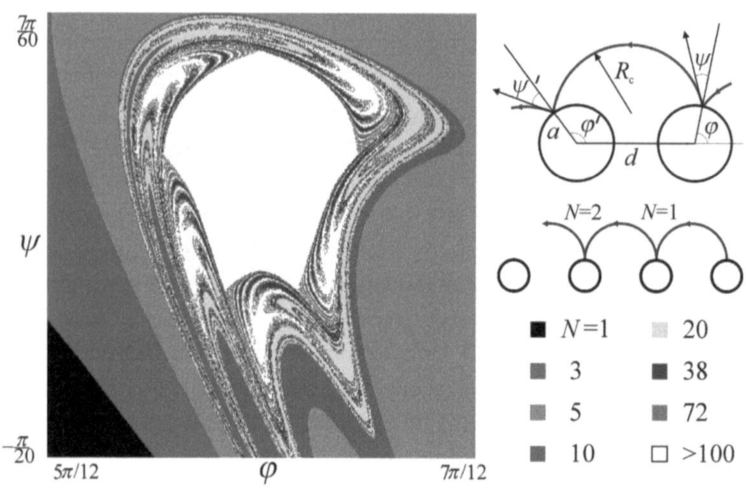

FIGURE 5.4 Poincare section of the trajectories "skipping" along the row of antidots presented in angular coordinates. The number of following jumps along the row of antidots occurring without dephasing is shown in color.

account a finite mean free path of 5–10 μm (and a lattice period of the order of 1 μm), one can classify trajectories performing at least 5–10 jumps along an array as the category of runaway trajectories. Such trajectories occupy a large volume in the phase space, and provide an essential anomalous contribution to conductivity.

It should be noted that another model was proposed in Ref. [6] to explain commensurability oscillations in square lattices by pinned electron orbits surrounding one (trajectory *h* in Fig. 5.3) or several (trajectories *i*, and *j* in Fig. 5.3) antidots, depending on the magnitude of the magnetic field. This approach is often considered as the main model in literature, whereas, in a series of our contributions [7,8] and investigations by other groups [9,10], it was shown that the runaway trajectories model is more consistent with a wide scope of experimental data, while the pinned-orbit model corresponds primarily to the case of dense antidot lattices.

In order to clarify how the symmetry of a lattice of antidots influences transport properties, the lattices of other types were investigated. Presented below are the results obtained on a hexagonal "honeycomb" lattice. The magnetoresistance of the honeycomb lattice of antidots is shown in Fig. 5.5A. It demonstrates four distinct peaks. Analysis of

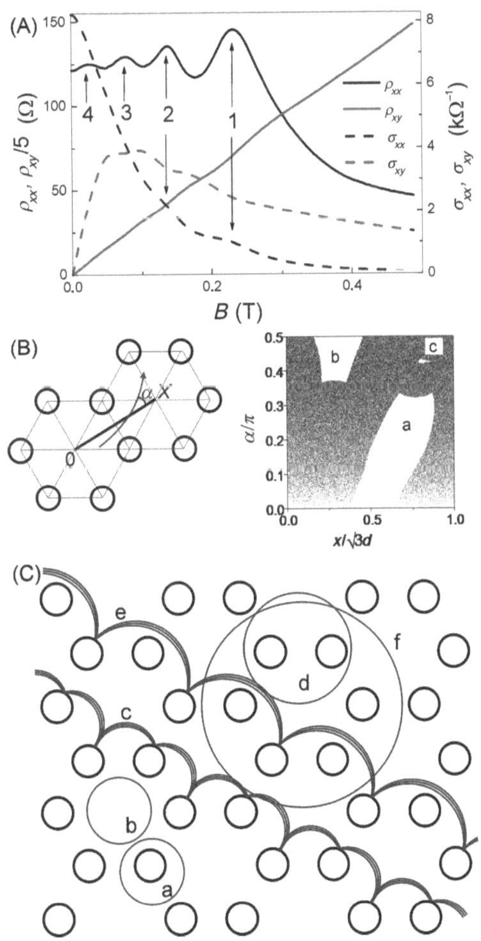

FIGURE 5.5 (A) Longitudinal (ρ_{xx}) and Hall (ρ_{xy}) magnetoresistivity of a honeycomb antidot lattice. Magnetoconductivity was calculated from the measured magnetoresistivity; (B) lattice structure (left) and Poincare section (right). *White areas* correspond to the special trajectories shown in (C).

the Poincare sections, together with the types of possible electron trajectories (Fig. 5.5B and C), shows that the main commensurability peak in the honeycomb lattice is caused by both runaway (c) and pinned (a, b) trajectories. Peak 2 is formed by trajectories (d) and (e), peak 3 is caused by pinned orbits, and peak 4 originates from the break in linear channeling trajectories by the applied magnetic field. Thus, depending on the lattice structure, the commensurability oscillations can appear as a result of both runaway and pinned trajectory formation.

5.3 Electron Transport in "Star" and "Caterpillar" Billiards

In antidot lattices the electron momentum relaxation length l_p limited by scattering on a random impurity potential exceeds the distance between artificial scatterers, but it is less than the whole lattice size (sample). With the samples size decreasing to below l_p, the description of their transport properties in terms of local conductivity tensor becomes no longer valid. A proper description of such systems is given by the Landauer–Buttiker formalism [11] in terms of the transmission probabilities from one sample terminal to another. One such system is the "star"-shaped billiards representing an example of so-called Sinai billiards (Fig. 5.6). The billiard shape is determined by six antidots placed in regular hexagonal vertices, with ballistic entrances (contacts) to the billiard cavity

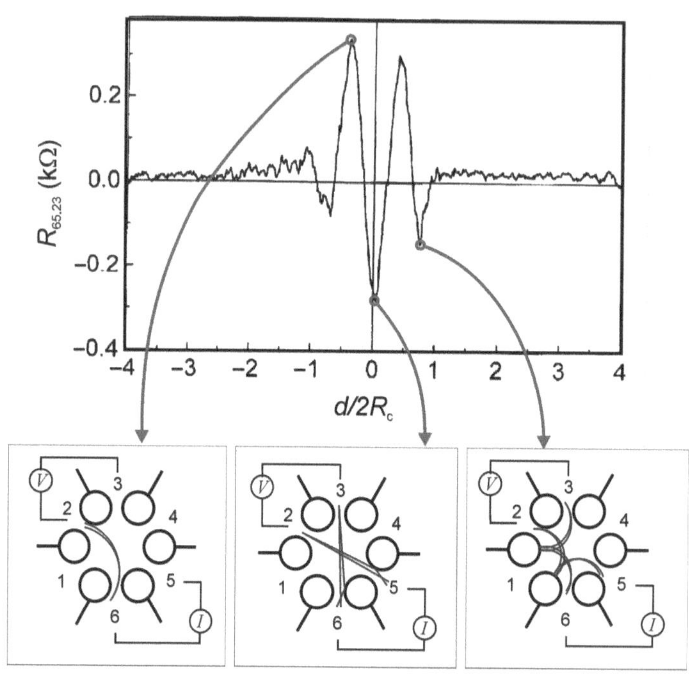

FIGURE 5.6 Magnetoresistance of "star"-like billiard and electron trajectories, which cause the features of magnetoresistance.

formed between them. The four-terminal resistance of the billiard can be obtained by passing the current between one pair of contacts (i, j), and measuring the voltage drop between the other pair (k, m), $R_{ij,km} = V_{km}/I_{ij}$. The magnetoresistance $R_{65.23}$ of the "star"-like billiard, measured at a temperature of 4.2K is shown in Fig. 5.6. The topology of the scheme for the measurement of $R_{65.23}$ is similar to the one used in van der Pauw measurements of the diagonal resistivity component ρ_{xx}. Thus, the magnetoresistance of a 2DEG system with an isotropic effective mass (which is true for the GaAs/AlGaAs heterojunction) measured by this method should be zero. One can see in Fig. 5.6 that the magnetoresistance of the "star"-billiard is not zero. It is not only large, it is also of the alternate sign.

This anomalous behavior is explained by the ballistic flythrough of an electron across the sample, driven by the magnetic field. In a zero magnetic field, a collimated flythrough from contact 5 to contact 2, and from 6 to 3, determines the resistance. The crossover transition of electrochemical potential leads to the inversion of electrical response compared to the case of an ordinary macroscopic device with diffusive transport. With the increasing magnetic field, flythrough 6–2 becomes possible due to the cyclotron curvature of the orbit, the crossover transition is suppressed, and the resistance becomes positive. With a further magnetic field increase, the crossover transition comes again into play due to colliding trajectories, and the resistance takes negative values again.

One more example of Sinai billiards is the "caterpillar"-like billiard (Fig. 5.7). We have fabricated a billiard of this type introducing two rows of periodically situated antidots, forming a waved channel connected to the terminals for longitudinal and Hall resistance measurements. Additional channels adjacent to the left and right side of the billiards are used to heat the 2DEG by the heating current and, thus, obtaining the temperature gradient necessary for measurements of the diffusive thermopower component. Typical billiard dimensions

FIGURE 5.7 Electron microscope image of the "caterpillar"-like billiard.

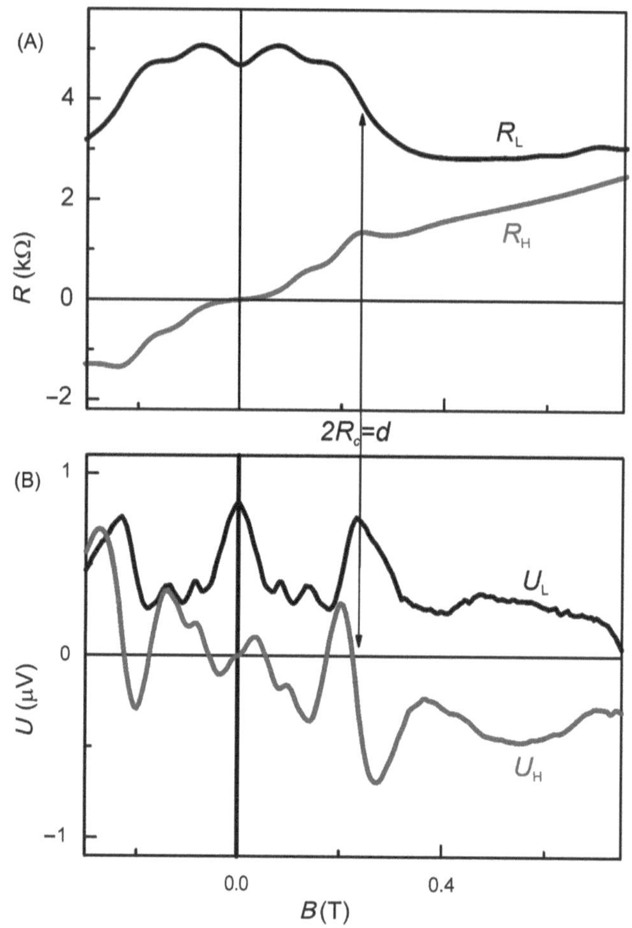

FIGURE 5.8 (A) Longitudinal and Hall magnetoresistance of "caterpillar"-like billiard; (B) Longitudinal and Hall thermopower.

do not exceed 3 μm, which is less than l_p. The magnetoresistance of the "caterpillar" is presented in Fig. 5.8A. It can be seen in the figure that the resistance increases with increasing magnetic field due to the channeling trajectories destruction. When approaching commensurability condition $2R_c = d$, the magnetoresistance turns to a dip caused by the formation of runaway trajectories skipping along the channel.

The magnetic field dependence of both the longitudinal component of thermopower U_L, and the transverse component U_H (Nernst–Ettingshausen effect), are shown in Fig. 5.8B. In the commensurability region $2R_c = d$, longitudinal thermovoltage U_L exhibits a maximum, and U_H has S-shaped behavior. One can see in Fig. 5.8 that thermopower demonstrates considerably more pronounced features compared to magnetoresistance. One can explain this by the fact that the thermopower measurement is, in fact, a differential technique, since the measured thermopower is determined by derivatives of transmission coefficients as shown in Ref. [12], devoted to the generalization of the Landauer–Buttiker formalism to thermoelectric phenomena. The thermopower magnitude is determined by the difference in transmission coefficients for quasielectrons above the Fermi level and quasiholes below it. This difference reverses the sign close to the region $2R_c = d$.

5.4 Weak Localization in a Square Antidot Lattice

Beside the effects related to classical electron dynamics, the billiards considered demonstrate quantum phenomena connected with the interference of electron waves on the artificial potential. With decreasing temperature, the phase coherence length L_φ of the electronic wave function becomes longer than the lattice period. This is the condition when quantum phenomena can come into play. One can observe the weak localization phenomenon that manifests itself by decreasing sample conductivity compared to the classical Drude value. The origin of the weak localization is interference of close electron trajectories propagating the closed loop in opposite directions. Applying a magnetic field leads to a phase difference between these trajectories which is proportional to the loop-encircled area. Therefore, the area distribution of closed trajectories determines the magnetic field dependence of the weak localization correction to classical conductivity. In ordinary conductors with disordered scatterers, one cannot point out any particular area. With increasing magnetic field, the contributions from a number of loops with different areas rapidly oscillate, leading to a monotonous suppression of the weak localization correction [13].

The situation is different in a periodic antidot lattice. The magnetic field dependence of the conductivity in a square antidot lattice is shown in Fig. 5.9A. Against the background of monotonous suppression of weak localization one can see a dip corresponding to the magnetic flux quantization (with flux quantum $\Phi_0 = hc/e$) through the area of the antidot lattice cell d^2. The self-crossing (closed) trajectories distribution over the encircled areas obtained by numerical calculations is presented in Fig. 5.9B. The distribution has a pronounced maximum at $S = d^2$. The presence of this particular area explains the observed features of weak localization in the periodical antidot lattice [14].

5.5 Mesoscopic Conductance Fluctuations in Sinai Billiards

In samples with dimensions less than the phase coherence length, the interference of electron waves leads to mesoscopic conductance fluctuations (MCF). Changing magnetic fields affect the phases and, thus, the interference conditions for quasiclassical trajectories, resulting in a change in conductance. These changes resemble chaotic fluctuations, while in contrast to them they are reproducible during successive magnetic field sweeps. In ordinary disordered conductors the mean square value of MCF is close to the universal value e^2/h independent of the sample material [15]. That is why these fluctuations are also called universal.

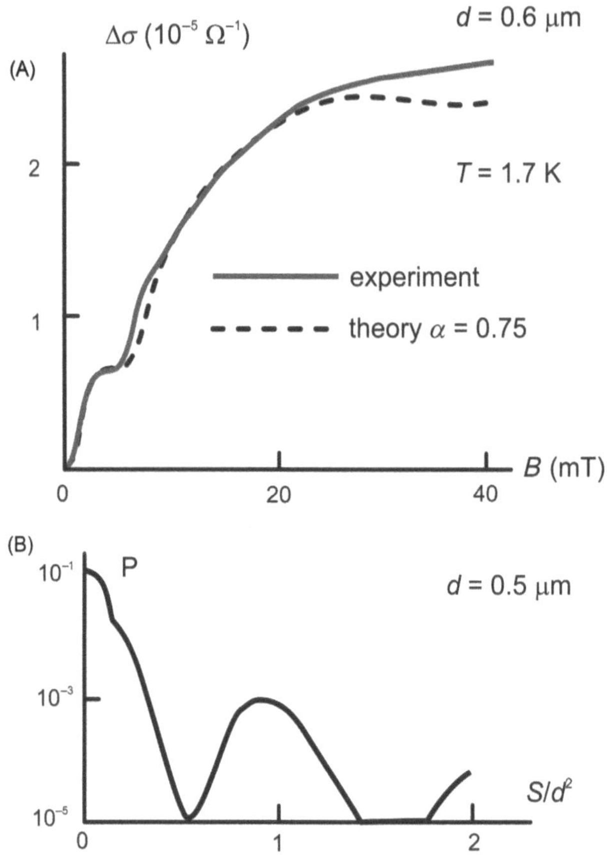

FIGURE 5.9 (A) The suppression of weak localization correction to the conductivity of the square antidot lattice by magnetic field; (B) The self-crossing trajectories distribution over the encircled areas.

Below are the results of experiments aimed at investigating the properties and features of MCF in antidot lattices [16]. Test samples were Hall bars with dimensions of $10\times100\ \mu m^2$ covered with square antidot lattices with a period of $d=500$ nm. The antidot diameter (including the surrounding depletion layer) was equal to $2a=220$ nm.

At a temperature of $T=4.2$K the magnetoresistance demonstrates a well-pronounced peak at a magnetic field of $B\approx0.53$ T (Fig. 5.10A). With the temperature decreasing below 1.7K, one can observe MCF against the background of commensurability oscillations. As one can see from Fig. 5.10B, MCF are observed both in low fields (below 0.3 T), and in the region of the main commensurability peak (around 0.5 T). The MCF amplitude presented in Fig. 5.10C remains practically unchanged, and is approximately equal to $10^{-3}e^2/h$ over a wide temperature range from 1.7K down to 10 mK. This indicates MCF nonuniversality in antidot lattices. The nonuniversal behavior of MCF can be understood from the already mentioned distribution $P(S)$ of closed trajectories over the encircled areas (Fig. 5.9B), and it has

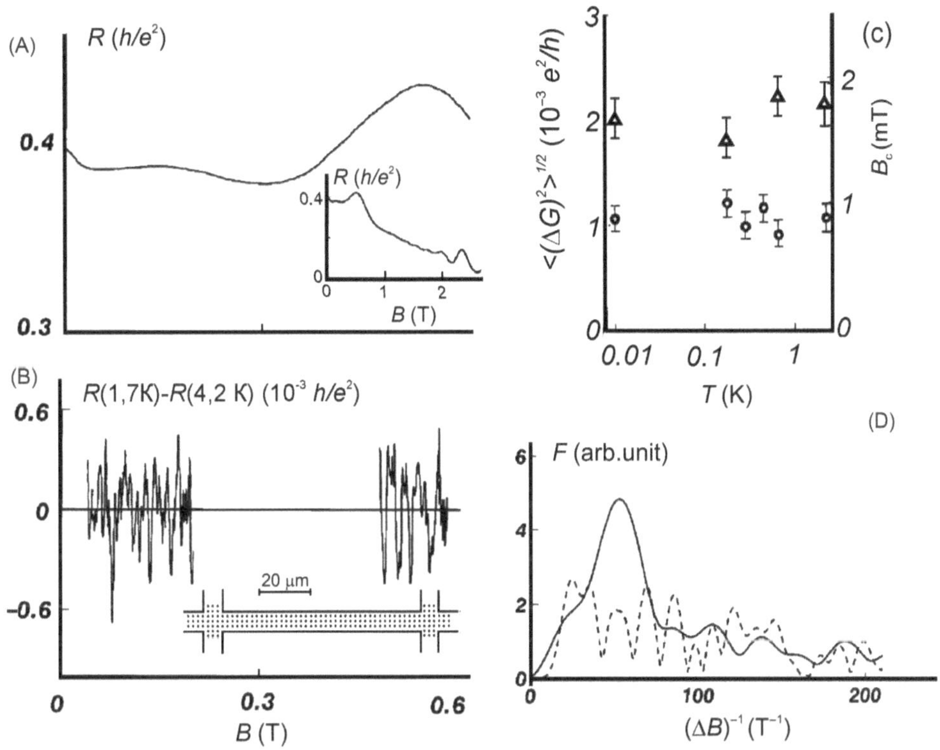

FIGURE 5.10 (A) Magnetoresistance of the square antidot lattice at $T=4.2$K; (B) Mesoscopic conductance fluctuations at $T=20$ mK; (C) Temperature dependence of root mean square amplitude of mesoscopic conductance fluctuations (*circles*) and correlation magnetic field (*triangles*); (D) Fourier spectrum of mesoscopic conductance fluctuations.

its maximum at $S=d^2$, in contrast to ordinary disordered conductors with monotonous exponential distribution $P(S)=\exp(-\alpha S)$. As a result, with the decreasing temperature, when phase coherence length L_φ becomes longer than d, the quantum interference correction is determined not by trajectories with length L_φ (as it would be for ordinary conductors), but by trajectories with characteristic length d (encircling area d^2), whose number is anomalously large and temperature-independent. The MCF spectrum in the region of the main commensurability peak presented in Fig. 5.10D demonstrates a maximum corresponding to the magnetic flux quantization through the area equal to the area of the lattice cell, minus the area of an antidot $d^2-\pi a^2$. This quasi-periodicity is also connected with anomalous distribution $P(S)$ remaining in higher magnetic fields. Phenomenologically, one can distinguish three particular areas in the lattice, which are d^2, $(d^2-\pi a^2)$, and πa^2. The oscillations already mentioned corresponding to the first two can be seen, while the last area πa^2 is too small in our case; it corresponds to oscillations with a large period comparable to the period of commensurability oscillations and, thus, is unobservable against their background.

5.6 Hysteretic Magnetoresistance of a Two-Dimensional Electron Gas in the Quantum Hall Effect Regime

One of the most important and interesting areas in the study of electron transport in nano- and mesoscopic structures is adiabatic transport in high magnetic fields, when the quantum Hall effect is observed. In quantizing magnetic fields, the electron spectrum splits into Landau levels and, at low temperatures, a 2DEG demonstrates the quantum Hall effect, which manifests itself in zero longitudinal resistance R_L and quantized Hall resistance $R_H = h/(2e^2 N)$, where N is the number of filled Landau levels. In this regime, N edge current states, drifting along the edge of the sample, are formed. In conventional macroscopic samples, electron backscattering between the current states propagating along the remote opposite edges is suppressed, thus leading to the disappearance of R_L.

In Ref. [17] we created a constriction in the macroscopic 2DEG, and thus made two edge current states propagating along the opposite edges of the sample closer to each other. It has been found that the magnetoresistance of the constricted 2DEG demonstrates a giant hysteresis near the integer filling factors (Fig. 5.11). The observed hysteresis is independent from the magnetic field sign, as well as from its parallel component [18]. This independence excludes the spin nature of the phenomenon. At filling factor $\nu = 1$, the hysteresis is observed against a background of zero magnetoresistance, and has an amplitude of 10 kΩ and a width of 1.2 T. The effect is observable over a wide temperature range; it is independent from

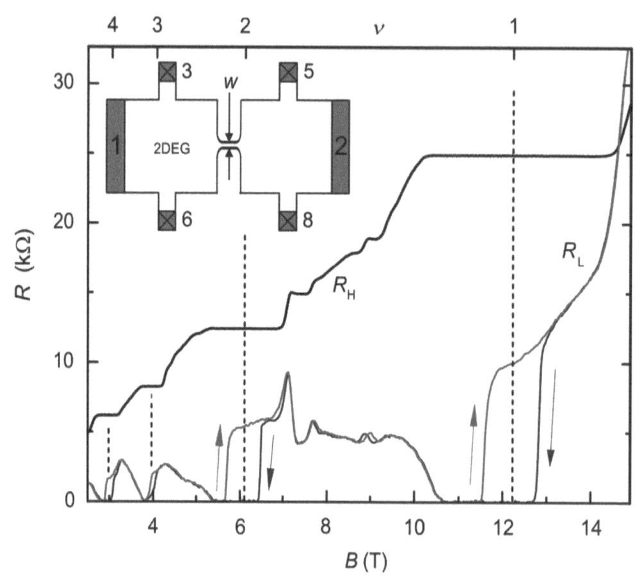

FIGURE 5.11 The magnetoresistance of a 2DEG with a constriction. The hysteresis of longitudinal magnetoresistance $R_L = U_{34}/I_{12}$ is observed at the centers of the plateaus of macroscopic 2DEG Hall resistance $R_H = U_{35}/I_{12}$. The geometry of the sample is shown in the inset.

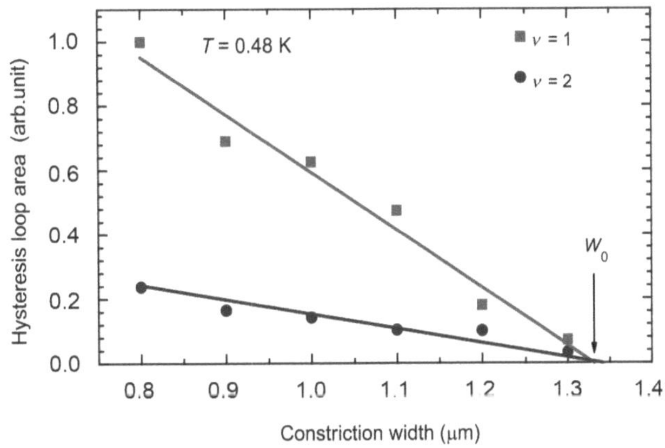

FIGURE 5.12 The dependence of the hysteresis loop area on the constriction width at filling factors $\nu = 1$ and 2 at a temperature of 0.48K.

temperature in the range of 60 mK to 400 mK, and is completely suppressed at temperatures higher than 1.4K [19].

It has been shown that hysteresis has significant phenomenological similarities to the magnetization hysteresis inherent in ferromagnetic materials showing multistability, jumps of relaxation, and the anhysteretic curve. The time relaxation of the hysteresis has fast and slow phases, similar to that of nonequilibrium magnetization of a 2DEG in the QHE regime, pointing to their common origin [20].

However, in contrast to the case of ferromagnets [21], the hysteresis observed in 2DEG is unusually inverted, i.e., the system response (magnetoresistance) advances the external excitation (a small change of external magnetic field), rather than retards it.

Studying the dependence of the hysteresis loop area on the lithographic constriction width (Fig. 5.12) showed that the hysteresis vanishes at widths bigger than the critical width $W_0 \approx 1.35$ μm. This points to the edge nature of the nonequilibrium currents, and allows us to determine the width of the area where these currents are located, which is about 0.5 μm [22].

The observed hysteresis can be explained by a nonequilibrium redistribution of electrons between the edge and the bulk states of Landau levels (Fig. 5.13A and B). This results from a considerable suppression of inter-level transitions in the quantum Hall effect regime, and leads to topological transitions in the electron liquid in the constricted 2DEG (Fig. 5.13C and D) [20].

5.7 Ballistic Effects in a Suspended 2DEG

The investigation results of electron transport in nanostructures containing 2DEG detached from a semiconductor substrate are discussed below. Such systems, as a new object of study, require detailed consistent research into their mechanical features and electron transport. The direct experimental comparison implying the performance of similar electrical

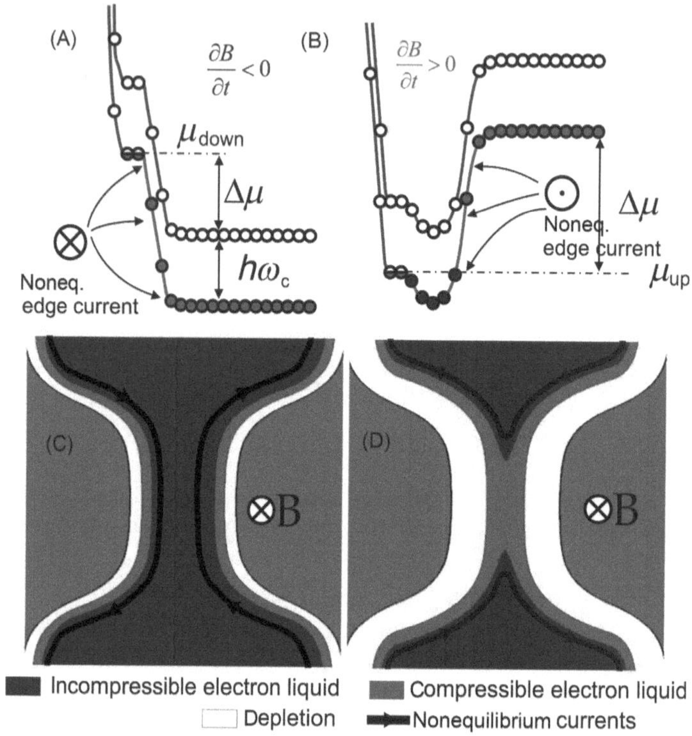

FIGURE 5.13 The structure of nonequilibrium currents and electron liquid in a 2DEG with the constriction at down (A and C), and up (B and D), sweep of the magnetic field in the quantum Hall effect regime.

measurements on the same samples before and after their suspension is a reasonable and simple method to reveal the electron transport features in suspended structures. We used this method to study suspended Hall bars, antidot lattices, quantum point contacts, and single-electron transistors.

The Hall bars (Fig. 5.14) were fabricated by photolithography from AlAs/GaAs heterostructures containing high-mobility 2DEG grown by molecular beam epitaxy. The 2DEG density was $n = 6 \times 10^{11}$ cm^{-2}, while the mobility was $\mu = 2 \times 10^{6}$ cm^2/V s measured at 4.2K. The 160 nm-thick heterostructures were grown above the 400 nm-thick $Al_{0.8}Ga_{0.2}As$ sacrificial layer which was afterwards selectively removed from under the Hall bars by means of isotropic wet etching in the HF-based etchant. A typical length of the suspended Hall bars was about 10 μm. The electron density remained practically unchanged after suspension, while the mobility was halved. Magnetoresistance measurements were performed over a wide temperature range from 0.48K to 40K.

Both nonsuspended and suspended Hall bars demonstrate negative longitudinal magnetoresistance in low magnetic fields ($B \leq 0.2$ T, Fig. 5.15). The absence of temperature dependence of magnetoresistance is evidence of the ballistic rather than quantum interference nature of the effect.

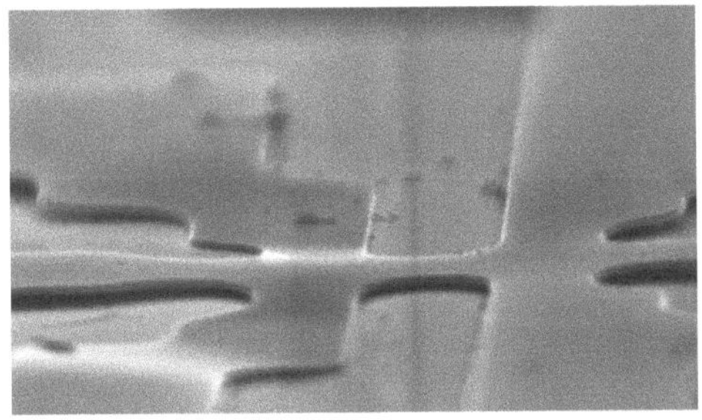

FIGURE 5.14 Electron microscope image of the suspended Hall bar.

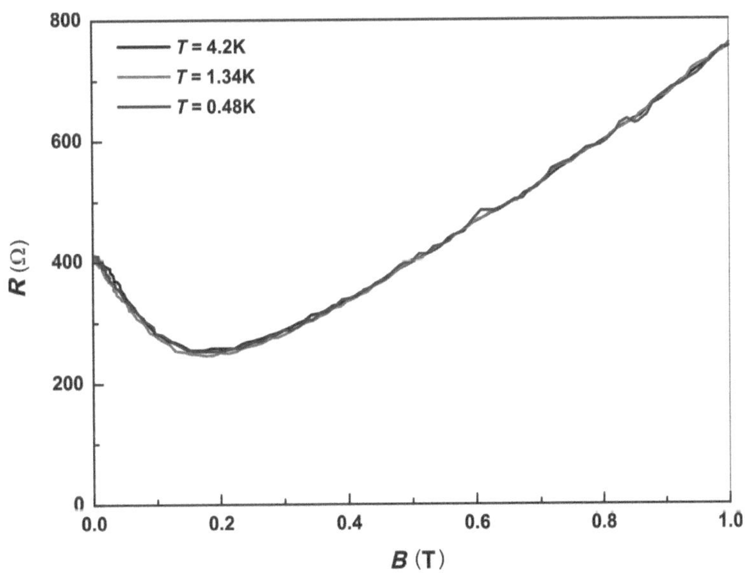

FIGURE 5.15 Magnetoresistance of the suspended Hall bar at low magnetic fields.

At a magnetic field corresponding to the minimum magnetoresistance ($B \sim 0.2$ T), the electron orbit cyclotron diameter is about 2 μm, which is equal to the effective width of the narrow part of the Hall bar (including the depletion regions). Taking into account the ballistic nature of the observed effect, we can conclude that the mean free path of electrons is at least larger than 2 μm [23].

5.8 Suspended 2DEG Structured With an Antidot Lattice

The next step was the creation of nanostructures from the ballistic suspended Hall bars. The Hall bars, together with voltage probes, were covered with a periodical square lattice of antidots by means of electron beam lithography followed by anisotropic plasmachemical etching

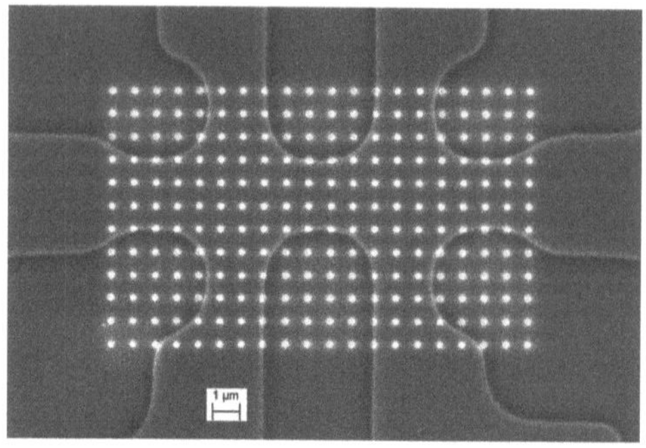

FIGURE 5.16 Hall bar covered with a periodic antidot lattice.

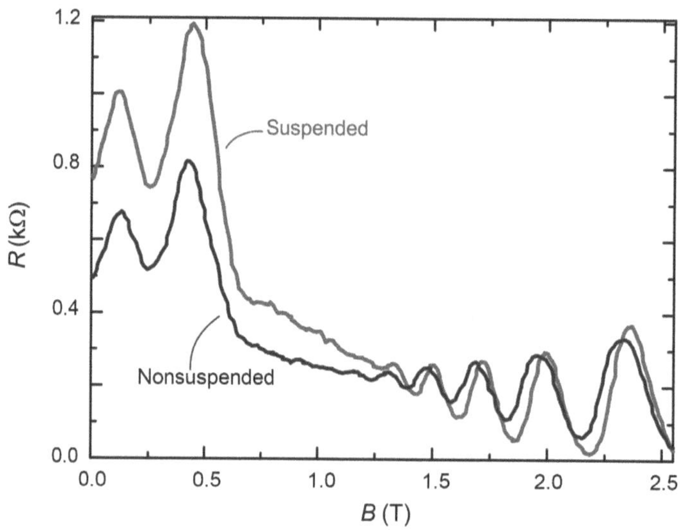

FIGURE 5.17 Longitudinal magnetoresistance of nonsuspended and suspended Hall bars with an antidot lattice.

(Fig. 5.16). The antidots' diameter varied from 100 to 200 nm, while the lattice period varied from 500 to 800 nm for different series of samples.

The longitudinal magnetoresistance of suspended and nonsuspended Hall bars with a lattice of antidots in low magnetic fields is shown in Fig. 5.17. The magnetoresistance demonstrates two peaks of commensurability oscillations and posterior Shubnikov–de Haas oscillations. Commensurability oscillations originate from geometrical resonances. For example, the main peak of these oscillations corresponds to the magnetic field when the cyclotron diameter coincides with the period of an antidot lattice [24]. Hall magnetoresistance also demonstrates ballistic features with the same origin. Temperature dependences of the commensurability oscillations are similar for nonsuspended and suspended samples (Fig. 5.18) and, for both cases, they are suppressed at the same temperature of about 50K.

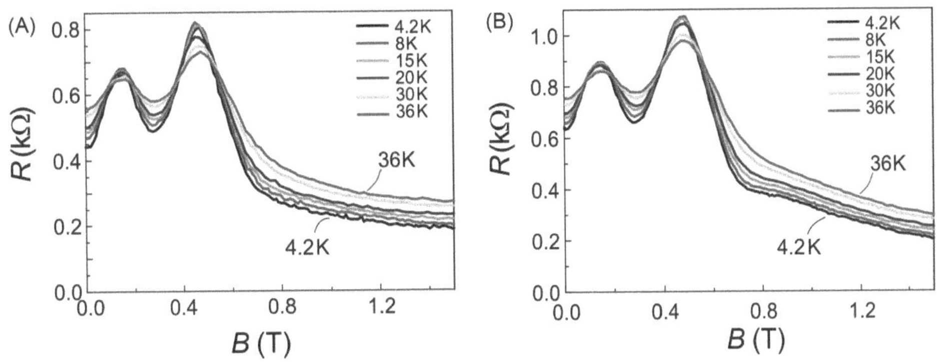

FIGURE 5.18 Temperature dependence of the commensurability oscillation of the magnetoresistance of: (A) nonsuspended; and (B) suspended Hall bars with an antidot lattice.

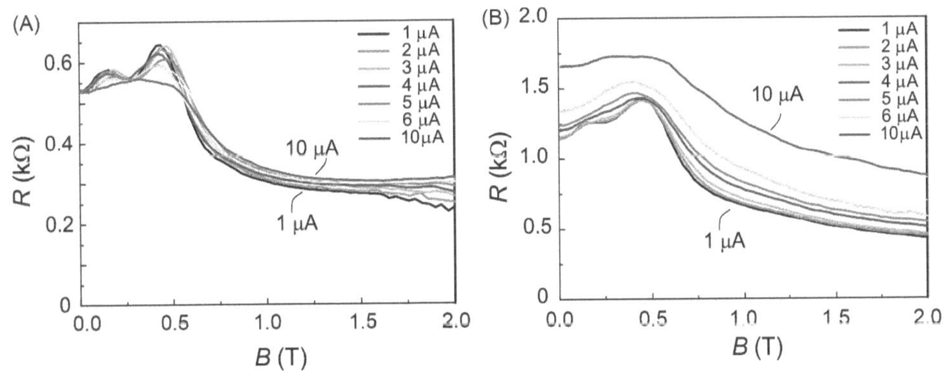

FIGURE 5.19 Dependence of longitudinal magnetoresistance of: (A) nonsuspended; and (B) suspended Hall bars on the applied dc current.

However, nonlinear effects essentially differ in nonsuspended and suspended samples. These effects have been studied by investigation of the commensurability oscillations dependence on the applied dc current. For this purpose, we passed a low-amplitude ac current together with a dc current through the sample, and measured their resistance on the ac signal. It has been found that the dc current suppresses the commensurability oscillations both in nonsuspended and suspended samples, but the behavior of this suppression is quite different (Fig. 5.19).

Moreover, the magnetic field dependence of ballistic thermopower in suspended Hall bars with an antidot lattice has been studied. For this purpose, ac current was passed through one pair of contacts, and the response appearing on the other pair of contacts was measured at doubled frequency. The results of these measurements are shown in Fig. 5.20. One can see that not only magnetoresistance, but also the magnetic field dependence of thermopower, demonstrate commensurability oscillations in suspended samples, which are similar to the oscillations in nonsuspended samples [25].

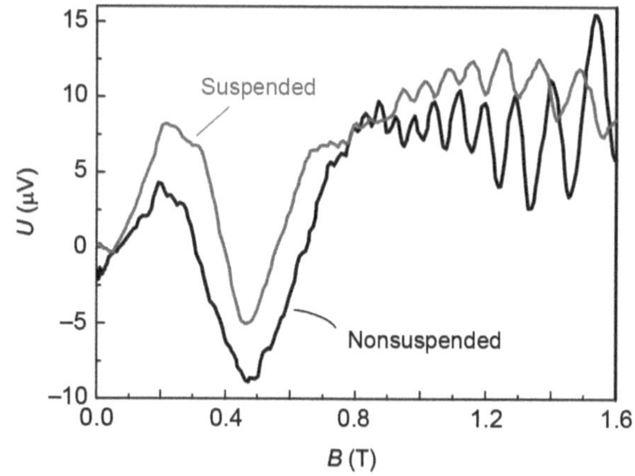

FIGURE 5.20 Magnetic field dependence of thermopower at $T = 1.5$K. The magnitude of the applied ac current is $I = 10\ \mu$A.

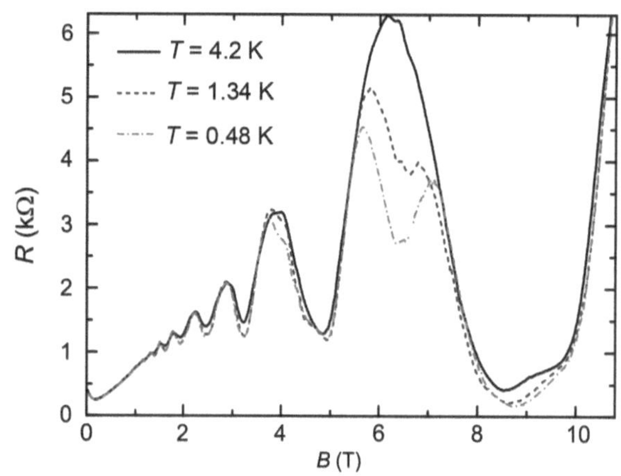

FIGURE 5.21 Temperature dependence of longitudinal magnetoresistance of the suspended Hall bar.

5.9 Quantum Hall Effect in Suspended Hall Bars

Longitudinal and Hall magnetoresistances of nonsuspended and suspended Hall bars without an antidot lattice have been studied in high magnetic fields. It has been found that the Hall magnetoresistance R_H in suspended samples has almost no difference in comparison with nonsuspended samples. The temperature dependences of Hall magnetoresistance are also similar for both cases. It has been demonstrated that longitudinal magnetoresistance R_L in nonsuspended samples vanishes at magnetic fields corresponding to the Hall plateaus, but it does not vanish in suspended samples even after the temperature is decreased by an order of magnitude (Fig. 5.21). We assume that such unusual behavior can be explained by the appearance of potential barriers between suspended and nonsuspended parts of the sample, which leads to a partial scattering of the edge current channels. To confirm this

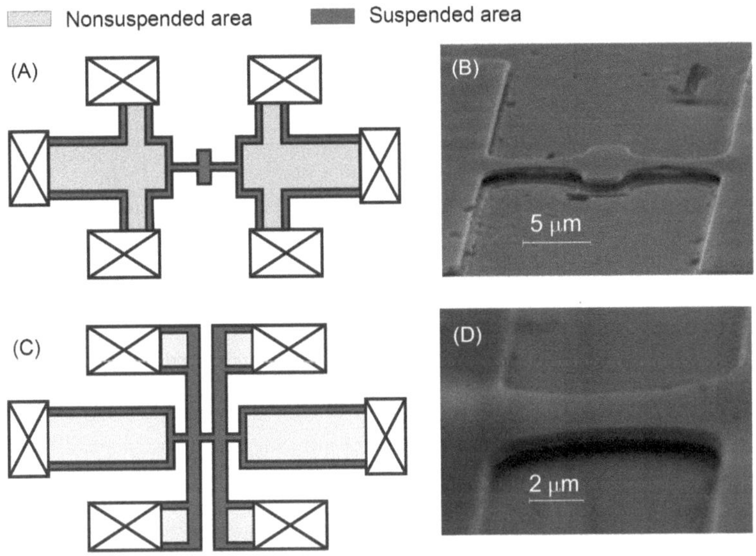

FIGURE 5.22 Diagrammatic representations and images made by electron microscopy of suspended Hall bars of two types: (A and B) Type I; and (C and D) Type II.

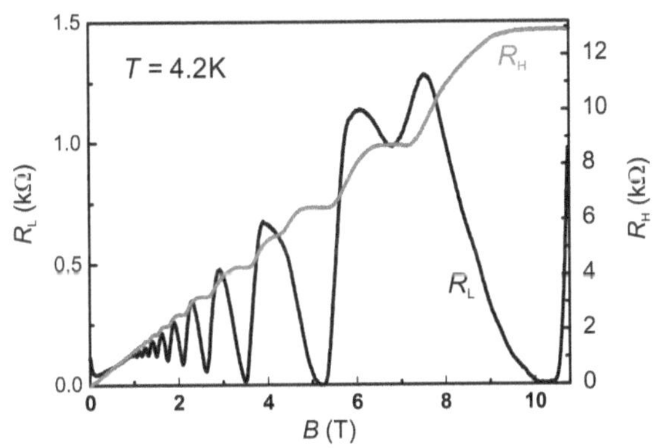

FIGURE 5.23 Quantum Hall effect in suspended samples at $T = 4.2K$.

assumption, we tried to avoid the effect by fabricating other Hall bars detached from the bulk together with the voltage probes (Fig. 5.22). The Hall magnetoresistance of these samples (Fig. 5.22C and D) demonstrates plateaus corresponding to the integer filling factors, and longitudinal magnetoresistance demonstrates vanishings which, in turn, correspond to the Hall plateaus (Fig. 5.23). Therefore, the experiments carried out demonstrate that the relocation of the suspension border out of the voltage probes reveals the typical quantum Hall effect regime and, thus, confirms our assumption.

We have also carried out an experimental study of the QHE breakdown by a dc current. First, we put the sample in a magnetic field corresponding to the disappearance of longitudinal magnetoresistance. Then we passed a dc current together with a low-amplitude ac

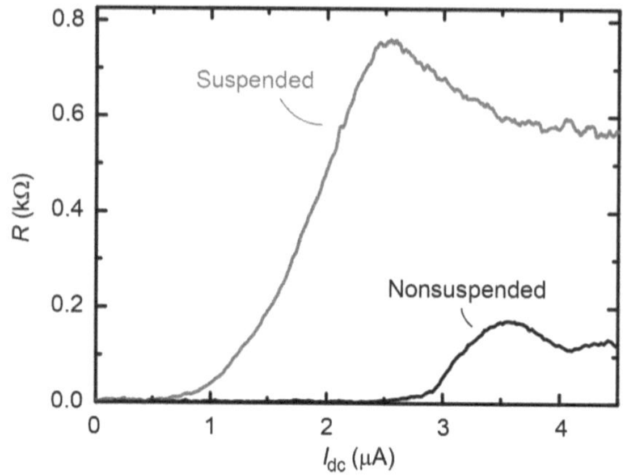

FIGURE 5.24 The dependence of the longitudinal resistance of suspended and nonsuspended Hall bars on applied dc current at a filling factor of $\nu = 4$ and a temperature of $T = 0.48\text{K}$.

current through the Hall bar, and measured the dependence of ac resistance on the magnitude of dc current for nonsuspended and suspended samples.

It has been shown that the critical current of the breakdown of the quantum Hall effect regime in suspended samples is several times lower than in nonsuspended samples (Fig. 5.24). This behavior can probably be explained by peculiarities of heat transport in suspended samples. It is obvious that, in the suspended sample, the heat loss rate is far less than in the nonsuspended sample, due to the poor heat coupling of the Hall bar to the substrate. Thus, the suspended sample is overheated at smaller heating currents, which leads to breakdown of the quantum Hall effect regime.

5.10 Suspended Quantum Point Contact

Two examples of the nanostructures demonstrating essentially quantum features of electron transport are a quantum point contact and a single-electron transistor. A quantum point contact is a 2DEG constriction with a width of an order of the Fermi wavelength. The electron states density in such a constriction is one-dimensional, and this fact leads to quantization of the constriction conductance in universal units: conductance quanta $2e^2/h$. Typically, such constrictions are formed by a metal split gate covering a surface. The suspended quantum point contacts were studied earlier in Refs. [26,27].

We created [28] a quantum point contact without using the split gate by means of electron beam lithography followed by anisotropic plasmachemical etching. The sample created represents a smooth constriction of a 2DEG with a lithographic width of 800 nm. The quantum point contact is equipped with two side gates separated from the constriction by narrow etched trenches (Fig. 5.25). The dependence of the constriction conductance on the equal voltages applied to the gates and the source-drain voltage demonstrated classical quantization in $2e^2/h$ units before the suspension. The quantization survived after the suspension, but it acquired some features (Fig. 5.26).

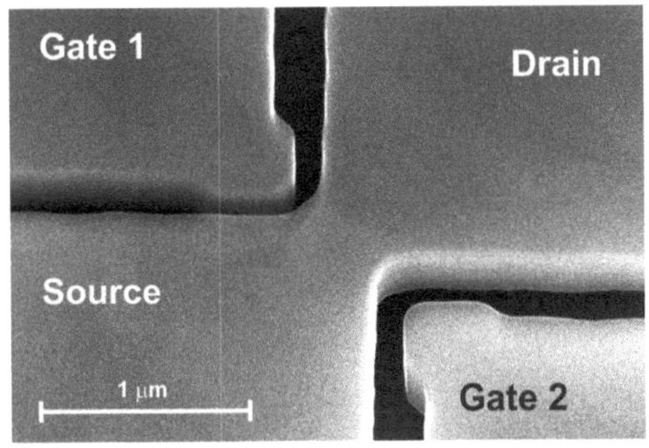

FIGURE 5.25 Scanning electron microscope image of the suspended quantum point contact obtained with the electron beam tilted at 56 degrees with respect to the surface.

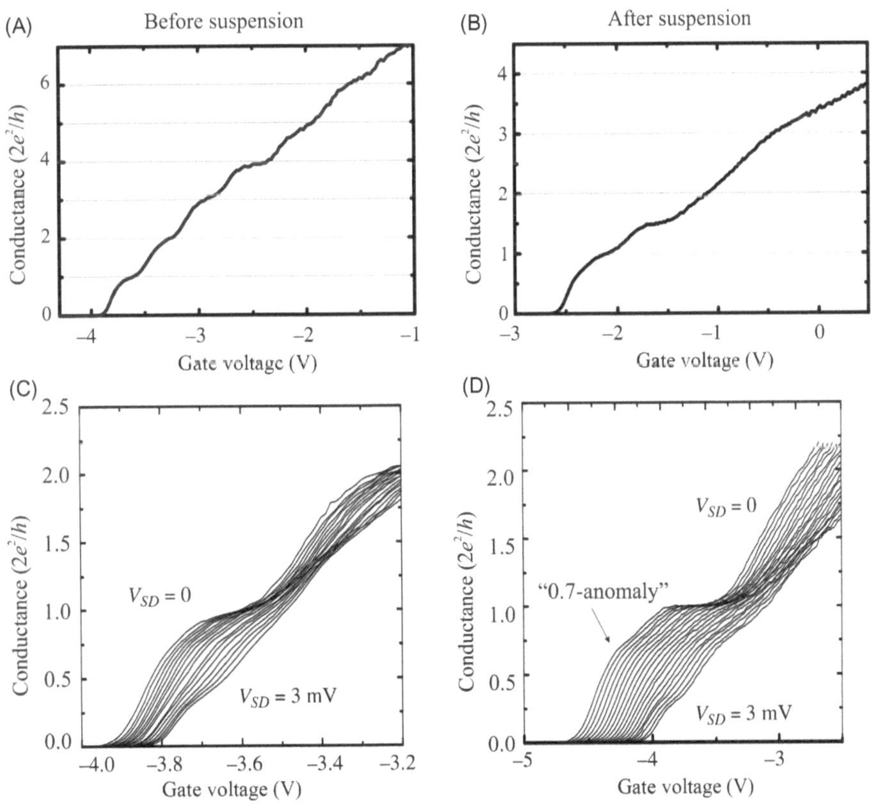

FIGURE 5.26 Conductance quantization before (A), and after (B), the suspension of the quantum point contact. The "0.7-anomaly" unobservable before the suspension (C), and appearing after it (D).

The most prominent of these features arose at conductance values close to $0.7 \times 2e^2/h$ (the so-called "0.7-anomaly"). The nature of this feature is the subject of most modern papers devoted to quantum point contacts. In spite of this, there is no generally accepted explanation for its appearance today. However, a consensus of the existing theories is that the "0.7-anomaly" is inexplicable by the single-particle model (unlike the integer plateaus appearing at conductance values being integer multiples of $2e^2/h$), and is related to inter-electron interaction.

In this context, the appearance of the "0.7-anomaly" after the suspension can be explained by the strengthening of electron–electron interaction due to electric field confinement in a membrane with a high dielectric constant $\varepsilon \approx 13$. This assumption does not contradict the rough numerical estimates obtained using the method of images. According to these estimates, the energy of interaction between two electrons distant by the length of the one-dimensional conductive channel (approximately 200 nm) should increase almost three times after the suspension.

5.11 Suspended Single-Electron Transistor

The increase in energy of the electron–electron interaction is especially interesting to study the Coulomb blockade observable in single-electron transistors. The main parameter characterizing these devices is the charging energy, i.e., the minimum energy an electron must have to overcome the electrostatic repulsion from a charged island, through which the current flows. The amplification in the electron–electron interaction should lead to an increase in the charging energy and, consequently, to an increase in the maximum working temperature of a single-electron transistor. Typically, the charging energy of single-electron transistors created from GaAs/AlGaAs heterostructures does not exceed several Kelvin in temperature units (Fig. 5.27).

The single-electron transistor created [29,30] represents a nanowire having two constrictions, with an island having a lateral size of about 600 nm placed between them (Fig. 5.28).

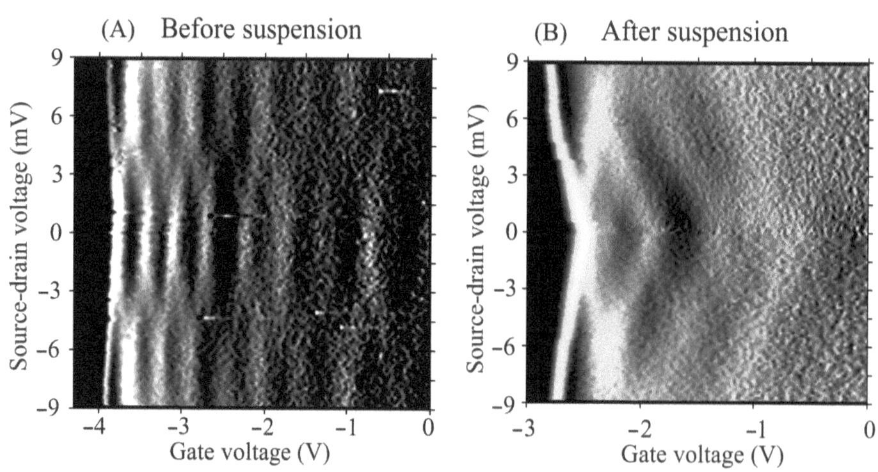

FIGURE 5.27 Evolution of the quantization plateaus upon application of a dc source-drain voltage. The conductance derivative on the gate voltages is shown in color. The *dark areas* correspond to the plateaus.

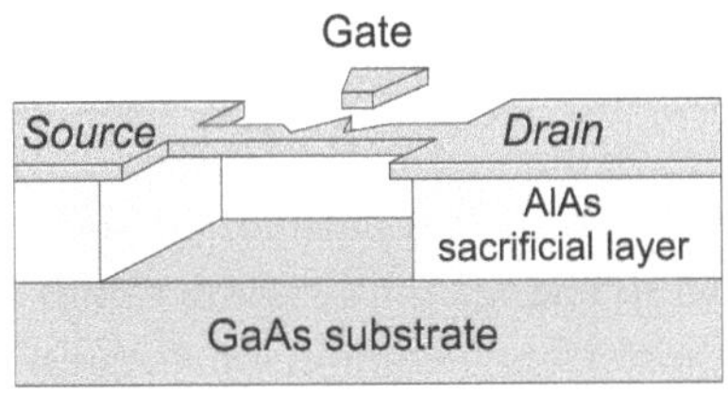

FIGURE 5.28 Schematic image of a suspended single-electron transistor.

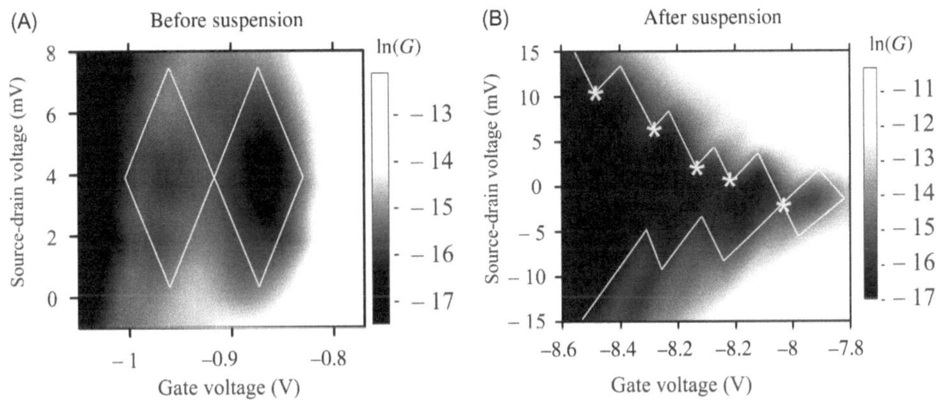

FIGURE 5.29 The conductance of a single-electron transistor measured as a function of gate voltage and source-drain voltage before (A), and after (B), the suspension.

The transistor is equipped with two gates on both sides of the nanowire. The transistor conductance is experimentally determined using ac signal measurements as a function of gate voltage V_G, and the dc component of source-drain voltage V_{SD}. The measurements performed before the suspension showed that the single-electron transistor created demonstrates the classical Coulomb blockade, manifesting itself as conductance lowering in definite areas in the (V_G, V_{SD}) plane, separated by regions of higher conductance (the left part of Fig. 5.29). The charging energy determined by the size of the lowered conductance areas along the V_{SD} axis is 40K in temperature units.

Analogous measurements performed on the same single-electron transistor after the suspension showed that the Coulomb blockade survives in suspended devices (the right side of Fig. 5.29), with the charging energy significantly increased (up to 170K in temperature units). The observed increase is a consequence of the amplification in electron–electron interaction – is caused by separation from the substrate with a high dielectric constant.

A feature of the dependence characterizing the suspended transistor (Fig. 5.29) is the overlap of the dark areas corresponding to the Coulomb blockade in the (V_G, V_{SD}) plane. This overlap indicates that, besides the Coulomb blockade, there is an additional tunneling blockade with a non-Coulomb nature in the suspended single-electron transistor. The results obtained agree with the theoretical paper [31] predicting such additional blockade in a suspended single-electron transistor. The appearance of the additional blockade caused by the ability of the suspended nanowire to bend, with an electron tunneling accompanied by a change in the bending value, as explained in this paper. This means that, in a suspended single-electron transistor, an electron should have an energy sufficient not only to overcome the Coulomb repulsion, but also to change the bending value. This phenomenon, first experimentally observed in the Institute of Semiconductor Physics, is called "elastic deformation blockade" in Ref. [31].

5.12 Euler Buckling Instability of Suspended Nanostructures

As shown above, suspended nanostructures not only have "electrical degrees of freedom," but also additional, mechanical degrees of freedom, with electron transport and mechanics being coupled via physical mechanisms of mutual influence. Nanostructures with the properties described can be classified as NEMS, the study of which has begun recently and which has become a rapidly developing research area.

However, not only electron-transport, but also the mechanical properties of the suspended nanostructures require a consistent and detailed study to implement their potential for practical applications. GaAs/AlGaAs heterostructures are often considered stress-free, because the layers comprising them are lattice-matched with an accuracy of up to a tenth of a percent. However, the nanostructures made from them often become deformed after suspension (Fig. 5.30).

The origin of these deformations was revealed by means of a comparison of the shapes and the bending values of suspended nanowires with a different geometry [32]. Bending was shown to occur when the nanowire length exceeded a certain critical value (Fig. 5.31). At the same time, nanowires with a subcritical length were shown to remain straight. The existence of critical length is characteristic for Euler buckling instability of the nanowires subject to the action of a longitudinal compressive force. The compressive force can arise due to the lattice

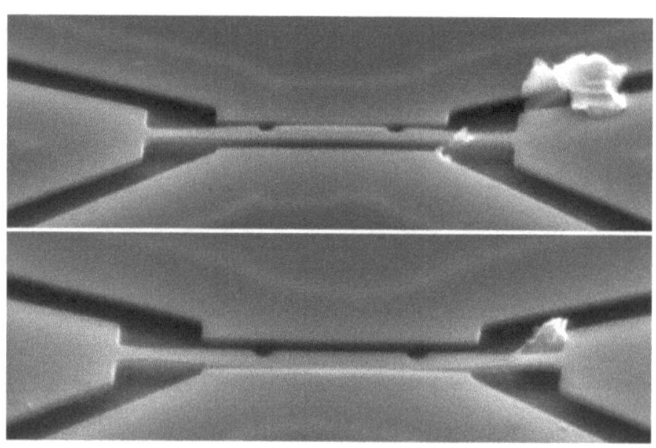

FIGURE 5.30 The deformed nanostructures.

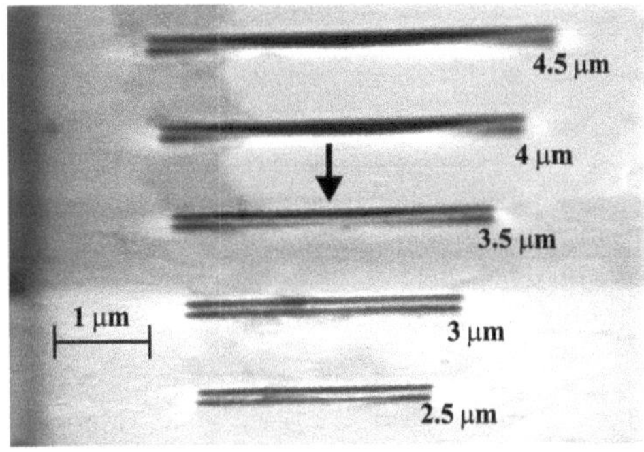

FIGURE 5.31 Scanning electron microscope image of 450 nm-wide nanowires obtained with the electron beam tilted at 80 degrees with respect to the surface. Exceeding the critical length equal to 3.5 μm leads to the nanowires bending, resulting from the Euler buckling instability.

constants mismatch between the different layers in GaAs/AlGaAs heterostructures, though the mismatch is small. However, wide (450 nm) nanowires were shown to become unstable, and bend at a length of 3.5 μm, which is almost three times smaller than the length of 9 μm that could be expected if the compressive force has the nature considered. This indicates that the nanowires are subject to an additional compressive stress much higher than that arising due to the lattice constants mismatch.

This mechanical stress can arise during the reaction of selective etching when the sacrificial layer residues are chemically transformed into the so-called rosenbergite $AlF_3 \cdot 3(H_2O)$. The appearance of a significant mechanical stress, in this case, does not contradict rough numerical estimates.

The appearance of the additional mechanical stress can be avoided if the nanowires are narrow enough (Fig. 5.32). It was shown experimentally that the maximal critical length preventing bending of the 50 nm-wide nanowires is equal to 5 μm. This critical length, as well as the dependence of the bending value on the nanowires length measured (Fig. 5.32), agree with numerical estimates implying that the mechanical stress is caused by the lattice mismatch only. Note, that, since the nanowires width is less than their thickness (90 nm), they are bent in the direction parallel to the surface, in contrast to the wide nanowires.

5.13 Nonlinear Dynamics of Suspended Nanostructures

The fact that elastic stress has a significant impact on the oscillatory properties of NEMS based on GaAs/AlGaAs heterostructures was demonstrated using a device representing a field-effect transistor with the conductivity channel having the shape of a suspended nanowire (Fig. 5.33) [33]. The nanowire is 6 μm long, 600 nm wide, and 90 nm thick. There are four side gates near the nanowire. A high-frequency electrical signal was applied to one of them (labeled G1 in Fig. 5.33) to drive the nanowire resonant vibrations. The opposite gate was grounded, and a dc voltage was applied to the two remaining gates to control nanowire

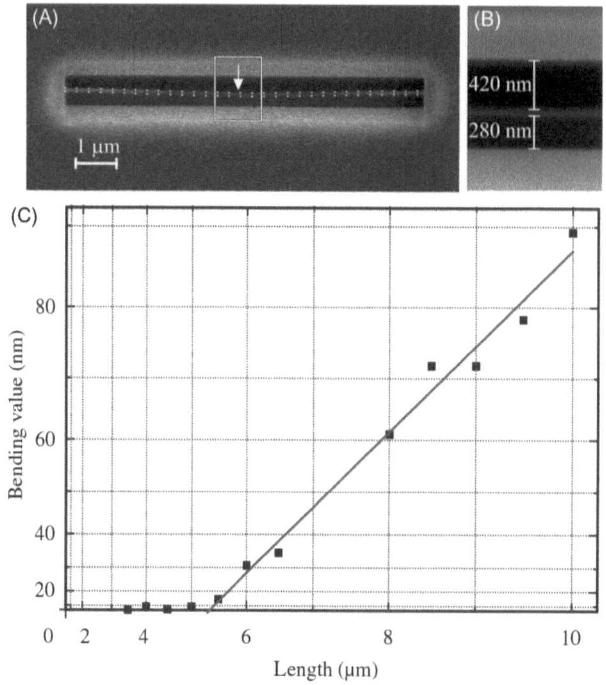

FIGURE 5.32 An example of an image of a 50 nm-wide nanowire bent due to Euler buckling instability (A and B). The bending value is shown as a function of the nanowires' length (C).

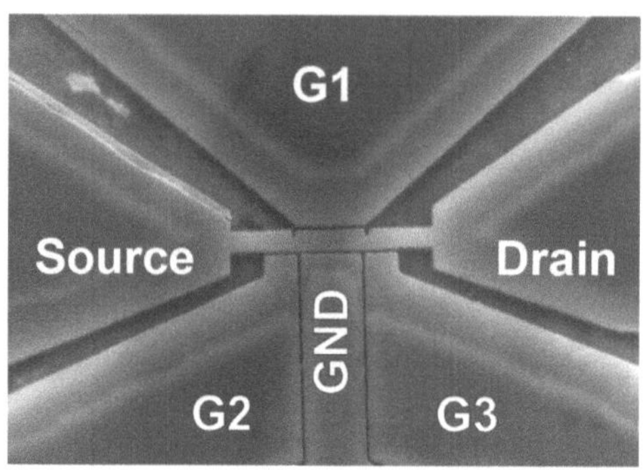

FIGURE 5.33 An image of the suspended nanowire connecting the source and the drain areas. The nanowire mechanical vibrations are driven when an rf driving signal is applied to gate G1. The gate GND is grounded, while gates G2 and G3 are used to control the conductance value by means of applying a dc voltage.

conductance. The sample was in a vacuum at a temperature of 4.2K during the measurements.

It was found that the dependence of nanowire conductance on driving signal frequency has a characteristic form, shown in Fig. 5.34. The frequency down-sweeping leads to a gradual increase in the conductance, starting from 7.7 MHz; the increase ended with an abrupt

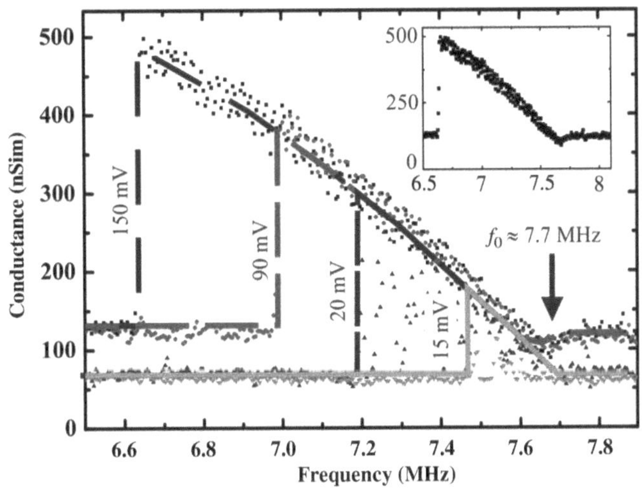

FIGURE 5.34 Nonlinear vibrations of the suspended nanowire detected by means of measuring its conductance.

FIGURE 5.35 The nanowire bending seen with the help of a scanning electron microscope.

drop to approximately the initial value. The increase in the driving signal amplitude results in a left shift of the frequency value corresponding to the drop. The specific form of the curves obtained is the same as the dependence of the vibration amplitude on the driving frequency that can be obtained for a classical nonlinear oscillator. The typical frequencies corresponding to the observed resonance agree to the numerical estimate that can be obtained for the fundamental flexural mode of the vibrations in a direction perpendicular to the surface.

The existing papers devoted to NEMS typically report on mechanical vibration detection by measuring a signal coherent to the driving signal, i.e., having a high frequency. These measurements require cumbersome devices, such as vector network analyzers. Our results show that this can be avoided by measuring the time-averaged conductance at a low frequency. Thus, NEMS with 2DEG represent multifunctional devices combining mechanical resonators and detectors in one device, and give an opportunity to detect the vibrations using a simple method of conductance measurement.

The influence of built-in mechanical stress on the oscillatory properties of suspended nanostructures is manifested in the type of nonlinearity shown in Fig. 5.34. If the nanowires were not prestressed, then the conductance (and the vibration amplitude) would demonstrate the abrupt drop on the right rather than on the left side of the curves shown in Fig. 5.34. Such behavior would be expected because of nonlinear elongation of the vibrating nanowire, resulting in the super-linear character of the restoring force. The abrupt drop appearing on the left side of the curves reflects the fact that the nanowire is initially bent (Fig. 5.35) due to longitudinal compressive stress. The system effectively represents an oscillator vibrating in a double-well potential near one of the minima. The value of the nanowire's static bending is measured using an atomic force microscope. Knowing this value, one can determine the vibrations' amplitude from the width of the resonant curves shown in Fig. 5.34. It is equal to 20 nm at 150 mV root mean square amplitude of the driving signal.

Acknowledgments

The work was funded by RFBR (projects No. 15-02-05774_a, 16-02-00579_a, 16-32-00405 mol_a, 16-32-60130 mol_a_dk).

References

[1] K. Ensslin, P.M. Petroff, Magnetotransport through an antidot lattice in GaAs-$Al_xGa_{1-x}As$ heterostructures, Phys. Rev. B41 (1990) 12307.

[2] D.R. Hofstadter, Energy levels and wave functions of Bloch electrons in rational and irrational magnetic fields, Phys. Rev. B14 (1976) 2239.

[3] E.M. Baskin, G.M. Gusev, Z.D. Kvon, A.G. Pogosov, M.V. Entin, Stochastic dynamics of 2D electrons in a periodic lattice of antidots, JETP Lett. 55 (1992) 678.

[4] E.M. Baskin, A.G. Pogosov, M.V. Entin, Classic chaotic dynamics of 2-dimensional electrons in the periodical lattice of antipoints, JETP 110 (1996) 2061–2086.

[5] Ya.G. Sinai, Dynamical systems with elastic reflections, Russ. Math. Surv. 25 (1970) 137–189.

[6] R. Fleischmann, T. Geisel, R. Ketzmerick, Magnetoresistance due to chaos and nonlinear resonances in lateral surface superlattices, Phys. Rev. Lett. 68 (1992) 1367–1370.

[7] M.V. Budantsev, Z.D. Kvon, A.G. Pogosov, A.E. Plotnikov, N.T. Moshegov, A.I. Toropov, Nonlocal effects in a two-dimensional electron gas with a periodic lattice of scatterers, JETP Lett. 63 (1996) 347–352.

[8] M.V. Budantsev, Z.D. Kvon, A.G. Pogosov, N.T. Moshegov, A.E. Plotnokov, A.I. Toropov, Transport in one-dimensional electron Sinai billiards, Surf. Sci. 361 (1996) 739–741.

[9] K. Tsukagoshi, T. Nagao, M. Haraguchi, S. Takaoka, K. Murase, K. Gamo, Commensurability oscillations by runaway and pinned electrons, Superlatt. Microstruct. 23 (1998) 493–496.

[10] S. Lüthi, T. Vancura, K. Ensslin, R. Schuster, G. Böhm, W. Klein, Electron trajectories in rectangular antidot superlattices, Phys. Rev. B55 (1997) 13088–13092.

[11] M. Büttiker, Four-terminal phase-coherent conductance, Phys. Rev. Lett. 57 (1986) 1761–1764.

[12] A.G. Pogosov, M.V. Budantsev, D. Uzur, A. Nogaret, A.E. Plotnikov, A.K. Bakarov, et al., Thermopower of a multiprobe ballistic conductor, Phys. Rev. B66 (2002) 201303.

[13] B.L. Altshuler, D.E. Khmelnitskii, A.I. Larkin, P.A. Lee, Magnetoresistance and Hall effect in a disordered two-dimensional electron gas, Phys. Rev. B 22 (1980) 5142–5153.

[14] G.M. Gusev, Z.D. Kvon, A.G. Pogosov, P. Basmaji, Weak localization effects in electron billiards, JETP 110 (1996) 696–702.

[15] P.A. Lee, A.D. Stone, H. Fukuyama, Universal conductance fluctuations in metals: effects of finite temperature, interactions, and magnetic field, Phys. Rev. B35 (1987) 1039–1070.

[16] M.V. Budantsev, Z.D. Kvon, A.G. Pogosov, L.V. Litvin, V.G. Mansurov, V.P. Migal, et al., Mesoscopic conductance fluctuations in an electronic billiard, JETP Lett. 59 (1994) 645–650.

[17] M.V. Budantsev, A.G. Pogosov, A.E. Plotnikov, A.K. Bakarov, A.I. Toropov, J.C. Portal, Giant hysteresis of magnetoresistance in the quantum hall effect regime, JETP Lett. 86 (2007) 264–267.

[18] M.V. Budantsev, A.G. Pogosov, A.K. Bakarov, A.I. Toropov, J.C. Portal, Effect of an in-plane magnetic field on magnetoresistance hysteresis of the two-dimensional electron gas in the integer quantum Hall effect regime, JETP Lett. 89 (2009) 92–95.

[19] M.V. Budantsev, A.G. Pogosov, A.E. Plotnikov, A.K. Bakarov, A.I. Toropov, J.C. Portal, Nonequilibrium state of the two-dimensional electron gas in the integer quantum Hall effect regime, JETP Lett. 89 (2009) 46–49.

[20] M.V. Budantsev, D.A. Pokhabov, A.G. Pogosov, E.Y. Zhdanov, A.K. Bakarov, A.I. Toropov, Hysteretic phenomena in a 2DEG in the quantum Hall effect regime studied in a transport experiment, Semiconductors 48 (2014) 1423–1431.

[21] M.V. Budantsev, A.G. Pogosov, D.A. Pokhabov, E.Y. Zhdanov, A.K. Bakarov, et al., Inverted hysteresis of magnetoresistance of a 2DEG at integer filling factors, AIP Conf. Proc. 1399 (2011) 601–602.

[22] M.V. Budantsev, A.G. Pogosov, D.A. Pokhabov, E.Y. Zhdanov, A.K. Bakarov, A.I. Toropov, Nonequilibrium currents in the quantum Hall effect regime spatially resolved by transport experiment, J. Phys. Conf. Ser. 456 (2013) 012005.

[23] A.G. Pogosov, M.V. Budantsev, E.Y. Zhdanov, D.A. Pokhabov, A.K. Bakarov, A.I. Toropov, Electron transport in suspended semiconductor structures with two-dimensional electron gas, Appl. Phys. Lett. 100 (2012) 241916.

[24] A.G. Pogosov, M.V. Budantsev, E.Y. Zhdanov, D.A. Pokhabov, Ballistic electron transport in structured suspended semiconductor membranes, AIP Conf. Proc. 1566 (2013) 211–212.

[25] E.Y. Zhdanov, A.G. Pogosov, M.V. Budantsev, D.A. Pokhabov, A.K. Bakarov, A.I. Toropov, Ballistic thermopower of suspended semiconductor Hall bars with two dimensional electron gas, J. Phys. Conf. Ser. 643 (2015) 012079.

[26] Y. Okazaki, I. Mahboob, K. Onomitsu, S. Sasaki, H. Yamaguchi, Quantum point contact displacement transducer for a mechanical resonator at sub-Kelvin temperatures, Appl. Phys. Lett. 103 (2013) 192105.

[27] C. Roessler, M. Herz, M. Bichler, S. Ludwig, Freely suspended quantum point contacts, Solid State Commun. 150 (2010) 861–864.

[28] A.A. Shevyrin, A.G. Pogosov, M.V. Budantsev, A.K. Bakarov, A.I. Toropov, S.V. Ishutkin, et al., The features of ballistic electron transport in a suspended quantum point contact, Appl. Phys. Lett. 104 (2014) 203102.

[29] A.G. Pogosov, M.V. Budantsev, A.A. Shevyrin, A.E. Plotnikov, A.K. Bakarov, A.I. Toropov, Blockade of tunneling in a suspended single-electron transistor, JETP Lett. 87 (2008) 150–153.

[30] A.G. Pogosov, M.V. Budantsev, R.A. Lavrov, A.E. Plotnikov, A.K. Bakarov, A.I. Toropov, et al., Coulomb blockade and the thermopower of a suspended quantum dot, JETP Lett. 83 (2006) 122–126.

[31] N. Nishiguchi, Elastic deformation blockade in a single-electron transistor, Phys. Rev. B 68 (2003) 121305.

[32] A.A. Shevyrin, A.G. Pogosov, M.V. Budantsev, A.K. Bakarov, A.I. Toropov, S.V. Ishutkin, et al., The role of Euler buckling instability in the fabrication of nanoelectromechanical systems on the basis of GaAs/AlGaAs heterostructures, Appl. Phys. Lett. 101 (2012) 241916.

[33] A.A. Shevyrin, A.G. Pogosov, M.V. Budantsev, A.K. Bakarov, A.I. Toropov, S.V. Ishutkin, et al., High-amplitude dynamics of nanoelectromechanical systems fabricated on the basis of GaAs/AlGaAs heterostructures, Appl. Phys. Lett. 103 (2013) 131905.

6

Modeling of Quantum Transport and Single-Electron Charging in GaAs/AlGaAs-Nanostructures

O.A. Tkachenko[1], V.A. Tkachenko[1], Z.D. Kvon[1,2], D.V. Sheglov[1], A.L. Aseev[1]

[1]RZHANOV INSTITUTE OF SEMICONDUCTOR PHYSICS OF THE SIBERIAN BRANCH OF THE RUSSIAN ACADEMY OF SCIENCES, NOVOSIBIRSK, RUSSIA [2]NOVOSIBIRSK STATE UNIVERSITY, NOVOSIBIRSK, RUSSIA

6.1 Introduction

The study of single-electron charging phenomena and quantum transport in man-made nanosystems embedded in classical electric circuits and containing a small number of electrons has become possible owing to advances in nanotechnology and modern experimental physics. In Coulomb blockade conditions, the resistance of a single nanosystem is determined by discrete recharging of ultra-small capacitances of tunnel-isolated Coulomb islands [1]. On the contrary, phase coherence and size quantization of electron energy play the main role in quantum transport through a nanoscale object. Observation of pronounced and regular effects are only possible in structures of high quality [2], which can be achieved in GaAs/AlGaAs heterojunctions with a two-dimensional electron gas (2DEG). The electronic system is formed by nanolithography and gate voltage control. Nonuniform electric fields are created by all the charged and neutral parts of the structure, including the metal gates on the surface, doping layers, and the 2DEG layer. Thus, the electrostatic potential confining electrons is formed at the depth of the 2DEG. At low temperatures, it is sensitive to accidental recharging of impurity centers. Modeling of 3D electrostatics allows us to calculate the capacitances and charge distribution in nanostructures, and to choose an appropriate mechanism of electron transport, which facilitates the interpretation of experiments (Ref. [3] and references therein).

Nanodevices presented here had diverse designs and were fabricated in different laboratories. A small three-terminal quantum dot and two different small ring interferometers were fabricated in the Institute of Semiconductor Physics, Novosibirsk, Russia (ISP SB RAS) [4–12]; a quasi-one-dimensional quantum dot and two types of quantum point contacts (QPCs) were created in the Cavendish Laboratory, England [13–18]. The 2DEG was placed at large [4–7,13–18] and small distances [8–12] from a semiconductor surface, and it was formed in different ways: using modulation or delta doping, and without any impurity doping, when electrons were drawn electrostatically by gate voltages [13]. Lateral 2DEG structuring was implemented by means of etching [4–9], local anodic oxidation [10–12], field

Advances in Semiconductor Nanostructures. DOI: http://dx.doi.org/10.1016/B978-0-12-810512-2.00006-8

control (total metal gate [4–7,13], in-plane side gates [5,8,9], stripe, and split metal gates [13–18]). Simple shapes of regularly arranged rectangles [13–18], concentric circles [4], and complex shapes [5–12] were used for nanoscale subsurface patterning. Measurements were performed over different temperature ranges. Construction of models and analysis of calculations were carried out in close cooperation with theoreticians, experimenters, and technologists (Ref. [3] and references therein).

Theoretical results were compared with experimental data during the course of numerical experiments. For this purpose, a set of computer programs were developed on the base of universal algorithms [19–30]. These programs solve: (1) the three-dimensional Poisson equation using the Thomas–Fermi approach; (2) the two-dimensional Schrödinger equation by scattering matrix method or the recursive Green's function technique; or (3) equations describing single-electron tunneling in systems of weakly coupled Coulomb islands and reservoirs based on the orthodox theory of the Coulomb blockade.

Numerical modeling allowed us to explain some of the observed effects: deviation from exact conductance quantization in a short QPC (Section 6.2); large-scale Fabry–Perot resonances and paradoxical penetration of the Coulomb oscillations into the region of large conductance of a quasi-one-dimensional quantum dot; backscattering resonances and redirections of electrons in triangular quantum dots (Section 6.3); single-electron charging of triangular quantum dots in a ring interferometer, phase drift, and temperature-induced ordering of Aharonov–Bohm oscillations in a small ring (Section 6.4). Graphene-like lattices of quantum dots and antidots are discussed in Section 6.5.

6.2 Quantum Point Contacts

These devices are field-effect transistors with a channel in the form of a short ballistic quantum wire, which is formed by a constriction in a high mobility two-dimensional electron gas. The conductance through the QPC as plotted against the split-gate voltage shows clear quantized plateaux. The quality of the plateaux is a measure of smoothness of the constriction, as defined by absence of impurities and significant potential fluctuations in the contact. Thus, electron scattering is determined by the geometry of the confining potential. Two kinds of QPCs with two-level metal gates, created and experimentally studied at the Cavendish Laboratory, were numerically modeled in Refs. [13,18]. One of the QPCs is implemented in a heterostructure without doping, i.e., the 2DEG is induced electrostatically at the GaAs/AlGaAs interface (100 nm below the surface) when a sufficient positive bias is applied to the metal top gate. The polyimide layer, about 500 nm thick, separates the top gate from the surface split gate. The top gate changes the carrier density (the Fermi level), while the split gate provides the confinement and controls the conductance of the channel [13]. For this system the calculations based on the structure design reproduce, in the main, measured dependence $G(V_{sg}, V_{tg})$ over a wide range of gate voltages, including the voltages at which the conductance steps occur, the shape and total number of the plateaux (Fig. 6.1).

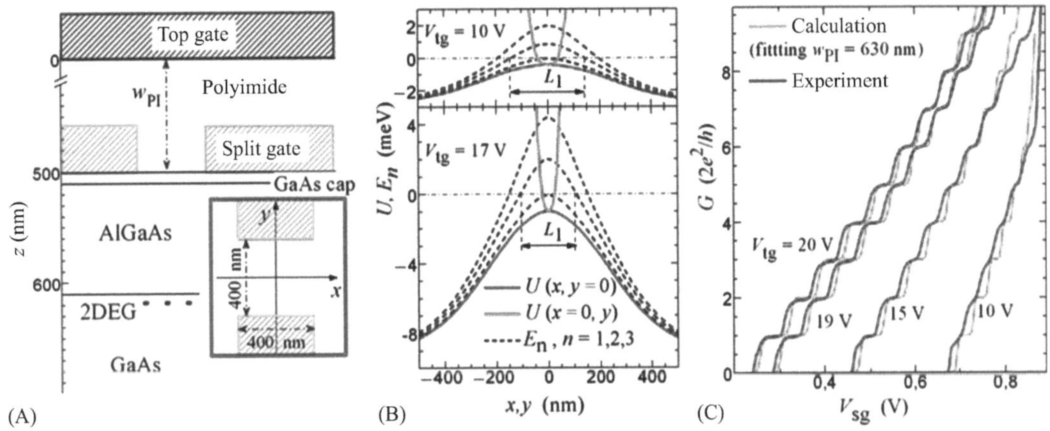

FIGURE 6.1 (A and B) Modeling of 3D electrostatics of a 1D channel in an induced 2DEG. (A) Schematic view of the structure; (B) The 1D-subbands $E_n(x)$, the cross-sections of the 2D effective potential U for two top gate voltages. The single-mode part of the channel is denoted by L_1. (C) Modeling of 2D quantum transport: the theoretical (temperature $T = 0$) and experimental ($T = 280$ mK) conductance curves (data from O.A. Tkachenko, V.A. Tkachenko, D.G. Baksheyev, K.S. Pyshkin, R.H. Harrell, E.H. Linfield, et al., Electrostatic potential and quantum transport in a one-dimensional channel of an induced two-dimensional electron gas, J. Appl. Phys. 89 (2001) 4993–5000). The thickness of polyimide w_{PI} was tuned to fit the experiment.

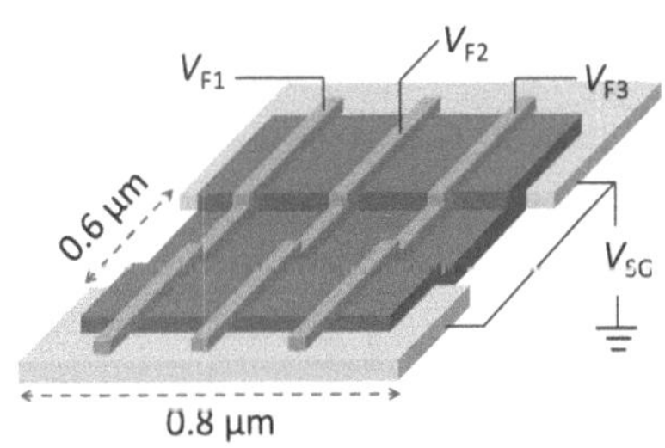

FIGURE 6.2 Schematic view of the metal gate system. *From C.-T. Liang, M.Y. Simmons, C.G. Smith, G.H. Kim, D.A. Ritchie, M. Pepper, Experimental evidence for Coulomb charging effects in an open quantum dot at zero magnetic field, Phys. Rev. Lett. 81 (1998) 3507–3510; C.-T. Liang, O.A. Tkachenko, V.A. Tkachenko, D.G. Baksheyev, M.Y. Simmons, D.A. Ritchie, et al., Gradual decrease of conductance of an adiabatic ballistic constriction below $2e^2/h$, Phys. Rev. B 70 (2004) 195324.*

In another configuration of the QPC, doping layers were situated at a large distance (80 nm) on both sides of the quantum well (20 nm thick), and scattering on impurities was suppressed [18]. Potentiometer contacts were deliberately brought close to the constriction to reduce the contribution of series resistance R_s of the 2DEG. Thus, the conductance of the single-mode constriction was measured precisely enough without resorting to subtraction of R_s. The constriction was electrostatically-shaped in the 2DEG by a pair of split gates lying on the semiconductor surface (Fig. 6.2). Three narrow finger gates separated from the lower gate by a dielectric layer and placed across the channel were used to modulate the electrostatic potential in the channel. The outer finger gates were grounded to the 2DEG to reduce the size of barrier along the constriction, while the center finger gate voltage was varied over a wide range and was used to control the height of barrier. In this setup the first conductance step shows a gradual decrease of the height (from $2e^2/h$ to $0.97 \times 2e^2/h$) with increasing negative voltage on the center finger gate. Realistic numerical modeling of electrostatics of the device and the transmission through the constriction produces results that are very similar to the experimental data.

Calculation of electrostatics shows that the tops of the first and the second subbands move with different rates in response to the change of the split gate voltage. This leads to a slower decrease of the reflection compared to a simultaneous increase in tunneling, and therefore, the onset of the conductance step occurs at a lower value than $G = 2e^2/h$. Interestingly, in the energy dependence of conductance the effect is absent. Thus, modeling shows the fundamental difference between changing the Fermi energy and changing the gate voltage [18].

6.3 Two-Terminal and Three-Terminal Quantum Dots

Conducting quantum dots defined electrostatically in a 2DEG operate as a single-electron transistor when the contacts are in a tunneling (closed) regime. In an open regime they can reveal gate-voltage interference oscillations or effects of resonant transmission. It is known from theory and experiment that interference effects can occur in a tunneling regime, but single-electron charging in an open regime is prohibited by the Coulomb blockade theory. In this respect, the paradoxical results of some experiments [14] require an explanation. So, the structure with a pair of split gates and three finger gates (Fig. 6.2) was used to create a novel type of quantum dot (Fig. 6.3; [14–17]). For this purpose, some depletion voltage was applied to the outermost fingers ($V_{F1} = V_{F3}$) to control the height of the entrance and exit barriers. The central finger gate was used to control the depth of the quantum dot (Figs. 6.3A and 6.4A). Calculations of 3D electrostatics and 2D quantum transport demonstrated that this novel type of dot had a unique geometric versatility, and in some voltage regimes essentially differed from the standard quantum dots by suppression of inter-1D subband scattering [15] (Fig. 6.3A and B). Thus, the device could reveal a simple quasi-one-dimensional behavior. This property of the quantum dot made it possible to observe simultaneously large-scale Fabry–Perot resonances of electron interference, and the Coulomb oscillations of ballistic conductance at $G > 2e^2/h$ [16] (Fig. 6.4). It also explained an amazing penetration of single-electron charging effects far into the open region.

Thus, in this device, phenomena of single-electron charging and coherent electron transmission at $G > 2e^2/h$ coexist. The orthodox theory of Coulomb blockade has been adapted to this case [16,17], and calculations of the voltage gate characteristics for different temperatures in the main reproduced the experimentally observed large-scale features, frequent Coulomb oscillations, and overall view of the curves (Fig. 6.4B).

Quantum dots with two inputs commonly have a convex boundary which makes it difficult to reduce their area. Small triangular quantum dots with a concave boundary are an exception. They naturally occur at a junction of quantum wires, e.g., at the entrance of a ring

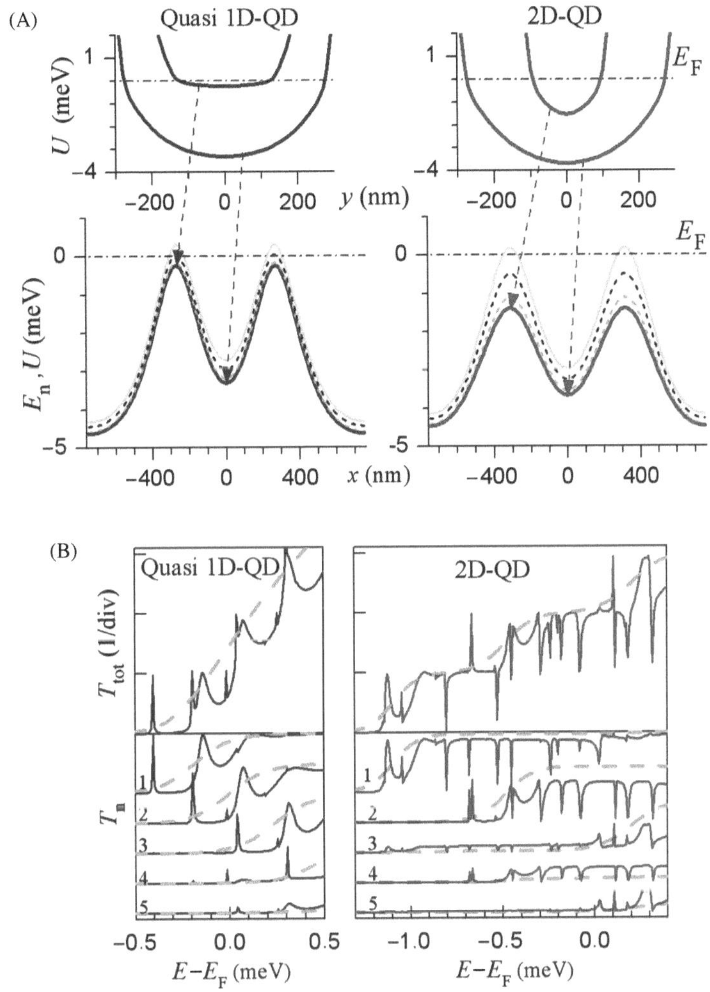

FIGURE 6.3 (A) Confining potential in different cross-sections (top), and the lowest three 1D-subbands $E_n(x)$ (bottom), for two different quantum dot devices: channel with overlying finger gates (left), and usual quantum dot device (right). One-dimensional subband spacing is almost constant inside the quasi-1D quantum dot, in contrast to the 2D quantum dot. (B) Total transmission coefficient and contributions of five 1D subbands for these two cases; *dashed lines* show corresponding transmission coefficients for a single constriction. Inter-1D subband mixing is suppressed in the quasi-1D quantum dot. *Data from D.G. Baksheyev, O.A. Tkachenko, V.A. Tkachenko, The role of intersubband mixing in single-electron charging of open quantum dot, Phys. E 6 (2000) 414–417.*

interferometer (Fig. 6.5). Their existence was predicted [4], and experimentally verified, for different nanostructures created in the Institute of Semiconductor Physics, Novosibirsk [5–12].

Triangular quantum dots with three inputs were created in structures with a 20 nm spacer layer (Fig. 6.5B). In the high-resistance regime, high-amplitude oscillations of single-electron charging of the dot were measured [5]. Calculation of the capacitance of the dot with respect to the side gate verified the single-electron origin of the oscillations.

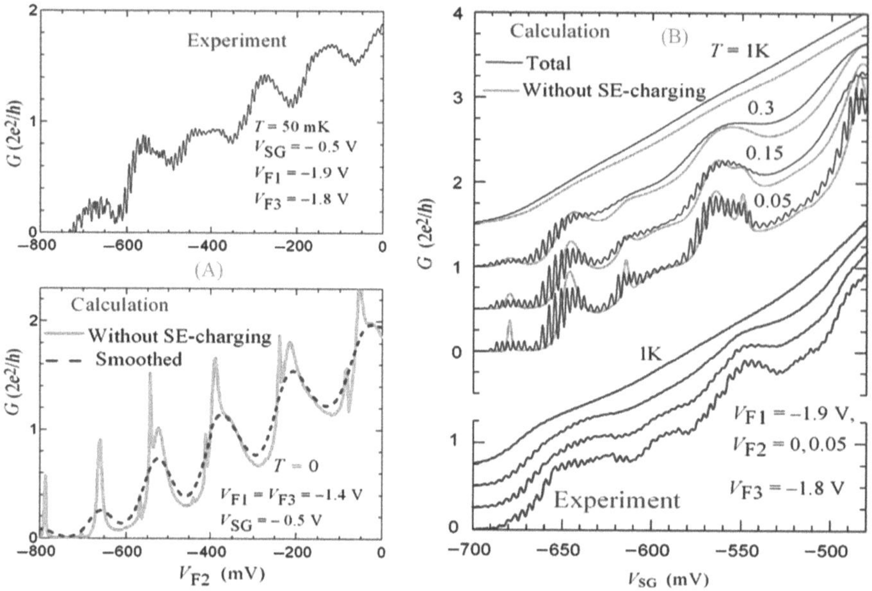

FIGURE 6.4 Calculated and measured conductance of the dot as a function of central finger gate voltage (A), and split-gate voltage (B). *Data from C.-T. Liang, M.Y. Simmons, C.G. Smith, G.H. Kim, D.A. Ritchie, M. Pepper, Experimental evidence for Coulomb charging effects in an open quantum dot at zero magnetic field, Phys. Rev. Lett. 81 (1998) 3507–3510; O.A. Tkachenko, D.G. Baksheyev, V.A. Tkachenko, C.-T. Liang, M.Y. Simmons, C.G. Smith, et al., Coulomb charging effect in an open quantum dot device, J. Phys. Condens. Matt. 13 (2001) 9515–9534; V.A. Tkachenko, D.G. Baksheyev, O.A. Tkachenko, C.-T. Liang, Coulomb oscillations of the ballistic conductance in a quasi-one-dimensional quantum dot, JETP Lett. 74 (2001) 209–212.*

Oscillations of another type were observed in the open regime (Fig. 6.6). They were explained by a resonant Fano interference [6]. To corroborate the interference nature of the conductance oscillations, calculations of 3D electrostatics of the device and coherent single-particle magnetotransport were performed. Calculations show that the characteristic distance between dips in conductance, ΔV_g and ΔE_F, corresponds to a change in the number of electrons in the quantum dot by about 10.

A slight asymmetry of the device was introduced by a displacement (10 nm) of antidots from their ideal positions, and a small variation in the depth of etched cylinders. This randomization simulated technology imperfections and different mesoscopic states of the sample which arise due to impurity recharging. Thus, slightly different confinement potentials were obtained, and they were used for calculations of transmission coefficients (Fig. 6.6). In the self-consistent calculation of the electrostatic potential and electron density, the Fermi level was set at zero. When conductance of the dot was calculated as a function of energy, the reference point of energy was also zero, and the confining potential did not change with energy.

The Coulomb oscillations and the Fano resonances can be observed in a two-terminal quantum dot and its asymmetry is not important. However, a weak asymmetry of a three-terminal triangular quantum dot gives rise to a new effect: redirection of the electron flow from one port of the dot to another with change of the Fermi level (Fig. 6.7) [7]. The whole device is covered

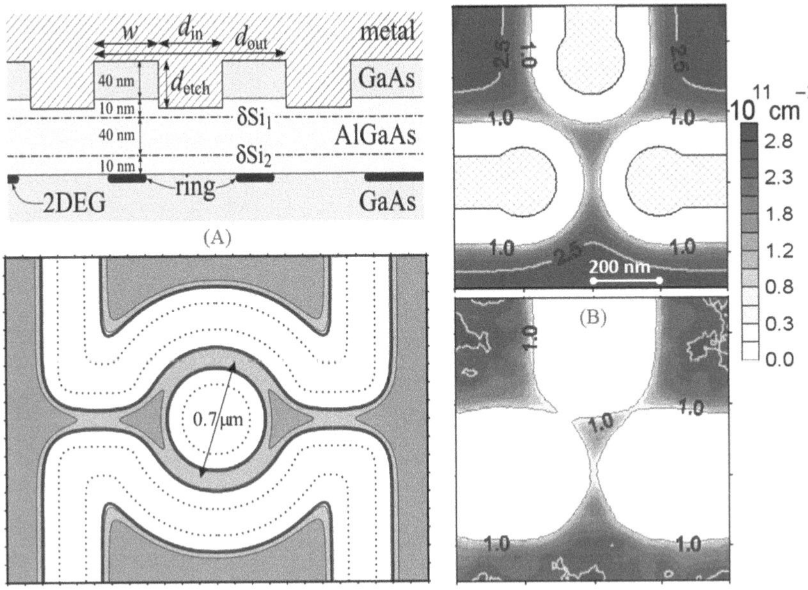

FIGURE 6.5 Two structures with triangular quantum dots. (A) Ring interferometer, vertical section of structure across the center of ring (data from O.A. Tkachenko, V.A. Tkachenko, D.G. Baksheev, Z.D. Kvon, J.C. Portal, Electrostatic potential, energy spectrum, and Fano resonances in a ballistic ring interferometer based on an AlGaAs/GaAs heterojunction, JETP Lett. 71 (2000) 255–258). (B) Three antidots, shown by the maps of electron density in 2DEG. Depletion regions surround etching regions; bottom right panel shows the influence of impurity disorder (data from V.A. Tkachenko, Z.D. Kvon, O.A. Tkachenko, D.G. Baksheev, O. Estibals, J.C. Portal, Coulomb blockade in a lateral triangular small quantum dot, JETP Lett. 76 (2002) 720–723).

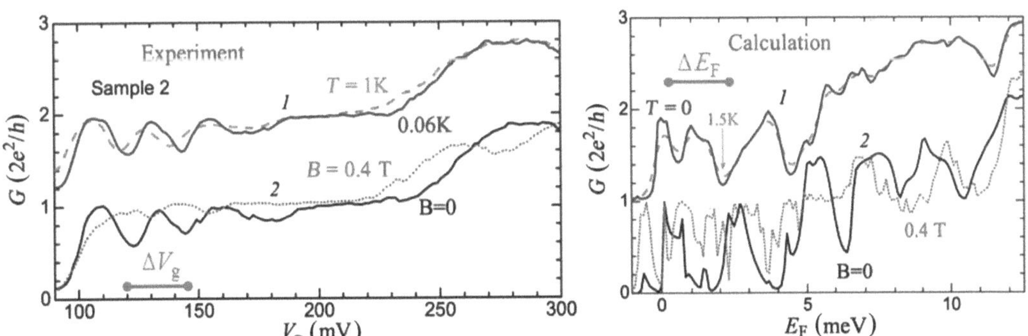

FIGURE 6.6 Calculated and measured conductance of triangular quantum dot. Cases 1 and 2 correspond to slightly different potentials (offset by $2e^2/h$ for clarity). *Data from V.A. Tkachenko, O.A. Tkachenko, Z.D. Kvon, D.G. Baksheev, A.L. Aseev, J.C. Portal, Coherent scattering in a small quantum dot, JETP Lett. 80 (2004) 588–592.*

by a TiAu gate. Applying a voltage on this gate allows control of the Fermi energy in the 2DEG. A total set of four-terminal resistances required to restore all conductance/transmission coefficients $G_{ik} = (2e^2/h)T_{ik}$ was measured in the temperature range of 50–100 mK. It was found that electrons flow from reservoir 1 to reservoir 2 at some gate voltage, and to reservoir 3 at another

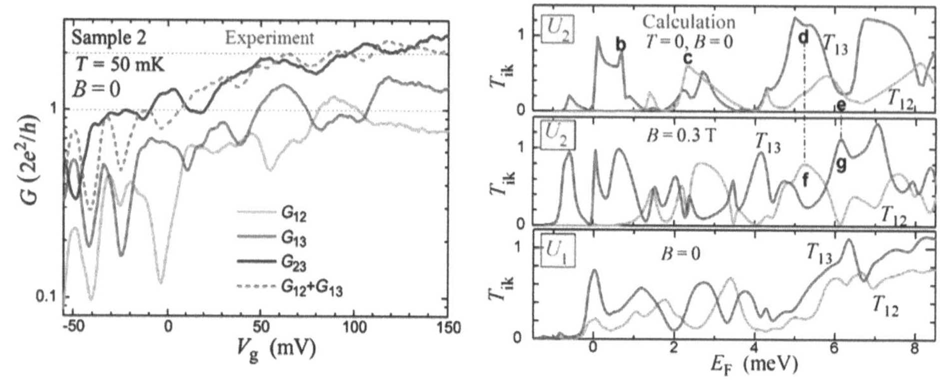

FIGURE 6.7 Redirections of electron flow in quantum dots: experiment and calculation for two slightly different potentials U_1 and U_2 at indicated values of magnetic field *B*. *Data from O.A. Tkachenko, V.A. Tkachenko, Z.D. Kvon, A.L. Aseev, J-C. Portal, Quantum interferential Y-junction switch, Nanotechnology 23 (2012) 095202.*

voltage. The polarization parameter $\gamma_{13} = (G_{21} - G_{23})/(G_{21} + G_{23})$ changes from -0.7 to $+0.4$, with addition to the dot of only 6–10 electrons. The effect was explained qualitatively by simple formulas, and numerically with the help of 3D electrostatics and 2D quantum transport calculations [7].

The reason for the effect was single-particle interference in conditions of weak asymmetry of the Y-branch (Fig. 6.8). The calculations show that the effect can occur in some other configurations, e.g., when the tip of the scanning gate microscopy is used as a movable local gate that splits electron waves going through a QPC. The effect becomes more noticeable in a weak perpendicular magnetic field [7]. It has been concluded that similar redirections are typical for waves of any origin (e.g., acoustic or electromagnetic waves) through splitters of wavelength size.

6.4 Small Ring Interferometers

Open and closed Aharonov–Bohm rings were created in the thin-spacer heterojunction structures at the ISP SB RAS. Their radius of 130–90 nm was as small as nanolithography and depletion effects would allow [8–12]. A thin 3 nm spacer allowed a high density of 2DEG. Originally rings were fabricated by electron lithography and reactive ion etching. Certain rings demonstrated resistance $R \gg h/e^2$ unusually high for quantum interferometers [8,9]. Besides Aharonov–Bohm magnetic field oscillations, the resistance of the ring displayed side gate voltage oscillations that were sinusoidal for a temperature of 4.2K, but at low temperature (100 mK) a fine structure of oscillations with dominating pairwise grouping of narrow peaks was observed (Fig. 6.9).

Calculations of 3D electrostatics based on heterostructure details, as well as images of control samples taken with scanning electron and atomic force microscopes, helped to build the picture of internal organization of the interferometer (Fig. 6.10). Triangular quantum dots were found to occupy almost all the area of the electron system of the interferometer

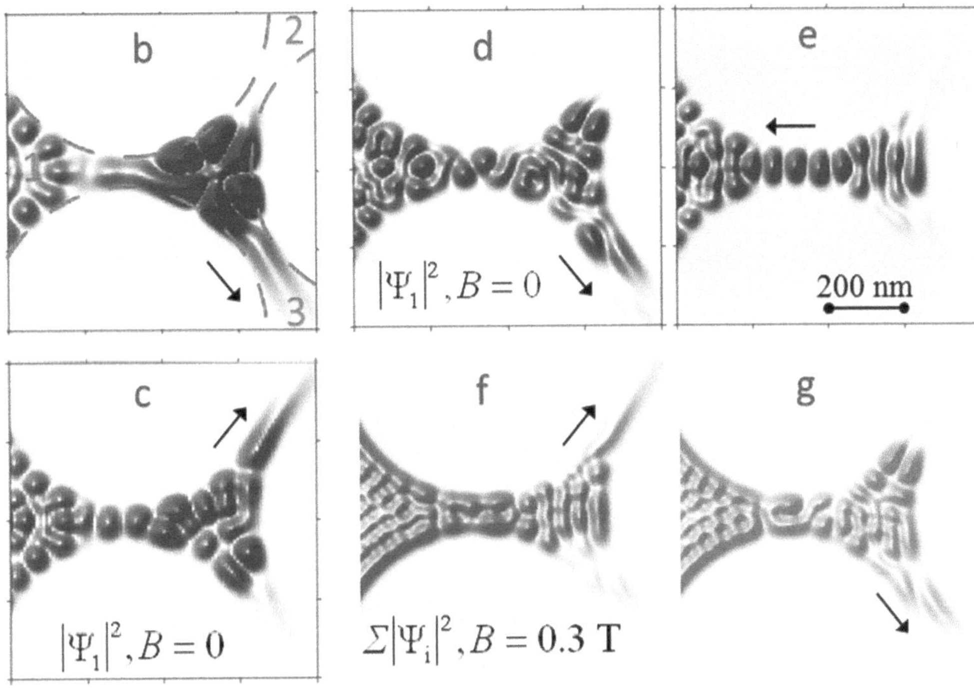

FIGURE 6.8 The probability density calculated for the states marked (b–g) on the transmission curves (Fig. 6.7, case U_2): contribution of the first transverse mode for $B = 0$ (b–e), and total probability density for $B = 0.3$ T (f and g). *Dashed lines* show three antidots. The electrons are incident from port 1 [7].

(excluding the central antidot). The dots were connected to the reservoirs, and to each other, with short and narrow channels (single-mode constrictions). The calculations took into account the nonideality of etched areas and potential fluctuations induced by random distribution of charged impurities.

Calculation of capacitances of triangular dots relative to the gates, to each other, and to reservoirs was performed. It was found that the gate oscillation period 6 mV corresponds to the gate capacitance of a single triangular quantum dot.

Consideration of this device as the system of Coulomb islands and calculation of the current using the Monte–Carlo method (Fig. 6.11) [9] shows that the fine oscillation structure at the lowest temperature can, in general, be explained by Coulomb interaction between quantum dots and a small (10%) difference between their gate capacitances. To the first approximation, the system has two Coulomb islands (Fig. 6.11A), that leads to splitting into doublets and frequency doubling at some gate voltages (Fig. 6.11B). In the second approximation, the ring movement states providing Aharonov–Bohm effect can be considered weakly connected with the states in reservoirs and triangular quantum dots. In this case, the system has three Coulomb islands (Fig. 6.11C), which provide the conductance triplets (Fig. 6.11D) observed in the experiment at $T = 0.1$K (Fig. 6.9D). The calculation also shows sinusoid oscillations at $T = 4.2$K, which corresponds to the measurements.

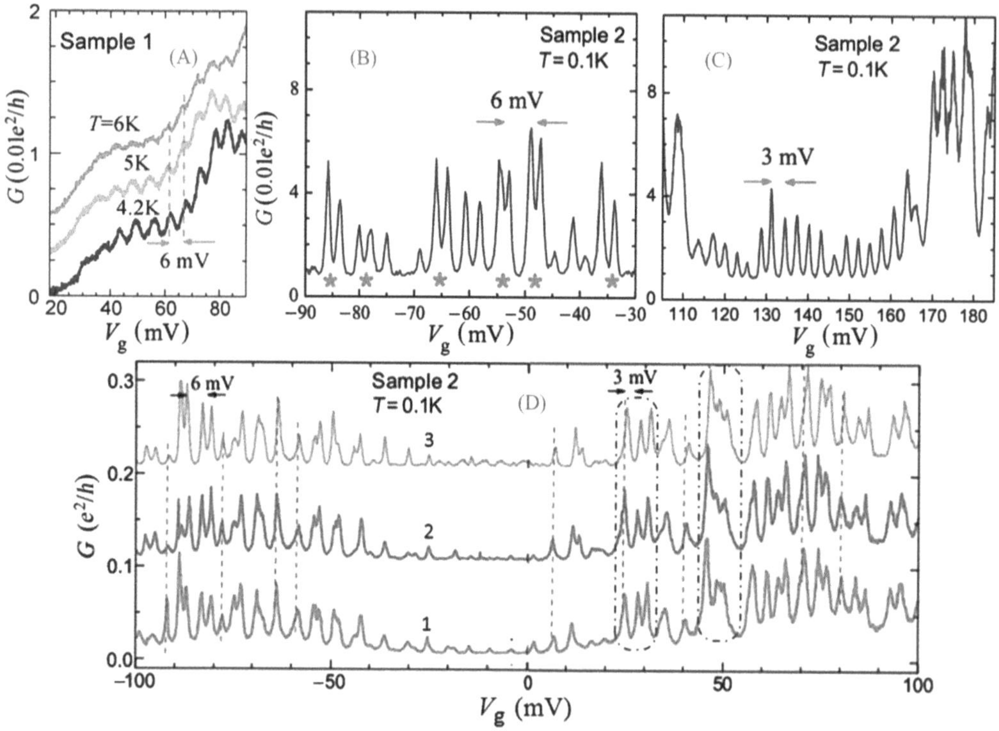

FIGURE 6.9 Measured conductance of small high-resistance rings. (A) Sinusoidal oscillations; (B) doublet splitting; (C) doubling of gate oscillation frequency; (D) fine structure with doublets and triplets on curves 1–3 for one mesoscopic state (curves 2 and 3 are shifted vertically for clarity). *Data from V.A. Tkachenko, A.A. Bykov, D.G. Baksheev, O.A. Tkachenko, L.V. Litvin, A.V. Latyshev, et al., Single-electron charging of triangular quantum dots in a ring interferometer, J. Exp. Theor. Phys. 97 (2003) 317–330; D.G. Baksheyev, A.A. Bykov, V.A. Tkachenko, O.A. Tkachenko, L.V. Litvin, A.V. Latyshev, et al., Coulomb interaction of triangular quantum dots in a small ring interferometer, Int. J. Nanosci. 2 (2003) 495–503.*

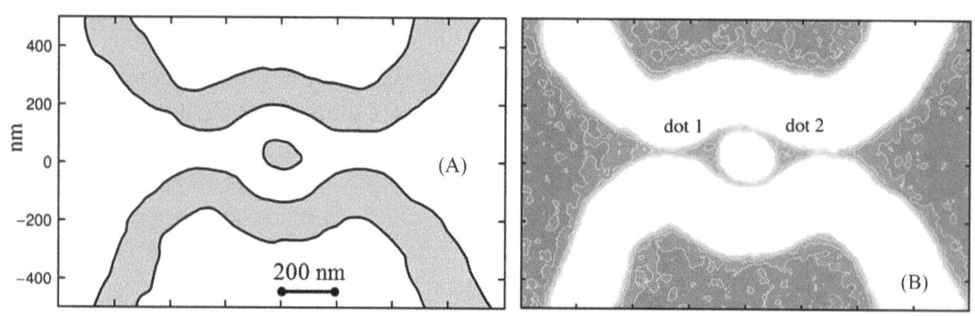

FIGURE 6.10 Lateral view of the model ring. (A) Etched areas, according to scanning electron microscopy. (B) Distribution of electron density up to the maximum $1.2 \cdot 10^{-12}$ cm^{-2} [9].

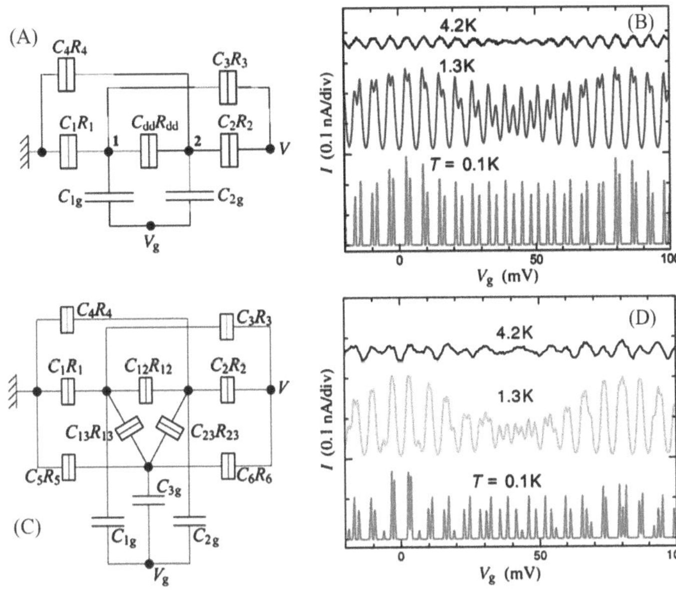

FIGURE 6.11 Two models of single-electron transport in the high-resistance ring interferometer (data from V.A. Tkachenko, A.A. Bykov, D.G. Baksheev, O.A. Tkachenko, L.V. Litvin, A.V. Latyshev, et al., Single-electron charging of triangular quantum dots in a ring interferometer, J. Exp. Theor. Phys. 97 (2003) 317–330; D.G. Baksheyev, A.A. Bykov, V.A. Tkachenko, O.A. Tkachenko, L.V. Litvin, A.V. Latyshev, et al., Coulomb interaction of triangular quantum dots in a small ring interferometer, Int. J. Nanosci. 2 (2003) 495–503). (A and C) Equivalent circuits of the device. (B and D) Current versus gate voltage V_g (curves for $T = 0.1K$ are shifted for clarity).

To overcome limitations of the "etched" rings, other rings were fabricated in the ISP SB RAS by local anodic oxidation [10,11]. The programmed anodic oxidation raster pattern appears on the surface of the semiconductor heterostructure at the first scanning under the conducting tip of atomic force microscope. The value and duration of the negative probe voltage define the size of the lithographic point and the depth of oxidization, i.e., the approach of the semiconductor surface to the plane of 2DEG located at a small 25 nm distance from the original surface. During local anodic oxidation deviations that appear from the pattern depend on many conditions. The samples made by the same programmed pattern are different, as with any technology. However, with this fabrication method, control scanning with tip voltage disabled provides the surface relief pattern where even the smallest technology-introduced geometric imperfections can be seen (Fig. 6.12). It is known that an anodic oxide penetrates below the heterostructure surface to a depth of h, which is almost proportional to the relief altitude. This allows us to determine h quite accurately, and hence the distance between the DEG and the semiconductor–oxide boundary. The two-dimensional map $h(x,y)$ (Fig. 6.12C and D) is used as the starting condition when calculating 3D electrostatics of the structure and 2D distribution of electron density [9]. Impurity disorder was neglected during calculations in order to better understand the role of disorder introduced by the patterning process. It can be seen that it was transferred to the electron

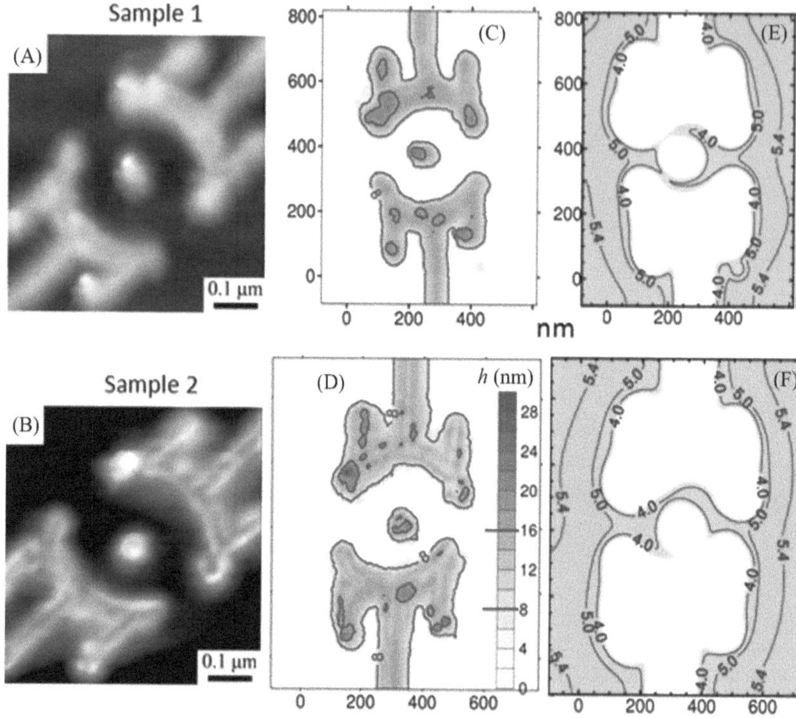

FIGURE 6.12 Illustration of the ring fabrication variations (data from V.A. Tkachenko, Z.D. Kvon, D.V. Sheglov, A.V. Latyshev, A.I. Toropov, O.A. Tkachenko, et al., Aharonov-Bohm oscillation amplitude in small ballistic interferometers, JETP Lett. 79 (2004) 136–140): (A and B) AFM images of two samples fabricated with the same template; (C and D) depths *h* of local anodic oxidization recovered by APM images; (E and F) electron density $N(x,y)$ in units 10^{-11} cm^{-2} calculated on the basis of the maps $h(x,y)$.

system in a smoothed form. The system has turned out to be asymmetric, but the ring remains connected in Sample 1 (Fig. 6.12). Sample 2 shows a broken electron ring. As the confining potential is calculated for each sample, taking into account their peculiarities, this calculation is much closer to the technology and experiments compared to calculations based on an idealized geometry [4] or geometry of control samples [8,9] that may differ considerably from the working samples. It should be noted that only the AFM image of the device does not allow us to reveal the electron ring fragmentation without a self-consistent calculation of 3D electrostatics and the following solution of the 2D Schrödinger equation with calculated 2D effective potential (2D scattering problem). Fig. 6.13 shows the result for Sample 1 in the zero magnetic field for fixed energy of ballistic electron.

Contributions to total probability density are shown for the first three transverse modes (25 important modes were taken into account, with all possible transitions among them). Comparing calculations and measurements of working samples allowed study of the influence of technological imperfections on transport properties of the ring for the first time [9]. Resistance measurement of nominally similar structures, as well as calculation of 3D

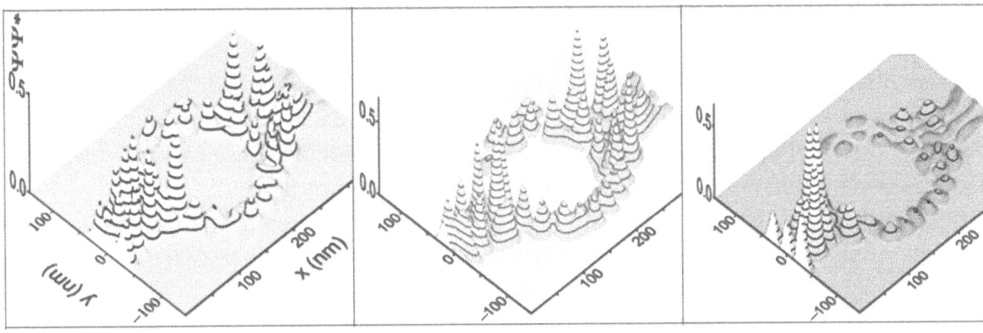

FIGURE 6.13 $|\Psi(x,y)|^2$ for waves with energy $E_F = -1$ meV, incoming to the ring from the left (modes 1–3 from left to right). Calculation corresponds to Sample 1 of Fig. 6.12E. High peaks of $|\Psi|^2$ indicate the interference in the input triangular quantum dots.

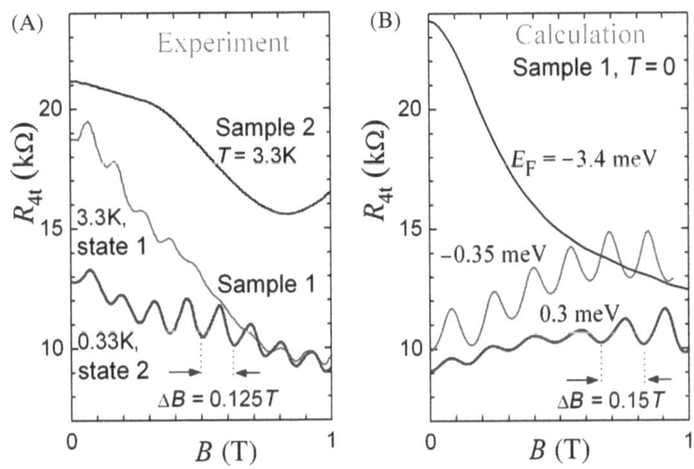

FIGURE 6.14 Resistance for the sample shown in Fig. 6.12: (A) Measured four-terminal resistance of Sample 2 and Sample 1; (B) Calculated two-terminal resistance curves for Sample 1 show the transition from a fragmented to a connected ring with increase of Fermi level E_F. Self-consistent calculation of potential and electron density E_F is fixed at zero. *Data from V.A. Tkachenko, Z.D. Kvon, D.V. Sheglov, A.V. Latyshev, A.I. Toropov, O.A. Tkachenko, et al., Aharonov-Bohm oscillation amplitude in small ballistic interferometers, JETP Lett. 79 (2004) 136–140.*

electrostatics and 2D quantum transport for zero temperature, shows that any weak asymmetry of electron rings will produce the Aharonov–Bohm (AB) oscillations of small amplitude (Fig. 6.14), with the broken ring excluding these oscillations at all.

In order to model the interferometer's behavior at different temperatures, calculation of electron transmission coefficient through the ring was made over a wide range of magnetic field B and electron energy E (Fig. 6.15). The 2D scattering problem, similar to the one shown in Fig. 6.13, was solved at each point (E,B) for confining 2D potential determined for $E_F = 0$. These massive calculations have been performed in parallel on a supercomputer cluster [12,31]. The transmission coefficient with an accuracy to multiplier $2e^2/h$ is the conductance at zero temperature, and E means the Fermi level E_F. Calculations were made for two close implementations of potential simulating different mesoscopic states of Sample 1. They were obtained by solving the 3D-electrostatic problem for slightly different (within 20%) values of the coefficient connecting observed height of surface bulging with the oxidation depth. Figs. 6.15 and 6.16 show that the conductance behavior for these cases is qualitatively similar, however it differs in details, as it should be in mesoscopics.

The period of the AB oscillations is defined by the effective area of the ring; however, the phase and amplitude of oscillations vary. The map of total conductance $G(E,B)$ clearly shows

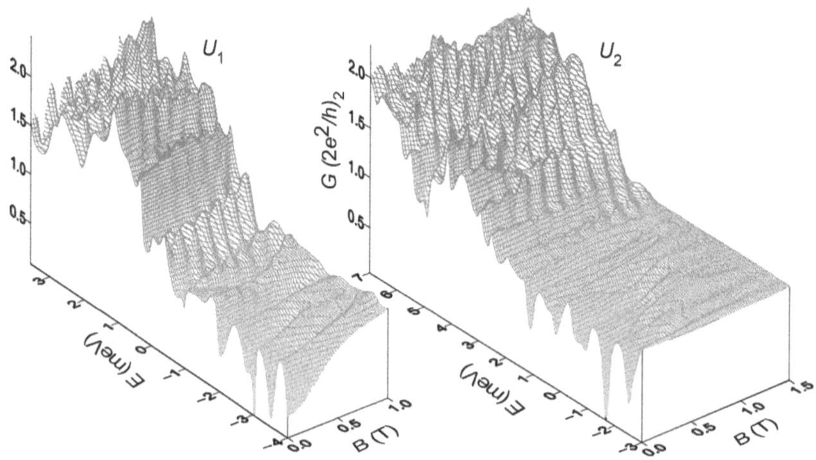

FIGURE 6.15 Calculated conductance $G(E,B)$ for U_1, U_2 (x,y) in Sample 1 (zero temperature limit). *Data from O.A. Tkachenko, V.A. Tkachenko, J.C. Portal, Phase drift and temperature-induced ordering of h/e oscillations in a small ring interferometer, JETP Lett. 97 (2013) 10–14.*

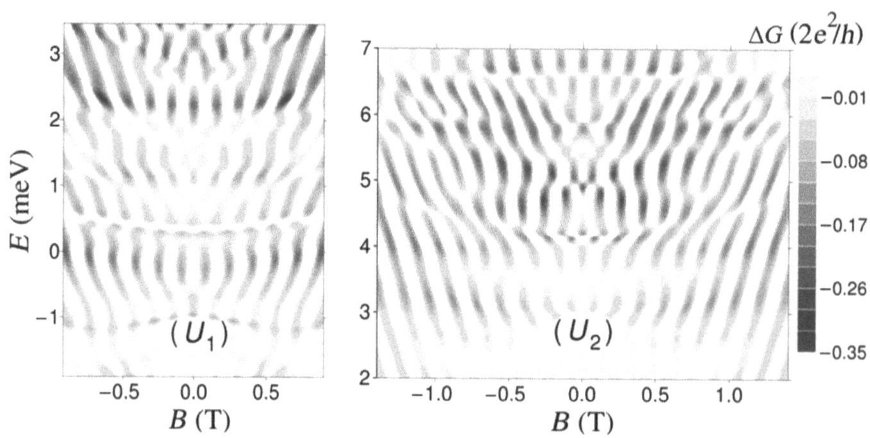

FIGURE 6.16 Behavior of the negative part of the difference between the total and background conductance for slightly different $U_1(x,y)$, $U_2(x,y)$. *Data from O.A. Tkachenko, V.A. Tkachenko, J.C. Portal, Phase drift and temperature-induced ordering of h/e oscillations in a small ring interferometer, JETP Lett. 97 (2013) 10–14.*

cutoff energy where the ring becomes connected and the AB oscillations appear (Fig. 6.15). For U_2, it is 3 meV higher than for U_1. The AB oscillations are suppressed below this boundary, but at $B=0$ the Fano dips can be seen; these are caused by interference in triangular quantum dots. Accordingly, similar to Fig. 6.6, with increasing B the dips quickly disappear when the conductance approaches $2e^2/h$. When energy is higher than the cutoff energy, the dips transform into more complicated fluctuations of background conductance explained by interference at the input quantum dots. If the background conductance is subtracted from the total value G, then the complicated behavior of the AB oscillations at $T=0$ becomes clearer (Fig. 6.16).

It is interesting that the amplitude of AB oscillations is small compared with $2e^2/h$ at any low temperature in any experiment with rings, while for simple 1D ring models it is $2e^2/h$. The loss of phase coherence was usually used to explain this low value. However, calculations for the working samples with natural asymmetry show that the AB oscillations are small even if both arms of the ring are open and the phase coherence is fully preserved.

The experiments with the rings have shown that the AB oscillation amplitude varies. We obtained similar amplitude variation in numerical modeling. Fig. 6.16 clearly demonstrates horizontal and slightly tilted strips of higher amplitude. The average interval between these strips is the same as for background conductance fluctuations and conductance dips, and is determined by the interference in input quantum dots. A small amplitude of the AB oscillations and their variations are explained by a residual asymmetry of the ring, and redirections of ballistic electron between the interferometer arms. Indeed, as seen above, the redirections occur with the same energy interval as the backscattering resonances in the triangular quantum dot (Fig. 6.6–6.8).

In order to clarify the role of these resonances in the ring interferometer, we calculated the nonequilibrium current distribution for two values of magnetic field, corresponding to a maximum and minimum of one of the AB oscillations at $T=0$ (Fig. 6.17). Modeling shows

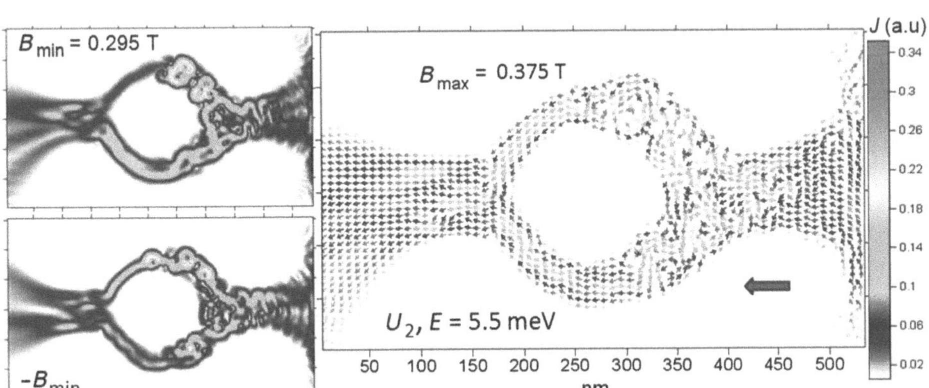

FIGURE 6.17 Contribution to nonequilibrium current distribution at $E=5.5$ meV for U_2. Plots correspond to minimum and maximum conductance in one of the AB oscillations. The electrons are incident from the right (the region of higher chemical potential).

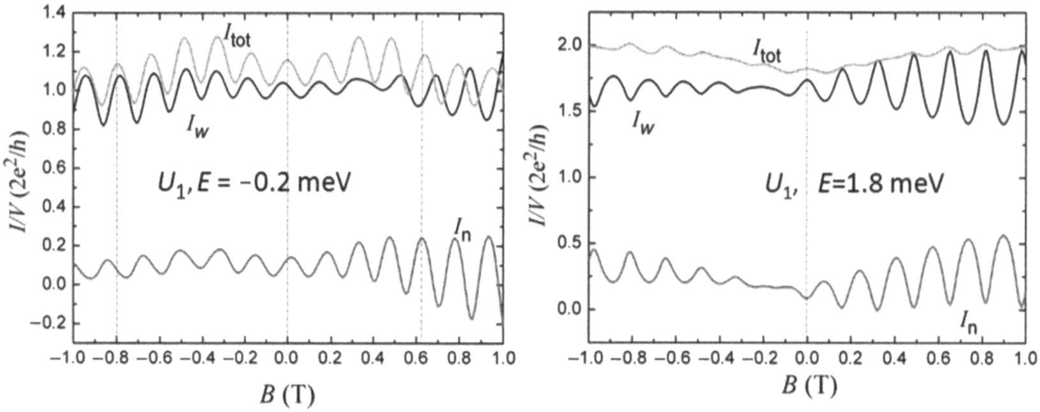

FIGURE 6.18 Two examples for h/e oscillations of nonequilibrium current running through the wide and narrow interferometer branches. Voltage V is applied between two reservoirs. Total current is $I = I_w + I_n$. Amplitude of the AB oscillations in total current is increased at $E = -0.2$ meV and almost vanishes at $E = 1.8$ meV.

that the sign reversal of magnetic field changes the currents in the arms of the ring, although the total conductance of the ring does not change. The sum of currents running through the arms is the total current. It is possible to determine the contributions of the "wide" and the "narrow" arms to the total conductance G. The example shows the polarization degree $(G_w - G_n)/G$ changing from 57% to 5.5% at $G_{min} = 0.86 \cdot 2e^2/h$, and from 37% to -2.4% at $G_{max} = 1.05 \cdot 2e^2/h$, when positive B is replaced with negative B. Similarly, in the three-terminal quantum dot it can be seen that switching of ballistic electrons between the interferometer arms is closely connected to changes in the interference pattern in the input triangular quantum dots and near the channel narrowing, where nanovortices occur. The Lorenz force pushes the electrons incident on the ring from the right downward at $B > 0$ and upward at $B < 0$.

Calculations of the current distributions show that magnetic field h/e oscillations are present in partial conductances G_w, G_n of the ring channels. Their amplitude depends on the magnetic field, and the phase is controlled by the redirection effect in triangular quantum dots of the interferometer. For example, if G_w, G_n oscillate in antiphase and the oscillation amplitudes are comparable, then the AB oscillations weaken in the total conductance $G = G_w + G_n$ (Fig. 6.18, case $E = 1.8$ meV). As the nonequilibrium currents I_w, I_n are not symmetrical by B, the phase of h/e oscillations of I_w, I_n cannot be equal to 0 or π at $B = 0$ (Fig. 6.18, case $E = -0.2$ meV).

Several experiments revealed an oddity that could not be explained: in two-terminal measurements the AB oscillation phase shows continuous drift with increasing energy. However, simple 1D models and many observations, e.g., [32], provide only phase jumps by π, in accordance with the constancy of oscillation period, and the requirement for symmetry by magnetic field (the Onsager relation). These phase π-jumps occur in Fig. 6.16 ($E = 0.5$ meV for U_1, $E = 6.7$ meV for U_2), but there is also continuous phase drift in the form of slanted wavy lines. In real rings with finite width of channels there is no contradiction between the AB oscillation phase drift, the Onsager relation, and the approximate constancy of the AB

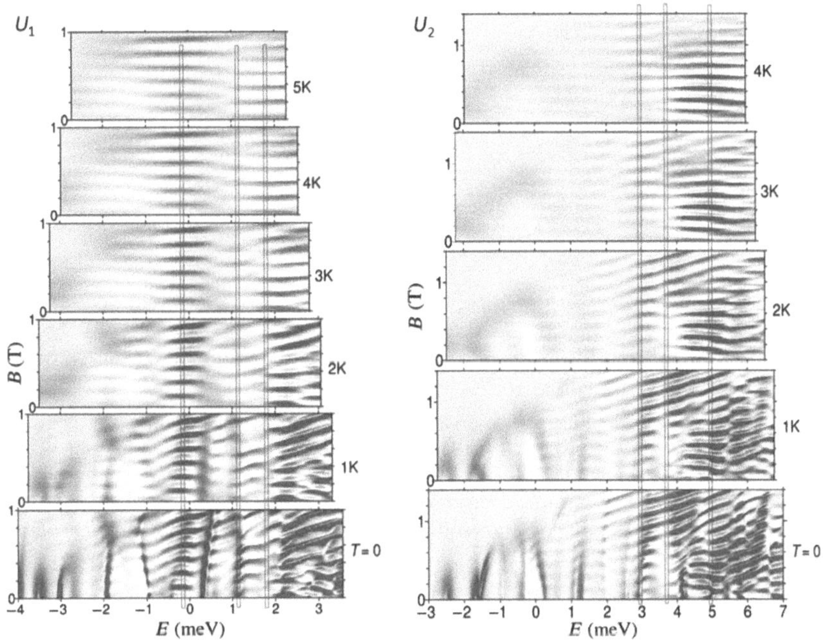

FIGURE 6.19 Maps of the calculated AB oscillation behavior with temperature growth for two close implementations of effective potential U_1 and U_2 in Sample 1 (data from V.A. Tkachenko, Z.D. Kvon, D.V. Sheglov, A.V. Latyshev, A.I. Toropov, O.A. Tkachenko, et al., Aharonov-Bohm oscillation amplitude in small ballistic interferometers, JETP Lett. 79 (2004) 136–140). *Vertical lines* mark ring states with varying thermal damping.

oscillation period. The AB oscillations can split or merge in some points of the plane (E,B) with increasing E at zero temperature (Fig. 6.16). Additional oscillations generally appear at $B = 0$, and they scatter symmetrically in the opposite directions $B > 0$ and $B < 0$. The rate of such scattering varies in E and B, but on average it preserves the AB oscillation period. Furthermore, the frequency of crossing by oscillations of the outermost verticals by B is approximately the same as for dips of the background conductance (Fig. 6.15), and for strips of the enhanced amplitude AB oscillations (Fig. 6.16). Hence, the AB oscillation phase drift is caused by the interference of electrons in the small triangular quantum dot.

For nonzero temperature conductance, calculations average the transmission coefficient over the energy range around the respective E_F corresponding to thermal smearing of Fermi function. Fig. 6.19 shows the results of such averaging [12]. The main result is that the AB oscillation phase is stabilized with increased T, and this takes place before the AB oscillation amplitude fluctuations are suppressed. The AB amplitude sometimes behaves oddly, and can even increase with T at certain intervals of E_F, instead of damping. This happens in the places where there are almost no oscillations at $T = 0$, but they migrate here at $T > 0$ from neighboring regions due to averaging by Fermi distribution. On the contrary, if a narrow range of E with strong oscillations at $T = 0$ is surrounded by "bald" regions, then these oscillations die out with increasing T faster than in wider energy strips filled by oscillations.

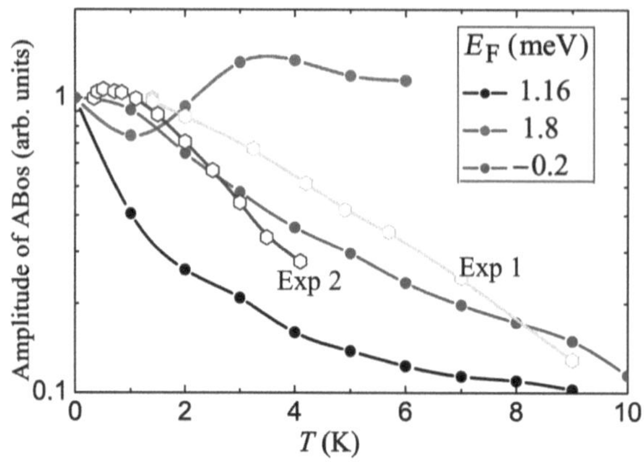

FIGURE 6.20 Calculated dependence of *h/e* oscillation amplitude on temperature for different energy states in potential U_1 and experimental curves exp1, exp2 for Sample 1. All curves are normalized to the first value point. *Data from V.A. Tkachenko, Z.D. Kvon, D.V. Sheglov, A.V. Latyshev, A.I. Toropov, O.A. Tkachenko, et al., Aharonov-Bohm oscillation amplitude in small ballistic interferometers, JETP Lett. 79 (2004) 136–140; O.A. Tkachenko, V.A. Tkachenko, J.C. Portal, Phase drift and temperature-induced ordering of h/e oscillations in a small ring interferometer, JETP Lett. 97 (2013) 10–14.*

The temperature evolution of the amplitude of the AB oscillations is shown in Fig. 6.20 for some interesting states denoted by vertical lines in Fig. 6.19 case U_1, after removing the background and processing of the h/e oscillations by Fourier transform. It has been previously thought that the thermal damping rate of the AB oscillations is stable because it is related to the phase coherence time which depends only on temperature. Nevertheless, the incoherence does not play any significant role in small rings at low temperature. In this case, the coherence, ring asymmetry, electron flow redirections, and averaging by Fermi distribution can produce diverse temperature dependencies, including nonmonotonic ones.

Note that measurements performed for the same sample in different cryostats [33] reveal different thermal damping rates of the AB oscillations (Fig. 6.20). Calculations of the temperature dependences did not consider impurity disorder, and were not tuned to fit the experiment, yet the modeling allowed us to understand why thermal damping of the oscillation amplitude can change in different measurements of the same sample.

Finally, the period of the AB oscillations in small rings cannot be easily estimated from the map of local anodic oxidization, because the interferometers have an asymmetric eye shape with wide connections to the reservoirs, and the area of the ring is not well defined. It turned out that the period of the AB oscillations in calculation changed from 0.14 to 0.16 T with energy, while the measured one was 0.125 T (Figs. 6.14, 6.16, and 6.18). The period of the AB oscillations in the experiment likely depended on a certain state of the sample. In our modeling, only two realizations of the electrostatic potential were tried. Modeling of the ring with 95 nm effective radius [11] shows that the AB oscillation period varies within 20% (Fig. 6.21C and D), and the minimum calculated period is close to the measured period 0.15 T. Therefore, the period of the AB oscillations (and the effective area of the ring), as well as the amplitude, phase, and thermal damping rate demonstrate mesoscopic behavior.

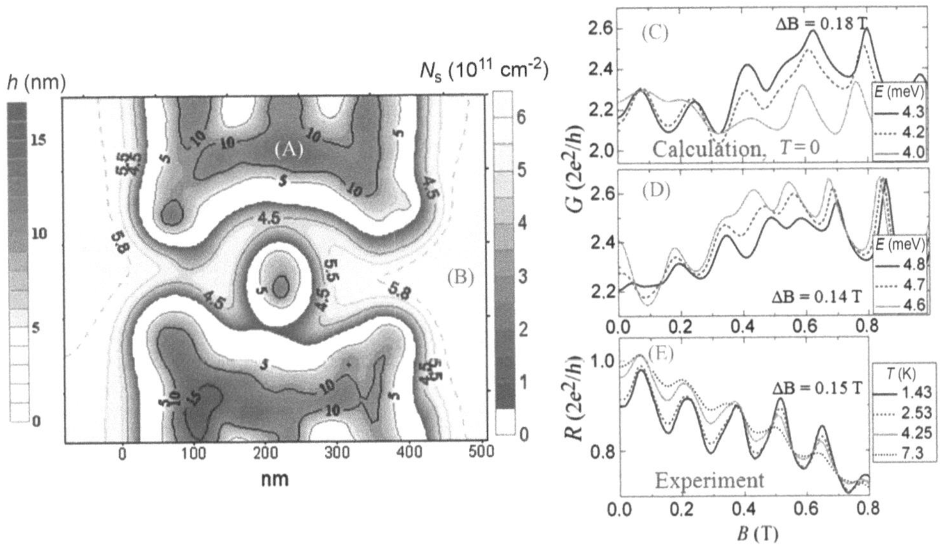

FIGURE 6.21 Modeling of the 95 nm ring fabricated by local anodic oxidation. (A) The map of local anodic oxidization depth h. (B) Calculated electron density N_s. (C D) Calculated conductance of the ring for different Fermi energy *E*. (E) Measured ring resistance *R*. *Data from V.A. Tkachenko, D.V. Sheglov, Z.D. Kvon, E.B. Olshanetsky, A.V. Latyshev, A.I. Toropov, et al., Smallest Aharonov-Bohm interferometer, fabricated by local anodic oxidation, in: Proc.14th Int. Symp. Nanostructures Physics and Technology, Ioffe Physico-Technical Institute. St. Petersburg, 2006, pp. 250–251.*

6.5 Graphene-Like Lattices of Quantum Antidots and Quantum Dots

Lattices of quantum dots/antidots with submicron periods have already been created over a long time by nanopatterning of ultra-high-mobility 2DEGs in GaAs/AlGaAs heterostructures. However, miniband spectra of these 2D artificial crystals have not been observed yet. The main reason is that the amplitude of random potential caused by impurities is much greater than the width of the miniband features. For example, one of the purposes of creating graphene-like lattices [34,35] in GaAs/AlGaAs structures is to obtain massless Dirac fermions of extremely low energy. The energy scale is three orders of magnitude lower than in graphene for lattices with a 100 nm period. Such quasiparticles have not yet been obtained in experimental conditions. However, recent calculations of 3D electrostatics and 2D quantum transport have determined the necessary parameters for such structures and acceptable levels of disorder (Figs. 6.22–6.24). Instead of using modulation-doped structures [34,35], it was proposed to form graphene-like lattices of quantum antidots [36] and quantum dots [37] in the gate voltage-induced 2DEG with two metal gates (Fig. 6.22A). The Schottky gate is perforated and separated from the top gate by a thin 5 nm insulator. Fig. 6.22 shows the dispersion laws for graphene-like lattices of quantum antidots and dots for optimized vertical and lateral layer sizes of nanostructures, and at optimal values of gate voltages.

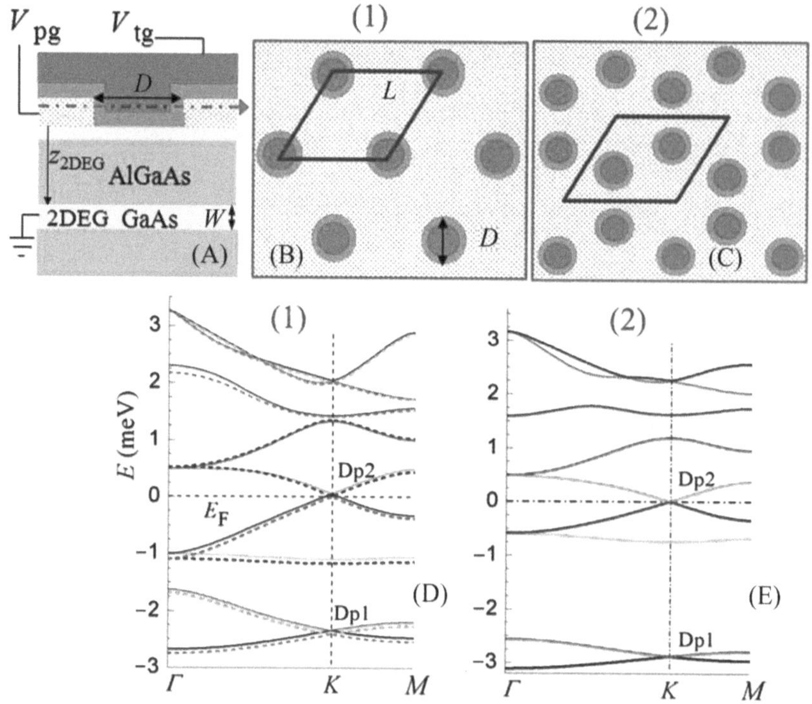

FIGURE 6.22 Schematic view of the structure: (A) Vertical cut; (B and C) lateral cut at the depth shown by *dash-dot arrow*. (1) Triangular lattice of antidots, attractive V_{pg}, repulsive V_{tg}. (2) Honeycomb lattice of open quantum dots, roles of V_{pg} and V_{tg} are exchanged. (D and E) Calculated miniband spectra at optimal parameters of periodic lattices (1) and (2): $z_{2DEG} = 37$ nm, $L = 130$ nm, $D = 60$ nm (B and D), and $D = 50$ nm (C and E). *Data from O.A. Tkachenko, V.A. Tkachenko, I.S. Terekhov, O.P. Sushkov, Effects of Coulomb screening and disorder on an artificial graphene based on nanopatterned semiconductor, 2D Mater. 2 (2015) 014010; O.A. Tkachenko, V.A. Tkachenko, Modeling of semiconductor graphene-like nanosystems, in: Proc. Int. Conf. "Advanced Mathematics, Computations and Applications 2015", Akademgorodok, Novosibirsk, 2015, pp. 764–770 (Electronic resource, in Russian).*

Calculations of 3D electrostatics in the Poisson–Thomas–Fermi approximation (*dashed line*) and the Poisson–Hartree approach (*solid line*) give very close results (Fig. 6.22D). The second Dirac feature (Dp2) in Fig. 6.22E, unlike the previous project of graphene-like lattice of quantum dots [34], is not overlapped by other miniband states, and has almost the same width $\Delta \approx 0.6$ meV as in the case of the lattice of antidots.

In connection with the well-known analogy between these quasiparticles and Weyl (massless) neutrinos, we note that Δ is 10–13 orders of magnitude lower than the energy of experimentally studied ultra-relativistic neutrinos, but is almost equal to the energy of the most abundant in the universe Big Bang neutrinos (0.4 meV).

It is interesting to know what level of errors in the fabrication of gates will not yet destroy the second Dirac feature in optimized structures. These errors were modeled by random variations within a few nanometers in position and radius δr of holes in the Schottky gate. Note that the level of $\delta r \leq 3$ nm is at the edge of current nanotechnology capability.

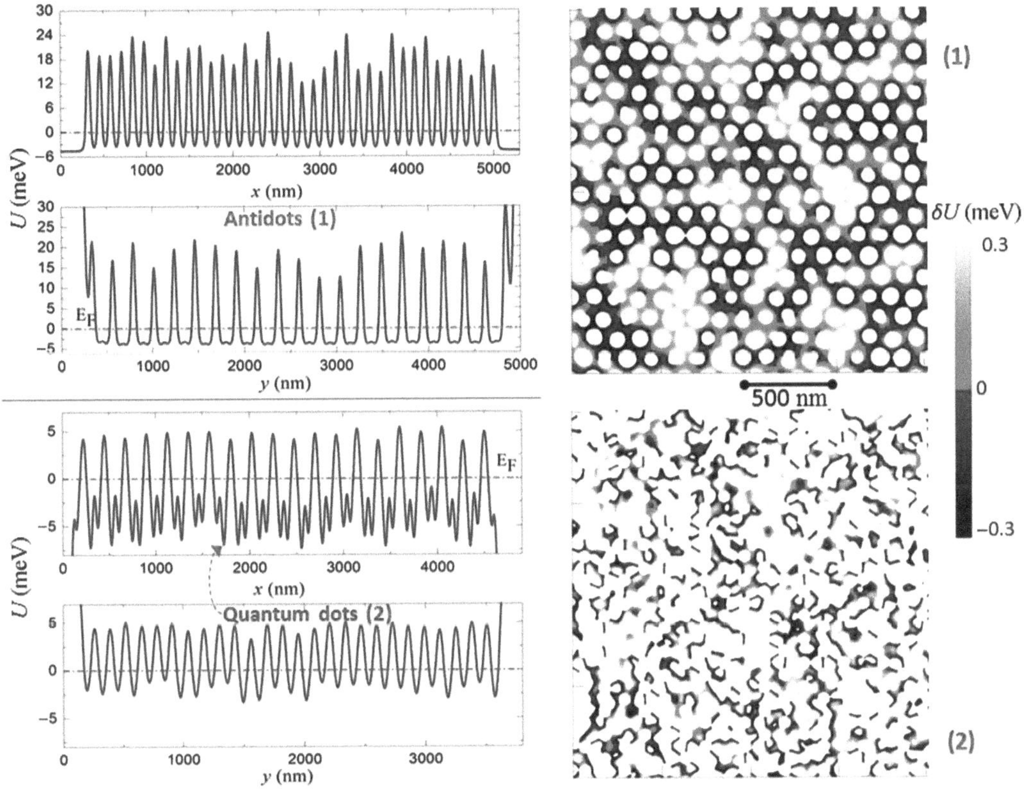

FIGURE 6.23 On the left: 2D potential in the cross-sections through the center of the quantum resonators with antidots or dots in a wide channel at disorder in the geometry of a perforated gate $\delta r \leq 5$ nm. On the right: Difference between ideal and disorder potentials $\delta U = U - U_{\text{ideal}}$ in a narrow range Δ:$|\delta U| < \Delta/2$ and $U < 0$, where $\Delta = 0.6$ meV.

Fig. 6.23 shows the behavior of the confining potential for 2D graphene lattices of quantum antidots and dots with a disorder $\delta r \leq 5$ nm. The basic difference between these lattices becomes evident from a comparison of potential fluctuations $\delta U = U - U_{\text{ideal}}$ with the width Δ of the second Dirac feature. An amplitude of δU in the region $U > E_F$ is much greater for the antidot lattice than for the dot lattice, but these fluctuations are not critical for quantum transport, since there are no electrons in the region. On the contrary, when $U < E_F$ the amplitude of δU is smaller than $\Delta/2$ almost everywhere for the antidot lattice, but only in narrow regions for the lattice of quantum dots. This happens because the sources of disorder are located directly above the quantum dots or antidots, and contribution to screening by the Schottky barrier is correspondingly suppressed or enhanced.

Finally, Fig. 6.24 shows how the disorder influences the conductance of quantum resonators with quantum antidot/dot lattices. One can see that disorder $\delta r \leq 2.5$ nm preserves miniband features, including the second Dirac point, in the antidot case, and destroys them in the case of quantum dots. The result agrees with the expected effect of screening by the

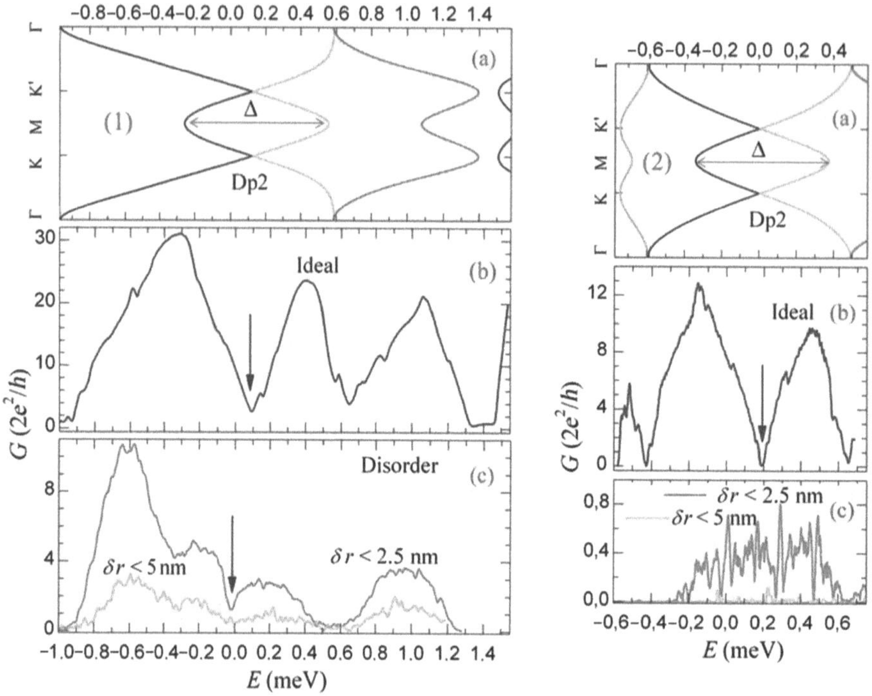

FIGURE 6.24 (a) The miniband spectrum of lattice of antidots (2) (data from O.A. Tkachenko, V.A. Tkachenko, Modeling of semiconductor graphene-like nanosystems, in: Proc. Int. Conf. "Advanced Mathematics, Computations and Applications 2015", Akademgorodok, Novosibirsk, 2015, pp. 764–770 (Electronic resource, in Russian), or quantum dots (1) near the second Dirac point (Dp2). (b and c) The conductance of the quantum resonator with lattices: the ideal case and disorder.

Schottky gate and the calculation of 2D potential fluctuations. If experiments proposed in Refs. [36,37] are undertaken, it will serve the technological progress of GaAs/AlGa as nanostructures, and reliable testing of their quality.

6.6 Conclusions

The research method described has many applications. It has great potential and can be used in other areas of nanophysics and nanotechnology. Specifically, manufacturing and characterization of experimental samples can be anticipated and corrected with realistic modeling based on trustworthy algorithms. For example, designing artificial graphene can be advanced.

Acknowledgments

The authors are grateful to D.G. Baksheev for help in setting up HPC computations, and to many colleagues for discussions and permission to use experimental data, particularly A.I. Toropov, A.A. Bykov, E.B. Olshanetsky,

L.V. Litvin, T.A. Gavrilova, A.V. Latyshev, C.-T. Liang, M.Y. Simmons, C.G. Smith, D.A. Ritchie, G.-H. Kim, M. Pepper, K.S. Pyshkin, R.H. Harrell, E.H. Linfield, C.J.B. Ford, O. Estibals, J.-C. Portal, I.S. Terekhov, O.P. Sushkov.

Computations were carried out at the Siberian Supercomputer Center (Novosibirsk, Russia), IDRIS (CNRS, France), and Joint Supercomputer Center (Moscow, Russia). The authors gratefully acknowledge partial support from Russian Science Foundation (grant #14-22-00143).

References

[1] K.K. Likharev, Correlated discrete transfer of single electrons in ultrasmall tunnel junctions, IBM J. Res. Dev. 32 (1988) 144–158.

[2] T. Ando, A.B. Fowler, F. Stern, Electronic properties of two-dimensional systems, Rev. Mod. Phys 54 (1982) 437–672.R. Landauer, Transport as a consequence of incident carrier flux, in: B. Kramer, G. Bergmann, Y. Bruynserade (Eds.), Localization Interactions and Transport Phenomena, Springer, Heidelberg, 1985, pp. 38–50.D.S. Fisher, P.A. Lee, Relation between conductivity and transmission matrix, Phys. Rev. B23 (1981) 6851–6854.

[3] O.A. Tkachenko, V.A. Tkachenko, Z.D. Kvon, A.V. Latyshev, A.L. Aseev, Introscopy of quantum nanoelectronic devices, Nanotechnol. Russ. 5 (2010) 676–695.

[4] O.A. Tkachenko, V.A. Tkachenko, D.G. Baksheev, Z.D. Kvon, J.C. Portal, Electrostatic potential, energy spectrum, and fano resonances in a ballistic ring interferometer based on an AlGaAs/GaAs heterojunction, JETP Lett. 71 (2000) 255–258.

[5] V.A. Tkachenko, Z.D. Kvon, O.A. Tkachenko, D.G. Baksheev, O. Estibals, J.C. Portal, Coulomb blockade in a lateral triangular small quantum dot, JETP Lett. 76 (2002) 720–723.

[6] V.A. Tkachenko, O.A. Tkachenko, Z.D. Kvon, D.G. Baksheev, A.L. Aseev, J.C. Portal, Coherent scattering in a small quantum dot, JETP Lett. 80 (2004) 588–592.

[7] O.A. Tkachenko, V.A. Tkachenko, Z.D. Kvon, A.L. Aseev, J.-C. Portal, Quantum interferential Y-junction switch, Nanotechnology 23 (2012) 095202.

[8] A.A. Bykov, D.G. Baksheev, L.V. Litvin, V.P. Migal, E.B. Ol'shanetskii, M. Cassé, et al., Transport properties of a GaAs/AlGaAs ring interferometer in the tunneling regime, JETP Lett. 71 (2000) 434–437.

[9] V.A. Tkachenko, A.A. Bykov, D.G. Baksheev, O.A. Tkachenko, L.V. Litvin, A.V. Latyshev, et al., Single-electron charging of triangular quantum dots in a ring interferometer, J. Exp. Theor. Phys. 97 (2003) 317–330. D.G. Baksheyev, A.A. Bykov, V.A. Tkachenko, O.A. Tkachenko, L.V. Litvin, A.V. Latyshev, et al., Coulomb interaction of triangular quantum dots in a small ring interferometer, Int. J. Nanosci. 2 (2003) 495–503.

[10] V.A. Tkachenko, Z.D. Kvon, D.V. Sheglov, A.V. Latyshev, A.I. Toropov, O.A. Tkachenko, et al., Aharonov-Bohm oscillation amplitude in small ballistic interferometers, JETP Lett. 79 (2004) 136–140.

[11] V.A. Tkachenko, D.V. Sheglov, Z.D. Kvon, E.B. Olshanetsky, A.V. Latyshev, A.I. Toropov, et al., Smallest Aharonov-Bohm interferometer, fabricated by local anodic oxidation, Proc.14th Int. Symp. Nanostructures Physics and Technology, Ioffe Physico-Technical Institute, St. Petersburg, 2006, pp. 250–251.

[12] O.A. Tkachenko, V.A. Tkachenko, J.C. Portal, Phase drift and temperature-induced ordering of h/e oscillations in a small ring interferometer, JETP Lett. 97 (2013) 10–14.

[13] O.A. Tkachenko, V.A. Tkachenko, D.G. Baksheyev, K.S. Pyshkin, R.H. Harrell, E.H. Linfield, et al., Electrostatic potential and quantum transport in a one-dimensional channel of an induced two-dimensional electron gas, J. Appl. Phys. 89 (2001) 4993–5000.

[14] C.-T. Liang, M.Y. Simmons, C.G. Smith, G.H. Kim, D.A. Ritchie, M. Pepper, Experimental evidence for Coulomb charging effects in an open quantum dot at zero magnetic field, Phys. Rev. Lett. 81 (1998) 3507–3510.

[15] D.G. Baksheyev, O.A. Tkachenko, V.A. Tkachenko, The role of intersubband mixing in single-electron charging of open quantum dot, Phys. E6 (2000) 414–417.

[16] O.A. Tkachenko, D.G. Baksheyev, V.A. Tkachenko, C.-T. Liang, M.Y. Simmons, C.G. Smith, et al., Coulomb charging effect in an open quantum dot device, J. Phys. Condens. Matt 13 (2001) 9515–9534.

[17] V.A. Tkachenko, D.G. Baksheyev, O.A. Tkachenko, C.-T. Liang, Coulomb oscillations of the ballistic conductance in a quasi-one-dimensional quantum dot, JETP Lett. 74 (2001) 209–212.

[18] C.-T. Liang, O.A. Tkachenko, V.A. Tkachenko, D.G. Baksheyev, M.Y. Simmons, D.A. Ritchie, et al., Gradual decrease of conductance of an adiabatic ballistic constriction below $2e^2/h$, Phys. Rev. B70 (2004) 195324.

[19] S.E. Laux, D.J. Frank, F. Stern, Quasi-one-dimensional electron states in a split-gate GaAs/AlGaAs heterostructure, Surf. Sci. 196 (1988) 101–106.

[20] J.A. Nixon, J.H. Davies, H.U. Baranger, Breakdown of quantized conductance in point contacts calculated using realistic potentials, Phys. Rev. B43 (1991) 12638–12641.

[21] A. Kumar, Self-consisted calculations on confined electrons in 3-dimensional geometries, Surf. Sci. 263 (1992) 335–340.

[22] A.J. Peck, S.J. Bending, A new approach to the modelling of GaAs/AlGa As nanostructures, Semicond. Sci. Technol. 9 (1994) 188–192.

[23] M. Stopa, Quantum dot self-consistent electronic structure and the Coulomb blockade, Phys. Rev. B54 (1996) 13767–13783.

[24] M. Cahay, M. McLennan, S. Datta, Conductance of an array of elastic scatterers: a scattering-matrix approach, Phys. Rev. B37 (1988) 10125–10136.

[25] Y. Takagaki, D.K. Ferry, Conductance of quantum waveguides with a rough boundary, J. Phys. Condens. Matter 4 (1992) 10421–10432.

[26] T. Ando, Quantum point contacts in magnetic fields, Phys. Rev. B44 (1991) 8017–8027.

[27] T. Usuki, M. Saito, M. Takatsu, R.A. Kiehl, N. Yokoyama, Numerical analysis of ballistic-electron transport in magnetic fields by using a quantum point contact and a quantum wire, Phys. Rev. B52 (1995) 8244–8255.

[28] A. Cresti, R. Farchioni, G. Grosso, G.P. Parravicini, Keldysh-Green function formalism for current profiles in mesoscopic systems, Phys. Rev. B68 (2003) 075306.

[29] H. Grabert, G.-L. Ingold, M.H. Devoret, D. Estève, H. Pothier, C. Urbina, Single electron tunneling rates in multijunction circuits, Z. Phys. B84 (1991) 143–155.

[30] L.R.C. Fonseca, A.N. Korotkov, K.K. Likharev, A.A. Odintsov, A numerical study of the dynamics and statistics of single electron systems, J. Appl. Phys. 78 (1995) 3238–3251.

[31] O.A. Tkachenko, V.A. Tkachenko, Supercomputer modeling of semiconductor quantum nanosystems, Vychisl. Metody Program. 13 (2012) 253–262 (in Russian).

[32] Z.D. Kvon, L.V. Litvin, V.A. Tkachenko, A.L. Aseev, One-electron transistors using the Coulomb blockade and quantum interference effects, Phys. Usp. 42 (1999) 402–405.

[33] E.B. Olshanetsky, Z.D. Kvon, D.V. Sheglov, A.V. Latyshev, A.I. Toropov, J.C. Portal, Temperature dependence of Aharonov-Bohm oscillations in small quasi-ballistic interferometers, JETP Lett. 81 (2005) 625–628.

[34] M. Gibertini, A. Singha, V. Pellegrini, M. Polini, G. Vignale, A. Pinczuk, et al., Engineering artificial graphene in a two-dimensional electron gas, Phys. Rev. B79 (2009) 241406.

[35] L. Nádvorník, M. Orlita, N.A. Goncharuk, L. Smrcka, V. Novak, V. Jurka, et al., From laterally modulated two-dimensional electron gas towards artificial graphene, New J. Phys. 14 (2012) 053002.

[36] O.A. Tkachenko, V.A. Tkachenko, I.S. Terekhov, O.P. Sushkov, Effects of Coulomb screening and disorder on an artificial graphene based on nanopatterned semiconductor, 2D Mater. 2 (2015) 014010.

[37] O.A. Tkachenko, V.A. Tkachenko, Modeling of semiconductor graphene-like nanosystems, Proc. Int. Conf. "Advanced Mathematics, Computations and Applications 2015", Akademgorodok, Novosibirsk, 2015, pp. 764–770 (Electronic resource, in Russian).

7

Spectroscopy of Vibrational States in Low-Dimensional Semiconductor Systems

A.G. Milekhin[1,2], D.R.T. Zahn[3]

[1]RZHANOV INSTITUTE OF SEMICONDUCTOR PHYSICS OF THE SIBERIAN BRANCH OF THE RUSSIAN ACADEMY OF SCIENCES, NOVOSIBIRSK, RUSSIA [2]NOVOSIBIRSK STATE UNIVERSITY, NOVOSIBIRSK, RUSSIA [3]TECHNICAL UNIVERSITY OF CHEMNITZ, CHEMNITZ, GERMANY

Progress in epitaxial growth of crystal films of nanometer thickness led to the appearance of a number of new objects of investigations in solid state physics, namely low-dimensional structures which include quantum wells, superlattices, quantum wires, quantum dots (QDs), etc. Theoretical investigations showed that the space confinement of free carriers and crystal lattice vibrations in low-dimensional structures drastically change their electron and phonon spectrum. A number of new, quantum size effects included quantizing the electron spectrum, confinement of optical and acoustic phonons, and folding the acoustic phonons. Coupled vibrations of optical phonons with intersubband and intrasubband plasmons (plasmon–phonon vibrations) and interface phonons propagating along the interface need to be further investigated. Determination of the phonon spectrum is highly important and actual from the point of view of practical applications of low-dimensional structures in micro- and optoelectronic devices while phonons influence on the relaxation processes of excited electrons and mobility of free carriers.

The aim of this chapter is to review the main regularities of formation and nature of phonon modes, coupled plasmon–phonon excitations, and vibrational states in low-dimensional semiconductor systems derived using the methods of vibrational spectroscopy, including Raman scattering and infrared (IR) spectroscopies.

7.1 Phonons in Semiconductor Superlattices

The confinement effect of optical phonons and folding acoustic phonons in periodical structures consisting of binary (such as III–V) compounds can be illustrated using the linear chain model. It allows the behavior of the phonons propagating along the superlattice (SL) growth (usually the [001] direction) [1] to be understood, and gives reasonable agreement with experimental data. This model is based on the assumption that atoms of a particular sort are located in the planes perpendicular to the [001] direction, and that these planes are

Advances in Semiconductor Nanostructures. DOI: http://dx.doi.org/10.1016/B978-0-12-810512-2.00007-X

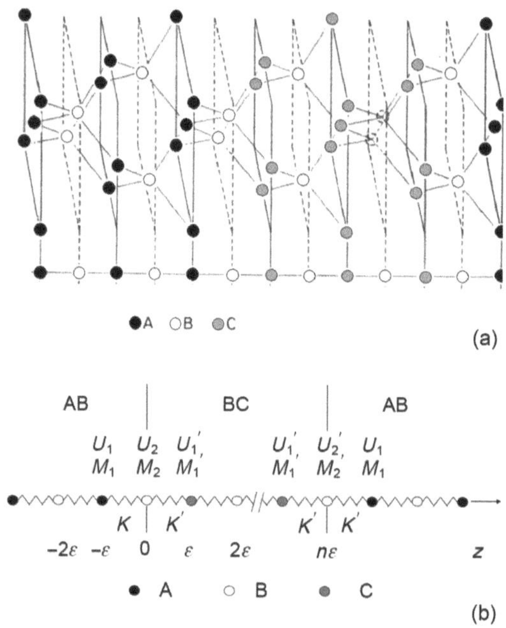

FIGURE 7.1 (a) A fragment of $(AB)_3/(BC)_3$ SL oriented in the [001] direction; (b) Schematic representation of the linear chain model.

moving as a whole (Fig. 7.1a). It allows the three-dimensional calculations to one-dimensional models to be reduced (Fig. 7.1). Here, the superlattice (SL) consists of material layers AB and BC. Taking into account the interaction between the nearest atoms, the motion equation inside each layer can be written as:

$$\begin{aligned} M_1\frac{d^2U_1(m\varepsilon)}{dt^2} &= -K\{2U_1(m\varepsilon)-U_2(m\varepsilon)-U_2[(m-1)\varepsilon]\},\\ M_2\frac{d^2U_2(m\varepsilon)}{dt^2} &= -K\{2U_2(m\varepsilon)-U_1[(m+1)\varepsilon]-U_1[m\varepsilon]\}, \end{aligned} \tag{7.1}$$

where $U(m\varepsilon)$ is the displacement of the atoms along direction z, located at $z=m\varepsilon$, ε is the thickness of a monolayer ($\varepsilon=a/2$), M is the mass of the corresponding atom, and K is the stiffness constant. Taking into account boundary conditions and the structure periodicity, the solution of Eq. (7.1) leads to the dispersion relation:

$$\cos(qd)=\cos(\alpha d_1)\cos(\beta d_2)-\frac{1}{2}\left[F+\frac{1}{F}\right]\sin(\alpha d_1)\sin(\beta d_2), \tag{7.2}$$

where $F=\dfrac{(2K-M_2\omega^2)}{(2K'-M_2'\omega^2)}tg(\alpha\varepsilon)ctg(\beta\varepsilon)$, α and ε are the phonon wave vectors, q is the superlattice wave vector, and $d=d_1+d_2$, d_1, d_2 are thicknesses of the layers consisting of materials AB and BC, respectively.

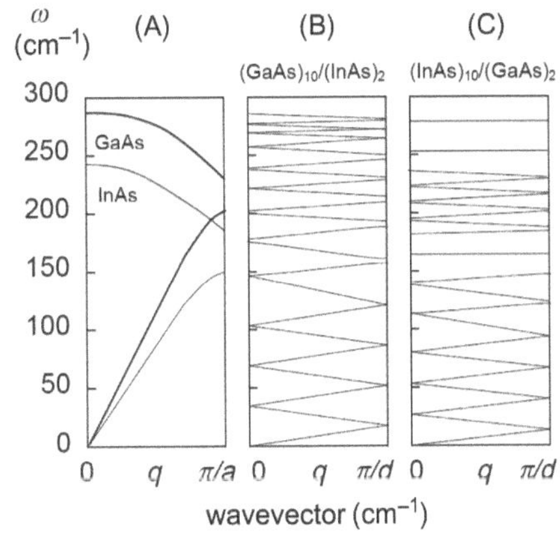

FIGURE 7.2 The calculated dispersion curves of LA and LO phonons: (A) in bulk GaAs and InAs; (B) for $(GaAs)_{10}/(InAs)_2$ SL; and (C) $(InAs)_{10}/(GaAs)_2$ SL [2].

Using this model, the dispersion relations for acoustic and optical phonons in GaAs/AlAs and InAs/GaAs SLs were calculated [1,2]. Fig. 7.2 shows dispersions of longitudinal optical (LO) and acoustic (LA) phonons for $(GaAs)_{10}/(InAs)_2$ and $(InAs)_{10}/(GaAs)_2$ SLs.

A new periodicity leads to folding of the phonon dispersion into a new Brillouin zone, and to appearance of dispersion gaps at the boundaries of the Brillouin zone (Fig. 7.2) from $q_{\min} = 0$ to $q_{\max} = \frac{\pi}{d}$ compared to the Brillouin zone in bulk InAs and GaAs (with $q_{\max} = \frac{\pi}{d}$).

Optical phonon dispersions in InAs and GaAs are located in different spectral regions. Therefore, they become localized within a particular layer. In this case, the phonons of one material cannot propagate inside the layers of the second material, and attenuate rapidly at a distance of 1–2 monolayers from the interface. These phonons do not reveal any dispersion in the direction perpendicular to SL layers, and can be represented as standing waves confined inside each layer. The quantized wave number of the confined mode corresponding to the integer number of half-wavelengths localized within the layer of thickness d_1 is:

$$\alpha = \frac{\pi m}{(d_1 + \delta)} \quad \text{or} \quad \beta = \frac{\pi m}{(d_2 + \delta)}, \tag{7.3}$$

where m is the quantum number of the confined modes, and $\delta \approx 1$ is the penetration depth of the confined mode into the neighboring layer expressed in monolayers.

For most materials with a diamond or zinc blend crystal lattice, the dispersion of optical phonons is negative. Hence, the spectral position of confined phonons is red shifted with respect to the frequencies of optical phonons from the center of the Brillouin zone. The effect of the spatial localization of phonons in semiconductor SL was investigated in detail using Raman spectroscopy. A set of LO phonon modes confined in SL layers was observed in Raman spectra of GaAs/AlGaAs [1–3], GaAs/InAs [2,4], Ge/Si [5,6], and GaSb/AlSb [7,8] SLs. It was shown [3] that odd modes are observed in the crossed scattering geometry, while even modes can be seen in parallel geometry. Raman spectroscopy was successfully used for establishing the dispersion of LO phonons in GaAs, AlAs, Ge, and Si [3,6,9].

Under certain conditions, IR spectroscopy allows both TO and LO odd phonons in semiconductor SLs to be studied. When this activity was initiated in the Institute of Semiconductor Physics (1990), the application of IR spectroscopy was limited to studying the phonon parameters of bulk materials, thick films, and periodic heterostructures composed of relatively thick layers for which the phonon confinement is negligible [10]. Later, new methods for determining the optical phonon and plasmon frequencies in A^3B^5 materials and their ternary solid solutions, composition of solid solutions, film and heterostructure thicknesses, concentration and effective mass of free carriers were developed [10,11]. Impurity states, as well as interband and intraband electronic excitations in heterostructures, were investigated using IR absorption [12,13].

7.1.1 GaAs/AlAs Superlattices

The main results regarding investigation of the phonon spectrum in SLs were obtained by Raman spectroscopy. A detailed review devoted to this subject is presented in the book of M. Cardona [3]. According to selection rules, only LO phonons are allowed in Raman spectra of GaAs/AlAs SLs grown along the (001) direction, while both TO and LO vibrations can be studied from the analysis of IR spectra. Therefore, application of both Raman and IR spectroscopies allows complete information about the vibrational spectrum of SLs to be obtained.

For a proper description of IR spectra of SLs it is necessary to take into account the confinement of optical phonons, and to consider the discrete phonon spectrum. In IR spectra the vibrational modes with wave numbers multiple to π/d, where d is the thickness of a SL layer, are allowed. TO modes confined in SL layers correspond to the poles of the dielectric function. Hence, their observation is possible in s-polarized light when the electric field vector is directed along the SL layers. Confined LO vibrations are located at zeros of dielectric functions that make them active in p-polarized light when the electric field vector is perpendicular to the plane of SL layers. In these conditions, the resonance absorption at LO frequencies can be observed. The confinement effect of LO and TO phonons leads to a shift of a fundamental mode (LO_1 or TO_1 mode) towards lower frequencies with respect to the corresponding values of the bulk constituents. Moreover, observation of the odd confined modes (with the quantum number higher than one) is also possible.

In our papers [14–21] the confinement effect of LO and TO phonons in the layers of GaAs/AlAs SLs were studied from analysis of IR reflection spectra. We were able to observe a number of TO modes confined in GaAs and AlAs layers in IR reflection spectra of $(GaAs)_n/(AlAs)_m$ (short n/m) SLs measured at the normal incidence of the light. Here, n and m are the thicknesses of the corresponding layers expressed in monolayers. Fig. 7.3 shows the IR reflection spectra with different layer thickness in the spectral range of TO phonons in GaAs. One can see a stair-like spectral dependency of IR reflection resulting from the TO phonon confinement. The frequencies of the observed modes were determined with good accuracy from the derivative of reflection $dR/d\nu$ (Fig. 7.3A and B). The minima of the derivative correspond to confined TO modes in GaAs and AlAs. However, a precise position of confined TO modes can be determined from the best fit of the calculated spectra to the experimental ones.

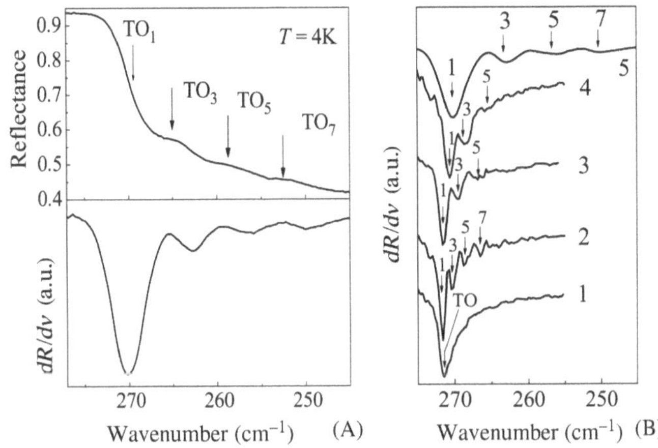

FIGURE 7.3 (A) IR reflection spectra and their derivatives for 10/10 SL in the frequency range of TO phonon of GaAs. (B) The derivative of IR reflection spectra: (1) monocrystal GaAs film ($d = 1.7\ \mu m$); (2) 30/30; (3) 17/17; (4) 14/10; (5) 10/10 SLs.

With increasing the quantum number of confined modes, their intensities decrease due to the decrease of their dipole momentum. Herewith, all possible odd confined modes (the fifth mode is not shown in the figure) are observed for 10/10 SL that show evidence of the high crystalline quality of the SLs. With increasing thickness of the GaAs layers, the frequency of the first confined TO_1 aspires to the value of the TO phonon in the bulk GaAs (curve 1, Fig. 7.3B).

The IR reflection spectra measured with p-polarized light reveal vibrational modes which give evidence of the confinement of LO phonons in GaAs and AlAs layers. IR spectra for a number of GaAs/AlAs SLs with different layer thickness measured at these conditions are presented in Fig. 7.4. As one can see from the figure, the frequency of the first confined LO mode decreases with decreasing layer thickness (TO modes show similar behaviour) from the LO phonon frequency in thick AlAs layer of long-period SLs (405 cm^{-1}) to the frequency of for the first confined mode in AlAs layers in 25/2 SL (392 cm^{-1}).

The wave number corresponding to each odd mth confined mode was determined from the relation $k_m = \frac{\pi m}{a(n+\delta)}$, where i is the quantum number of the confined mode, n is the number of monolayers (MLs) in the corresponding layer, and a is the lattice constant of a layer. The penetration depth of confined mode GaAs in the neighboring AlAs layers was taken as $\delta = 0.8$ [3].

From analysis of the complex dispersion of optical phonons of GaAs [1] one can conclude that the penetration depth of AlAs vibrations into a GaAs layer is less than 0.1 nm $(\mathrm{Im}(k))^{-1} < 0.1$ nm. Therefore, the penetration depth of AlAs vibration into a GaAs layer was taken as $\delta = 0$. The validity of this penetration depth is confirmed by our further experiments investigating the confinement effect of TO phonons in the GaAs layer separated by ultra-thin $Al_xGa_{1-x}As$ layers [22]. It was shown that the 1 ML-thick AlAs layer represents a good barrier for GaAs phonons, while $Al_xGa_{1-x}As$ with $x < 0.5$ is penetrating. More detailed analysis of

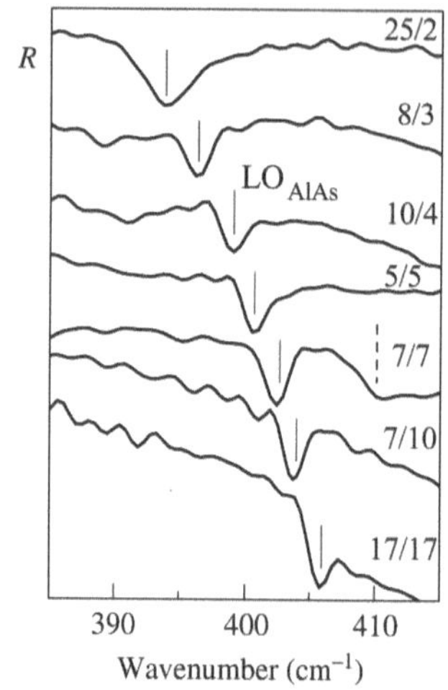

FIGURE 7.4 IR reflection spectra of GaAs/AlAs SLs with different layer thickness measured in the spectral range of LO phonons in AlAs at the incidence angle of 70 degrees in the p-polarized light at $T = 4K$. The *vertical lines* indicate the LO_1 modes.

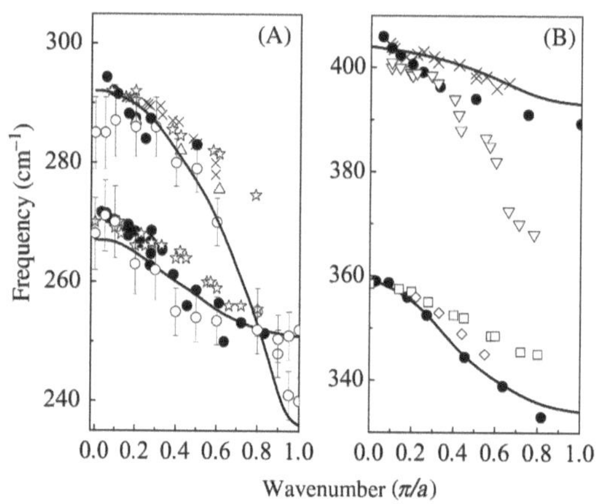

FIGURE 7.5 The dispersion of TO and LO phonons in GaAs and AlAs. Our IR and Raman data are shown by *solid circles* and *stars*, respectively. The rest of the symbols are the literature data [2,4,26]. The *solid lines* represent the results of the first principle microscopic calculations.

vibrational modes requires calculations where the whole atomic structure of the SL is taken into account. These calculations were carried out for structures with ultra-thin $Al_xGa_{1-x}As$ layers [14,23,24]. Using the values of the confined mode frequencies and corresponding wave numbers, the dispersion relations for TO and LO phonons in GaAs and AlAs (Fig. 7.5) were determined. They agree well with the calculations (curve 4), Raman [25], and neutron scattering [26] data.

It is obvious that, due to the anisotropic structure of SLs, one can expect anisotropy effects in their vibrational spectrum. The most pronounced evidence of anisotropy is the

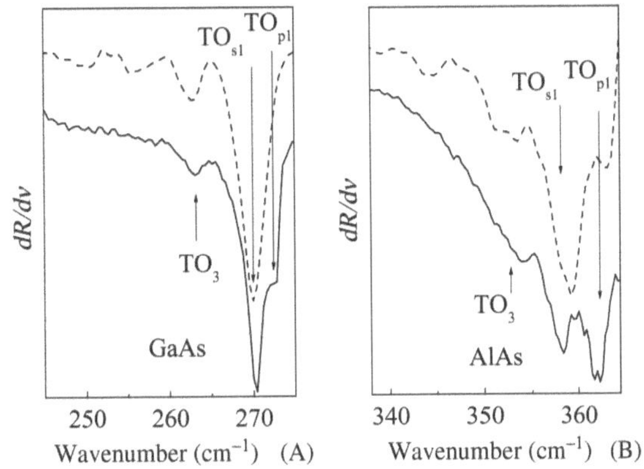

FIGURE 7.6 The derivative of IR reflection spectra of GaAs/AlAs 10/10 SL measured in the spectral range of TO phonons of: (A) GaAs; and (B) AlAs at the normal (*dashed lines*) and incline incidence in the p-polarized light (*sold lines*) at $T = 4.2$K.

splitting of the first confined (fundamental) mode into two modes with different directions of the polarization vector. The vibrational mode with the polarization vector directed along the layer plane is independent on the propagation angle (s-polarized mode). The p-polarized mode, with the polarization vector normal to the layer plane, possesses a strong angular dispersion. The frequencies of s- and p-polarized modes with higher quantum numbers are degenerated and remain constant with increasing propagation angle. This is due to the decrease of their dipole momentum responsible for anisotropy with increasing quantum number of confined modes. The study of anisotropy of TO phonons in GaAs and AlAs layers of GaAs/AlAs SLs was carried out using IR spectroscopy and was published in Refs. [18,21].

For investigation of anisotropy of TO phonons, IR reflection spectra were recorded at the angles of incidence of $\theta = 10$ and 70 degrees. Typical derivative IR spectra of $(GaAs)_{10}/(AlAs)_{10}$ SLs are presented in Fig. 7.6. The spectra of reflection derivative recorded at normal incidence reveals distinct modes corresponding to confined TO_{s1} phonons propagating normal to the layers. At off-normal incidence, the contribution of TO phonons propagating along the layers and having the polarization vector directed normal to the layers (TO_{pi} modes) should be considered. Since the frequency of TO_{s1} and TO_{p1} modes are degenerated, the splitting of TO_1 phonon into two modes is observed in the derivative of reflection spectra measured at $\theta = 70$ degrees (Fig. 7.6).

As was shown in our papers [27,28], direct excitation of interface phonons in SLs and their observation in IR reflection spectra is also possible using a diffraction grating. For this, the metal diffraction grating with period of $d < \lambda$, where λ is the wavelength of the light, is deposited on the SL surface. To achieve effective interconnection, the diffraction grating is located at distances $L \ll \lambda$ from the layer surface. The external planar electromagnetic wave entering the heterostructure normal to the surface with the diffraction grating, excites 2D plasmons in the heterostructure with wave vectors $k_i = 2\pi i/d$, $(i = 1, 2, 3, \ldots)$.

Lowering the symmetry of SLs grown along the (311) direction with respect to SLs oriented along the [001] direction complicates the identification of vibrational modes confined in SL layers. It leads to the appearance of optical modes in (311)-oriented SLs having either pure transverse (A″ modes with the polarization vector directed along the $[1\bar{1}0]$ direction), or mixed longitudinal/transverse (A′ modes) character [29]. Moreover, surface corrugation in (311)A-oriented SLs with a period of 3.2 nm can cause the splitting of confined modes due to phonon confinement in narrow and broad parts of the corrugated layers [30].

Systematic study of the vibrational spectrum of $(GaAs)_n/(AlAs)_m$ SLs over a wide range of layer thicknesses ($n = 7$, 10, 12, 28, and $m = 7$, 16, 17, 24 MLs) [31–34] grown simultaneously on (311) A- and (311) B-oriented GaAs surfaces was carried out. The behavior of confined optical phonons in the long-period (311)-oriented SLs is very similar to the SLs grown along the (001) direction. However, in the short-period GaAs/AlAs SLs for which the formation of quantum wires due to surface corrugation was possible, we were able to observe anisotropy of vibrational spectrum of the SLs from the analysis of their IR reflection spectra [33,34]. The splitting of the fundamental TO_1 modes into two with different directions of the polarization vector was observed and interpreted in terms of formation of GaAs and AlAs quantum wire arrays. The study of conductivity anisotropy [35] shows the principal possibility of the quantum wire array formation in GaAs/AlAs SL grown on (311)A surfaces. It was established that the maximal splitting of TO confined modes is observed for the structures for which the maximal conductivity anisotropy was established [36]. This confirms the validity of the conclusions made.

7.1.2 GaSb/AlSb Superlattices

In contrast to GaAs/AlAs system, GaSb/AlSb SLs are characterized by a large lattice parameter mismatch in GaSb and AlSb (0.65%). As in the case of GaAs/AlAs SLs, their optical phonons are localized in the corresponding layers.

We have performed the first systematic investigation of the confinement effect of optical phonons in layers of the short-period GaSb/AlSb SLs using IR spectroscopy and Raman scattering [37]. GaSb/AlSb SLs with a superior crystalline quality were grown in the A.V. Rzhanov Institute of Semiconductor Physics (ISP) by means of molecular beam epitaxy. The high resolution transmission electron microscopy (HRTEM) images of 7/7 and 4/4 SLs shown in Fig. 7.7A and B indicate a well-controlled periodicity and a layer thickness homogeneity of the SL with an accuracy of 2 MLs. The control of the average SL period was performed from the comparison of folded acoustic mode frequencies derived from Raman scattering data and the dispersion of folded acoustic phonons calculated with the Rytov model [38].

$$\cos(qd) = \cos\left(\frac{\omega d_1}{v_1}\right)\cos\left(\frac{\omega d_2}{v_2}\right) - \frac{k^2+1}{2k}\sin\left(\frac{\omega d_1}{v_1}\right)\sin\left(\frac{\omega d_2}{v_2}\right), \tag{7.4}$$

where $k = \dfrac{v_1\rho_1}{v_2\rho_2}$; $v_{1,2}$ and $\rho_{1,2}$ are the sound velocities and density in GaSb and AlSb, respectively.

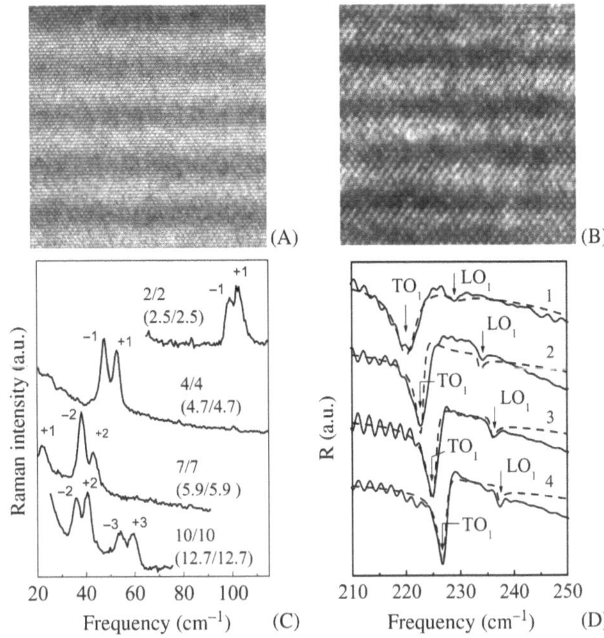

FIGURE 7.7 HRTEM images of GaSb/AlSb SLs: (A) 7/7; and (B) 4/4. Dark and bright areas correspond to GaSb and AlSb layers, respectively. (C) Raman spectra of the folded LA phonons in GaSb/AlSb SLs. The nominal and calculated (in brackets) layer thicknesses are given in the figure. (D) IR reflection spectra of GaSb/AlSb, 2/2, 4/4, 7/7, and 10/10 SLs (curves 1–4, respectively) measured in polarized light (at $\Theta \sim 70$ degrees). Calculated spectra are shown by *dashed lines*.

Using Eq. (7.4) and the frequencies of folded LA phonons observed in the Raman spectra (Fig. 7.7C), the average SL periods were calculated. As one can see for the figure, they agree well with the electron microscopy data (Fig. 7.7).

The confinement effect of optical phonons was discovered in short-period GaAsb/AlSb SLs from the analysis of IR reflection spectra. Fig. 7.7D shows the IR reflection spectra of GaSb/AlSa which reveal features corresponding to the first confined LO_1 and TO_1 modes in GaSb layers. As in the case of GaAs/AlAs SLs, the low-frequency shift of the modes is observed that provides evidence of a strong confinement of LO and TO phonons in GaSb/AlSb SLs. The confinement effect of even LO phonons was also demonstrated in the resonant Raman scattering experiments [37]. Using the experimental frequency positions of confined modes and the corresponding wave numbers, the dispersion relations were obtained for TO and LO phonons in GaSb and AlSb [37].

7.2 Plasmon–Phonon Modes in GaAs/AlAs Superlattices

Since SLs refer as anisotropic structures, the propagation of free carriers in SLs depends on the direction. It remains free along the SL layers, while in the direction of SL growth it is determined by the thickness of the SL layers. Hence, the interaction of free carriers with phonons depends strongly on the direction, as well as the dielectric response of SLs, which describes their optical properties as also anisotropic. The proximity of the frequencies of plasmons and optical vibrations of crystal lattices in ionic semiconductors leads to the appearance of coupled plasmon–phonon modes. One can expect plasmon–phonon interaction in the doped SLs composed of ionic semiconductors.

We considered the influence of electronic excitations on the spectrum of lattice vibrations in GaAs/AlAs SLs [19,20]. Two limited cases were chosen when optical phonons interact with the plasmon excitations propagating along (2D plasmons) and perpendicular to the SL layers. In the case of relatively thick barriers, when the electron spectrum along the SL axis consists of a set of discrete states E_i, intersubband plasmon excitations can occur. Herewith, it is necessary to have the lower (higher) electronic states occupied (unoccupied) with electrons. With decreasing the SL period or barrier thickness, the electrons in the neighboring layers interact with each other. As a result, the discrete state E_i broadens, forming minizones. The frequency of intrasubband plasmon excitations will be determined as:

$$\Omega_{pz} = (4\pi e^2 N/m_z)^{1/2} \tag{7.5}$$

where m_z is the effective mass in a partially occupied minizone, and N is the 3D electron density.

The energy spectrum of such SLs was calculated using the effective mass method in the envelope function approximation [39]. If the Fermi level is located in the minizone, then the electrons at the Fermi level can propagate along the SL axis. Again, if their energy approaches the energy of LO phonons, the coupled modes of intrasubband plasmons with optical phonons occur.

For example, the lowest minizone of $(GaAs)_{17}/(AlAs)_2$ SL is composed of the Γ electronic states in GaAs and AlAs. A large width of the minizone (up to 75 meV) allowed us to change the occupation of the minizone by increasing the doping level in a controllable way [40]. As one can see from Fig. 7.8, full occupation of the minizone is reached at $n = 2.5 \cdot 10^{18}$ cm^{-1}. Variation of the minizone occupation causes shifts of plasmon–phonon modes observed in IR spectra. We carried out analysis of the IR reflection spectra and determined the intrasubband plasmon frequencies Ω_{pz} for various doping levels (Fig. 7.8). It is interesting that Ω_{pz} approaches a maximum for the concentration of $n = 1.4 \cdot 10^{18}$ cm^{-3}, and decreases to zero at $n = 5.6 \cdot 10^{18}$ cm^{-3}. The latter fact indicates that the lowest minizone is completely filled with electrons, and the electrons cannot move along SL axis anymore and, hence, interact with LO phonons polarized along SL axis.

7.3 Phonons in QD Structures

As will be shown below, the built-in mechanical strain leads to significant changes of the optical phonon spectrum. Hence, let us begin consideration of the phonon spectrum in structures with relaxed or nonstrained QDs. QDs composed of metal (cadmium, zinc, lead,

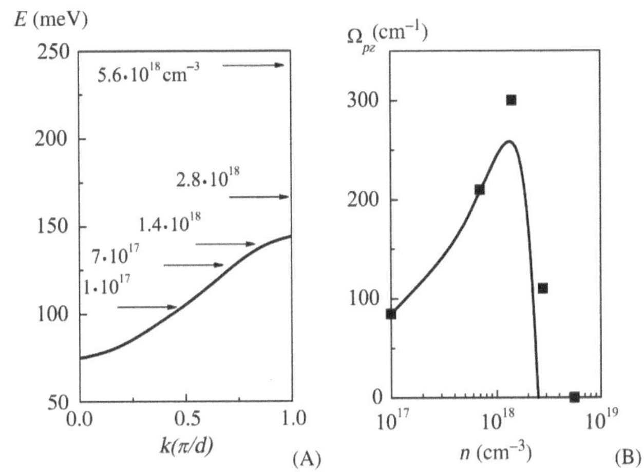

FIGURE 7.8 (A) The structure of the lowest minizone in $(GaAs)_{17}/(AlAs)_2$ SL. The occupation of the minizone at the given electron concentration is indicated by arrows; (B) The dependence of the intrasubband plasmon frequency on the electron concentration in the SL. The solid line represents the dependence calculated using the envelope function model.

etc.), chalcogenides synthesized by means of the Langmuir–Blodgett (LB) method, as well as Ge QDs grown by molecular beam epitaxy in the Volmer–Weber growth mode on oxidized Si surfaces are typical representatives of such objects.

7.3.1 Relaxed QDs Fabricated by Langmuir–Blodgett Technology

Relaxed QDs of metal chalcogenides (CdS, ZnS, CdZnS, PbS, CuS, Ag_2S, and ZnO) were formed using LB technology successfully developed in ISP [41–49].

The scheme of QD formation processed by LB technology is presented in Fig. 7.9. At the first stage (Fig. 7.9A), multimolecular layers (metal behenates $C_{21}H_{43}COO{-}Me$ (Me = Cd, Pb, Cu, Ag, Zn)) with different thicknesses (from 1 to 440 MLs) are deposited on a semiconductor substrate. As a result of chemical reaction of gaseous H_2S with behenate films, semiconductor QDs of metal sulfides are formed in an organic matrix (Fig. 7.9B). Further annealing of the structures at $T = 200°C$ leads to the removal of organic matrix and the formation of free-standing QDs on a solid substrate (Fig. 7.9C). Variation of monolayer number in a multimolecular matrix allows the QD areal density to be changed over a wide range. QDs with an ultra-low areal density were fabricated by decreasing the concentration of cadmium salt by 1–2 orders of magnitude in a monolayer-thick film of behenic acid. Free-standing ZnO QDs were fabricated as a result of annealing of either multimolecular films of zinc behenate, or ZnS QDs at temperatures of 400–600°C on Si substrates covered by a 100 nm-thick Pt layer. Formation of QDs in a matrix, as well as the formation process of free-standing QDs, was monitored by means of IR spectroscopy judging from a change in vibrational spectra of the LB films after their reaction with H_2S and the matrix removal.

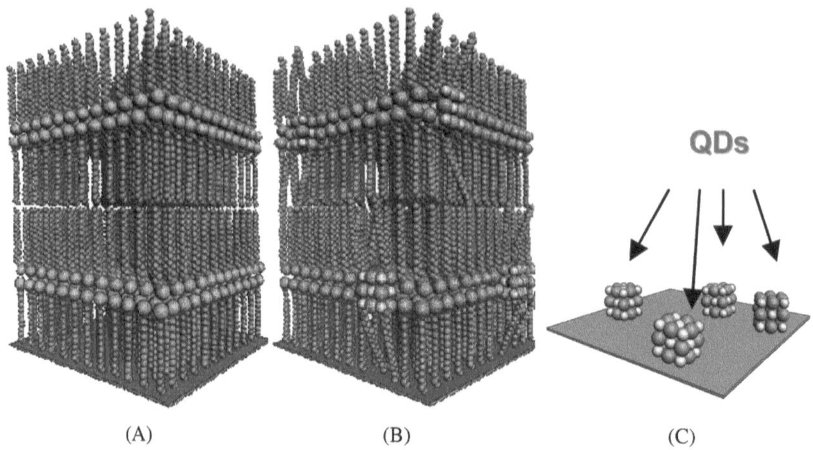

FIGURE 7.9 The scheme of QD formation process by LB technology. (A) Multimolecular layers of the matrix; (B) QDs in an organic matrix; (C) Free-standing QDs.

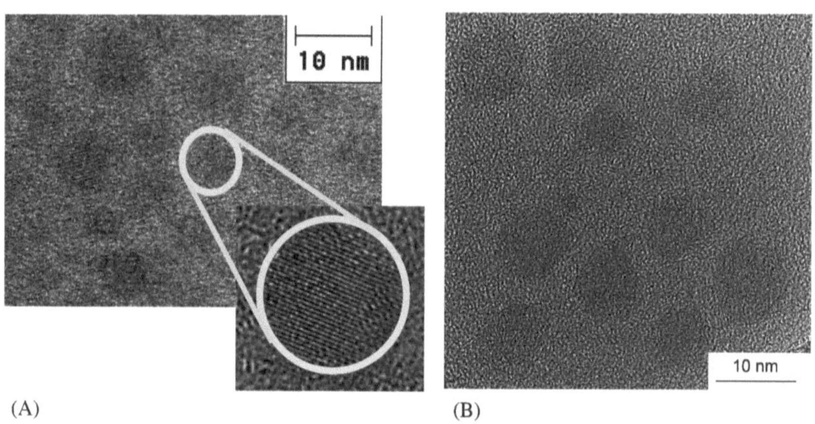

FIGURE 7.10 HRTEM images of: (A) $Cd_{0.5}Zn_{0.5}S$ QDs; and (B) CuS QDs annealed at 200°C.

The structural parameters of QDs such as size, shape, areal density, and crystalline structure were determined by direct structural methods including scanning electron microscopy (SEM) and atomic force microscopy (AFM). The size and crystalline structures of HK CdS, ZnS, PbS, Ag_2S, and ZnO were established using X-ray diffraction [41,43,47–49]. PbS and ZnS QDs reveal a cubic crystal structure, CdS and ZnO QDs have a hexagonal structure, and Ag_2S QDs are monoclinic. QDs of metal sulfides predominantly have a spherical shape, and ZnO QDs are represented as domes. The QD size amounts to 3–10 nm (Fig. 7.10), the lateral size and the height are 40 and 5 nm [50], respectively.

QD structures were investigated by Raman and IR spectroscopies [42–50].

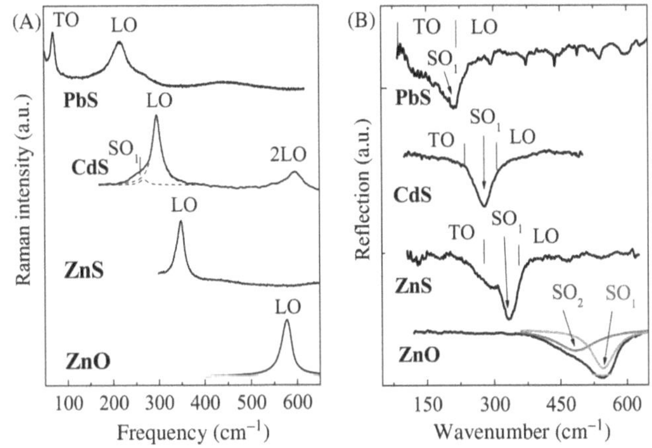

FIGURE 7.11 (A) Raman; and (B) IR reflection spectra of ZnS, CdS, PbS, and ZnO QDs. The *vertical lines* indicate the position of LO and TO phonon frequencies in the bulk materials.

Fig. 7.11A shows the Raman spectra of the QDs fabricated utilizing the LB method in the spectral range of lattice vibrations of materials composing the QDs. As one can see, the Raman spectra reveal predominantly the vibrational modes related with LO and TO phonons in QDs (Fig. 7.11A), while surface optical (SO) modes are clearly observed in the IR spectra (Fig. 7.11B). The frequency positions of the observed SO modes (ω_{SO}) in spherical QDs correspond to the fundamental modes of SO phonons (SO_1) calculated in the dielectric continuum model [51] using the equation:

$$\frac{\omega_{SO}^2}{\omega_{TO}^2} = \frac{\varepsilon_0 + \varepsilon_m(l+1)/l}{\varepsilon_\infty + \varepsilon_m(l+1)/l} \tag{7.6}$$

where $l = 1.2.3.$, and ε_m, ε_0, ε_∞ are the dielectric constant of the matrix material, the low- and high-frequency dielectric constants of QD materials, respectively. The observation of two SO modes for ZnO QDs is due to their anisotropic crystal structure (wurtzite) [50].

7.3.2 Relaxed Ge/SiO_2 QDs

The introduction of a SiO_2 spacer layer of nanometer thickness before growth was proposed for fabrication of small-sized Ge QDs during epitaxial growth [52,53]. The formation of QDs was confirmed by HRTEM data [53,54]. Fig. 7.12 represents the planar and cross-sectional HRTEM images of relaxed Ge QDs. Ge QDs have a shape of a half-sphere with a base size of 4–6 nm and the height of 3–4 nm. The areal density amounts to about $2 \cdot 10^{12}$ cm^{-2}.

Typical Raman spectra of QD structures grown at different nominal Ge thickness (0.6, 0.9, 1.2, 1.5 nm) on a silicon oxide layer measured in the $z(x,y)\bar{z}$ scattering geometry at an excitation energy of 2.47 eV are presented in Fig. 7.12B. The peak observed at 302 cm^{-1}, corresponds to the optical phonon modes localized in Ge QDs. Its intensity decreases with

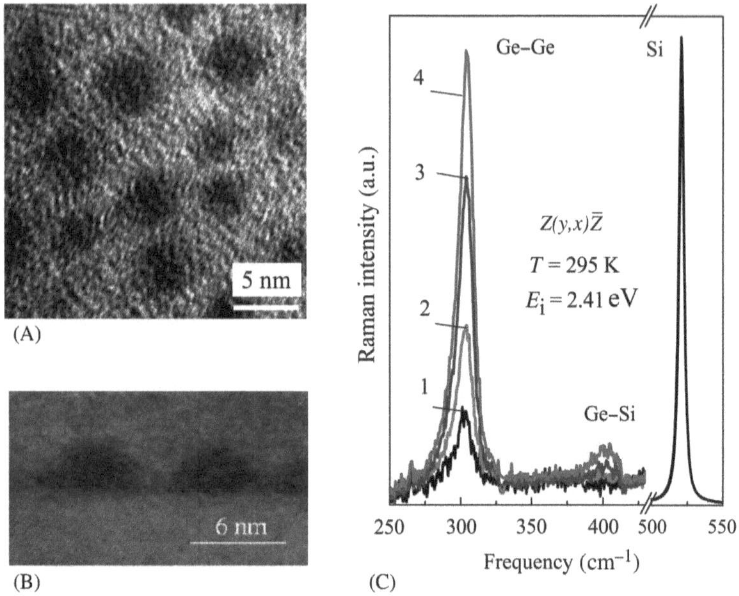

FIGURE 7.12 (A) The cross-sectional; and (B) plane-view HRTEM images of relaxed Ge QDs. (C) Raman spectra of relaxed Ge QDs (nominal thickness of Ge 0.6, 0.9, 1.2, and 1.5 nm) (lines 1–4, respectively).

decreasing nominal thickness of the Ge layer, and can be explained by decreasing the areal density of Ge QDs. The feature near 400 cm^{-1} is assigned to the Ge–Si vibrational modes and manifests intermixing of the Ge and Si atoms at heterointerfaces [55,56]. The phonon mode at 520 cm^{-1} corresponds to the Raman scattering by optical phonons from a Si substrate.

We carried out the first investigation of Raman scattering of Ge/SiO_2 QDs at different excitation energies. It was shown that the frequency position of optical phonons in the relaxed Ge QDs (298 cm^{-1}), even in nonresonant conditions (1.83 eV), is located below the corresponding value in Ge bulk (303 cm^{-1}) (Fig. 7.13A). With increasing excitation energy from 2.5 to 2.7 eV, the position of the Ge phonon mode is shifted towards lower frequencies by 10 cm^{-1}. The shift is accomplished by decreasing intensity. The dependence of Raman intensity on the excitation energy has a maximum near 2.0–2.4 eV. The observed phonon mode frequency and intensity dependencies on the excitation energy (Fig. 7.13) are explained by the confinement effect of optical phonons in the arrays of QDs inhomogeneous in size.

The intensity maximum at 2.35 eV (at Ge nominal thickness of 1 nm) corresponds to the E_0 resonance in Ge which reaches a value of 2.5 eV due to the confinement effect [57]. For a thicker Ge layer (nominal thickness of 2 nm), the maximum energy position is shifted upward (2.42 eV); this indicates the formation of smaller Ge QDs. The feature observed at 2.2 eV most probably is due to the E_1 transition in Ge QDs. The Raman intensity for smaller QDs for which the electronic transitions have larger energy increases approaching excitation energy of E_0. Since QDs are inhomogeneous in size, larger QDs for which the confinement

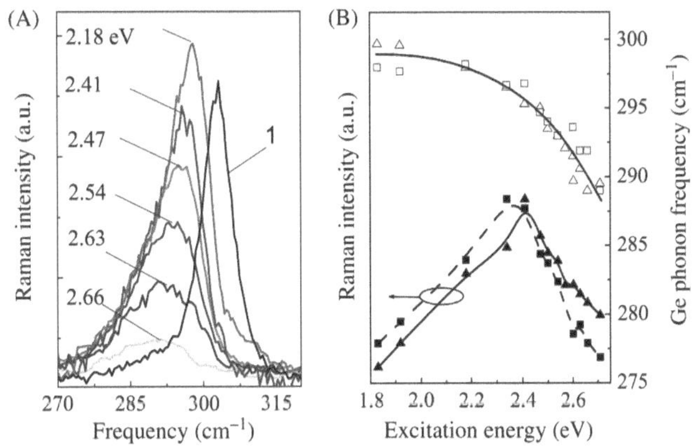

FIGURE 7.13 (A) Raman spectra of relaxed Ge QDs (Ge layer thickness is 1 nm) and bulk Ge (curve 1), measured in $z(x,y)\bar{z}$ scattering geometry with different excitation energy. (B) Shows the dependencies of the Raman intensity and the frequency of Ge optical phonons on the excitation energy for Ge QDs with the nominal thickness of 1 nm (*squares*) and 2 nm (*triangles*).

effect is negligible contribute predominantly to the Raman process in nonresonant conditions. With increasing excitation energy (above 2.3 eV), smaller QDs are involved in the Raman process that reveals itself in decreasing the optical phonon frequency observed in the experiment.

Using the phonon confinement model [58] we determined the average Ge QD size taking part in the Raman process. According to this model, the spectral shape of the signal due to the Raman scattering by optical phonon in spherical QDs with the size d is determined as:

$$I(\omega)\propto \int_0^1 \frac{4\pi k^2 \cdot e^{-k^2 d^2/4}}{[\omega-\omega(k)]^2+\left[\frac{\Gamma_0}{2}\right]^2} dk, \tag{7.7}$$

where Γ_0 is the spectral width in a bulk crystal, $\omega(k)$ is the phonon dispersion, and k is the phonon wave vector. From a comparison of the frequency and the half-width of the optical phonon modes in the calculated and experimental Raman spectra, the average size of QDs contributing to the Raman process was determined as 7.5; 5; 4; 3; 2.2; and 2 nm at an excitation energy of 2.18; 2.41; 2.47; 2.54; 2.63; and 2.66 eV, respectively.

Thus, resonant size-selective Raman scattering allows the optical phonon modes in Ge QDs to be identified, and the influence of confinement effect on vibrational spectrum of QDs to be studied.

7.3.3 Strained GeSi/Si QDs

The optical phonons from the center of the Brillouin zone in diamond-like semiconductors is degenerated three-fold. Biaxial mechanical strain leads to lowering of the crystal symmetry, and causes the mode splitting into a doublet and a singlet for the modes with atom displacements along and perpendicular to the (001) plane, respectively. The shift of the optical phonon modes propagating along the [001] direction as a function of mechanical strain can be derived from the following equations [59]:

$$\Delta\omega(LO) = \frac{1}{2\omega_0}\left[p\varepsilon_{zz} + 2q\varepsilon_{xx}\right] \tag{7.8}$$

for LO phonons, and:

$$\Delta\omega(TO) = \frac{1}{2\omega_0}\left[p\varepsilon_{xx} + q(\varepsilon_{xx} + \varepsilon_{zz})\right] \tag{7.9}$$

for TO phonons. Here, ω_0 is the phonon frequency in the absence of strain, p, q, and r are the phonon deformation potentials, $\varepsilon_{xx} = \varepsilon_{yy}$ and ε_{zz} are the deformation tensor components for the directions perpendicular and parallel to the [001] direction. Using Eqs. (7.8) and (7.9), the deformation tensor components were calculated for Ge/GeSi [60,61], GaSb/AlSb [7,8], InAs/InGaAs [2,62], and InGaAs/InP SLs [63].

We investigated Raman scattering by Ge/Si QD structures grown by MBE over a wide range of temperatures (300–650°C), and determined the deformation tensor components [54,64–73]. It was established, that for low growth temperature (300°C) the QDs with pyramid-like shape (or hut-clusters) are predominantly formed, while larger islands with complicated faceting (or dome-clusters) are created at elevated temperatures. The average base size amounts to 15 and 20–100 nm and the height is 2 and 4–6 nm for hut- and dome-clusters, respectively.

Fig. 7.14 shows the Raman spectra of the samples under investigation measured in the $z(y,x)\bar{z}$ and $y'(z,x')\overline{y'}$ scattering geometries which allow, respectively, LO and TO phonons in Ge QDs to be observed [65]. Two intensive peaks corresponding to the scattering by TO and LO phonons are observed at 308 and 315 cm^{-1}, respectively. Weaker features at 404 and 415 cm^{-1} are due to transverse and longitudinal Ge–Si vibrational modes (not shown here). The frequency position of TO and LO phonons exceeds the corresponding value of the phonons in Ge bulk (300 cm^{-1} at $T = 300K$), this indicates the presence of compressive strain in the QDs. With increasing growth temperature and decreasing Si spacer thickness, the frequency of optical phonons is decreasing (Fig. 7.14).

For QDs composed of Ge_xSi_{1-x} solid solution, the frequency ω_0 should be replaced by the expression $\omega_0(x) = 280.8 + 19.37x$ [74]. The frequencies of LO and TO phonons in Ge_xSi_{1-x} QDs are predominantly determined by QD composition, while the splitting value depends only on the deformation tensor components. Simultaneous measurement of these two frequencies guarantees the determination of mechanical strain and composition in strained GeSi/Si QD structures. Using Eqs. (7.8) and (7.9), the tensor component ε_{xx} for the structure grown at a low temperature ($T_g = 300°C$) was determined as -0.034. Ge content in the QDs amounts to 85%. Significant changes take place in the Raman spectra of GeSi QD SLs grown at elevated temperatures (Fig. 7.14). With increasing growth temperature (from 300°C to 650°C), the frequency positions of Ge optical phonon and LO–TO splitting

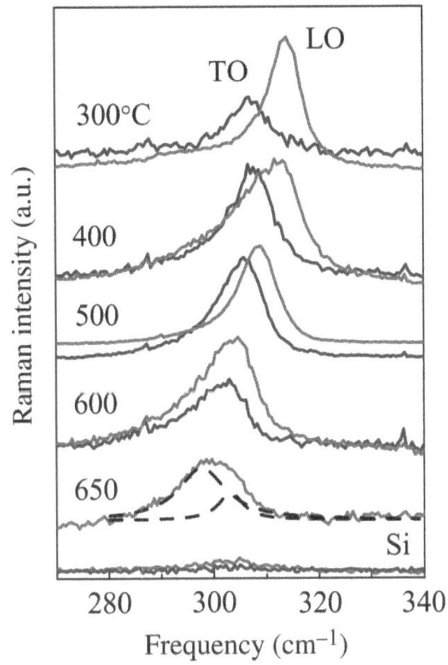

FIGURE 7.14 Raman spectra of Ge/Si QDs grown at a temperature of $T = 300-650°C$ and the spectra of Si substrate measured in $z(y,x)\bar{z}$ and $y'(z,x')\overline{y'}$ scattering geometries.

are decreasing. This manifests in the relaxation of elastic strain and increasing Si content in GeSi QDs. For the structure grown at $T = 600-650°C$, Ge composition reaches a value of 50%. These values agree well with the data obtained later by other research groups for the Ge/Si structures grown in similar conditions [75,76].

It was shown that the dependence of Ge–Ge mode frequency in GeSi/Si QD structures on the excitation energy reveals a nonmonotonic behavior [71]. At low growth temperatures (300–400°C) the phonon frequencies are determined by the built-in mechanical strain and the phonon confinement effect. Herewith, the dependence of LO phonons in Ge QDs corresponds qualitatively with the dependency for relaxed Ge/SiO_2 QDs shown in Fig. 7.13A. However, with increasing the temperature this dependence changes qualitatively. It was found that two mode distribution of QDs in size at a growth temperature of 500°C plays a significant role. At low excitation energies, the phonon frequencies are determined by the contribution of relaxed GeSi QDs of a large size, while at higher excitation energies, strained QDs of smaller size contribute resonantly to the Raman process. This causes increasing LO phonon frequency at an excitation energy above 2.4 eV. At further temperature increases, the areal QD density decreases, and the phonon frequencies are described in terms of atom intermixing effect in QDs and are independent of the laser excitation energy.

While the dependence of optical phonons confined in Ge QDs on the excitation energy evidences the zero-dimensional character of Ge/Si QD structures, the folding effect of acoustic phonons observed in multilayered Ge/Si structures is typical for two-dimensional systems. Fig. 7.15 shows Raman spectra of Ge QD multilayered structures with a nominal

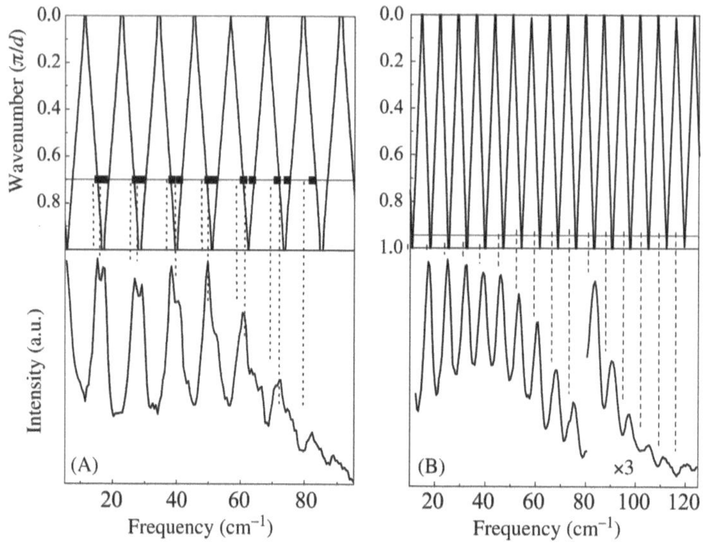

FIGURE 7.15 Raman spectra of samples: (A) 1.1/20; and (B) 1.4/30 measured in $z(xx)\bar{z}$ scattering geometry (below). In the upper part of the figure, the dispersion of LA phonons calculated within the Rytov theory [39] is shown. The *horizontal line* corresponds to the wave vector value used in the Raman experiments.

thickness of Ge of 1.1 and 1.4 nm, separated by 20 and 30 nm Si (samples 1.1/20 and 1.4/30, correspondingly), measured in the $z(xx)\bar{z}$ scattering geometry. The doublets of folded LO phonons up to 15th order are observed in the Raman spectra that evidences sharp interfaces in the Ge/Si QD SLs. Using the Rytov model [39], dispersions of optical phonons were calculated as shown in Fig. 7.15. The frequency positions of the folded LA phonons were determined using a wave vector of $q_s \approx 4\pi n/\lambda_L = 1.054 \times 10^6\ \text{cm}^{-1}$, where λ_L is the laser wavelength (514.5 nm), and n is the refraction coefficient ($n \approx 4.33$ for Ge and Si at 514.5 nm).

The structure periods derived from the calculations amounts to 22.9 and 37.9 nm for samples 1.1/20 and 1.4/30 which exceed the nominal thickness by 10–20% [67,77]. HRTEM experiments allowed the structure period to be independently measured, and confirmed the validity of conclusions based on the Raman data. According to Raman data, the frequencies of folded acoustic phonons are defined by the average structure period and do not depend on internal structure of the layers. The last conclusion was confirmed for all investigated periodical InAs/AlAs, InGaAs/AlAs, and InAs/AlGaAs QD structures [78–81].

7.3.4 A^3B^5 QDs

A^3B^5 QD nanostructures offer a unique opportunity to form QDs with various mechanical built-in strains. The lattice constant in InAs (0.606 nm) exceeds the corresponding values in GaAs (0.565 nm) and AlAs (0.566 nm); therefore, InAs QDs in matrices AlAs and GaAs undergo a compressive strain along QD layers, and a tensile strain along the growth direction [78].

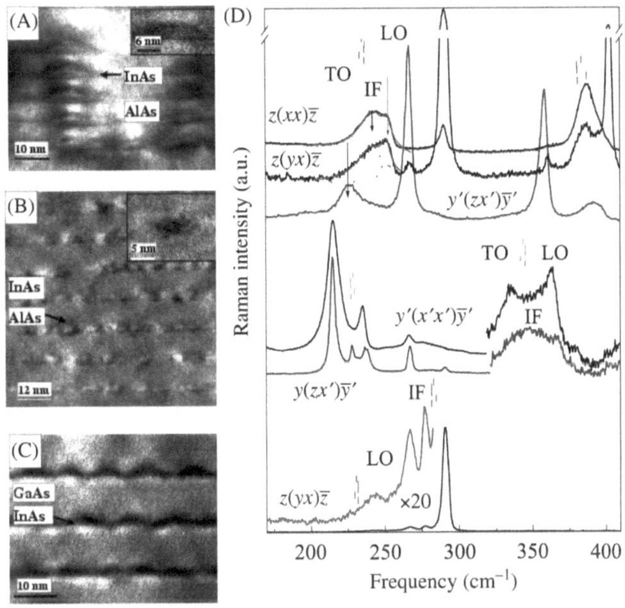

FIGURE 7.16 HRTEM cross-sectional images of the structures with InAs QDs in: (A) AlAs; and (B) GaAs matrices; (C) AlAs QDs in InAs matrix; and (D) Corresponding Raman spectra. The insets show HRTEM images QDs with a higher resolution. The *vertical lines* indicate the calculated IF phonon frequencies.

In AlAs QDs in an InAs matrix the situation is opposite: QDs are compressed in the growth and subject to tensile strain along the layers.

We investigated the phonon spectra of InAs QDs in an AlGaAs matrix and AlAs nanocrystals in an InAs matrix grown in ISP using MBE in the Stranski–Krastanov growth mode [64].

The cross-sectional HRTEM images shown in Fig. 7.16A indicate the lens-like shape of InAs QDs grown in AlAs and GaAs matrices (samples A and C, respectively), and the ellipsoidal shape of AlAs QDs in InAs (sample B). InAs QDs in sample A (C) have a base size of 20 nm (10 nm), and a height of about 4 nm (1.5 nm). The average size of AlAs QDs determined from the HRTEM images amounts to 4–5 nm for the base and 2–4 nm in height.

Fig. 7.16 represents the Raman spectra of samples A, B, and C measured in different scattering geometries which reveal confined TO, LO, and interface (IF) phonons. According to selection rules for SLs composed of materials having a zinc blend crystal structure [82], only LO phonons can be observed in the $z(yx)\bar{z}$ and $y'(x'x')\overline{y'}$ scattering geometries, while TO phonons are active in the $y'(zx')\overline{y'}$ geometry. In resonant conditions IF phonons can be activated. As one can see from the figure, these selection rules are fulfilled also for the QD SLs under investigation.

As one can see from Fig. 7.16, TO and LO phonon modes in matrix materials of samples A, B, and C are observed in the allowed geometries at frequencies close to their bulk values. The frequencies of the optical phonons localized in QDs differ from their bulk analogs. The frequency positions of TO and LO phonons in samples A and C in InAs QDs are shifted towards higher frequencies by 10–15 cm^{-1} with respect to the bulk values in InAs due to mechanical strain in QDs. The frequency of LO phonon mode in InAs QDs of sample C is (243 cm^{-1}) that is lower than the corresponding value in sample A (251 cm^{-1}). The possible reason for the difference is a partial strain relaxation and/or atomic intermixing at the interfaces between QDs and matrix. The signs of mechanical deformations in sample B are opposite to that which causes a low-frequency shift (30–40 cm^{-1}) of optical phonons localized in AlAs QDs of sample C [82].

The strain-induced shifts of optical phonons were calculated for InAs and AlAs QDs in accordance with Eqs. (7.8) and (7.9) [73,83]. The values of phonon deformation potentials p and q in AlAs are not known. However, since the force and elastic constants in GaAs and AlAs are close [84], the values p/ω_0^2 and q/ω_0^2 should be the same. The averaged tensor deformation components were calculated for the distribution of deformation in QDs [85]. For the calculation of the phonon mode shift in InAs QDs in an AlAs matrix the signs of deformation tensor components were changed to their opposite. Herewith, a good agreement was observed with Raman and resonant PL experimental data [86], and those calculated by other authors [83].

We have investigated the spectrum of interface phonons in A^3B^5 QDs [64,87–91]. It was shown that the frequency position of IF phonons is well described by the dielectric continuum model [92]. In the calculations we supposed [88] that AlAs QDs in an InAs matrix have a shape of oblate ellipsoids with ellipsoid axis ratio 2:1, which corresponds well to the value derived from HRTEM measurements. According to this model, each IF mode is determined by two quantum numbers $l(l \geq 0)$ and $m, |m| \leq l$. The IF modes with the lowest quantum numbers contribute predominantly to the Raman process, while the modes with higher l are strongly localized at the heterointerfaces and interact weakly with charge carriers [88]. Therefore, the first three IF modes (with quantum numbers $l=1, m=0$, $l=1, m=1$, and $l=2, m=0$) were calculated (Fig. 7.16). One can see from the figure, the calculated frequencies of IF modes in ellipsoidal QDs correspond well to the value experimentally observed in Raman spectra.

If the matrix is composed of an $Al_xGa_{1-x}As$ ternary solid solution which demonstrates a two mode behavior of its phonon spectrum, then Raman spectra reveal three sets of IF phonons in the range of InAs optical phonon modes originated from GaAs- and AlAs-like phonons (Fig. 7.17) with quantum numbers $l=1$ and $m=0,1$.

Obviously, InAs QD structures grown on GaAs and AlAs surfaces are characterized by two types of IF phonons, namely the IF phonons propagating along a planar interface wetting layer (WL)/matrix and along a corrugated interface QD/matrix. We have shown that it is possible to distinguish these two contributions by growing the structures GaAs/InAs QDs/AlAs with asymmetric barriers (Fig. 7.18a), consequently excluding one of the IF phonon types [90].

Fig. 7.18b and c show the Raman spectra of InAs QDs measured in the $z(x,x)\bar{z}$ and $z(y,x)\bar{z}$ scattering geometries and are thematically described in Fig. 7.18a. Raman spectrum of sample A′, measured in the $z(x,x)\bar{z}$ geometry reveals a broad asymmetric feature located between TO and LO phonon mode frequencies of AlAs. The frequency position and the shape of the feature can be explained by two effects. First, AlAs-like IF phonons from the

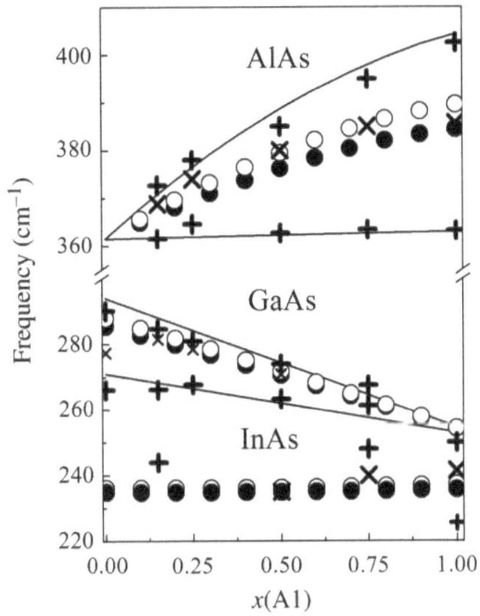

FIGURE 7.17 The dependence of IF phonon frequencies on the composition $Al_xGa_{1-x}As$ matrix. Signs (+) and (×) correspond to the frequency position of AlAs- and GaAs-like IF phonons, derived from Raman data. The *solid lines* indicate the frequency positions of AlAs- and GaAs-like phonons in a bulk AlGaAs. Signs • and ○ correspond to the frequency positions of the (1,0) and (1,1) IF phonon modes.

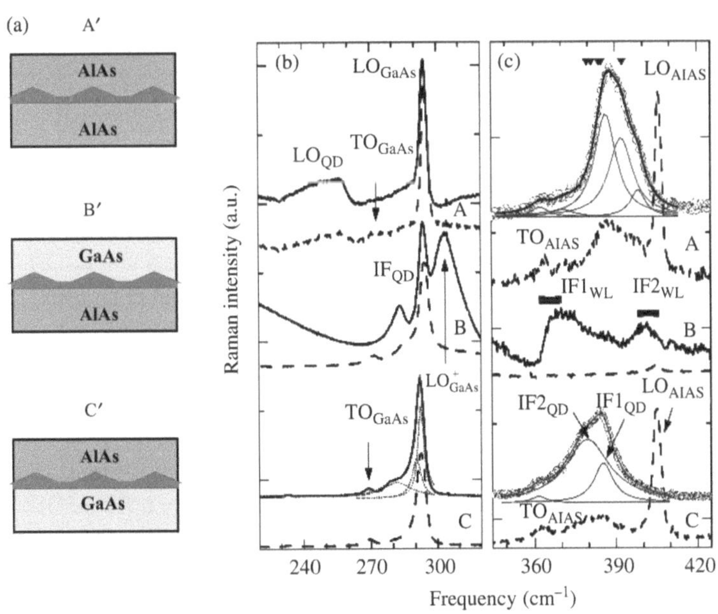

FIGURE 7.18 (a) The scheme of the period of A′, B′, and C′ structures; (b) and (c) The corresponding Raman spectra measured in the $z(xx)\bar{z}$ (*solid lines*) and $z(yx)\bar{z}$ (*dashed lines*) scattering geometries with 647.1 nm. The calculated IF phonon frequencies are shown by *triangles*, and the frequency intervals of the IF phonons propagating in the vicinity of planar InAs/AlAs interfaces are indicated by the *horizontal lines*.

corrugated interface InAs QDs/AlAs-matrix contribute to the Raman scattering at the frequencies at the middle between TO and LO phonons of AlAs. Second, the observed asymmetry of the broad feature can be due to Raman scattering by AlAs-like IF phonons from a lower interface WL/matrix, with the frequencies close to those of TO and LO phonons in AlAs [26].

Sample B′ (Fig. 7.18a) has only a planar InAs WL/AlAs matrix interface, while sample C′ has a corrugated QD/matrix interface, because the second interface is InAs/GaAs. The Raman spectrum of sample B′ does not reveal any features in the middle between the frequencies of TO and LO phonons of AlAs. Instead, two phonon modes near 370 and 400 cm^{-1} are assigned to AlAs-like IF phonons from the planar interfaces between InAs WL and AlAs matrix. These modes are indicated in Fig. 7.18c as IF1WL and IF2WL. IF phonon frequencies were calculated using the model of dielectric continuum [92], supposing that InAs QDs have a spheroidal shape. The first five calculated IF modes ($l \leq 2, m = 0, 1$) are presented in Fig. 7.18c by *triangles*. As one can see from Fig. 7.18c, a good agreement between the calculated and experimental IF phonon frequencies for sample A′ is observed. The position of experimental phonon modes was determined from the best fit by Lorentzian curves. The horizontal lines in Fig. 7.18a indicate the positions of IF phonons calculated using the dielectric continuum model [93] for planar SLs with a large thickness ratio of AlAs and InAs layers that corresponds to the case of sample A′.

Thus, the analysis of Raman spectra of asymmetric GaAs/InAs/AlAs QD structures allows the contributions of IF phonons originated from a planar WL/matrix interface to be distinguished, and the frequency positions of IF phonons in QD structures observed in experiment correspond well to the calculated ones within the dielectric continuum model.

7.4 Phonons in Nanostructures: From Array Toward a Single Nanostructure

Typically, QD arrays are composed of QDs which are nonhomogeneous in size and shape. This complicates investigation of a fine structure of phonon modes of a single QD, while the vibrational spectrum of an array represents a superposition of spectra of single QDs with different size and shape. QDs have a nanometer size that leads to an extremely low Raman cross-section and, as a consequence, to a tiny Raman signal from a single QD. Herewith, principal questions of phonon spectrum formation remain unanswered. Therefore, one of the approaches for enhancing the Raman signal developed in laboratory №5 (ISP) is the synthesis and investigation of QD arrays with an ultra-low areal density by resonant and surface-enhanced Raman scattering (SERS) aiming at the determination of the phonon spectrum of a single QD.

7.4.1 Resonant Raman Scattering

Resonant Raman scattering requires the coincidence of laser excitation energy with the energy of some electronic transitions (such as interband transitions), and leads to increasing Raman response by optical phonons in QDs (approximately by three orders of magnitude). Fig. 7.19 demonstrates the comparison of resonant and nonresonant Raman scattering by a dense array of ZnS and ZnO QDs formed by LB technology [43,50]. As is seen from the figure, enhancement of Raman scattering by LO phonon modes in resonant conditions amounts to three orders of magnitude. Even larger enhancement was obtained for ZnO QDs. The appearance of the overtones up to the ninth order evidences the high crystalline quality of the synthesized QDs.

Utilizing the micro-Raman setup in resonant Raman scattering experiments allowed us to probe the phonon response of PbSe arrays with an areal density up to 10 QDs/μm^2 [46]. Fig. 7.20 demonstrates the Raman spectrum of dense PbSe QDs which reveals the LO

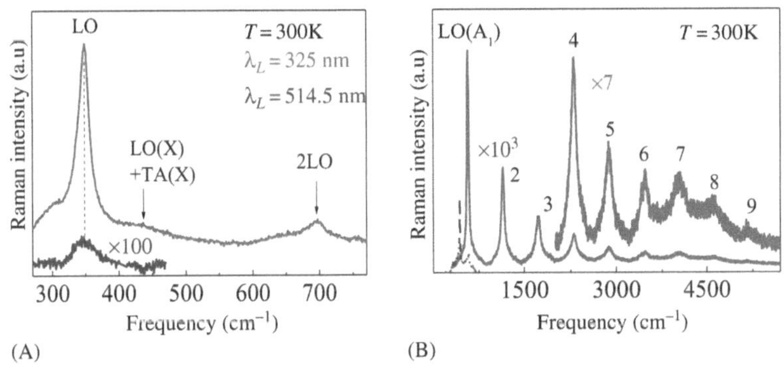

FIGURE 7.19 Comparison of resonant ($\lambda_L = 325$ nm) and nonresonant ($\lambda_L = 514.5$ nm) Raman spectra of: (A) ZnS; and (B) ZnO QDs.

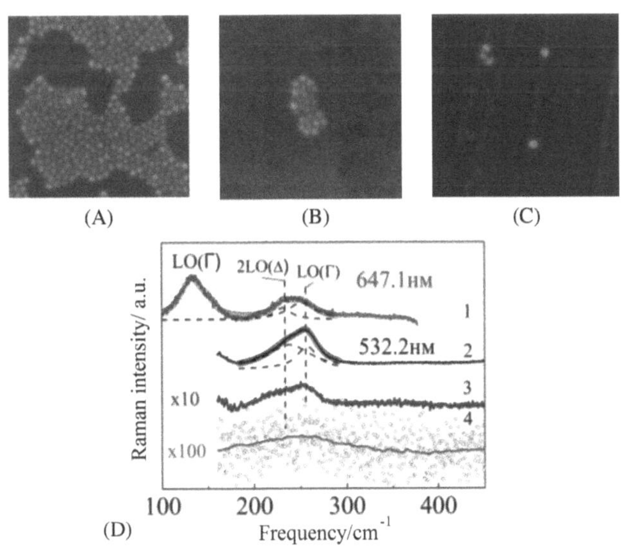

FIGURE 7.20 (A)–(C) SEM images of PbSe QDs (400 × 400 nm^2) with different areal density (bright objects). (D) Raman spectra of QD arrays with different areal density: 1 and 2 are from the arrays shown in (A), 3 and 4 are from the arrays shown in (B) and (C).

phonon mode and its overtone. In resonant conditions (at 532.2 nm) the overtone mode appears clearly for the structure with a dozen QDs (curve 4).

7.4.2 Surface-Enhanced Raman Scattering

Another possibility for enhancing the Raman signal is the utilization of SERS phenomenon. SERS has been widely used for investigation of a low number of materials, and allows the Raman signal to be enhanced by a factor of 10^5-10^6. Despite of the absence of physical limitations on the observation of the SERS effect in semiconductor QDs, no investigations in this field were carried out.

Recently, we have reported on the observation of SERS by confined optical and SO phonons in dense arrays of CdS [94] and ZnO [50] QDs after deposition of Ag nanoclusters on the QD surface. It was shown that Raman enhancement in QDs reaches the value of 150 for optical phonons in CdS QDs [44,46,95] and about 10^3 for SO phonons in ZnO QDs [50]. The comparison of the Raman and SERS spectra of CdS QDs is shown in Fig. 7.21, and demonstrates the advantage of SERS for studying the QD phonon spectrum. As one can see from the figure, the intensive modes of LO phonons and their overtones appear in the SERS spectrum. Probing the SERS in double resonant conditions (approximity of the excitation energy with the interband transition and plasmon energy in QDs and metal nanoclusters, respectively) leads to further enhancement of the effective scattering and Raman scattering (factor of 700) [46]. This allowed us to investigate a single QD layer with an areal density of about $4 \cdot 10^{11}$ cm^{-2}. Deposition of Ag nanoclusters on the surface of ZnO QDs leads to the appearance of intensive modes in Raman spectra which were interpreted as SO phonon modes in ZnO QDs, similar to those considered above and observed in IR spectra (Fig. 7.11).

Formation of metal nanoclusters with LSPR energy varied in a controllable way allowed us to investigate the SERS phenomenon. The arrays of gold nanoclusters with a size from a dozen to hundreds of nanometers (Fig. 7.22A) were fabricated in the ISP. The LSPR energy of such arrays is located in the red–infrared spectral region [47]. These arrays provide a significant enhancement of the Raman signal by phonons in CuS [47,49,94] and CdSe [96,97] QDs formed on Au arrays using LB technology. Fig. 7.22B shows the Raman spectra of CuS

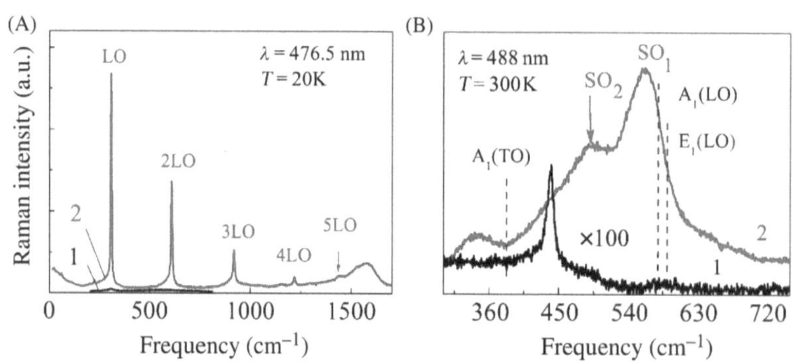

FIGURE 7.21 Comparison of Raman (curve 1) and SERS (curve 2) spectra of: (A) CdS; and (B) ZnO QD arrays.

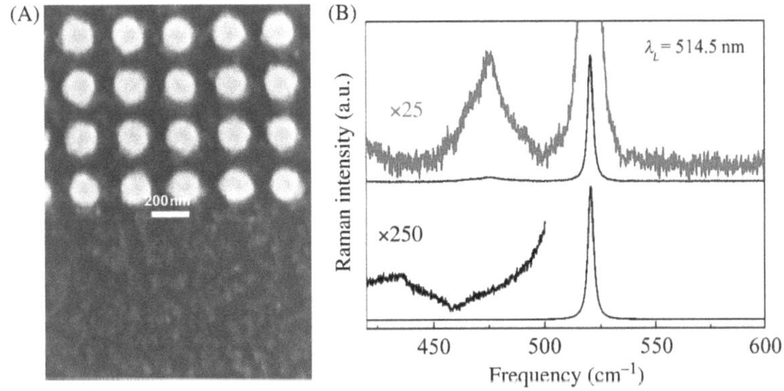

FIGURE 7.22 (A) SEM images of ordered gold arrays with deposited CuS QDs. (B) Raman spectra measured from the area with CuS QDs deposited on a Si surface (below) and on the gold nanocluster array (above).

QDs homogeneously deposited on Au nanocluster arrays and a Si surface in the vicinity of the arrays. The Raman spectra of QDs deposited on Si reveals the features corresponding to a bulk Si, while the Raman spectra of QDs deposited on the arrays of a gold nanocluster demonstrate the optical phonon mode in CuS QDs.

Acknowledgments

I would like to express my thanks to Yu.A. Pusep, my PhD supervisor, for my initiation as a researcher and for his major contribution to investigation of GaAs/AlAs SLs, A.I. Topopov, D.V. Dmitiev, A.K. Barakov, V.V. Preobrazhenskii, B.R. Semyagin, A.I. Nikiforov, O.P. Pchelyakov for the growth of InGaAs/AlGaAs, GaSb/AlSb, and Ge/Si samples, L.L. Sveshnikova, T.A. Duda for fabrication of QDs using the LB technology, N.A. Yeryukov for assistance in IR and Raman experiments, E. Rodyakina, A.K. Gutakovskii, S.S. Kosolobov, A.V. Latyshev for the samples with metal arrays and structural analysis carried out using electron microscopy methods, A.S. Medvedev for technical assistance.

Investigation of SERS by optical phonons in QDs is carried out with a financial support of the Russian Science Foundation (project 14-12-01037) and the Ministry of Education and Science of the Russian Federation.

References

[1] C. Colvard, T. Gant, M. Klein, R. Merlin, R. Fisher, H. Morcoc, et al., Folded acoustic and quantized phonons in (GaAl)As superlattices, Phys. Rev. B31 (1985) 2080–2091.

[2] V.A. Gaisler, A.O. Govorov, T.V. Kurochkina, N.T. Moshegov, S.I. Stenin, A.I. Toropov, et al., Phonon spectrum of GaAs-InAs superlattices, JETP 71 (3) (1990) 603–609.

[3] M. Cardona, G. Güntherodt (Eds.), Light Scattering in Solids V, Springer-Verlag, Berlin, 1989.

[4] V.A. Gaisler, Raman scattering of layered semiconductor structures (Dr. Sci. thesis's) (in Russ), Novosibirsk, 1996.

[5] R. Schorer, G. Abstreiter, S. Gironcoli, E. Molinari, H. Kibbel, H. Presting, In-plane Raman scattering of (001)-Si/Ge superlattices: theory and experiment, Phys. Rev. B49 (1994) 5406–5414.

[6] J. Zi, K. Zhang, X. Xie, Vibrational properties of SiGe superlattices, Prog. Surf. Sci. 54 (1997) 69–113.

[7] G.P. Schwartz, G.J. Gualtieri, W.A. Sunder, L.A. Farrow, Light scattering from quantum confined and interface optical vibrational modes in strained-layer GaSb/AlSb superlattices, Phys. Rev. B36 (1987) 4868–4877.

[8] P.V. Santos, A.K. Sood, M. Cardona, K. Ploog, Y. Ohmori, H. Okamoto, Raman scattering from GaSb/AlSb superlattices: acoustic, optical, and interface vibrational modes, Phys. Rev. B37 (1988) 6381–6392.

[9] Z.P. Wang, H.X. Han, G.H. Li, Raman-scattering studies of LO phonons in GaAs/$Al_xGa_{1-x}As$ superlattices, Phys. Rev. B42 (1990) 9693–9696.

[10] S. Perkowitz, Infrared and Millimeter Waves, vol. 8, Academic Press, New York, NY, 1983.

[11] S. Perkowitz, Optical Characterization of Semiconductors: Infrared, Raman and Photoluminescence Spectroscopy, Academic Press, New York, NY, 1993.

[12] R. Dingle, A.C. Gossard, W. Wiegmann, Direct observation of superlattice formation in a semiconductor heterostructure, Phys. Rev. Lett. 34 (1975) 1327–1330.

[13] L.C. West, S.J. Eglash, First observation of an extremely large dipole infrared transition within the conduction band of a GaAs quantum well, Appl. Phys. Lett. 46 (1985) 1156–1158.

[14] B. Levine, R. Malik, K. Choi, C. Bethea, D. Kleinman, J. Vandenberg, Strong 8.2 μm infrared intersubband absorption in doped GaAs/AlAs quantum well waveguides, Appl. Phys. Lett. 50 (1987) 273–275.

[15] Yu Pusep, S.W. da Silva, J.C. Galzerani, D.I. Lubyshev, P. Basmaji, A. Milekhin, et al., Spectroscopy of the optical vibrational modes in GaAs/$Al_xGa_{1-x}As$ heterostructures with monolayer-wide $Al_xGa_{1-x}As$ barriers, Phys. Rev. B52 (1995) 2610–2618.

[16] Yu Pusep, A. Milekhin, M. Sinyukov, K. Ploog, A. Toropov, Trapped phonons in the reflection spectra of GaAs/AIGaAs superlattices, JETP Lett. 52 (9) (1990) 464–469.

[17] Yu Pusep, A. Milekhin, N.T. Moshegov, V. Tikhomirov, A. Toropov, Penetration of optical vibrations localized in the layers of GaAs/AIAs periodic structures into the barrier according to reflection spectra, JETP Lett. 54 (1) (1991) 43–47.

[18] Yu Pusep, A. Milekhin, A. Toropov, Anharmonicity of GaAs optical vibrations in GaAs/AIAs superlattices, JETP Lett. 55 (10) (1992) 586–588.

[19] Yu Pusep, A. Milekhin, A. Toropov, FTIR-spectroscopy of $(GaAS)_n/(AlAs)_m$ superlattices, Superlattices Microstruct. 13 (1) (1993) 115–123.

[20] Yu Pusep, A. Milekhin, A. Toropov, FTIR spectroscopy of longitudinal confined phonons and plasmon-phonon vibrational modes in GaAs n/AlAs m superlattices, Solid State Electron. 37 (4–6) (1994) 613–616.

[21] Yu Pusep, A. Milekhin, N.T. Moshegov, A. Toropov, A study of the vertical transport of electrons in $(GaAs)_n(AlAs)_m$ superlattices by Fourier transform infrared spectroscopy, J. Phys. Cond. Matter 6 (1994) 93–100.

[22] Yu Pusep, A. Milekhin, A. Toropov, Anisotropy of zone-centre optical phonons in $(GaAs)_n/(AlAs)_m$ superlattices, J. Phys. Cond. Matter 7 (1995) 1493–1498.

[23] A.G. Milekhin, Yu.A. Pusep, V.V. Preobrazhenskii, B.R. Semyagin, D.I. Lubyshev, Localization of transverse optical phonons of GaAs in GaAs/$Al_xGa_{1-x}As$ periodic structures with paired quantum wells, JETP Lett. 59 (7) (1994) 493–496.

[24] A. Milekhin, Yu Pusep, Yu Yanovskii, I. Marakhovka, IR Fourier spectroscopy of double-quantum-well GaAs/AlAs heterostructures, Sov. Sol. St. Phys. 38 (5) (1996) 1605–1609.

[25] Yu Pusep, S.W. da Silva, J.C. Galzerani, D.I. Lubyshev, P. Basmaji, A. Milekhin, et al., Atomic-scale characterization of interfaces in the GaAs/AlGaAs superlattices, Mater. Sci. Eng. B35 (1995) 180–183.

[26] A.K. Sood, J. Menendez, M. Cardona, K. Ploog, Resonance Raman scattering by confined LO and TO phonons in GaAs-AlAs superlattices, Phys. Rev. Lett. 54 (1985) 2111–2114.

[27] J.L.T. Waugh, G. Dolling, Crystal dynamics of gallium arsenide, Phys. Rev. 132 (1963) 2410–2412.

[28] A. Milekhin, M. Roesh, E. Batke, K. Köhler, P. Ganser, Grating-coupler excited interface phonons in GaAs/AlAs superlattices, Solid State Commun. 112 (7) (1999) 387–390.

[29] A. Milekhin, M. Roesh, E. Batke, D.R.T. Zahn, K. Köhler, P. Ganser, et al., Infrared and Raman studies of confined and interface optical phonons in short-period GaAs/AlAs superlattices with a grating coupler, J. Vac. Sci. Technol. B17 (4) (1999) 1738–1741.

[30] Z.V. Popovich, E. Richter, J. Spitzer, M. Cardona, A.J. Shield, R. Notzel, et al., Phonon properties of (311) GaAs/AlAs superlattices, Phys. Rev. B49 (1994) 7577–7583.

[31] S.W. da Silva, Yu.A. Pusep, J.C. Galzerani, M.A. Pimenta, D.I. Lubyshev, P.P. Gonzalez Borrero, et al., Raman study of interface roughness in. $GaAs._n/AlAs._n$ superlattices grown on tilted surfaces: evidence of corrugation of the (113) interface, Phys. Rev. B53 (1996) 1927–1932.

[32] A. Milekhin, Yu Pusep, D. Lubyshev, V. Preobrazhenskii, B. Semyagin, Confined optical vibrations of (311) GaAs/AlAs superlattices, IOP Conf. Ser. 145 (1996) 437–442.

[33] A.G. Milekhin, Yu. A. Pusep, Yu Yanovskii, V.V. Preobrazhenskii, B.R. Semyagin, Localized optical phonons in GaAs/AlAs superlattices grown on (311)*A* and (311)*B* surfaces, Sov. Sol. St. Phys. 40 (1998) 507–509.

[34] A. Milekhin, Yu Yanovskii, V. Preobrazhenskii, B. Semyagin, Yu Pusep, Optical properties of the (311) oriented GaAs/AlAs superlattices and quantum wire-like structures, Phys. E 2 (1998) 368–371.

[35] A.G. Milekhin, Yu.A. Pusep, Yu Yanovskii, V.V. Preobrazhenskii, B.R. Semyagin, Optical phonons in quantum-wire structures, JETP Lett. 67 (2) (1998) 120–124.

[36] V. Ya Prints, I.A. Panaev, V.V. Preobrazhenskii, B.R. Semyagin, High-temperature anisotropy of the conductivity of superlattices of GaAs quantum wires grown on faceted 311A surfaces, JETP Lett. 60 (3) (1994) 217–220.

[37] A.G. Milekhin, Yu. A. Pusep, Yu Yanovskii, V.V. Preobrazhenskii, B.R. Semyagin, D.I. Lubyshev, Localized optical vibrational modes in GaSb/AlSb, superlattices, JETP Lett. 64 (5) (1996) 393–397.

[38] S.W. da Silva, Yu.A. Pusep, J.C. Galzerani, D.I. Lubyshev, A.G. Milekhin, V.V. Preobrazhenskii, et al., Optical phonon spectra of GaSb/AlSb superlattices: influence of strain and interface roughnesses, J. Appl. Phys. 80 (1) (1996) 597–599.

[39] S.M. Rytov, Acoustic properties of fine layered media, Sov. Phys. Acoust. 2 (1956) 71–83.

[40] G. Bastard, Superlattice band structure in the envelope function approximation, Phys. Rev. B24 (1981) 5693–5697.

[41] A. Milekhin, L. Sveshnikova, T. Duda, N. Surovtsev, S. Adichtchev, D.R.T. Zahn, Vibrational spectra of quantum dots formed by Langmuir–Blodgett technique, J. Vac. Sci. Technol. B28 (2010) C5E22–C5E24.

[42] D. Yu Protasov, W.-B. Jian, K.A. Svit, T.A. Duda, S.A. Teys, A.S. Kozhuhov, et al., Formation of arrays of free-standing CdS quantum dots using the Langmuir–Blodgett technique, J. Phys. Chem. C 115 (41) (2011) 20148–20252.

[43] A.G. Milekhin, L.L. Sveshnikova, T.A. Duda, N.V. Surovtsev, S.V. Adichtchev, Yu.M. Azhniuk, et al., Resonance effects in Raman scattering of quantum dots formed by the Langmuir-Blodgett method, J. Phys. Conf. Ser. 245 (2010)012045-1-4.

[44] A. Milekhin, L. Sveshnikova, T. Duda, N. Surovtsev, S. Adichtchev, D.R.T. Zahn, Optical phonons in nanoclusters formed by the Langmuir-Blodgett technique, Chin. J. Phys. 49 (1) (2011) 63–70.

[45] A.G. Milekhin, N.A. Yeryukov, L.L. Sveshnikova, T.A. Duda, C. Himcinschi, E.I. Zenkevich, Resonant Raman scattering of ZnS, ZnO, and ZnS/ZnO core/shell quantum dots, Appl. Phys. A 107 (2012) 275–278.

[46] A. Milekhin, N. Yeryukov, L. Sveshnikova, T. Duda, S. Kosolobov, A. Latyshev, et al., Raman scattering for probing semiconductor nanocrystal arrays with a low areal density, J. Phys. Chem. C 116 (2012) 17164–17168.

[47] N.A. Yeryukov, A.G. Milekhin, L.L. Sveshnikova, T.A. Duda, E.E. Rodyakina, E.S. Sheremet, et al., Surface enhanced Raman scattering by organic and inorganic semiconductors formed on laterally ordered arrays of Au nanoclusters, Thin Solid Films 543 (2013) 35–40.

[48] A. Milekhin, N. Yeryukov, L. Sveshnikova, T. Duda, D. Protasov, A. Gutakovskii, et al., CdZnS quantum dots formed by the Langmuir–Blodgett technique, J. Vac. Sci. Technol. B31 (2013)04D109-1-7.

[49] A.G. Milekhin, L.L. Sveshnikova, T.A. Duda, N.A. Yeryukov, N.V. Surovtsev, S.V. Adichtchev, et al., Surface-enhanced Raman scattering by semiconductor nanostructures, Optoelectron. Instrum. Data Process. 49 (5) (2013) 504–513.

[50] A.G. Milekhin, N.A. Yeryukov, L.L. Sveshnikova, T.A. Duda, E.I. Zenkevich, S.S. Kosolobov, et al., Surface enhanced Raman scattering of light by ZnO nanostructures, J. Exp. Theor. Phys. 113 (6) (2011) 983–991.

[51] R. Ruppin, R. Englman, Optical phonons of small crystals, Rep. Prog. Phys. 33 (1970) 149–196.

[52] A.I. Nikiforov, V.V. Ulyanov, A.G. Milekhin, O.P. Pchelyakov, S.A. Teys, S. Schulze, et al., Formation of Ge nanoislands on pure and oxidized Si surfaces by MBE, Phys. State Sol. (c) 1 (2) (2004) 360–363.

[53] A.A. Shklyaev, M. Shibata, M. Ichikawa, High-density ultrasmall epitaxial Ge islands on Si(111) surfaces with a SiO_2 coverage, Phys. Rev. B62 (2000) 1540–1543.

[54] A.G. Milekhin, A.I. Nikiforov, M. Yu Ladanov, O.P. Pchelyakov, S. Schulze, D.R.T. Zahn, Resonant Raman scattering by strained and relaxed germanium quantum dots, Phys. Solid State 46 (2004) 92–96.

[55] A.V. Kolobov, Y. Maeda, K. Tanaka, Raman spectra of Ge nanocrystals embedded into SiO_2, J. Appl. Phys. 88 (2000) 3285–3289.

[56] M. Seon, M. Holtz, T.-R. Park, O. Brafman, J.C. Bean, Effect of hydrostatic pressure on the Raman spectrum of Ge_nSi_m multiple quantum wells with $n < 4$ and $m < 7$, Phys. Rev. B58 (1998) 4779–4784.

[57] A.B. Talochkin, BA Markov, C.P. Suprun, A.I. Nikiforov, Raman scattering of light by optical phonons in Si-Ge-Si structures with quantum dots, JETP Lett. 64 (1996) 219–224.

[58] I.H. Campbell, P.M. Fauchet, The effects of microcrystal size and shape on the one phonon Raman spectra of crystalline semiconductors, Solid State Commun. 58 (1986) 739–741.

[59] F. Cerdeira, C.J. Buchenauer, F.H. Pollak, M. Cardona, Stress-induced shifts of first-order Raman frequencies of diamond-and zinc-blende-type semiconductors, Phys. Rev. B5 (1972) 580–593.

[60] F. Cerdeira, A. Pinczuk, J.C. Bean, B. Batlogg, B.A. Wilson, Raman scattering from Ge_xSi_{1-x}/Si strained-layer superlattices, Appl. Phys. Lett. 45 (1984) 1138–1140.

[61] G. Abstreiter, H. Brugger, T. Wolf, H. Jorke, H.J. Herzog, Strain-induced two-dimensional electron gas in selectively doped Si/Si_xGe_{1-x} superlattices, Phys. Rev. Lett. 54 (1985) 2441–2444.

[62] M. Nakayama, K. Kubota, H. Kato, N. Sano, Raman scattering from GaAs-$In_xGa_{1-x}As$ strained-layer superllatices, Solid State Commun. 51 (1984) 343–345.

[63] J. Groenen, G. Landa, R. Carles, P.S. Pizani, M. Gendry, Tensile and compressive strain relief in $In_xGa_{1-x}As$ epilayers grown on InP probed by Raman scattering, J. Appl. Phys. 82 (2) (1997) 803–809.

[64] A. Milekhin, D. Tenne, D.R.T. Zahn, in: S. Bandyopadhyay, H.S. Nalwa (Eds.), Quantum Dots and Nanowires, American Scientific Publishers, Los Angeles, 2003, pp. 75–419.

[65] A.G. Milekhin, A.I. Nikiforov, O.P. Pchelyakov, S. Schulze, D.R.T. Zahn, Phonons in self-assembled Ge/Si structures, Phys. E13 (2002) 982–985.

[66] A. Milekhin, A. Nikiforov, M. Ladanov, O. Pchelyakov, D. Tenne, S. Schulze, et al., Resonant Raman scattering by strained and relaxed Ge quantum dots, Mat. Res. Soc. Symp. Proc. 737 (2003) 1–6E.13.7.

[67] A. Milekhin, S. Schulze, D.R.T. Zahn, N. Stepina, A. Yakimov, A. Nikiforov, Raman scattering study of Ge dot superlattices, Appl. Surf. Sci. 175 (2001) 629–635.

[68] A. Milekhin, A.I. Nikiforov, O.P. Pchelyakov, S. Schulze, D.R.T. Zahn, Phonons in Ge/Si superlattices with Ge quantum dots, JETP Lett. 73 (2001) 461–464.

[69] A. Milekhin, A. Nikiforov, O. Pchelyakov, S. Schulze, D.R.T. Zahn, Size-selective Raman scattering in self-assembled Ge/Si quantum dot superlattices, Nanotechnology 13 (2002) 55–58.

[70] A.G. Milekhin, A.I. Nikiforov, M.Yu Ladanov, O.P. Pchelyakov, D.N. Lobanov, A.V. Novikov, et al., Phonons in Ge/Si quantum dot structures: influence of growth temperature, Phys. E 21 (2–4) (2004) 464–468.

[71] A.G. Milekhin, A.I. Nikiforov, O.P. Pchelyakov, A.G. Rodrigues, J.C. Galzerani, D.R.T. Zahn, Resonant Raman scattering in GeSi/Si superlattices with GeSi quantum dots, JETP Lett. 81 (2005) 30–33.

[72] M. Ya Valakh, R. Yu Holiney, V.N. Dzhagan, Z.F. Krasil'nik, O.S. Lytvyn, D.N. Lobanov, et al., Raman spectroscopy and electroreflectance studies of self-assembled SiGe nanoislands grown at various temperatures, Phys. Solid State 47 (2005) 54–57.

[73] M.Ya Valakh, V.O. Yukhymchuk, V.M. Dzhagan, O.S. Lytvyn, A.G. Milekhin, A.I. Nikiforov, et al., Raman study of self-assembled SiGe nanoislands grown at low temperatures, Nanotechnology 16 (2005) 1464–1468.

[74] H.K. Shin, D.J. Lockwood, J.-M. Baribeau, Strain in coherent-wave SiGe/Si superlattices, Solid State Commun. 114 (2000) 505–510.

[75] P.H. Tan, D. Bougeard, G. Abstreiter, K. Brunner, Raman characterization of strain and composition in small-sized self-assembled Si/Ge dots, Phys. Rev. B68 (2003)125302-1-6.

[76] A.V. Baranov, A.V. Fedorov, T.S. Perova, R.A. Moore, V. Yam, D. Bouchier, et al., Analysis of strain and intermixing in single-layer Ge/Si quantum dots using polarized Raman spectroscopy, Phys. Rev. B73 (2006)075322-1-6.

[77] A. Milekhin, N. Stepina, A. Yakimov, A. Nikiforov, S. Schulze, D.R.T. Zahn, Raman scattering of Ge dot superlattices, Eur. Phys. J. B16 (2000) 355–359.

[78] D.A. Tenne, V.A. Haisler, A.I. Toropov, A.K. Bakarov, A.K. Gutakovsky, D.R.T. Zahn, et al., Raman study of self-assembled GaAs and AlAs islands embedded in InAs, Phys. Rev. B61 (2000) 13785–13790.

[79] A. Milekhin, A. Toropov, A. Bakarov, S. Schulze, D.R.T. Zahn, Resonant Raman scattering in nanostructures with InGaAs/AlAs quantum dots, JETP Lett. 83 (11) (2006) 505–508.

[80] A. Milekhin, N. Yeryukov, A. Toropov, D. Dmitriev, E. Sheremet, D.R.T. Zahn, Raman scattering of InAs/AlAs quantum dot superlattices grown on (001) and (311)B GaAs, surfaces, Nanoscale Res. Lett. 7 (2012) 476-1-5.

[81] Y. Kamikawa-Shimizu, R. Patterson, A. Milekhin, K. Akahane, Y. Shoji, D.R.T. Zahn, et al., Study of phonons in self-assembled InAs quantum dots embedded in an InGaAlAs matrix, Phys. E 57 (2014) 1–5.

[82] A. Fainstein, P. Etchegoin, M. Chamberlain, M. Cardona, K. Totemeyer, K. Eberl, Selection rules and dispersion of GaAs/AlAs multiple-quantum-well optical phonons studied by Raman scattering in right-angle, forward, and backscattering in-plane geometries, Phys. Rev. B51 (1995) 14448–14458.

[83] M. Grundmann, O. Stier, D. Bimberg, InAs/GaAs pyramidal quantum dots: strain distribution, optical phonons, and electronic structure, Phys. Rev. B52 (1995) 11969–11981.

[84] P. Giannozzi, S. de Gironcoli, P. Pavone, S. Baroni, Ab initio calculation of phonon dispersions in semiconductors, Phys. Rev. B43 (1991) 7231–7242.

[85] A.D. Andreev, J.R. Downes, D.A. Faux, E.P. O'Reilly, Strain distributions in quantum dots of arbitrary shape, J. Appl. Phys. 86 (1999) 297–305.

[86] R. Heitz, M. Grundmann, N.N. Ledentsov, L. Eckey, M. Veit, D. Bimberg, et al., Multiphonon-relaxation processes in self-organized InAs/GaAs quantum dots, Appl. Phys. Lett. 68 (1996) 361–363.

[87] A.G. Milekhin, A.I. Toropov, A.K. Bakarov, M.Yu Ladanov, G. Zanelatto, J.C. Galzerani, et al., Vibrational spectroscopy of InAs and AlAs quantum dot structures, Phys. E 21 (2004) 241–246.

[88] M.Yu Ladanov, A.G. Milekhin, A.I. Toropov, A.K. Bakarov, A.V. Latyshev, D.A. Tenne, et al., Interface phonons in semiconductor nanostructures with quantum dots, JETP 101 (2005) 554–561.

[89] A.G. Milekhin, A.I. Toropov, A.K. Bakarov, M.Yu Ladanov, G. Zanelatto, J.C. Galzerani, et al., Optical phonons in InAs and AlAs quantum dot structures, Appl. Surf. Sci. 234 (2004) 45–49.

[90] A.G. Milekhin, A.I. Toropov, A.K. Bakarov, D.A. Tenne, G. Zanelatto, J.C. Galzerani, et al., Interface phonons in InAs and AlAs quantum dot structures, Phys. Rev. B70 (2004)085313-1-5.

[91] A. Milekhin, D. Tenne, A. Toropov, A. Bakarov, S. Schulze, D.R.T. Zahn, Raman study of interface phonons in InAs quantum dot structures, Phys. State Sol. (c) 1 (2004) 2629–2633.

[92] P.A. Knipp, T.L. Reinecke, Classical interface modes of quantum dots, Phys. Rev. B46 (1992) 10310–10320.

[93] R.E. Camley, D.L. Mills, Collective excitations of semi-infinite superlattice structure: surface plasmons, bulk plasmons, and the electron-energy-loss spectrum, Phys. Rev. B29 (1984) 1695–1706.

[94] A. Milekhin, L.L. Sveshnikova, T.A. Duda, N.A. Yeryukov, E.E. Rodyakina, A.K. Gutakovskii, et al., Surface-enhanced Raman spectroscopy of semiconductor nanostructures, Phys. E 75 (2016) 210–222.

[95] A.G. Milekhin, L.L. Sveshnikova, T.A. Duda, N.V. Surovtsev, S.V. Adichtchev, D.R.T. Zahn, Surface enhanced Raman scattering by CdS quantum dots, JETP Lett. 88 (12) (2008) 799–801.

[96] E. Sheremet, A.G. Milekhin, R.D. Rodriguez, T. Weiss, E.E. Rodyakina, O.D. Gordan, et al., Surface- and tip-enhanced resonant Raman scattering from CdSe nanocrystals, Phys. Chem. Chem. Phys. 17 (2015) 21198–21203.

[97] A.G. Milekhin, L.L. Sveshnikova, T.A. Duda, E.E. Rodyakina, V.M. Dzhagan, O.D. Gordan, et al., Surface-enhanced Raman scattering by colloidal CdSe nanocrystal submonolayers fabricated by the Langmuir–Blodgett technique, Beilstein J. Nanotechnol. 6 (2015) 2388–2395.

[98] Yu Pusep, M. Silva, J. Galzerani, A. Milekhin, N. Moshegov, A. Toropov, Fourier-transform infrared and Raman spectroscopies of plasmon anisotropy in heavily doped GaAs/AlAs superlattices, J. Appl. Phys. 79 (1996) 8024–8029.

PART
II

Surface, Interface, Epitaxy

8

Atomic Processes on the Silicon Surface

A.V. Latyshev[1,2], L.I. Fedina[1], S.S. Kosolobov[1], S.V. Sitnikov[1], D.I. Rogilo[1], E.E. Rodyakina[1,2], D.A. Nasimov[1], D.V. Sheglov[1], A.L. Aseev[1]

[1]RZHANOV INSTITUTE OF SEMICONDUCTOR PHYSICS OF THE SIBERIAN BRANCH OF THE RUSSIAN ACADEMY OF SCIENCES, NOVOSIBIRSK, RUSSIA [2]NOVOSIBIRSK STATE UNIVERSITY

8.1 Introduction

The fundamental knowledge accumulated in the area of surface science provides a foundation for the development of many industrial technologies, including epitaxial ones. At present, the comprehension of atomic processes on the surface is a crucial point for advanced manufacturing of low-dimensional systems based on superlattices, quantum wires, quantum dots, etc. Because of the high sensitivity of their electronic and optical properties to any structural inhomogeneity, high demand for smooth thin film surfaces and heterointerfaces, as well as for high structural quality arise. However, vicinal surfaces (formed due to a wafer miscut) commonly used in semiconductor technologies exhibit various morphological instabilities, and step pattern formation can be important in directed growth and nanofabrication. Therefore, one of the main subjects of current surface science is to study surface instabilities and their evolution under various external driving forces (e.g., growth, etching (sublimation), and electric current), since the patterns generated may provide useful templates for nanoscale fabrication [1−3]. According to the classical approach, sublimation and growth processes include mass transport of adatoms and their attachment/detachment to/from monatomic steps, so that the role of surface vacancies in step kinetics becomes unclear. It is expected that at high sublimation temperatures vacancies can be created to take part in mass transport and step motion. However, to visualize sublimation vacancies, extremely large step-free terraces significantly exceeding the diffusion length of adatoms are necessary; as we will show here the step-free terrace should be about 120 μm in size. We also show that such large atomically flat terraces turn out to be useful for overall investigations of both sublimation and growth processes, allowing determination of the main kinetic parameters for adatom and vacancy interactions with steps (Schwöbel barriers), and their 2D island nucleation and growth (2DNG).

Despite the numerous structural investigations of atomically clean semiconductor surfaces carried out, there was no information about the instabilities of elementary steps until our ultrahigh vacuum reflection electron microscopy (UHV REM) development in 1985 [4−6]. This was substantially because of the difficulty of visualization of a surface dynamic process with high spatial and time resolution [5,6].

Advances in Semiconductor Nanostructures. DOI: http://dx.doi.org/10.1016/B978-0-12-810512-2.00008-1

8.2 Experimental Method

8.2.1 In Situ Ultrahigh Vacuum Reflection Electron Microscopy

A unique technique called in situ UHV REM (Fig. 8.1) has been developed in ISP SB RAS to study various technological processes on an atomically clean Si surface [4]. We designed a small quasiclosed box with cryogenic differential pumping to install into the column of a JEOL microscope to form ultrahigh vacuum conditions near a sample's surface. Inside the box, the possibility of resistive sample heating by passing alternating or direct electric current was provided. This heating method was the best to be adapted in the differential pumping chamber by providing minimum heater dimensions, and minimal perturbations of the electron beam and sample contaminations from the heater components. Additionally, to carry out in situ growth experiments, we built evaporation cells for metal and semiconductors into the box, as well as a controllable gas inlet.

Because of the small angle of the electron beam reflected from a crystal surface, UHV REM is highly sensitive to any elements of surface structure. Depending on surface orientation, it provides a spatial resolution high enough to visualize individual monatomic steps (0.31 nm-high on Si(111), and 0.14 nm-high on Si(001)) over a wide temperature range, up to the melting point of silicon. First of all, the in situ UHV REM technique was applied to visualize monatomic step dynamics under different surface treatments, and second it was used to develop technologies related to the utilization of atomically clean semiconductor surfaces [7–9].

8.2.2 Preparation of an Atomically Clean Si Surface

Annealing of a Si specimen at sublimation temperatures in the UHV chamber of the REM leads to the formation of an atomically clean Si surface. One can assess whether a complete surface cleanout has been achieved by the absence of additional reflections related to

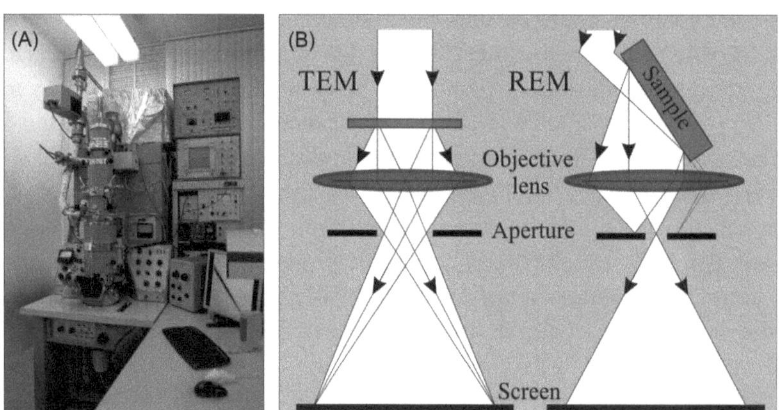

FIGURE 8.1 (A) Exterior view of the ultrahigh vacuum reflection electron microscope; and (B) schematic representation of the electron beam path in transmission and reflection mode.

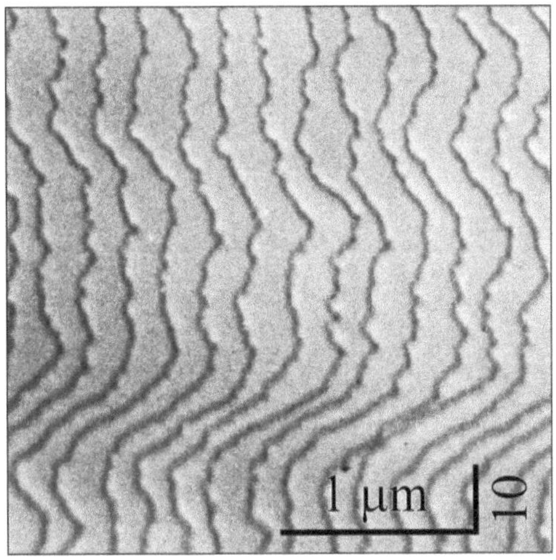

FIGURE 8.2 Typical REM image of the atomically clean Si(111) surface with a train of 0.31 nm-high monatomic steps.

carbide contaminations or the presence of reversible $(7\times7)\Leftrightarrow(1\times1)$ and $(1\times2)\Leftrightarrow(1\times1)$ phase transitions on the Si(111) and Si(001) surfaces [10]. The adequacy of these criteria is validated by UHV molecular beam epitaxy techniques equipped with devices for chemical surface analysis [11]. Fig. 8.2 shows a typical REM image of an atomically clean Si(111) surface containing a train of elementary steps with a height equal to one interplanar space, 0.31 nm [12]. Because of the small angle (1–5 degrees) of surface observation, the REM image is compressed by a factor of 30–50 along the incident electron beam. This leads to the appearance of different magnification scales in parallel and perpendicular to the electron beam direction in the REM image. Dark wavy lines in the REM image (Fig. 8.2) represent elementary atomic steps, which are a superposition of two contrast types: diffraction and phase. The diffraction (Bragg) contrast is induced by the areas of lattice deformation near monatomic steps. The phase (Fresnel) contrast is linked with the phase shift of electron waves reflected from adjacent terraces divided by the monatomic step. The monatomic step contrast has been analyzed in comparison with the contrast of lattice deformation around a gliding 60-degree dislocation emerging at the surface. Since the deformation field of a single dislocation is well known, this allows estimation of the lattice deformation near monatomic steps on the Si(111) surface to be about $2\cdot10^{-4}$ [13].

8.3 Step Motion on Vicinal and Step-Free Si(111) Surfaces During Sublimation

Sublimation is the transition of a substance from the solid phase directly to the gas phase without passing through an intermediate liquid phase. REM visualizes the sublimation process when the sample temperature reaches 900°C and steps start to move toward upper

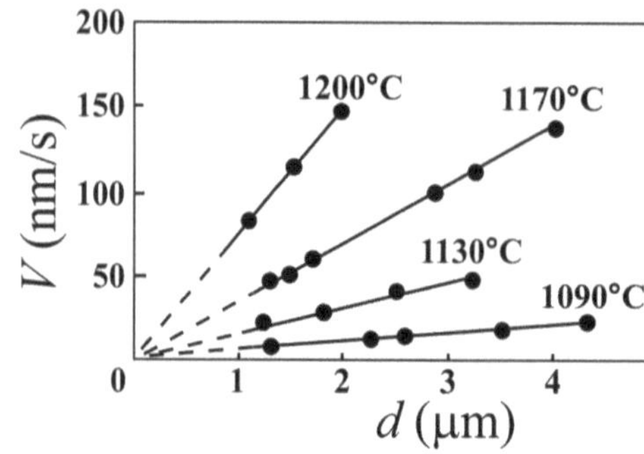

FIGURE 8.3 Dependence of monatomic step velocity V on the interstep distance d during sublimation.

terraces that correspond to adatom detachment from steps. According to classical Barton, Cabrera, and Frank (BCF) theory [14], the rate of parallel step train growth depends on the relation between the interstep distance d and the adatom diffusion length λ_s. If $d > \lambda_s$ then the step rate does not depend on d, but if $d < \lambda_s$ then the step rate is proportional to d and equals $V = d\nu\sigma \exp(-W/kT)$, where ν is atomic frequency, σ is adatom supersaturation on terraces, W is total evaporation energy, k is Boltzmann's constant, and T is the sample temperature.

The first in situ REM experiments on vicinal Si(111) surfaces had shown that the step rate during sublimation was, indeed, proportional to the distance between steps, d, in the temperature range between 900°C and 1400°C, if d was below 5 μm. This proved that the step motion occurs under the $d < \lambda_s$ condition, implying step adatom exchange during sublimation (Fig. 8.3). A value of $\lambda_s \geq 5$ μm is the lower limit of adatom diffusion length at high T. The total evaporation energy $W = 4.2 \pm 0.2$ eV was determined to be close to sublimation enthalpy [15]. Thus, the step motion on the vicinal Si(111) surface did not display any signs of vacancy involvement in step kinetics, in accordance with BCF theory.

To clarify the role of vacancies in step dynamics, we have undertaken comparative investigations of step motion on the Si(111) surface with extremely low miscut angle (below 0.001 degrees), and on commonly used ($\sim$0.1 degrees) substrates during pure sublimation assisted with oxygen etching at $T = 970-1300$°C, which is known to induce surface vacancies [16]. In the first case, we used substrates with dimples prepared by lithography followed by high temperature annealing. This resulted in the formation of 120 μm-wide step-free Si(111) terraces in the dimple's center, and allowed study of the nucleation and growth of vacancy two-dimensional (2D) islands. Additionally, widely spaced monatomic steps close to the extra-large terrace could be used for studies of step dynamics to compare with those on vicinal substrates.

The REM images presented in Fig. 8.4A and B show the Si(111) surface in a dimple's center at 1100°C during pure sublimation and oxygen etching, respectively. The *thin dark lines* are monatomic steps separated by singular terraces of gray contrast. The step density strongly increases during oxygen etching, because of fast nucleation and the growth of

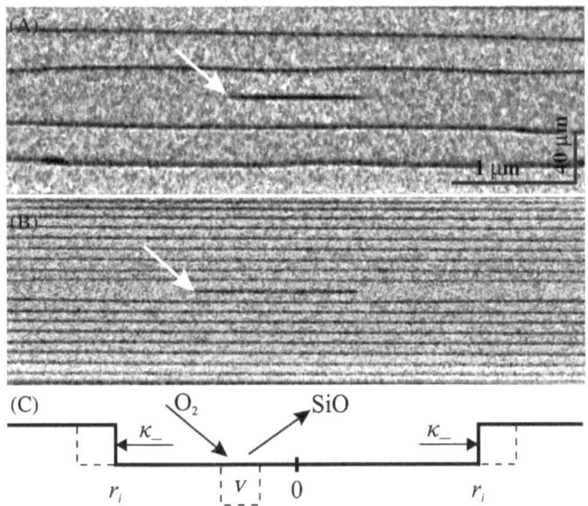

FIGURE 8.4 REM images of vacancy 2D island formation on 120 μm-wide atomically flat Si(111) terraces at 1100°C during: (A) sublimation, and (B) oxygen exposure (etching rate 0.9 ML/s). (C) Schematic representation of surface vacancy formation.

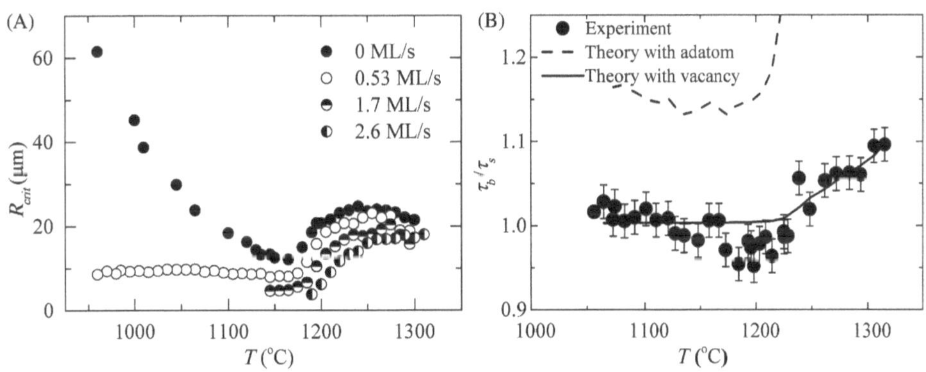

FIGURE 8.5 (A) Temperature dependences of R_{crit} during sublimation and oxygen etching; and (B) τ_b/τ_s ratio for pure sublimation.

vacancy 2D islands (marked by a *white arrow*). Due to the above-mentioned REM foreshortening, the round vacancy 2D island appears in the image as a strongly elongated ellipse (see Fig. 8.4).

When the 2D island diameter becomes larger than the critical value D_{crit} ($D_{crit} = 2R_{crit}$), a new vacancy island nucleates in the center of the initial one. It is clear that straightforward desorption of silicon atoms reserving surface vacancies in the terrace's center takes place during both sublimation and oxygen etching, but their $R_{crit}(T)$ dependences become essentially different in the temperature range between 970°C and 1180°C (Fig. 8.5A). Forced generation of vacancies by fixed oxygen flow provides invariable R_{crit}; however, the sublimation process displays *decreasing* values of R_{crit} from $\sim$60 to $\sim$11 μm when T increases from 970°C to 1180°C. The activation energy for $R_{crit}(T)$ decay is found to be 2.7 ± 0.1 eV. This suggests that R_{crit} is determined mainly by desorption of adatoms detached from the vacancy 2D island edges; nevertheless, their diffusion length λ_s depends on the recombination rate with sublimation vacancies. When the growing terrace size exceeds twice λ_s, adatoms detached from terrace edges cannot reach its center, and a new vacancy 2D island nucleates. This allows determination of the upper estimates of $\lambda_s < 60$ and $\lambda_s < 10$ μm at $T = 970$°C and $T = 1180$°C, respectively. At the same time, a small $R_{crit} < 10$ μm implying a low λ_s is observed during oxygen etching up to $T_{crit} = 1180$°C. This is a critical temperature where sublimation and oxygen etching $R_{crit}(T)$ dependences meet (Fig. 8.5A). Surprisingly, R_{crit} begins to increase in both cases with a further T increase. Homma et al. earlier observed a similar temperature dependence of step dynamics during sublimation with 2.4 eV of activation energy [17]. However, because of the smaller terrace sizes ($\sim$100 μm) employed, they did not observe the formation of vacancy 2D islands. Taking into account these results, we can propose a lower limit of adatom diffusion length of $\lambda_s > 50$ μm at $T \sim 1000$°C.

Nonlinear $R_{crit}(T)$ dependences (Fig. 8.5A) suggest that different atomic processes dominate sublimation at temperatures below and above $T_{crit} = 1180$°C. Oxygen etching at $T < T_{crit}$ just sets up a nearly constant concentration of surface vacancies determining invariable small R_{crit} and λ_s. On the contrary, the formation of vacancies during sublimation is an activated process, but it affects *mainly* λ_s. Therefore, the detachment of adatoms from steps, their diffusion, and their desorption dominate below T_{crit}. In this case, the observed $R_{crit}(T)$ decay corresponds to the difference between adatom desorption energy, E_{des}, and adatom diffusion energy, E_D, (1.3 eV, see Section 8.5.1): $E_{des} - E_D \sim 2.7$ eV. It results in $E_{des} \sim 4.0$ eV. Thus, the sublimation mechanism at $T_{crit} < 1180$°C is dominated by desorption of adatoms detached from steps. If the difference between sublimation and oxygen etching $R_{crit}(T)$ dependences at $T_{crit} > 1180$°C disappears, this suggests that adatom detachment from steps becomes less effective in comparison with straightforward evaporation of surface atoms.

To understand the evaporation process at $T_{crit} > 1180$°C in more detail, we have analyzed the dynamics of widely spaced steps close to 120-μm flat terraces and steps on vicinal substrates. We measured times τ_b and τ_s for the atomic step passage of the distance between neighboring steps on large ($W_b \sim 10$–60 μm at different T) and small ($W_s \sim 0.2$ μm) terraces, respectively. Fig. 8.5B shows the measurements of the τ_b/τ_s ratio as a function of T. We have found that the slope of experimental $\tau_b/\tau_s(T)$ also dramatically changes at $T_{crit} > 1180$°C, and cannot only be described within the scope of classical BCF theory considering desorption of adatoms detached from steps. However, assuming surface vacancies (advacancies) as a main species for providing mass transport at $T_{crit} > 1180$°C, it makes BCF theory capable of describing our experimental data (see Fig. 8.5B) [16]. Fast desorption of surface atoms reserving surface vacancies at $T_{crit} > 1180$°C leads to a strong increase in the density of vacancies and their interaction with steps. We will show below (see Section 8.6.2) that the energy barrier of vacancy interaction with the step $\sim$1.5 eV can be determined from vacancy 2DNG kinetics during oxygen etching. Using this value, we have estimated the activation energy of step motion to be $E_\sigma = 4.30 \pm 0.05$ eV at $T_{crit} > 1180°C$, and to be associated with the generation of surface vacancies that may lead to surface-bulk defect exchange and incomplete melting layer formation.

In general, the step distribution on a vicinal Si surface during sublimation depends on step–step interaction [18] and step orientation [19]. The presence of thermal fluctuations causes more complex step behavior. Equilibrium step distance is defined by the balance between entropic forces initiating step fluctuation and step–step repulsion forces [20,21]. The term describing thermal step fluctuation $\eta(y, t)$ is introduced by Langevin formalism [22] and obeys line stochastic differential equations. In most cases, a step fluctuation is less than the interstep distance and is neglected by theory. There are several special cases where these fluctuations must be taken into account [23]. The analysis of a balance between attachment/detachment processes, and taking the requirement for minimum step stiffness into account allowed evaluation of the stiffness of steps on the Si(111) surface in thermodynamic equilibrium as $\beta \approx 8 \cdot 10^{-10}$ J/m.

We have applied in situ REM to study single fluctuations caused by step pinning through strain fields of gliding dislocation emerging at the surface [24]. These points on the surface have been shown to act as centers of step inhibition. Step detachment from the exit point of the dislocation occurs when the step curvature reaches a critical value. Several time dependences of step perturbation amplitude h at fixed T are presented in Fig. 8.6. The interaction between monatomic steps and dislocations gliding under external stress has been studied in more detail in Ref. [25]. We have revealed the asymmetry of interaction between steps and dislocations gliding in step-up and step-down directions. This asymmetry is explained by the nature of the lattice distortion around steps within the framework of a simplified atomistic model.

We have found that dynamic step edge stiffness on the Si(111) surface increases by two orders of magnitude when T increases from 1250°C to 1400°C [26]. This fact is highly unexpected from the viewpoint of the theoretical framework [22]. When elevating crystal temperature, one should expect an increase in step fluctuation amplitude due to kink density rising. This should, in turn, increase the effective width of atomic steps, which corresponds to step line stiffness reduction. We imply here that the effective step width is the average statistical

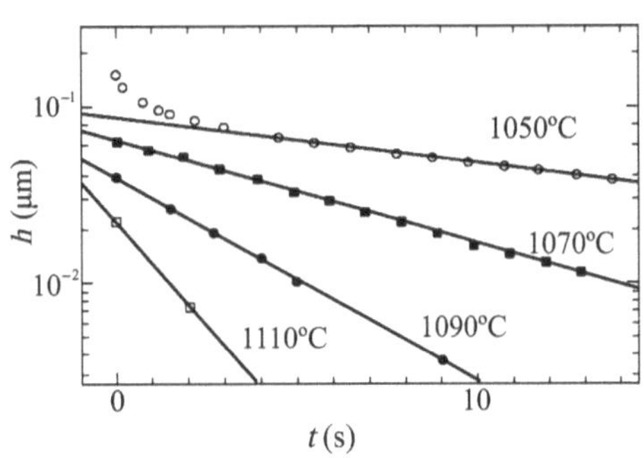

FIGURE 8.6 Time dependences of step perturbation amplitude h after interaction with a glide dislocation.

deviation of step segments from its average position. When the sample temperature increases above a certain critical point, this effect should trigger "incomplete surface melting"; a morphological transition to infinite effective step width. This effect has been discovered on the surface of many metals and semiconductors, including the Si(001) surface [27], but not on the Si(111) surface. Dynamic step stiffness has been theoretically considered in Ref. [28] by analysis of nonequilibrium step fluctuations on the basis of extended BCF theory with the application of linear stochastic differential equations. This analysis predicts the increase of step roughness during epitaxial growth, and the smoothing of steps during sublimation, which agrees with the experimental results obtained.

8.4 Instability of the Regular Step Train

8.4.1 Instability of Step Train During Sublimation and the Effect of Adatom Electromigration

When studying the surface morphology of samples heated resistively using direct current (DC) by in situ REM, we have discovered that depending on *T* and current direction, reversible instabilities in the elementary step train emerge during sublimation on both vicinal Si (111) [15] and Si(001) surfaces [29]. These instabilities cause redistribution of steps from the regular train to step-bunches: regions of high step density divided by low step density areas. Fig. 8.7A and B display REM images of the Si(111) surface with regular steps (RSs) and step-bunches, respectively. We have studied heating conditions for step-bunching, and defined factors affecting temperature ranges in this transformation [15]. Both the direction and magnitude of DC influence step-bunching temperature ranges so that the transition of RSs to step-bunches is observed when reversing DC direction or magnitude. The reversibility of step-bunching consists of step-bunch disintegration into RSs when modifying sample

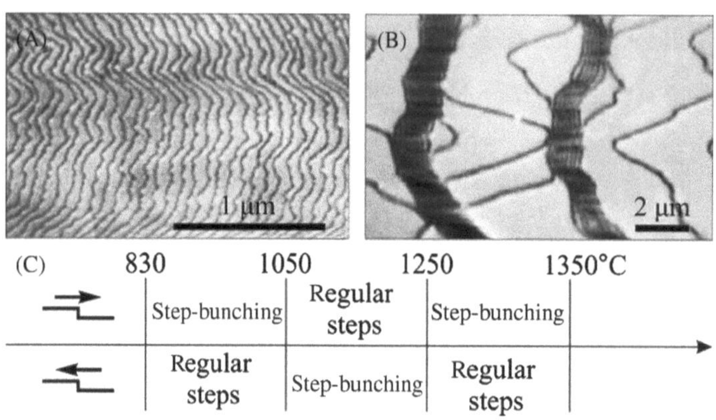

FIGURE 8.7 REM images of Si(111) surface with: (A) regular monatomic steps; and (B) step-bunches; (C) temperature ranges of step-bunching for different DC directions. *Arrows* show DC directions relatively to step front.

temperature to values lying outside step-bunching ranges, or reversing DC direction at fixed T (Fig. 8.7C). The last fact authenticates the existence of the effective charge of adatoms caused by their partial ionization and entrainment with charge carriers [30]. Heating of samples with alternating current never leads to step-bunching.

One can count the monatomic steps and interstep distance in a bunch by measuring its width. This distance appears to be about a few nanometers; the step-bunch does not therefore represent atomic planes with high indexes (facets). Prolonged sample annealing leads to the coalescence of step-bunches, and formation of large atomically flat terraces (up to $\sim 10\ \mu m$). These features of step-bunching have been further confirmed by UHV scanning electron microscopy [31], atomic force microscopy (AFM) [32], scanning tunneling microscopy [33], and synchrotron-radiation X-ray scattering [34].

The effect of electromigration on the Si(001) surface is more complicated, due to anisotropic adatom diffusion on the surfaces with (1×2) and (2×1) superstructures. The evidence for such anisotropy is based on the asymmetrical shape of 2D islands during sublimation and growth [29]. It has been found that DC heating the Si(001) surface with atomic steps affects the ratio between areas occupied by (1×2) and (2×1) superstructures [35]; at $T > 800°C$, the (1×2) or (2×1) superstructure dominates at the surface depending on DC direction, which is the reason for atomic step pairing. At $T < 800°C$, or at any temperature when the sample is heated by alternating current, both (1×2) or the (2×1) domains occupy equal areas. Again, the atomic step train on the Si(001) surface is not stable by sample heating with DC only. We have also found a critical interstep distance below which atomic steps do not form bunches.

Using in situ REM we have determined that step-bunching on the Si(111) surface becomes more complicated during prolonged sample annealing when so-called "antibands" form (Fig. 8.8A–C) [36,37]. Every antiband is a band consisting of steps with the opposite front direction formed at some distance from the initial step-bunch, which generally reproduces its shape. Since the step-bunch significantly modifies adatom distribution in its vicinity, the velocity of step segments varies with the distance from the bunch which leads to the generation of macrokinks on monatomic steps, and eventually to the formation of an antiband parallel to the step-bunch. The formation of antibands and step-bunches is a reversible process; antiband disintegration occurs, therefore, when the DC direction is reversed or when the sample temperature is modified out of the step-bunching temperature interval.

The discovery of adatom electromigration on the Si surface has promoted the development of new atomic process physics on semiconductor surfaces related to the effective charge of adsorbed atoms. The problem of the theoretical description of the current-induced step-bunching phenomenon has been addressed by numerous scientific groups (see Refs. [7,38]). The results obtained by our group have practical importance because they allow the distribution of monatomic steps to be controlled: they allow the formation of a surface with RSs, step-bunches divided by large atomically flat terraces or regions with a small step density, and even more complicated surface morphology consisting of step-bunches and antibands.

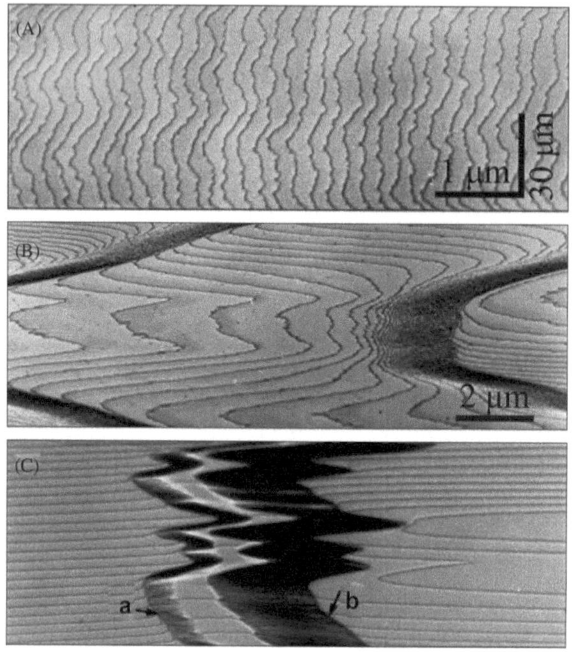

FIGURE 8.8 REM images of the Si (111) surface with: (A) regular step train; (B) step-bunches; (C) a step-bunch (b), and an antiband (a).

8.4.2 Effective Charge of Si Adatom on the Si(111) Surface

Using in situ REM and ex situ AFM techniques, we have shown the existence of an adatom concentration gradient on terraces between step-bunches during sublimation, epitaxy, and near-equilibrium conditions on the Si(111) surface at temperatures above 900°C [39]. Adatoms shift toward lower terraces within the $T = 1050-1240°C$ range, and toward the opposite direction at $T = 1250-1350°C$. The reason for the adatom drift is proposed to be an effective adatom charge that is negative at 1100°C, and positive at 1300°C. We have found that the sign of the effective charge, as well as adatom drift during DC sample heating, does not depend on the supersaturation on the surface. Based on experimental data obtained during Si sublimation at 1280°C, the effective charge of an adatom has to be 0.07 ± 0.01 of an electron's elementary charge.

In order to realize a smooth distribution of adatom concentration on the Si surface, we used a special sample holder called a "flat capillary" (Fig. 8.9). It consists of two Si samples (1) placed side-by-side at a small distance (0.1 mm), divided by insulators (2), and having electrical contacts (3) that allow independent heating of these samples by an electric current. This sample holder makes Si surface supersaturation control at $T > 900°C$ possible on one sample by evaporation from another sample. Near-equilibrium (net atom flux from the surface is about zero) or epitaxial growth conditions (net atom flux onto the surface is greater than zero) can be realized using this sample holder at temperatures up to the melting point of silicon.

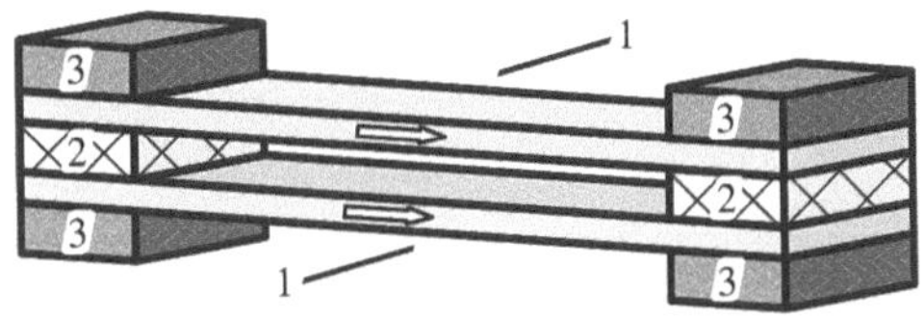

FIGURE 8.9 Schematic drawing of a flat capillary consisting of Si samples (1), insulators (2), and electrical contacts (3).

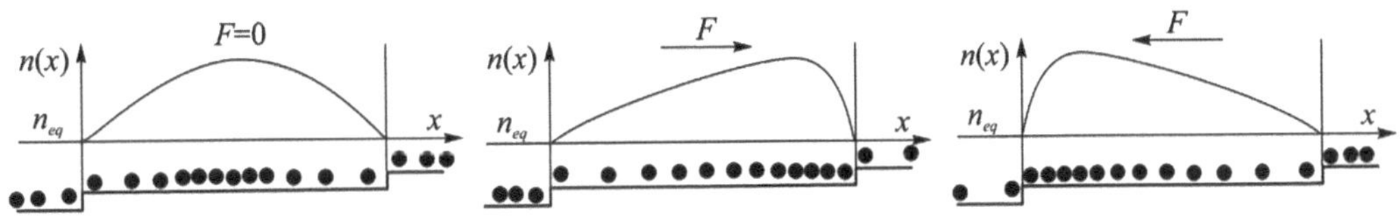

FIGURE 8.10 Schematic representation of adatom concentration redistribution on a terrace between atomic steps under external force.

The velocity of a monatomic step during epitaxial growth is suggested to be proportional to the sum of adatom concentration gradients on two adjacent terraces near the step. The higher the gradients are, the faster is the step velocity. The schematic representation of adatom concentration on a terrace under the action of external force F is shown in Fig. 8.10. When a sample is heated resistively, electric field E interacts with adatoms having effective charge q_{eff}, and electromigration force $F = q_{eff}E$ influences every adatom [15]. Unfortunately, there are no direct methods for effective adatom charge measurement at high temperatures. However, if q_{eff} is positive, the direction of adatom drift coincides with the electric field. Adatom drift creates the asymmetry of adatom distribution on terraces (Fig. 8.10). To investigate this asymmetry, we have analyzed the step velocity aligned to DC direction with reoriented steps on the Si surface moving perpendicularly to the applied electric field. The step velocity depends on local adatom supersaturation near each segment of a reoriented step, which therefore allows evaluation of the adatom concentration gradient caused by the electromigration force.

8.4.3 The Behavior of Monatomic Steps During $(1 \times 1) \Leftrightarrow (7 \times 7)$ Phase Transition

The superstructural $(1 \times 1) \Leftrightarrow (7 \times 7)$ transition has been studied by in situ REM to reveal atomic processes induced by this transition [40]. We have found that the upper edges of monatomic steps serve as nucleation centers for (7×7) domains (Fig. 8.11), and that steps displace in a step-down direction when the temperature decreases below the transition point (about 830°C). The step displaces back in the step-up direction during the opposite $(7 \times 7) \Rightarrow (1 \times 1)$ transition [41]. The magnitude of both step-down and step-up motions equals about 0.2 of the terrace width between steps, which corresponds to the absorption or generation of about $4 \cdot 10^{14}$ atoms/cm^2 by steps. Hence, it follows that the nonreconstructed Si surface is not atomically flat at $T > 830°C$ due to an anomalously high concentration of

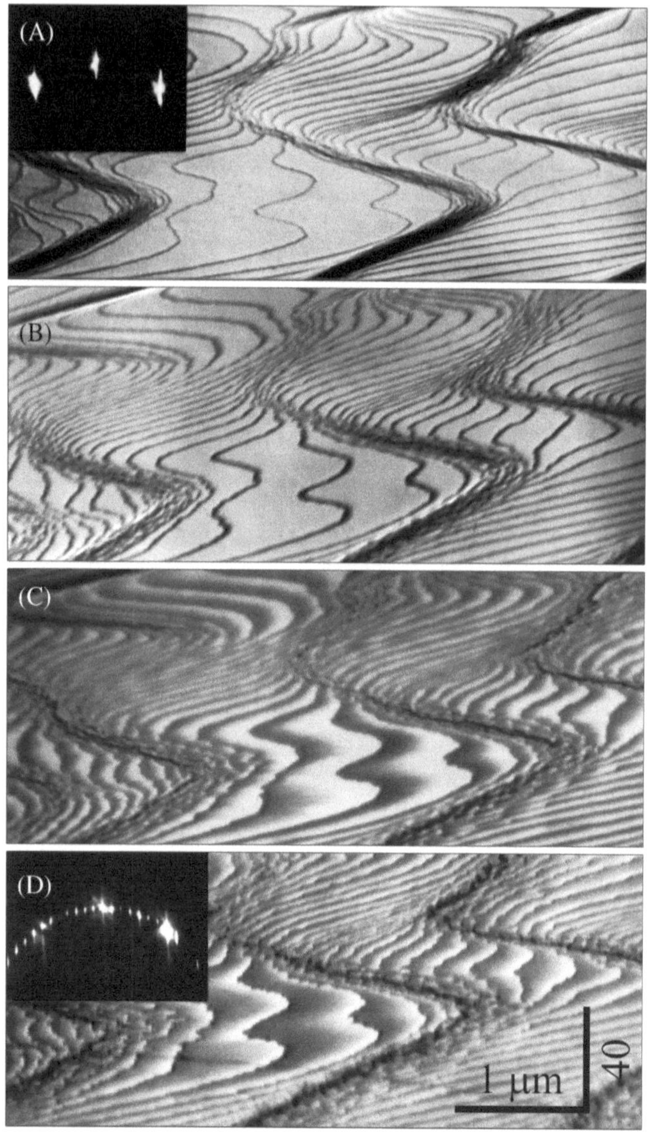

FIGURE 8.11 Consecutive stages of superstructural (1 × 1)⇒(7 × 7) transition obtained in a matrix reflection in the same location on the Si(111) surface with monatomic steps and step-bunches. (A) before the transition, and (B-D) under a slow temperature decrease from 830°C within the range of 25 ± 5°C. Dark contrast between steps corresponds to (7 × 7) reconstructed areas of the surface.

adatoms (about 0.2 per monolayer). The disappearance of the (7 × 7) structure is accompanied by a strong reduction in specular reflected electron beam intensity, more than is predicted by the Debye–Waller factor. During the opposite transition, the intensity rises again. This conclusion has been confirmed by calculations of electron diffraction from the nonreconstructed Si(111) surface in Ref. [42], where the best fit with experimental data is achieved for a surface with high adatom concentration. The existence of adatom excess on the nonreconstructed Si(111) surface has been confirmed by quenching experiments [43]. In this case, (7×7) and (1×1) domains form simultaneously with adatom clusters due to impeded adatom absorption by monatomic steps at a high cooling rate.

8.5 Step Motion During Si Growth

8.5.1 Si Growth on the Nonreconstructed Si(111) Surface

Based on in situ REM, we have first analyzed Si growth on a vicinal Si(111) surface at higher temperatures. Before Si deposition, at $T > 870^\circ$C the step train moves slowly in a step-up direction due to sublimation. When the Si flux from the evaporator comes onto the surface, a deceleration of step-up motion takes place until steps and their further motion in a step-down direction come to a complete halt [44]. We have not observed the formation of 2D islands on terraces between steps on a vicinal Si surface that corresponded to an adatom diffusion length of λ_s much greater than step distances [45].

The problem of step flow to 2DNG transition is widely studied both theoretically and experimentally (see, e.g., Ref. [46]). Among many fundamental parameters of crystal growth, λ_s is one of the most significant because it defines the growth mode. According to the classical Einstein relation, $\lambda_s^2 = 2D_s\tau_s$, where $D_s = D_0\exp(-E_s/kT)$ is the coefficient of adatom diffusion, E_s is the activation energy, D_0 is a preexponential factor, τ_s is the lifetime of an adatom before its desorption from the surface, and k is Boltzmann's constant. The diffusion length measured in the first experiments was of the order of a few microns, which demanded the D_0 factor to be corrected, as was discussed in Ref. [47]. The most likely reason for this is the need to account for the effect of long adatom jumps on surface diffusion [48].

During step-flow growth, λ_s exceeds half of a step distance, and atoms deposited onto the surface are able to migrate on a terrace and to reach steps. However, a critical distance exists between steps (critical terrace width, λ) above which the nucleation of 2D islands occurs [45]. The classical treatment of crystal growth [49,50] implies that the nucleation of a 2D island starts with the formation of a stable nucleus consisting of more than i atoms. This approach also implies that adatom attachment to a step takes place without an energy barrier and that growth kinetics are limited only by adatom diffusion on the surface (diffusion limited (DL) kinetics) [51–53]. We can therefore use the classical rate equation theory to determine the critical nucleus size during Si homoepitaxy. In this case, the dependence of λ, as well as typical island separation L_s on an infinite terrace, turn into the laws:

$$N_{2D}^{-1} \equiv L_s^2 \propto R^{-\chi} \exp\frac{-E_{2D}}{kT} \tag{8.1}$$

and

$$\lambda^2 \propto R^{-\chi} \exp\frac{-E_{2D}}{kT}, \tag{8.2}$$

where N_{2D} is the concentration of 2D islands, R is deposition rate, χ is the scaling exponent, and E_{2D} is the effective activation energy of 2DNG:

$$E_{2D} = \chi^{DL}\left(E_D + \frac{E_i}{i}\right) = E_{2D}^{DL} \tag{8.3}$$

and

$$\chi = \frac{i}{i+2} = \chi^{DL} \tag{8.4}$$

where E_D is the activation energy of adatom diffusion on the Si(111) surface, i is the nucleus size, and E_i/i is the dissociation energy of a critical nucleus per adatom. Thus, the classical theory relates χ and E_{2D} parameters to the surface diffusion activation energy E_D, the nucleus size i, and its dissociation energy E_i. Consequently, the analysis of experimental $N_{2D}(R, T)$ and $\lambda(R, T)$ dependences can be used to estimate these fundamental parameters.

Using in situ REM, we have studied 2DNG on extra-large ($\sim$100 μm) step free Si(111) terraces at $T > 830°C$. Fig. 8.12A–F show REM images of Si growth at $T = 950°C$. Due to REM image compression, the initial round terrace is visualized in the image as a horizontal stripe of bright contrast bordered by two atomic steps (marked by numbers 1 and 2 in Fig. 8.12A–E). After initiation of Si deposition, the nucleation of 2D islands in the shape of short dark lines is observed (Fig. 8.12B). One can see that each 2D island is easily distinguishable, and we are able to calculate 2D island concentration N_{2D} at this growth stage. As Si coverage $\Theta = Rt$ increases, no nucleation of new islands occurs (Fig. 8.12C–E), and 2D islands grow until a complete Si(111) layer is formed (Fig. 8.12F). We regarded this instant of time as $\Theta = 1$ ML, and used it to calculate Si deposition rate, R. Fig. 8.12G displays a topographical AFM image of 2D islands at $\Theta \approx 0.35$ ML. One can see that their real shape is smooth and rounded, which is typical [51,52] for 2D islands on the Si(111)-(1 × 1) surface. We have plotted experimental 2D island concentration, N_{2D} versus R and T using log–log (Fig. 8.12H) and Arrhenius (Fig. 8.12I) coordinates, respectively. Note that all experimental N_{2D} values were measured at $\Theta \sim 0.05$ ML. The slopes of $N_{2D} \propto R^{\chi}$ dependences first obtained at $T > T_c$ (Fig. 8.12H) are equal to $\chi = 0.81$–0.82, and correspond to the critical nucleus consisting of $i = 9$ particles.

The classical Eqs. (8.1–8.4) have been derived under the assumption that a stable nucleus appears after the association of more than i mobile particles. Such nuclei are allowed to grow only at the expense of new particle attachment, while smaller nuclei may also dissociate by particle detachment. However, according to contemporary crystal growth conceptions [54–56], $i \gg 1$ values can be obtained at a reasonably high T when the probability of atom detachment is notably large for nuclei of all sizes. In this case, in spite of $\chi \xrightarrow{i \gg 1} 1$ predicted by theory (Eq. 8.4), χ goes to $\chi^* < 1$ at $i \gg 1$ [54–57]. Therefore, our experimental χ^{DL} values of about 0.82 correspond to larger critical nucleus sizes than is in accordance with Eq. (8.4), i.e., $i > 9$.

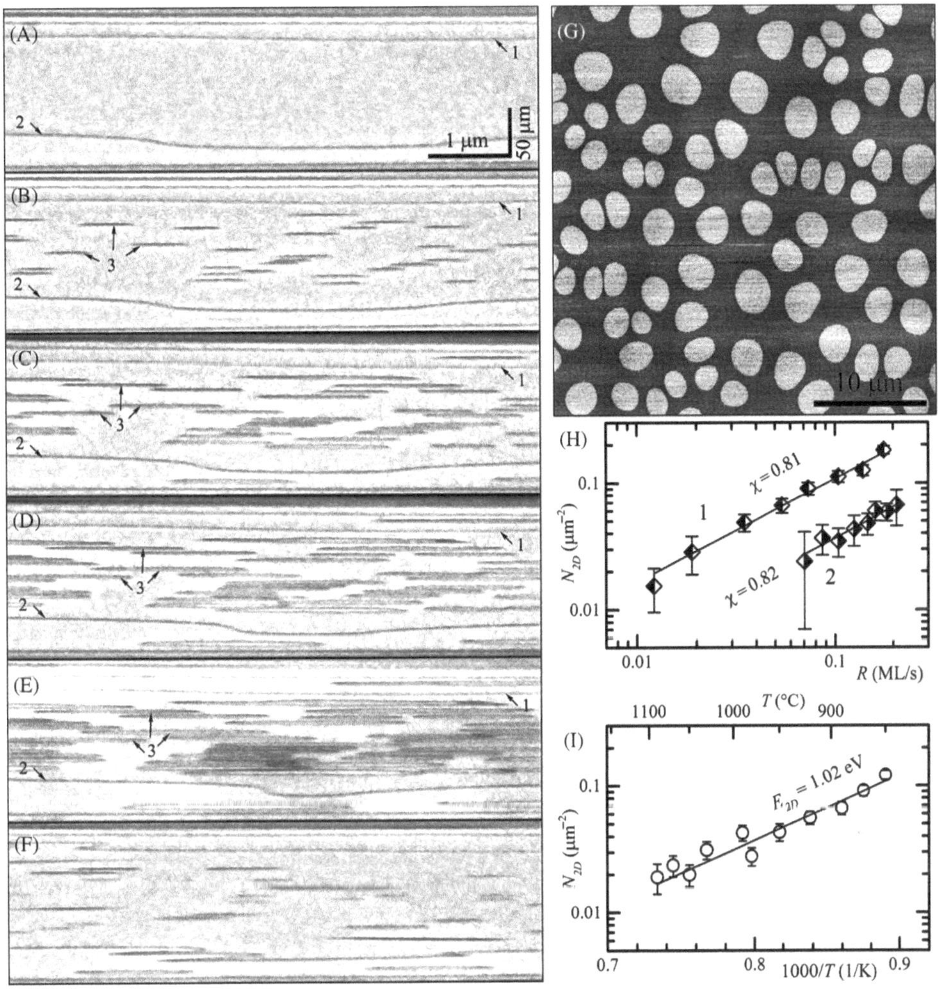

FIGURE 8.12 REM images of initial stages of Si growth on a 100 μm-Si(111) terrace at $T = 950°C$ and $R = 0.16$ ML/s: (A) 0 s, (B) 0.2 s, (C) 0.8 s, (D) 1.2 s, (E) 1.8 s, and (F) 6.2 s. In (A–E) monatomic steps initially bordering the extra-large terrace are labeled by 1 and 2; growing 2D islands nucleated in (B) are labeled 3. (G) Topographical AFM image of 2D islands after $\Theta \approx 0.35$ ML Si deposition at $T = 850°C$. (H) Experimental dependences $N_{2D}(R)$ at fixed T: (1) 850°C, (2) 950°C. (I) Experimental dependence $N_{2D}(T)$ at fixed $R = 0.125$ ML/s.

The slope of experimental $N_{2D} \propto \exp\frac{E_{2D}}{kT}$ obtained at $T = 850-1090°C$ corresponds to $E_{2D} \approx 1.02$ eV (Fig. 8.12I). Since a large ($i > 9$) critical nucleus represents, in fact, an unstable 2D island on the Si(111)-(1 × 1) surface, the atom detachment from the nucleus can be considered as the atom detachment from the monatomic step bordering the unstable island. Assuming that the interaction of adatoms with steps on the Si(111) surface (attachment/detachment) involves adatom pairs [51], one can estimate the dissociation energy of a critical nucleus per adatom E_i/i as one-half of the adatom pair formation energy

E_{ad} by detachment from the monatomic step: $E_i/i \approx E_{ad}/2$. Next, using Eq. (8.3) and substituting $E_{ad} = 0.23$ eV [51], and our experimental values $\chi^{DL} = 0.82$ and $E_{2D} = 1.02$ eV, we can estimate the activation energy of adatom diffusion on the Si(111)-(1 × 1) surface:

$$E_D \approx \frac{E_{2D}^{DL}}{\chi^{DL}} - \frac{E_{ad}}{2} = 1.13\text{eV}.$$

This estimate is less than the 1.30 eV obtained in Ref. [51], but the sum $E_D + E_{ad} = 1.36$ eV is consistent with the 1.3 eV [52] and 1.2 ± 0.2 eV [58] obtained previously.

8.5.2 Si Growth on the Step-Bunched Si(111)-(7 × 7) Surface

To study the impact of atomic steps on 2DNG more closely, we have analyzed 2D island nucleation on flat step-free terraces of the step-bunched Si(111)-(7 × 7) surface in the initial stage (first layer) and late growth stages when a pyramid-like structure bordered with a number of monatomic steps forms [59]. Nucleation in the first layer allows use of the classical analysis of 2D island concentration (Eq. 8.1) not affected by atomic steps. Analysis of late growth stages is based on critical terrace width scaling analysis of 2DNG (Eq. 8.2), that can be applied to information obtained about competition between 2DNG and adatom sink to steps.

The Si(111)-(7 × 7) surface with step-bunches divided by variable atomically flat terraces of width $W_{in} \sim 1-10$ μm has been prepared by using step-bunching phenomenon (Fig. 8.13A). At the initial stage of Si growth shown in Fig. 8.13B, the nucleation of high-density 2D islands takes place in the terrace's center. Two asymmetrical denuded zones of bright contrast are also clearly seen in the REM image near step-bunches. The denuded zone near the descending step-bunch (on the left) is about 30% wider than the one near the ascending step-bunch (on the right). This authenticates a preferential adatom attachment to a descending step-bunch rather than to an ascending one. Because the 2D island concentration N_{2D} in 2DNG zone is too high, they are not resolved separately in the REM image.

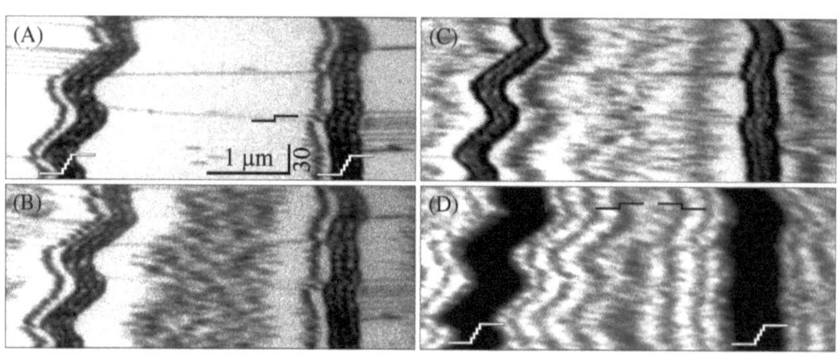

FIGURE 8.13 REM images of Si growth on the step-bunched Si(111)-(7 × 7) surface: (A) before, and after (B) 0.12 ML, (C) 1 ML, and (D) prolonged Si deposition at 600°C.

However, in situ REM easily detects the complete Si layer formation, with a few 2D islands atop after 1 ML deposition (Fig. 8.13C).

Fig. 8.14A shows an AFM image of the Si(111)-(7 × 7) surface after $\Theta \approx 0.15$ ML Si deposition at 650°C. *White arrows* denote chains of 2D islands nucleated at domain walls. These walls, as well as monatomic steps, serve as a network of effective sinks for adatoms and create denuded zones adjoining them. Two-dimensional islands near denuded zones have larger sizes and are a greater distance from each other compared to the islands in the central regions of superstructural domains. We used only central domain regions to calculate the mean island separation $L_s^2 \equiv N_{2D}$ dependences on R presented in Fig. 8.15A. One can see from Fig. 8.14A that the islands nucleated within (7 × 7) superstructural domains have clear triangular faceting with $[\bar{1}\bar{1}2]$-type step edges that are known to be related to the hampered nucleation of the double kink [60–63]. As we have shown first in Ref. [64], this is

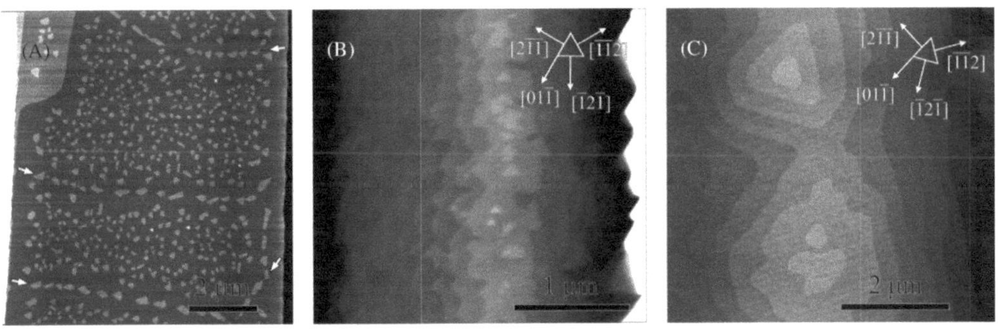

FIGURE 8.14 AFM images of 2D islands in: (A) the initial; and (B and C) late growth stages on flat terraces of the step-bunched Si(111) surface at (A and B) 650°C and (C) 750°C.

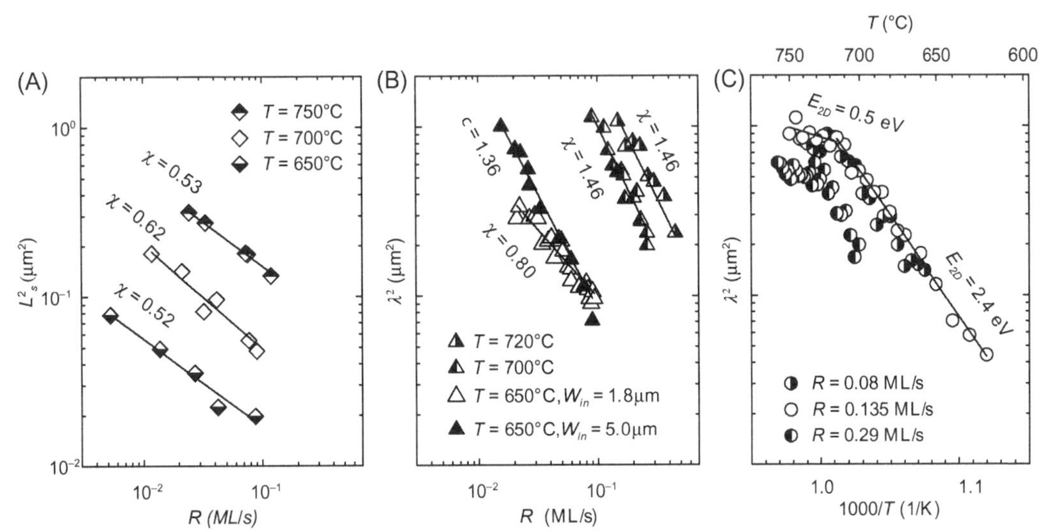

FIGURE 8.15 Experimental: (A) $L_s^2(R)$; (B) $\lambda^2(R)$; and (C) $\lambda^2(T)$ dependences.

due to the existence of the Schwöbel barrier ($\sim$0.9 eV) for adatom attachment to descending steps with $[\bar{1}12]$-type step edges that cause AL growth kinetics and step permeability. In order to analyze experimental $L_s^2(R)$ obtained under AL conditions, a modified rate equation theory can be used [65]. Under AL conditions, $L_s^2(R,T)$ retains the form of Eq. (8.1), but E_{2D} and χ expressions differ from Eqs. (8.3) and (8.4):

$$\chi^{AL} = \frac{2i}{i+3}, \tag{8.5}$$

$$E_{2D}^{AL} = \chi^{AL}\left(E_D + \frac{E_i}{i} + \frac{i+1}{i}E_{ES}\right), \tag{8.6}$$

where E_{ES} is the Schwöbel barrier ($E_{ES} \sim 0.9$ eV [64]). The $\chi = 0.52-0.62$ values presented in Fig. 8.15A signify that the critical nucleus consists of $i=1$ particle when 2DNG takes place apart from any sinks.

Continuous Si growth on flat terraces of the step-bunched Si(111) surface at $T=600-830°C$ leads to the formation of the multilayer structures shown in Fig. 8.13D. During Si deposition the height of this structure increases, but the uppermost layer width gradually decreases to a critical value λ, then periodic 2DNG occurs on the λ-width terrace. AFM images of pyramid-like structures grown at 650°C and 750°C are presented in Fig. 8.14B and C. In both cases, the structures are bordered by zigzagged or straight $[\bar{1}12]$-type step edges suggesting AL kinetics, however, their shape is different. In the last case, triangular pyramids rather crown a multilayer structure. The time evolution of these structures is periodic at fixed T and R, and is repeated in each Si monolayer. During this period, all layers of the multilayer structure widen owing to a step-flow caused by adatom attachment. At the same time, there is no observable displacement of the step-bunches due to the large number of steps they contain. We have found that the pyramid-like shape depends not only on R and T, but also on the initial terrace size between bunches (W_{in}) which controls the number of 2D layers. The lower T, the greater R, and the larger W_{in} are, the higher is the pyramid-like structure formed.

It is obvious that the 2DNG on a λ-width terrace competes with adatom attachment to descending steps bordering this structure. To characterize that quantitatively, we can compare $\lambda^2(R)$ dependences (Fig. 8.15B) with $L_s^2(R)$ dependence not affected by adatom sinks to steps (Fig. 8.15A). Despite identical fitting formulae $\lambda^2 \propto R^{-\chi}$ and $L_s^2 \propto R^{-\chi}$, experimental λ^2 and L_s^2 scaling are different within the same T and R range (Fig. 8.15). We have found that at $T=650°C$ the critical nucleus size increases from $i=1$ for 2DNG far from steps (Fig. 8.15A), to $i=2$ and $i=8$ for 2DNG on λ-width terraces with a W_{in} increase (Fig. 8.15B) when the step number of pyramid-like structures rises. This testifies that the zigzagged steps of growing pyramid-like structures are permeable. We assume that the step permeability assigns increasing nucleus size until the downhill adatom current dominates pyramid-like growth. It is believed that step-up adatom attachment is limited by a slightly higher energy barrier than we have found for step-down interaction ($E_{ES} \sim 0.9$ eV) that specifies Si growth kinetics at $T > 720°C$. At present, this energy barrier has not yet been determined. However, at

$T > 720°C$, step faceting by straight $[\bar{1}\bar{1}2]$-type edges becomes stronger (see Fig. 8.14B and C), which suggests the enhancement of step permeability, and the appearance of a dominating uphill adatom current.

We have found that $\lambda(T)$ dependences plotted in Arrhenius coordinates at several fixed R display the change of their slope E_{2D} at T around 720°C (Fig. 8.15C). In particular, $\lambda^2 \propto \exp(-E_{2D}/kT)$ fitting of the plot obtained at $R = 0.135$ ML/s gives $E_{2D} \approx 2.4$ eV within the 620–720°C range. This value is close to the 2.0 and 2.2 eV reported previously at $T < 600°C$ in Refs. [66,67], respectively. According to modified rate equation theory [65], experimental values of $\chi > 1$ clearly indicate AL kinetics up to $T \approx 720°C$. Further, E_{2D} drops to 0.5 eV at $T > 720°C$ and suggests that we observe a AL⇔DL kinetic transition [68]. However, this transition coincides with a notable change in pyramid-like structure morphology (Fig. 8.14C), i.e., as we assume, linked to the enhancement of the uphill adatom current. As a result, the dispersion of λ values at $T > 720°C$ increases (Fig. 8.15C). It is clear that 2DNG kinetics on a λ-width terrace at $T > 720°C$ well deserves ampler treatment.

8.5.3 Si Growth on an Extra-Large Step-Free Si(111)-(7 × 7) Surface

Additionally, we have studied prolonged Si growth (~300 ML) on ~100 μm-wide atomically flat terraces at $T = 600–750°C$. We observed again the multilayer growth at Si deposition rates of $R \geq 3 \cdot 10^{-2}$ ML/s, while planar layer-by-layer growth took place at lower R. Figs 8.16A and B display AFM images of multilayer Si growth at 600°C and 750°C, respectively. In both cases, the growth surfaces seem to be similar to the pyramid-like growth on flat terraces of the step-bunched surface. According to recent theoretical concepts [69], the very fact of multilayer structure formation (multilayer kinetic roughening) arises due to the predominance of a destabilizing uphill adatom current over a stabilizing downhill adatom current. To characterize the multilayer growth on extra-large terraces, we have analyzed layer coverage distributions obtained from the AFM images (Fig. 8.16C). Both plots in Fig. 8.16C display the linear slopes and small height of multilayer growth after ~300 ML Si deposition. These linear slopes and the "slope selection" have been theoretically attributed to the permeability of

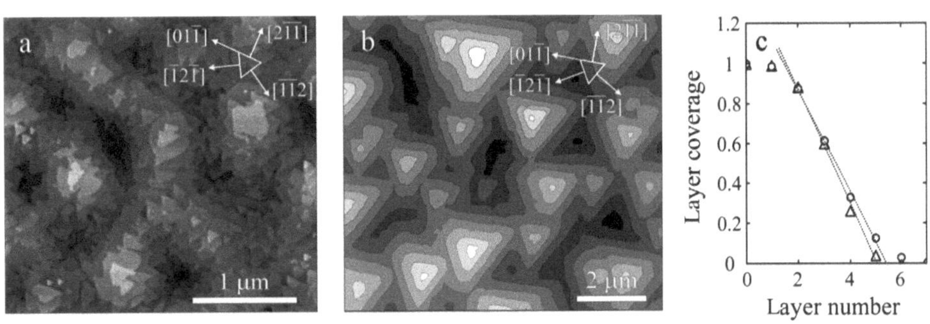

FIGURE 8.16 (A) AFM; and (B) processed AFM image of the Si(111)-(7 × 7) surface after 300 ML Si deposition onto 100 μm-wide atomically flat terraces at (A) $T = 600°C$, $R = 3 \cdot 10^{-2}$ ML/s and (B) $T = 750°C$, $R = 8 \cdot 10^{-2}$ ML/s. (C) Layer coverage distributions: *triangles* and *circles* for (A) and (B), respectively, display linear slopes.

monatomic steps [70]. Moreover, the step-up permeability has been recently shown to trigger multilayer structure formation by early second layer nucleation atop 2D islands [71]. However, in this case the critical size decreases. The crucial role of step permeability for multilayer Si/Si(111)-(7 × 7) growth is authenticated by strong faceting of steps by [$\bar{1}\bar{1}$2]-type edges (Fig. 8.16A and B). It is of particular importance that multilayer structures do not appear during low-*R* Si growth. This fact suggests that the balance between destabilizing uphill and stabilizing downhill adatom currents depends strongly on growth parameters *T* and *R*. In other words, this balance is defined by step properties (Schwöbel barriers, kink density, step orientation, etc.), and can be controlled during technological treatments by *T* and *R* variation.

8.6 Step Motion During Oxygen Etching

8.6.1 Interaction of Oxygen With the Si Surface

To optimize the oxidation processes of nanoscale structures, we have investigated the interaction of oxygen with an atomically clean Si(111) surface within the $T = 500–900°C$ range [72]. When partial oxygen pressure is increased to a critical value P_{crit}, we have not registered any modification of intensity in basic and superstructural diffraction reflections. At the same time, in situ UHV REM shows the motion of monatomic steps at the Si(111) surface toward ascending terraces, corresponding to surface etching by oxygen [73].

It is well known that adsorbed oxygen interacts with silicon and forms volatile silicon monoxide (SiO): Si + O(ads)→SiO(gas)↑. This reaction describes silicon etching by oxygen at $T > 500°C$ by means of SiO removal from the surface. This process is similar to sublimation, and explains why in situ REM displays step motion toward ascending terraces. However, the temperatures for oxygen etching are substantially lower than for pure thermal silicon sublimation. Fig. 8.17 shows the dependences of monatomic step velocity *V* on the width of adjacent terraces by oxygen etching of the vicinal Si(111) surface. The wider the adjacent terraces are, the faster monatomic steps move.

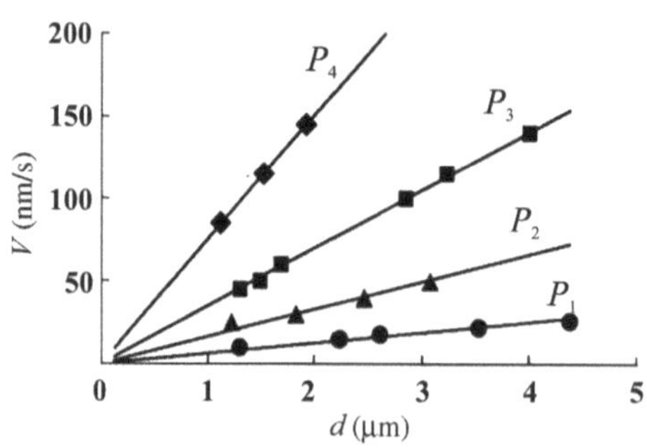

FIGURE 8.17 Experimental dependence of step velocity *V* on the width of adjacent terraces *d* at different oxygen pressure ($P_1 < P_2 < P_3 < P_4 < P_5$).

Additional REM contrast has been observed on the wide terrace between monatomic steps at an oxygen pressure above P_{crit}. This contrast corresponds to the formation of negative 2D islands with one interplanar spacing depth. This finding is consistent with the results obtained by Shimuzu et al. [74]. Further exposure of the Si surface to oxygen atmosphere leads to an increase in island sizes. Real oxidation (oxide formation) of the Si surface begins above a certain oxygen pressure, when homogeneous ripple contrast, as well as the termination of step motion, appear in REM images.

Based on these results, we may conclude that oxygen etching of the Si surface proceeds either by the step-flow mechanism, or by nucleation of 2D negative islands, depending on temperature and oxygen pressure. Thus, generation and migration of surface vacancies followed by their 2D island nucleation or interaction with steps can describe both these mechanisms. The 2DNG mechanism, including periodic formation and coalescence of 2D negative islands, describes surface etching at lower temperatures or higher oxygen pressures. Step-flow etching takes place at higher temperatures or lower oxygen pressures. It is important to note that, similar to the growth process, the shape of monatomic steps during etching is changed from smooth to zigzag due to the formation of macrokinks. This strongly suggests that kinks form due to the interaction of vacancies with atoms located on the steps, rather than due to the interaction of moving monatomic steps with 2D negative islands. A thorough analysis of the REM images taken from the same area at different azimuthal angles shows that macrokinks consist of linear segments oriented parallel to the $\langle 110 \rangle$ crystallographic direction. The oscillations of specular electron beam intensity corresponding to nucleation and coalescence of 2D negative islands have been observed during oxygen etching of the Si surface at high temperatures. We have estimated the activation energy of surface vacancy diffusion to be 1.35 ± 0.05 eV [75].

8.6.2 Kinetics of Negative 2D Island Nucleation and Growth During Oxygen Etching

In situ REM have recently been used to study the nucleation and growth kinetics of negative 2D islands on extremely wide (up to 120 μm in diameter) atomically flat terraces of the Si(111) surface during oxygen etching at $T > 1000°C$ in more detail [76]. To create such a large terrace, the sample was first milled by a low energy Ar^+ beam to form a round 2 μm-deep crater on the surface. Then the sample was annealed in UHV REM at $T \sim 1300°C$ to remove natural oxide and contamination related to the preparation technique. During annealing (sublimation) a wide terrace bordered with concentric monatomic steps appears in the dimple's center due to multiple negative 2D island formation and growth (see Fig. 8.4A). After that, the sample temperature was decreased and molecular oxygen was supplied from a barium peroxide source placed in front of the sample installed in the UHV chamber of the REM. The solid state source was heated to temperatures above 500°C by passing an electrical current.

When oxygen pressure at fixed T rises, the number of atomic steps inside the dimple increases, and the terrace size between steps decreases (see Fig. 8.4B). The frequency of island nucleation is found to rise with oxygen flux at fixed T. The interaction between oxygen and silicon atoms on the surface is therefore believed to be the dominating mechanism for vacancy formation. Previously reported experimental [72,77,78] and theoretical [79] results suggest that surface vacancies are mobile, and their diffusion should be considered as well as adatom diffusion.

We have measured a critical size (in our case it is a diameter) D_{crit} of the central atomically flat terrace on a REM image recorded just before the nucleation of a new island at different oxygen fluxes. In our experiments, the width of an atomically flat terrace can reach up to 120 μm. Similar to the Si growth process, D_{crit} can be measured in dependences on island nucleation frequency ν (recalculated with etching rate) and sample temperature, to apply the classical approach for the analysis of nucleation and growth kinetics of negative 2D islands. Since we have not observed in our experiments a significant difference in spatial distribution or kinetics of the atomic step motion related to the direction of heating current, the isotropic model is used. We consider vacancy diffusion on a circular terrace with a radius r_i bounded by an ascending atomic step (Fig. 8.4C). The solution of the theoretical problem is an equation describing $D_{crit}(\nu)$ dependence in two regimes of kinetics: DL and attachment/detachment limited (ADL).

$$D_{crit}^2 \approx a^2 \left(\frac{\nu_0}{\nu}\right)^\chi \exp\frac{-E}{kT}$$

Here, parameters χ and E depend on a kinetic regime and are defined according to Table 8.1.

The dependence of D_{crit}^2 on ν in the $T = 1000-1300°C$ range is given in Fig. 8.18. We have obtained the scaling exponent χ by extracting slopes from $D_{crit}^2(\nu)$ measurements taken over a range of temperatures. Then the experimental data were approximated by function $D_{crit}^2 \propto A(T)(\nu_0/\nu)^\chi$, where $A(T)$ is a prefactor depending on T. Both χ and prefactor dependences on temperature are shown in Figs. 8.19A and B. One can see that two different kinetics regimes with $\chi = 1.15 \pm 0.10$ at temperatures exceeding $T_{crit} = 1180°C$, and $\chi = 0.95 \pm 0.03$ at lower temperatures are observed. According to the classical approach, if $\chi > 1$ the 2DNG kinetics is described by the ADL mode, and at $\chi < 1$ with the DL one.

At temperatures above T_{crit}, based on classical theory, we have estimated the critical size of the nucleus as $i = 4$. The prefactor $A(T)$ follows the Arrhenius form with an activation energy of 3.9–4.1 eV within this T range. When substituting the vacancy diffusion

Table 8.1 Scaling Exponents χ and Activation Energies E for DL (Diffusion Limited) and ADL (Attachment/Detachment Limited) Kinetic Regimes

Regime	χ	E
DL	$\frac{i}{i+2}$	$\frac{E_i + iE_D}{i+2}$
ADL	$\frac{2i}{i+3}$	$\frac{2(E_i + iE_D + (i+1)E_{AD})}{i+3}$

activation energy $E_D = 1.3$ eV and $i = 4$ into the equation for effective activation energy in the ADL regime:

$$E = 2[E_i + iE_D + (i+1)E_{AD}]/(i+3),$$

where E_{AD} is the energy barrier of vacancy interaction with the step, we can determine $E_{AD} = 1.50 \pm 0.15$ eV. The obtained E_{AD} is greater than the diffusion energy of the surface vacancy. Thus, we conclude that the vacancy 2D island nucleation on the Si(111) surface during oxygen etching at temperatures above $T_{crit} = 1180°C$ is realized in the ADL regime. However, at temperatures below T_{crit}, an annihilation between adatoms detached from steps and vacancies becomes significant, and the classical approach should be expanded to include both adatoms and vacancies. If we assume, as discussed in Section 8.3, that desorption through adatom detachment from a step dominates below $T_{crit} = 1180°C$, then DL kinetics in this temperature range can be explained. In this case, the energy of adatom diffusion ($\sim$1.13 eV) is always greater than the energy of adatom formation, 0.23 eV [51].

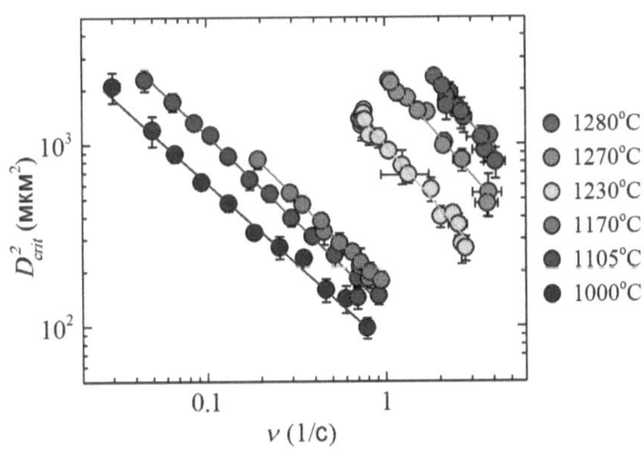

FIGURE 8.18 Dependence of squared critical terrace diameter D^2_{crit} of vacancy 2D island on their nucleation frequency ν at different fixed T.

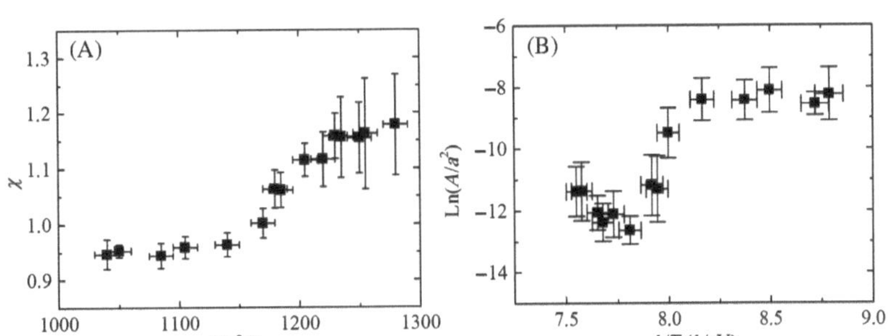

FIGURE 8.19 Temperature dependences of: (A) Scaling exponent χ; and (B) prefactor A.

8.7 Initial Stages of Heteroepitaxial Growth on the Si(111) Surface

The investigation of the initial stages of Ge epitaxy on the stepped Si(111) surface by using in situ UHV REM techniques has shown that the growth of the first monolayers obeys the same laws as during Si homoepitaxy. According to well-known data, Ge adatoms attach to the substrate's monatomic steps if growth takes place by the step-flow mechanism, and the adatoms nucleate 2D islands if terraces are wider than a critical terrace width of λ that scales with temperature as $\lambda^2 \propto \exp(-1.2\text{eV}/kT)$ [80].

The control of fractional Ge coverage deposition onto the stepped Si(111) surface allows realization of the conditions for the formation of lateral superlattices consisting of Ge stripes alongside monatomic steps that alternate with silicon stripes. The main impediment to lateral superlattice formation during Ge deposition is the effect of monatomic step clustering caused by polycentric nucleation of 2D Ge islands induced in superstructural domains on the Si(111) surface.

When studying the nucleation of new phase islands during surfactant heteroepitaxy by in situ UHV REM, we have also revealed variable instabilities of RS trains on the vicinal Si(111) surface induced by submonolayer adsorption of impurity atoms that trigger surface reconstructions. The formation of superstructural domains during reversible phase transition is accompanied not only by reversible monatomic step displacement, but also by step clustering. In particular, we have found that macrosteps (bunches) form on the clean Si(111) surface during the $(1 \times 1) \Leftrightarrow (7 \times 7)$ superstructural transition [41] and impurity-induced Si (7×7)Ge, Si(5×5)Ge, Si(5×2)Au, and Si(3×1)Ca phase transitions [33,81]. The polycentric domain nucleation of a new surface structure and adatom diffusion between the domains are the key conditions of macrostep formation. The density of macrosteps depends on the initial monatomic step separation, substrate temperature, and superstructural transition speed; it is also determined by kinetic limitations related to the requirement for significant mass transport on the surface [82,83].

During the formation of calcium induced (3×1) domains (Fig. 8.20), we have detected the influence of heating electric current direction on domain shape, and on the configuration of monatomic step clusters, which is caused by the electromigration of adsorbed atoms [81]. The presence of kinetic limitations for step aggregation during superstructural transitions allows control of the final step density by the selection of submonolayer coating deposition conditions.

Using UHV REM, the peculiarities of pseudomorphous Ge film formation by step flow and 2D island nucleation have been studied [84]. We have found two types of denuded zones around a three-dimensional (3D) Ge island. The first one represents the zone free from small 2D Ge islands around a large 3D island. The second one is provided by the appearance of a closed atomic step surrounding a 3D island that moves off the 3D island in a step-up direction. This allows us to arrive at the conclusion that the generation of Si atoms from the step takes place during the attachment of Ge atoms to the 3D island. When the crystal temperature increased, we observed significant enhancement of Si atom generation by a monatomic step during 3D Ge island growth.

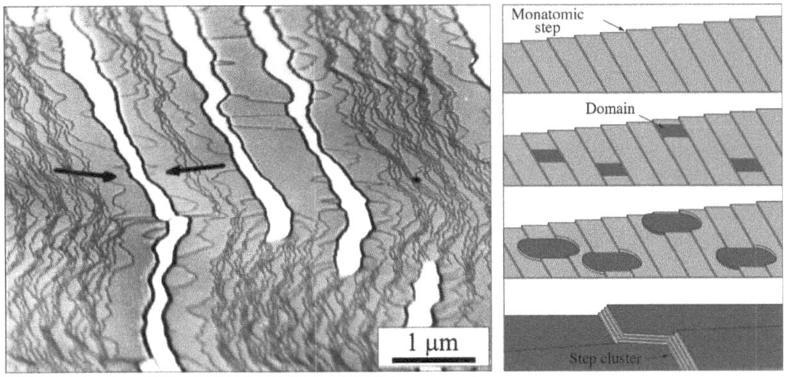

FIGURE 8.20 (At the left) REM image of submonolayer calcium induced nanostructures: extended superstructural (3 × 1) domains distributed periodically on the surface (minimum domain width is 20 nm, and maximum length is up to 1 mm). (At the right) The schematic representation of a step cluster formation.

8.8 Step-Bunching Induced by Gold Adsorption on the Si(111) Surface

As we have shown in previous sections, the movement of RSs depends on surface treatments: they move in a step-down direction during growth, and they move toward higher situated terraces during sublimation due to the detachment of atoms from step kinks followed by adatom desorption from terraces. The rate of step movement depends on the mean step distance and the crystal temperature [12]. In accordance with Ref. [9], the system of monatomic steps is unstable with respect to fluctuations in the distance between them, which leads to the formation of step-bunches (SBs). The RS ⇔ SB transition is reversible upon a change in the polarity of the applied voltage required for Joule heating of the crystal. The system of RSs is stable at 900°C upon DC heating directed toward higher situated terraces, and step-bunches form after DC direction alternation.

We used in situ REM to study the impact of gold adsorption on the behavior of monatomic steps on the Si(111) surface within the $T = 850-1260$°C range. While Si crystal was resistively heated, even submonolayer gold deposition on the Si surface lead to fast redistribution of RSs to step-bunches (SB) and *vice versa*. These RS ⇔ SB morphological transitions on the Si surface have been investigated as a function of gold coverage and the direction of heating direct electrical current (DC). The successive back and forward RS ⇔ SB transitions on the Si surface after deposition of 0.75 ML of gold take place during further isothermal annealing at 900°C: RS (0) ⇒ SB (0.72−0.42) ⇒ RS (0.42−0.24) ⇒ SB (0.24−0.07) ⇒ RS (0.07−0). The numbers given in parentheses are the intervals of gold monolayer coverage where a step distribution remains invariant. The reversal of DC current direction triggers RS ⇒ SB and SB ⇒ RS changes within the same gold coverage ranges [85].

The deposition of gold on the silicon surface was realized from a small-volume evaporator representing a crucible of a tungsten wire through which an electric current was passed. The deposition rate of gold atoms was precisely determined by the time necessary to fill the

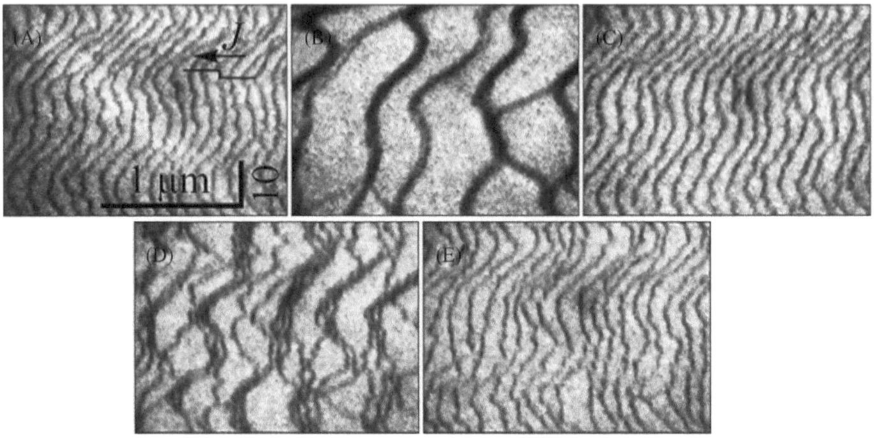

FIGURE 8.21 REM images of the Si(111) surface after: (A) ~0.75 ML initial gold deposition and subsequent thermal annealing at 900°C for (B) 4 min, (C) 20 min, (D) 43 min, and (E) 50 min. An *arrow* in (A) shows DC current direction.

entire surface of the Si(111)-(5 × 2)Au superstructure at 500°C, which corresponded to a 0.42 ML gold coverage. In this case, the superstructure type can be identified by microdiffraction patterns obtained from the chosen surface area.

Let us consider the behavior of monatomic steps upon gold adsorption at 900°C during heating by a DC current directed toward the upper terraces [86]. Under these conditions, atomic steps retained a regular distribution for a long period, e.g., several hours. The temperature and the amount of adsorbed gold ($\Theta \sim 0.75$ ML) in the experimental studies were selected so as to exclude any faceting or gold-induced reconstructions, as well as 3D gold island formation at a given temperature. A typical sequence of REM images of the same silicon surface area taken from fragments of video during gold deposition and subsequent thermal annealing is shown in Fig. 8.21. After the deposition of $\Theta \sim 0.75$ ML gold on the silicon surface, heating of the gold source was switched off and the crystal was isothermally annealed at 900°C for 50 min. During gold deposition, the intensity of basic diffraction reflections decreased, but additional reflections pointing to reconstruction were not observed. Since the surface before and immediately after gold deposition was completely identical, a REM image of only the Si(111) surface before gold deposition is shown in Fig. 8.21A. To prove the existence of gold deposited on the surface (which could fully dissolve in the bulk of crystal or reevaporate), the substrate temperature was sharply decreased to 500°C immediately after adsorption. At this temperature, a Si(111)-($\sqrt{3} \times \sqrt{3}$)Au superstructure formed on the surface. This superstructure covered the entire surface (with $5 \times 200\ \mu m^2$ dimensions) visualized by REM. The formation of ($\sqrt{3} \times \sqrt{3}$) superstructure domains at this temperature started when the gold coverage of the silicon surface was greater than 0.68 ML [13]. The occurrence of this superstructure indicated that about 0.7 ML gold was retained on the surface, despite the high substrate temperature (900°C). At lower amounts of gold, a partial coverage of the surface by the ($\sqrt{3} \times \sqrt{3}$) superstructure was observed at 500°C. It is clear that the surface concentration of gold decreases during thermal annealing at $T = 900$°C, because of its partial dissolution in the crystal bulk, and after 4-min annealing RSs transform to step-bunches separated by broad (from 0.5 μm to several micrometers) singular terraces (see Fig. 8.21B).

Note that the step-bunching induced by gold deposition differs from that initiated by electromigration during sublimation. In the latter case, the rearrangement of steps to step-bunches proceeds through a change of step velocities toward the step-up direction. In the former case, bunches are formed due to movement of atomic steps toward both step-up (adatom detachment from steps) and step-down (adatom attachment to steps) directions. Moreover, some fragments of the same monatomic step moved in both directions, corresponding to sublimation and epitaxial growth. Such strange behavior of steps cannot be explained by gold adsorption on steps. Analysis of the diffraction step contrast has not revealed any features associated with the increase of local step deformation that would provide evidence of gold incorporation into the steps. However, after 20-min annealing at 900°C the morphology of the Si(111) surface returned to RSs (Fig. 8.21C).

This observation suggests that a short-term step-bunching effect induced by gold deposition provides the increase in silicon adatom density, which initiates the step movement in a step-down direction. A possible source of excessive adatoms on the surface may be gold diffusion into the crystal's bulk by a kick-out mechanism. In this case, a gold atom substitutes a silicon atom in a lattice position and creates a self-interstitial atom that can drain to the surface. However, because a small amount of gold diffused into the bulk, the step-bunching lasts for a short time. In order to estimate the quantity of gold atoms on the surface at this annealing stage, the crystal temperature was decreased to 500°C and additional reflections corresponding to Si(111)-(5 × 2)Au reconstruction were detected. In accordance with the phase diagram, this indicated that the gold coverage decreased to $\Theta \sim 0.44$ ML.

Surprisingly, subsequent isothermal annealing at 900°C for 40 min again led to a change from RSs to step-bunches (Fig. 8.21D). In this case, the formation of step-bunches differed from the initial one shown in Fig. 8.21B; the large terraces contained zigzagged clustered steps. This appears to be an incomplete step-bunching process. A REM image of such a step--bunched surface is in agreement with a previously observed step-bunch morphology induced by a very small amount of gold ($\Theta < 0.005$ ML) [87]. We may carefully assume that the second short-term step-bunching is triggered again by a small number of self-interstitials generated in subsurface layers through the well-known process of oxide precipitate formation in Cz-Si during long duration (40 min) annealing. However, the impact of bulk self-interstitials on surface instabilities is a new phenomenon and needs more investigation. Finally, the formation of regularly arranged steps is detected after 47-min annealing (Fig. 8.21E). We cannot exclude that the effective charge of adatoms and their sign depends on the density of gold adatoms [87]. As the crystal temperature decreased to room temperature, no additional reflections related to gold deposition were observed in the diffraction pattern, except for the reflections typical of the Si (111)-(7 × 7) superstructure. This allows the conclusion that the 0.7 ML gold adsorbed on the silicon surface completely dissolved in the crystal bulk during the thermal annealing at 900°C that lasted for approximately 47 min. Additional annealing for several hours had not revealed any change in step distribution. It is clear that these transformations of the Si(111) surface morphology cannot be explained by the formation of a new surface reconstruction, as in case of low temperature gold adsorption [88].

8.9 Conclusion

In situ UHV REM is a powerful technique that supplies fundamental knowledge concerning atomic processes on the Si surface under various surface treatments necessary to develop new technologies. The main advantage of this technique is the possibility for carrying out in situ observations of surface dynamics at temperatures up to the melting point of silicon. The adoption of advanced surface morphology design (large-scale step-free terraces, step-bunches, etc.) moves in situ REM toward quantitative characterization of growth and sublimation processes, and evaluation of such fundamental parameters as adatom diffusion length, adsorbed atom concentration, critical terrace width, nucleus size, activation energy of diffusion, migration energies of adatoms and vacancies, energy barriers for adatom, and vacancy interaction with steps, etc.

Acknowledgment

The authors gratefully acknowledge partial financial support from RSF (project No. 14-22-00143).

References

[1] J.-N. Aqua, I. Berbezier, L. Favre, T. Frisch, A. Ronda, Growth and self-organization of SiGe nanostructures, Phys. Rep. 522 (2013) 59–189. http://dx.doi.org/10.1016/j.physrep.2012.09.006.

[2] M. Michailov, Nanophenomena at Surfaces: Fundamentals of Exotic Condensed Matter Properties, Springer Berlin Heidelberg, Berlin, Heidelberg, 2011. Available from: http://dx.doi.org/10.1007/978-3-642-16510-8.

[3] C. Dupas, P. Houdy, M. Lahmani, Nanoscience: Nanotechnologies and Nanophysics, first ed., Springer-Verlag Berlin Heidelberg, Berlin, Heidelberg, 2007. Available from: http://dx.doi.org/10.1007/978-3-540-28617-2.

[4] A.A. Kroshkov, A.L. Aseev, E.A. Baranova, A.V. Latyshev, O.A. Yakushenko, Differential ultrahigh-vacuum pump for electron microscope, Instrum. Exp. Tech. 28 (1985) 220–224. http://inis.iaea.org/Search/search.aspx?orig_q = RN:19024051 (accessed 29.02.16).

[5] N. Osakabe, Y. Tanishiro, K. Yagi, G. Honjo, Reflection electron microscopy of clean and gold deposited (111) silicon surfaces, Surf. Sci. 97 (1980) 393–408. http://dx.doi.org/10.1016/0039-6028(80)90675-5.

[6] K. Yagi, Reflection electron microscopy: studies of surface structures and surface dynamic processes, Surf. Sci. Rep. 17 (1993) 307–362. http://dx.doi.org/10.1016/0167-5729(93)90002-7.

[7] K. Yagi, H. Minoda, M. Degawa, Step bunching, step wandering and faceting: self-organization at Si surfaces, Surf. Sci. Rep. 43 (2001) 45–126. http://dx.doi.org/10.1016/S0167-5729(01)00013-9.

[8] S.S. Kosolobov, A.V. Latyshev, Step bunching on silicon surface under electromigration, in: M. Michailov (Ed.), Nanophenomena Surfaces Fundam. Exot. Condens. Matter Prop, Springer Berlin Heidelberg, Berlin, Heidelberg, 2011, pp. 239–258. Available from: http://dx.doi.org/10.1007/978-3-642-16510-8_11.

[9] A.V. Latyshev, A.B. Krasilnikov, A.L. Aseev, Application of ultrahigh vacuum reflection electron microscopy for the study of clean silicon surfaces in sublimation, epitaxy, and phase transitions, Microsc. Res. Tech. 20 (1992) 341–351. Available from: http://dx.doi.org/10.1002/jemt.1070200405.

[10] A.V. Latyshev, Reflection electron microscopy of the stepped silicon surfaces, Phys. Low-Dimens. Struct. 3–4 (1999) 169–180.

[11] A.V. Latyshev, In-situ UHV reflection electron microscopy of silicon surface dynamic processes, in: M. Michailov (Ed.), Proc. EWSSW'96 "Thin Film. Phase Transitions Surface,", Bulgarian Academy Science, Sofia, 1996, pp. 142–152.

[12] L.I. Fedina, D.V. Sheglov, S.S. Kosolobov, A.K. Gutakovskii, A.V. Latyshev, Precise surface measurements at the nanoscale, Meas. Sci. Technol. 21 (2010) 54004. http://stacks.iop.org/0957-0233/21/i = 5/a = 054004.

[13] A.L. Aseev, A.V. Latyshev, A.B. Krasilnikov, S.I. Stenin, Reflection electron microscopy study of the structure of atomic clean silicon surface, in: E. Mizera (Ed.), Defects Cryst., World Scientific Publishing Co., Singapore, 1987, pp. 231–237.

[14] W.K. Burton, N. Cabrera, F.C. Frank, The growth of crystals and the equilibrium structure of their surfaces, Philos. Trans. R. Soc. Lond. A (1951) 299–358.

[15] A.V. Latyshev, A.L. Aseev, A.B. Krasilnikov, S.I. Stenin, Transformations on clean Si(111) stepped surface during sublimation, Surf. Sci. 213 (1989) 157–169. http://dx.doi.org/10.1016/0039-6028(89)90256-2.

[16] S.V. Sitnikov, A.V. Latyshev, S.S. Kosolobov, Advacancy-mediated atomic steps kinetics and two-dimensional negative island nucleation on ultra-flat Si(111) surface, J. Cryst. Growth, http://dx.doi.org/10.1016/j.jcrysgro.2016.05.048 in press.

[17] Y. Homma, H. Hibino, T. Ogino, N. Aizawa, Sublimation of the Si (111) surface in ultrahigh vacuum, Phys. Rev. B 55 (1997) R10237.

[18] C. Alfonso, J.M. Bermond, J.C. Heyraud, J.J. Métois, The meandering of steps and the terrace width distribution on clean Si (111): an in-situ experiment using reflection electron microscopy, Surf. Sci. 262 (1992) 371–381.

[19] J.C. Girard, S. Gauthier, S. Rousset, W. Sacks, S. de Cheveigné, J. Klein, Dynamics of high index step equilibrium fluctuations as observed by scanning tunneling microscopy, Surf. Sci. 301 (1994) 245–252. http://dx.doi.org/10.1016/0039-6028(94)91304-8.

[20] G.S. Bales, A. Zangwill, Morphological instability of a terrace edge during step-flow growth, Phys. Rev. B 41 (1990) 5500–5508. Available from: http://dx.doi.org/10.1103/PhysRevB.41.5500.

[21] N.C. Bartelt, J.L. Goldberg, T.L. Einstein, E.D. Williams, J.C. Heyraud, J.J. Métois, Brownian motion of steps on Si (111), Phys. Rev. B 48 (1993) 15453.

[22] P. Nozières, Shape and growth of crystals in solids far from equilibrium, in: C. Godrèche (Ed.), Solids Far From Equilib., Cambridge University Press, Cambridge, 1992, p. 1.

[23] R. Kariotis, M.G. Lagally, Calculation of terrace-edge structure distributions for vicinal surfaces, Surf. Sci. 248 (1991) 295–305. http://dx.doi.org/10.1016/0039-6028(91)91175-W.

[24] A. Latyshev, H. Minoda, Y. Tanishiro, K. Yagi, Ultra high vacuum reflection electron microscopy study of step-dislocation interaction on Si(111) surface, Jpn. J. Appl. Phys. 34 (1995) 5768. Available from: http://dx.doi.org/10.1143/JJAP.34.5768.

[25] A.V. Latyshev, H. Minoda, Y. Tanishiro, K. Yagi, UHV REM investigation of the interaction between steps and dislocation on silicon (111) surface II, Surf. Sci. 357-358 (1996) 550–554. http://dx.doi.org/10.1016/0039-6028(96)00221-X.

[26] A.V. Latyshev, H. Minoda, Y. Tanishiro, K. Yagi, Dynamical step edge stiffness on the Si (111) surface, Phys. Rev. Lett. 76 (1996) 94.

[27] N.C. Bartelt, R.M. Tromp, E.D. Williams, Step capillary waves and equilibrium island shapes on Si(001), Phys. Rev. Lett. 73 (1994) 1656–1659. Available from: http://dx.doi.org/10.1103/PhysRevLett.73.1656.

[28] M. Uwaha, Evolution of a train of interacting steps in a surface diffusion field, J. Cryst. Growth 128 (1993) 92–96. http://dx.doi.org/10.1016/0022-0248(93)90301-C.

[29] L.V. Litvin, A.B. Krasilnikov, A.V. Latyshev, Transformations of the stepped Si(001) surface structure induced by heating the specimen by a current, Surf. Sci. 244 (1991) L121–L124. http://dx.doi.org/10.1016/0039-6028(91)90483-9.

[30] A.V. Latyshev, A.L. Aseev, Monatomic steps on silicon surfaces, Phys. Usp. 41 (1998) 1015–1023. Available from: http://dx.doi.org/10.1070/PU1998v041n10ABEH000462.

[31] Y. Homma, M. Suzuki, H. Hibino, Step band structures on vicinal Si(111) surfaces created by DC resistive heating, Appl. Surf. Sci. 60–61 (1992) 479–484. http://dx.doi.org/10.1016/0169-4332(92)90463-8.

[32] M. Suzuki, Y. Homma, Y. Kudoh, R. Kaneko, Step bunching structure on vicinal Si(111) surfaces studied by scanning force microscopy, Appl. Surf. Sci. 60-61 (1992) 460–465. http://dx.doi.org/10.1016/0169-4332(92)90460-F.

[33] H. Tokumoto, M. Iwatsuki, Scanning tunneling microscopy of clean silicon surfaces at elevated temperatures, Jpn. J. Appl. Phys. 32 (1993) 1368–1378. Available from: http://dx.doi.org/10.1143/JJAP.32.1368.

[34] M.J. Ramstad, R.J. Birgeneau, K.I. Blum, D.Y. Noh, B.O. Wells, M.J. Young, D.C. current-dependent faceting of vicinal Si(111), Europhys. Lett. 24 (1993) 653–658. Available from: http://dx.doi.org/10.1209/0295-5075/24/8/006.

[35] A.V. Latyshev, A.B. Krasil'nikov, A.L. Aseev, S.I. Stenin, Effect of electric current on the ration of the areas of the (2 × 1) and (1 × 2) domains at the clean (001) surface of silicon during sublimation, JETP Lett. 48 (1988) 526–529.

[36] A.V. Latyshev, A.B. Krasilnikov, A.L. Aseev, In situ REM study of monatomic step behaviour on Si (111) surface during sublimation, Ultramicroscopy 48 (1993) 377–380.

[37] A.V. Latyshev, A.B. Krasilnikov, A.L. Aseev, UHV REM study of the anti-band structure on the vicinal Si (111) surface under heating by a direct electric current, Surf. Sci. 311 (1994) 395–403.

[38] C. Misbah, O. Pierre-Louis, Y. Saito, Crystal surfaces in and out of equilibrium: a modern view, Rev. Mod. Phys. 82 (2010) 981–1040. Available from: http://dx.doi.org/10.1103/RevModPhys.82.981.

[39] E.E. Rodyakina, S.S. Kosolobov, A.V. Latyshev, Drift of adatoms on the (111) silicon surface under electromigration conditions, JETP Lett. 94 (2011) 147–151. Available from: http://dx.doi.org/10.1134/S0021364011140128.

[40] A.V. Latyshev, A.B. Krasilnikov, A.L. Aseev, L.V. Sokolov, S.I. Stenin, Reflection electron microscopy study of clean Si(111) surface reconstruction during the (7 × 7)-(1 × 1) phase transition, Surf. Sci. 254 (1991) 90–96. http://dx.doi.org/10.1016/0039-6028(91)90641-5.

[41] A.V. Latyshev, A.L. Aseev, S.I. Stenin, Anomalous behavior of monatomic steps in (1 × 1)⇔(7 × 7) reconstruction at an atomically clean silicon (111) surface, JETP Lett. 47 (1988) 530–532.

[42] S. Kohmoto, A. Ichimiya, Determination of the Si(111) "1 × 1" structure at high temperature by reflection high-energy electron diffraction, Surf. Sci. 223 (1989) 400–412. http://dx.doi.org/10.1016/0039-6028(89)90669-9.

[43] Y.-N. Yang, E.D. Williams, High atom density in the "1 × 1" phase and origin of the metastable reconstructions on Si(111), Phys. Rev. Lett. 72 (1994) 1862–1865. Available from: http://dx.doi.org/10.1103/PhysRevLett.72.1862.

[44] A.V. Latyshev, A.L. Aseev, A.B. Krasilnikov, S.I. Stenin, Initial stages of silicon homoepitaxy studied by in situ reflection electron microscopy, Phys. Stat. Sol. 113 (1989) 421–430. Available from: http://dx.doi.org/10.1002/pssa.2211130220.

[45] A.V. Latyshev, A.B. Krasilnikov, A.L. Aseev, Self-diffusion on Si(111) surfaces, Phys. Rev. B 54 (1996) 2586–2589. Available from: http://dx.doi.org/10.1103/PhysRevB.54.2586.

[46] T. Nishinaga, Handbook of Crystal Growth, Thin Films and Epitaxy, North-Holland, Amsterdam, 1994.

[47] C. Alfonso, J.C. Heyraud, J.J. Métois, About the sublimation of Si surfaces vicinal of {111}, Surf. Sci. 291 (1993) L745–L749. http://dx.doi.org/10.1016/0167-2584(93)90280-V.

[48] D.C. Senft, G. Ehrlich, Long jumps in surface diffusion: one-dimensional migration of isolated adatoms, Phys. Rev. Lett. 74 (1995) 294–297. Available from: http://dx.doi.org/10.1103/PhysRevLett.74.294.

[49] J.A. Venables, Rate equation approaches to thin film nucleation kinetics, Philos. Mag. 27 (1973) 697–738. Available from: http://dx.doi.org/10.1080/14786437308219242.

[50] J.A. Venables, Atomic processes in crystal growth, Surf. Sci. 299-300 (1994) 798–817. http://dx.doi.org/10.1016/0039-6028(94)90698-X.

[51] A.B. Pang, K.L. Man, M.S. Altman, T.J. Stasevich, F. Szalma, T.L. Einstein, Step line tension and step morphological evolution on the Si(111) (1 × 1) surface, Phys. Rev. B 77 (2008) 115424. Available from: http://dx.doi.org/10.1103/physrevb.77.115424.

[52] H. Hibino, C.-W. Hu, T. Ogino, I.S.T. Tsong, Decay kinetics of two-dimensional islands and holes on Si (111) studied by low-energy electron microscopy, Phys. Rev. B 63 (2001) 245402. Available from: http://dx.doi.org/10.1103/PhysRevB.63.245402.

[53] B.J. Gibbons, S. Schaepe, J.P. Pelz, Evidence for diffusion-limited kinetics during electromigration-induced step bunching on Si(111), Surf. Sci. 600 (2006) 2417–2424. http://dx.doi.org/10.1016/j.susc.2006.03.046.

[54] C. Ratsch, P. Šmilauer, A. Zangwill, D.D. Vvedensky, Submonolayer epitaxy without a critical nucleus, Surf. Sci. 329 (1995) L599–L604. http://dx.doi.org/10.1016/0039-6028(95)00353-3.

[55] I.G. Neizvestny, N.L. Shwartz, Z.S. Yanovitskaja, 2D island system formation during initial stage of MBE growth with large critical island size, Phys. Low-Dimens. Struct. 11 (2000) 137–145.

[56] I.G. Neizvestny, N.L. Shwartz, Z.S. Yanovitskaja, Two-dimensional nucleation during epitaxy with large critical nucleus size (translation), Mikroelectronika 31 (2002) 84–92.

[57] C. Ratsch, A. Zangwill, P. Šmilauer, D.D. Vvedensky, Saturation and scaling of epitaxial island densities, Phys. Rev. Lett. 72 (1994) 3194–3197. Available from: http://dx.doi.org/10.1103/PhysRevLett.72.3194.

[58] A. Pimpinelli, J. Villain, D.E. Wolf, J.J. Métois, J.C. Heyraud, I. Elkinani, et al., Equilibrium step dynamics on vicinal surfaces, Surf. Sci. 295 (1993) 143–153. http://dx.doi.org/10.1016/0039-6028(93)90190-U.

[59] D.I. Rogilo, L.I. Fedina, S.S. Kosolobov, B. Ranguelov, A.V. Latyshev, 2D Si island nucleation on the Si (111) surface at initial and late growth stages: on the role of step permeability in pyramidlike growth, J. Cryst. Growth, http://dx.doi.org/10.1016/j.jcrysgro.2016.06.028 in press.

[60] S.A. Teys, A.B. Talochkin, B.Z. Olshanetsky, Formation of Ge nanoislands before the completion of a wetting layer in the Ge/Si(111)system, J. Cryst. Growth 311 (2009) 3898–3903. http://dx.doi.org/10.1016/j.jcrysgro.2009.06.021.

[61] L. Andersohn, T. Berke, U. Köhler, B. Voigtländer, Nucleation behavior in molecular beam and chemical vapor deposition of silicon on Si(111)(7 × 7), J. Vac. Sci. Technol. A 14 (1996) 312–318. http://dx.doi.org/10.1116/1.579894.

[62] B. Voigtländer, M. Kästner, P. Šmilauer, Magic islands in Si/Si(111) homoepitaxy, Phys. Rev. Lett. 81 (1998) 858–861. Available from: http://dx.doi.org/10.1103/PhysRevLett.81.858.

[63] J. Mysliveček, T. Jarolímek, P. Šmilauer, B. Voigtländer, M. Kästner, Magic islands and barriers to attachment: a Si/Si(111)7 × 7 growth model, Phys. Rev. B 60 (1999) 13869–13873. Available from: http://dx.doi.org/10.1103/PhysRevB.60.13869.

[64] D.I. Rogilo, L.I. Fedina, S.S. Kosolobov, B.S. Ranguelov, A.V. Latyshev, Critical terrace width for two-dimensional nucleation during Si growth on Si(111)-(7 × 7) surface, Phys. Rev. Lett. 111 (2013) 36105. Available from: http://dx.doi.org/10.1103/PhysRevLett.111.036105.

[65] D. Kandel, Initial stages of thin film growth in the presence of island-edge barriers, Phys. Rev. Lett. 78 (1997) 499–502. Available from: http://dx.doi.org/10.1103/PhysRevLett.78.499.

[66] W.F. Chung, K. Bromann, M.S. Altman, The transition to step flow growth on the clean and surfactant covered Si(111) Surface studied by in-situ LEEM, Int. J. Mod. Phys. B 16 (2002) 4353–4362. Available from: http://dx.doi.org/10.1142/S0217979202015431.

[67] K.L. Man, A.B. Pang, M.S. Altman, Kinetic length and step permeability on the Si(111) (1×1) surface, Surf. Sci. 601 (2007) 4669–4674. Available from: http://dx.doi.org/10.1016/j.susc.2007.05.039.

[68] B. Ranguelov, M.S. Altman, I. Markov, Critical terrace width for step flow growth: effect of attachment-detachment asymmetry and step permeability, Phys. Rev. B 75 (2007) 245419. Available from: http://dx.doi.org/10.1103/PhysRevB.75.245419.

[69] J.W. Evans, P.A. Thiel, M.C. Bartelt, Morphological evolution during epitaxial thin film growth: formation of 2D islands and 3D mounds, Surf. Sci. Rep. 61 (2006) 1–128. http://dx.doi.org/10.1016/j.surfrep.2005.08.004.

[70] E. Korutcheva, K. Koroutchev, I. Markov, Slope selection of mounds with permeable steps in homoepitaxy, Eur. Phys. J. B 86 (2013) 60. Available from: http://dx.doi.org/10.1140/epjb/e2012-30884-4.

[71] Y.Y. Hervieu, I. Markov, Kinetics of second layer nucleation with permeable steps, Surf. Sci. 628 (2014) 76–81. Available from: http://dx.doi.org/10.1016/j.susc.2014.05.016.

[72] S.S. Kosolobov, A.L. Aseev, A.V. Latyshev, In situ study of the interaction of oxygen with the Si(111) surface by ultrahigh-vacuum reflection electron microscopy, Semiconductors 35 (2001) 1038–1044. Available from: doi:10.1134/1.1403568.

[73] R.E. Schlier, H.E. Farnsworth, Structure and adsorption characteristics of clean surfaces of germanium and silicon, J. Chem. Phys. 30 (1959) 917. Available from: http://dx.doi.org/10.1063/1.1730126.

[74] N. Shimizu, Y. Tanishiro, K. Kobayashi, K. Takayanagi, K. Yagi, Reflection electron microscope study of the initial stages of oxidation of Si (111)-7×7 surfaces, Ultramicroscopy 18 (1985) 453–461.

[75] A.L. Aseev, S.S. Kosolobov, A.V. Latyshev, S.A. Song, A.A. Saranin, A.V. Zotov, et al., In situ REM and ex situ SPM studies of silicon (111) surface, Phys. Stat. Sol. 202 (2005) 2344–2354. Available from: http://dx.doi.org/10.1002/pssa.200521232.

[76] S. Sitnikov, S. Kosolobov, A. Latyshev, Attachment-detachment limited kinetics on ultra-flat Si(111) surface under etching with molecular oxygen at elevated temperatures, Surf. Sci. 633 (2015) L1–L5. http://dx.doi.org/10.1016/j.susc.2014.12.004.

[77] J.B. Hannon, M.C. Bartelt, N.C. Bartelt, G.L. Kellogg, Etching of the Si(001) surface with molecular oxygen, Phys. Rev. Lett. 81 (1998) 4676–4679. Available from: http://dx.doi.org/10.1103/PhysRevLett.81.4676.

[78] A. Pimpinelli, J.-J. Métois, Macrovacancy nucleation on evaporating Si(001), Phys. Rev. Lett. 72 (1994) 3566–3569. Available from: http://dx.doi.org/10.1103/PhysRevLett.72.3566.

[79] C. Misbah, O. Pierre-Louis, A. Pimpinelli, Advacancy-induced step bunching on vicinal surfaces, Phys. Rev. B 51 (1995) 17283–17286. Available from: http://dx.doi.org/10.1103/PhysRevB.51.17283.

[80] A.V. Latyshev, A.B. Krasilnikov, A.L. Aseev, In situ reflection electron microscope observation of two-dimensional nucleation on Si(111) during epitaxial growth, Thin Solid Films 281–282 (1996) 20–23. http://dx.doi.org/10.1016/0040-6090(96)08564-1.

[81] A.B. Krasilnikov, A.V. Latyshev, A.L. Aseev, Transformation of a Si(111) stepped surface during the formation of a Si(3×1)Ca superstructure, Surf. Sci. 290 (1993) 232–238. http://dx.doi.org/10.1016/0039-6028(93)90706-P.

[82] A.V. Latyshev, A.B. Krasilnikov, A.L. Aseev, Direct REM observation of structural processes on clean silicon surfaces during sublimation, phase transition and epitaxy, Appl. Surf. Sci. 60–61 (1992) 397–404. http://dx.doi.org/10.1016/0169-4332(92)90450-C.

[83] A.B. Krasilnikov, A.V. Latyshev, A.L. Seev, S.I. Stenin, Monoatomic step clustering during superstructural transitions on Si (111) surface, J. Cryst. Growth 116 (1992) 178–184.

[84] A.V. Latyshev, A.B. Krasilnikov, A.L. Aseev, UHV reflection electron microscopy investigation of the monoatomic steps on the silicon (111) surface at homo-and heteroepitaxial growth, Thin Solid Films 306 (1997) 205–213.

[85] S.S. Kosolobov, S.A. Song, E.E. Rodyakina, A.V. Latyshev, Initial stages of gold adsorption on silicon stepped surface at elevated temperatures, Semiconductors 41 (2007) 448–452. Available from: http://dx.doi.org/10.1134/S1063782607040173.

[86] S.S. Kosolobov, S.A. Song, L.I. Fedina, A.K. Gutakovskii, A.V. Latyshev, Instability of the distribution of atomic steps on Si(111) upon submonolayer gold adsorption at high temperatures, JETP Lett. 81 (2005) 117–121. Available from: doi:10.1134/1.1898002.

[87] A.V. Latyshev, H. Minoda, Y. Tanishiro, K. Yagi, Adatom effective charge in morphology evolution on Si (111) surface, Appl. Surf. Sci. 130 (1998) 60–66. http://dx.doi.org/10.1016/S0169-4332(98)00025-7.

[88] K. Aoki, H. Minoda, Y. Tanishiro, K. Yagi, REM studies of adsorption-induced phase transitions and faceting in the Si(111)-Au system, Surf. Rev. Lett. 05 (1998) 653–663. Available from: http://dx.doi.org/10.1142/S0218625X98001018.

9

Atomic Structure of Semiconductor Low-Dimensional Heterosystems

A.K. Gutakovskii[1], A.V. Latyshev[1,2], A.L. Aseev[1]

[1]RZHANOV INSTITUTE OF SEMICONDUCTOR PHYSICS OF THE SIBERIAN BRANCH OF RUSSIAN ACADEMY OF SCIENCES, NOVOSIBIRSK, RUSSIA [2]NOVOSIBIRSK STATE UNIVERSITY, NOVOSIBIRSK, RUSSIA

A qualitatively new approach, based on the use of size effects in so-called quantum heterosystems (HS) or low-dimensional semiconductor systems, has been actively developed in recent years to advance semiconductor properties. Application of quantum HS allows one to control the fundamental semiconductor parameters such as band gap width, effective mass and charge-carrier mobility, energy spectrum of electron subsystem, refraction index, etc., and enables the creation of materials for nanoelectronics with unique electronic and optical characteristics not typical of a bulk material.

The realization of all HS quantum advantages is possible only in the case when a material being used has quite high structural perfection. Therefore, establishing of the HS formation mechanisms, an interface morphology, and the main defect-formation regularities in them, are important scientific–technological tasks, the solution of which implies, first of all, knowing the real structure of a material. The greatest progress has been made in transmission electron microscopy, using electrons with an energy in the range of 200–400 keV, among methods for investigating the bulk structure of crystal objects. In recent decades, with progress in device-manufacturing technology, and emergence of high-resolution electron microscopes with a resolution better than 0.1 nm, the potential of this method has considerably grown—both in the field of observation of materials' atomic structure, and in the field of electron microscope image processing and interpretation. The method of transmission high-resolution electron microscopy (HREM) and the new techniques integrated with it, such as scanning transmission electron microscopy (STEM), the filtration method of nonelastic-scattered electrons, diffraction in convergent-beam, Z-contrast method, and methods of local chemical analysis, are nowhighly demanded. Constitutiveparts of HREM investigations are the methods of computer simulation and digital processing of HREM images. The use of these methods not only provides unambiguity and correct interpretation of experimental images of an object's atomic structure, but also allows quantitative maps for distribution of deformations and lattice distortions over the area of an object under study to be obtained.

The results of using HREM to study the real structure of low-dimensional heterosystems at the atomic level, obtained in ISP SB RAS in recent decades, are the main content of the present chapter. A brief description of the analytical HREM method, is considered in the first part of the chapter.

Advances in Semiconductor Nanostructures. DOI: http://dx.doi.org/10.1016/B978-0-12-810512-2.00009-3

9.1 Analytical HREM

Quite detailed descriptions of the peculiarities of analytical HREM are presented in a number of monographs and reviews (see, e.g., Refs. [1–3]). Modern transmission-type microscopes can be equipped with lens aberration correctors of the electron-optical system, electron beam monochromators, energy filters, and detectors for operation in the scanning transmission and local chemical analysis modes. The main possible modes of a modern high-resolution analytical electron microscope and their designations are the following:

- High-resolution transmission electron microscopy (HREM);
- Scanning transmission electron microscopy (STEM);
- Electron energy loss spectroscopy (EELS);
- Energy dispersive X-ray analysis (EDXA);
- Energy-filtered transmission electron microscopy (EFTEM).

HREM and STEM are the main modes for imaging the atomic structure of an object under study. In the first case, an object is illuminated with a parallel coherent electron beam (planewave) which scatters (diffracts) on the atom potential of an object, forming secondary electron waves. The HREM image is the result of such wave's interference in the image plane of a microscope lens, and it reflects the atomic potential projection on the image plane. In the second case, the object is illuminated with an acute-focused coherent electron beam, which is scanned from point-to-point over the sample surface. The diameter of the focused beam on the sample surface determines the microscopic resolution in this regime, and may be less than 0.1 nm. The transmitted electron waves are registered either by bright-field (BF) or dark-field (DF), or high-angle annular dark-field (HAADF) detectors. The STEM image, as well as the HREM image of a very thin crystal, show the atom potential projection of an object on an image plane.

The main characteristic of the HREM method is the spatial resolution value determined by the position of the first zero of transfer contrast function (TCF) in the microscope's optical system under Scherzer optimal lens defocus (Δf_{sh}) [3]. This value is also called resolution on "point to point" and is determined by the following expression:

$$\delta_p = 0.66 C_S^{1/4} \lambda^{3/4}, \tag{9.1}$$

where λ, is the electron wavelength; and C_S is the lenses spherical aberration coefficient. The optimal defocus value is determined by the expression:

$$\Delta f_{sh} = -(4/3 C_S \lambda)^{1/2}. \tag{9.2}$$

The TCF for the Scherzer defocus is given in Fig. 9.1A. It is important to note that, in this case, the spatial frequencies carrying the information about the studied object with a value

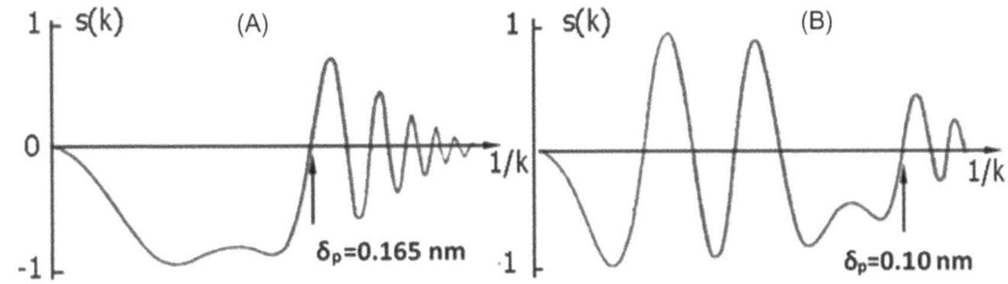

FIGURE 9.1 Trasfer function of contrast $s(k)$ of a high-resolution electron microscope at accelerating voltage 400 kV and $C_S = 0.85$ mm for $\Delta f = -42.5$ (A); and -114 nm (B); k = spatial frequency.

to $1/\delta_p$, are transmitted with minimal phase changes; therefore, simple image interpretation is possible.

There is also an information limit of resolution δ_l (resolution "on lines"). Its value is determined by the smallest image peculiarities which can be revealed by the given electron-optical system. However, in this case, the possibility of unambiguous image interpretation is missing, and a comparison of experimental images with theoretical ones is required. Fig. 9.1B shows the TCF for the defocus value $\Delta f = -114$ nm, at which resolution of "lines" at 0.11 nm should be expected.

Line resolution is determined by objective lens chromatic aberration and it also depends on the stability of the microscope's electromagnetic parameters (accelerating voltage, currents of lenses, deviating coils, etc.), which affect defocus fluctuation (ΔF). Defocus fluctuation in the first approximation is determined as:

$$\Delta F = C_c \Delta E / E, \tag{9.3}$$

where C_c is the objective lens chromatic aberration coefficient, and $\Delta E/E$ is the microscope electromagnetic parameter instability. The line resolution (the information limit of microscope resolution) is determined as:

$$\delta_i = (\pi \lambda \Delta F / 2)^{1/2}. \tag{9.4}$$

As it follows from Eq. (9.1), the main factor which determines electron microscope resolution in the HREM mode is the objective lens spherical aberration. In the STEM regime, the microscope resolution is determined by the diameter of a focused beam on a sample surface. As the beam focuses by a condenser system of lenses, it is by this system that spherical aberration will mainly influence resolution.

To eradicate lens spherical aberration, modern microscopes, e.g., TITAN 80-300(FEI), JEM-2200FS (JEOL), are furnished with spherical aberration correctors, which enable the setting of any C_S value, including that equal to zero, which allows an appreciable increase in resolution. In this case, point and line resolutions coincide and are determined by Eq. (9.4).

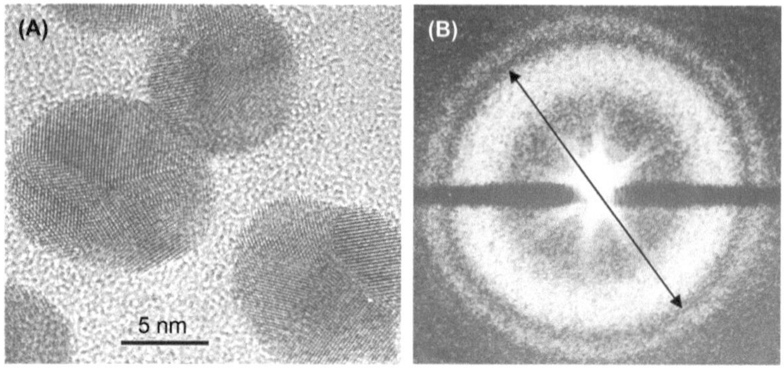

FIGURE 9.2 Experimental HREM image of gold crystal particles on amorphous germanium film (A); and the corresponding diffractogram from the amorphous film site (B). *Arrows* indicate the transmission band marked on TFC (Fig. 9.1A) from 0 to $1/\delta_p$. ($\delta_p = 0,16$ nm).

As follows from Eq. (9.3), reaching a high-resolution value requires stable conditions for the microscope's electronic device operation, including measures to reduce vibrations, temperature fluctuations, and external electromagnetic fields.

We carried out experiments to determine the resolution of electron microscopes JEM-4000EX (JEOL) and TITAN 80-300(FEI) installed in ISP SB RAS, Novosibirsk.

The JEM-4000EX microscope is supplied with a pole-piece of an ultra high resolution objective lens UHP-40H with passport spherical and chromatic aberration values of 1.0 and 1.6 mm, respectively. The experimental determination of spherical aberration coefficient, point to point and line resolutions was carried out according to the methods of [2], and resulted in values of 0.85 mm, 0.16, and 0.10 nm, respectively (see Figs. 9.1 and 9.2).

The TITAN 80-300(FEI) microscope is equipped with an objective lens spherical aberration corrector and a monochromator. As mentioned earlier, the theoretical value of instrumental resolution for such devices is determined by Eq. (9.4). The real instrumental resolution measurements in the HREM mode were carried out using the Young fringes method. A slight shift in the lighting beam is made during one HREM image frame registration. It is equivalent to the lighting of an object from two closely located coherent point sources. The following Fourier transformation of this image allows Young pole imaging. The radius of the reciprocal space region, where Young poles are clearly resolved, determines the maximal spatial frequency resolvable in the HREM image, i.e., microscope instrumental resolution δ (Fig. 9.3). In our case, $\delta = \delta_p = \delta_l = 0.08$ nm.

Determination of the instrumental resolution in the STEM mode was carried out with monocrystalline gold and silicon objects. Typical STEM images of these objects are illustrated in Fig. 9.4.

A very important addition to HREM and/or STEM data on atomic material structure is information about the chemical composition of an object under study. This information, in certain cases, can be obtained from the contrast peculiarities in HREM and/or STEM images.

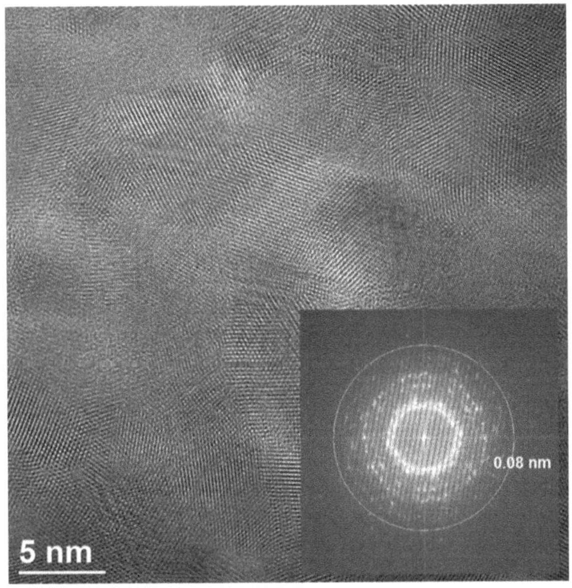

FIGURE 9.3 Experimental HREM image of crystalline gold particles and the corresponding Fourier spectrum with Young fringes. The diameter of the reciprocal space region, where Young fringes are clearly resolved, corresponds to the spatial frequency, resolvable in the HREM image, of 1/0.08 nm.

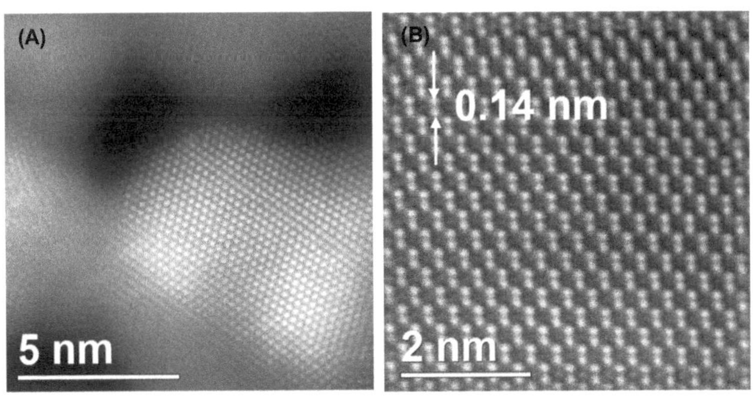

FIGURE 9.4 HAADF STEM images of the crystalline Au(110) (A) and Si(110) (B). TITAN 80-300, 300 kV.

But, in most cases, the solution of these tasks requires specialized methods of local chemical analysis. There are two basic methods applied in carrying out a chemical analysis, and those integrated in transmission-type electron microscopes are EDAX and EELS.

Energy-dispersive X-ray analysis (EDAX) is based on the registration of characteristic X-ray radiation (in the energy range to 40 keV) excited by an incident electron beam, and it

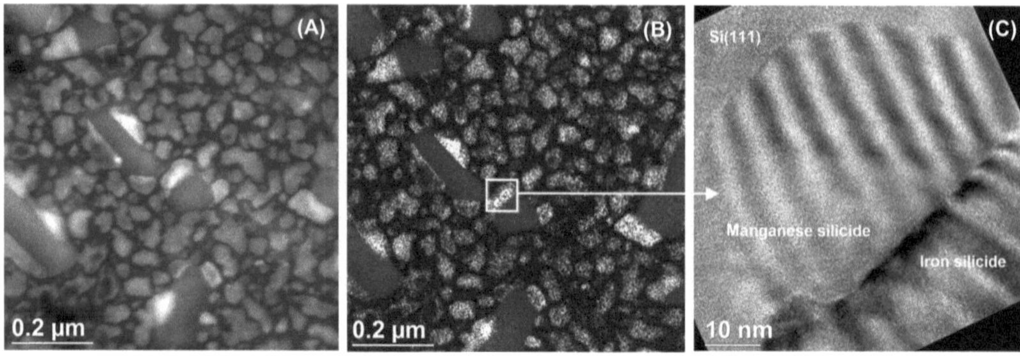

FIGURE 9.5 EFTEM (A and B) and HREM (C) images of iron silicide crystals (*red* (dark gray in print versions) in Fig. 9.5B), and of manganese (*green* (light gray in print versions) in Fig. 9.5B) on Si(111): (A) Image in elastic-scattered electrons (zero loss of image) without energy loss; (B) Composite image of Fe, Mn, and Si distribution maps obtained in nonelastic-scattered electrons with losses 708 eV (L-edge Fe), 640 eV (L-edge Mn), and 99.2 eV (M-edge Si); (C) HREM image of the nanocrystal conjugation border from the site marked in the *white square* in Fig. 9.5B.

allows a local chemical analysis to be carried out from carbon to uranium; however, it is more convenient for studying chemical elements with high atomic numbers.

Electron energy loss spectroscopy (EELS) is based on the analysis of the energy spectrum of nonelastic-scattered electrons that pass through a sample, and it enables a qualitative and quantitative local chemical analysis to be carried out on elements with small atomic numbers: e.g., boron, carbon, nitrogen, oxygen.

Modern microscopes allow a simultaneous use of STEM, EDAX, and EELS regimes to obtain maps of chemical element distributions over the area of an object under study (chemical mapping). Herein, spatial resolution can reach parts of a nanometer.

One more method for chemical analysis and mapping is obtaining *HREM images with energy filtration: EFTEM* (energy-filtered transmission electron microscopy). This method is based on the use of precision energy filters built into an electron microscope [4,5]. The precision of energy filtration over energy in such filters is approximately 1 eV. By positioning the filter energy gap in different EELS spectral regions, it is possible to form an image in nonelastic-scattered electrons on any chemical element, as they are part of elements of the object under study. Fig. 9.5 is an example of the combined use of HREM and EFTEM modes to study the structural–chemical peculiarities of the crystals obtained by iron and manganese deposition in an ultrahigh vacuum on the Si(111) surface. The experiments in deposition were undertaken in IACP FEB RAS, Vladivostok, the structural–chemical analysis was carried out by electron microscope TITAN 80-300 in ISP SB RAS, Novosibirsk.

It can be clearly seen in Fig. 9.5 that the iron silicide nanocrystal shape is anisotropic: its longitudinal size is 100–300 nm, and transverse size is 50–150 nm. The shape of manganese silicide nanocrystals is isotropic, and they are evenly distributed over the silicon surface. Their characteristic size varies from 20 to 100 nm.

The importance of image simulation is dependent on the specific peculiarities of the HREM method. The dynamic character of electron scattering in a sample, and the presence of wave aberrations in an electron microscope optical system, lead to a distortion of diffracted electron wave phases and amplitudes. Therefore, in a general case, electron–microscopic contrast is intricately connected with sample structure, so the interferential picture of the atomic planes observed in HREM does not always correctly reflect the atomic structure of an object, and image simulation is inevitably involved to interpret it. The coincidence of calculated and experimental contrast allows one to evidence of the correspondence of the suggested model to a real structure. Visual comparison of calculation and experiment is believed to be common, due to insufficient development of quantitative comparison of experimental and calculated images. The multislice method, first proposed by Cowley and Moodie in 1957, is based on the principles of physical optics [3], and is analyzed in detail by the authors of [6]; it has become the most widely used method for HREM image simulation. In our studies we used the "MUSLI" program package developed by A.L. Chuvilin in the Institute of Catalysis, SB RASB [7], and then it was tested and adjusted in ISP SB RAS.

9.2 Automodulation of Chemical Composition in $Cd_xHg_{1-x}Te$ Films

Automodulation of the chemical composition in solid solutions of $Cd_xHg_{1\text{-}x}Te$ grown by MBE was first found by the authors of [8]. It can be seen in Fig. 9.6 that bands of different contrast are observed over the whole solid solution $Cd_xHg_{1\text{-}x}Te$ layer of constant composition. The distance between the bands of different contrasts changes from 20 nm in the lower region of constant composition (at a distance about 1 μm from the border with a CdTe layer), to 60 nm near the heterosystem surface. Bands of different contrasts are also present in the film region with a variable composition; however, the distance between them is considerably smaller and is equal to 5 nm near the heteroborder CdHgTe/CdTe. The angle between the bands and the growth surface (301) is within 20–40 degrees, increasing toward the film surface. In the insert to Fig. 9.6B there is an HREM image of the film structure, in which there is a clearly seen alternation of the layers differing in contrast, several interplanar distances thick, lying in plane (100). Bright layers are of an order of 3–7 interplanar distances (100) thick. Different layer contrasts may be conditioned by the difference in their chemical composition. HREM image simulation was carried out to check this assumption and to determine the depth of composition modulation. HREM simulation with a following comparison of calculated and experimental data is the only method for determining the depth of composition modulation in the neighboring layers of such small thickness in the absence of local methods of chemical analysis.

The models of $Cd_xHg_{1-x}Te$ film crystalline structure with a 5–10% difference in the composition of "x" were built to determine the modulation depth.

HREM image calculations were made for standard operation parameters of electron microscope JEOL-4000EX: accelerating voltage $U = 400$ keV, and spherical aberration

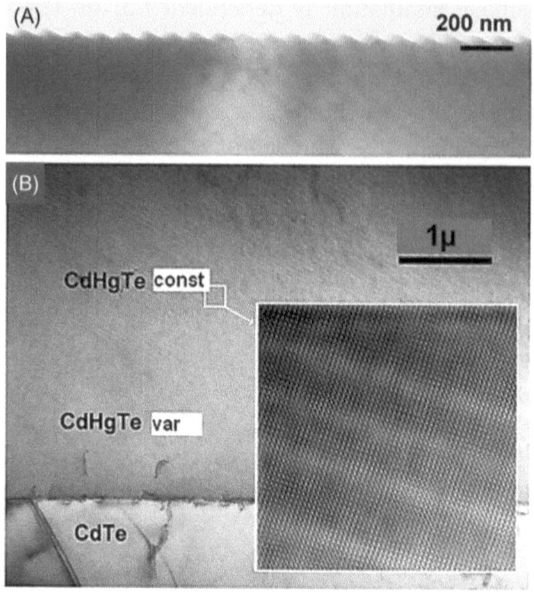

FIGURE 9.6 HREM images of cross-section (001) for heterosystem $Cd_{0.22}Hg_{0.78}Te$-$Cd_xHg_{1-x}Te$-CdTe-ZnTe-GaAs (301): (A) Image of the near-surface $Cd_{0.22}Hg_{0.78}Te$ layer region; (B) Image of the region containing part of the CdTe layer, variable composition layer $Cd_xHg_{1-x}Te$(var), and part of the constant composition layer $Cd_{0.22}Hg_{0.78}Te$ (const).

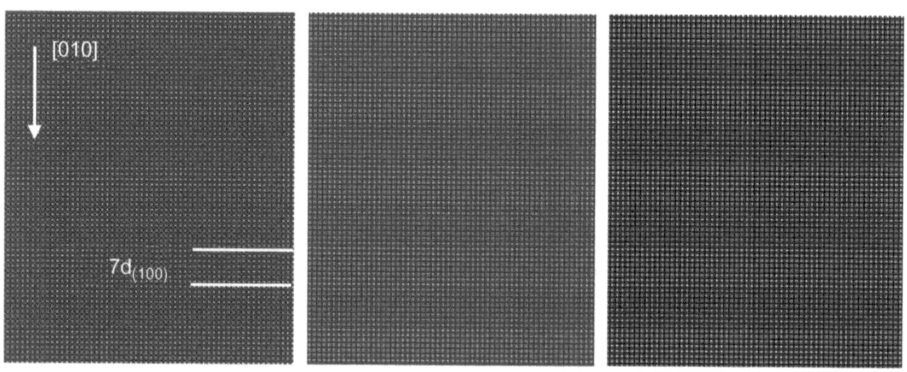

FIGURE 9.7 Theoretical HREM cross-section (001) image for the heterosystem of alternating $Cd_{(0.22+x)}Hg_{(0.78-x)}Te$ layers. Left – right "x" = 0.05, 0.07, 0.10.

coefficient $C_s = 1$ mm. The analysis and comparison of experimental and calculated images on the relative intensity of the neighboring layers with the composition variation showed that the modulation depth of the composition in these systems is ~7% (Fig. 9.7).

The obtained result agrees well with electrophysical measurements. After the modulated film was annealed for several hours at $T = 220-240°C$, the modulation of the composition disappeared in the whole film bulk. Alignment of the composition after annealing is confirmed by a change in the photoconductivity spectrum. It was shown in Ref. [8] that, after annealing, the long wave photoconductivity edge is shifted to the short wave region at 8 μm. Evaluations showed that, for such a shift, the compositions in the neighboring layers should have values of $x = 0.195$ and $x = 0.270$ after growth, and the same composition with $x = 0.233$ after annealing, i.e., the depth of composition modulation before film annealing is 7.5%, this agrees well with the HREM image simulation results.

9.3 Nanocrystal Visibility Limits in an Amorphous Matrix

The analysis of nanocrystalline cluster contrast in an amorphous matrix in HREM images is required, first, to get correct information about their shape and dimensions and, second, to determine their optimal imaging conditions (accelerating voltage, defocus, etc.). Simulation of HREM images includes creation of models for crystalline cluster atomic structure in an amorphous matrix, i.e., a mathematical description of all model atom coordinates and the image's proper calculation for these models. Some theoretical CdS nanocrystal HREM images in the amorphous matrix and results of their analysis are given below as an example.

In Fig. 9.8 there are calculated models of quasi-spherical CdS nanocrystals with a cubic crystalline lattice type on the amorphous substrate and corresponding theoretical HREM images for two substrate thicknesses.

FIGURE 9.8 Models of quasi-spherical CdS nanocrystals with a cubic crystalline lattice type on the amorphous substrate (left), and the corresponding theoretical HREM images for different substrate thicknesses. Crystal mid-size is 1.7 nm, amorphous matrix thickness 3 nm (middle), and 12 nm (right). Orientation of surface nanocrystals relatively to the electron beam corresponds to (100) (upper row) and (111) (lower row) habit planes.

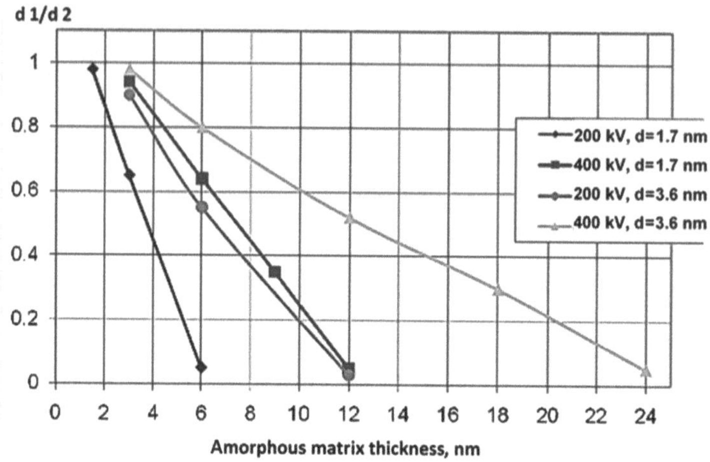

FIGURE 9.9 Dependences of the ratio of CdS nanocrystal size, measured in the theoretical HREM image (d1), to the real size (d2), set by the model established in calculations, on the amorphous matrix thickness for various accelerating voltages.

It can be seen from the example presented that, first, nanocrystal imaging at HREM investigations is possible in cases when the ratio of crystal mid-size and the amorphous matrix thickness exceeds some critical value. Second, nanocrystal sizes in HREM images are smaller than their real size. Using HREM image simulation for the system under study, it is possible to determine this critical value, and also to evaluate the error in geometrical size measurements of the analyzed nanocrystals on an amorphous substrate. The results of the analysis of theoretical HREM images for CdS nanocrystals on the carbon substrate are presented in Fig. 9.9. In this figure, there are dependences of nanocrystal size in relation to the theoretical HREM image to the real size (simulation-set size) on amorphous matrix thickness.

It follows from the analysis of these dependences that the error in determining the size grows with an increase in amorphous matrix thickness and a decrease of accelerating voltage value at other equal conditions. The crystallite sizes established in experimental HREM images are smaller than those of the simulated sizes. For instance, at an amorphous matrix thickness of the samples under study ~30–60 nm, the error in determining the nanocrystal size in the HREM image is ~20%.

9.4 The Geometrical Phase Method for Quantitative Analysis of Crystalline Lattice Deformations

As was pointed out in the section 9.1, the dynamic character of electron scattering in the sample and the presence of wave aberrations in the electron microscope optical system lead to distortion of diffracted electron waves and amplitudes, which considerably complicate HREM image interpretation in terms of atom locations, and require the involvement of theoretical image simulation. At the same time, neither dynamic scattering nor lens aberrations affect the vectors of scattered waves (Bragg scattering angles), which carry information about interplanar distances in a crystal. This information can be obtained directly from HREM images interpreted without involving simulation. Absolute precision of interplanar distance measurements with such a method is not high, due to systematic errors and the necessity for numerous calibrations. However, relative precision may reach 10^{-3}–10^{-4}, and it is limited, mainly, by the quality of sample preparation. Therefore, the HREM method is ideal for measuring local variations in lattice parameters, in particular, for building local stress fields with a resolution better than 1 nm.

There are two groups of methods for measuring local lattice distortions on HREM images. The so-called "peak-finding" methods, based on the analysis of positions for contrast maxima and minima in the image [9,10], belong to the first group. A shift of extremums positions (bright and dark spots in the lattice image) relative to the reference lattice are interpreted in terms of deformation.

The methods using "geometrical phase analysis" [11] belong to the second group. These methods operate in the reciprocal space and analyze a shift of the spatial frequency phase in relation to the reference frequency in the HREM image of an ideal crystal, expressing lattice distortions through it.

The geometrical phase method (GPM) is very sensitive to the smallest disturbances of crystal lattice periodicity and, therefore, it is used for imaging local disturbance sites of material structure, e.g., emergence of dislocations.

The classical GPM was created to analyze crystal lattice distortions in a single-component system. As an image of a crystalline object is periodic and is described by a set of spatial frequencies, it is possible to apply Fourier transformation to it and to obtain a Fourier transform of this image, i.e., the whole spectrum of spatial frequencies in the reciprocal space. Analyzing a spatial frequency phase shift, in other words, the region near a maximum of interest in the Fourier spectrum, it is possible to express crystal lattice distortions through it. This is the gist of the GPM.

The mathematical base of the GPM is considered in detail in Refs. [12,13] and, herein, there are only main equations on which the HREM image digital processing mechanism is based for building phase images and obtaining quantitative information about crystalline lattice dilatations and deformations. The contrast in an HREM image is periodic and can be presented as a Fourier series:

$$I(\mathbf{r}) = \sum_{\mathbf{g}} H_{\mathbf{g}} \exp(2\pi i \mathbf{g} * \mathbf{r}), \tag{9.5}$$

where $I(\mathbf{r})$ is image intensity in point $\mathbf{r}$, $\mathbf{g}$ are spatial frequencies corresponding to Bragg reflexes, and $H_{\mathbf{g}}$ are Fourier coefficients.

Taking into account the fact that symmetrical Fourier components from the real function (image) are complex-conjugated, the contrast for each separate system of **g** planes can be written as:

$$I_{\mathbf{g}}(\mathbf{r}) = 2A_{\mathbf{g}}(\mathbf{r})\cos(2\pi\mathbf{g} * \mathbf{r} + P_{\mathbf{g}}(\mathbf{r})), \tag{9.6}$$

where $A_{\mathbf{g}}$ is the amplitude of sinusoidal periodic contrast corresponding to the system of **g** planes, and $P_{\mathbf{g}}$ determines a shift of this sinusoid related to image borders.

Function $A_{\mathbf{g}}(\mathbf{r})$ in the reciprocal space is presented by intensity distribution of the corresponding reflex of **g**, and can be obtained by inverse Fourier transformation of the region around the corresponding reflex:

$$A_{\mathbf{g}}(\mathbf{r}) = A_{\mathbf{g}}(\mathbf{k}) \otimes \mathrm{M}(\mathbf{k} - \mathbf{g}). \tag{9.7}$$

Here **k** is the current coordinate in the reciprocal space, and M(**k**-**g**) is the mask in the reciprocal space in the vicinity of the chosen inverse lattice vector **g**. The mask size does not exceed the Brillouin zone, and the form can be different: Gaussian Lorentzian, etc. For an ideal crystal, when there are no **g**-variations on the coordinate, this mask is delta-function $\delta(\mathbf{k}\text{-}\mathbf{g})$.

Phase shift field $P_{\mathbf{g}}(\mathbf{r})$ can be interpreted variously [7]: either as a field of shifts ($\Delta\mathbf{r}(\mathbf{r})$):

$$P_{\mathbf{g}}(\mathbf{r}) = \mathrm{P}_0 - 2\pi\mathbf{g} * \Delta\mathbf{r}(\mathbf{r}), \tag{9.8}$$

or as a local variation of lattice parameters ($\Delta\mathbf{g}(\mathbf{r})$):

$$P_{\mathbf{g}}(\mathbf{r}) = P_0 + 2\pi\Delta\mathbf{g}(\mathbf{r}) * \mathbf{r}. \tag{9.9}$$

Having the phase shift field and neglecting the constant term P_0, it is possible to determine the shift field for the given system of planes directly:

$$\Delta\mathbf{r}(\mathbf{r}) = -P_{\mathbf{g}}(\mathbf{r})/2\pi\mathbf{g} \tag{9.10}$$

For small variations of **g**, the gradient from the phase shift field will result in the field of variations (**g**) in the reciprocal space, which will reflect the field of interplanar distance variations in the direct space, i.e., in the HREM image:

$$\Delta\mathbf{g}(\mathbf{r}) = \nabla P_{\mathbf{g}}(\mathbf{r})/2\pi. \tag{9.11}$$

A sequence of operations for crystalline lattice distortion imaging is presented in Fig. 9.10 with an example of HREM cross-section digital image processing for heterosystem GaN-Al_2O_3(0001) toward $[11\bar{2}0]$. Each point intensity in Fig. 9.10D corresponds to this or that

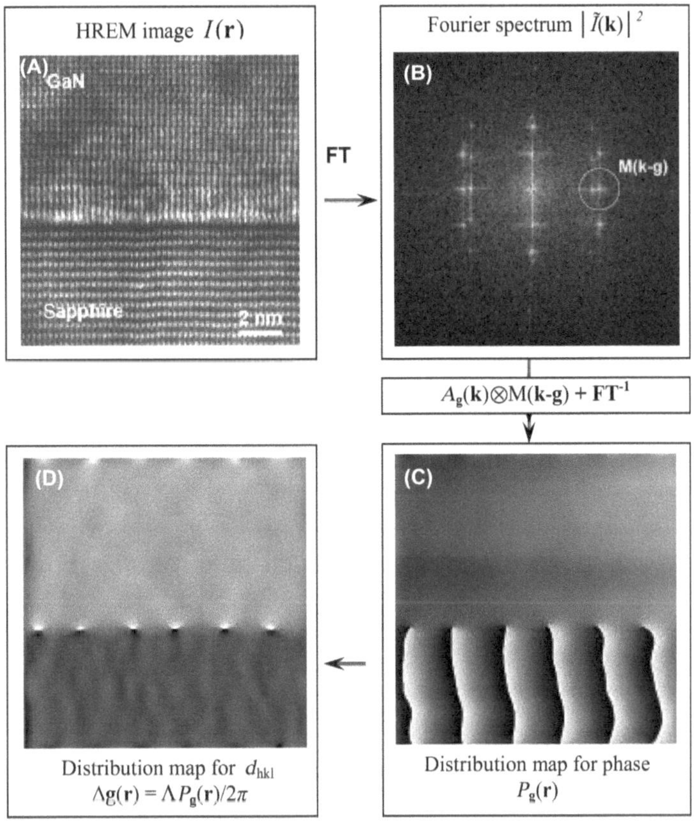

FIGURE 9.10 (A–D) Sequence of HREM image digital processing for crystal lattice distortions imaging close to the misfit dislocation cores.

interplanar distance value (**d**) for planes [$\bar{1}$100]. It is seen that (**d**) has its constant value in the film region equal to ~0.276 nm, and it decreases in a hop-like manner to ~0.242 nm at the heteroborder. These values are exactly equal to the corresponding interplanar distances in bulk crystals GaN and Al_2O_3, i.e., there are no elastic deformations either in the film or in the substrate.

A characteristic feature of the image in Fig. 9.10D is a specific contrast as black-and-white "sockets" periodically located along the interface. Such contrast is conditioned by the presence of crystalline lattice compression and extension regions. The lattice parameter decreases in the compression region, and this region looks darker. The lattice parameter increases in the extension region, and this region looks lighter. Such kinds of distortions correspond to crystal deformations in the vicinity of misfit dislocation (MD) core. On the symmetrical character of distortions of the atomic planes around the MD cores, one can assert that these are purely edge dislocations with Burgers vectors ($\bar{\mathbf{b}}$) $a/2$ [$\bar{1}$100] lying in the

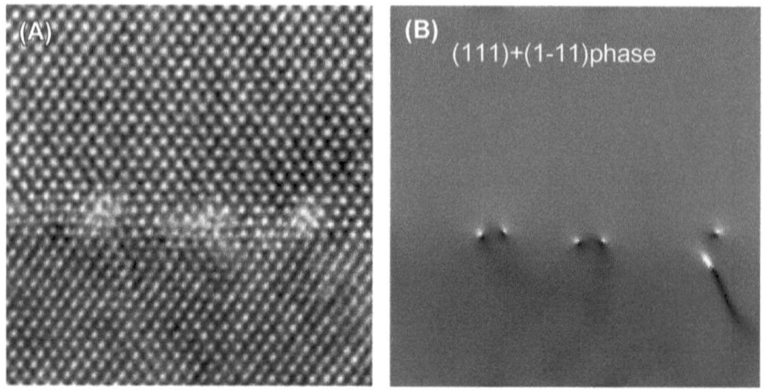

FIGURE 9.11 Misfit dislocations in the interface InSb–GaAs(001): (A) Experimental HREM image; (B) Total map of $\mathbf{d}_{111}$ and $\mathbf{d}_{1\text{-}11}$ distributions.

interface plane. The period of MD sequence determined in experimental images is ~2.1 nm. These dislocations are a thermodynamically equilibrium heterostructural element, which provide misfit accommodation of film crystalline lattices and the substrate. Considering that, for heterosystem GaN-Al_2O_3(0001) $\overline{\mathbf{b}} = 0.276$ nm and lattice parameters misfit (**f**) is 13%, it is possible to determine the period of the MD (**L**) sequence as $\mathbf{L} = \overline{\mathbf{b}}/\mathbf{f} = 2.123$ nm. The coincidence of experimental and theoretical parameters of value **L** indicates full misfit accommodation for film lattice parameters and the substrate, i.e., full relaxation of the heteroepitaxial stress in this system.

In Fig. 9.11 there is an example of using GPM for analyzing the structure of InSb-GaAs (001) interface, for which $\mathbf{f} = 13\%$ and the average distance among MDs is about seven interplanar spacing in the interface. It can be clearly seen in Fig. 9.2B, which is two superimposed maps of interplanar spacing distributions for planes (111) and $(1\bar{1}1)$, that MD cores are splitted. Each MD core consists of two closely located 60-degree dislocations with Burgers vectors $a/2\ [0\bar{1}1]$ and $a/2\ [10\bar{1}]$ tilted toward the interface plane. It is worth noticing that the total Burgers vector of these dislocations is $a/2$ [1–10] and lies in the interface plane, i.e., each pair of 60-degree dislocations is equivalent to one purely edge Lomer dislocation. Such a situation is indicative of the mechanism of heteroepitaxial stress relaxation in this system, which is realized by correlated gliding of two different 60-degree dislocations from the growing surface along the inclined crossing planes {111}. Such a mechanism of heterosystem plastic deformation was first observed by us for system InAs-GaAs(001) [14], and was theoretically grounded in the contributions by Kvam [15] and Gosling [16].

The authors of Ref. [17] developed a new algorithm for HREM image digital processing with GPM for the imaging and quantitative analysis of ultra-small ($\sim 10^{-4}$) crystalline lattice dilatations and deformations in the objects under study. This algorithm (generalized GPM) was first extended into multicomponent systems (superlattices, etc.) and tested in detail assisted by special test models. These models included all data of a real object, such as thickness; presence of heteroepitaxial stresses; crystal lattice temperature

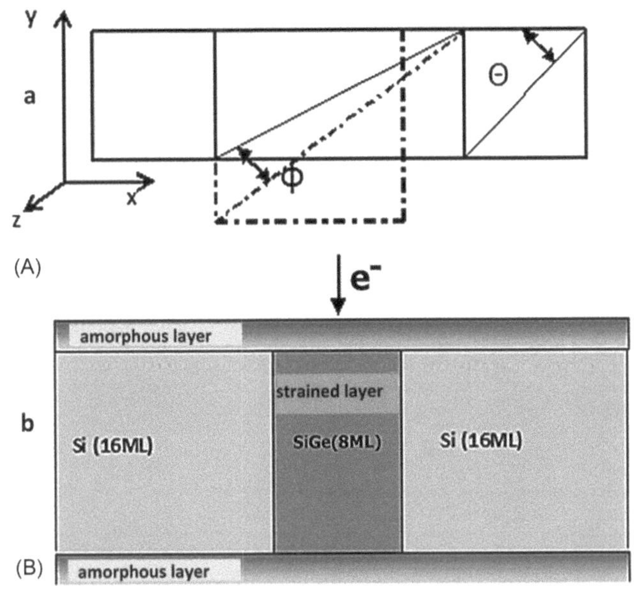

FIGURE 9.12 Schematic views of the tetragonally deformed crystal lattice (A), and the test model of strained heterostructure Si-GeSi-Si (B); $\phi = (\sin^2\Theta_{hkl})/2 \times (1+\nu)/(1-\nu) \times f$.

oscillations (DW: Debye–Waller factor); presence of a surface amorphous or disordered layer, which is frequently formed during object preparation for transmission electron microscopy; presence of stress relaxations on thin foil free surfaces; etc. The test models of stressed systems were a pseudomorphous elastically-deformed Ge_xSi_{1-x} film, 8 monolayers thick, coherently conjugated, over the upper and lower heterointerfaces, with a nondeformed Si matrix, 32 monolayers thick (Fig. 9.12). Such a model corresponds to real stressed superlattices, which are used in many device structures. The Ge concentration was set as 0.1% and 5% in the solid solution. The test model elementary cell dimensions were $21.5 \times 21.7 \times 0.384$ nm, and the number of cell atoms was 10,752. Program "MUSLI" was used for HREM image simulation. Value DW = 0.01 corresponds to minimal atomic oscillations in the crystal (ideal crystal lattice). In the real sample, DW ≈ 0.46.

After the simulation, digital processing of theoretical HREM images was made by GPM using a commercial Digital Micrograph program package and the scripts we specially developed. Unlike the commonly used algorithm, which employs only one averaged reference grid, our algorithm allows a preliminary identification of two reference grids, corresponding to the nondeformed Si and GeSi layers, and, then, calculation of the absolute values of all deformation tensor components for each of the layers that form the heterosystem.

The distribution maps of deformations ε_{xx} and ε_{yy} and the profiles of deformation distribution, registered top-down, are shown in Fig. 9.13. According to these maps, layer

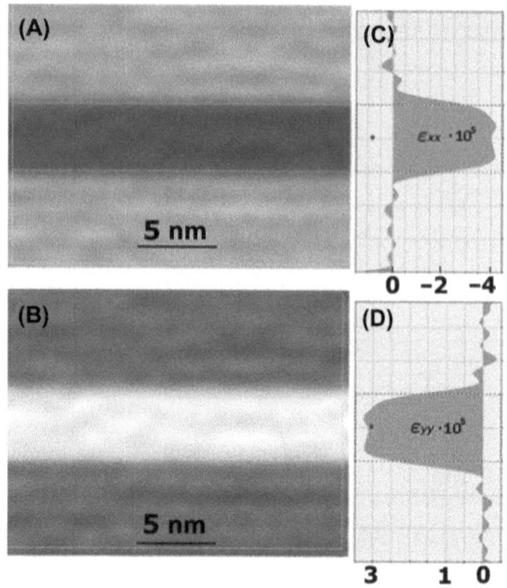

FIGURE 9.13 Distribution maps of deformations ε_{xx} and ε_{yy} for strained heterosystem Si-$Ge_{0.001}Si_{0.999}$-Si (A and B) corresponding to the HREM image in Fig. 9.6A. Right (C and D) are the corresponding distribution profiles ε_{xx} and ε_{yy}.

$Ge_{0.001}Si_{0.999}$ is tetragonally distorted, i.e., compressed in the horizontal direction at the value of 4×10^{-5}, and extended in the vertical direction at 3×10^{-5}. There are no deformations in Si layers.

In the real sample (DW $\approx$0.46, "frozen phonons" model), the lattice distortions visibility limit depends on the "ideality" of the thin foil under study and, first of all, on the disordered surface layer thickness. The influence of scattering on phonons and the amorphous surface layer was studied with the example of Si-$Ge_{0.05}Si_{0.95}$-Si. The results of HREM image simulation and processing for models with different amorphous surface layer thicknesses are given in Fig. 9.14. It was shown that considering the scattering on phonons leads to an inconsiderable weakening of contrast on the map of deformation distribution, and practically does not affect the interface blurring (compare Fig. 9.14A and B). At the same time, the presence of the amorphous surface layer can very strongly weaken the contrast in phase images (Fig. 9.14D) and considerably complicate the imaging and quantitative measurements of lattice distortions and deformations in strained systems. Because of this, the thinner the foil under study is, the stronger the influence of the amorphous layer is. At the equal ratio of amorphous and crystalline layer thicknesses (equal to, in our case, 1/5), the smaller the crystalline phase thickness (compare Fig. 9.14C and D) is, the weaker the contrast on deformation distribution maps is.

Samples with a thickness of about 10 nm toward the electron beam are needed for HREM investigations. It is known that the deformation field near a free surface is different from the deformation field in the bulk, due to partial stress relaxation on a free surface: the thinner an object is, the stronger the difference in the deformation field near a free surface is. The

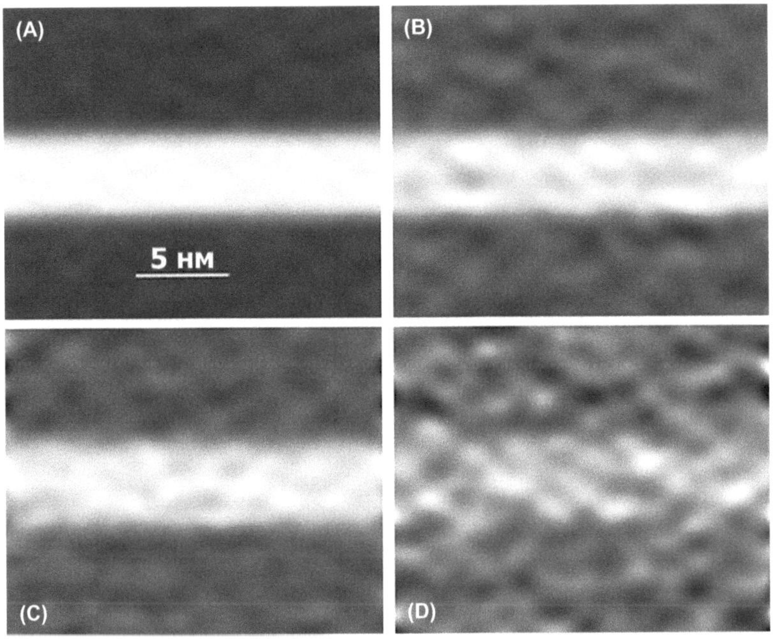

FIGURE 9.14 Distribution maps of deformation ε_{yy} for strained heterosystem Si-$Ge_{0.05}Si_{0.95}$-Si: (A) Ideal crystal; (B–D) Considering the scattering on phonons and the amorphous surface layer. The amorphous surface layer thickness was 1 nm (D) and 2 nm (C), the thickness of the crystalline part of the object was 5 nm (D) and 10 nm (A–C).

influence of surface relaxation was studied in thin foil HREM image contrast and lattice distortions distribution on test models of stressed systems using GPM.

The structure of the test model resembled a pseudomorphous tetragonally deformed Ge film (8 monolayers thick), coherently conjugated with a nondeformed Si substrate (20 monolayers thick). The parameter misfit of the conjugated lattices was $\mathbf{f} = 0.038$. The test model elementary cell dimensions were $\sim 13 \times 8 \times 6$ nm, and the number of atoms was 28,560 (cell thickness toward electron beam incidence was 6 nm, i.e., the thickness of the foil under study assisted with HREM).

The method of molecular mechanics, realized in the commercial HyperChem program package (version 7.5) was used for our model relaxation. The following boundary conditions were used: all atoms are fixed (fixed position) on all lateral surfaces, and atoms have broken bonds on the lower and upper surfaces. The atomic model of the stressed system before and after relaxation is shown in Fig. 9.15.

The theoretical HREM images (Fig. 9.16) were then calculated for the electron microscope operating at an accelerating voltage of 250 kV ($C_s = 1.2$ mm, beam angular divergence: 0.5 mrad, focus blurring: 50 nm, diameter of objective aperture: 6.0 nm^{-1}). Digital processing of theoretical HREM images was carried out at the final stage with GPM using the special scripts we developed for the commercial Digital Micrograph program package.

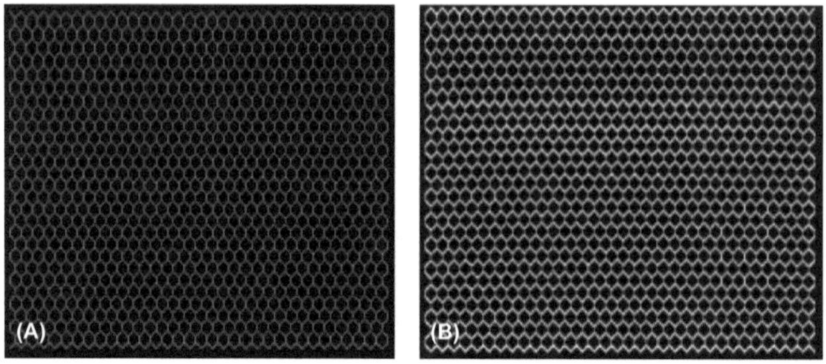

FIGURE 9.15 Atomic model of strained system Ge × Si with **f** = 0.038 before (A), and after (B), surface relaxation with the method of molecular dynamics.

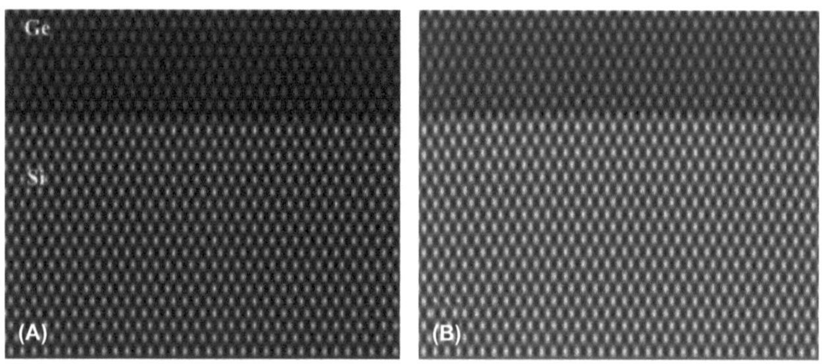

FIGURE 9.16 The simulated HREM images of strained Ge-Si system before (A) and after (B) surface relaxation.

The aperture diameter in the reciprocal space near inverse lattice vector **g** (111) was 5 pics. The profiles of interplanar distance changes (**d**) were registered in a top-down direction in the corresponding images. The phase image parameters and the corresponding profiles are shown in Fig. 9.17.

The analysis of phase pictures showed that the HREM image contrast and lattice distortion distribution in the strained system are modified by the surface relaxation effect. Due to this effect, the interplanar spacing for planes {111} changes from 0.32 to 0.317 nm in the Ge film, and from 0.314 to 0.3145 nm in the Si substrate. It is seen that surface relaxation distorts the monotonous character of interplanar spacing dependences on the thickness of the strained structure in phase HREM images, and leads to a seeming blurring of the interfaces. It may lead to incorrect extrapolation of the results of HREM analysis over the initial (massive) samples, where the surface relaxation effect is considerably weaker. Nevertheless, at optimal observation conditions, the average deformation value of

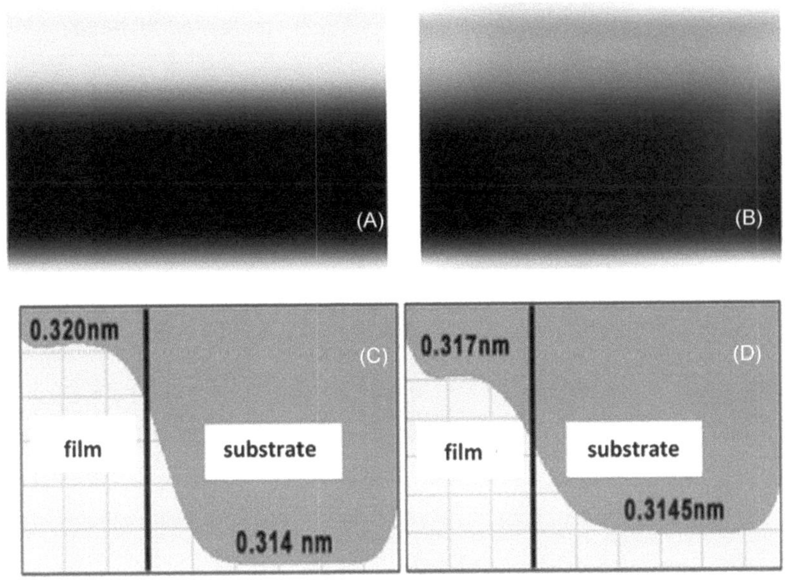

FIGURE 9.17 Phase images and the corresponding profiles of a change of value "**d**" for (111) planes: (A and C) Initial model; (B and D) Relaxed model.

each layer of the heterosystem is determined by GPM with a precision not worse than hundredths of parts of a percentage.

These conclusions are confirmed by the use of GPM for the HREM image processing of heterosystem GeSi × Si. The HREM cross-section (110) image of heterosystem Si-$Ge_{0.20}Si_{0.80}$-Si (A) and the distribution map corresponding to it (B), and the distribution profile (C) of interplanar spacing (111) over the heterosystem's thickness are presented in Fig. 9.18. It is seen in the phase image and the intensity profile, corresponding to it, that both interfaces (lower and upper) seem blurred at the distance of an order of 5 nm, and the character of a change of interplanar spacing $\mathbf{d}_{111}$ is evidently not monotonous. However, the average value d_{111} of the solid solution thickness, equal to 0.341 nm, corresponds to a tetragonally deformed crystal lattice $Ge_{0.20}Si_{0.80}$ with an accuracy of 0.01%. Thus, surface relaxation causes the seeming interface blurring, and decreases quantitative measurement precision on HREM images.

Summing up the results of testing the generalized GPM, one can say that this method is a powerful instrument for imaging and quantitative measurement of crystal lattice dilatations and deformations in HREM investigations. It is shown that the presence of an amorphous surface or disordered layer and surface stress relaxation on free thin foil surfaces, analyzed in the microscope, are the main factors that limit imaging sensibility and decrease quantitative measurement precision of lattice distortions on HREM images at approximately two orders of magnitude.

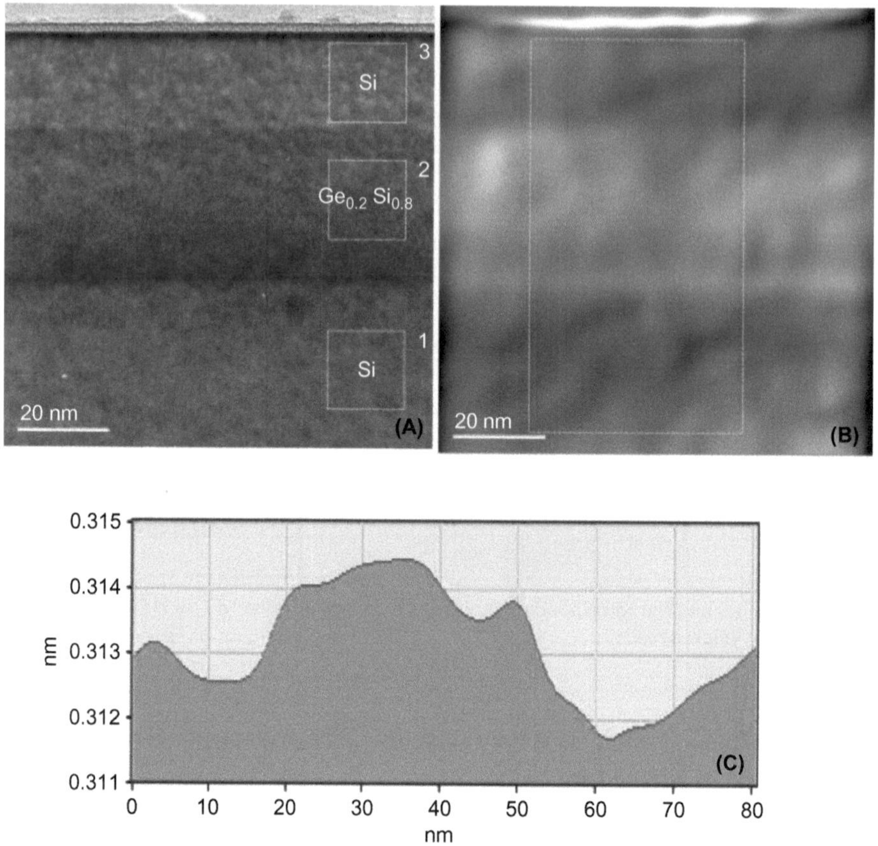

FIGURE 9.18 Experimental HREM image of cross-section (110) of heterosystem Si-$Ge_{0.20}Si_{0.80}$-Si (A) and the distribution map (B), corresponding to it, and the distribution profile of interplanar spacing (111) (C) over the thickness of the heterosystem. The profile is imaged top-down.

9.5 Atomic Structures of Multicomponent Systems

9.5.1 Strained (InGaAsSb)-(GaSb) and (InAs)-(GaSb) Superlattices

The peculiarities of the energy system in strained short-period superlattices (SSPSL) (InAs)-(GaSb) lead to a considerable suppression of Auger recombination velocity, and an increase in minority carrier lifetimes. The consequence of this phenomenon is an increase in photodetectivity temperature limit based on SSPSL, compared to photodetectors (PDs) based on CdHgTe solid solutions. The key role in SSPSL optical properties realization occurs in GaSb/InAs heteroborders. The composition and thickness of transition layers, bordering the interface of materials determine residual mechanical stresses in the grown SSPSL, and should be such that they could partially compensate the stresses occurring in SSPSL due to the incoincidence of GaSb and InAs lattices. A wide spectrum of photoelectron devices, including two-colored PDs on one crystal, functional in the range of 3–5 μm and 8–12 μm [18], have now been realized on the second-type of GaSb/InAs interface. At the same time, the main factors determining the type of interface: InSb or GaAs, their blurring and roughness, etc., have not been discovered so far. All these parameters depend considerably on a concrete technology of the molecular sources used, etc., and, hence, the task of establishing the influence of synthesis conditions on SSPSL structural–morphological parameters is always topical.

Growth experiments were carried out in ISP SB RAS using modernized installations "Katun-S" and "Shtat" by the V.V. Preobrazhensky group. The installations are equipped with a gate source of arsenic and antimony of an original design. They are also equipped with a fast electron diffractometer, and devices for precision control and measurements of molecular flow density directly in the growth region. Very clean materials (Ga, In, As, Sb) were used for loading molecular sources, and monocrystal wafers of GaSb(001) were used as substrates. Two types of SLs were grown and studied: (InGaAsSb)-(GaSb) (10 periods) and (InAs)-(GaSb) (~300 periods). [19,20].

The investigation of structural–morphological peculiarities of these objects was investigated using the HREM method with microscopes JEM-4000EX (JEOL) and TITAN 80-300 (FEI). The samples for HREM investigations were prepared as cross-sections on planes {110}, normal to the growth surface, by argon ion etching with energy 5 keV (initial stage) and 1 keV (final stage). The ion beam incidence angle was 5–10 degrees; sample rotation velocity was 5 r/min. A HREM cross-section (110) image of heterosystem (InGaAsSb)-(GaSb)-GaSb (001) is shown in Fig. 9.19. It can be seen that all SLs layers are homogeneous in thickness and contrast. There are no structural defects in the bulk of the heteroepitaxial layer or on the interfaces. This fact is indicative of coherent conjugation of the alternating crystal lattice layers. The conjugation interfaces in Fig. 9.19 look sharp and flat-parallel.

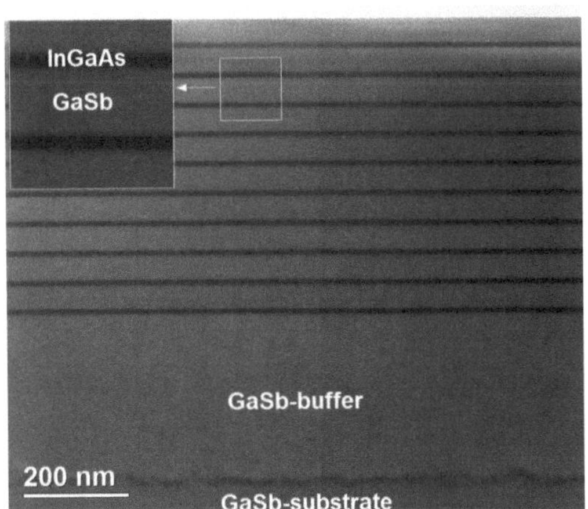

FIGURE 9.19 Bright-field cross-section (110) TEM image of heterosystem $[(InGaAsSb)\text{-}(GaSb)]_{10}$–GaSb(001). JEM-4000EX, 400 keV.

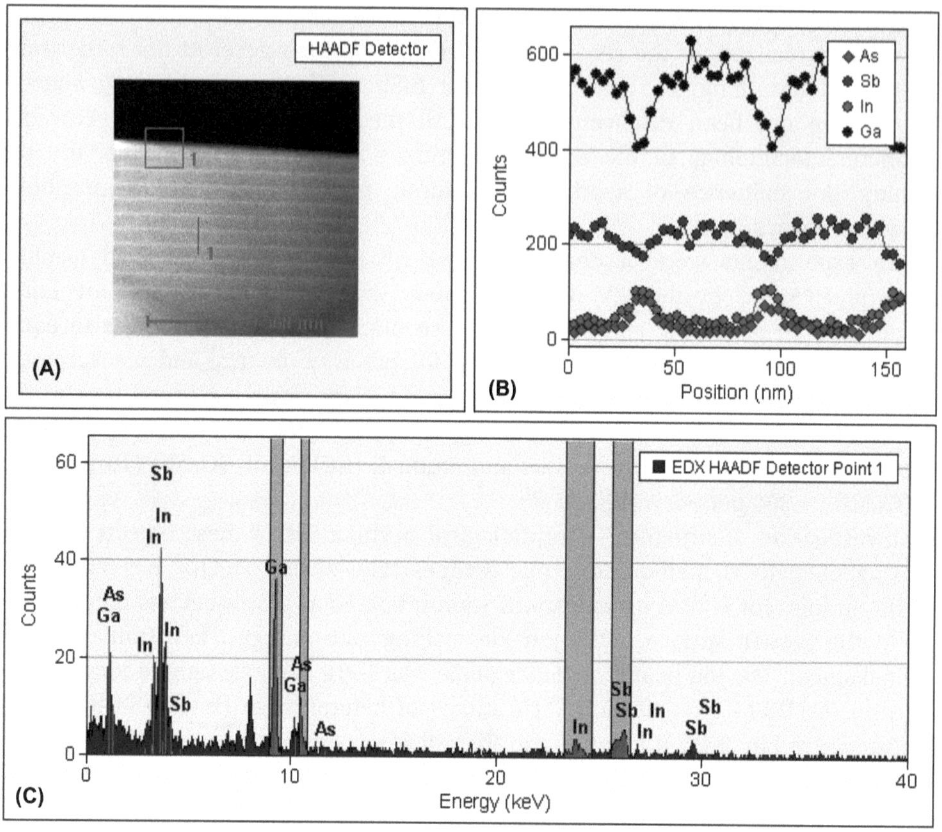

FIGURE 9.20 HAADF cross-section (110) image of heterosystem (InGaAsSb)-$(GaSb)_{10}$-GaSb(001) (A); profiles of In, Ga, As, and Sb distribution along line 1 (B), and 1 spectrum of characteristic X-ray radiation (C) over the area of the square. TITAN 80-300, 300 keV, STEM regime.

However, the analysis of chemical elements distribution carried out over the heterosystem's thickness reveals a considerable intermixing of chemical elements on the conjugation interfaces (Fig. 9.20).

Bright field cross-section (110) TEM images of SSPSL fragments $[(GaSb)_{10}\text{-}(InAs)_6]_{320}$ (A) and $[(GaSb)_{10}\text{-}(InAs)_9]_{290}$ (B) are presented in Fig. 9.21. The electron microdiffraction picture, corresponding to the image of Fig. 9.21A, is shown in Fig. 9.21C. The presence of the strongly periodic structure formed by alternating layers clearly manifests itself in the electron microdiffraction image as satellite reflexes along the growth direction [001] close to the main reflections. This satellite periodicity precisely corresponds to the SSPSL period, which was ~4.8 nm for the sample. The TEM images in Fig. 9.21A and B were obtained under strong two-wave diffraction from planes (200), and their peculiarity is the layering of contrast in the region of InAs layers. These layers appear as three alternating dark-light-dark regions. Under two-wave dynamic diffraction contrast, formed on crystal lattice distortions, two observable crystal structural peculiarities are possible.

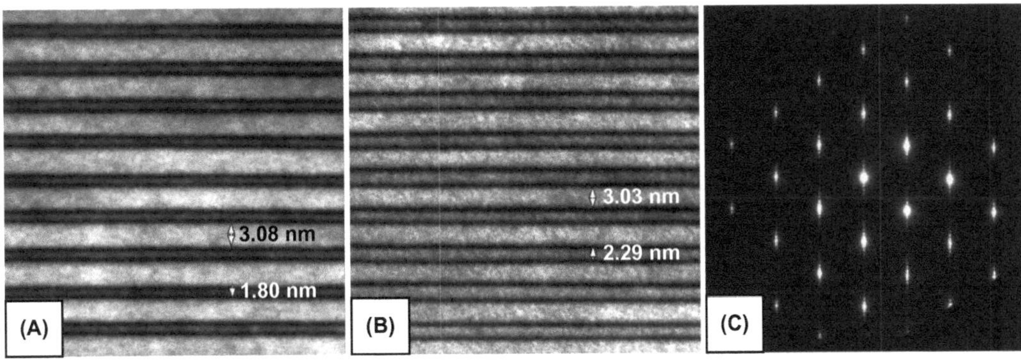

FIGURE 9.21 Bright-field TEM cross-section (110) images of SSPSL fragments $[(GaSb)_{10}\text{-}(InAs)_6]_{320}$ (A), and $[(GaSb)_{10}\text{-}(InAs)_9]_{290}$ (B). (C) is the picture of electron microdiffraction corresponding to (A).

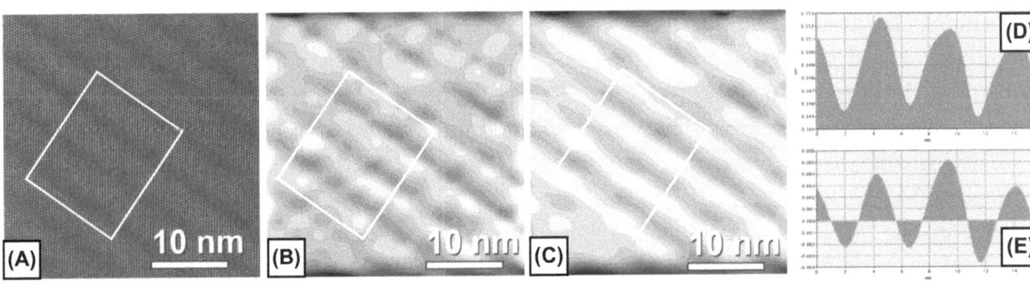

FIGURE 9.22 HREM cross-section (110) image of SSPSL fragment $[(GaSb)_{10}\text{-}(InAs)_6]_{320}$ (A); distribution maps of interplanar spacing $d_{(111)}$ (B); and elastic deformation ε_{xx} (C). The distribution profiles of $d_{(111)}$ and ε_{xx}, imaged along the long side of the *marked rectangles* in (B) and (C), respectively, are presented in (D and E).

First, the presence of elastic deformation occurs due to not a large ($\sim$0.6%) disagreement of the periods of conjugated material crystal lattices. This deformation value is insufficient to break interface coherence (pseudomorphism) and formation of MDs, but it is quite sufficient for the formation of diffraction contrast on elastic-deformed layers. The second structural peculiarity consists of the formation of transition regions between GaSb and InAs layers, in which chemical composition is different from the nominal composition of conjugated layers. These can be GaAs and InSb layers or their triple (fourfold) solid solutions.

The analysis of crystal lattice parameter changes of GaSb and InAs layers alternating over the SSPSL thickness was carried out using digital HREM image processing. In Fig. 9.22A there is a typical experimental HREM cross-section (110) image of SSPSL fragment $[(GaSb)_{10}\text{-}(InAs)_6]_{320}$, and in Fig. 9.22B are the results of this digital image processing with GPM.

The following conclusions were made according to the results obtained.

- Alternating SSPSL layers are coherently coordinated over the interfaces, GaSb layers being elastic-compressed, InAs layers are elastic-extended.

- The sign-variable mechanic stress field (elastic deformations) is formed over all the superlattice thickness. The deformation value ε_{xx} of each layer is about 0.4% for GaSb and 0.2% for InAs. The average SSPSL deformation value toward growth is about 0.2%, and it agrees with the X-ray diffractometry data obtained for these SSPSLs.
- Interplanar spacing $d_{(111)}$ in the transition layers between GaSb and InAs is about 0.3455 nm, which is less than the corresponding value for bulky InAs (0.3500 nm). It can be indicative of GaAs-type interfaces and formation of solid InGaAs solution in these regions.

9.5.2 Film and Interface Structures in Heterosystem GaAs-GaP-Si (001)

Integration of silicon electronics and the $A^{III}B^{V}$-based element base is of interest due to perspectives about various device developments: monolithic electron devices with silicon control circuits, or multicascade solar cells with an active silicon layer, etc. The whole spectrum of MW and optoelectronics based on $A^{III}B^{V}$ compounds and their solutions can be obtained with molecular-beam epitaxy (MBE). The results of our investigation of the atomic structure of interfaces and structural defects in the GaAs-GaP-Si heterosystem grown by MBE on Si substrates deviated from plane (001) at 6 degrees toward [110] [21]. The choice of substrate vicinal orientation allows one to avoid antiphase domain formation in an epitaxial layer.

The investigations were carried out on electron microscope JEOL-4000EX at accelerating voltage 400 kV. A typical Dark field TEM image under "weak beam" conditions (004) is presented in Fig. 9.23A. It can be seen in the figure that the threading dislocation density is maximal near the film-substrate border and, when distant from it, the threading dislocation density decreases at more than two orders of magnitude. The decrease of threading dislocation density is connected with annihilation of dislocations of a different sign during gliding.

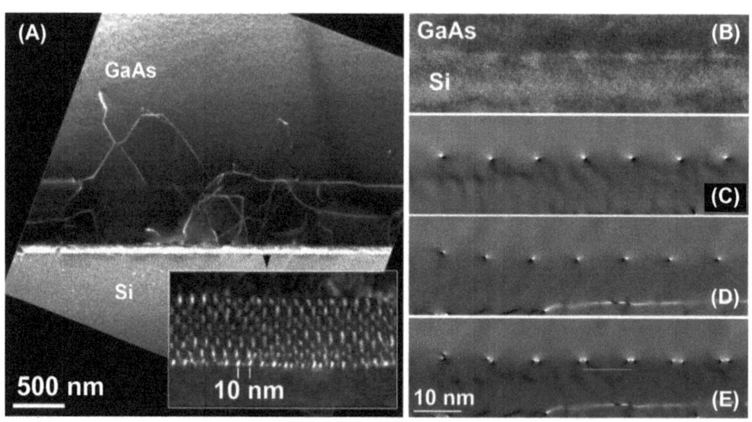

FIGURE 9.23 Dark-field TEM cross-section (110) images of heterosystem GaAs/Si(00) under "weak beam" (2-20) conditions. In the insert to (A) there is the fragment of the interface image obtained by tilting the sample around the axis [1-10] at 18 degrees. (B) shows the experimental HREM image of the GaAs-Si interface and the results of its digital processing (C-E): (C) Distribution map of interplanar spacing (111); (D) Distribution map of interplanar spacing (1-11); (E) Total (C) + (D).

In the insert to Fig. 9.23A there is the DF-TEM image of the GaAs-Si interface obtained by tilting the sample around the [1-10] direction parallel to the interface. The ordered misfit dislocation (MDs) network is clearly seen. The alternation period of MDs is 10 nm, which indicates full relaxation of heteroepitaxial strains in this system. The atomic structure of MDs was studied in the high-resolution mode. An example of the GaAs-Si interface HREM image is shown in Fig. 9.23B. Digital processing of the such image with the GPM allows extraction of quantitative information about the crystal structure of the material: interplanar spacing, deformation of layer lattices, etc. and, also the local sites of strong crystal structure distortions (dislocation cores location). In Fig. 9.23C–E, there are the so-called phase images corresponding to the experimental HREM image in Fig. 9.23B. Figs. 9.23C and D are the phase images for planes (111) and (1-11), respectively; and Fig. 9.23E is the result of summing the previous images. The MDs in these images appear as black-and-white sockets corresponding to crystalline lattice extension and compression regions close to the MD cores. In Fig. 9.23E, the MDs appear as double-sockets typical of split dislocations. During heteroepitaxial strain relaxation 60-degree MDs are formed over adjacent glide planes {111}, forming an atom configuration like that of pure edge Lomer MDs in the interface. We observed the analogous picture earlier in heterosystem Ge-Si (001) [22]. The alternation period of MDs is 10 nm, which corresponds to full strains relaxation exactly by edge MDs.

It was established that the use of the buffer GaP(001) layer on Si and application of cyclic annealing during gallium arsenide deposition allows the growth of 3–5 μm thick GaAs layers with a threading dislocation density less than 1×10^6 cm^{-2} and mean square surface roughness of not more than 1 nm over an area of 100 μm × 100μm on Si substrates. Quantum wells $In_{0.17}Ga_{0.83}As$ grown on such layers are characterized by an intensive peak of photoluminescence at 77K, which is provided by the high structural perfection of the material.

9.6 Iron Disilicide in a Silicon Matrix

At present, during the engineering of silicon-based integrated circuits (IC), developers more and more often come across the problem of the fast operation of metal interconnections, integrating various blocks of complex IC, such as central processors from various computational devices. The solution to the problem is seen in the transition from electric to optical connections. However, there is no efficient light diode compatible with existing VLSI technology. It is connected to the fact that silicon proper, as a nondirect band semiconductor with band gap 1.12 eV, cannot provide efficient light diodes of 1.3–1.5 μm. Therefore, creating such diodes based on silicon planar technology is a topical task which developers try to solve in different ways, including both fabricating thin-film heterostructures and quantum wells based on direct band A_3B_5 semiconductors.

Iron disilicide (β-$FeSi_2$) is a quasidirect band semiconductor with a band gap width of 0.86 eV [23] which, in the case of crystal lattice compression [24], becomes direct. Within the β-$FeSi_2$/Si system, the film-β-$FeSi_2$/Si substrate-like [25] and-Si/β-$FeSi_2$/n-Si-like [26] structures have the best device properties. The first heterostructure allowed creation of an efficient photodetector operating at small inverse shift values and demonstrating a photoresponse of 0.2 A/W at a wavelength of 1300 nm; the second was a light diode with a 1 μm thick β-$FeSi_2$ film with output optical radiation power of 0.4 mW and 1550–1600 nm at room temperature. At the film approach, efficient electroluminescence is observed at quite high pump current density values only , due to high film defectiveness which can be reduced only with long high-temperature annealing (850°C and higher), which leads to the intensive diffusion of Fe atoms [27] and the appearance of high nonradiative recombination centers density. Hence, the film approach to creating silicon-based light diodes is limited by high defectiveness in both analyzed systems, and it cannot be used in integrated performance for building light and photodiodes operating in a wavelength range of 1.3–1.5 μm. The alternative approach, which consists of in-building direct band and quasidirect band semiconductor nanocrystals with a band gap width of c. 0.8 eV (such as β-$FeSi_2$) in the active region of the silicon *p–n* junction, allows one to overcome these limitations.

It was shown in Refs. [28,29] that the formation of β-$FeSi_2$ by both ion implantation and epitaxial methods—such as reactive (RE) and molecular-beam epitaxy (MBE)—is accompanied by the formation of high point and line defect concentrations. Therefore, long time (up to 20 h) high-temperature (to 900°C) annealing was required to decrease the layers' defectiveness. Thermal treatment leads to impurity diffusion, which does not allow the use of silicon irradiated structures in the technology of creating silicon-based integrated microcircuits. On the other hand, at low temperatures, the light emission of the devices obtained with this method decreased several hundreds, even thousands of times at room temperature.

A light diode quantum efficiency increase with β-$FeSi_2$ nanocrystal multilayers built in the silicon matrix can be reached by optimization of growth procedures, the aim of which is to reduce structural defects at nanocrystal in-building and increase nanocrystal bulk concentration in the operation region. An investigation into the peculiarities of β-$FeSi_2$ nanocrystal atomic structure, morphology, and spatial distribution in a crystalline silicon matrix, and also the structure of nanocrystal–matrix interfaces, depending on synthesis conditions, was carried out using HREM to achieve this aim.

Structures with β-$FeSi_2$ nanocrystals (NCs) were grown on substrates cut out of *n*-type conductivity silicon wafers with orientation (001) and (111) in a VARIAN ultrahigh vacuum chamber [30]. To obtain an atomically-clean Si surface, the substrate was subjected to short-time heating (c. 1 s) at 1250°C; the pressure in the chamber was not more than 10^{-9} torr. Layers with nanocrystals were formed using two methods: (1) a combination of reactive and solid-phase epitaxy (RE + SPE) by iron layer deposition (nominal thickness 0.1 – 0.4 nm, temperature 375°C, with a further annealing over 2 min, at 475 °C); (2) SPE method by iron layer deposition (thickness 0.1–0.4 nm at room temperature) with a further annealing over 20 min. Multilayer heterostructures were grown by repeated NCs formation cycles and epitaxial overgrowth by *p*-type conductivity silicon layers at 700°C. It was shown with the slow electron diffraction method that silicon grew epitaxially over the nanocrystals. The obtained structures with nanocrystals were overgrown at 700°C by epitaxial contact layer $p^+ - Si$ with an impurity concentration of 10^{18} cm^{-3} and thickness of 100 nm. The growth experiments were fulfilled by the staff of the Optics and Electrophysics Laboratory, IACP FEB RAS, who

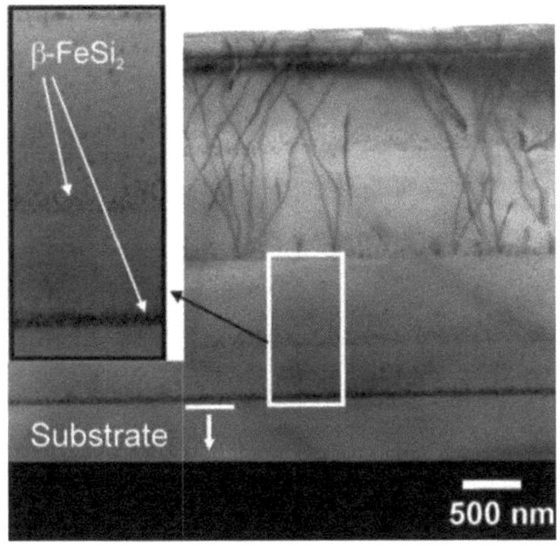

FIGURE 9.24 Bright-field TEM cross-section image of the four-layer heterosystem Si/-NCs- β-$FeSi_2$/Si(001) grown with the combined method.

have wide experience (1966–2015) in carrying out experiments of "silicide – silicon" heterosystem synthesis using MBE, RE, and SPE methods. The investigations of β $FeSi_2$ nanocrystal structure and morphology were carried out with HREM on microscopes JEOL-4000EX and FEI-TITAN 80-300 at accelerating voltages of 400 and 300 kV, respectively [31,32] .

According to the analysis of experimental HREM images, a high density of threading dislocations, nucleating on the biggest nanocrystals (Fig. 9.24), appear at multilayer NCs system formation with the combined method (Fig. 9.25).

Using the SPE method, it is possible to suppress the process of threading dislocation formation. In this case, two nanocrystal types are formed in a silicon matrix: small quasi-spherical NCs (type 1) and large oblong NCs (type 2). The atomic structure of small NCs was first imaged and it was established that they have an orthorhombic crystal lattice (space group CmCa) with interplanar spacing parameters 9.879, 7.799, and 7.839 Å. Beside the diffraction maxima from the Si lattice, clearly imaged are four maxima, corresponding to interplanar spacing parameters 7.72, 2.06, 1.97, and 2.36 Å, in the images obtained by Fourier transform of experimental HREM image.

These values are characteristic for interplanar spacing of planes (010) or (001); (201); (331) and (312) in β-$FeSi_2$, respectively. The angle between planes (010) or (001) β-$FeSi_2$ and Si (111) is equal to 1.5 degrees, which is indicative of their epitaxial coherent conjugation with a small-angled grain boundary formation. The use of Fourier filtration of the HREM experimental image (Fig. 9.26A) allows to separate clear imaging of either the crystal lattice of the Si matrix (Fig. 9.26B), or the crystal lattice of NC-β-$FeSi_2$ (Fig. 9.26C).

The analysis of type 2 NCs reveals interplanar spacing parameters 4.91, 4.80, and 3.89 Å, which coincide with the data for planes (100), (111), and (010) or (001) β-$FeSi_2$, respectively. The angle between β-$FeSi_2$ (100) and Si (110) is 3 degrees, which also confirms, just as for small NCs (type 1), the formation of an epitaxial coherent interface (Fig. 9.27).

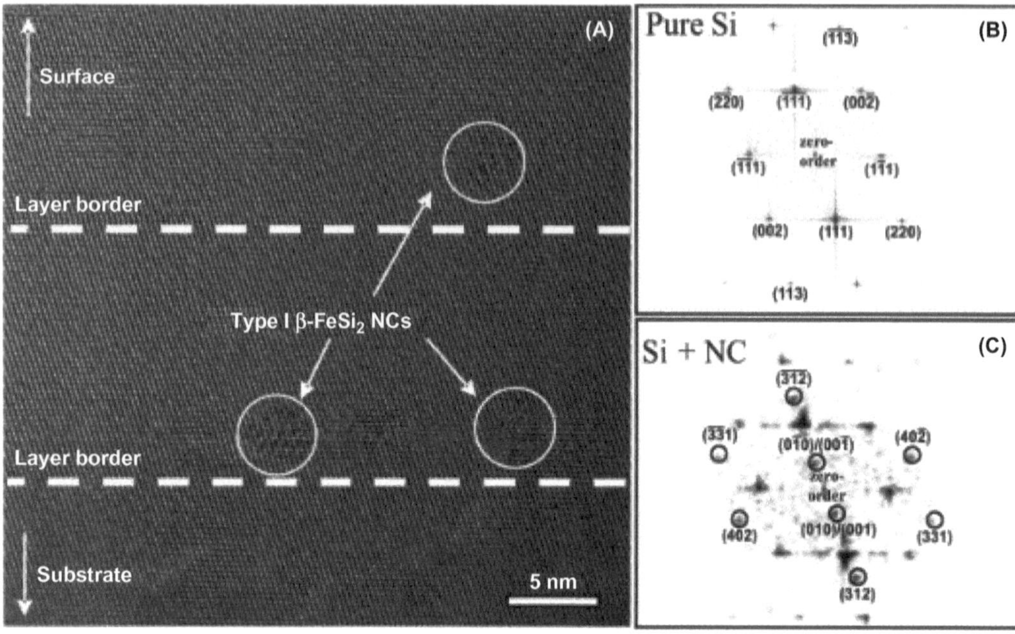

FIGURE 9.25 HREM image of small β-$FeSi_2$ -NCs (type I) in the Si matrix (A), and the corresponding Fourier spectra from the regions of clean silicon (B), and Si + NCs (C). The reflexes from β-$FeSi_2$ -NCs are marked in *black circles*.

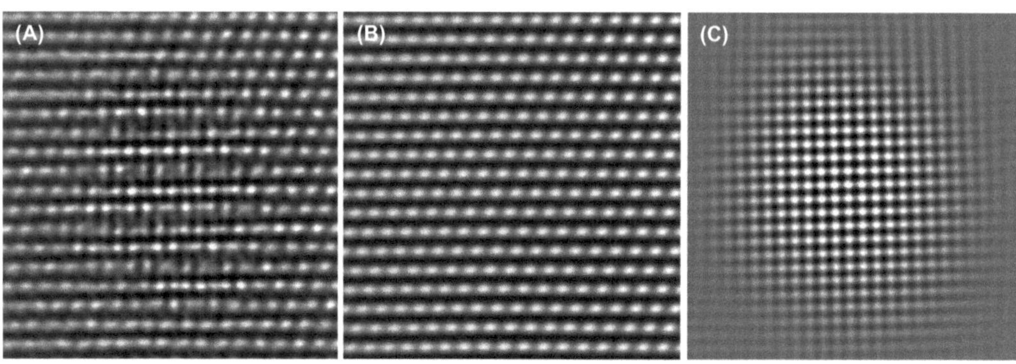

FIGURE 9.26 Experimental (A), and filtered (B and C) HREM images of small β-$FeSi_2$ -NC (type I) in the Si matrix.

It was first found that, when using silicon substrates with surface orientation (111) at multilayer β-$FeSi_2$-Si heterosystems growth in the first layer, counting from the substrate, small crystals (less than 5 nm) extending along the interface (Fig. 9.28A and C) are formed. The bulk of nanocrystals are found in the last layer near the samples' surface (Fig. 9.28A and B), i.e., the effect shown of nanocrystals "floating" toward the growth surface. These NCs have a shape extended toward the growth surface. Their size in this direction depends on the number of iron layers deposited first, and it increases from approximately 20–60 nm for systems

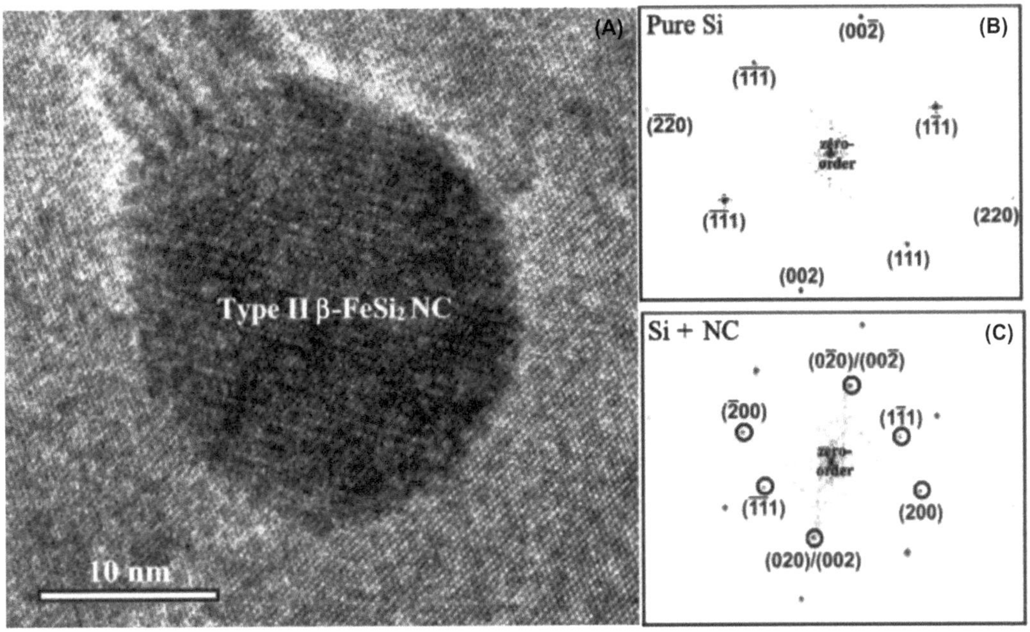

FIGURE 9.27 HREM image of big β-$FeSi_2$ NCs (type II) in the Si matrix (A), and the corresponding Fourier spectra from the regions of clean silicon (B), and Si + NCs (C). Reflexes from β-$FeSi_2$ NCs are marked in *black circles*.

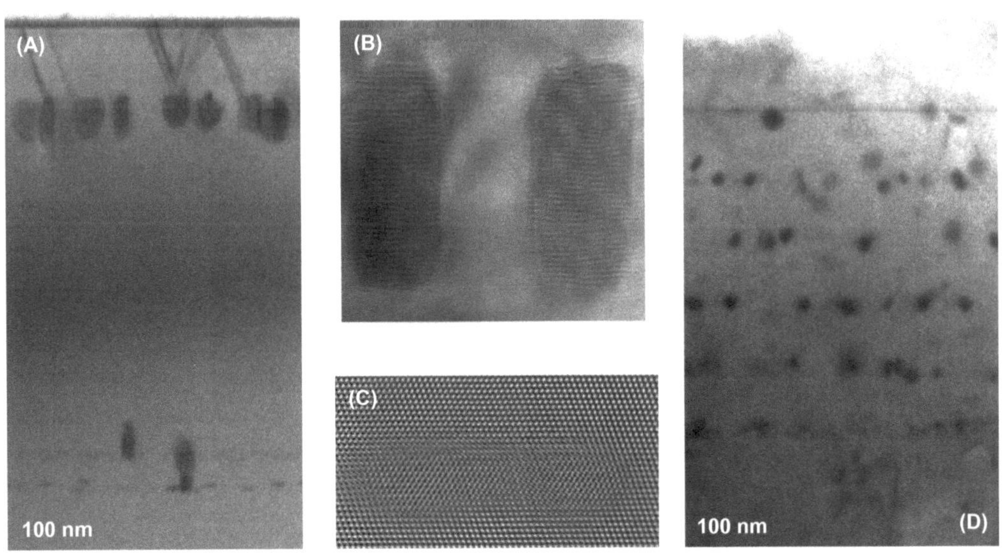

FIGURE 9.28 Electron-microscope cross-section images of the multilayer β-$FeSi_2$-Si heterosystems grown with MBE on substrates Si(111) (A), and Si(001) (D). (B) and (C) HREM image of the β-$FeSi_2$NCs formed in the first and eighth iron layers (counting from the substrate). The thickness of all deposited iron layers was ~0.1 nm. The number of iron layers was eight (A–C) and four (D).

with two and eight iron layers, respectively. The size in the perpendicular direction is c. 10–15 nm. Using silicon substrates with orientation (001), nanocrystals are formed in each layer (Fig. 9.28D), and the NC shape anisotropy expressed is far weaker.

Electroluminescence (EL) investigations in the systems grown with the combined method showed that the EL-line connected with luminescence on dislocations, dominates in the spectra recorded at a temperature of 5K. A temperature increase to 77K leads to full EL quenching. Unlike the structures obtained with the combined method, integrated EL intensity for the structures formed with the SPE method decreases not more than four times with a temperature increase from 7K to 300K. Therein, the strong EL-line, responsive to energy of 0.84 eV, which is connected with luminescence on β-$FeSi_2$ -NCs, is observed in the spectra. It was noticed that EL intensity of the structures grown in similar conditions with the SPE method appreciably increases at the transition from substrates with orientation (110) to substrates of orientation (111), for which β-$FeSi_2$ NC dimensions are considerably higher.

Acknowledgments

The authors gratefully acknowledge partial financial support from RSF (project No. 14-22-00143).

References

[1] L. Reimer, Transmission Electron Microscopy: Springer Series in Optical Science, 36, Spinger-Verlag, 1984.

[2] H. Bakker, A. Bleeker, P. Mul, Design and Performance of an Ultra-High-Resolution 300kV Microscope, Ultramicroscopy 64 (1996) 17–34.

[3] J.M. Cowley, Diffraction physics, Third Edition, Elsevier, Amsterdam—Lausanne—New York—Oxford—Shannon—Tokyo, 1995, p. 477pages

[4] R.F. Egerton, A revised expression for signal/noise ratio in EELS, Ultramicroscopy 9 (1982) 387.

[5] L. Krivanek, A.J. Gubbens, N. Dellby, C.E. Meyer, Design and first applications of a post-column imaging filter, Microsc. Microanal. Microstruct 3 (1992) 187.

[6] P. Goodman, A.F. Moodie, Acta Crystalogr A30 (1974) 280.

[7] A. Chuvilin, U. Kaiser, On the peculiarities of CBED pattern formation revealed by multislice simulation, Ultramicroscopy 104 (2005) 73–82.

[8] I.V. Sabinina, A.K. Gutakovskii, Y.G. Sidorov, and A.V. Latyshev, Automodulation of the chemical composition in epitaxial CdxHg1-xTe(301) films grown by MBE JETP Letters (in Russian) 2011, v.94, No4, 348–352.

[9] O. Brandt, K. Ploog, R. Bierwolf, M. Hohenstein, Phys. Rev. Lett 68 (1992) 1339.

[10] S. Paciornik, R. Kilaas, U. Dahmen, Ultramicroscopy 56 (1994) 94.

[11] M.J. Hytch, M. Gandais, Phil.Mag A 72 (1995) 619.M. Takeda, J.J. Suzuki, Opt.Soc.Am A 13 (1996) 1495. M.J. Hytch, Microsc. Microanal. Microstruct 8 (1997) 41.G. Ade, R. Lauer, Ultramicroscopy 77 (1999) 177–185.

[12] S. Kret, P. Ruterana, A. Rosenauer, D. Gerthsen, Extracting quantitative information from high resolution electron microscopy, Phys.Stat. Sol. (B) 227 (2001) 247.

[13] M.J. Hytch, E. Snoeck, R. Kilaas, Quantitative measurement of displacement and strain fields from HREM micrographs, Ultramicroscopy 74 (1998) 131.

[14] V.Y. Karasev, N.A. Kiselev, E.V. Orlova, M.A. Gribelyuk, A.K. Gutakovsky, Y.O. Kanter, et al., HREM of epitaxial layers in the InAs/GaAs system, Ultramicroscopy 35 (1) (1991) 11.

[15] E.P. Kvam, D.M. Maher, C.J. Humpreys, Variation of dislocation morphology with strain in GexSi1 − x epilayers on (100)Si, J. Mater. Res 5 (1990) 1900.

[16] T.J. Gosling, Mechanism for the formation of 90° dislocations in high-mismatch (100) semiconductor strained-layer systems, J. Appl. Phys. 74 (1993) 5415.

[17] A.K. Gutakovsky, A.L. Chuvilin, Se Ahn Song, Application of high-resolution electron microscopy for imaging and quantitative analysis of deformation fields in heterosystems, Bull. Russ. Acad. Sci. Phys. 71 (10) (2007) 1464−1470 (in Russian); ISSN 1062-8738, Bull. Russ. Acad. Sci. Phys., 2007, Vol. 71, No. 10, pp. 1426−1432.

[18] R. Robert, W. Martin, S. Johannes, R. Frank, F. Joachim, S. Ralf, et al. Status of mid-infrared superlattice technology in Germany, in: Proc. of SPIE, 2009, 7222, pp. 72220T-1−72220T-13.

[19] A.V. Vasev, E.A. Emelyanov, D.F. Feklin, M.A. Putyato, B.R. Semyagin, V.V. Preobrazhenskii, The exchanging mechanisms in anion sublattice during formation of InAs/GaSb heterointerfaces by MBE (Thesis of E-MRS 2014 FALL MEETING) (15-19 September, 2014, Warsaw, Poland), p. 12.

[20] E.A. Emelyyanov, A.V. Vasev, B.R. Semyagin, A.P. Vasilenko, A.A. Komanov, A.K. Gutakovskii, et al., InAsSb on GaAs (001): influence of the arsenic molecules form on composition and crystalline properties of MBE layers, J. Phys. Conf. Ser. 643 (2015) 012006 (6).

[21] E.A. Emelyanov, D.F. Feklin, A.V. Vasev, M.A. Putyato, B.R. Semyagin, A.K. Gutakovskii, et al., Heteroepitaxy of $A^{III}B^{V}$ films on vicinal substartes Si(001), Autometry 50 (3) (2014) 13−22 (in Russian).

[22] Y.B. Bolkhovityanov, A.S. Deryabin, A.K. Gutakovskii, L.V. Sokolov, Mechanism of induced nucleation of misfit dislocations in the Ge-on-Si(001) system and its role in the formation of the core structure of edge misfit dislocations, Acta Mater. 61 (2013) 617−621.

[23] M.C. Bost, J. Mahan, A clarification of the index of refraction of betta-iron disilicide, J. Appl. Phys. 64 (1988) 2034.

[24] L. Miglio, G. Malegori, Origin and nature of the band gap in β-$FeSi_2$, Phys. Rev. B 52 (3) (1995) 1448−1451.

[25] S. Izumi, M. Shaban, N. Promos, K. Nomoto, T. Yoshitake, Near-infrared photodetection of β-$FeSi_2$/Si heterojunction photodiodes at low temperatures, Appl. Phys. Lett. 102 (2013) 032107.

[26] M. Suzuno, T. Koizumi, T. Suemasu, p-Si/β-$FeSi_2$/n-Si double-heterostructure light-emitting diodes achieving 1.6 μm electroluminescence of 0.4 mW at room temperature, Appl. Phys. Lett. 94 (2009) 213509.

[27] K. Takakura, H. Ohyama, K. Takarabe, T. Suemasu, F. Hasegawa, Hole mobility of p-type beta-FeSi2 thin films grown from Si/Fe multilayers, J. Appl. Phys. 97 (2005) 093716.

[28] T. Suemasu, Y. Ugajin, S. Murase, T. Sunohara, M. Suzuno, Photoluminescence decay time and electroluminescence of p-Si/β-FeSi2 particles/n-Si and p-Si/β-FeSi2 film/n-Si double-heterostructures light-emitting diodes grown by molecular-beam epitaxy, J. Appl. Phys. 101 (124) (2007) 506.

[29] C. Spinella, S. Coffa, C. Bongiorno, S. Pannitteri, M.G. Grimaldi, Origin and perspectives of the 1.54 μm luminescence from ion-beam-synthesized β-$FeSi_2$ precipitates in Si, Appl. Phys. Lett. 76 (2000) 173.

[30] S.A. Dotsenko, N.G. Galkin, L.V. Koval, V.O. Polyarnyi, J. Surf, In situ differential reflectance spectroscopy study of solid phase epitaxy in Si(111)Fe and Si(111)Cr systems, Sci. Nanotechnol. 4 (2006) 319.

[31] A.V. Shevlyagin, D.L. Goroshko, E.A. Chusovitin, K.N. Galkin, N.G. Galkin, A.K. Gutakovskii, Enhancement of the Si p-n diode NIR photoresponse by embedding β-FeSi2 nanocrystallites, Sci. Rep. 5 (2015) 14795. http://dx.doi.org/10.1038/srep14795.

[32] T.S. Shamirzaev, N.G. Galkin, E.A. Chusovitin, D.L. Goroshko, A.V. Shevlyagin, A.K. Gutakovski, et al., Electroluminescent 1.5-μm light-emitting diodes based on p + -Si/NC β-FeSi2/n-Si structures, Semiconductors 49 (4) (2015) 508−512. Available from: http://dx.doi.org/10.1134/S1063782615040211.

10

Formation of GaAs Step-Terraced Surfaces by Annealing in Equilibrium Conditions

V.L. Alperovich[1,2], I.O. Akhundov[1,2], D.M. Kazantsev[1,2], N.S. Rudaya[1], E.E. Rodyakina[1,2], A.S. Kozhukhov[1], D.V. Sheglov[1], A.N. Karpov[1], N.L. Shwartz[1,3], A.S. Terekhov[1], A.V. Latyshev[1,2]

[1]RZHANOV INSTITUTE OF SEMICONDUCTOR PHYSICS OF THE SIBERIAN BRANCH OF RUSSIAN ACADEMY OF SCIENCES, NOVOSIBIRSK, RUSSIA [2]NOVOSIBIRSK STATE UNIVERSITY, NOVOSIBIRSK, RUSSIA [3]NOVOSIBIRSK STATE TECHNICAL UNIVERSITY, NOVOSIBIRSK, RUSSIA

10.1 Introduction

Atomically flat crystal surfaces are necessary for fundamental surface science and various applications. Only such surfaces yield opportunities for reproducible fabrication of nanoscale structures, due to self-assembling phenomena at crystal growth [1], or by using modern atomic probe techniques [2]. Nowadays, such nanostructures are the objects of intense fundamental studies and, in the future, they may become the basis for nanoelectronic and nanophotonic devices. The practical importance of atomically flat III–V semiconductor surfaces is determined by the opportunity to use them to improve the performance of modern optoelectronic devices. In particular, atomically flat substrates are used for the growth of epitaxial III–V heterostructures, which can be employed in various optoelectronic devices.

A common technique for preparing solid flat surfaces is chemo-mechanical polishing (CMP), which can provide very small root mean square (rms) roughness $R_q \sim 0.1-0.2$ nm, comparable to the interatomic distance. However, due to mechanical action during polishing, CMP leads to the formation of a disordered near-surface layer with distorted coordination of surface atoms and, as a consequence, a high concentration of structural defects. Thus, despite small values of rms roughness, the surface prepared by CMP is disordered on the atomic scale. This disorder can be effectively reduced by allowing surface migration at elevated temperatures. Thus, by means of annealing, a crystal surface can be brought to an ordered step-terraced morphology, which corresponds to a regular set of atomically flat terraces separated by monatomic steps. The terrace width is determined by the angle of misorientation from the singular crystal face. Almost perfect step-terraced surfaces of silicon are obtained by annealing in a vacuum [3,4]. However, the application of this method to III–V semiconductors is problematic because, due to the high and varying evaporation rates of the

Advances in Semiconductor Nanostructures. DOI: http://dx.doi.org/10.1016/B978-0-12-810512-2.00010-X

III and V components, it is difficult to find the vacuum annealing temperature range in which surface diffusion is sufficiently effective for surface smoothing, while sublimation is still negligible [5]. It is generally claimed that an effective method for preparing smooth crystal surfaces is epitaxial growth. However, growth is performed in nonequilibrium conditions and, therefore, does not always yield smooth step-terraced surfaces, due to possible kinetic instabilities.

To avoid deterioration of the surface relief due to kinetic instabilities, thermal smoothing of crystal surfaces should be performed in conditions close to equilibrium, when the atomic fluxes are equilibrated so neither surface sublimation, nor growth occurs [6]. If such equilibrium is violated, there is a net atomic flux toward the surface which leads to crystal growth, or atomic flux from the surface, which leads to sublimation. Under both growth or sublimation, the roughness of an initially flat surface will, in general, increase with annealing duration. It should be noted that, as a rule, a global thermodynamic equilibrium cannot be reached in a complex heterogenous system, such as a crystal surface with adatoms, vacancies, atomic steps, kinks, and other morphological elements. Therefore, one can search for local equilibrium (quasi-equilibrium) conditions, which provide effective smoothing of a crystal surface.

It was shown by Ding et al. [7] that in situ annealing of MBE-grown GaAs(001) films at sufficiently high As_4 vapor pressure and moderate temperature led to the formation of atomically flat terraces separated by monatomic steps, while lowering As_4 pressure and increasing temperature led to surface roughening. Annealing GaAs(001) in metal-organic chemical vapor deposition (MOCVD) setups in a mixture of arsine and hydrogen may lead to the formation of terraced surfaces with monatomic steps [8], or to step bunching [9], depending on the crystallographic orientation and other experimental conditions.

Thermal smoothing experiments in the growth chambers of MBE or MOCVD setups are expensive and time-consuming. A more efficient technique of GaAs surface thermal smoothing was proposed in Ref. [10]. This technique consists of annealing in conditions close to equilibrium, in the quartz tube of a liquid phase epitaxy setup, under a flow of pure molecular hydrogen. The quasi-equilibrium conditions are achieved in a narrow slit (two-dimensional capillary) between two adjacent GaAs samples. Although the opportunity to obtain a step-terraced GaAs surface by annealing in equilibrium conditions was experimentally demonstrated [10], a number of important questions related to this method remained open. First, the closeness of annealing conditions to equilibrium was not determined. Deviations in the conditions toward growth or sublimation may lead to kinetic instabilities, which hinder ordered step-terraced surface formation. Second, adequate methods should be developed for characterizing step-terraced surface morphology formation. Such methods will open new opportunities for comparing experiment with theory, elucidating the surface smoothing microscopic mechanisms, and determining the respective parameters. To our knowledge, an analytical microscopic theory, which could adequately describe the kinetics of the step-terraced morphology formation, is not yet available. In the absence of an analytical theory, direct Monte Carlo (MC) simulation, which consists of modeling atom random walk with an account of the interaction between atoms [4,11] could be used. MC simulation is widely used for modeling crystal growth and sublimation, and it can also be explored to describe surface smoothing and the formation of step-terraced morphology. Third, it is known that, at high temperatures, surface smoothing changes to disordering (roughening) of the surface relief. However, questions about the conditions and mechanisms of the roughening transition are still not well understood. Along with scientific interest, these questions are of practical importance. Indeed, to facilitate surface diffusion, and thus, to increase smoothing efficiency, the annealing temperature should be as high as possible, but well below the roughening transition temperature. Finally, the thermal smoothing of GaAs/AlGaAs glass-bonded transmission-mode photocathode structures and a possible influence of dislocations on the surface morphology, which are introduced under thermo-mechanical stress relaxation, were not studied earlier. In this chapter, the results of the developments in GaAs surface smoothing techniques and study of the mechanisms of step-terraced surface formation are described [12–20].

10.2 GaAs Surface Thermal Smoothing Technique

Thermal smoothing experiments were carried out on samples which were cut from commercially available "epi-ready" GaAs(001) substrates with small rms roughness $R_q \leq 0.15$ nm prepared by the CMP technique. The anneals were performed in the quartz tube of a liquid phase epitaxy setup, under a flow of pure molecular hydrogen at temperatures in the range of $T = 400–800°C$. Prior to annealing, the surface oxides were removed in a solution of HCl in isopropyl alcohol [21]. Conditions close to equilibrium were provided by two methods. In the first one, the GaAs substrate was covered with the other GaAs sample. Thus, during the annealing, in a narrow slit (two-dimensional capillary) between the adjacent samples a quasi-equilibrium between the solid and vapor phases was reached [10]. The width of the capillary was determined by the macroscopic flatness of the samples, and was estimated to be below ~ 1 μm. In the second method, the equilibrium between the GaAs surface and the vapors of arsenic and gallium was provided by the presence of a saturated Ga-GaAs melt. The melt was put in a special graphite cassette, together with the GaAs(001) sample meant for smoothing. The morphology of the initial and annealed GaAs surfaces was studied ex situ using an atomic force microscope (AFM) Solver P-47H (NT-MDT), assuming that the oxide layer formed in the air replicates the surface relief with submonolayer accuracy [22].

To determine the proximity of the annealing conditions to equilibrium, we performed experiments on the sample with a lithographic mark. To prepare such a mark, a 10 nm deep and 100 nm wide trench was made along the perimeter of a 5×5 μm^2 square by AFM tip-induced oxidation [23], and subsequently the oxide was removed [21] (Fig. 10.1A). This square trench enabled us to identify specific monatomic steps after the ex situ wet treatment and annealing. It was found that oxidation in the air, and removal of oxides in the solution of HCl in isopropyl alcohol, led to the formation of dips of monatomic depth (~ 0.3 nm) ~ 30 nm in diameter on terraces, while atomic steps kept their position and shape within experimental accuracy [24].

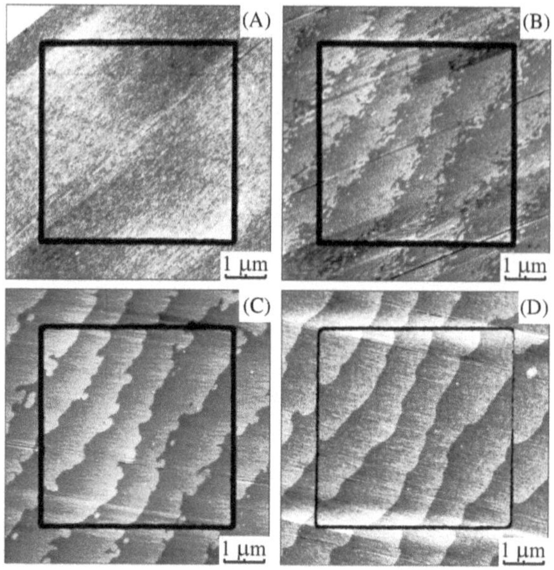

FIGURE 10.1 AFM images of the epi-ready GaAs(001) substrate with a square trench lithographic mark before (A), and after thermal smoothing in quasi-equilibrium conditions at $T=625°C$ for 10 min (B), 40 min (C), and 100 min (D).

Shown in Fig. 10.1 are the AFM images of a GaAs(001) substrate area with the square mark before (A), and after (B–D), consecutive anneals at $T=625°C$. No traces of "step-and-terrace" morphology are observed before the anneals (Fig. 10.1A). Distinct terraces bounded by jagged steps are seen in the AFM image after annealing for 10 min (Fig. 10.1B). On the terraces there are substantial concentrations of two-dimensional "positive" and "negative" islands of monatomic height and depth, respectively. An increase in the annealing time led to step-smoothing, to the increase of the average lateral size of the islands, and to a decrease in their concentration (Fig. 10.1C). As a result of annealing for 100 min at $T=625°C$, the surface morphology approached that of an "ideal" vicinal surface with a regular set of atomically flat terraces of ~ 1 μm wide separated by nearly straight equidistant steps of monatomic height (~ 0.3 nm). The width of the terraces corresponds to the substrate miscut angle of ~ 0.02.

To detect deviations in the annealing conditions from equilibrium toward growth or sublimation, we analyzed the position of monatomic steps after consecutive anneals. In the case of growth or sublimation one would expect the movement of the steps toward lower or higher lying terraces, respectively. A detailed comparison of the AFM images measured on patterned GaAs(001) substrates after anneals for various durations did not reveal any significant shift of the step mean positions. The accuracy of the step position measurements allowed us to estimate the upper limit of the growth (or sublimation) rate to be well below 3×10^{-5} monolayers per second. It should also be noted that the step-flow growth (or sublimation) would lead to flattening of mesas into step-free facets, similar to the phenomenon observed during epitaxial growth on mesa-structured GaAs substrates [25]. However, as is seen from

Fig. 10.1B–D, each step inside the mesa is a continuous extension of the step outside the mesa, with the uppermost and lowest terraces on the mesa having approximately the same width compared to other terraces. Thus, within experimental accuracy, the annealing conditions are close to equilibrium.

10.3 Thermal Smoothing GaAs(001) Surface: Isochronal Anneals at Various Temperatures

To study the surface relief evolution under thermal smoothing by the "capillary" technique, and in the presence of a Ga-GaAs melt, a set of isochronal anneals at various temperatures were performed [13]. Compared to studying the smoothing kinetics, isochronal anneals are less time-consuming and, thus, more efficient for the determination of optimal annealing conditions, because the diffusion process rates depend exponentially on temperature.

Fig. 10.2A–C shows the AFM topographic images of the as-received chemo-mechanically polished GaAs(001) substrate and images after isochronal 2-h anneals in a capillary at various temperatures. Starting from $T = 520°C$, vicinal terraces with two-dimensional islands and pits of monatomic height and depth, respectively, appear on the surface. Similar to the results presented in Fig. 10.1, the increase in temperature leads to step shape smoothing, to an increase in the mean lateral island and pit sizes, and to a decrease in their concentration (Fig. 10.2B). A further temperature increase up to $T = 650°C$ leads to the islands incorporation into steps, and to step straightening (Fig. 10.2C). A zoomed image of a 900×900 nm^2 area of the smoothed surface with a single step separating two flat terraces is shown in Fig. 10.2D. On these terraces, there are no islands and pits, at least with lateral sizes larger than the instrumental resolution of the AFM microscope, which is about 5–7 nm. Shown in Figs. 10.2E and F are the z–x profile across a single step, with the surface height distribution calculated over the image shown in Fig. 10.2D. These data yield a step height of approximately 0.3 nm, which coincides with the GaAs one unit cell step height within the calibration accuracy of our measurements. Thus, the surface morphology approached that of an "ideal" vicinal surface with a regular set of atomically flat terraces of ~ 1 μm in width separated by monatomic steps.

The microscopic mechanisms of step-terraced surface formation consist in the recombination of adjacent adatoms and advacancies, small islands and pits, the growth of large islands (or pits) at the expense of small ones (Ostwald ripening), incorporation of islands and pits into steps, and step straightening. At higher annealing temperatures, starting from $T \approx 675°C$, morphological ordering is changed to disordering, which revealed itself in step meandering, nucleation of islands and pits on the terraces, and complete destruction of regular terrace arrays at $T \geq 700°C$ (not shown in Fig. 10.2). This disordering may be caused by the deviation from equilibrium due to a gradual depletion of the near-surface regions with arsenic, rather than by the fundamental entropy-driven roughening transition [26]. Possible reasons for the depletion are the incongruent evaporation of GaAs, vapor flow through the open sides of the capillary, and restricted diffusion flow of arsenic from the GaAs bulk to the surface.

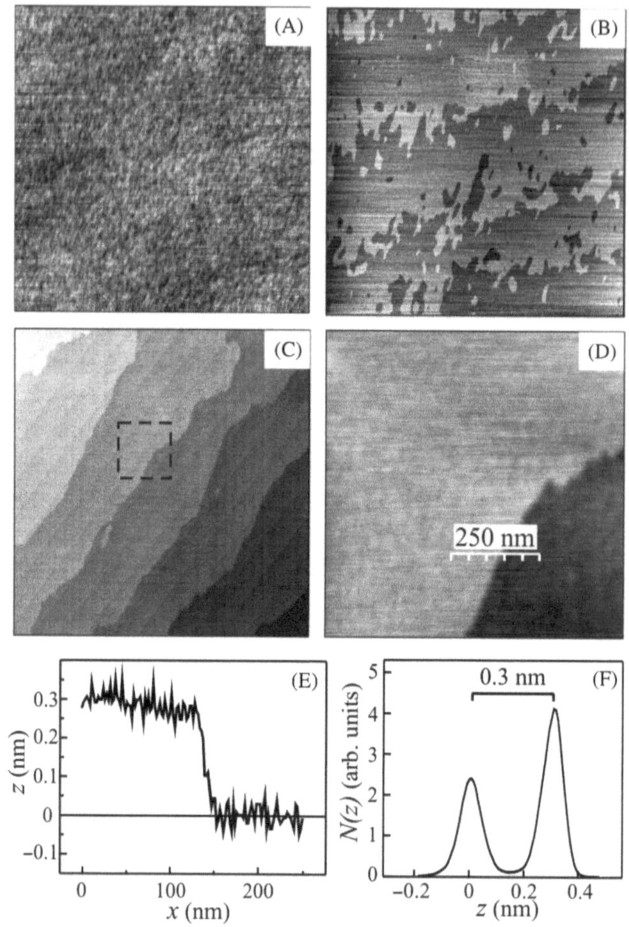

FIGURE 10.2 (A) A typical $5 \times 5\ \mu m^2$ AFM image of the GaAs(001) substrate initial surface; (B and C) $5 \times 5\ \mu m^2$ AFM images of this substrate after 2-h anneals in the capillary at $T = 575°C$ (B) and 650°C (C); (D) A zoomed image of a $900 \times 900\ nm^2$ area marked by the *dashed square* on (C); (E) A 250 nm z–x profile across the monatomic step marked in (D); (F) The height distribution calculated over the image (D). *Reprinted with permission from V.L. Alperovich, I.O. Akhundov, N.S. Rudaya, D.V. Sheglov, E.E. Rodyakina, A.V. Latyshev, et al., Step-terraced morphology of GaAs(001) substrates prepared at quasiequilibrium conditions, Appl. Phys. Lett. 94 (2009) 101908. Copyright 2009, American Institute of Physics.*

To provide annealing conditions closer to equilibrium, in the second method the anneals were performed in a quasi-closed graphite cassette containing a saturated Ga-GaAs melt. The AFM images of GaAs(001) substrates annealed in the cassette at various temperatures are shown in Fig. 10.3. It can be seen that the formation of a regular array of terraces occurred in the temperature range $T \leq 700°C$. As for the "capillary" method, an increase in the annealing temperature up to 650°C leads to an increase in the mean size of the islands, a decrease in their concentration, and step straightening. It can be seen from Fig. 10.3 that,

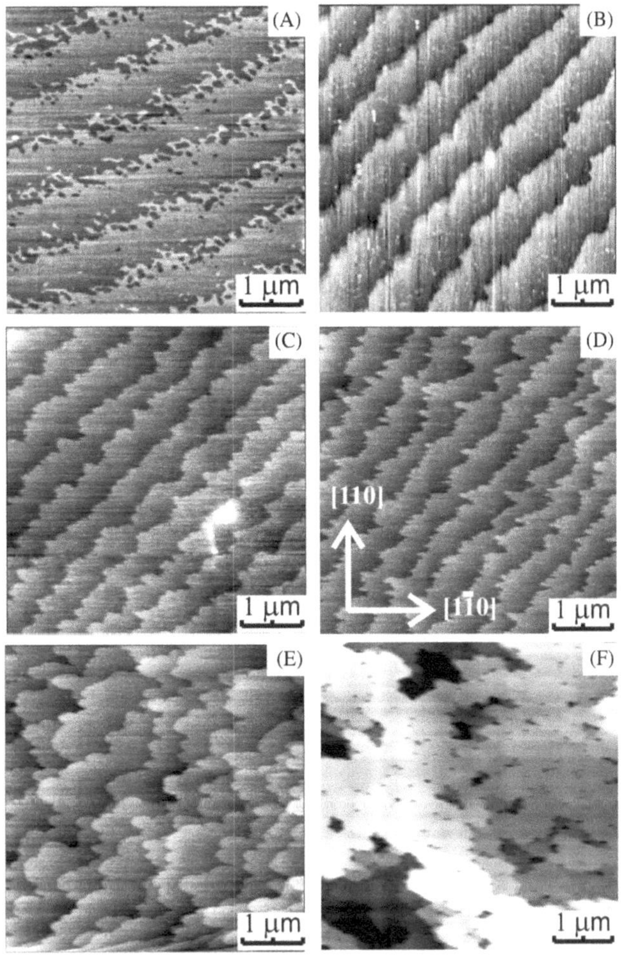

FIGURE 10.3 AFM images of a GaAs(001) substrate after 1-h anneals in equilibrium with the Ga-GaAs melt at various temperatures *T*: (A) 575°C; (B) 650°C; (C) 675°C; (D) 700°C; (E)725°C; (F) 775°C. *Reprinted with permission from V.L. Alperovich, I.O. Akhundov, N.S. Rudaya, D.V. Sheglov, E.E. Rodyakina, A.V. Latyshev, et al., Step-terraced morphology of GaAs(001) substrates prepared at quasiequilibrium conditions, Appl. Phys. Lett. 94 (2009) 101908. Copyright 2009, American Institute of Physics.*

although the system of terraces and monatomic steps is preserved up to $T = 725°C$, the steps become wavier at $T > 650°C$. It can also be seen that the steps meander preferentially along the [1$\bar{1}$0] direction, presumably due to the anisotropy of the adatom diffusion coefficients and incorporation probabilities on the GaAs(001) surface. A complete destruction of the terraces accompanied by the formation of deep pits took place at $T \geq 775°C$.

To characterize the surface morphology evolution during thermal flattening and roughening quantitatively, along with the rms roughness of the terraces, we determined the full length of monatomic steps, including the island perimeters, L_s, normalized to straight steps

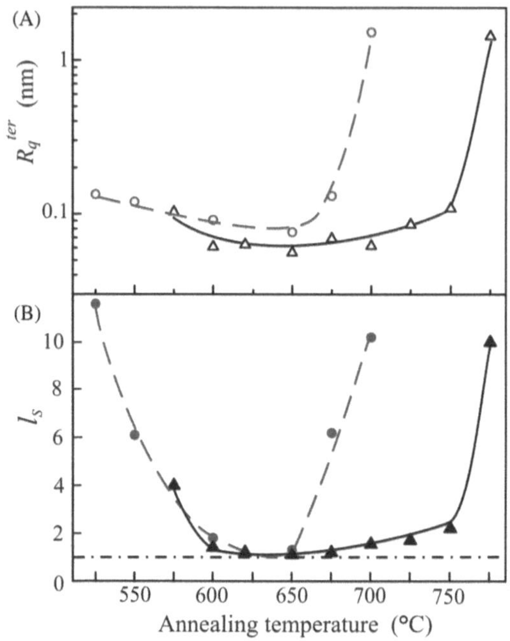

FIGURE 10.4 Temperature evolution of terrace rms roughness R_q^{ter} (A), and normalized step length l_s (B), for the GaAs(001) substrates annealed in the capillary (*circles, dashed lines*) and in equilibrium with the Ga-GaAs melt (*triangles, solid lines*). The *dashed and solid lines* are drawn as a visual guide. The *dashed-dotted line* denotes the lower limit for step length $l_s = 1$. *Reprinted with permission from V.L. Alperovich, I.O. Akhundov, N.S. Rudaya, D.V. Sheglov, E.E. Rodyakina, A.V. Latyshev, et al., Step-terraced morphology of GaAs(001) substrates prepared at quasiequilibrium conditions, Appl. Phys. Lett. 94 (2009) 101908. Copyright 2009, American Institute of Physics.*

of length L_0 on the ideal vicinal surface with the same area and the same terrace width: $l_s = L_s/L_0$ [13,17]. The full length of steps has a clear thermodynamic meaning, because the step contribution to the surface free energy is proportional to the step length [17].

Shown in Fig. 10.4A and B are the rms roughness of terraces R_q^{ter}, and the dimensionless step length l_s, respectively, as a function of the isochronal annealing temperature. It is seen that at $T < 650°C$, rms roughness weakly depends on temperature, while the step length decreases by an order of magnitude and approaches its limiting value $l_s = 1$. Consequently, the step length is a more useful characteristic of step-terraced morphology formation as compared to the rms roughness. It can be seen from Fig. 10.4B that the optimal temperatures of anneals with durations of about 1 h are in the range from 600°C to 650°C for annealing, both in the capillary and in the presence of the Ga-GaAs melt.

At high temperatures, both rms roughness and step length increase sharply due to surface roughening. It can be seen from Fig. 10.4 that, under annealing in the cassette with the Ga-GaAs melt, the threshold roughening temperature is $\Delta T \sim 100°C$ higher than the threshold temperature characteristic for annealing in the capillary. This fact confirms that, under annealing in the capillary, the observed step-terraced morphology destruction is not related to the fundamental "roughening transition" [26], but is due to deviations from equilibrium conditions, presumably caused by an arsenic vapor flux through the capillary edges. Additional annealing experiments in the cassette with the Ga-GaAs melt proved that, in this case, surface roughening was also caused by deviations of the annealing conditions from equilibrium toward sublimation or growth. Thus, under GaAs annealing we were not able to observe the fundamental roughening transition. It should be noted that indirect evidence of such transitions were obtained by Bartelt et al. for the Si(001) surface [27], although, to the best of our knowledge, there is still a lack of direct experimental proof. The observation of roughening transition is hindered by difficulties in keeping the exact balance between the sublimation and deposition atom fluxes at elevated temperatures, because the evaporation rate exponentially increases with temperature.

10.4 GaAs(001) Surface Smoothing Kinetics

To further elucidate the mechanisms of step-terraced morphology formation, we studied the kinetics of surface smoothing at a fixed temperature $T = 625°C$ and various annealing durations in the range from 15 min to 8 h [17]. The anneals were performed in a graphite cassette with the Ga-GaAs melt. Each anneal was carried out on a new sample cut from one GaAs(001) substrate. AFM images of the sample surfaces after selected anneals are shown in the top row of Fig. 10.5. The variations in the terrace width and vicinal step direction from sample to sample are due to smooth variations in the misorientation angle caused by the macroscopic surface relief of the substrate. The surface relief evolution under increasing anneal duration at a fixed temperature is qualitatively similar to that observed in the set of isochronal anneals up to 650°C. An array of terraces separated by jagged steps and a high concentration of islands and dips was formed after the first 15-min anneal (Fig. 10.5A). By increasing the annealing duration, the island and dip concentration decreased, and the steps straightened. Thus, the surface morphology approached that of the ideal vicinal surface (Fig. 10.5B and C).

To characterize the step-terraced morphology formation process quantitatively, we used Fourier and autocorrelation analyses, the rms roughness R_q [28], and the total length of monatomic steps L_s [13,17]. Along with the rms roughness R_q calculated over surface areas with several terraces, we used the rms roughness R_q^{ter} calculated over a single terrace, which does not contain a contribution from steps, and was determined by the island and pit concentration on the terrace. Shown in Fig. 10.6 in double log scales are the annealing duration dependences of rms roughness values R_q and R_q^{ter} (Fig. 10.6A), and excess step length $\Delta l_s = (l_s - 1)$. It can be seen that the rms roughness R_q, denoted by diamonds, is constant within the experimental scatter ~20%, due to a small roughness of the initial epi-ready substrates, and to the step contribution, which restricts the minimal rms roughness at the ideal vicinal GaAs surface by a value of 0.08 nm. This minimal value is denoted in Fig. 10.6A by the horizontal *dashed-dotted line*.

Unlike R_q, terrace rms roughness R_q^{ter} and normalized excess step length Δl_s decrease with increasing annealing duration. However, R_q^{ter} decreases relatively weakly, by approximately a factor of 2. This decrease is comparable to the experimental error (~0.05 nm),

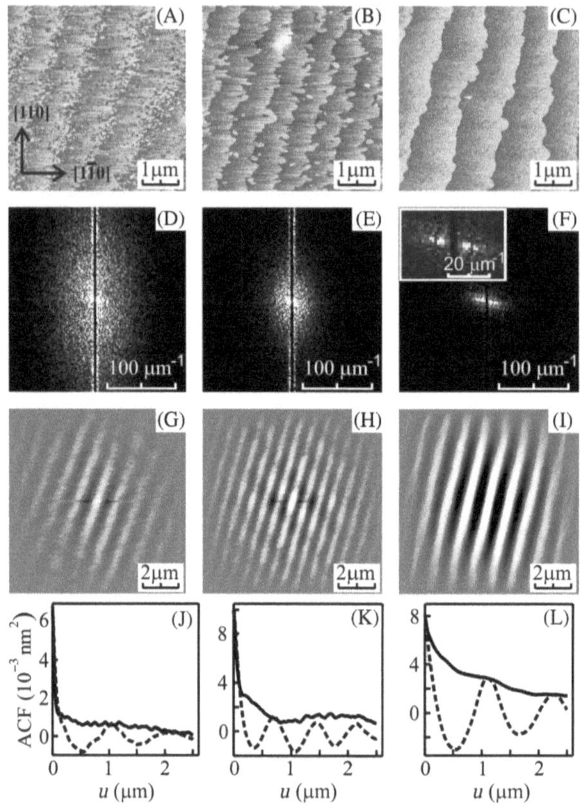

FIGURE 10.5 AFM surface relief images (top row, A–C), their Fourier transforms (second row, D–F), autocorrelation functions (ACFs, third row, G–I), and cross-sections of the ACFs in the directions perpendicular (*dashed lines*) and parallel (*solid lines*) to the steps (bottom row, J–L) obtained after anneals at 625°C for 15 min (A, D, G, and J), 40 min (B, E, H, K), and 240 min (C, F, I, and L). A zoomed image of the FT with two peaks, which arise due to the periodicity of the surface steps, is shown in the inset of Fig. 10.1F. *Reprinted from I.O. Akhundov, V.L. Alperovich, A.V. Latyshev, A.S. Terekhov, Kinetics of atomic smoothing GaAs(001) surface in equilibrium conditions, Appl. Surf. Sci. 269 (2013) 2–6, Copyright (2013), with permission from Elsevier.*

which is restricted by the AFM noise. On such smooth surfaces, the variations of surface relief measurement precision, which are determined mainly by the AFM tip quality, can also contribute substantially to the rms roughness determination error.

As compared to rms roughness, the decrease in the excess length of steps Δl_s is much more pronounced (by a factor of ≈ 30). It can be seen from Fig. 10.6 that, in the double logarithmic scale, the measured dependences $R_q(t)$ and $\Delta l_s(t)$ can be approximated by *straight lines*, indicating power dependences on the annealing time:

$$R_q^{ter}(t) = a_0 \cdot (t/t_0)^{\alpha}$$

$$\Delta l_s(t) = (t/t_1)^{\beta}$$

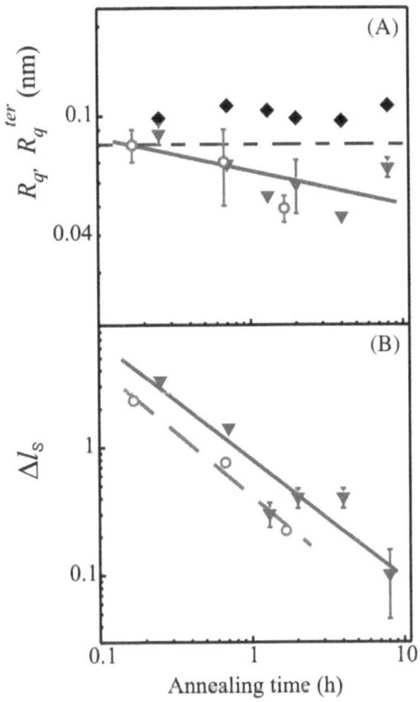

FIGURE 10.6 Kinetics of root mean square roughness $R_q(t)$ and $R_q^{ter}(t)$ (A), and the dimensionless excess of step length $\Delta l_s(t)$ (B). The kinetics corresponds to the annealing temperature $T = 625°C$. The double logarithmic scale is used in order to approximate the experiment by power dependences (*straight lines*, see the text).

where $a_0 = 0.28$ nm is the period of the GaAs lattice in the [001] direction. For the step length the fitting parameters were determined to be equal to $\alpha = -0.16 \pm 0.04$, $t_0 = 8 \times 10^{-4}$ h, $\beta = -1 \pm 0.15$, and $t_1 = 0.8$ h. Thus, the excess of the step length is inversely proportional to the annealing time *t*, while the temporal dependence of the rms roughness is much weaker $|\alpha| << 1$. These observations confirm the conclusion made from the experiments with isochronal anneals [13] that for characterizing step-terraced morphology formation on the epi-ready substrate surfaces with small initial values of R_q, the step length is a more adequate parameter compared to the rms roughness.

The substrate surface relief nonuniformity yields a substantial contribution to the step length determination error on different samples. The sample with the lithographic mark allowed us to exclude this contribution, and also to track in detail the shape evolution of the particular atomic steps. The step length kinetics measured on the sample with the square lithographic trench (Fig. 10.1) are shown by *circles* in Fig. 10.6B. It can be seen that these kinetics are also described by the inversely proportional dependence on the annealing duration.

To further characterize the step-terraced surface formation under annealing we used two-dimensional Fourier transforms (FTs) $F(\vec{k}) = \int h(\vec{r})\exp(-i\vec{k}\,\vec{r})d\vec{r}$ and autocorrelation functions (ACFs) $ACF(\vec{u}) = < h(\vec{r}) \cdot h(\vec{r} - \vec{u}) >$ calculated over the AFM images. Here, $h(\vec{r})$ is the height of the surface relief as a function of coordinate $\vec{r}$ in the surface plane, the brackets denote averaging over the surface. Shown in the second row of Fig. 10.5 are the FT power spectra images calculated for the respective AFM images shown in the first row. For all images in Fig. 10.5D–F, the FTs contain a doublet of narrow peaks, which are symmetrically placed with respect to zero wave vector $k = 0$. These peaks arise due to the formation of a periodic set of terraces, separated by monatomic steps and situated along the line perpendicular to the mean direction of the vicinal steps at wave vectors corresponding to relief periodicity $k = 2\pi/b_t$. Here, b_t is the mean width of terraces. The peaks are superimposed on a broad diffuse "halo" originating from the random distribution of islands, pits, and fluctuations of the step shape. The intensity and width of the halo characterize the surface morphology deviation from that of an ideal vicinal surface. Therefore, in the gray scale FT images shown in Fig. 10.5D–F, we intentionally selected the low amplitude range of FT power spectra in order to make the halo clearly visible, with the doublet of peaks being out of scale. The exception is the inset of Fig. 10.5F, which shows a zoomed image of the distinctly resolved doublet of peaks for the surface with mature step-terraced morphology.

It can be seen from Fig. 10.5D and E that the halo has an elliptical shape and, at the initial stage of smoothing, is elongated along the [110] direction. The anisotropy of the halo reflects preferential orientation of jagged step segments along the $[1\bar{1}0]$ direction which, in turn, arises supposedly due to the anisotropy of the adatom diffusion coefficients and incorporation probabilities on the GaAs(001) surface [29,30]. A similar anisotropy is observed in the morphology of films grown by molecular beam epitaxy [30]. The increase in annealing time leads to a distinct reduction in the size of the FT halo, which signals narrowing of the spectrum of spatial frequencies as the surface approaches the step-terrace morphology. At the final stage of smoothing, after the recombination of islands and pits, the shape of FT depends mainly on step meandering, which determines the terrace width fluctuations and the shape of the doublet of peaks. In this case, the FT image is elongated along the doublet, i.e., in a direction perpendicular to the steps rather than along the crystal axes (see Fig. 10.5F).

The ACFs calculated over respective AFM images are shown in Fig. 10.5G–I. ACFs provide the information which is complementary to the results of Fourier analysis. The periodic strips in the ACF images reflect the periodicity of step-terraced surfaces. The strips become more distinct with the increase in annealing time, as the surface approaches ideal step-terrace morphology. The cross-sections of the ACFs in the directions parallel and perpendicular to the vicinal steps are shown in the bottom row of Fig. 10.5. While the presence of oscillations in the ACF cross-sections perpendicular to the steps is a trivial consequence of the step-and-terrace periodicity, the observed increase in typical damping lengths of the cross-sections parallel to the steps inherently characterizes the smoothing process, i.e., the increase in the mean size and the decrease in the concentration of islands and pits at the initial stages of step-terrace morphology formation, and also straightening of the step at the final stage. Approximation of a parallel cross-section by an exponent can yield a correlation length λ. It was found that the dependences of the FT halo width and correlation length on the annealing duration can also be described by power functions: the halo width decreases approximately as $t^{-1/2}$, while the correlation length increases as $t^{1/2}$ [17].

10.5 MC Simulation of GaAs Step-Terraced Surface Formation

To clarify thermal smoothing mechanisms at a microscopic level, to determine the parameters that govern the process of step-terraced morphology formation, and to find optimal conditions for effective surface smoothing, one should compare the experimental results on smoothing kinetics to a microscopic theory. However, to our knowledge, an analytical microscopic theory which could adequately describe the kinetics of step-terraced morphology formation is not yet available. The existing theory describes island nucleation and the growth of large islands at the expense of small ones [31], but does not take into account the influence of vicinal steps on this process. It is reasonable to assume that the temporal dependence of the island mean size should be changed by the presence of steps, which act as sinks for adatoms. In the absence of an analytical theory, experimental results should be compared to numerical simulations. The MC method, which consists of the random walk simulation of atoms along the surface with an account of their interaction, is widely used for simulation of various crystal growth and sublimation processes [32]. This method can also be used for surface smoothing simulation [33–35]. Kawamura and Ishii [33] simulated GaAs(001) surface smoothing under growth interruptions in a molecular beam epitaxy setup. The diffusion activation energy and the lateral bond energies of gallium and arsenic were determined from the comparison of experiment with the simulation. However, in Ref. [33] vicinal steps were present at the initial surface, unlike the substrate surface after CMP. Irisawa et al. [34] described the formation of a step-terraced morphology on an initially rough vicinal $SrTiO_3(001)$ surface under vacuum anneals by MC simulation in the Kossel crystal model. Zhdanov [35] simulated single monatomic step formation on a flat crystal surface, although the results were not compared to the experiment. To the best of our knowledge, MC simulations of smoothing rough vicinal semiconductor surfaces were not performed earlier. In this section, experimental results on the GaAs(001) surface smoothing kinetics are compared to MC simulations [19,20]. This comparison allowed us to evaluate microscopic parameters which determine the smoothing process: the diffusion activation energy, lateral bond energy, and adatom desorption energy.

MC simulations of the surface smoothing and step-terrace morphology formation were performed on the Kossel crystal in the standard solid-on-solid model [32]. In this model, the complicated processes of III and V components mass transfer and the formation of GaAs unit cells are replaced by a random walk and attachment/detachment of one kind of "atoms" without taking into account a real atomic structure of the crystal unit cell and surface reconstructions. Surface atoms are allowed to hop to adjacent free sites on the terrace, incorporate into a step, leave the step, or desorb from the surface. The initial rough surface was prepared by the addition of a random relief to a flat inclined plane, as described in Ref. [19].

The experimental and simulated kinetics of surface relief under the step-terraced morphology formation as a result of annealing in equilibrium conditions are shown in Fig. 10.7. The initial rough surfaces are shown in the left column. The annealing time increases from left to right. The experimental AFM images of the GaAs(001) surface [17] are shown in the

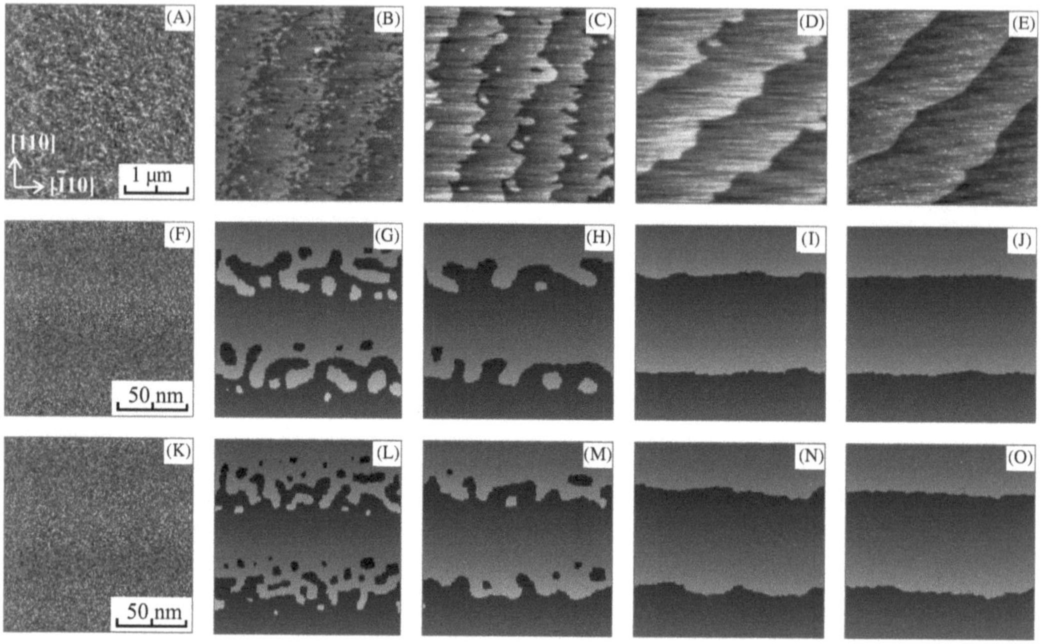

FIGURE 10.7 Experimental (A–E) and simulated (F–O) surface images before (A, F, and K), and after, annealing at the temperature of 900K during 15 min (B, G, and L), 40 min (C, H, and M), 120 min (D, I, and N), and 480 min (E, J, and O) without (F–J), and with (K–O), atom migration through the gas phase taken into account. *Reprinted from D.M. Kazantsev, I.O. Akhundov, A.N. Karpov, N.L. Shwartz, V.L. Alperovich, A.S. Terekhov, et al., Monte Carlo simulation of GaAs(001) surface smoothing in equilibrium conditions, Appl. Surf. Sci. 333 (2015) 141–146, Copyright (2015), with permission from Elsevier.*

top panels. Surface smoothing simulations without and with atom migration through the gas phase are shown in the middle and bottom panels, respectively. The linear size of the simulated surface is smaller by a factor of 20 than the size of the experimental surface due to limited computational resources. The procedure of simulation parameters selection is described below. It can be seen from Fig. 10.7 that, despite the simplicity of the model and different scales of the simulated and experimental surfaces, the simulation qualitatively reproduces the evolution of the experimental surface morphology and enables us to trace the main stages of surface smoothing. These stages consist of: (1) fast recombination of adjacent adatoms and advacancies, small islands, and pits; (2) the growth of large islands (or pits) at the expense of small ones ("Ostwald ripening," Fig. 10.2B, G, and L); (3) incorporation of islands and pits into steps (Fig. 10.2C, H, and M), and (4) straightening of the steps (Fig. 10.2D–E, I, J, N–O). Comparison of the simulated surface images in the middle and bottom panels shows that taking into account atom migration through the gas phase leads to a decrease in the mean size of islands. A possible reason could be due to the decrease of adatom diffusion length due to desorption.

To describe step-terrace morphology formation quantitatively, and to determine the model parameters which govern this process, we compared the experimental and simulated

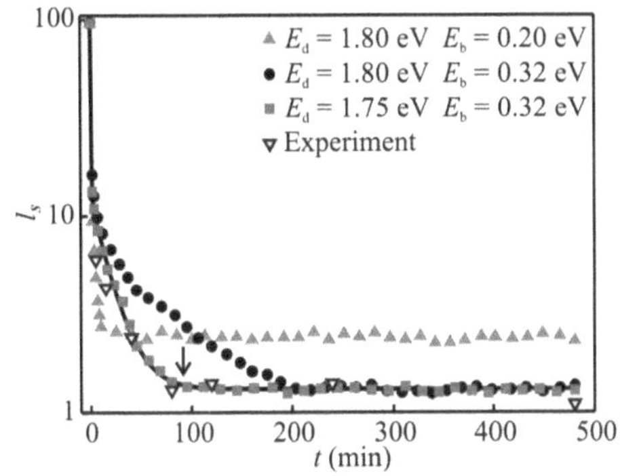

FIGURE 10.8 Experimental and simulated step length kinetics. The simulated kinetics are calculated for various diffusion activation energies E_d and lateral bond energies E_b. The approximation of the experimental and simulated kinetics by Eq. (10.1) for $E_d = 1.75$ eV, $E_b = 0.32$ eV is shown with the *solid line*. Experimental surface smoothing time $t_s = 90$ min is indicated by the *arrow. Reprinted from D.M. Kazantsev, I.O. Akhundov, A.N. Karpov, N.L. Shwartz, V.L. Alperovich, A.S. Terekhov, et al., Monte Carlo simulation of GaAs(001) surface smoothing in equilibrium conditions, Appl. Surf. Sci. 333 (2015) 141–146, Copyright (2015), with permission from Elsevier.*

kinetics of monatomic step length L_s. As was discussed in Sections 10.3 and 10.4, the dimensionless step length L_s/L_0 normalized by the length of straight steps on an ideal vicinal surface L_0 is a more adequate parameter for characterization of step-terrace morphology formation than the rms roughness [13,17]. We start by comparing experimental step length kinetics $l_s(t)$ to simulations, taking into account the atom migration only by means of surface diffusion. Experimental kinetics $l_s(t)$ are shown in Fig. 10.8, along with the simulated kinetics calculated for various values of diffusion activation energy E_d and lateral bond energy E_b. It can be seen that the shapes of all graphs are similar in general, although they depend on the values of E_d and E_b. A quick decrease in the step length at the initial stage of annealing corresponds to recombination of small adjacent islands and pits. At the second stage, the decrease in step length slows down. This stage corresponds to Ostwald ripening and incorporation of the islands into forming vicinal steps. Finally, the step length saturates to the value corresponding to equilibrium step-terraced morphology. At this stage, the undulations of the step length are caused by thermal fluctuations, while the mean value of l_s is independent of time.

To characterize different stages of step-terraced morphology formation and to define the surface smoothing time, we approximated the step length kinetics using the following semi-empirical expression:

$$l_s = l_{s1}e^{-t/t_1} + l_{s2}e^{-t/t_2} + l_s^{eq} \tag{10.1}$$

where t_1 is the island and pit recombination time, t_2 is the Ostwald ripening time, l_{s1} and l_{s2} are respective pre-exponential factors, and l_s^{eq} is the equilibrium step length. For our experimental conditions and model parameters, typical values of t_1 are in the range of 0.1–0.5 min, and the values of t_2 are in the range of 5–40 min. The approximation in

Eq. (10.1) for the simulation which gives the best fit to the experiment is shown in Fig. 10.8 by the *solid line*. The surface smoothing time t_s is defined as the time when the second term in Eq. (10.1) is smaller by a factor of 10 than the third term:

$$t_s = t_2 \ln \frac{10 l_{s2}}{l_s^{eq}} \tag{10.2}$$

The numerical factor 10 is selected in such a way that smoothing time t_s corresponds to a clearly visible kink in kinetics $l_s(t)$. This kink is indicated by the *arrow* in Fig. 10.8.

As expected, a small decrease in diffusion activation energy E_d yields a significant acceleration of surface smoothing because, in the model [32], the time between two consecutive jumps exponentially increases with E_d:

$$\tau \sim \exp(E_d/kT) \tag{10.3}$$

The equilibrium step length does not depend on E_d, but it increases with a decrease in lateral bond energy E_b. The best fit of the experimental kinetics by the simulation yields $E_d = 1.75 \pm 0.03$ eV, and $E_b = 0.32 \pm 0.02$ eV. To take into account a significant difference in the sizes of experimental and simulated surface areas, we obtained the simulated dependence of smoothing time on terrace width. An assumption can be made that the smoothing time is proportional to the terrace width squared because, in accordance with the random walk diffusion model, the time t necessary for an atom traveling at a distance L is proportional to squared L: $t_s \sim L^2$. In fact, the simulated smoothing time increases with increasing L faster than L^2: $t_s \sim L^{2.4}$. A possible reason for the distinction may consist in the fact that, in the course of step-terrace morphology formation, adatoms "waste time" for the islands formation on terraces. Extrapolation of the obtained dependence $t_s(L)$ to the experimental terrace width yields the expected experimental smoothing time for the fitted E_d and E_b values. Then corrected diffusion activation energy value $E_d = 1.20 \pm 0.03$ eV was calculated using Eq. (10.3).

The previous estimation of E_d was made neglecting a mass transfer through the gas phase. To elucidate the contribution of atom migration through the gas phase into step-terrace morphology formation, and to estimate adatom desorption energy E_{des}, we performed simulations which take into account atom desorption and readsorption. To simulate quasi-equilibrium between the surface and the gas phase, the atom flux to the surface was kept equal to the desorption flux from the surface. The simulation results for various E_{des} are shown in Fig. 10.9. Atom migration through the gas phase leads to a decrease in the surface smoothing time compared to the simulation of surface diffusion only. To keep the smoothing time constant while the mass transfer through the gas phase increases with decreasing E_{des}, it is necessary to slow down surface diffusion by increasing E_d. Fig. 10.9A shows the relation between diffusion activation energy E_d and desorption energy E_{des} calculated for a fixed smoothing time $t_s = 90$ min. For a sufficiently big $E_{des} \geq 2.3$ eV, mass transfer through the gas phase does not affect surface smoothing because the atom diffusion length becomes comparable to the terrace width. The calculated dependence of the mean island diameter d on

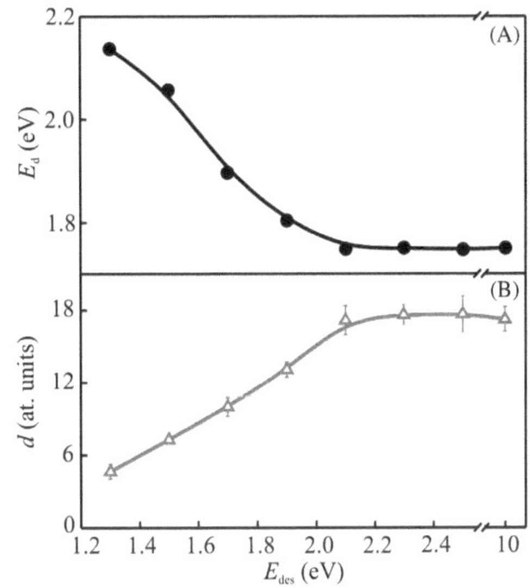

FIGURE 10.9 (A) Calculated relation between diffusion activation energy E_d and desorption energy E_{des} for the constant smoothing time equal to 90 min; (B) Calculated dependence of mean island diameter *d* (in atomic units) on E_{des} after annealing during 15 min. In the calculation, the value of E_d was selected according to the relation shown in (A). *Reprinted from D.M. Kazantsev, I.O. Akhundov, A.N. Karpov, N.L. Shwartz, V.L. Alperovich, A.S. Terekhov, et al., Monte Carlo simulation of GaAs(001) surface smoothing in equilibrium conditions, Appl. Surf. Sci. 333 (2015) 141–146, Copyright (2015), with permission from Elsevier.*

desorption energy for a fixed annealing time of 15 min is shown in Fig. 10.9B. The values of diffusion activation energy used in this calculation correspond to the relation between E_d and E_{des} shown in Fig. 10.9A. It can be seen that the increase in E_{des} yields an increase in mean island diameter due to the increase in adatom diffusion length. Comparing the sizes of islands obtained in the simulation and in the experiment allowed us to determine desorption energy $E_{des} = 2.1 \pm 0.05$ eV and diffusion activation energy $E_d = 1.3 \pm 0.05$ eV. To take into account the difference in the sizes of experimental and simulated surface areas, we obtained the simulated dependences of the mean island size and the smoothing time on the terrace width, and extrapolated these dependences to the experimental terrace width value in the same way as stated above.

It is interesting to compare the value of E_d obtained in the present work to those known from literature. Independent experiments and first-principle calculations yield a broad range of E_d, from 0.85 to 4.0 eV for various surface reconstructions of GaAs. The most reliable values were obtained from experiments on GaAs molecular beam epitaxy [36,37]. Our diffusion activation energy value is close to the value $E_d = 1.3$ eV determined for the GaAs(001)-$\beta(2 \times 4)$ surface [36].

Here we determined the surface energies without taking into account GaAs(001) surface anisotropy due to reconstructions [19]. More details on anisotropy in Ostwald ripening and step-terraced surface formation on the GaAs(001) surface are given in Ref. [20]. Fourier and autocorrelation analyses are used to reveal surface relief anisotropy and provide information about island and pit shape, and their size distribution. Two origins of surface anisotropy are revealed. At the initial stage of surface smoothing, crystallographic anisotropy is observed, and it is caused, presumably, by the anisotropy of surface diffusion at GaAs(001). A difference in diffusion activation energies along the [110] and $[1\bar{1}0]$ axes of the (001) surface is estimated as $\Delta E_d \approx 0.1$ eV, from the comparison of experimental results and simulation. At later stages of surface smoothing, the surface relief anisotropy is determined by the vicinal steps direction [20].

10.6 Step-Terraced GaAs(001) Surfaces With Straight Monatomic Steps Induced by Dislocations

The preparation of nearly ideal vicinal surfaces is hindered by the fact that, at real surfaces, vicinal steps are not perfectly straight because their shape is determined by long-period undulations of the surface relief, which cannot be reduced at reasonable temperatures and periods of annealing. In the present study, we experimentally proved the possibility of forming a rectangular grid of straight monatomic slip steps induced by generation of dislocations due to thermo-mechanical stress relaxation in GaAs/AlGaAs heterostructures bonded to glass substrates [15].

AlGaAs/p-GaAs/AlGaAs double heterostructures were epitaxially grown on GaAs(001) substrates, and were bonded to glass plates by means of thermocompression bonding. The GaAs substrate and AlGaAs layer were removed by selective wet etching. Such glass-bonded structures are used as GaAs(C_s,O) transmission-mode photocathodes [12,38]. In the glass-bonded heterostructures there are thermo-mechanical stresses due to bonding and subsequent anneals. At sufficiently high temperatures, stress relaxation causes generation of dislocations in the heterostructure. These dislocations, in turn, lead to an increasing nonradiative recombination rate of photoelectrons, to a decreasing photoemission quantum yield, to the formation of "dark line defects," and to changes in the surface morphology. The present work is aimed at experimental investigation of the impact of dislocation formation in glass-bonded AlGaAs/p-GaAs heterostructures on GaAs surface morphology. To achieve this aim, isochronous 1-h anneals of such heterostructures were performed under equilibrium conditions provided by the presence of a saturated Ga-GaAs melt in the vicinity of the GaAs surface [13,15].

The AFM images of the GaAs(001) surface of the GaAs/AlGaAs glass-bonded structure before and after anneals at various temperatures are shown in Fig. 10.10. Annealing the initially rough surface eventually leads to step-terraced morphology formation. The atomically flat terraces separated by curved monatomic steps were observed earlier as a result of annealing "epi-ready" GaAs(001) substrates [13]. It can be seen that, along with curved "vicinal" steps, a rectangular grid of straight steps is formed on the surface. Such "cross-hatch" patterns are typical for epitaxial films grown on lattice mismatched substrates and, apparently, have the same origin arising from the plastic relaxation of mechanical stresses by

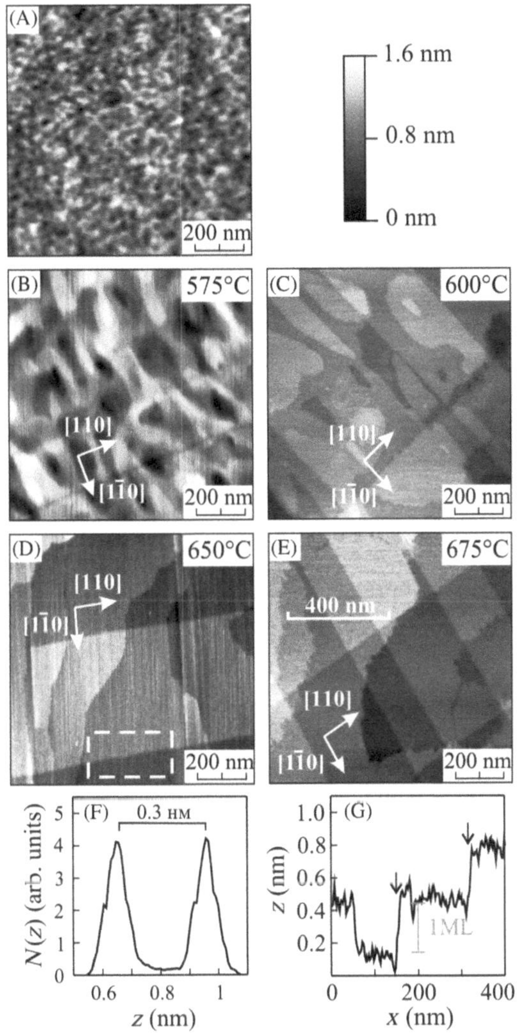

FIGURE 10.10 AFM images of the GaAs(001) surface of the GaAs/AlGaAs heterostructure bonded to glass before (A), and after 1-h anneals at increasing temperatures (B–E). (F) Height distribution function calculated over the rectangular region shown by the *dashed line* in (D). (G) *z–x* cross-section of the 400 nm segment shown in (E). The slight angle distortions are due to the incomplete correction of AFM scanner nonlinearities. *Reprinted from I.O. Akhundov, D.M. Kazantsev, V.L. Alperovich, N.S. Rudaya, E.E. Rodyakina, A.V. Latyshev, Formation and interaction of dislocation-induced and vicinal monatomic steps on a GaAs(001) surface under stress relaxation, Scr. Mater., 114 (2016) 125–128, Copyright (2016), with permission from Elsevier.*

generation of misfit dislocations [39–41]. In our case, the most probable reason for the straight step formation is the generation of 60-degree misfit dislocations. The density of the dislocation grid, which is estimated using surface step density, is about $\sim 3 \times 10^4\ cm^{-1}$. However, unlike typical cross-hatch patterns, which are usually rough and visible with an optical microscope or even with the naked eye, we observe here the dislocation-induced slip steps of monatomic height due to the low roughness of GaAs/AlGaAs structure surfaces. This is confirmed by the height distribution function (Fig. 10.10F) determined for the surface area which contains flat terraces on both sides of a slip step. Also, the cross-section shown in Fig. 10.10G proves that, within experimental accuracy, both the vicinal and slip steps have the same height of approximately 0.3 nm, corresponding to the period of the GaAs lattice in the [001] direction.

It would be interesting to discuss here the qualitative differences between vicinal and slip steps. The first apparent difference is that the dislocation steps are straight and have definite crystallographic directions, while the vicinal steps are curved, with their shape and direction are determined by uncontrollable smooth surface relief undulations. Second, there are no intersections within the set of vicinal steps, while dislocation steps apparently intersect with each other and with vicinal steps. Third, in the surface fragment with the certain sign of a miscut angle, the vicinal steps have the same sign and, thus, form a monotonically descending (or ascending) staircase. On the contrary, as is seen from Fig. 10.10, adjacent dislocation steps may have opposite signs. Therefore, the dislocation steps form a set of orthogonal stripes with nonmonotonically interleaving lower and higher lying terraces. Finally, the dislocation steps are more stable with respect to thermal shape fluctuations compared to the vicinal steps. The higher thermal stability of slip steps is confirmed by an AFM image after annealing at high temperature $T = 675°C$ (Fig. 10.10E): it is seen that slip steps remain straight, while vicinal steps acquire small-scale, but clearly visible, undulations, caused, presumably, by kink bunching. The higher thermal stability of slip steps is probably due to the anisotropy of linear step tension: the steps along the singular [110] directions correspond to a minimum of step line tension.

More detailed results of studying the formation and interaction of curved vicinal and straight dislocation-induced monatomic slip steps on the smooth GaAs(001) surface are presented in Ref. [42]. Typical dislocation phenomena, like transverse glide, are revealed in the slip step patterns. Annihilation of steps with opposite signs and anticrossing of slip steps with each other and with vicinal steps are studied [42]. The observation of straight slip steps may be of interest for reproducible fabrication of quasi-1D structures. Also, this observation may form the basis for a nondestructive control method of thermo-mechanical stress relaxation in GaAs/AlGaAs heterostructures bonded to glass.

10.7 Conclusions

In conclusion, in this chapter, the techniques for GaAs thermal smoothing in equilibrium conditions, and for the characterization of step-terraced surface morphology formation are further developed. The kinetics of smoothing GaAs(001) are measured, and the results are characterized by temporal evolution of the rms roughness, total length of monatomic steps, FTs, and ACFs. Experiments on samples with lithographic marks allowed us to trace the evolution of individual atomic steps' shape, and evaluate the closeness of annealing conditions to equilibrium. The comparison of the experiment with MC simulation enabled us to elucidate the main stages and microscopic mechanisms of step-terraced morphology formation, and to evaluate the effective parameters which govern this process. The formation and interaction of curved vicinal steps and straight dislocation-induced steps of monatomic height were studied under thermo-mechanical stress relaxation in GaAs/AlGaAs heterostructures bonded to glass.

Acknowledgments

This work was partly supported by RSF (Grant No. 14-22-00143), by RFBR (Grant No. 14-02-00352) and by the Russian Academy of Sciences (Project No. 0306-2015-0012). The experiments were partly carried out with the equipment and supported by CKP "NANOSTRUKTURY,".

References

[1] C. Teichert, Self-organization of nanostructures in semiconductor heteroepitaxy, Phys. Rep. 365 (2002) 335–432.

[2] M.F. Crommie, C.P. Lutz, D.M. Eigler, Confinement of electrons to quantum corrals on a metal-surface, Science 262 (1993) 218–220.

[3] A.V. Latyshev, A.L. Aseev, A.B. Krasilnikov, S.I. Stenin, Transformations on clean Si(111) stepped surface during sublimation, Surf. Sci. 213 (1989) 157–169.

[4] H.-C. Jeong, E.D. Williams, Steps on surfaces: experiment and theory, Surf. Sci. Rep. 34 (1999) 171–294.

[5] Y. Fan, I. Karpov, G. Bratina, L. Sorba, W. Gladfelter, A. Franciosi, Atomic scale roughness of GaAs(001) 2×4 surfaces, J. Vac. Sci. Technol. B 14 (1996) 623–631.

[6] A. Pimpinelli, J. Villain, Physics of Crystal Growth, Cambridge University Press, Cambridge, 1998, p. 373.

[7] Z. Ding, D.W. Bullock, P.M. Thibado, V.P. LaBella, K. Mullen, Atomic-scale observation of temperature and pressure driven preroughening and roughening, Phys. Rev. Lett. 90 (2003) 216109.

[8] J.E. Epler, T.A. Jung, H.P. Schweizer, Evolution of monolayer terrace topography on (100)GaAs annealed under an arsine/hydrogen ambient, Appl. Phys. Lett. 62 (1993) 143–145.

[9] K. Ohkuri, J. Ishizaki, S. Hara, T. Fukui, Multiatomic step formation on GaAs(001) vicinal surfaces during thermal treatment, J. Cryst. Growth 160 (1996) 235–240.

[10] V.L. Alperovich, O.E. Tereshchenko, N.S. Rudaya, D.V. Sheglov, A.V. Latyshev, A.S. Terekhov, Atomically flat GaAs(001) surfaces with regular arrays of monatomic steps for nanotechnology, in: Proc. 12th Int. Symp. Nanostructures: Physics and Technology, St Petersburg, Russia, 2004, pp. 203–204.

[11] K. Sudoh, M. Okano, T. Irisawa, K. Matsumoto, M. Uwaha, Relaxation dynamics of labyrinthine submonolayer films, Surf. Sci. 609 (2013) L1–L4.

[12] V.V. Bakin, A.A. Pakhnevich, A.G. Zhuravlev, A.N. Shornikov, I.O. Akhundov, O.E. Tereshchenko, et al., Semiconductor surfaces with negative electron affinity, e-J, Surf. Sci. Nanotechnol. 5 (2007) 80–88.

[13] V.L. Alperovich, I.O. Akhundov, N.S. Rudaya, D.V. Sheglov, E.E. Rodyakina, A.V. Latyshev, et al., Step-terraced morphology of GaAs(001) substrates prepared at quasiequilibrium conditions, Appl. Phys. Lett. 94 (2009) 101908.

[14] I.O. Akhundov, D.V. Sheglov, A.S. Kozhukhov, N.S. Rudaya, V.L. Alperovich, A.V. Latyshev, et al., Proc. 18th Int. Symp. Nanostructures: Physics and Technology, St. Petersburg, Russia, 2010, pp. 327–328.

[15] I.O. Akhundov, N.S. Rudaya, V.L. Alperovich, S.N. Kosolobov, A.V. Latyshev, A.S. Terekhov, Step-terraced morphology formation on patterned GaAs(001) substrates, in: Proc. 20th Int. Symp. Nanostructures: Physics and Technology, Nizhny Novgorod, Russia, 2012, pp. 122–123.

[16] I.O. Akhundov, A.S. Kozhukhov, V.L. Alperovich, Characterization of GaAs(001) step-terraced morphology formation, in: Proc. XIII International Conference and Seminar on Micro/Nanotechnologies and Electron Devices, EDM, Erlagol-Altai, Russia, 2012, pp. 9–12.

[17] I.O. Akhundov, V.L. Alperovich, A.V. Latyshev, A.S. Terekhov, Kinetics of atomic smoothing GaAs(001) surface in equilibrium conditions, Appl. Surf. Sci. 269 (2013) 2–6.

[18] D.M. Kazantsev, I.O. Akhundov, A.N. Karpov, N.L. Shwartz, V.L. Alperovich, A.V. Latyshev, Monte Carlo simulation of GaAs smoothing in equilibrium conditions, in: Proc. 21st Int. Symp. Nanostructures: Physics and Technology, Saint Petersburg, Russia, 2013, pp. 279–280.

[19] D.M. Kazantsev, I.O. Akhundov, A.N. Karpov, N.L. Shwartz, V.L. Alperovich, A.S. Terekhov, et al., Monte Carlo simulation of GaAs(001) surface smoothing in equilibrium conditions, Appl. Surf. Sci. 333 (2015) 141–146.

[20] D.M. Kazantsev, I.O. Akhundov, N.L. Shwartz, V.L. Alperovich, A.V. Latyshev, Anisotropy in Ostwald ripening and step-terraced surface formation on GaAs(001): experiment and Monte Carlo simulation, Appl. Surf. Sci. 359 (2015) 372–379.

[21] V.L. Alperovich, O.E. Tereshchenko, N.S. Rudaya, D.V. Sheglov, A.V. Latyshev, A.S. Terekhov, Surface passivation and morphology of GaAs(100) treated in HCl-isopropanol solution, Appl. Surf. Sci. 235 (2004) 249–259.

[22] D.A. Nasimov, D.V. Sheglov, E.E. Rodyakina, S.S. Kosolobov, L.I. Fedina, S.A. Teys, et al., AFM and STM studies of quenched Si(111) surface, Phys. Low-Dimens. Semicond. Struct. 3–4 (2003) 157–165.

[23] D.V. Sheglov, A.V. Prozorov, D.A. Nasimov, A.V. Latyshev, A.L. Aseev, Peculiarities of nanooxidation on flat surface, Phys. Low-Dimens. Struct. 5–6 (2002) 239–246.

[24] I.O. Akhundov, N.S. Rudaya1, V.L. Alperovich, A.S. Kozhukhov, A.V. Latyshev, A.S. Terekhov, Influence of oxidation and oxide removal on the morphology of step-and-terraced GaAs surfaces, in: Proc. 22nd Int. Symp. Nanostructures: Physics and Technology, Saint Petersburg, Russia, 2014, pp. 257–258.

[25] H. Weishart, E. Bauser, M. Konuma, H.-J. Queisser, Monomolecular steps of ultra-low density on (100) growth faces of liquid phase epitaxial GaAs, J. Cryst. Growth 137 (1994) 335–336.

[26] J. Lapujoulade, The roughening of metal surfaces, Surf. Sci. Rep. 20 (1994) 195–249.

[27] N.C. Bartelt, R.M. Tromp, E.D. Williams, Step capillary waves and equilibrium island shapes on Si(001), Phys. Rev. Lett. 73 (1994) 1656–1659.

[28] P.A. Arutyunov, A.L. Tolstikhina, Phenomenological description of surface characteristics measured by the method of atomic force microscopy, Crystallogr. Rep. 43 (1998) 483–492.

[29] Y.N. Yang, Y.S. Luo, J.H. Weaver, L.T. Florez, Effects of annealing on the surface morphology of decapped GaAs(001), Appl. Phys. Lett. 61 (1992) 1930–1932.

[30] G.R. Bell, T.S. Jones, B.A. Joyce, Direct observation of anisotropic step activity on GaAs(001), Surf. Sci. 429 (1999) L492–L496.

[31] S.A. Kukushkin, V.V. Slezov, Disperse Systems on the Surface of Solids (Evolution Approach): Mechanisms of Thin Film Formation, Nauka, St. Peresburg, 1996, p. 309.

[32] D.V. Brunev, I.G. Neizvestny, N.L. Shwartz, Z.Sh Yanovitskaya, Influence of Schwoebel barriers and surface diffusion anisotropy on surface relief evolution during epitaxial growth: simulation, Phys. Low-Dimens. Struct. 5–6 (2001) 173–183.

[33] T. Kawamura, A. Ishii, Monte Carlo simulation of recovery process after MBE growth on GaAs(100), Surf. Sci. 493 (2001) 438–446.

[34] T. Irisawa, K. Matsumoto, K. Sudoh, H. Iwasaki, M. Uwaha, Smoothing of anatomically rough vicinal surface—STM observation and MC simulation, Surf. Sci. 602 (2008) 2880–2885.

[35] V.P. Zhdanov, Diffusion of adsorbed particles with attractive lateral interactions at low temperature, Surf. Sci. 617 (2013) 199–206.

[36] J.H. Neave, P.J. Dobson, B.A. Joyce, J. Zhang, Reflection high-energy electron diffraction oscillations from vicinal surfaces—a new approach to surface diffusion measurements, Appl. Phys. Lett. 47 (1985) 100–102.

[37] J.M. Van Hove, P.I. Cohen, Reflection high energy electron diffraction measurement of surface diffusion during the growth of gallium arsenide by MBE, J. Cryst. Growth 81 (1987) 13–18.

[38] R.L. Bell, Negative Electron Affinity Devices, Clarendon, Oxford, 1973, p. 148.

[39] A.M. Andrews, A.E. Romanov, J.S. Speck, M. Bobeth, W. Pompe, Development of cross-hatch morphology during growth of lattice mismatched layers, Appl. Phys. Lett. 77 (2000) 3740–3742.

[40] M.A. Lutz, R.M. Feenstra, F.K. LeGoues, P.M. Mooney, J.O. Chu, Influence of misfit dislocations on the surface morphology of $Si_{1-x}Ge_x$ films, Appl. Phys. Lett. 66 (1995) 724–726.

[41] Yu.B. Bolkhovityanov, A.S. Deryabin, A.K. Gutakovskii, M.A. Revenko, L.V. Sokolov, Dominating nucleation of misfit dislocations from the surface in GeSi/Si(001) films with a stepwise composition grown by means of molecular-beam epitaxy, J. Cryst. Growth 293 (2006) 247–252.

[42] I.O. Akhundov, D.M. Kazantsev, V.L. Alperovich, N.S. Rudaya, E.E. Rodyakina, A.V. Latyshev, Formation and interaction of dislocation-induced and vicinal monatomic steps on a GaAs(001) surface under stress relaxation, Scr. Mater. 114 (2016) 125–128.

11

Atomic Processes in the Formation of Strained Ge Layers on Si(111) and (001) Substrates Within the Stransky–Krastanov Growth Mechanism

S.A. Teys

RZHANOV INSTITUTE OF SEMICONDUCTIOR PHYSICS OF THE SIBERIAN BRANCH OF RUSSIAN ACADEMY OF SCIENCES, NOVOSIBIRSK, RUSSIA

11.1 Introduction

The level of strain due to misfit arising in heterosystems with different lattice parameters is determined by the value of misfit between the mating lattices. A popular system composed of Ge on Si substrate with a 4% difference in lattice parameters is a well known strained structure. The growth according to the Stransky–Krastanov mechanism has been considered for a long time a characteristic feature of Ge epitaxy on Si [1–4]. This classic mechanism implies layer-by-layer growth of the so-called wetting layer (WL), followed by transition to the formation of three-dimensional islands. A widespread explanation of the transition from the layer-by-layer growth of WL to three-dimensional growth due to the change in the energy of growing film with the accumulation of misfit strain [5] is far from being a complete depiction of the nature of this phenomenon. The development of scanning tunneling microscopy (STM) [6] promoted the accumulation of additional information on different stages of growth in the Ge-Si heterosystem at the atomic level. In particular, this method provided new data on the mechanisms of formation of the pseudomorphous state of Ge phase on Si substrates with different orientations, as well as about the changes in the surface structure of the growing Ge layer [7–13]. The processes of Ge phase formation are accompanied by elastic and plastic deformations, surface phase transitions, and periodic atomic mass transfer.

The purpose of the present work was to demonstrate the mechanism of transition from WL formation to the start of three-dimensional growth, and the features of atomic processes during the growth of Ge film on Si(111) and (001) surfaces.

Advances in Semiconductor Nanostructures. DOI: http://dx.doi.org/10.1016/B978-0-12-810512-2.00011-1

11.2 Surface Preparation, Growth of Ge and STM Measurements

An atomically clean Si surface was obtained and epitaxial growth of Ge was carried out under ultra-high vacuum conditions in a STM manufactured by OMICRON. Silicon samples $12 \times 3 \times 0.4\ mm^3$ in size were cut out of n-type silicon wafers with a resistance of 0.5 ohm cm. Deviation of the surface from (111) and (001) planes did not exceed 0.5 degrees. Before mounting in the STM chamber, the samples were treated in an aqueous solution of a mixture of nitric acid and hydrogen peroxide to form a thin protective oxide layer. Subsequent removal of the protective oxide by heating in a vacuum allowed us to obtain a reproduceable atomically clean Si surface. The samples were preliminarily degassed in the vacuum chamber for not less than 4 h at a temperature of about 600°C, then the silicon surface was cleaned by heating at 1250°C for several seconds at a pressure not higher than $(1-2) \times 10^{-10}$ torr. The sample was heated by a direct current passing through. The sample temperature was calibrated with the help of an optical pyrometer. Sample cooling was usually carried out by a step-wise decrease in the heating current at a rate of 50°C/min within a temperature range of 900–350°C. Sample cooling after the growth stage occurred at a rate of several tens of degrees per second, due to interruption of the heating current. This allowed us to "freeze" the instantaneous growth picture, in contrast to inertial systems of indirect sample heating.

The source of Ge was a tungsten evaporator in the form of a metal strip with a fixed weighted portion of Ge. Due to the specific features of the lattice of diamond-like semiconductors, epitaxial growth on the Si(111) surface proceeds through double atomic layers: bilayers (BL), while on Si(001) it occurs through monolayers (ML). One Ge bilayer: BL_{111}, contains 1.44×10^{15} at/cm^2, which makes 2.3 ML_{001} because of the differences in surface density of atoms at the surfaces with different orientations. Deposition rate varied within the range 10^{-3}–10^{-1} BL/min. Ge coverage value was estimated from total Ge volume in two-dimensional three-bilayer islands [10] formed on Si(111) terraces far from the steps and interdomain boundaries (IB). Growth was carried out at a substrate temperature of 300–700°C; the optimal growth temperature for the majority of experiments was 400°C.

The STM images of the surface were taken at room temperature. For all STM images, light regions correspond to the prominent parts of surface relief, while dark regions correspond to deepenings. Tungsten tips obtained by means of electrolytic etching in alkaline solution were used as a STM probe. Analysis and processing of the STM images were carried out using the free software WSxM 4.0 [14,15].

11.3 Formation Ge Wetting Layer on Si(111)

Six stages of the structural state of Ge phase at the initial stage of growth on Si(111) in the mode of molecular beam epitaxy (MBE) can be demonstrated (Fig. 11.1).

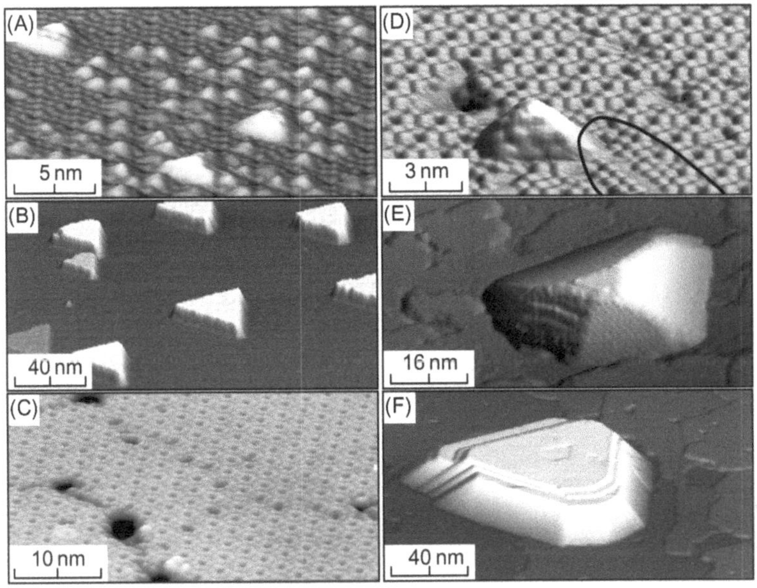

FIGURE 11.1 STM images of six successive stages of the formation of Ge phase on Si(111) at 400°C with germanium flux 9×10^{-3} BL/min. (A) Nucleation of Ge nanoclusters in 7×7 half-cells and 2D triangular Ge nanoislands of minimal size into two 7×7 cells; (B) formation of mainly three-bilayer Ge islands on Si(111)-7×7; (C) formation of the wetting layer with the surface structure 5×5 and numerous defects on the surface; (D) formation of a sharp-peaked nucleus of 3D Ge islands on the wetting layer; the region of the 2×2 structure characteristic of nonstrained Ge is outlined with a black curve; (E) 3D Ge island as a truncated pyramid at the moment of formation of a flat (111) top with the structures $2 \times$ [8; 2; $\sqrt{3}$] of nonstrained Ge (see Fig. 11.5A); (F) expanded 3D Ge island with subbilayer steps at the flat top from the emergence of partial dislocations along sloping planes {111}.

The first feature of WL formation on the Si(111) surface is accumulation of Ge in the form of nanoclusters (Fig. 11.1A) in the triangular halves of the unit cells of Si(111)-7×7 surface structure [7–10]. It follows from the experimental data [10,16] that half-cells 7×7 are intermediate shallow potential wells in which the primary accumulation of Ge in the form of nanoclusters occurs. The formation of nanoclusters on terraces may be observed below the growth temperature of 500°C, while the diffusion length of Ge adatoms remains less than the characteristic terrace width of 150–200 nm. Ge atoms do not stay in the shallow potential wells E_c of 7×7 half-cells at the growth temperatures $>500°C$. They overcome long distances and are embedded into deeper potential wells at the edges of steps forming a saw-shaped edge. Or they stay at the defects of the interdomain boundaries forming triangular islands (Fig. 11.2). An increase in Ge flux and a decrease in growth temperature cause an increase in the probability of nanocluster formation in 7×7 half-cells, up to their maximal concentration. Knowing the dimensions of the triangular half-cell 7×7, one can easily find that the maximal possible concentration of nanoclusters cannot exceed 3.2×10^{13} cm^{-2}. Typical MBE modes are far from equilibrium growth conditions because they are realized at relatively high deposition rates. In this situation, observation of the oscillations of the

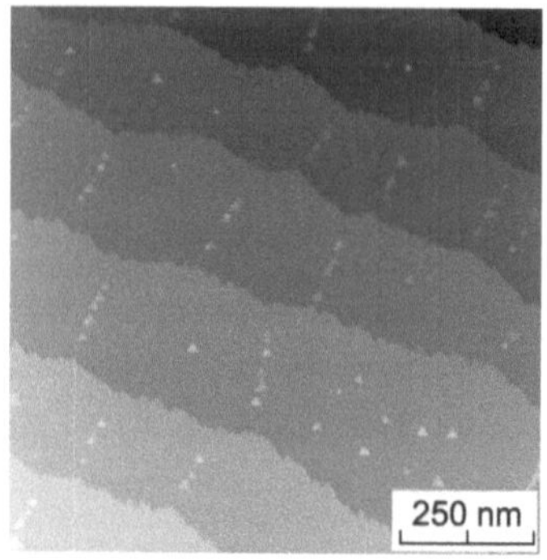

FIGURE 11.2 STM image of the Si(111) surface after the deposition of 0.05 BL of Ge at a temperature of 450°C at a rate of 9×10^{-3} BL/min. The chains of triangular Ge islands are located along the interdomain boundaries of the 7×7 structure.

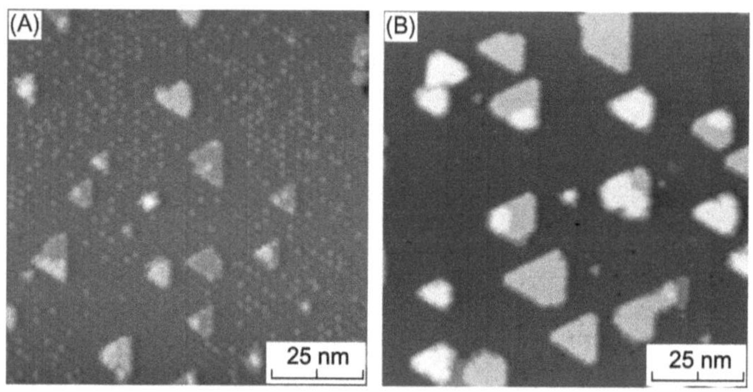

FIGURE 11.3 STM images of the Si(111) surface after Ge epitaxy at a temperature of 380°C at a rate of 4×10^{-3} BL/min. (A) Nanoclusters and islands after the deposition of 0.14 BL Ge; (B) only 2D islands are seen after continued Ge deposition up to 0.5 BL.

mirrored electron beam occurs under conditions of a high concentration of growth centers. Therefore, it is highly probable that deep-modulated oscillations of the intensity of a mirrored electron beam are provided directly by the formation and fusion of nanoclusters in 7×7 half-cells without the formation of two-dimensional islands [13].

Under quasi-equilibrium growth conditions, at first Ge nanoclusters are formed intensely in 7×7 half-cells (Fig. 11.3A), then they disappear completely (Fig. 11.3B) after nucleation and expansion of 2D Ge islands on even terraces [16,17]. An intermediate period of growth with the density of nanoclusters and islands equal to 3.7×10^{12} cm^{-2} and 2.5×10^{11} cm^{-2}, respectively, is shown in Fig. 11.3A. One can clearly see in Fig. 11.3A that there are many nanoclusters around the islands. While nanoclusters are not observed in Fig. 11.3B after tripling the amount of deposited Ge, the density of islands remains almost unchanged at 2.4×10^{11} cm^{-2}. Cluster decomposition in the presence of Ge islands also occurs during additional annealing of the samples at the growth temperature in the absence of Ge flux [10,18].

Decomposition of nanoclusters starts due to redistribution of Ge atoms into deeper potential wells E_s corresponding to atomic positions at the edges of steps on islands one bilayer high. These embedding sites are also located along initial bilayer Si steps, because surface regions free from nanoclusters and islands are formed along the steps during growth; these regions are called depletion zones [19]. The prevailing sink of Ge adatoms during the growth process to step edges prevents accumulation of the atoms in 7×7 half-cells, nucleation of nanoclusters, and formation of islands near 1 BL steps.

Returning to the formation of WL under quasi-equilibrium growth conditions, it is necessary to stress the second feature of WL formation. At the stage of completion of Ge redistribution from nanoclusters into 2D pseudomorphous islands (Fig. 11.1B), the islands increased not only in lateral size, but also in height, according to the multilayer growth mechanism [7,20]. Therefore, under quasi-equilibrium growth conditions, island growth according to the Vollmer–Weber mechanism is observed in the Si(111) surface.

The third feature is the limitation of Ge island height to three bilayers [9]. The height of the majority of islands in Fig. 11.1B is 3 BL; triangular islands with the height of 1 and 2 BL can be seen in the left part of Fig. 11.1B. The limitation of island height was recorded experimentally in Refs. [7,9,10], even before formation of the continuous strained layer. It is necessary to stress that exactly the same height limitation occurs during homoepitaxial growth of nonstrained Si islands on Si(111) [9,21]. So, height limitation is determined not by the accumulation of mismatch strain, but by the properties of the diamond-like lattice. Integration of sole steps into a configuration three BL high is also characteristic of the pure vicinal Si(111) surface deflected by several degrees toward a direction of the type of $\langle\bar{1}\bar{1}2\rangle$ [22]. It was established by us [19] that the depletion zone for three-bilayer steps during Ge growth on Si(111) is larger than that for steps 1 BL high. The increased depletion zone for three-bilayer steps is evidence of an even deeper potential well, E_{3s}, for the sink of adsorbed atoms, in comparison with a bilayer step.

An increased in "absorbing capacity" of three-bilayer steps is likely connected with the specific atomic structure of the edges of these steps. The presence of longitudinal and transverse dimers, and the high concentration of dangling bonds at the edge of a three-bilayer step [23] provide a quite logical explanation for the increased probability of the capture of diffusing Ge adatoms. In addition, it explains the limitation of island height to three bilayers. The increased sinking of Ge adatoms to the edges of three-bilayer steps, not only from the substrate surface, but also from small upper planes of the islands, prevents the formation of new upper layers on the islands and at the same time accelerates their lateral growth. Successive formation of deeper and deeper potential wells ($E_c < E_s < E_{3s}$) during Ge growth defines the direction of adsorbate redistribution, and the current morphology of the growth surface.

The fourth feature of Ge growth on Si(111) is the change in the surface structure of 2D Ge islands. It should be noted that the tightly packed surface structure $c2 \times 8$ is characteristic of the pure nonstrained Ge(111) surface [24,25]. The regions with 2×2 and $2 \times \sqrt{3}$ structures can also be detected with the help of STM on the Ge growth surface [12]. The cells of 2×2 and $2 \times \sqrt{3}$ phases are fragments of the $c2 \times 8$ phase, so below we will designate this group of related germanium structures as $2 \times [8;\ 2;\ \sqrt{3}]$. The density of $2 \times [8;\ 2;\ \sqrt{3}]$

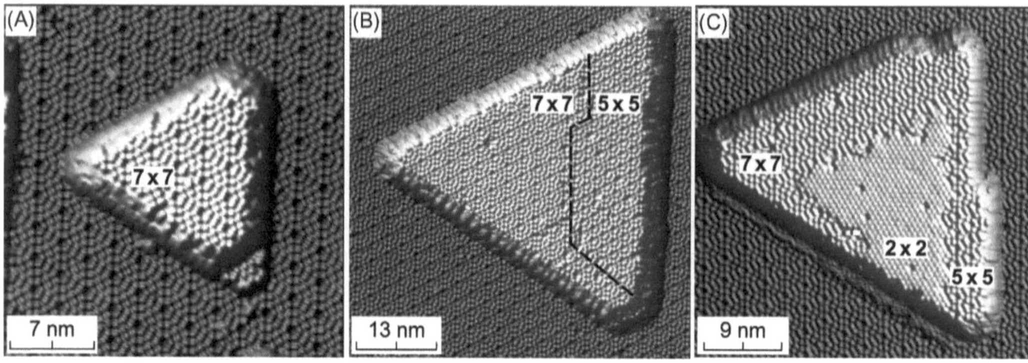

FIGURE 11.4 STM images of three-bilayer Ge islands. (A) With the 7×7 structure; (B) with 7×7 and 5×5 structures; (C) with 7×7, 5×5, and 2×2 structures.

structures 1 BL thick and the incomplete adatom layer is 2.25 at per primitive 1×1 unit cell. During Ge growth on Si(111), the surface of an overwhelming majority of Ge islands repeats the structure of a pure silicon surface, 7×7 [7–10] (Fig. 11.4A). This configuration of Ge atoms is quite logical for partial removal of strain in the islands, due to the arrangement of a less dense (2.08 at/1×1) surface structure (7×7)-Ge. Further lateral growth of the islands is accompanied by a gradual transition to the surface structure (5×5)-Ge with even lower density: 2.00 at/1×1 (Fig. 11.4B). According to our experimental data, after the deposition of 0.8 BL of Ge at a temperature of 400°C and a deposition rate of 2×10^{-3} BL/min, about 70% of the surface of Ge islands is occupied by the 7×7 structure, while 30% is occupied by the 5×5 structure. Only separate small islands contained tightly packed 2×2 (Fig. 11.4C) and $\sqrt{3} \times \sqrt{3}$ structures; at the moment of coalescence of three-bilayer islands into a continuous WL, the 5×5 structure occupies 100% of the surface [7,8,10] (Fig. 11.1C). A similar sequence of changes in surface structures $c2 \times 8 \rightarrow 7 \times 7 \rightarrow 5 \times 5$ was also observed in another strained heterosystem Sn-Ge(111) with an increase in Sn concentration [26]. So, changes in the surface structures during growth is a peculiar indicator of changes in the strained state of the heterosystem. A similar sequence of surface phase transitions depending on the strained state of the lattice is also confirmed by calculations relying on the first principles with using the density functional theory [27].

At increased growth temperatures (500–600°C) and increased adatom path length, the probability of Ge integration into bilayer steps dominates, three-bilayer growth is suppressed, and the transition to the surface structure 5×5 occurs when Ge thickness is less than 3 BL.

11.4 The Transition from the Wetting Layer to Three-Dimensional Growth of Ge on Si(111)

The fifth feature of Ge growth on Si(111) is an increase of several times in the rate of surface diffusion of Ge atoms over the WL at a constant growth temperature. In this situation, the

WL surface is composed of Ge atoms with the new structure 5 × 5 [16,17,28]. It should be noted that island nucleation on numerous IB, and the formation of nanoclusters in the unit cells of the 5 × 5 structure, do not occur on the WL surface, compared to the initial Si(111)-7 × 7 surface at the same growth temperature [10,16]. Under constant growth conditions, the density of nuclei of 3D islands on the surface of the WL (Fig. 11.1D) is several tens of times lower than the density of 2D Ge islands on the initial Si(111)-7 × 7 substrate (Fig. 11.3). Potential wells that are too shallow for the nucleation of 3D islands are formed on the surface of the WL with the Ge(111)-5 × 5 structure. For the sink of Ge atoms, deeper potential wells appear in the regions with increased defect content in the sites where several IB intersect, or near surface regions with 2 × [8; 2; $\sqrt{3}$] structures (Fig. 11.1D) characteristic of a nonstrained Ge(111) surface [11]. Rapid subsequent growth of 3D islands with {113} facets [4,7,29] also means the formation of deeper potential wells for the sink of atoms at the {113} surface in comparison with the potential relief of the Ge(111) surface with 5 × 5 structure. The region in which atoms get assembled around 3D islands increases up to several hundred nanometers. In addition, the growth of 3D islands is accompanied by partial destruction of elastically strained WLs near them [30], as well as the disappearance of some islands on the WL [9]. The transition of Ge atoms from the strained WL to 3D islands is evidence of the new redistribution of atoms from shallow potential wells of the 5 × 5 structure into deeper minima on the {113}-3 × 1 surfaces of 3D Ge islands.

The sixth feature, according to our data, is the nucleation of 3D Ge islands, at first in the form of pointed pyramids (Fig. 11.1D), in the assumed sites of the smallest deformation of the Ge lattice. Several facts provide indirect evidence of this. First, surface regions with a tightly packed 2 × 2 structure are observed near the nuclei of 3D islands (Fig. 11.1D, outlined with a *black curve*); this structure is characteristic of a nonstrained Ge(111) surface. Second, 3D Ge islands in the form of a trihedral pyramid are facetted by the {113} planes (Fig. 11.5A) with a surface structure of 1 × 3 [7]. It should be noted that the surface structure Ge(113)-1 × 3 was discovered on nonstrained germanium by Olshanetsky and coauthors [31]. Third,

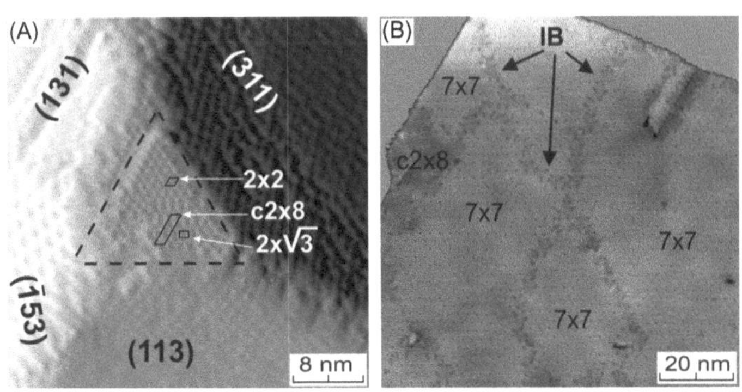

FIGURE 11.5 STM images of 3D Ge islands on the wetting layer. (A) An island 9 nm high immediately after the formation of the flat top (outlined with *dash line*), orientations of side faces are indicated in parentheses, *arrows* indicate the unit cells of surface structures of nonstrained Ge(111); (B) the image of the flat top of an island 6 nm high with several domains of the 7 × 7 structure; *arrows* indicate interdomain boundaries (IB).

at the moment of formation of flat tops of 3D Ge islands (Fig. 11.1E), to a height of 7 ± 2 nm [11,12], the surface structure of upper (111) surfaces contains only tightly packed unit cells of $2 \times [8; 2; \sqrt{3}]$ structures peculiar to the nonstrained Ge(111) surface (Fig. 11.5A).

A further increase in the size of 3D islands and, therefore, elastic deformation, again causes rearrangement of the family of 2×2, $2 \times \sqrt{3}$, and $c2 \times 8$ structures into the 7×7 structure at the tops of 3D islands (Fig. 11.5B) and back. The first transition from $2 \times [8; 2; \sqrt{3}]$ to a less dense 7×7 structure is due to an increase in the elastic deformations of the 3D island. The inverse transition to the structure of nonstrained germanium $2 \times [8; 2; \sqrt{3}]$ is connected with the plastic relaxation of 3D islands due to the introduction of mismatch dislocations into the Ge-Si(111) interface [11]. In Fig. 11.5B, one can see several domains of the 7×7 phase separated by domain boundaries, the regions of which contain mainly the 2×2 phase. The presence of the domains of 7×7 structure, shifted with respect to each other by a noninteger value of the translation vector of the 7×7 surface structure, is evidence of independent nucleation of the 7×7 phase at different regions of the flat top of Ge islands. The phase transition involves partial strain relaxation, due to the formation of less dense packing of the adatoms in the upper layer, and the formation of deep corner vacancies in the 7×7 unit cell. A wave-like deformation of the upper atomic plane of the island is clearly seen in Fig. 11.5B. This kind of Ge surface modulation on Si(111) was discovered for the first time with the help of analysis of the profiles of diffraction reflections of slow electrons (SPA LEED) in Ref. [32], which was interpreted as a result of the formation of a net of mismatch dislocations sliding in the Ge-Si interface. Later on, surface modulation was also visualized by means of STM on Ge islands [33]. Theoretical calculation of strain field arising within the volume of Ge film as a result of the introduction of mismatch dislocations into the interface showed that the surface modulation from the network of dislocations can be observed with the help of STM only to a Ge film thickness of <8 nm [34].

At the same time as the phase transitions of $7 \times 7 \rightarrow 2 \times [8; 2; \sqrt{3}]$, during plastic relaxation, steps appear on island surfaces due to the emerging dislocations; their height is less than 1 BL (Fig. 11.6). Dislocations are introduced into elastically strained Ge islands

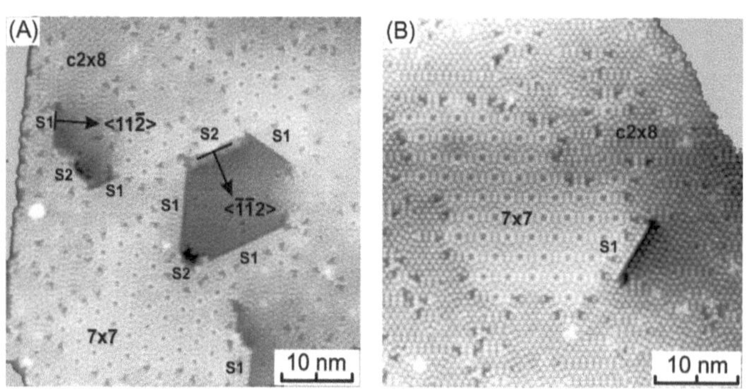

FIGURE 11.6 STM images of the tops of Ge islands after the deposition of three germanium bilayers on the Si(111) surface at a temperature of 400°C. (A) On an island 12 nm high, S_1 steps are 0.13 ± 0.02 nm high, S_2 steps are 0.24 ± 0.02 nm high; *arrows* mark the fronts of the steps in the direction of the lower-lying terrace; (B) an island 15 nm high, with nucleating S_1 step and stacking fault.

both through sliding in the plane of the interface (111) and along the sloping slip planes {111}. Both complete 60-degree dislocations, and partial Shockley dislocations with a 30 or 90 degrees (edge) orientation of Burgers vector can be introduced at the sloping planes. During the formation of Shockley dislocations, stacking faults arise in the sloping planes; they were discovered in some islands with the help of TEM [11]. The output of stacking faults to island surfaces causes the formation of steps with a height less than the thickness of one bilayer. We discovered two types of subbilayer steps: S_1 and S_2, with a height of 1/3 and 2/3 BL, respectively (Fig. 11.6A). The S_1 steps are known from literature [7], while S_2 were discovered for the first time by us.

The steps are always arranged along the directions of the type of $\langle 110 \rangle$. The normal to the front of the S_1 step is the direction of the type of $\langle 11\overline{2} \rangle$, while the normal to the front of S_2 is the direction $\langle \overline{1}\overline{1}2 \rangle$. A STM image of the initial stage of the nucleation of an S_1 step and stacking fault from the surface of a strained Ge island is shown in Fig. 11.6B.

The statistical processing of the images of flat tops of Ge islands with the orientation (111) showed that under growth conditions at 400°C, Ge flux 5×10^{-3} BL/min, and coverage value 3 BL, the S_1 type steps are mainly observed on island tops (Fig. 11.6B). The fraction of S_2 type steps does not exceed 6%. An increase in Ge coverage by only 0.5 BL causes a noticeable increase in island heights and in the number of S_2 steps (up to 40%). In addition, as a rule, S_1 steps arise in the central part of island tops and at its ends, while S_2 steps arise only in the central part of the island top, provided that there are one or two common (extreme, or angular) points with S_1 steps (Fig. 11.6A). A possible reason of the formation of these steps is the nucleation of partial Shockley dislocations with Burger's vectors of the type $(a/6)\langle 112 \rangle$ on the surface of strained islands; they are inclined to the step line at an angle of 30 or 90 degrees; as a result, the heights of the steps differ by a factor of 2. Sloping planes {111} passing through S_1 steps contain subtraction stacking faults, so partial dislocations extend from step ends to the depth of the island. It should be pointed out that the fraction of partial dislocations sliding in the sloping slip planes {111}, and forming the steps of S_1 and S_2 types, is insignificant with respect to the major number of mismatch dislocations into the Ge-Si interface under quasi-equilibrium growth conditions.

11.5 Formation of a Wetting Layer and Hut-clusters of Ge on Si(001)

The above features of Ge growth on Si(111) do not manifest themselves for Ge epitaxy on Si (001) substrates. Unlike for the (111) surface, atomically clean nonstrained (001)Si and Ge surfaces have identical atomic structure of surface phases with a periodicity of 2×1 at room temperature [25]. As a result of reconstruction, surface atoms form pairs, or dimers, which get lined up in even rows along the $\langle 110 \rangle$ direction (insert in Fig. 11.7A). This surface structure contains only half of the unsaturated dangling bonds in comparison to the nonrearranged surface. Unlike for the (111) plane, the (001) surface rearrangement involves only one upper monolayer

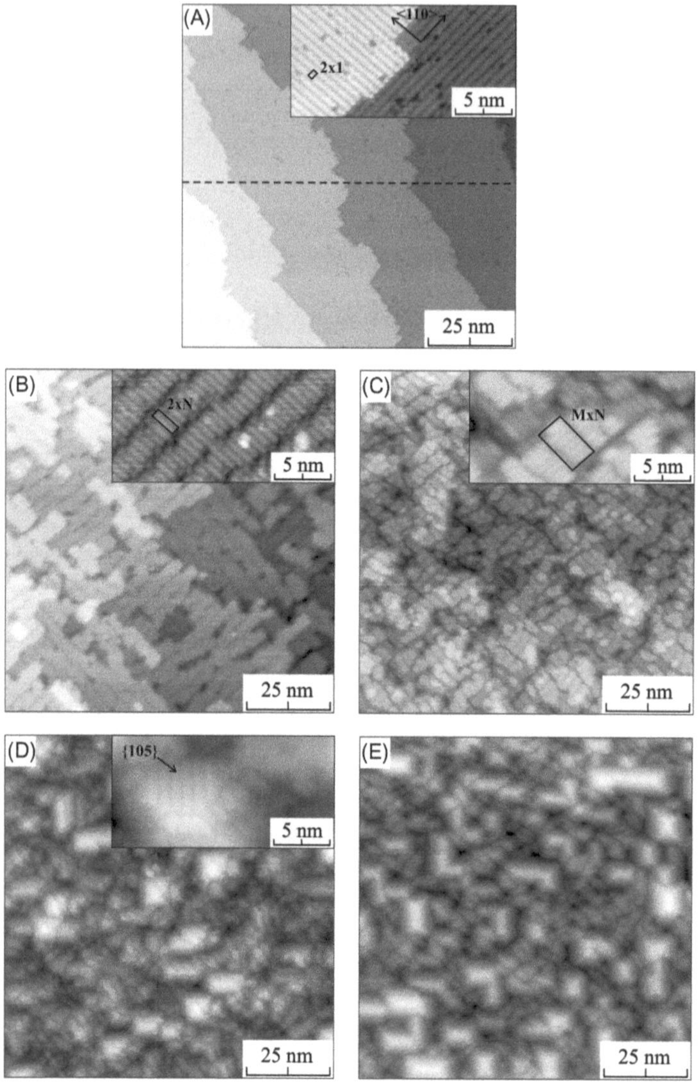

FIGURE 11.7 STM images of successive changes in the surface morphology of the wetting layer of Ge on Si(001) during growth at a rate of 1.2×10^{-2} ML/min at a temperature of 400°C. (A) Initial clean Si(001) surface; insert shows the rows of dimers and the unit cell of the 2×1 surface structure; (B) Ge $2 \times N$ surface structure after the deposition of 1 ML of Ge; insert shows the $2 \times N$ unit cell; (C) the surface after deposition of 2 ML of Ge; insert shows the $M \times N$ structure; (D) the surface after deposition of 3 ML of Ge; insert shows a separate {105} facet; (E) the surface with hut-clusters after the deposition of 4 ML of Ge.

of atoms without the participation of additional adatoms. The rows of dimers on neighboring terraces are located mutually perpendicularly, as shown in the insert in Fig. 11.7A.

During the growth process, Ge atoms mainly get built into step edges, and they also form 2D islands 1 ML high on extended terraces, similar to the growth on the Si(111) surface.

However, unlike for the (111) orientation, the formation of nanoclusters does not occur. The formation of nanoclusters is impossible in unit cells of small size 2×1. A meeting of two adatoms on the Si(001)-2×1 surface leads to the formation of a stable dimer on the upper-lying layer. According to the data of electron diffraction, the unit cells of the surface structure 2×1 based on dimers remain stable up to a temperature of $\sim 1200^\circ$C [35].

Alternating on the surface sections of dimers and vacancies in the form of absent dimers is the reason for the observation of different surface structures on Si(001) [36–38], including those connected with the presence of admixture atoms. For Ge growth on Si(001), many researchers [9,39–43] observed the formation of so-called $2\times N$ structures (Fig. 11.7B). The unit cell 2×1 remains almost unchanged, but new periodicity with N decreasing from 16 to 8 with an increase in Ge coverage appears in the diffraction patterns, due to the formation of regular vacancies and their alignment perpendicular to dimer rows. This is the method by which the Ge-Si(001) heterosystem removes part of the mismatch strain [9]. It must be stressed that the absence of each eighth dimer on the surface is evidence that the upper layer is filled with Ge adatoms only to 88%. This means that after the deposition of exactly 1.0 ML Ge, after independent calibration of Ge flux, the real thickness of the Ge layer on the surface reaches approximately 1.14 ML. Formally, the appearance of the $2\times N$ structure indicates a strain relaxation process through surface rearrangement, similar to the case of Si(111). However, the unit cells of the 2×1 structure do not change; instead, accumulation of vacancies in the growing layer takes place; therefore, the density of the upper layers decreases.

An increase in Ge coverage above 1 ML leads to the formation of the rows of vacancies perpendicular to already existing ones (Fig. 11.7C). The formation of a so-called $M\times N$ structure is observed on the surface [42–44]. These $M\times N$ unit cells are limited by the grooves formed by vacancies (insert in Fig. 11.7C). Surface roughness increases so rapidly (Figs. 11.7B and C) that it is impossible to distinguish the initial system of regular steps on the STM image (Fig. 11.7A).

The development of roughness leads to several interesting effects. First of all, the surface diffusion of the adsorbate slows down, because the vacancies lined up are in fact the edges of steps. Anticipating, it should be stressed, that the density of future hut-clusters [5] remains almost unchanged during low-temperature growth. According to our data, the densities of hut-clusters for Ge flux equal to 1.2×10^{-2} ML/min with a 4 ML coating, differ only insignificantly and are equal to 5.7×10^{11} cm^{-2} and 5.2×10^{11} cm^{-2}, for growth temperatures of 300°C and 400°C, respectively. Second, the actual thickness of the WL and surface roughness increase due to the accumulation of vacancies in the growing layer (Fig. 11.8). In Fig. 11.8, the lower line with small undulations corresponds to the surface profile of the initial clean Si (001) surface, with regular monoatomic steps and rows of dimers. The average amplitude of surface roughness does not exceed 2 ML. In the case of ideal uniform distribution of 1 ML of deposited Ge over the surface (*shaded area* in Fig. 11.8), the amplitude of surface roughness would remain unchanged. However, measured amplitudes of surface roughness (Fig. 11.8) increase monotonously with the deposition of Ge layers. Third, small film regions surrounded by the grooves composed of vacancies turn into isolated islands possessing the potential for free relaxation of the strained lattice. Indeed, an increase in the lattice

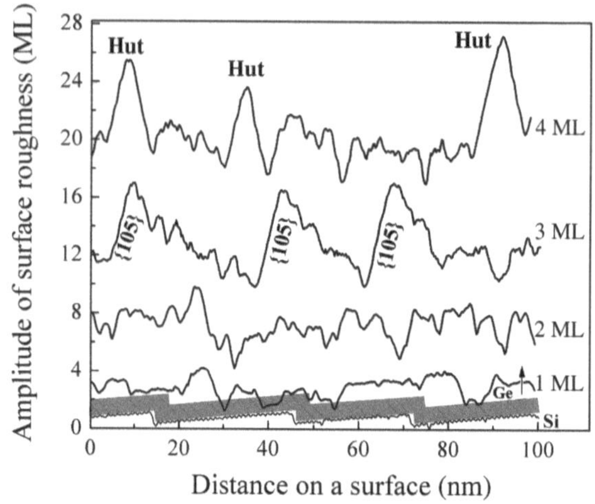

FIGURE 11.8 Experimental profiles of the roughness of Ge film on the growth surface depending on the thickness of the deposited Ge layer at 400°C at a rate of 1.2×10^{-2} ML/min. The lower line with small goffer corresponds to the profile of initial clean Si(001) surface measured along the *dashed line* on the STM image shown in Fig. 11.7A. A uniformly distributed 1 ML of Ge is shown schematically as a *shaded area*.

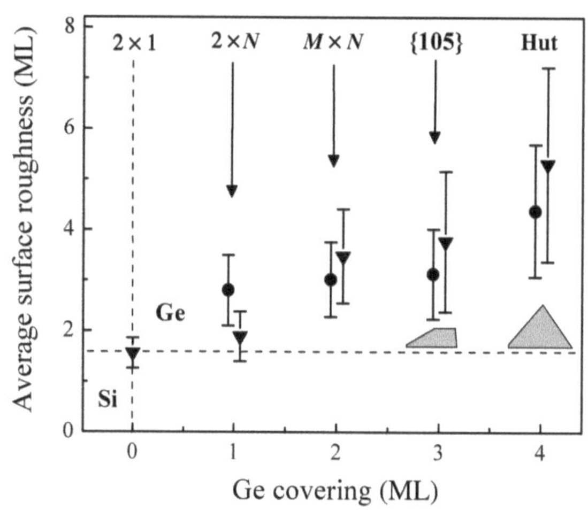

FIGURE 11.9 Dependence of the average roughness of Si(001) surface on coverage with Ge at a growth rate of 1.2×10^{-2} ML/min. *Circles* mark growth at 300°C, *triangles* mark growth at 400°C. *Arrows* show surface structures corresponding to Ge coating.

parameter of the WL surface to values larger than the value for bulk Ge at this stage of growth was detected by means of fast electron diffraction [45]. So, with a smooth increase in average surface roughness (Fig. 11.9), the dimer is conserved as the basic structural element of the surface, and the unit cell 2×1 is also conserved. This is the first distinguishing feature of the rearrangement of surface structure on Si(001) during Ge epitaxy.

For facets {105} and hut-clusters to be formed (Figs. 11.7D and E), the appearance of a 10-degree deviation from the (001) plane is necessary. Such a deviation from the substrate orientation is seen first of all in deepenings and grooves of the $M \times N$ structure or near the steps. The initial formation of {105} facets in pits was demonstrated for the first time with the help of STM by Goldfarb for Ge growth by means of gas-phase epitaxy [42,46]. The direct experimental confirmation of the formation of separate {105} facets became possible as a result of the observation of the same surface region with the help of STM. However, the presence of hydrogen in gas-phase epitaxy [42,46] cast doubt on the total mechanism of the formation of separate {105} facets. However, the existence of faceting in pits [43] and the formation of hut-clusters of Ge near deepenings also demonstrated this mechanism for the MBE mode. The {105} faceting in pits is also present in STM images after MBE, as reported by other authors [9,47,48].

Our experimental data obtained in the MBE mode also confirm the formation of separate {105} facets in deepenings on the rough surface, as shown in Fig. 11.7D. A single {105} face surrounded by an $M \times N$ structure is clearly seen in the insert to Fig. 11.7D. The image was obtained owing to experiments carried out under quasi-equilibrium growth conditions, with small Ge fluxes and a high rate of sample cooling. This allowed us to detect the intermediate stage of the formation of a single facet: the nucleus of a future tetrahedral pyramid, or a hut-cluster. So, the second distinguishing feature of Ge growth on Si(001) is the formation of single {105} facets in the region of vacancy grooves and WL faceting by {105}, mainly to the depth of the WL, occurring at low growth temperatures under the conditions of limited surface diffusion. The size of separate {105} facets is determined by the expansion of neighboring hut-clusters or their facets, as well as by the number of Ge atoms of the WL near these facets. As a result of local {105} faceting of the WL, many researchers had the opportunity of observing a dense array of hut-clusters. Chaotic WL faceting was demonstrated most expressively for low-temperature (300°C) Ge epitaxy [44]. The array of uneven merged hut-clusters is evidence of the irregular formation and growth of {105} facets.

The formation of separate {105} facets in pits was not detected in direct observation of the surface in the MBE mode for Ge deposition rates equal to 6×10^{-2} ML/min [49,50]. One can see that the formation of fully-fledged tetrahedral pyramids (hut-clusters) occurred rather rapidly within the time necessary to record one STM image frame (approximately within 1 min). Therefore, it is an even greater problem to detect the intermediate stages of the growth of hut-clusters at higher growth rates: 4.2 ML/min and relatively slow subsequent cooling $\sim$0.4 °C/s [51]. In addition, it was demonstrated experimentally by Goldfarb [42] that a hut-cluster can be formed directly at the site of deepening, while more often the formation of hut-clusters is observed near steps and deepenings [42,46,52]. Therefore, some pits may be enclosed by hut-clusters. It is also necessary to state that the video [50] clearly demonstrates the transformation of hut-clusters with a square base into wedge-shaped hut-clusters [49,50]. The transition of pyramids into wedge-shaped hut-clusters is also confirmed by the symmetrical plots showing the changes in the concentration of hut-clusters of different kinds, though the authors deny the transformation of the pyramids with a square base into wedge-shaped pyramids with a rectangular base [48,51].

The deviation of the surface from (001) is also possible due to ascending diffusion of the adsorbate and mass transfer to energy-attractive growth centers in the upper layers, for example, to surface regions with lattice parameters close to those of bulk Ge [45], or to impurity centers [53]. As a rule, these mechanisms dominate at increased growth temperatures. An increase in temperature causes acceleration of the diffusion processes, Ge atoms pass over diffusion barriers on the steps and on the $M \times N$ structure, and become concentrated in deeper potential wells of growth centers. For example, Ge epitaxy at $500-650°C$ demonstrates the formation of "prepyramids" [54], and the generation of separate hut-clusters [5,44]. The concentration of hut-clusters decreases sharply because separate pyramids do not adjoin each other through {105} facets, and do not hinder the growth of each other. According to our experimental data, at a Ge flux equal to 1.2×10^{-2} ML/min the concentration of hut-clusters decreases to 1×10^{11} cm^{-2} at the growth temperature of 500°C and coverage of 4 ML.

The high rate of {105} facet growth is explained by low diffusion barriers at this surface [55,56]. In addition, the formation of {105} facets is accompanied by the local transition of Ge atoms from initial {001} planes into {105} planes, which is the evidence of the presence of deeper potential wells for Ge building in {105} facets.

11.6 Conclusion

To conclude, it must be noted that we considered the features of Ge phase formation of Si (111) and (001) substrates under quasi-equilibrium conditions of Ge growth within the framework of the Stransky–Krastanov mechanism. Successive formation of deeper and deeper potential wells during Ge growth was demonstrated. An increase and partial relaxation of strain determines the direction of Ge redistribution, and the current morphology of the growth surface. The mechanism of the transition of Ge film from WL to three-dimensional growth was demonstrated.

The consistent change of surface structures $2 \times [8;\ 2;\ \sqrt{3}] \rightarrow 7 \times 7 \rightarrow 5 \times 5$ on (111) substrates and $2 \times 1 \rightarrow 2 \times N \rightarrow M \times N$ on (001) during the formation of the WL is accompanied by a smooth decrease in the density of the growing film, and is a particular indicator of the change in the strained state of the Ge-Si heterosystem. The transition from the WL to three-dimensional growth due to the formation of new facets provides the next stage of the relaxation of elastic strain on the side facets of three-dimensional islands.

During Ge epitaxy on Si(001), before the formation of {105} facets, a smooth increase in surface roughness occurs; the basic structural element of the surface, i.e., dimer, is conserved in this process, as well as the unit cell of the surface structure 2×1. The formation of single {105} facets in vacancy grooves of the rough surface leads to faceting of the WL mainly to the depth, and to the formation (at low growth temperature, 300–400°C) of an array of hut-clusters with the density only weakly dependent on growth temperature.

Acknowledgments

The work was carried out with the financial support from the Russian Foundation for Basic Research and the Integration Project of the Joint Programme of the National Academy of Sciences of Ukraine and the Russian Academy of Sciences.

References

[1] T. Narusawa, W.M. Gibson, Pseudomorphic structure at the interface of Ge on Si(111) studied by high-energy-ion scattering, Phys. Rev. Lett. 47 (1981) 1459–1462.

[2] T. Ichikawa, S. Ino, RHEED study on the Ge/Si(111) and Si/Ge(111) systems: reaction of Ge with the Si (111)(7 × 7) surface, Surf. Sci. 136 (1984) 267–284.

[3] M. Asai, H. Ueba, C. Tatsuyama, Heteroepitaxial growth of Ge films on the Si(100)-2 × 1 surface, J. Appl. Phys. 58 (1985) 2577–2583.

[4] S.M. Pintus, S.I. Stenin, A.I. Toropov, E.M. Truhanov, V.Yu Karasyov, Morphological transformations of thin heteroepitaxial films, Thin Solid Films 151 (1987) 275–288.

[5] Y.-W. Mo, D.V. Savage, B.S. Swartzentruber, M.G. Lagally, Kinetic pathway in Stranski-Krastanov growth of Ge on Si(001), Phys. Rev. Lett. 65 (1990) 1020–1023.

[6] G. Binning, H. Rohrer, Ch Gerber, E. Weibel, Surface studies by scanning tunneling microscopy, Phys. Rev. Lett. 49 (1982) 57–60.

[7] U. Kohler, O. Jusko, G. Pietsch, B. Muller, M. Henzler, Strained-layer growth and islanding of germanium on Si(111)-(7 × 7) studied with STM, Surf. Sci. 248 (1991) 321–331.

[8] N. Motta, A. Sgarlata, R. Calarco, Q. Nguyen, J. Castro Cal, F. Patella, et al., Growth of Ge-Si(111) epitaxial layers: intermixing, strain relaxation and island formation, Surf. Sci. 406 (1998) 254–263.

[9] B. Voigtlander, Fundamental processes in Si/Si and Ge/Si epitaxy studied by scanning tunneling microscopy during growth, Surf. Sci. Rep. 43 (2001) 127–254.

[10] S.A. Teys, B.Z. Olshanetsky, Formation of the wetting layer in Ge/Si(111) epitaxy at low growth rates studied with STM, Phys. Low-Dimens. Struct. 1/2 (2002) 37–46.

[11] S.A. Teys, E.M. Trukhanov, A.S. Ilyin, A.K. Gutakovskii, A.V. Kolesnikov, Initial stages of Ge epitaxy on Si (111) under quasi-equilibrium growth conditions, JETP Lett. 92 (2010) 388–395.

[12] S.A. Teys, E.M. Trukhanov, A.S. Ilyin, A.V. Kolesnikov, Superstructure phase transitions in Ge on Si(111) at the initial stage of epitaxial growth, Bull. Russ. Acad. Sci. Phys. 75 (2011) 1055–1058.

[13] S.A. Teys, Features of atomic processes at the formation of a wetting layer and nucleation of three-dimensional Ge Islands on Si(111) and Si(100) surfaces, JETP Lett. 96 (2012) 794–802.

[14] WSxM—free software at: <www.wsxmsolutions.com>.

[15] I. Horcas, R. Fernandez, J.M. Gomez-Rodriguez, J. Colchero, J. Gómez-Herrero, A.M. Baro, WSXM: a software for scanning probe microscopy and a tool for nanotechnology, Rev. Sci. Instrum. 78 (2007) 013705–013708.

[16] S.A. Teys, A.B. Talochkin, B.Z. Olshanetsky, Formation of Ge nanoislands before the completion of a wetting layer in the Ge/Si(111) system, J. Cryst. Growth 311 (2009) 3898–3903.

[17] S.A. Teys, A.B. Talochkin, K.N. Romanyuk, B.Z. Olshanetsky, Growth of germanium nanoislands and nanowiers on singular and vicinal Si(111) surfaces prior to the formation of a wetting layer, Phys. Solid State 46 (2004) 80–84.

[18] Yu.B. Bolkhovityanov, S.Ts Krivoschapov, A.I. Nikiforov, B.Z. Ol'shanetskii, O.P. Pchelyakov, L.V. Sokolov, et al., Synthesis of ordered Ge–Si heterostructures containing ultrasmall germanium nanoclusters, Phys. Solid State 46 (2004) 64–66.

[19] K.N. Romanyuk, S.A. Teys, B.Z. Olshanetsky, Scanning tunneling microscopy study of the growth and self-organization of Ge nanostructures on vicinal Si(111) surfaces, Phys. Solid State 48 (2006) 1820–1826.

[20] B. Voigtlander, A. Zinner, T. Weber, High temperature scanning tunneling microscopy during molecular beam epitaxy, Rev. Sci. Instrum. 67 (1996) 2568–2572.

[21] B. Voigtlander, T. Weber, Nucleation and growth of Si/Si(111) observed by scanning tunneling microscopy during epitaxy, Phys. Rev. B54 (1996) 7709–7712.

[22] B.Z. Olshanetsky, S.A. Teys, Phase transitions on clean and nickel containing vicinal Si(111) surfaces, Surf. Sci. 230 (1990) 184–196.

[23] S.A. Teys, K.N. Romanyuk, R.A. Zhachuk, B.Z. Olshanetsky, Orientation and structure of triple step staircase on vicinal Si(111) surfaces, Surf. Sci. 600 (2006) 4878–4882.

[24] R.M. Feenstra, A.J. Slavin, Scanning tunneling microscopy and spectroscopy of cleaved and annealed Ge(111) surfaces, Surf. Sci. 251/252 (1991) 401–407.

[25] K. Oura, V.G. Lifshits, A.A. Saranin, A.V. Zotov, M. Katayama, Surface Science An Introduction, Springer-Verlag, Berlin Heidelberg, 2003.

[26] T. Ichikawa, S. Ino, Structural study of Sn-induced superstructures on Ge(111) surfaces by RHEED, Surf. Sci. 105 (1981) 395–428.

[27] R. Zhachuk, S. Teys, J. Coutinho, Strain-induced structure transformations on Si(111) and Ge(111) surfaces: a combined density-functional and scanning tunneling microscopy study, J. Chem. Phys. 138 (2013) 224702–224709.

[28] V. Cherepanov, B. Voigtlander, Influence of material, surface reconstruction, and strain on diffusion at the Ge(111) surface, Phys. Rev. B69 (2004) 125331–125338.

[29] P.M.J. Maree, K. Nakagawa, F.M. Mulders, J.F. Van der Veen, K.L. Kavanagh, Thin epitaxial Ge-Si(111) films: study and control of morphology, Surf. Sci. 191 (1987) 305–328.

[30] A.A. Shklyaev, M. Shibata, M. Ichikawa, Instability of two-dimensional layers in the Stranski-Krastanov growth mode of Ge on Si(111), Phys. Rev. B58 (1998) 15647–15651.

[31] B.Z. Olshanetsky, V.I. Mashanov, A.I. Nikiforov, LEED studies of clean high Miller index surfaces of germanium, Surf. Sci. 111 (1981) 429–440.

[32] M. Horn-von Hoegen, A. Al-Falou, H. Pietsch, B.H. Muller, M. Henzler, Formation of interfacial dislocation network in surfactant mediated growth of Ge on Si(111) investigated by SPA-LEED, Surf. Sci. 298 (1993) 29–42.

[33] S.N. Filimonov, V. Cherepanov, N. Paul, H. Asaoka, J. Brona, B. Voigtlander, Dislocation networks in conventional and surfactant-mediated Ge/Si(111) epitaxy, Surf. Sci. 599 (2005) 76–84.

[34] A.S. Il'in, E.M. Trukhanov, S.A. Tiis, A.K. Gutakovskii, Analysis of the dislocation structure at the Ge/Si (111) heterointerface, J. Surf. Investig. 8 (2014) 787–793.

[35] B.Z. Olshanetsky, A.V. Rzhanov, F.L. Edelman, Structure of the silicon surface (100) after oxidation at high temperatures, Semiconductors 7 (1973) 2312–2315.

[36] A.E. Dolbak, B.Z. Olshanetsky, S.I. Stenin, S.A. Teys, T.A. Gavrilova, Effect of nickel on clean silicon surfaces: transport and structure, Surf. Sci. 218 (1989) 37–54.

[37] P.W. Murray, R. Lindsay, F.M. Leibsle, P.L. Wincott, G. Thornton, Direct observation of the c(8×8) defect structure on Si(001) using scanning tunneling microscopy, Phys. Rev. B54 (1996) 13468–13471.

[38] W. Kim, H. Kim, G. Lee, Ja-Yong Koo, Initial stage of carbon incorporation into Si(001) and one-dimensional ordering of embedded carbon, Phys. Rev. Lett. 89 (2002) 106102–106104.

[39] Y.-W. Mo, M.G. Lagally, Scanning tunneling microscopy studies of the growth process of Ge on Si(001), J. Cryst. Growth 111 (1991) 876–881.

[40] X. Chen, F. Wu, Z. Zhang, M.G. Lagally, Vacancy-vacancy interaction on Ge-covered Si(001), Phys. Rev. Lett. 73 (1994) 850–853.

[41] F. Wu, X. Chen, Z. Zhang, M.G. Lagally, Reversal of step roughness on Ge-covered vicinal Si(001), Phys. Rev. Lett. 74 (1995) 574–577.

[42] I. Goldfarb, P.T. Hayden, J.H.G. Owen, G.A.D. Briggs, Nucleation of "Hut" pits and clusters during gas--source molecular-beam epitaxy of Ge/Si(001) in situ scanning tunneling microscopy, Phys. Rev. Lett. 78 (1997) 3959–3962.

[43] A. Rastelli, H. von Kanel, G. Albini, P. Raiteri, D.B. Migas, L. Miglio, Morphological and compositional evolution of the Ge/Si(001) surface during exposure to a Si flux, Phys. Rev. Lett. 90 (2003) 216104.

[44] M. Tomitori, K. Watanabe, M. Kobayashi, O. Nishikawa, STM study of the Ge growth mode on Si(001) substrates, Appl. Surf. Sci. 76/77 (1994) 322–328.

[45] A.I. Nikiforov, V.A. Cherepanov, O.P. Pchelyakov, Growth of Ge islands on the Si(100) Surface by MBE, J. Surf. Investig. 6 (2002) 74–78.

[46] I. Goldfarb, P.T. Hayden, J.H.G. Owen, G.A.D. Briggs, Competing growth mechanisms of Ge/Si(001) coherent clusters, Phys. Rev. B56 (1997) 10459–10468.

[47] D.E. Jesson, M. Kastner, B. Voigtlander, Direct observation of subcritical fluctuations during the formation of strained semiconductor Islands, Phys. Rev. Lett. 84 (2000) 330–333.

[48] L.V. Arapkina, V.A. Yuryev, Ge quantum dot arrays grown by ultrahigh vacuum molecular-beam epitaxy on the Si(001) surface: nucleation, morphology, and CMOS compatibility, Nanoscale Res. Lett. 6 (2011) 522–615.

[49] M. Kastner, B. Voigtlander, Kinetically self-limiting growth of Ge islands on Si(001), Phys. Rev. Lett. 82 (1999) 2745–2748.

[50] <http://www.fz-juelich.de/video/voigtlaender>.

[51] L.V. Arapkina, V.A. Yuryev, Nucleation of Ge quantum dots on the Si(001) surface, Phys. Rev. B82 (2010) 045315.

[52] I. Goldfarb, Effect of strain on the appearance of subcritical nuclei of Ge nanohuts on Si(001), Phys. Rev. Lett. 95 (2005) 025501–025504.

[53] R. Butz, H. Luth, The influence of carbon on the surface morphology of Si(100) and on subsequent Ge island formation, Thin Solid Films 336 (1998) 69–72.

[54] A. Vailionis, B. Cho, G. Glass, P. Desjardins, D.G. Cahill, J.E. Greene, Pathway for the strain-driven two-dimensional to three-dimensional transition during growth of Ge on Si(001), Phys. Rev. Lett. 85 (2000) 3672–3675.

[55] F. Montalenti, D.B. Migas, F. Gamba, L. Miglio, Fast isotropic adatom diffusion on Ge(105) dot facets, Phys. Rev. B70 (2004) 245315–245318.

[56] L. Huang, G.-H. Lu, F. Liu, X.G. Gong, First-principles study of adsorption and diffusion on Ge/Si(001)-(2 × 8) and Ge/Si(105)-(1 × 2) surfaces, Surf. Sci. 601 (2007) 3067–3072.

12

Molecular Beam Epitaxy of $Cd_xHg_{1-x}Te$

Yu.G. Sidorov[1], A.P. Anciferov[1], V.S. Varavin[1], S.A. Dvoretsky[1], N.N. Mikhailov[1], M.V. Yakushev[1], I.V. Sabinina[1], V.G. Remesnik[1], D.G. Ikusov[1], I.N. Uzhakov[1], G.Yu. Sidorov[1], V.D. Kuzmin[1], S.V. Rihlicky[1], V.A. Shvets[1], A.S. Mardezov[1], E.V. Spesivcev[1], A.K. Gutakovskii[1], A.V. Latyshev[1,2]

[1]*RZHANOV INSTITUTE OF SEMICONDUCTOR PHYSICS OF THE SIBERIAN BRANCHOF THE RUSSIAN ACADEMY OF SCIENCES, NOVOSIBIRSK, RUSSIA* [2]*NOVOSIBIRSK STATE UNIVERSITY, NOVOSIBIRSK, RUSSIA*

Infrared imaging systems are required for applications in medicine, agriculture, the chemical industry, ferrous and nonferrous metals metallurgy, the fuel industry, and other areas of the economy. Due to high quantum efficiency, tunable absorption wavelength, and a wide operating temperature range, HgCdTe (mercury cadmium telluride, MCT)-based infrared focal plane arrays are a good choice for the fabrication of high-performance infrared imaging systems. Creating arrays and matrices of photodetectors with a large number of elements imposes stringent requirements for narrow-band MCT. What is wanted is a large area plate MCT with highly uniform properties and low cost.

MCT technology over the past 20 years has evolved from the production of relatively small (diameter <10 mm) volume bars to large (up to 150 mm in diameter) epitaxial layers on alternative substrates. The electrical properties of solid solutions MCT are largely determined by the intrinsic point defects and residual impurities associated with the growing method. MCT is characterized by a high rate of diffusion of point defects. Therefore, a low growth temperature is need. These requirements are best suited to epitaxial techniques of growth. Molecular beam epitaxy (MBE) MCT technology has reached the level required for the preparation of material for practical applications. The material quality of the instrument grows with the composition of the alloys required for the application for all wavelengths;—from short-wave to a very long wavelength. In the MCT films grown by MBE, a concentration of conduction electrons (1−5) of 10^{14} cm^{-3} is routinely obtained, which is not inferior to the results obtained by liquid phase epitaxy (LPE), and developed methods of doping donor and acceptor impurities. Growth rates in the map layers using MBE currently account for 3−5 μm/h, with working layers of 10−15 μm thick grown in no more than 3 h, which is comparable to the times for growing layers using the LPE method. MBE is the most promising method for growing MCT layers on substrates of CdZnTe, gallium arsenide, and silicon. MBE is superior to other methods for growing MCT films because of a number of circumstances.

Advances in Semiconductor Nanostructures. DOI: http://dx.doi.org/10.1016/B978-0-12-810512-2.00012-3

12.1 Advantages and Problems of MBE MCT

1. MCT photodiodes with a heterojunction currently display the best characteristics. Heterostructures grown by LPE are usually limited to two or three layers, because different melts are required for the preparation of layers differing in composition and doping. MBE is ideally suitable for the growth of layers differing in composition and doping. MBE is also ideally suitable for deposition of multilayer structures used in two-color IR-videosystems and monolithic IR-arrays. MBE is the only technique well adapted to the fabrication of superlattice detectors for the very long-wave infrared range consisting of alternating CdTe and HgTe layers of 100–200 Å thickness.
2. In MBE, MCT film growth is carried out at the lowest temperatures (180°C in MBE, compared with 450°C for LPE and 300–400°C for metal organic vapor phase epitaxy). The reduction of growth temperature precludes impurity diffusion from the substrate, thus reducing background doping in the epilayer.
3. Unlike in LPE, no aggressive medium is present in MBE. The absence of an aggressive medium combined with low temperature and analytical facilities for surface control make the use of alternative substrates for growth of epitaxial MCT films possible. Alternative substrates (CdTe or CdZnTe epilayers grown on Al_2O_3, GaAs, or Si substrates) show a definite advantage over bulk CdZnTe substrates in terms of larger possible size, attainability, lower price, and improved matching of thermal expansion coefficient with the read out systems.
4. The main advantage of MBE when using alternative substrates is in the unlimited possibilities of growing different compounds in a single growth system. The advantage of growing multilayer structures in one device is determined by the fact that in MBE no contamination of interfaces occurs. Another advantage of MBE over LPE is that one can control the composition at the interfaces; this makes possible the creation of heterojunctions exhibiting the highest quality.
5. Increased productivity installations for growing MCT and reducing the cost of heterostructures can only be achieved with a high reproducibility of the process. The technologies which are controlled by facilities of built-in analytical processes of growth and preparation of substrates have the advantage, and in this respect, MBE has no competitors.

The implementation of the advantages of MBE, however, require overcoming extremely complex scientific and technological problems, which probably explains why MBE has not become the leading technique in MCT fabrication. High composition uniformity is needed over large areas of the substrate, with a high precision of composition specification and maintenance (<0.005). These are contradictory requirements. Typically, to achieve high uniformity in size the method of MBE substrate rotation is used. However, the rotation of the substrate eliminates continuous use of analytical methods for controlling the composition of the growing layer of MCT, so it is necessary to be guided only by the temperature of the molecular sources. Maintaining the composition method based on temperature control of the molecular sources is not sufficiently precise and reliable, due to the change over time in the amount of evaporated material sources which make it impossible to change the composition of a controlled thickness heteroepitaxial structure. In connection with this, we developed a MBE chamber design that ensures content uniformity on large area substrates without rotation, which allows continuous monitoring of the analytical methods of the growing process. Ellipsometry is the most suitable method for measuring MCT films composition. The successful application of in situ ellipsometry can uniquely specify the composition of the growing film MCT and can receive structures with a given profile of compositional distribution in thickness.

12.2 Defects Caused by the Use of Substrates From Nonisovalent Compounds

There are two objective factors leading to intensive defect formation in MBE grown A^2B^6 films. The first is due to the low growth temperature for compounds containing mercury. The second is due to the use of substrates from nonisovalent compounds, because substrates of the required size and quality made from the deposited materials have several disadvantages. It is known that intermediate compounds of the $A^3{}_2B^6{}_3$ type are formed at the A^2B^6–A^3B^5 interface [1]. These compounds may influence the growth mechanism and defect formation. In the research on epitaxy, however, one usually studies the influence of the lattice mismatch on the structure of fabricated films. The dependence of the structure of the grown heterosystems on technological growth conditions is also studied (see e.g., Ref. [2]). At the same time, little attention is paid to questions related to the influence of the mutual chemical interaction of the film and substrate components on the film growth and defect formation mechanisms.

We have studied these processes at the initial stage of MBE growth of A^2B^6 compounds on GaAs substrates taking the heteropair ZnSe-GaAs as an example. The ZnSe-GaAs system may be regarded as a model, as ZnSe and GaAs have nearly the same lattice parameters, the tendency to form chemical compounds film components, and the substrate is stronger than that of other systems, such as CdTe-GaAs. The stereochemistry of the surface compounds created from A^2B^6 and A^3B^5 compounds can be successfully studied when the number of valence electrons participating in the formation of surface compounds is taken into account. An excess of valence electrons can lead to a distortion in the tetrahedral configuration of the compounds at the interface (e.g., to an atom i.e., forming two or three, rather than four bonds), and as a result to differing orientations of the film and substrate.

Selenium interacts with GaAs, substitutes arsenic in the GaAs lattice, and forms a chemical bond with gallium [3]. In a lattice with a tetrahedral bond configuration, which is the case for GaAs, there should be four valence electrons per atom. An A^3 atom gives three valence electrons, and a B^6 atom six. As a result, for an A^3B^6 compound there is one extra electron. This is the reason why for A^3B^6 compounds, the tetrahedral lattice turns out to be unstable. Nevertheless, the formation of such a lattice for A^3B^6 compounds is still possible if one third of all the sites in the metal sublattice are vacant. The chemical formula for a compound corresponding to this type of lattice is $A_2{}^3B_3{}^6$. Indeed, in an $A_2{}^3B_3{}^6$ compound there will be $(3 \times 2 + 6 \times 3)/6 = 4$ electrons per site, with vacant sites included.

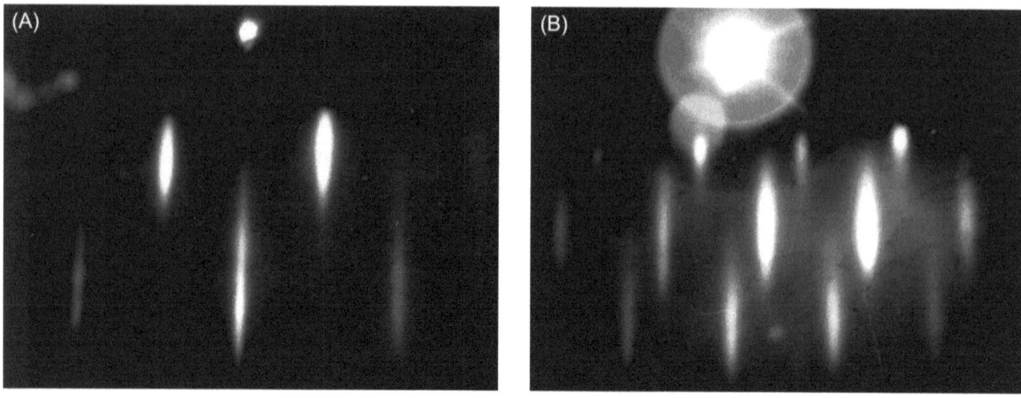

FIGURE 12.1 The diffraction pattern from the surface of GaAs(112)B after preparation with a high content of Se. (A) Azimuth ⟨111⟩, and (B) Azimuth ⟨110⟩.

It should be pointed out that in a crystal lattice with defects, the bonds of group six atoms are not equivalent. There are bonds to the metal atoms and saturated bonds oriented toward the vacant sites. Let us consider the surface of such a crystal formed by a group of six atoms whose saturated bonds have an outward orientation. This surface will have minimal excess energy and will be the most stable, because there are no broken unsaturated bonds on it. Obviously, the (111)B surface corresponds to this configuration of bonds. Therefore, one should expect that crystals of $A_2{}^3B_3{}^6$ compounds with a tetrahedral configuration of bonds will favor being faced by the (111)B surface. The formation of a surface with saturated bonds is possible not only for a compact $A_2{}^3B_3{}^6$ compound. If a (111)B GaAs substrate is treated with the vapor of a group six element, chalcogen atoms will substitute the arsenic atoms in the surface layer, and also form a surface with saturated bonds.

The epitaxy of A^2B^6 compound films on such surfaces will pose some problems connected with the saturated character of the surface bonds. A possible solution to this situation is the formation of a halogen atom with a six bond octahedral configuration. A hexavalent state is typical for chalcogen. In this case, three bonds of halogen atom direct to the three gallium atoms of substrate, and three bonds direct to the atoms of the second group element of the deposited film. Note that the configuration corresponds to (111)A. It follows that the excess valence electrons on the (111)B GaAs, selenium treated, can lead to a change of polarity (111) on the B(111) to A(111), and the formation of twins during film deposition A^2B^6 compounds.

Thus, the crystal–chemical model heterojunction A^2B^6/GaAs, based on an analysis of the average number of valence electrons per formula unit, predicts that the formation of the lattice with tetrahedral coordination bonds of Ga–Se leads to disruption of the balance of the valence electrons in the heterojunction. An excess of valence electrons on the heterojunction leads to faceting, twinning, and violation of the stoichiometry of the growing structure.

Experiments have shown [4] that steps on the GaAs(112)B surface are formed at sufficiently intensive treatment (temperature and pressure of selenium vapor, see Fig. 12.1); Se peaks in XPS spectra are observed from a surface other than Ga and As peaks.

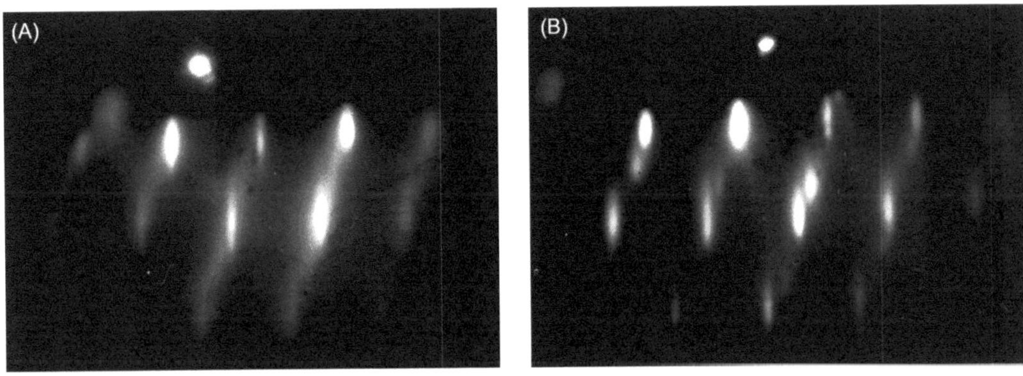

FIGURE 12.2 The diffraction patterns of growth in the initial moment of ZnSe on GaAs(112)B azimuth ⟨110⟩. (A) The growth at 30 s, (B) Growth at 90 s.

This type of pattern is characteristic of the diffraction system of equidistant steps on the surface of the sample along the direction of incidence [4]. It is evident that the splitting reflex multiple of the distance between the reflexes is in the ratio of 1–3. From this, it follows that the ratio between the step width and the identity parameter for (112)GaAs is 3.

Long-term exposure to Se vapor at temperatures exceeding 400°C leads to facets (111). Twinning is observed when A^2B^6 film is grown on such surface [5] (see Fig. 12.2).

The experimentally observed behavior of the surface treated with the molecular flow of selenium, is consistent with the findings of stereochemistry.

12.3 Processes in the Adsorption Layer at MBE $Cd_xHg_{1-x}Te$ and CdTe

Thermodynamic analysis of elemental Te and HgTe crystallization during MBE growth was carried out. The supersaturation was expressed through the Gibbs free energy of the Te and HgTe crystallization [6]:

$$\Delta G_{\mathrm{Te}} = -RT \ln\left[P_{\mathrm{Te}_2}^{1/2} / P_{\mathrm{Te}_2}^{01/2}\right]$$

$$\Delta G_{\mathrm{HgTe}} = -RT \ln\left[\left(P_{\mathrm{Te}_2}^{1/2} \times P_{\mathrm{Hg}}\right) / K_{\mathrm{HgTe}}\right]$$

where ΔG is the Gibbs free energy of the crystallization process, $P^0_{\mathrm{Te}_2}$ is the pressure vapor of tellurium over pure tellurium, K_{HgTe} is the HgTe dissociation constant [7], P_{Hg} is the pressure of Hg at the growth, and P_{Te_2} is the equivalent beam pressure of Te at the growth.

The dependence of ΔG on temperature is shown in Fig. 12.3 [8]. P_{Te_2} is the pressure which is needed for a 1 μm/h growth rate of the HgTe film, $P_{\mathrm{Hg}} = 10^{-3}$ Pa, *solid line*: ΔG_{HgTe} (crystallization of HgTe is possible at $T < T_2$); *dotted line*: ΔG_{Te} (crystallization of Te is possible at $T < T_1$).

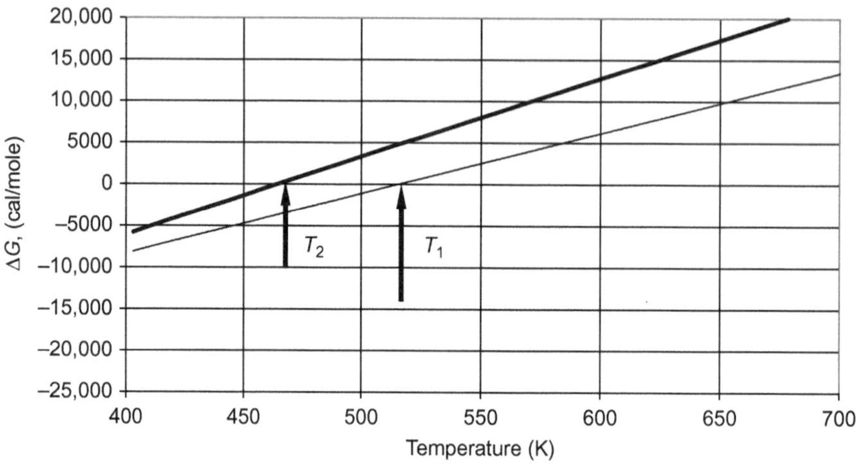

FIGURE 12.3 Calculated dependences of supersaturation on temperature: *solid line*: ΔG_{HgTe} (crystallization of HgTe at $T < T_2$); *dotted line*: ΔG_{Te} (crystallization of Te at $T < T_1$).

At $T_1 > T > T_2$, the single stable solid phase is elemental tellurium. Under these conditions the polycrystalline growth of Te is experimentally observed by RHEED. At $T < T_2$, the stable solid phases are Te and HgTe. The crystallization of these phases will be determined by kinetic circumstances.

A sufficient quantity of mercury on the surface facilitates the formation phase of mercury telluride. When the crystallization of HgTe is complicated (e.g., if mercury adsorption is difficult) the probability of elemental Te crystallization increases.

There are molecules of Te_2 in the molecular beam incident on the growing surface. The dissociation energy of the Te_2 molecule is 54.2 kcal. The diatomic molecule must dissociate into atoms before tellurium can be arranged in a crystal lattice.

As follows from the calculations, the dissociation energy of the Te_2 molecule on a singular surface and near the step decreases to 30 and 10 kcal, respectively. It is clear that in MCT MBE (low growth temperature) a good way of decreasing the Te_2 dissociation energy is to use the vicinal planes of substrates.

12.4 MBE System for Growing Narrow-Gap Solid Solutions Containing Mercury

When designing and building MBE installations for growing narrow-gap solid solutions of mercury, the following requirements were taken into account [9,10]:

1. According to the research into heteroepitaxy $A^{II}B^{VI}$ compounds on GaAs (Section 12.2), a necessary condition for reducing defects in heterojunctions is the absence of vapors of elements of the sixth group at the thermal cleaning of the GaAs substrate from the oxide. Satisfaction of this condition requires the introduction of a special technological installation chamber in which there is no sputtering of elements of the sixth group.

2. Due to the high pressure of mercury vapor, the vacuum system should provide an appropriate means of pumping. The use of a MCT technological module for growing buffer layers is impractical.
3. An automated control system (ACS) for technological processes should ensure the maintenance of precise conditions in the MCT process of growth.
4. The requirement for high composition uniformity layers in the area, together with the requirement for high precision of composition, necessitates the creation of a special tooling chamber for growing MCT, and the introduction of ellipsometric control growth in situ.

Ellipsometric control of the composition is based on the dependence of the MCT optical constants on X_{CdTe}. Thereby, at the angle of incidence $\varphi = 70$ degrees parameter Ψ varies from 4 degrees for CdTe to 15 degrees for HgTe, providing a composition sensitivity of $\delta X_{CdTe} \sim 0.001$.

To get the calibration curve $\Psi(X_{CdTe})$ accurate measurements of MCT films of high structural perfection were obtained by MBE and vapor phase epitaxy [11]. Quadratic approximation of these results (see Fig. 12.4) gives $\Psi(X_{CdTe}) = 14.90 - 13.14X_{CdTe} + 2.36X_{CdTe}^2$. The calibration curve can be recalculated in relation to the measurement conditions in the chamber (angle of incidence and temperature), and allows determination of the composition with an accuracy of $\delta X_{CdTe} = \pm 0.01$, and its relative change during film growth; is no worse than 0.001.

MCT heterostructures were grown on an MBE installation "Ob-M" equipped for operation with mercury [12]. The schematic layout of the system is shown in Fig. 12.5 It includes load-lock and unload-lock chambers. The loading is done in a laminar box. In

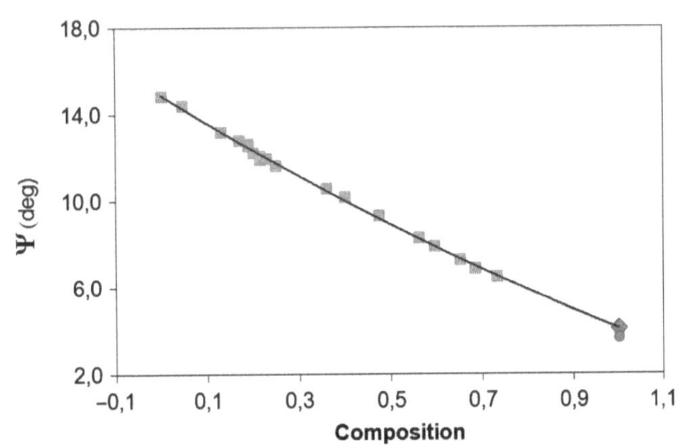

FIGURE 12.4 Calibration curve $\Psi(X_{CdTe})$ for composition determination. *Angle of light incidence is equal to 70°. Symbols are experimental data corresponding to the samples obtained by vapor phase epitaxy (squares), MBE (circles) and cleavage (rhomb);* the *solid curve* is an approximation of experiment by quadratic dependence $\Psi(X_{CdTe}) = 14.90 - 13.14X_{CdTe} + 2.36X_{CdTe}^2$.

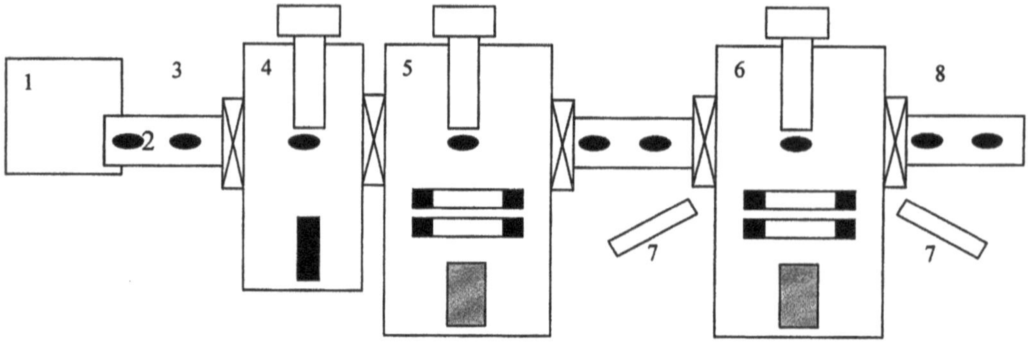

FIGURE 12.5 The scheme of MBE equipment for growing HS MCT. (1) Laminar box; (2) substrate; (3) load chamber; (4) chamber of preepitaxial treatment of substrates; (5) chamber of buffer layers growing; (6) chamber of MCT layers growing; (7) automatic ellipsometer; (8) unload chamber.

this MBE system, three technological chambers are used, with different compositions of the residual gases.

The preepitaxial substrate preparation chamber has an As molecular source. Only thermal treatment of the substrates at a required temperature in the arsenic vapor is performed there. The process is controlled with the help of RHEED.

The chamber for growing buffer layers composed of chalcogenides of group two elements contains vapor of tellurium at a level of 10^{-6} Pa and is, therefore, as has been shown, not suitable for carrying out the procedures of preepitaxial substrate preparation. The chamber is equipped with RHEED and ellipsometer for buffer layer thickness measurements. The chamber for growing narrow-gap solid solutions containing mercury is equipped with a special system of molecular sources providing a sufficiently uniform distribution of solid solution composition without substrate rotation. These sources provide a high uniformity of molecular flows over a large area at a minimum consumption of evaporated materials. Calculations show that the deviation of the composition (x_{CdTe}) should not exceed the value 0.0002 cm^{-1} at the exact alignment. The high uniformity of flow eliminates substrate rotation and implements continuous monitoring of the growth process.

The vapor pressure in the chamber is of the order of 10^{-4} Pa and, therefore, the chamber cannot be used for procedures of preepitaxial substrate surface preparation and buffer layer growth. The control of growth processes was carried out by in situ ellipsometry. An automatic ellipsometer was used for measurements of composition.

The MBE technology installation process is controlled by an automated system. The automated process control system equipment allows you to adjust the temperature of the molecular source and the substrate, and also to control ellipsometry film composition, and manage the growth process. A distinctive feature of the control system is the use of a built-in automated ellipsometer to determine the parameters of the film, with the introduction of feedback on the temperature of molecular beam sources, which allows you to fully automate the process of film growth.

A single-wavelength high-speed ellipsometer (LEF-755) has been developed which can provide control of the MBE MCT processes with its use [13]. No moving parts, as well as an absence of signal modulation during measuring provide a high-speed, which is limited only by digitization. In this device, a single measurement time is 200 ms, with a measurement accuracy $\delta\Psi = \delta\Delta = 0.01$ degrees.

12.5 Growing Heteroepitaxial MCT Structures on a GaAs Substrate

As shown in Section 12.2, surface GaAs(112)B restructuring is possible with the formation of facets (011), (111), and (113) at heteroepitaxy $A^{II}B^{VI}$ compounds on GaAs. ZnTe-CdTe buffer layers are grown on the substrate GaAs to match the crystal lattices of GaAs and MCT. Introduction of a layer of zinc telluride uniquely prevents the change of orientation that almost inevitably occurs during the deposition of cadmium telluride film on GaAs.

Technology developed in ISP SB RAS of MBE MCT used the GaAs substrate (310) [9,10,13]. The surface (310) is not widespread in MBE, and therefore the literature contains no information about its properties. Plane (310) belongs to the ⟨001⟩ zone formed planes (100) and (110). The surface (310) can be obtained by rotating the (100) plane to 19.11 degrees to the plane (110) relative to the [001] axis. It is believed that a system of equidistant steps is formed on the surface at these corners. For the plane (310) azimuth [001] correspond to the direction along the steps, and the azimuth $[\bar{1}30]$ corresponds to the direction across the steps.

In his handbook, Nichols [11] presented an elementary cell of the (310) surface of crystals with a diamond structure. This cell is a parallelogram with an angle forming 122.31 degrees. The handbook provides binding of the cell to two perpendicular directions, $[\bar{1}35]$ and $[\bar{1}3\bar{2}]$. Using the equation of vector geometry it is not difficult to see that the sides of the parallelogram are parallel to [001] and $[\bar{1}3\bar{2}]$, and its small diagonal is parallel to the direction $[0\bar{1}3]$ (see Fig. 12.6).

It was found that, during thermal removal of passivating oxide with GaAs(310) substrate, the most notable changes are observed in the azimuths [001] and $[\bar{1}3\bar{2}]$. As desorption of residual contaminants from the surface of the annealed substrate occurs, diffraction reflexes

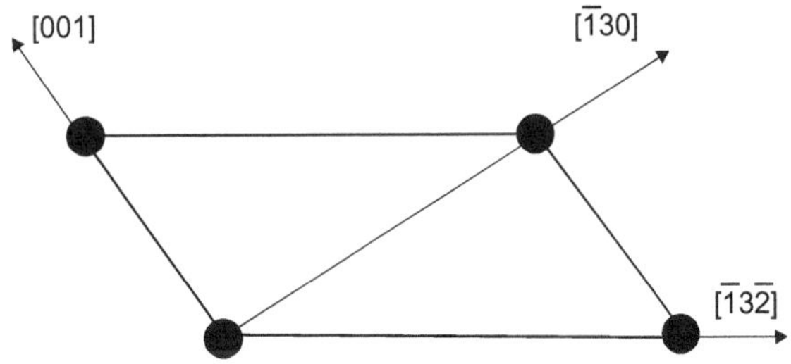

FIGURE 12.6 Elementary cell of the (310) surface of the crystals with diamond structure.

are narrower and stretched. The azimuth $[\bar{1}3\bar{2}]$ fractional reflexes appear, dividing the distance between the main reflexes in half. The azimuth [001] reflexes split, and the diffraction pattern takes the characteristic form of a stepped surface (see Fig. 12.7).

Auger analysis revealed that the thus reconstructed surface comprises O and C in amounts below Auger spectrometer detectivity. This morphological reconstruction surface was chosen by us as a reference point, indicating the receipt of an atomically clean surface of GaAs(310).

Preepitaxial processing of samples was carried out in two ways. The first method was to heat "epi-ready" substrate under a vacuum to remove the passivating oxide, and obtain the appearance of the superstructure (1 × 2). The second way is to remove the passivating oxide in a hydrochloric acid solution in isopropyl alcohol [14], followed by heating in vacuo until the superstructure appears (1 × 2). The second method rather than the first allowed us to lower the preepitaxial annealing temperature to 70°C. For a reference point for preepitaxial annealing, temperatures were selected for removal of passivating oxide with "epi-ready" substrate, which was considered equal to 580–600°C, and the temperature of GaAs destruction which is assumed to be 630–650°C. Relative changes in preepitaxial annealing temperature were determined by a thermocouple of the substrate manipulator.

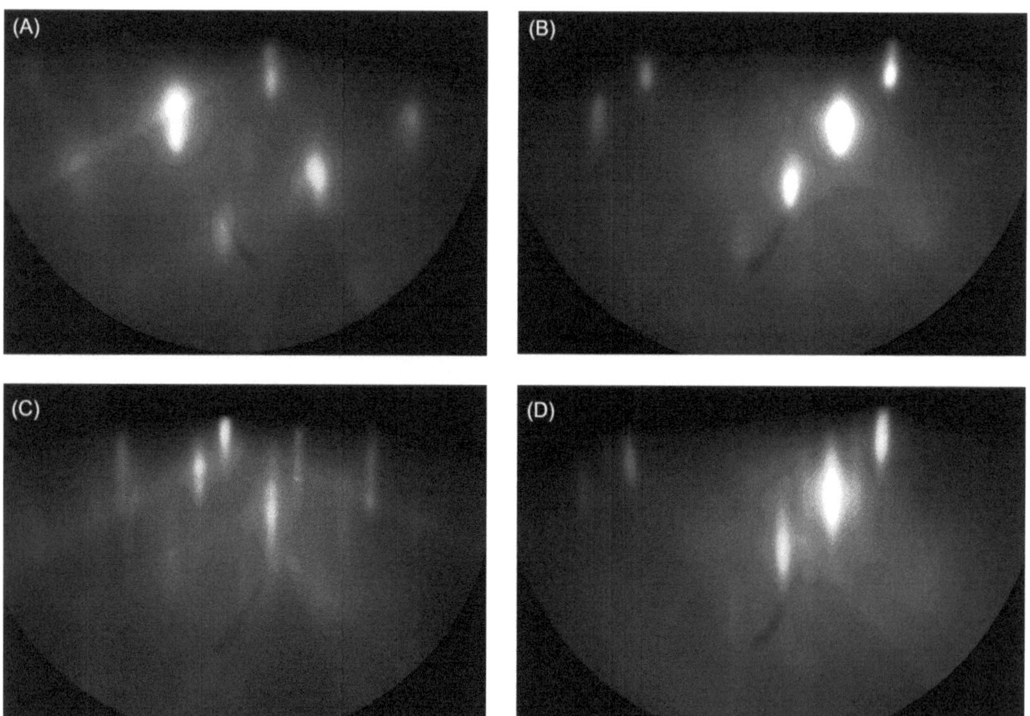

FIGURE 12.7 The diffraction pattern from the surface of GaAs(310) in the azimuth [001] and $[\bar{1}3\bar{2}]$. (A and B) Before annealing; (C and D) after annealing.

Since further considerations of the stoichiometry of the GaAs surface after annealing plays an important role, preepitaxial annealing regimes were tested on the substrate GaAs(100). Annealing at 600°C, and a background As pressure in the chamber of $3.2 \cdot 10^{-4}$ Pa leads to the appearance of an arsenic stabilized superstructure (2 × 4), which when cooled to 50°C passed to the superstructure (2 × 3). Therefore, it can be assumed that having the annealing temperature at 500–520°C and a background As pressure in the chamber of $5.6 \cdot 10^{-4}$ Pa should not introduce elemental Ga on the substrate surface GaAs(310).

12.5.1 Zinc Telluride Epitaxy

The pressure ratio of evaporated components has a decisive influence on the morphology of the growing film in MBEMBE of binary semiconductors. The optimum ratio of vapor pressures of groups II and VI components are unknown for ZnTe(310) and CdTe(310) films. Studies have established the following patterns of morphological transformations during zinc telluride and cadmium telluride epitaxy.

If, during growth, Te_2 vapor pressure is greater than the pressure of Zn or Cd vapor, by increasing the film thickness surface roughening occurs. In the diffraction pattern, this leads to a reflex constriction point and lower intensity (see Fig. 12.8B). If, during the growth, the vapor pressure of Zn and Cd is more than the Te_2 vapor pressure, an atomically smooth surface will be formed during epitaxy. The diffraction pattern from such a surface is characterized by narrow reflexes, sharp Kikuchi lines, and low diffusion background (see Fig. 12.8A).

12.5.2 Epitaxial Growth of Cadmium Telluride

A converging spiral-like curve in the $\Psi-\Delta$ plane is observed with the growth of cadmium telluride films on a GaAs substrate. The start point corresponds to GaAs, and the end point corresponds to bulk cadmium telluride. There is good agreement between the theoretical curve

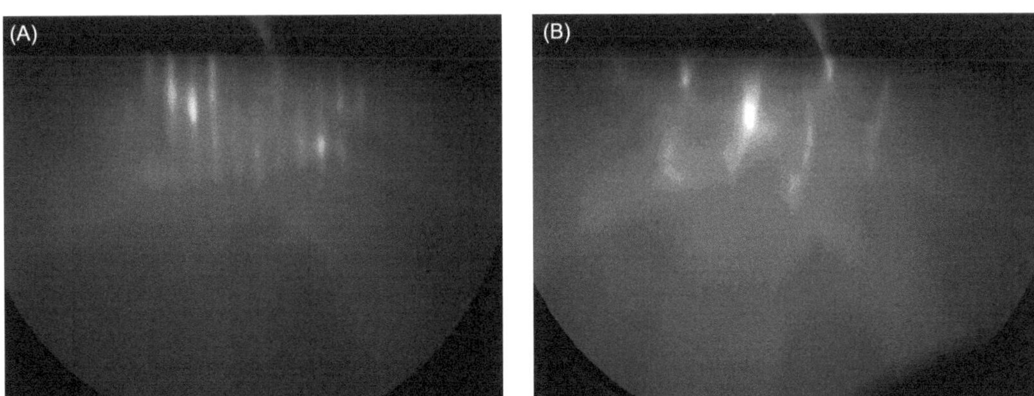

FIGURE 12.8 Diffraction patterns in azimuth [001] from the surface of CdTe(310) 5 μm thickness film. (A) Film grown under conditions $P_{Cd} > P_{Te}$, $T_S = 280°C$. (B) Restructuring of the surface shown in Fig. 5.3a at Te adsorption, $P_{Te} = 1 \cdot 10^{-5}$ Pa, $T_S = 210°C$.

and the experimentally measured Ψ, Δ. This fact allows the determination of the epitaxial film growth rate. However, in the case of nonoptimal growth conditions, a deviation of the measured ellipsometric parameters from the theoretical curve takes place [15].

The same as for zinc telluride, there is an optimal ratio of cadmium and tellurium flows, providing epitaxial cadmium telluride growth with the best structural perfection. For these growth conditions, the surface state must not be changed, i.e., no surface morphology should be developed. As in the previous case (zinc telluride growth), surface relief is equivalent to a layer with different optical constants. The angle delta (Δ) is the most sensitive of ellipsometric parameters to relief. Fig. 12.9 shows the evolution of ellipsometric parameter Δ for three cases of CdTe epitaxial growth. The conditions of surface preparation were identical in all cases, as well as the ZnTe buffer layer growth conditions. In these experiments, the CdTe growth rate was 2.1 μm/h, and the growth temperature $T = 290°C$. In the case of the optimal flow ratio $J_{Cd}/J_{Te_2} = 3.5$, changes in the ellipsometric parameters Ψ and Δ do not occur. When CdTe film is growing with a large excess $J_{Cd}/J_{Te_2} = 27$, a weak monotonic decrease of ellipsometric parameter Δ occurs. While growing CdTe film in conditions close to tellurium enrichment, $J_{Cd}/J_{Te_2} = 1$is accompanied by a sharp decrease of ellipsometric angle Δ by a value of ≈30 degrees at an early growth stage. Using this method, optimal process conditions for growing CdTe films were determined.

Epitaxial layers of CdTe, grown on ZnSe/GaAs(013) substrates with optimal technological parameters ($J_{Cd}/J_{Te_2} = 5-7$, $T_{sub} = 280–290°C$, growth rate ≈2 μm/h) have the best structural perfection (half-width of the X-ray rocking curve ≈2 angular minutes for 6 μm film thickness). The surface topography of such films, as shown by AFM studies, did not exceed 4 nm.

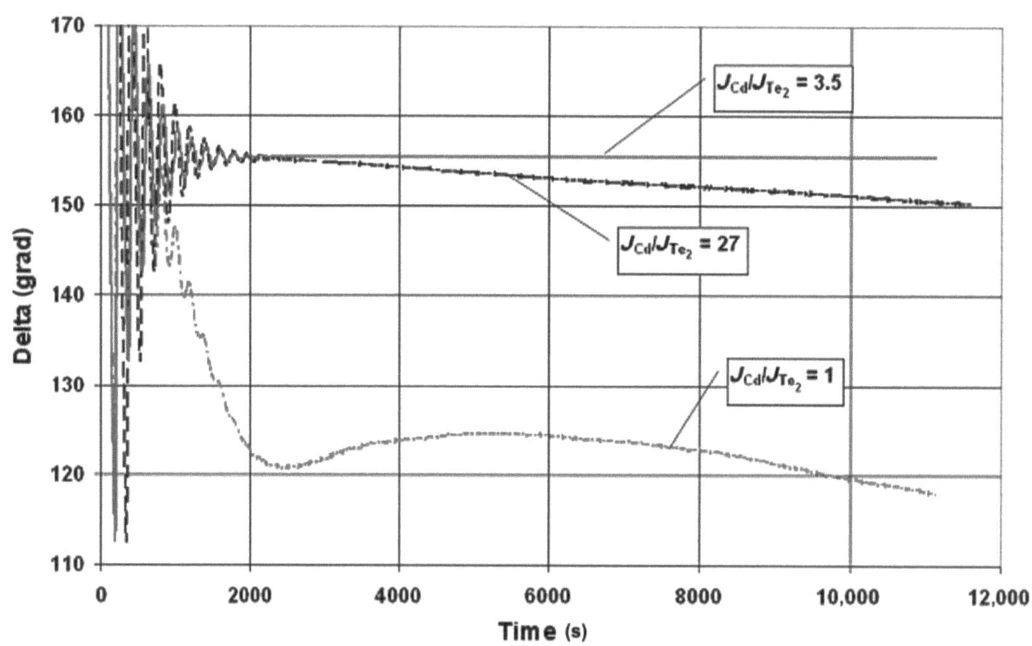

FIGURE 12.9 Changing the ellipsometric parameters during CdTe film growth.

12.5.3 MCT Growth

Growing MCT epitaxial layers were produced from separate sources of Cd, Te, and Hg. The special design of the sources, and their location in the growth chamber is described in Section 12.2. Monitoring the work of molecular sources, of the mercury flow sensor, of the substrate temperature, and of an automatic ellipsometer LEF-755, as well as process control, was carried out with the help of ACS [16].

Mode component flows in growing MCT differ substantially from the molecular flows. Molecular fluxes are considered when the atoms or molecules are not experiencing collisions when moving from the source to the growing surface. In a typical substrate, a temperature of 165–200°C is desired, and mercury vapor pressure above the growth surface should be no less than 10^{-3} Pa. Under these pressures the mercury flow influences on the value of tellurium flux coming onto the substrate. Mercury flow change also affects the amount of cadmium flow. Deposition of tellurium film on a cold substrate in the presence of mercury, and in the absence thereof, showed a significant change in the growth rate of the film of tellurium. Growth rates are reduced by 2–4 times with increasing mercury vapor pressure. Therefore, cadmium and tellurium flows arriving at the substrate surface substantially change when the mercury flow changes.

As noted in Section 12.3, tellurium falls on the surface in the form of diatomic molecules. Tellurium vapor pressure at the substrate temperature is small. If the diatomic tellurium does not react with mercury and cadmium, it cannot evaporate, and remains on the surface to form a tellurium solid phase. Thermodynamic analysis shows that at temperatures less than 190°C two stable phases may exist at the same time, crystalline phase tellurium and MCT [9]. In this case, the preferential formation of one or the other phase will be determined exclusively by the kinetics of formation of the corresponding phase.

Fig. 12.10 schematically shows the main possible processes occurring on the surface with tellurium. Tellurium molecules on the surface are involved in two processes: the dissociation

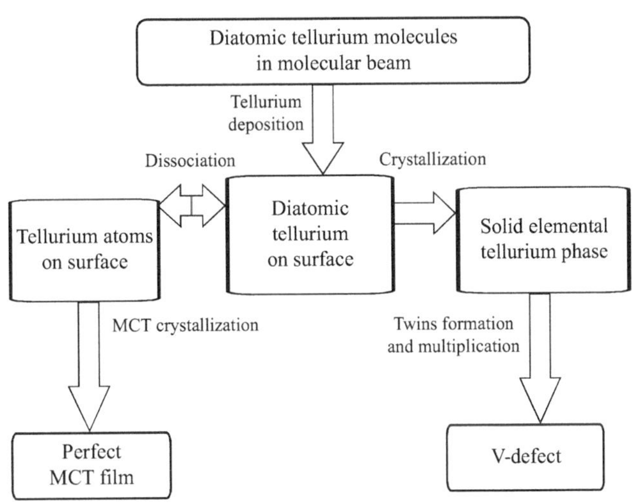

FIGURE 12.10 Processes involving tellurium occurring on the MCT surface.

of tellurium molecules and crystallization of perfect film MCT, and tellurium crystallization as a separate phase when the dissociation process does not have time to take place. In the latter case, the formation of a tellurium phase on the surface suppresses the crystal growth of MCT, and leads to an avalanche of defects. As a result, under optimum conditions, it is possible to observe irreversible deterioration of the crystal structure and HgCdTe surface relief development.

The MCT growth process is controlled in a "Ob-M" vacuum chamber with the automatic ellipsometer LEF-755 only. As in the case of buffer layers, at the initial stage of MCT growth one can observe the trace of ellipsometric parameters in the form of a convergent spiral (in the $\Psi-\Delta$ plane) due to the existence of a CdTe–HgCdTe interface. The results of these measurements are shown in Fig. 12.11 [17].

These results at the initial stage of growth allow determination of the film growth rate when the thickness and growth time are known. The point of the twisting "snail" allows us to determine the composition of the growing layer from the calibration curve (see Fig. 12.4). In accordance with the thermodynamic concepts discussed in Section 12.3, there is an opportunity to grow MCT films with different composition without substantially changing the technological conditions. Therefore, to reduce the impact of heterojunction and surface recombination of charge carriers (see Section 12.5 for more detail), it is typical to grow the structure represented in Fig. 12.12. On the heterojunction

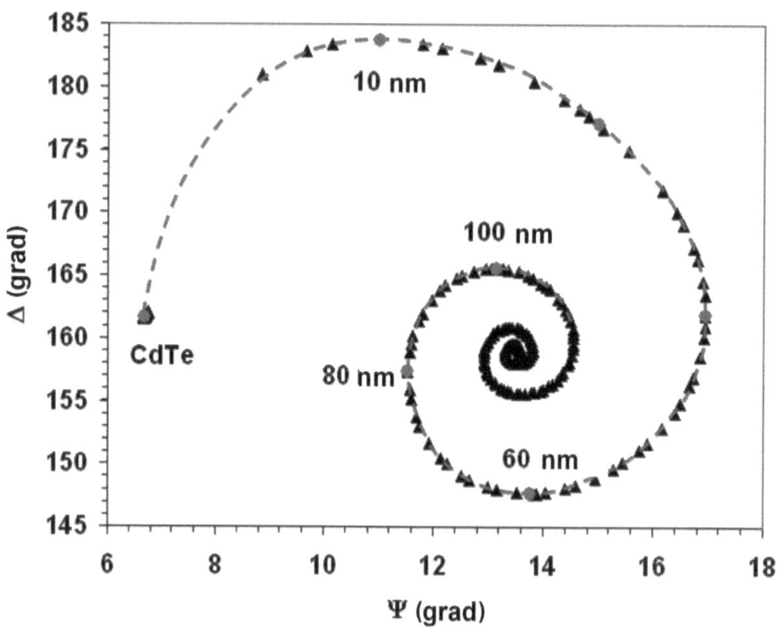

FIGURE 12.11 Variation of the ellipsometric parameters during growth of a MCT film on a CdTe substrate at an initial stage. The *dashed line* shows the calculated curve. Numbers near the circles indicate film thickness in nanometer. *Triangles* are experimental results.

with the buffer layer and near the surface more wide-band layers are grown. The inset in the figure shows the temperature change measured during the growth process by a polarizing pyrometer. It would be seen that during growth the substrate temperature is able to keep within ±1.5°C. Therefore, the composition accuracy during growth will not be worse than $\Delta X_{CdTe} = 0.001$. This fact is also confirmed by the following transmission spectra measurements [13], with step-by-step etching of the surface layer. It can be seen that there is complete agreement between the results of in situ ellipsometric measurements and stratified etching.

Thus, when the film grows in optimal conditions, we can precisely control the composition of the growing film. Ellipsometric in situ monitoring of the growth process allows us to grow potential barriers precisely, as well as potential wells, by growing built layers with different compositions. When growing the MCT nanostructures a high growth rate should be maintained in order to reduce growth time, and thereby reduce the influence of elements of interdiffusion. For precise control of composition and thickness, using the ellipsometric technique during real rates growth of MCT (3–5 μm/h) is necessary to ensure high speed and threshold sensitivity of the measurements [18].

When growing thin nanolayers (barriers or quantum wells) there are two interfaces formed, so the trajectory of the ellipsometric parameters will consist of two spiral curves. The shape of these curves is determined by the composition, and their extension by the layers' thickness. Fig. 12.13 [18] shows the variation in the ellipsometric parameters measured with a sharp decrease of the tellurium flux in a short time (~1 s.) by the partial overlap of the molecular tellurium source. Experimental results are shown by the symbols.

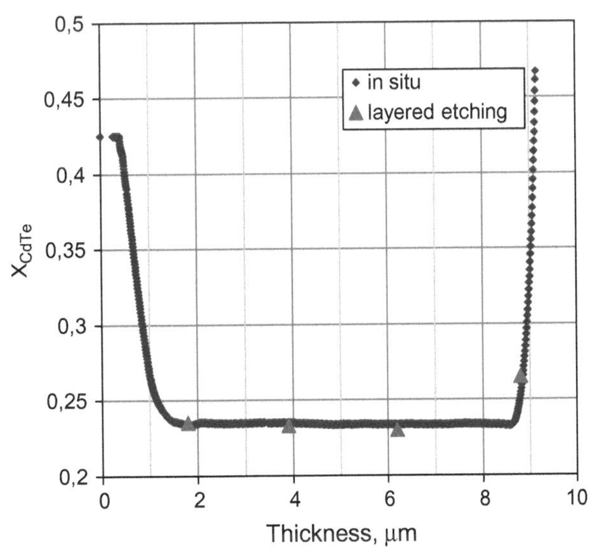

FIGURE 12.12 Composition distribution across thickness of HgCdTe MBE film measured by in situ ellipsometer. Triangles are composition measurements during step-by-step etching the epitaxial film.

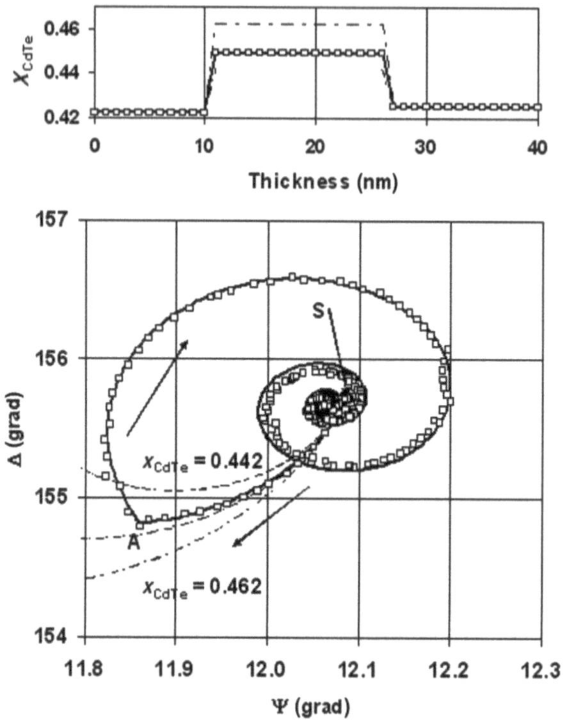

FIGURE 12.13 Ellipsometric parameter variation during growth of potential barrier of composition $X_{CdTe} = 0.449$ on the MCT substrate of $X_{CdTe} = 0.22$. The upper part of the picture: Dependence of the growing films composition on thickness; Lower part: Ellipsometric parameter variation. *Squares* are the experimental measurements, the line is a single-layer model. *Dashed curves* are modeling the potential barriers of composition $X_{CdTe} = 0.442$, 0.452, and 0.462. The *arrows* show the sequence of ellipsometric parameter variation. Point S corresponds to the initial composition. Curve SA corresponds to composition of the barrier layer.

Intervals between measurements are 1 s. The start variation (point S) corresponds to the steady growth of MCT with the composition $X_{CdTe0} = 0.422$. After shutting-off Te flow, there is a rapid change of parameters Ψ and Δ along the line SA. In accordance with the modeling conducted above, the area SA represents part of a spiral curve corresponding to a growing layer with a higher content of CdTe. After Te flow restoration, there is a break at point A of the ellipsometric parameters trajectory, which continue to change by drawing another spiral curve convolving to the starting point S.

The coincidence of the start and the end points of the process means that, after restoration of the Te flow, the composition of the growing MCT film returns to its initial value X_{CdTe0}. The composition profile is shown in the insert, and the calculated curve describing variation of ellipsometric parameters for this profile is represented by a *solid line* in the figure. The good agreement between the theoretical curve with the experimental data which occurs after Te flow restoration (part AS) proves the growth of a uniform layer with constant composition. A slight discrepancy corresponds to a composition change of a few thousandths mole fraction. The break of the experimental trajectory at point A shows a stepped composition change; the transition layer thickness is no more than 0.5 nm. The thickness of the barrier layer is determined by the length SA, and in this case it is 16.2 ± 0.1 nm.

In Fig. 12.13 *dotted lines* show parts of the curves calculated for barrier layer composition exceeding X_{CdTe0} at 0.02, 0.04, and 0.06 mole fraction. Comparing these curves with the experimental results one can see that the composition of the barrier layer is $X_{CdTe} = X_{CdTe0} + 0.027$, and accuracy is $\delta X_{CdTe} = \pm 0.002$.

It should be noted that the accuracy of determining the thickness and composition of grown nanostructures strongly depends on the absolute value of the layer thickness. From the results we can conclude that, for structures having practical application with a thickness of 10 nm, the ellipsometric technique provides the necessary accuracy of nanolayer thickness and composition, and also allows you to control the quality of the interface.

The above analysis shows the ability for controlled growth of the potential barriers and wells on the basis of solid solutions $Cd_xHg_{1-x}Te$. In the report of the Academy of Sciences of the USSR in 1987 a manuscript was published by our colleagues A.V. Rzhanov, K.K. Svitashev, A.S. Mardezhov, and V.A. Shvets about the possibility of ellipsometric control during the preparation of superlattices based on alternating layers of CdTe and HgTe [19].

The experiments showed [20] that ellipsometric control really enables the necessary accuracy for both the composition and thickness in growing embedded potential barriers and wells (when composition varies at any desired interval, and layer thickness is from a few to hundreds of nanometers), as well as when growing smooth varying band structures with the required variation of the composition gradient. An example of the grown structures with 30 HgTe quantum wells and wide-doped electrodes is shown in Fig 12.14.

Fig. 12.14 on the left side shows the structure of the quantum wells, the diagram on the right side is of the ellipsometric parameter variation during the growth of 30 HgTe QW. One can see that there is a piece-wise smooth curve during growth. The blue color shows the variation of the ellipsometric parameters during growth of wide-doped electrodes, and the red curves show HgTe QW.

12.5.4 The Nature of V-Shaped Defects in HgCdTe Epilayers Grown by MBE

Typical threading defects, which are called voids [20–22] or surface-craters, are commonly observed on the surfaces of HgCdTe epilayers grown by MBE. The cross-sizes of these defects extend to the top surface, so we called them V-shaped defects [23,24]. A typical density of V-shape defects is about $10^3\ cm^{-2}$. When a V-shaped defect is located within the active area of a photodiode it leads to a deterioration in diode parameters. Because of this, the density of V-shaped defects must be minimized, especially for fabrication of focal plane infrared detector arrays. In spite of the importance of this problem, the mechanism of nucleation and multiplication of defects during MBE have not been well investigated. The important role of elemental Te in the formation of V-shaped defects was shown in Ref. [24]. Several groups [20,25] demonstrated a reduction in defect density to 100–300 cm^{-2} by

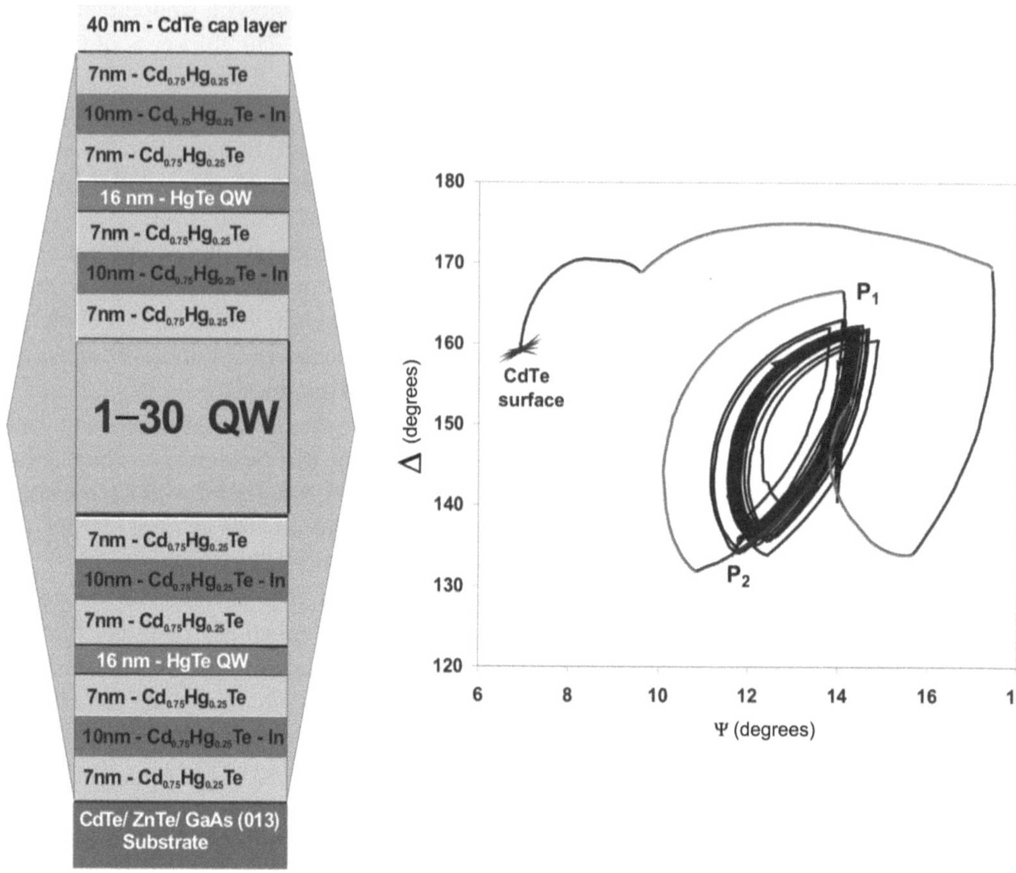

FIGURE 12.14 On the left side is a cross-sectional view of the structure, and on the right side are the ellipsometric parameter variations during the growth of 30 HgTe QW.

optimization of growth conditions and careful preparation of substrates. For reproducibility and improvement of this result, in our opinion, it is necessary to investigate the microstructure and mechanism of V-shaped defect formation in detail.

AFM and TEM investigations show [26] that V-shaped defects are complicated defects composed of twin lamellas, stacking faults, and areas with disordered structure with a high concentration of elemental Te, that can be up to 3% according to electron probe microanalysis [8] (see Fig. 12.15). Close to V-shaped defects, typical surface irregularities are observed.

Fig. 12.15 shows the typical AFM (A) and TEM (B) images of V-shaped defects observed on (301)HgCdTe surface. A comparison and analysis of these images allows us to find the correlation between the peculiarities of micromorphology in the defect area and its microstructure. Fig. 12.16 shows the scheme for the analysis of V-shaped defect images. In this scheme the pyramid formed by {111} planes inclined to the (301) surface (A) and the quadrangle trace of intersection of this pyramid with the (301) surface (B) are represented.

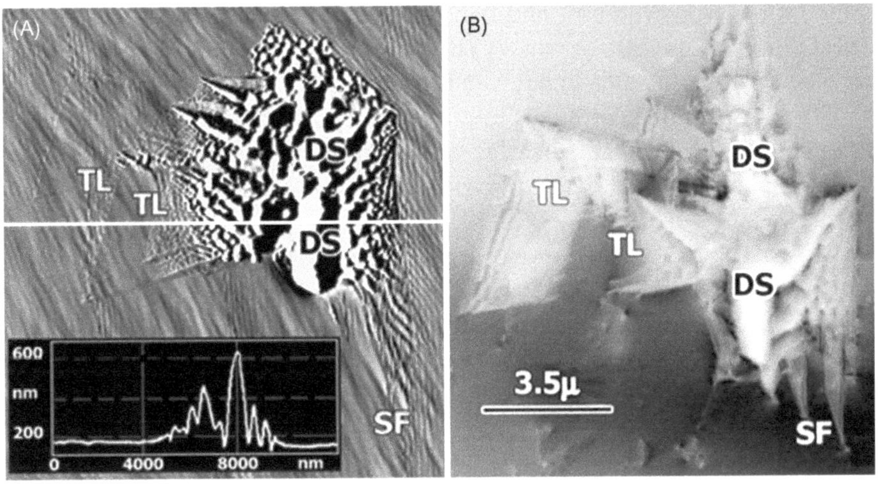

FIGURE 12.15 Typical AFM-image, $12 \times 12\ \mu m^2$ scan (A), and TEM-image (B), of V-shaped defects composed of twin lamellas (TL), stacking faults (SF), and disordered structure areas (DS) on the surface of (301)HgCdTe film. The insert in (A) shows the relief profile along the *white line*.

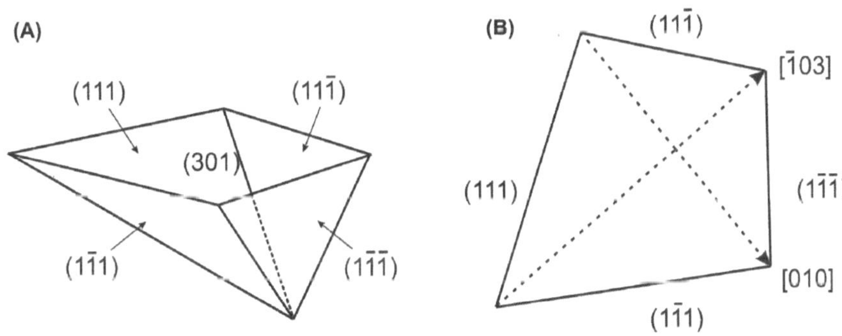

FIGURE 12.16 The pyramid formed by {111} planes inclined to (301) surface (A), and the trace of intersection of this pyramid with (301) surface (B).

In Fig. 12.15 we see the typical elements of V-shaped defects: SFs, TLs, and DS areas. The height profile along the white line in (A) is shown in the inset. The height of the DS-area is over 500 nm. The SFs and TLs grow along {111} planes from the V-shaped defect center. When a V-shaped defect image is compared with the scheme shown in Fig. 12.16, it is evident that the defect is bounded to the right by the $(1\bar{1}\bar{1})$ plane which is inclined at 68.58 degrees to the (301) plane, and to the left by a set of (111) planes which are inclined at 43.09 degrees to the (301) plane. TEM investigation of cross-sections shows that essentially all SFs observed in (301) HgCdTe layers are located in $(1\bar{1}\bar{1})$ planes, growing from the CdTe buffer layer. SFs which lie in the (111) planes are observed only in the area of V-shaped defects. The areas with disordered structure in the center part of the V-shaped defect are the most hazardous to photodiodes.

Thus, the formation mechanism and microstructure of these areas were closely studied. At low defect density (less than 10^3 cm^{-2}) TEM investigation is difficult because of the small size of the e-transparent foil area and low defect density. To follow the consecutive stages of V-shaped defect formation by TEM, it is best to use the growth of HgCdTe layers under non-optimal conditions: elevated temperatures and reduced Hg–vapor pressure. Under these conditions, continual V-shaped defect nucleation takes place, and defect density may be as great as 10^6 cm^{-2}.

The analysis of this TEM image reveals the consecutive stages of V-shaped defect formation, beginning with the initial ones shown in Fig. 12.17A. The lines (line-shape areas) of disordered structure are seen along the macrostep bends as dark lines with diffraction contrast. The microstructure along these lines was observed with atomic resolution. The HREM image shows the presence of precipitates with the distance between atomic layers precisely the same as in elemental Te (0.59 nm) (see insert in Fig. 12.17B).

Fig. 12.17C shows the presence of misoriented fine HgCdTe grains. Since Te activity during HgCdTe growth by MBE is sufficient for elemental Te formation [9], it would appear reasonable to assume that under nonoptimal conditions elemental Te can be accumulated along the growth front (along the macrosteps in the case of steady state growth) and can initiate the nucleation of misoriented HgCdTe grains and Te precipitates, at a size of 10 nm.

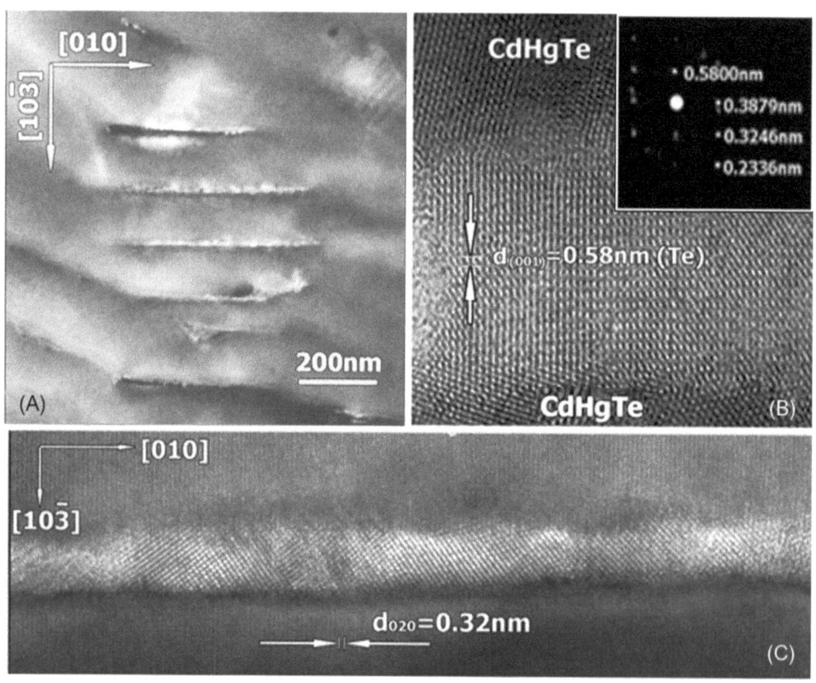

FIGURE 12.17 Initial stage of V-shape defect formation. (A) Bright field TEM-image of disordered structure that is formed along [010]-macrosteps, $g = 040$; (B) HREM-image and corresponding microdiffraction pattern of Te precipitates; (C) HREM-image of misoriented fine HgCdTe grains that are formed along [010]-macrosteps; the orientation of the HgCdTe grains is close to (001). The incident e-beam direction is [301] (A and C) and [110] (B).

In these places the macrosteps bend because its flow is pinned. As a result, short lengths of macrosteps can assume an exact [010] direction, as follows from the TEM data.

HgCdTe grains and Te precipitates can trigger the formation of dislocations, which are also observed along the macrostep bends (Fig. 12.17B). At this stage, the formation of walls with HgCdTe grains, excess Te, and dislocations begins, while HgCdTe layer growth in these walls with disordered structure extends along the growth direction.

Hereafter, we shall use the term " DS-walls" to mean the walls with disordered structure. Stereophotography was used to locate these DS-walls. The DS-wall images shown in Fig. 12.17A have been obtained under imaging conditions close to the [301] zone axis (when the incident e-beam direction is perpendicular to the (301) surface). Under these conditions, the width of the DS-wall image is minimum. When the sample is tilted 18 degrees about the [010] axis to the left (toward the (100) pole), and to the right (toward the (101) pole) the DS-wall image becomes wider and appears to the right and to the left from the [010] direction respectively. This indicates that DS-walls are normal to the (301) growth surface. The DS-walls observed by TEM intersect the (301) growth surface in the [010] direction. From analysis of a stereographic projection for the (301) plane, it follows that DS-walls are located in vertical (103) planes, which are perpendicular to the (301) plane and intersect it along the [010] direction. The formation of DS-walls which extended along the growth direction and contained HgCdTe grains, Te precipitates, and dislocations, and may be considered the initial stage of V-shaped defect formation. In the initial stages of HgCdTe layer growth the situation is complicated by the fact that the optimal growth conditions are not attained yet, whereas the relief perturbations on the substrate surface are present. In the experiments carried out the CdTe buffer layer plays the role of a substrate. Our investigation of CdTe buffer layers by TEM and AFM revealed that surface irregularities with a height of more than 10–20 nm are the result of threading defects.

Thus, the study of the microstructure and micromorphology of V-shaped defects in films $Cd_xHg_{1-x}Te(301)$ in successive stages of their formation by TEM, HRTEM, and AFM allowed elucidation of the mechanism of formation of V-defects in MBE $Cd_xHg_{1-x}Te$ [26]. The formation of the V-shaped defect begins from the capture of excess Te along the growth front at steady state stages of HgCdTe growth, and along the irregularities on the surface of the CdTe buffer layer at the initial stages. The captured Te initiates nucleation of misoriented HgCdTe grains, new Te precipitates, and dislocations in the vertical walls that are formed along the growth front under nonoptimal growth conditions. It was established that the formation of these walls with disordered structure (DS-walls) extended along the growth direction in the initial stage of V-shaped defect formation. Under the effect of strains, the splitting of perfect dislocations into partial ones with SF between them takes place. DS-walls bound the gliding of partial dislocations and lateral SF growth. While growing, the vertical columns packed with SFs and bounded by DS-walls are formed. These columns represent the basic part of the V-shaped defect—DS-area. Due to the continual Te nucleation and twinning, the polycrystalline growth in DS-area can occur. The extended SFs and TLs grow from the DS-areas along {111} planes. With film thickness growth, the size of DS-areas increases as well as the size of extended SFs and TLs. As a result, the lateral size of the V-shaped defect increases with increasing HgCdTe film thickness.

12.6 The Homogeneity of the Composition and Electrical Properties of Heteroepitaxial of MCT on a GaAs Substrate

MCT growth is highly sensitive to the orientation of the substrate surface. It was established experimentally that the orientation of gallium arsenide (103) is optimum for growing MCT. It helps to prevent the formation of twins, twin lamellae with the growth of II–VI compounds. This orientation is less sensitive to changes in the conditions of the growing process, and allows a single process to carry out the growth of MCT films of various compositions.

The composition distribution on the area of the films grown on GaAs substrates with a diameter of 76.2 mm, was measured using the spectra of transmission on a Bruker Fourier spectrometer with local measurements of 0.5 mm^2. We obtained the following composition distribution characteristics: X_{CdTe} mean = 0.20171, standard deviation $\Delta X = 0.000164$, and relative deviation $\Delta X/X = \pm 0.654\%$. This uniformity in the HS diameter of 76.2 mm compared to the best published data obtained by rotating the substrate [27].

Thanks to *in situ* ellipsometric control the MCT layers may have a predetermined composition profile, e.g., the working layer of constant composition and graded-gap layers. Fig. 12.12 illustrates the change in the composition of the thickness of typical HS MBE MCT layers with graded-gap measured by ellipsometer during the growth process. When the composition of the working layer of $X_{CdTe} = 0.22$ at the boundaries of the working layer created variband layers, the CdTe content rises to the surface and to the border with the buffer layer. Graded-gap layers with a high content of CdTe can be used to passivate the surface [28]. The increase in the band gap on the surface of the MCT and heteroboundaries film creates built-in fields, brushing aside nonequilibrium carriers from surfaces which can have high recombination rate. There is reason to believe that in this way you can increase the effective lifetime of nonequilibrium carriers [29]. Fig. 12.18 shows an example of the composition

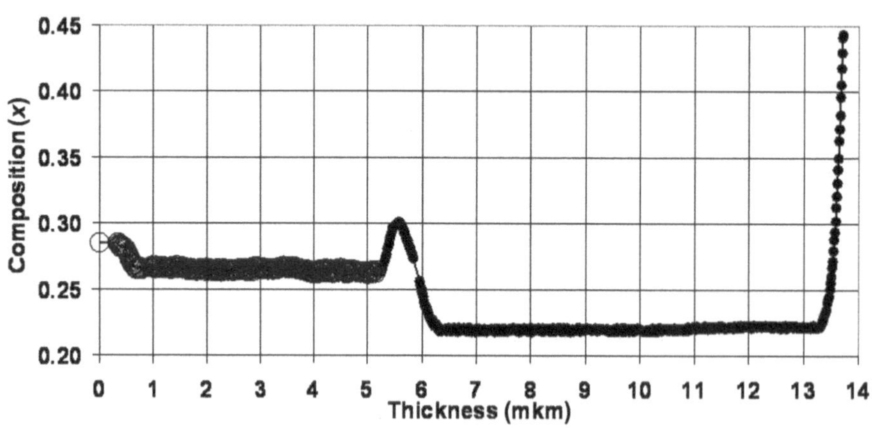

FIGURE 12.18 The results of in situ ellipsometry measurements of the film thickness: a sublayer 5 μm thick, a wide-layer between the working layer and sublayer, and the graded-gap layer on the surface.

distribution in a structure with a sublayer ($X_{CdTe} = 0.27$) with high conductivity, which can be used to decrease the series resistance in large format arrays and diode heterodyne. To prevent recombination of nonequilibrium carriers, the sublayer is separated from the active layer ($X_{CdTe} = 0.22$) by a wide-gap layer ($X_{CdTe} = 0.30$).

As-grown the MCT films are of n-type conductivity. The concentration of carriers in n-type films are in the range of $1 \cdot 10^{14}\ cm^{-3}$ to $1 \cdot 10^{15}\ cm^{-3}$ at mobilities of 50,000–150,000 cm^2/V s. Fig. 12.19 shows the dependence of carrier mobility in n-type material versus layer composition. The experimental values (*dots* on Fig. 12.19) agree well at liquid nitrogen and room temperatures with the theoretical values (*solid lines* on Fig. 12.19) from Ref. [30], calculated under the assumption that the dominant scattering by optical phonons. Mobility for the composition $X_{CdTe} = 0.158$ reaches 710,000 cm^2/V s at a temperature of 77K, and is comparable with the best mobility in the bulk material.

One of the most important parameters of the materials used for the manufacture of photodetectors is the lifetime of photoexcited carriers whose value is largely determined by the presence of recombination centers in the films. Layers of HgCdTe n-type conductivity with graded-gap layers at 77K have a record high minority carrier lifetime for material grown on a GaAs substrate. The temperature dependence of the lifetime of minority carriers in the film with $X_{CdTe} = 0.22$ is shown in Fig. 12.20. After etching, the upper variable-gap layer relaxation time of nonequilibrium carriers at a temperature of 77K is reduced by 4–5 times, indicating the positive effect of graded-gap layers. It should be noted that the carrier concentration in MCT structures with graded-gap layers does not differ from one of the structures without graded-gap layers. Perhaps dislocation does not affect the lifetime as much as it is thought.

n-type films are transferred to p-type conduction at a concentration ranging from $5 \cdot 10^{15}\ cm^{-3}$ to $2 \cdot 10^{16}\ cm^{-3}$ at mobilities of 300 – 600 cm^2/Vs of annealing temperature

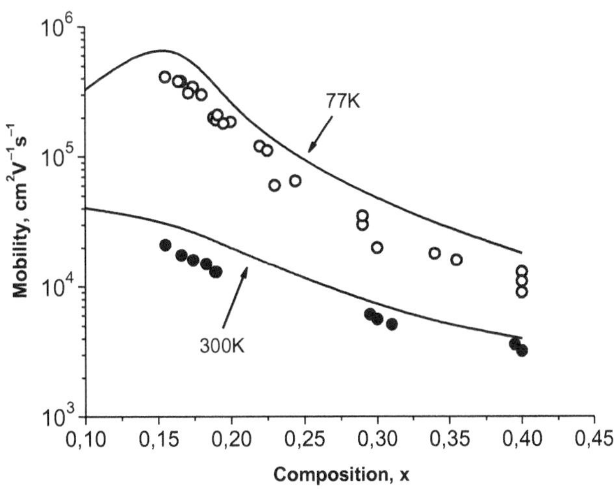

FIGURE 12.19 Mobility dependence on the composition of the MBE MCT films.

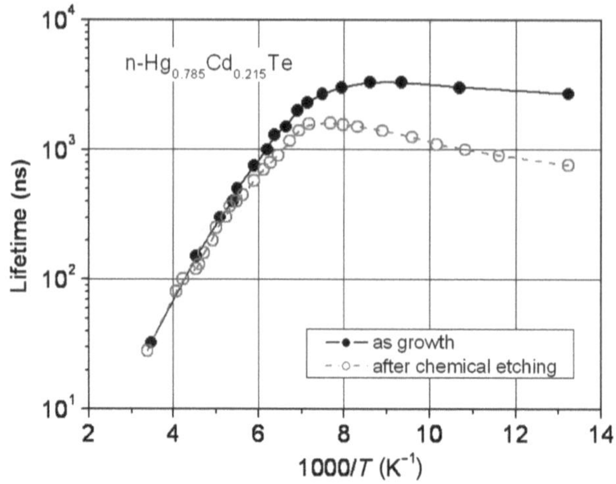

FIGURE 12.20 Temperature dependence of the minority carrier lifetime in HgCdTe, $X_{CdTe} = 0.22$.

230°C, mercury temperature 30°C, and an annealing duration of 20–40 h. Heat treatments used an ampoule filled with gas (hydrogen or helium). Dependence of the results of annealing on the type of gas were not observed. The ampoule was placed in a two-zone furnace. One zone is intended for heating the reservoir with mercury, and the second is used to heat the sample. Changing the conductivity type is reversible, annealing at a temperature of 230°C and mercury temperature above 180°C again gives n-type conductivity. The calculation of the temperature dependence of the concentration of vacancies in the composition range $X_{CdTe} = 0.2–1.0$ [31] makes it possible to determine the optimal parameters of heat treatment in graded-gap and multilayer structure MCT in order to obtain the desired profile of the concentration of vacancies.

The experimental evidence on the effect of annealing conditions on the properties of the films MCT suggests the presence of MCT films using GaAs substrates, in addition to moving acceptor centers with variable concentrations and also donor centers, the concentration of which depends on the growing conditions. The main donor centers in MCT films grown by MBE are apparently tellurium atoms in antistructural positions [32]. Experimental data on the change in the concentration of donor centers in the films MCT growing temperature confirm this assumption. Upon doping with indium in the MCT MBE method, formation of electrically neutral complexes In_2Te_3 does not occur, despite the high tellurium activity [33]. When indium doping is used during film growth of MCT MBE, the concentration of electrons in the films (in the range of $5 \cdot 10^{14}$ to $1.3 \cdot 10^{17}$ cm^{-3}) is directly proportional to the indium concentration, and is not changed by annealing to fill vacancies.

At temperatures above 700°C the As_4 cracking degree reaches saturation [34]. The efficiency of doping by arsenic diatomic molecules is two orders of magnitude higher than by As_4 but activation of arsenic, in any case, requires a high-temperature treatment of MCT films.

12.7 Conclusion

A automated multichamber installation for the growth of HS MBE MCT structures with quality control of the process of growing in real time was created. A set of technological chambers in the installation were determined on the basis of studies of the mechanisms of formation of defects in heterostructures. A MCT growing chamber is equipped with an automatic ellipsometer and molecular sources with ring diffusers. This solves MCT technology problems associated with high demands for homogeneity of composition and control of composition in the process of growing CMT layers.

The equipment allows growth of layers of MCT with a high uniformity of composition across the substrate area, and gives almost unlimited possibilities for modifying the parameters of the layer thickness of MCT, which is necessary for the development of new generations of photodetectors. The technological process is designed based on ideas about the mechanisms of formation and the nature of defects in HgCdTe layers, and allows preparation of the epitaxial structure of MCT with the necessary concentration, carrier mobility, and relaxation time of nonequilibrium photoconductivity. Photoelectric characteristics of HS MBE grown MCT meet the requirements in the manufacture of photosensitive element-based photoconductors and photodiodes.

The quantum well made MBE-based mercury telluride surface orientation (013) found a two-dimensional electron–hole system, consisting of light highly mobile electrons with concentration $N_s = (4-7) \cdot 10^{10}\ \mathrm{cm}^{-2}$ and mobility $\mu_n = (4-6)\ 10^5\ \mathrm{cm^2/V\,s}$, and heavier holes with a concentration of $P_s = (0.7-1.6)\ 10^{11}\ \mathrm{cm}^{-2}$ and mobility of $\mu_p = (3-7)\ 10^4\ \mathrm{cm^2/V\,s}$. The electron mobility recorded and the observed quantum effects confirm the high quality of the HgCdTe grown.

References

[1] A. Krost, W. Richter, D.R.T. Zahn, et al., Chemical reaction at the ZnSe/GaAs interface detected by Raman spectroscopy, Appl. Phys. Lett. 57 (17) (1990) 1981–1983.

[2] T. Scimeca, Y. Watanabe, F. Maeda, et al., Controlled passivation of GaAs by Se treatment, Appl. Phys. Lett. 62 (14) (1993) 1667–1669.

[3] M.V. Yakushev, V.G. Kesler, L.M. Logvinsky, Yu.G. Sidorov, XPS analysis of the ZnSe/GaAs(112)B hetero-systems grown by MBE, Poverhnost 2 (1997) 58–67 (in Russian).

[4] P.K. Larsen, P.L. Dobson (Eds.), Reflection high-energy electron diffraction and reflection electron imaging of surface, NATOASI Ser. B (1988) 188–194.

[5] M.V. Yakushev, Yu.G. Sidorov, L.V. Sokolov, Twinning at the initial stages of MBE growth of A2B6 compounds on GaAs substrates, Poverhnost 10 (1996) 35–46 (in Russian).

[6] F.A. Kuznetsov, Yu.G. Sidorov, I.E. Maronchuk, To the question of quantity description the transport reactions, Fiz. Tverd. Tela 6 (10) (1964) 2981–2983 (in Russian).

[7] C.H. Su, P.-K. Liao, R.F. Brebrick, Partial pressures over the pseudobinary solid solution $Hg_{1-x}Cd_xTe(s)$ for $x = 0.70$ and 0.95 and over four Te-rich ternary melts, J. Electrochem. Soc. 132 (1985) 942–949.

[8] V.S. Varavin, S.A. Dvoretsky, V.I. Liberman, N.N. Mikhailov, Yu.G. Sidorov, The controlled growth of high-quality mercury cadmium telluride, Thin Solid Films 267 (1995) 121–125.

[9] Yu.G. Sidorov, S.A. Dvoretsky, M.V. Yakushev, N.N. Mikhhailov, V.S. Varavin, V.I. Liberman, Pecularities of the MBE growth physics and technology of narrow-gap II-VI compounds, Thin Solid Films 311 (1997) 253–266.

[10] K.K. Svitashev, S.A. Dvoretsky, Yu.G. Sidorov, V.A. Shvets, A.S. Mardezhov, I.E. Nis, et al., The growth of high-quality MCT films by MBE using in situ ellipsometry, Cryst. Res. Technol. 29 (1994) 931–936.

[11] Yu.G. Sidorov, S.A. Dvoretsky, N.N. Mikhhailov, M.V. Yakushev, V.S. Varavin, A.P. Anciferov, Molecular beam epitaxy of compounds $Cd_xHg_{1-x}Te$, Equip. Technol. Opticheskу J. 67 (2000) 39–45 (in Russian).

[12] Yu.G. Sidorov, S.A. Dvoretsky, V.S. Varavin, N.N. Mikhhailov, M.V. Yakushev, I.V. Sabinina, Molecular beam epitaxy of solid solutions of cadmium-mercury-tellurium on "alternative" substrates, Fiz. Tech. Poluprovodn. 35 (2001) 1092–1101 (in Russian).

[13] J.F. Nicholas, An Atlas of Models of Crystal Surfaces, Gordon and Breach Science Publishers, New York, NY, 1965.

[14] Yu.G. Galicin, V.G. Mansurov, V.I. Poshevnev, A.S. Terekhov, Surface passivation of GaAs in alcoholic solutions of HCl, Poverhnost 10 (1989) 140–144 (in Russian).

[15] N.N. Mikhailov, V.A. Svets, S.A. Dvoretsky, E.V. Spesivtsev, Yu.G. Sidorov, S.V. Rykhlitski, et al., Ellipsometric control of the growth of nanostructures on the basis of $Cd_xHg_{1-x}Te$, Avtometriya 2 (2003) 71–80 (in Russian).

[16] V.S. Varavin, V.V. Vasiliev, S.A. Dvoretsky, N.N. Mikhailov, V.N. Ovsyuk, Yu.G. Sidorov, et al., HgCdTe epilayers on GaAs: growth and devices, Opto-electronics Rev. 11 (2) (2003) 99–111.

[17] V.A. Svets, S.V. Rykhlitski, E.V. Spesivtsev, N.A. Aulchenko, N.N. Mikhailov, S.A. Dvoretsky, et al., In situ ellipsometry for control of $Hg_{1-x}Cd_xTe$ nanolayer structures and inhomogeneous during MBE growth, Thin Solid Films 455–456 (2004) 688–694.

[18] A.V. Rzanov, K.K. Svitashev, A.S. Mardezhov, V.A. Svets, The control of parameters of the superlattices by the method of ellipsometry in the growing process, Dokl. Akad. Nauk 297 (1987) 604–609 (in Russian).

[19] S. Dvoretsky, N. Mikhailov, Yu Sidorov, V. Shvets, S. Danilov, B. Wittman, et al., Growth of HgTe quantum wells for IR to THz detectors, J. Electron. Mater. 39 (2010) 918–923.

[20] M. Zandian, E. Goo, TEM investigation of defects in arsenic doped layers grown in-situ by MBE, J. Electron. Mater. 30 (2001) 623–627.

[21] D. Chandra, F. Aqariden, J. Frazier, S. Gutzler, T. Orent, H.D. Shih, Isolation and control of voids and void-hillocks during molecular beam epitaxial growth of HgCdTe, J. Electron. Mater. 29 (2000) 887–892.

[22] E.C. Piquette, M. Zandian, D.D. Edwall, J.M. Arias, MBE growth of HgCdTe epilayers with reduced visible defect densities: kinetics considerations and substrate limitations, J. Electron. Mater. 30 (2001) 627–631.

[23] I.V. Sabinina, A.K. Gutakovsky, Preparation of TEM samples from compound semiconductors by chemomechanical polishing, Ultramicroscopy 45 (1992) 411–415.

[24] Yu.G. Sidorov, V.S. Varavin, S.A. Dvoretsky, V.I. Liberman, N.N. Mikhailov, I.V. Sabinina, et al., Growth and defect formation in $Cd_xHg_{1-x}Te$ films during molecular beam epitaxy, Growth Cryst. 20 (1996) 35–45.

[25] L. He, Y. Wu, L. Chen, S.L. Wang, M.F. Yu, Y.M. Qiao, et al., Composition control and surface defects of MBE-grown HgCdTe, J. Cryst. Growth 227–228 (2001) 627–631.

[26] I.V. Sabinina, A.K. Gutakovsky, Yu.G. Sidorov, A.V. Latyshev, Nature of V-shaped defects in HgCdTe epilayers grown by molecular beam epitaxy, J. Cryst. Growth 274 (2005) 339–346.

[27] P.S. Wijewarnasuriya, M. Zandian, D.D. Edwall, et al., MBE P-on-n $Hg_{1-x}Cd_xTe$ heterostructure detectors on silicon substrates, J. Electron. Mater. 27 (1998) 546–549.

[28] R.K. Bhan, V. Dhar, P.K. Chaudhury, et al., Comment on “Electrical properties of epitaxially grown CdTe passivation for long-wavelength HgCdTe photodiodes” [Appl. Phys. Lett. 65 (1994) 2725] Appl. Phys. Lett. 68 (1996) 2453–2455.

[29] V.G. Remesnik, A.M. Mischenko, N.N. Mikhailov, Patent RF №2022402, Bulleten izobreteniy №20 (1994).

[30] A. Rogalski, J. Piotrowski, Intrinsic infrared detectors, Prog. Quant. Electron. 12 (2/3) (1988) 87–289.

[31] V.S. Varavin, G.Yu Sidorov, Yu.G. Sidorov, The concentration of vacancies in the metal sublattice of the solid solutions of tellurides of cadmium and mercury, depending on the composition, J. Fiz. Himii 84 (9) (2010) 1605–1612 (in Russian).

[32] V.S. Varavin, S.A. Dvoretsky, N.N. Mikhailov, Yu.G. Sidorov, The donor defects in epitaxial layers of CdHgTe you-for example, by the method of molecular-polar beam epitaxy, Avtometriya 3 (2001) 9–19 (in Russian).

[33] V.S. Varavin, S.A. Dvoretsky, D.G. Ikusov, N.N. Mikhhailov, Yu.G. Sidorov, G.Yu Sidorov, et al., The dependence of electrophysical parameters of the films $Cd_xHg_{1-x}Te$ grown by molecular beam epitaxy, the doping level of indium, Fiz. Tech. Poluprovodn. 42 (6) (2008) 664–667 (in Russian).

[34] G.Yu Sidorov, N.N. Mikhailov, V.S. Varavin, D.G. Ikusov, Yu.G. Sidorov, S.A. Dvoretsky, The study of influence of temperature of arsenic cracking on the efficiency of its incorporation into the CdHgTe film in the process of molecular beam epitaxy, Fiz. Tehn. Poluprovodn. 42 (6) (2008) 668–671 (in Russian).

13

Surface Morphologies Obtained by Ge Deposition on Bare and Oxidized Silicon Surfaces at Different Temperatures

A.A. Shklyaev[1,2], K.N. Romanyuk[1], S.S. Kosolobov[1], A.V. Latyshev[1,2]

[1]RZHANOV INSTITUTE OF SEMICONDUCTOR PHYSICS OF THE SIBERIAN BRANCH OF RUSSIAN ACADEMY OF SCIENCES, NOVOSIBIRSK, RUSSIA [2]NOVOSIBIRSK STATE UNIVERSITY, NOVOSIBIRSK, RUSSIA

13.1 Introduction

Thin-layer Ge depositions on Si surfaces lead to the formation of dense arrays of different islands that are intensively used for the fabrication of Si–Ge nanostructures with unique physical properties [1]. These properties depend on the Si–Ge composition of the islands and their shape. The shape of the islands that appear under Ge deposition on bare Si(001) and Si(111) surfaces is well studied and exhibits a strong dependence on the growth temperature [2–22]. The island formation occurs under a dramatic strain caused by the 4.2% Si–Ge lattice mismatch. The Ge deposition on the Si(001) and Si(111) surfaces covered with an ultrathin, about 0.3–0.5 nm thick, SiO_2 film also results in Ge island formation. In spite of the use of amorphous SiO_2 substrate coverage, the islands can grow epitaxially with respect to the crystalline Si substrates. This occurs at relatively high temperatures at which bare Si nanoareas in the SiO_2 film appear, due to the SiO_2 decomposition through the formation of volatile GeO and SiO molecule at the initial growth stage [23,24]. At small Ge coverages, up to about 2 nm, the Ge islands grown on the SiO_2-covered Si surfaces are round-shaped, and they form extremely dense arrays with a concentration of $\sim 10^{12}$ cm^{-2} [23–26]. The islands arrays grown at temperatures below 400°C, when they form relatively weak bonds with the SiO_2 film, can be modified with a scanning tunneling microscope (STM) probe [27,28].

Due to the Si–Ge lattice mismatch, Ge growth on bare Si surfaces occurs through the Stranski–Krastanov (S–K) growth mode, in which the Ge wetting layer formation is followed by three-dimensional (3D) island nucleation and growth. The Si–Ge lattice strain causes Si–Ge intermixing during growth, which results in the appearance of a significant number of Si atoms in the wetting layers and islands [6,11,12,15,16]. Moreover, the 3D island formation initiates

Advances in Semiconductor Nanostructures. DOI: http://dx.doi.org/10.1016/B978-0-12-810512-2.00013-5

atom transfer from the wetting layers to the islands [7,14,17], and the deep trench appearance around the islands [6,12]. In contrast, Ge growth on SiO_2-covered Si surfaces begins with inhomogeneous SiO_2 film decomposition that produces bare Si nanoareas [23–26,29,30]. The nanoareas serve as places for the Ge islands epitaxial growth. The growth occurs without Ge wetting layer formation, and produces rounded islands in dense arrays [23,31].

We use STM and scanning electron microscopy to study the surface morphologies obtained by Ge depositions on the bare and SiO_2-covered Si(001) and Si(111) surfaces over a wide temperature range from 450°C to 950°C. The layers grown on the SiO_2-covered surfaces are composed of islands whose shapes appear to be different from the shapes of islands grown by the S–K growth mode. Moreover, the surface morphology indicates that the presence of the SiO_2 film residuals drastically reduces Si–Ge intermixing during island growth. At the same time, in the high-temperature growth range, Si–Ge intermixing becomes a dominating process which governs Si–Ge composition, shape, and special distribution of islands, and even leads to lateral nanowire (NW) formation.

13.2 Experimental Details

The experiments were carried out in an ultrahigh-vacuum chamber with a base pressure of about 1×10^{-10} torr. The chamber was equipped with an STM manufactured by Omicron. A Knudsen cell with a BN crucible was used for Ge deposition at a rate up to 1.2 nm/min, which was calibrated with the STM for the initial stage of Ge growth on the Si(111) surface at the low temperature of 400°C. After electrochemical etching, the W STM tips were sharpened by cutting their apexes from several sides using the 30 kV focused Ga ion beam of a 1540 XB Carl Zeiss microscope. As a result, the typical opening angle of the STM tip apexes, estimated from the STM images, was about 20 degrees. The STM images of the surfaces covered with Ge were usually obtained with a sample bias voltage of 2.4–3.0 V, and a constant current between 3 and 30 pA. The STM image processing with free software Gwyddion was used for generation of stereographic projection maps of faceting planes relative to the sample surface plane. This allowed nano-goniometry of facet angles and the identification of their Miller indices to be obtained [18].

After the samples were removed from the growth chamber, their scanning electron microscope (SEM) images were obtained using a Pioneer microscope manufactured by Raith. The use of SEM allows imaging of the surface areas in a wide range of sizes and with a velocity which is much higher than in methods such as STM and atomic force microscopy. This permits a more complete picture of the grown islands distribution in size and shape, and over the sample surface, to be obtained.

A $10 \times 2 \times 0.3\ mm^3$ sample was cut from an n-type Si(001) or Si(111) wafer with a miscut angle of about 10 degrees and a resistivity of 5–20 Ω cm. Clean Si surfaces were prepared by flash direct-current heating at ~1250°C. The sample temperature was measured using an IMPAC IGA 12 pyrometer. To grow the ultrathin SiO_2 films, the sample temperature was set to 400°C and then raised to 550°C for 10 min after oxygen had been introduced into the chamber at a pressure of 2×10^{-6} torr. The SiO_2 film, prepared under similar conditions, was previously shown to be 0.3–0.5 nm thick [23]. After the SiO_2 film growth, the chamber was pumped to the pressure of $\sim 1 \times 10^{-9}$ torr for 10 min, and then the Ge crucible temperature was raised up to 1210°C, depending on the required Ge deposition rate. The sample heating was turned off right after finishing the Ge deposition, or it was maintained for several minutes when the deposition was performed at relatively low temperatures.

13.3 Epitaxial Ge Growth on Bare and Oxidized Si(001) Substrates in the Middle Temperature Range

The Ge deposition on the bare Si(001) surface at low temperatures of 400–550°C leads to the formation of rectangular {510}-faceted islands called "hut-clusters" [2], as shown in Fig. 13.1A. At a temperature of 500°C or higher, square-based {510} faceted pyramids appear [2,18,32,33]. Several of these pyramids, at the initial stage of their formation, along with the hut-clusters, can be seen in Fig. 13.1A.

The Ge deposition on the Si(001) surface covered with ultrathin SiO_2 film results in island formation [23,24], as shown in Fig. 13.1B. The distinctive feature of the growth process is that the islands start to appear immediately after the beginning of the deposition, i.e., without the Ge wetting layer formation. The island array concentration reaches a value in the order of 10^{12} cm^{-2}, which is approximately one order of magnitude higher than that of Ge hut-clusters and pyramids that form on bare Si(001) surfaces. At such a high concentration, the average distance between the islands is about 10 nm. The shape of the small Ge islands looks rounded. However, the stereographic projection map (inset in Fig. 13.1B) shows the existence of faceting planes, in particular, with {311} orientation. One can note that the largest rounded island inclination angles, measured with the STM tip, reach 78 degrees, whereas the shape is close to the hemisphere according to the transmission electron microscopy (TEM) data [23,24,30,34].

Despite the fact that Ge is deposited on the amorphous SiO_2 film surface, it has been found, using high-energy electron diffraction and high resolution TEM, that the Ge islands grow epitaxially with respect to the crystalline structure of the Si substrate at relatively high growth temperatures (T_{gr}) [23,34]. The possibility of epitaxial growth appears due to the interaction between the deposited Ge atoms with SiO_2 which produces volatile GeO and SiO molecules [23,35]. This reaction leads to a partial SiO_2 film removal, and to the appearance of bare Si nanoareas. The temperature range in which the bare Si nanoareas are formed depends on the SiO_2 film thickness and the Ge deposition rate. Under our experimental conditions this occurs at $T_{gr} > 430$°C. At lower T_{gr}, SiO_2 decomposition does not occur, Ge islands grow over the SiO_2 film surface, and their crystal lattices have a random orientation with respect to the crystal lattice of the Si substrate. In the temperature range of 430–500°C, the Ge islands array is composed of a mixture of epitaxial and nonepitaxial islands. At temperatures above 500°C, all the Ge islands grow epitaxially on Si. Further Ge deposition at T_{gr} in the range from 430°C to 550°C leads to the neighboring islands coalescing. The epitaxial

FIGURE 13.1 STM images of: (A) the bare, and (B–C) SiO_2-covered Si(001) surfaces after Ge depositions in amounts of (A) 2 nm at 500°C, and (B) 1.2 nm and (C) 30 nm at 450°C. Insets show maps of stereographic projections obtained from the corresponding STM images [21].

island growth becomes dominating, and the islands take the shape with {113} facets on the side walls, as shown in Fig. 13.1C.

Ge deposition on the bare Si(001) surface at 600°C leads to the formation of islands of two shapes. These are the pyramids which are similar to that observed at low temperatures, and the large islands called multifaceted domes [32] with different crystallographic planes on their side walls, as shown in Figs. 13.2A and 13.3A. During further Ge deposition, the

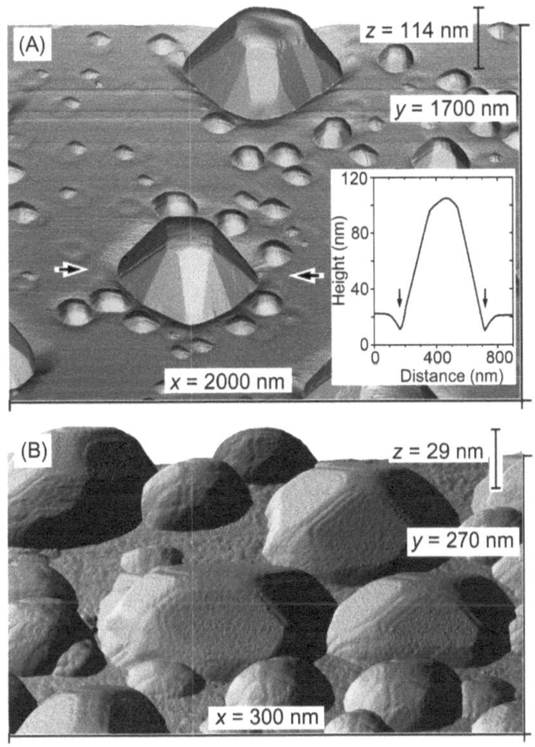

FIGURE 13.2 STM images of the surfaces prepared by the 10 nm Ge deposition at 600°C on: (A) the bare, and (B) SiO_2-covered Si(001) substrates. Inset shows the height profile along the line between arrows shown in (A) [21].

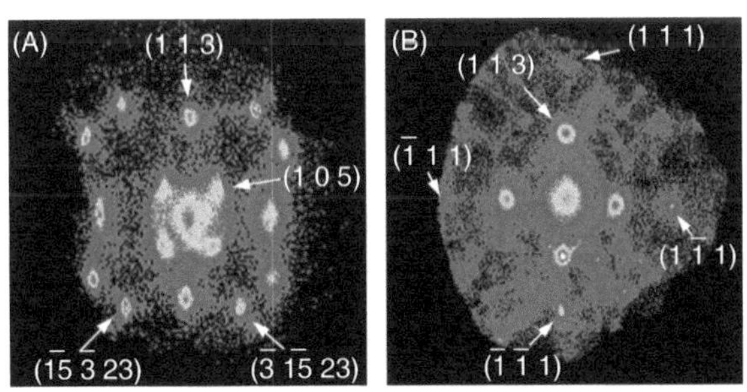

FIGURE 13.3 (A) and (B) Maps of stereographic projections obtained from the STM images shown in Fig. 13.2, respectively. To obtain the maps, fragments of the STM images, containing only one large island, were used [21].

multifaceted dome islands grow rapidly, while the pyramid islands can decrease in size and disappear. The same result is observed under subsequent annealing. This occurs due to the Ge and Si atoms transferring from the Ge wetting layer, pyramid islands, and Si substrate into the dome islands [22,32]. The surface morphology image in the intermediate stage, when a part of the pyramid islands still remains on the surface, is shown in Fig. 13.2A.

The Ge deposition on the SiO_2-covered Si(001) surface at 600°C results in the formation of islands with a higher concentration and smaller size in comparison to those grown on the bare Si(001) surface, as shown in Fig. 13.2. Besides, the islands on the bare and SiO_2-covered Si(001) surface are different in shape. Another essential distinction consists of trenches around the islands which are only formed around the islands grown on the bare Si surfaces [6,12]. The trenches appear under the strain caused by the Si–Ge lattice mismatch. The appearance of the trenches reduces the strain, not only due to the Si removal from the areas around the islands, but also due to changes in the composition of the islands which contain Si atoms transferred from the substrate. The absence of trenches, when the SiO_2-covered Si (001) surface is used, proves that the islands do not suffer high strain, and they do not contain a considerable amount of Si.

Analysis of the stereographic projection maps obtained from the STM images shows that the large multifaceted dome islands grown on the bare Si(001) surface at 600°C contain {510}, {311}, and {23 15 3} facets on the side walls, as shown in Fig. 13.3A, that are in agreement with [18,34]. The intensity of spots on the map is proportional to the total area occupied by the corresponding facets on the surface. The approximately equal spot intensities in Fig. 13.3A indicate that all of the facets revealed cover nearly equal areas on the island side walls.

The spot intensity in the map (Fig. 13.3B) for the surface obtained by Ge deposition on the SiO_2-covered Si(001) surface shows that {311} facets dominate over others on the island side walls, whereas {23 15 3} facets occupy much smaller surface areas. Besides, the presence of the very steep {111} side facets that are inclined at an angle of 54.7 degrees from the (001) plane is observed in the map. In the case of the bare Si(001) surface, such steep side facets have been observed only for the islands grown by Ge deposition at very high temperatures of 720–800°C [18,36].

13.4 Surface Morphologies Obtained by Ge Deposition on Si(001) at High Temperatures

Ge deposition on Si(001) at temperatures above 800°C leads to the formation of islands which are rather homogeneously distributed along the surface. This indicates that Ostwald ripening does not play any essential role. Surface morphology formation at such high temperatures can be imagined as follows. The island nucleation on the initially homogeneous surface occurs spontaneously, and forms a random spatial island distribution. The islands produce the lattice strain in the substrate. If dislocations, which reduce the lattice strain, are absent or their influence is not significant, larger islands induce a stronger lattice strain in the surrounding substrate area and, therefore, they are less favorable for further lateral growth in comparison with smaller islands. At the same time, the strain, being strongest at the island perimeter, decreases with distance [37,38]. This makes the closely located islands also less favorable for lateral growth. As a result, the islands become more uniform in size, and are homogeneously distributed over the surface during Ge deposition. At growth temperatures from about 850°C to 900°C, the island size distribution becomes monomodal [39].

The islands grown at 900°C exhibit ordering along the surface in the direction such as ⟨001⟩ (Fig. 13.4), as is well confirmed by the magnitude profiles in different directions of the FFT pattern obtained from the SEM image (Fig. 13.4C). That the distance between islands becomes adjusted is also confirmed by the presence of only one intense circle in the FFT pattern (Fig. 13.4B and C). Moreover, the island shape becomes uniform and symmetrical. The STM images showed that the islands have a cone shape with shallow side walls [39]. It should be emphasized that the islands ordering at 900°C (Fig. 13.4) required 60 nm of Ge deposition with a relatively low rate of 0.4 nm/min. In spite of a rather large deposited Ge coverage, the grown island sizes are relatively small. They are about the same as those of islands grown by 10 nm Ge deposition with a higher Ge deposition rate. This indicates that the Ge diffusion in the substrate is the dominating process which governs the surface morphology formation at such high temperatures.

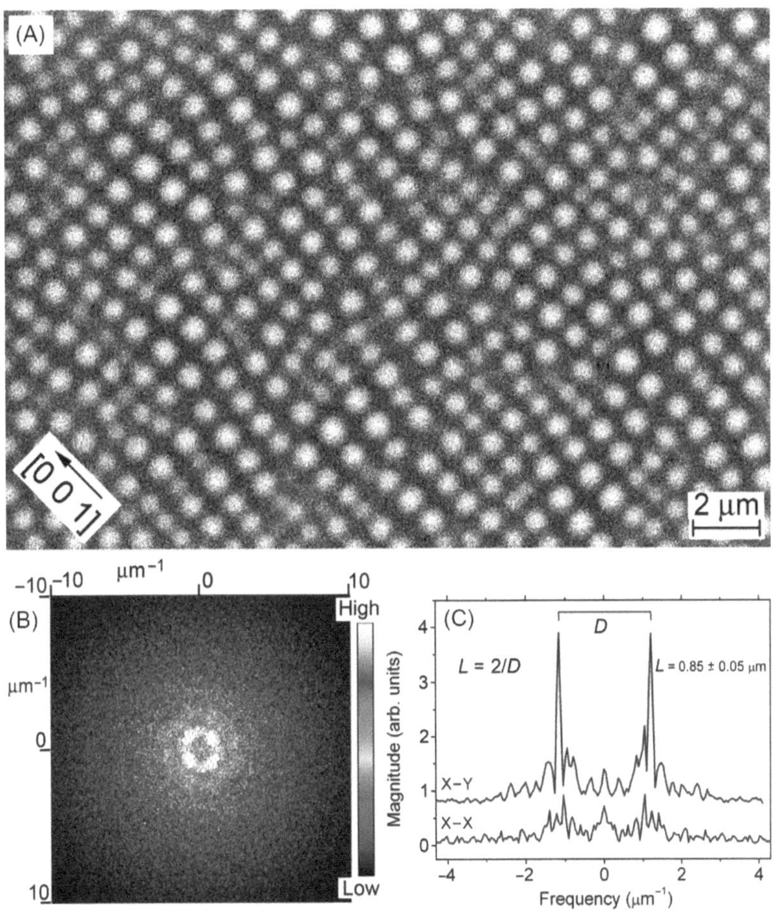

FIGURE 13.4 (A) SEM image of the Si(001) substrate covered with 60 nm Ge at about 900°C using the Ge deposition rate of 0.4 nm/min. (B) The FFT pattern obtained for the SEM image shown in (A). (C) Magnitude profiles along the lines passing through the center of the FFT pattern in the directions along its side (marked as X–X) and diagonal (X–Y) [39].

13.5 Ge Epitaxial Growth on Si(111) Substrates in the Middle Temperature Range

The morphology of Ge islands that are formed on bare Si(111) surfaces is studied in detail only for small Ge coverage [4,7–9,13,21]. Ge depositions in amounts of about 2 nm on the bare Si(111) surface at T_{gr} in the range of 320–440°C lead to the appearance of tall three-dimensional islands after Ge wetting layer formation. The mixture of tall and flat islands is observed at T_{gr} from about 440°C to 550°C, and only large flat islands are formed on the surface at $T_{gr} > 550$°C [9]. The tall island shape is a frustum of a tetrahedron with {311} facets on the side walls [4]. The surface morphology obtained after deposition of larger Ge coverage of 5 nm at 450°C shows that the tall and flat Ge islands preferably grow in the lateral dimension (Fig. 13.5A). Other facets, in addition to {311}, appear on the island side walls. The STM images show that their angular divergence $\Delta\varphi$ with {311} facets on the (111) plane is 36–38 degrees. This corresponds well with the {23 15 3} facets which have $\Delta\varphi = 36.5$ degrees. The {311} and {23 15 3} facets were previously classified as MAJOR for Ge [40]. Ge island nucleation and growth cause a partial decomposition of the Ge wetting layer, making it thinner [7]. This also leads to the formation of trenches around the islands [6,12], which start to appear at a temperature as low as 450°C, as can be seen in Fig. 13.5A [21]. The trenches become deeper at higher growth temperatures, reaching 4 nm at 500°C, as shown in Fig. 13.5B.

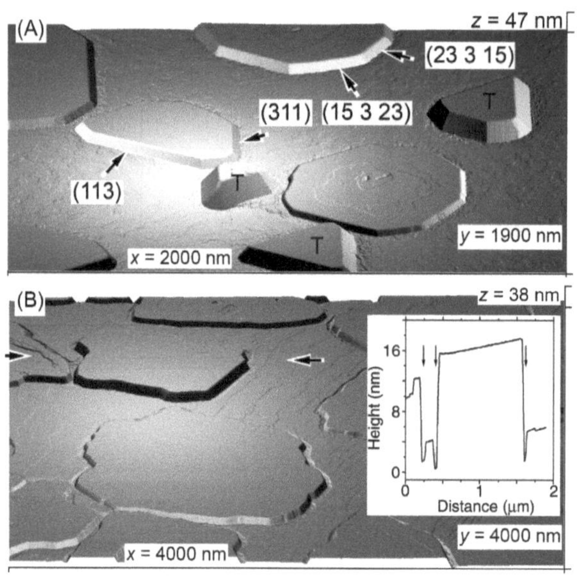

FIGURE 13.5 STM images of the surfaces prepared by 5 nm Ge deposition on the bare Si(111) surface (A) at 450°C, and (B) 500°C. The tall islands are marked by T in (A). Inset shows the height profile along the line between arrows shown in (B) [21].

Ge growth on the SiO_2-covered Si(111) surfaces at 340–550°C shows that its mechanism is similar to that on the SiO_2-covered Si(001) surfaces [24,31]. The difference is found in island shapes. The small epitaxial islands appear to be more rounded and do not exhibit any trace of {311} facets on their side walls, in contrast to that found for the islands grown on the SiO_2-covered Si(001) surface (Fig. 13.1B). The island shape remains rounded at Ge coverages of 4 nm, as shown in Fig. 13.6A. With increasing coverage, the size of the Ge islands increases due to their coalescence, and {311} facets are formed on the side walls (Fig. 13.6B). At a Ge coverage above 2 nm and temperature up to 550°C, the Ge islands form a dense array which completely covers the whole surface. At higher growth temperatures, approximately in the range of 550–640°C, larger islands grow faster, consuming smaller islands similar to that which occurs on bare Si(100) surfaces [32] owing to the Ostwald ripening mechanism. Such behavior is not observed for Ge on bare Si(111). The large island shape, shown in Fig. 13.7A, in addition to {311} facets, develops to contain {23 20 4}, {23 15 3}, and {331} facets (Fig. 13.8A and B), among which the dominating facets are {311} and {23 20 4}.

The surface morphology significantly changes at $T_{gr} \geq 640°C$, as shown in Fig. 13.7B. The shape of relatively flat and strongly truncated islands becomes {331}-facetless (Fig. 13.8C), whereas the tall islands with an aspect ratio of 0.2 additionally include groups of steep facets, such as (23 20 $\bar{4}$) and (23 15 $\bar{3}$), oriented around the {110} plane which have their inclination angles of 43 degrees from (111) (Fig. 13.8D). The relative spot intensities in the map of stereographic projections show that {23 15 3} and {23 20 4} facets occupy approximately equal island side wall areas. The drastic changes in the surface morphology, when T_{gr} becomes

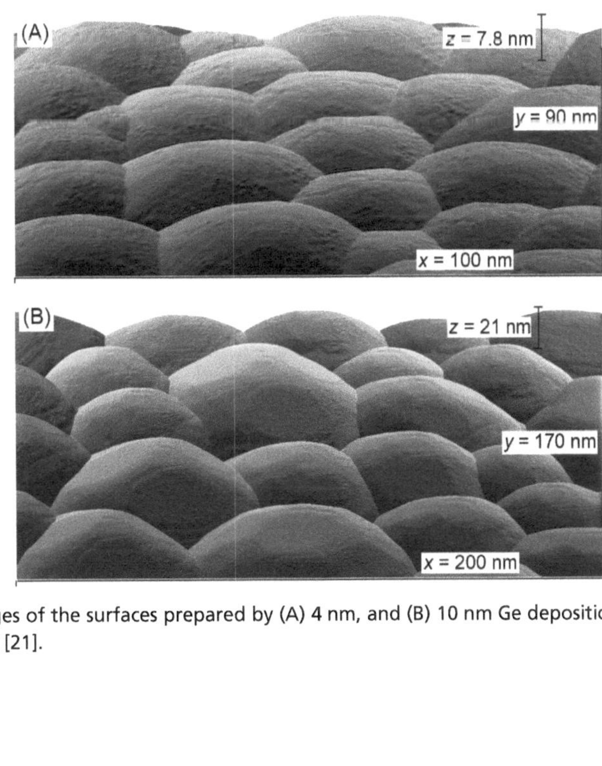

FIGURE 13.6 STM images of the surfaces prepared by (A) 4 nm, and (B) 10 nm Ge deposition on the SiO_2-covered Si (111) surfaces at 500°C [21].

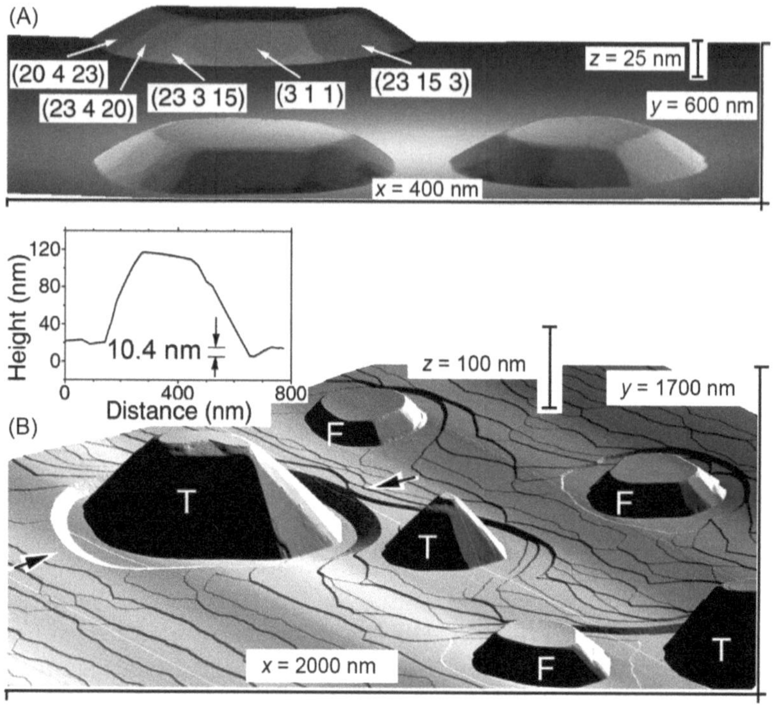

FIGURE 13.7 STM images of the surfaces prepared by 4 nm Ge depositions on the SiO_2-covered Si(111) surfaces at (A) 630°C, and (B) 650°C. Inset shows the height profile along the line between arrows shown in (B). The relatively flat and tall islands are marked by F and T in (B), respectively [21].

higher than 640°C, include the formation of deep trenches around the islands, whose depth reaches up to about 10 nm (Fig. 13.7B). The trenches formation indicates that the SiO_2 film is completely decomposed during Ge deposition [41]. This leads to the appearance of a strong Si–Ge lattice strain, and the growth conditions become similar to that on bare Si surfaces. However, the shapes of islands and their concentrations remain different, due to the differences in conditions at the initial growth stages.

13.6 High-Temperature Structures of SiGe on Si(111)

At temperatures around 800°C, the mechanism of surface morphology formation during Ge deposition on Si(111) appears to be essentially different from that in the middle temperature range of 400–600°C. Very large 3D islands are formed, whose shape appears to be formed by means of the dendritic growth mechanism (Fig. 13.9) [42,43]. Dendritic shapes of bulk and surface crystals usually appear as a result of strong differences in growth rates of different crystallographic planes, which often leads to fractal formation. However, the difference in growth rates for Ge on Si(111) are not significant in the middle temperature range.

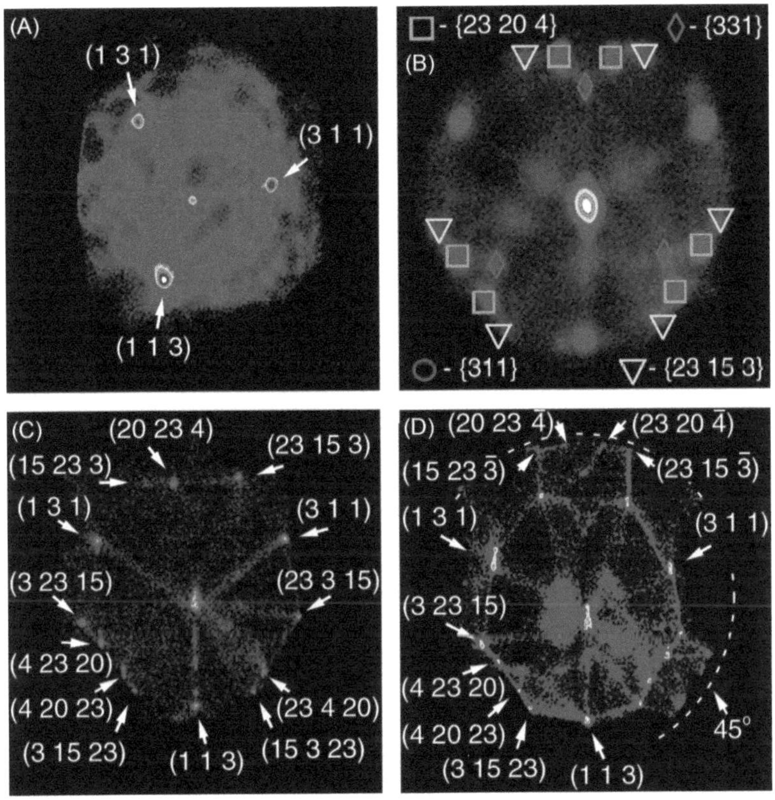

FIGURE 13.8 Maps of stereographic projections obtained from the STM images shown respectively in Fig. 13.6B for (A), in Fig. 13.7A for (B), and in Fig. 13.7B for (C) and (D) for which only fragments of the STM image were used with a relatively flat island for (C) and the tall island for (D) [21].

The surface morphology formation becomes highly anisotropic at 800°C. The anisotropy is realized through the breakdown of the growth front into a group of quasi-parallel ridges. When the large 3D islands are nucleated close to each other, they form the ridges nets by ridges intersections [43]. Similar ridges nets appear during high-temperature annealing of Ge layers deposited on Si(111) at relatively low temperatures. In this case, the morphology transformation occurs simultaneously over the entire surface and is caused only by the lattice strain and surface energy minimization without the participation of growth anisotropy [37].

Along the very large 3D islands, different small and flat 3D SiGe structures were observed after Ge depositions on Si(111) at temperatures around 850°C [44]. Their formation proceeds through the nucleation of islands, which then develop into long straight and winding NWs (Fig. 13.10). The NWs are formed in spite of their larger surface area size, in comparison with disk-like islands of the same base size, and longer perimeters at which the lattice strain is known to be the strongest. The general reason for NW formation is the minimization of the strain energy by means of elongation, instead of the flat 3D structures isotropic-like

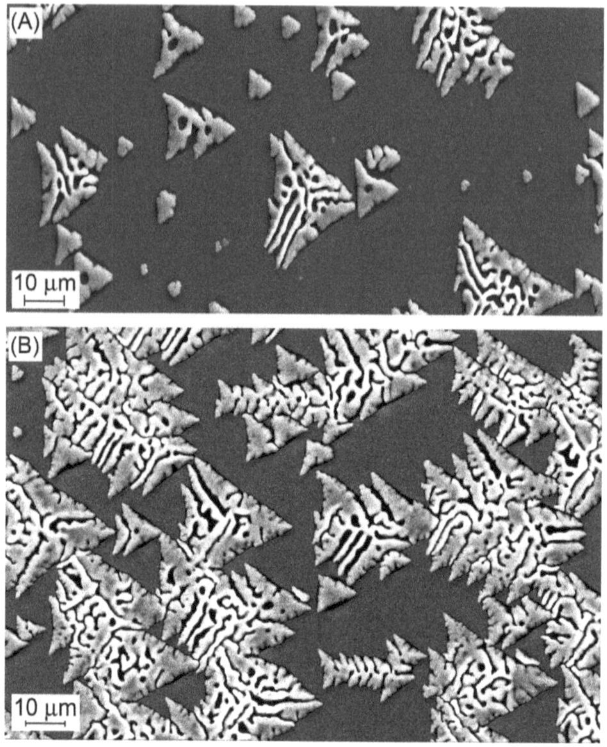

FIGURE 13.9 SEM images of Si(111) surfaces after (A) 21 nm, and (B) 65 nm Ge depositions at 800°C [42].

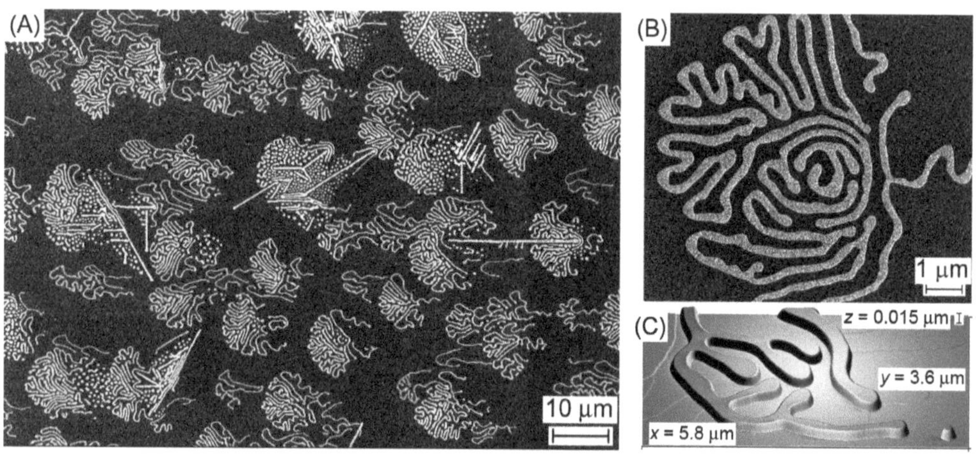

FIGURE 13.10 (A) and (B) SEM images of the NWs grown by 5 nm Ge deposition; and (C) the STM image of the surface morphology after 10 nm Ge deposition on Si(111) at 840°C [44].

growth, as has been theoretically demonstrated by Tersoff and Tromp [45]. Additional factors that govern the NW shape formation are different for straight and winding NWs. The crystallographic orientations of NWs on the Si(111) surface and their probable SiGe composition suggest that the energy barrier for the dislocation network formation at the NW/substrate interface is responsible for the straight shape of the NWs, and for their certain width. As for the winding NWs, however, they grow in the direction in which the Ge adatom concentration is favorable for NW formation with a required SiGe composition.

13.7 Surface Morphologies After Deposition of Large Ge Coverages

The conditions for Ge island epitaxial growth on SiO_2-covered Si surfaces are provided by the formation of bare Si areas at the initial Ge deposition stage. The bare Si area size increases with increasing T_{gr}, and can vary from several atoms through several nanometers up to the whole surface, when SiO_2 is completely decomposed. The surfaces with nanoareas of bare Si, or those slightly covered with Ge, were offered for use as substrates for nanocontact epitaxy [29,41,46–48]. The Ge island growth on bare Si areas and then the islands coalescence under further Ge deposition occurs over the SiO_2 film residuals. TEM data reveal shifts of the Ge lattice with respect to the Si lattice at the Si–Ge interface in places where Ge is separated from Si by SiO_2 film residuals [38,46]. In order to compensate for the 4.2% lattice mismatch between Ge and Si, the lattice shift should occur with a period of about 24 atoms (1/0.042) in the plane of the Si–Ge interface. Since the distance between the atoms on the (100) plane is about 0.4 nm, this corresponds to 9.6 nm. It coincides well with the average separation between the bare Si areas that are formed when Ge is deposited at T_{gr} in the range of 450–600°C [19,31]. It is essential that the Si–Ge lattice shifts provide the epitaxial Ge islands on the Si surfaces with the same lattice constant as in bulk Ge, indication that the islands appear to be completely relaxed.

Complete island relaxation with respect to the substrate is confirmed in structures obtained by high-coverage Ge depositions on oxidized Si(111) surfaces, which lead to a significant increase in the flat (111) areas on the islands top (Fig. 13.11). These areas undergo c(2×4) and c(2×8) reconstructions which are typical of the (111) surfaces of bulk Ge [49–51]. The observation of the reconstructions thus demonstrates that the grown epitaxial Ge layers on Si(111) are unstrained, since the strained islands prepared by Ge deposition on bare Si(111) surfaces exhibit the (5×5) reconstruction.

The STM data show that the ratio of the top flat areas to the areas of other facets, inclined from (111), increases with a decrease in the growth temperature from 550°C to 360°C. This occurs simultaneously with a decrease in the depth of depressions between the flat areas. This tendency does not extend to lower temperature range. Fig. 13.12 shows that the Ge layers grown at 250–300°C indeed exhibit a relatively low surface roughness; however, the surface morphology is mainly composed of small-sized stepped facets such as {433}, {755}, and {775}, whereas the flat (111) areas occupy a relatively small portion of the surface. The

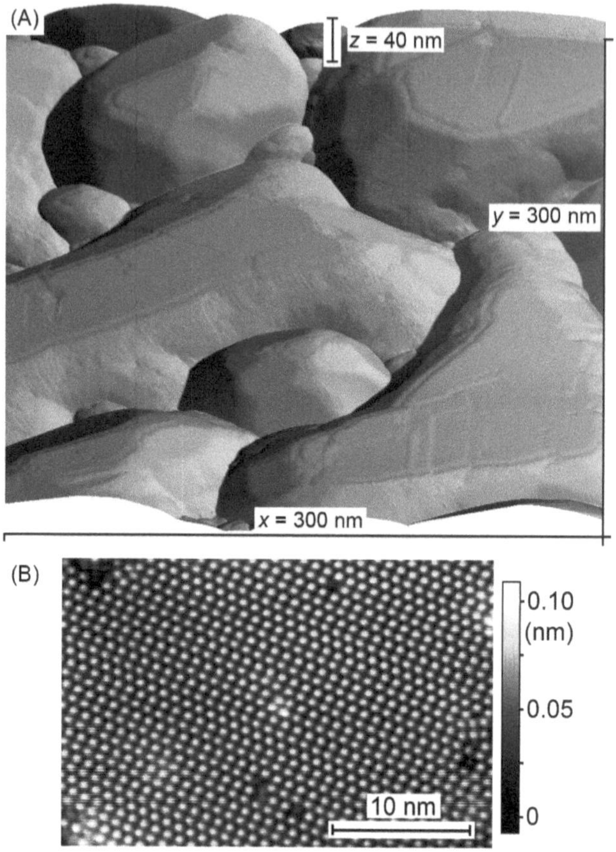

FIGURE 13.11 (A) STM image of the oxidized Si(111) surface after Ge deposition in two stages. Ge was deposited for a coverage of 1 BL at 620°C, and then at 360°C for 85 BL. (B) STM image of the c(2 × 4), and c(2 × 8) reconstructions observed on flat areas of the Ge layer [41].

root-mean-square roughness was estimated from the STM image, shown in Fig. 13.12, to be about 2.8 nm for the 140 nm thick Ge layer [41]. The most prevalent {433} and {755} facets consist of atomic steps with edges toward $[2\bar{1}\bar{1}]$, which are the sort of stepped surfaces that were observed on the Ge surfaces cleaved in vacuum [52] and found to be stable on the vicinal Ge surfaces inclined from (111) [53].

The islands grown by Ge deposition on Si surfaces covered with ultrathin SiO_2 films have improved crystalline quality with a reduced concentration of stacking faults [26]. The island coalescence under a further Ge deposition leads to the formation of continuous thin films with a low threading dislocations concentration [46,47]. The Si plates covered with such Ge films can serve as substrates which are appropriate for the growth of GaAs epitaxial layers.

The Ge growth on bare Si surfaces at elevated temperatures is accompanied by Si–Ge intermixing [6,12]. The Si content in the grown Ge islands exhibits a complicated spatial

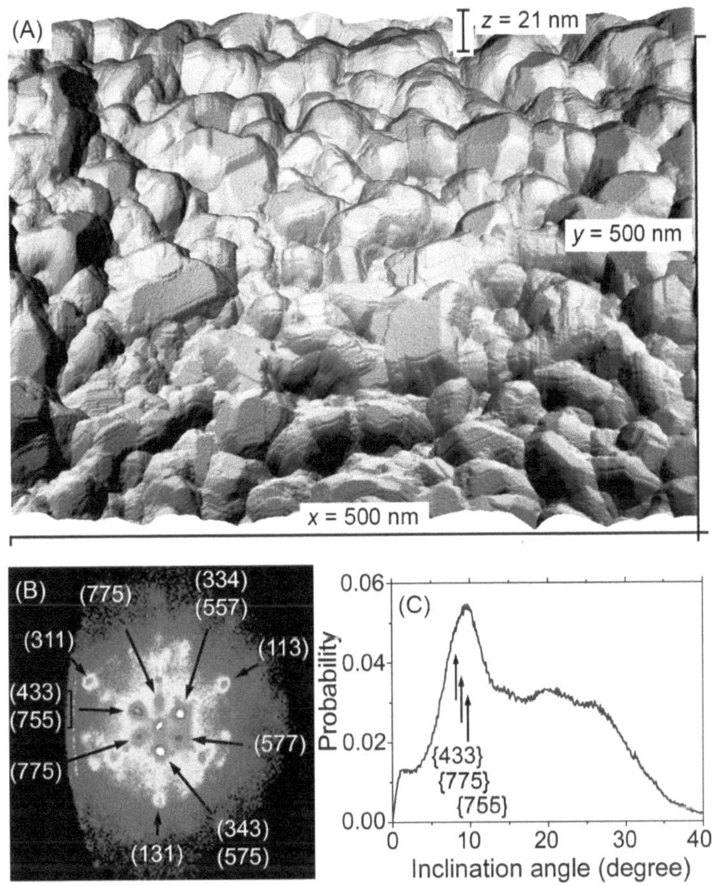

FIGURE 13.12 (A) STM image of the oxidized Si(111) surface after Ge deposition at two stages. Ge was deposited for a coverage of 2 BL at 590°C and then at temperatures in the range between 250°C and 300°C for the total coverage of 140 nm. (B) The map of stereographic projections of surface areas, and (C) portion of the areas as a function of their inclination angle from the sample surface obtained from the STM image [41].

distribution, and reaches up to 50% depending on the growth temperature [15,16,22,34]. The results of this work show the absence of any feature of Si–Ge intermixing during Ge growth on SiO_2-covered Si surfaces. The grown islands can be, therefore, referred to as relaxed pure Ge islands epitaxially grown on Si. The shape of pure Ge islands essentially differs from that of the islands prepared by the S–K growth mode. The shape of pure Ge islands on Si(001) is mostly composed of {311} and {111} facets, whereas the set of {501}, {311}, and {23 15 3} facets determines the shape of islands grown using the S–K growth mode.

The tall islands (Fig. 13.7B) that appear on the SiO_2-covered Si(111) surface at 650°C are not formed when Ge is deposited on bare Si(111) surfaces at high T_{gr}; so, their shape has not been described. The shape of such islands can be compared to the shape of the islands grown on bare Si(001) surfaces at similar T_{gr}, which contain a group of facets whose orientations lie around the {110} planes. However, the shape of the islands grown on the SiO_2-covered Si(111) surface contains {23 20 4} facets which are not observed in the shape of islands grown on bare Si(001) surfaces. It is interesting to note that the (001) plane, usually being regarded as energetically favorable, is not observed in the shapes of any island grown on Si(001) substrates.

Since the size of the areas that are covered with the SiO_2 film residuals can be varied, this allows one to prepare Si substrates with only small SiO_2 film fractions. Such substrates form the intermediate initial growth conditions between the bare and SiO_2-covered Si surfaces. It can be suggested that the use of such substrates allows one to control the shape of the islands and their composition and, thereby, to vary the Si–Ge nanostructure properties.

Besides the shape and chemical composition, Ge islands grown using ultrathin SiO_2 films form arrays with an island concentration of about 10^{12} cm^{-2}, which is about one order of magnitude greater than that provided by the S–K growth mode. Moreover, the corresponding Si–Ge quantum dot structures do not contain Ge wetting layers. Such structures can be used as light emitters in the 1.3–1.5 μm range [54,55]. Photodiodes, fabricated on the basis of the structures, exhibit an enhanced quantum efficiency at wavelength 1.3 μm [56]. It has been recently found that the separation distances between holes confined in the Ge quantum dots, and electrons located in the surrounding Si, are significantly larger than those in the structures grown by the S–K growth mode [57,58]. Longer distances cause a reduction in the electron–hole recombination rate, thereby increasing the lifetime of excited carriers. Such structures produce a photovoltage which is higher than that of S–K grown structures [59]. The results obtained here can serve as basis for the Si–Ge heterostructure parameter optimization that is in great demand for various applications.

Photoconductivity (PC) of Si–Ge structures with Ge quantum dots (QDs) grown on Si (100) surfaces covered with ultrathin SiO_2 films, measured as a function of the applied voltage, contain steps and a step with a peak at the step edge for low- and high-temperature grown structures, respectively [60]. These PC features are associated with the impact ionization of QD-related excitons. The PC at step edges increases by several orders of magnitude for a certain value which is governed by the balance between rates of photo-generation, recombination, and impact ionization of excitons. The electron localization deeper in Si from the Ge QD layer, in conjunction with a narrow binding-energy distribution of excitons, is suggested to be the main reason for the sharpness of PC steps. The PC appears to be very sensitive to impact ionization and QD preparation conditions. This allows the specific characteristics of QD structures, related to their electronic and structural properties, to be revealed.

In conclusion, the comparison of surface morphologies of Ge layers deposited on bare and SiO_2-covered Si(001) and Si(111) surfaces shows that the presence of SiO_2 film residues significantly reduces Si–Ge lattice strain, and prevents Si–Ge intermixing. This permits pure Ge islands and layers on Si substrates to be obtained. The pure Ge island shapes appear to be different from those of the islands grown by the S–K growth mode. The islands lateral ordering on Si(001) during the Ge deposition was found to occur at 900°C. This is associated with the realization of the dynamic equilibrium between the growth of islands and their decay through Si–Ge intermixing. Since the lattice strain and chemical composition essentially influence the properties of quantum dot structures, the use of ultrathin SiO_2 films permits fabrication of Si–Ge nanostructures with novel optoelectronic characteristics that are important for device development.

Acknowledgments

The partial financial support by RSF grant 14-22-00143 and 14-12-01037 is gratefully acknowledged. The experiments were partly carried out using the equipment of and supported by CKP "NANOSTRUKTURY," Ministry of Education and Science of the Russian Federation.

References

[1] O.P. Pchelyakov, Yu.B. Bolkhovityanov, A.V. Dvurechenskii, L.V. Sokolov, A.I. Nikiforov, A.I. Yakimov, et al., Silicon-germanium nanostructures with quantum dots: formation mechanisms and electrical properties, Semiconductors 34 (2000) 1229–1247.

[2] Y.W. Mo, D.E. Savage, B.S. Swartzentruber, M.G. Lagally, Kinetic pathway in Stranski-Krastanov growth of Ge on Si (001), Phys. Rev. Lett. 65 (1990) 1020–1023.

[3] U. Köhler, O. Jusko, G. Pietsch, B. Müller, M. Henzler, Strained-layer growth and islanding of germanium on Si(111)-(7 × 7) studied with STM, Surf. Sci. 248 (1991) 321–331.

[4] B. Voigtländer, A. Zinner, Simultaneous molecular beam epitaxy growth and scanning tunneling microscopy imaging during Ge/Si epitaxy, Appl. Phys. Lett. 63 (1993) 3055–3057.

[5] R. Gunnella, P. Castrucci, N. Pinto, I. Davoli, D. Sébilleau, M. De Crescenzi, X-ray photoelectron-diffraction study of intermixing and morphology at the Ge/Si(001) and Ge/Sb/Si (001) interface, Phys. Rev. B 54 (1996) 8882–8891.

[6] T.I. Kamins, E.C. Carr, R.S. Williams, S.J. Rosner, Deposition of three-dimensional Ge islands on Si (001) by chemical vapor deposition at atmospheric and reduced pressures, J. Appl. Phys. 81 (1997) 211–219.

[7] A.A. Shklyaev, M. Shibata, M. Ichikawa, Instability of two-dimensional layers in the Stranski-Krastanov growth mode of Ge on Si (111), Phys. Rev. B 58 (1998) 15647.

[8] N. Motta, A. Sgarlata, R. Calarco, Q. Nguyen, J.C. Castro, F. Patella, et al., Growth of Ge-Si(111) epitaxial layers: intermixing, strain relaxation and island formation, Surf. Sci. 406 (1998) 254–263.

[9] A.A. Shklyaev, M. Shibata, M. Ichikawa, Ge islands on Si(111) at coverages near the transition from two-dimensional to three-dimensional growth, Surf. Sci. 416 (1998) 192–199.

[10] A.B. Talochkin, A.A. Shklyaev, V.I. Mashanov, Super-dense array of Ge quantum dots grown on Si(100) by low-temperature molecular beam epitaxy, J. Appl. Phys. 115 (2014) 144306.

[11] X.R. Qin, B.S. Swartzentruber, M.G. Lagally, Diffusional kinetics of SiGe dimers on Si (100) using atom-tracking scanning tunneling microscopy, Phys. Rev. Lett. 85 (2000) 3660.

[12] F. Boscherini, G. Capellini, L. Gaspare, M. Di Seta, F. De Rosei, A. Sgarlata, et al., Ge–Si intermixing in Ge quantum dots on Si, Thin Solid Films 380 (2000) 173–175.

[13] B. Voigtländer, Fundamental processes in Si/Si and Ge/Si epitaxy studied by scanning tunneling microscopy during growth, Surf. Sci. Rep. 43 (2001) 127–254.

[14] A.V. Osipov, F. Schmitt, S.A. Kukushkin, P. Hess, Stress-driven nucleation of coherent islands: theory and experiment, Appl. Surf. Sci. 188 (2002) 156–162.

[15] F. Ratto, F. Rosei, A. Locatelli, S. Cherifi, S. Fontana, S. Heun, et al., Composition of Ge(Si) islands in the growth of Ge on Si(111) by x-ray spectromicroscopy, J. Appl. Phys. 97 (2005) 043516.

[16] A. Rastelli, M. Stoffel, A. Malachias, T. Merdzhanova, G. Katsaros, K. Kern, et al., Three-dimensional composition profiles of single quantum dots determined by scanning-probe-microscopy-based nanotomography, Nano Lett. 8 (2008) 1405–1409.

[17] R. Bergamaschini, M. Brehm, M. Grydlik, T. Fromherz, G. Bauer, F. Montalenti, Temperature-dependent evolution of the wetting layer thickness during Ge deposition on Si(001), Nanotechnology 22 (2011) 285704.

[18] M. Brehm, H. Lichtenberger, T. Fromherz, G. Springholz, Ultra-steep side facets in multi-faceted SiGe/Si (001) Stranski-Krastanov islands, Nanoscale Res. Lett. 6 (2011) 70.

[19] L.V. Arapkina, V.A. Yuryev, Nucleation of Ge clusters at high temperatures on Ge/Si(001) wetting layer, J. Appl. Phys. 111 (2012) 094307.

[20] K.N. Romanyuk, A.A. Shklyaev, B.Z. Olshanetsky, Structure and stability of Ge cluster on Si (111) surface in the presence of Bi surfactant, Sur. Sci. 617 (2013) 68–72.

[21] A.A. Shklyaev, K.N. Romanyuk, S.S. Kosolobov, Surface morphology of Ge layers epitaxially grown on bare and oxidized Si(001) and Si(111) substrates, Surf. Sci. 625 (2014) 50–58.

[22] J.N. Aqua, I. Berbezier, L. Favre, T. Frisch, A. Ronda, Growth and self-organization of SiGe nanostructures, Phys. Rep. 522 (2013) 59–189.

[23] A.A. Shklyaev, M. Shibata, M. Ichikawa, High-density ultrasmall epitaxial Ge islands on Si (111) surfaces with a SiO_2 coverage, Phys. Rev. B 62 (2000) 1540.

[24] A.A. Shklyaev, M. Ichikawa, Extremely dense arrays of germanium and silicon nanostructures, Physics—Uspekhi 51 (2008) 133–161.

[25] A.I. Nikiforov, V.V. Ul'yanov, O.P. Pchelyakov, S.A. Teys, A.K. Gutakovskii, Growth and structure of Ge nanoislands on an atomically clean silicon oxide surface, Phys. Solid State 46 (2004) 77–79.

[26] D. Leonhardt, J. Sheng, J.G. Cederberg, Q. Li, M.S. Carroll, S.M. Han, Nanoscale interfacial engineering to grow Ge on Si as virtual substrates and subsequent integration of GaAs, Thin Solid Films 518 (2010) 5920–5927.

[27] A.A. Shklyaev, M. Ichikawa, Nanostructures on oxidized Si surfaces fabricated with the scanning tunneling microscope tip under electron-beam irradiation, J. Vac. Sci. Technol. B 24 (2006) 739–743.

[28] A.A. Shklyaev, M. Ichikawa, Fabrication of germanium and silicon nanostructures using a scanning tunneling microscope, Physics—Uspekhi 49 (2006) 887.

[29] Y. Nakamura, T. Miwa, M. Ichikawa, Nanocontact heteroepitaxy of thin GaSb and AlGaSb films on Si substrates using ultrahigh-density nanodot seeds, Nanotechnology 22 (2011) 265301.

[30] A. Roy, T. Bagarti, K. Bhattacharjee, K. Kundu, B.N. Dev, Patterns in Ge cluster growth on clean and oxidized Si(111)-(7×7) surfaces, Surf. Sci. 606 (2012) 777–783.

[31] A.A. Shklyaev, M. Ichikawa, Effect of interfaces on quantum confinement in Ge dots grown on Si surfaces with a SiO_2 coverage, Surf. Sci. 514 (2002) 19.

[32] G. Medeiros-Ribeiro, A.M. Bratkovski, T.I. Kamins, D.A.A. Ohlberg, R.S. Williams, Shape transition of germanium nanocrystals on a silicon (001) surface from pyramids to domes, Science 279 (1998) 353–355.

[33] G. Medeiros-Ribeiro, T.I. Kamins, D.A.A. Ohlberg, R.S. Williams, Annealing of Ge nanocrystals on Si (001) at 550°C: metastability of huts and the stability of pyramids and domes, Phys. Rev. B 58 (1998) 3533–3536.

[34] N. Tanaka, S.P. Cho, A.A. Shklyaev, J. Yamasaki, E. Okunishi, M. Ichikawa, Spherical aberration corrected STEM studies of Ge nanodots grown on Si(001) surfaces with an ultrathin SiO_2 coverage, Appl. Surf. Sci. 254 (2008) 7569–7572.

[35] A.A. Shklyaev, M. Shibata, M. Ichikawa, Nanometer-scale germanium islands on Si (111) surface windows formed in an ultrathin silicon dioxide film, Appl. Phys. Lett. 72 (1998) 320–322.

[36] E. Sutter, P. Sutter, J.E. Bernard, Extended shape evolution of low mismatch $Si_{1-x}Ge_x$ alloy islands on Si (100), Appl. Phys. Lett. 84 (2004) 2262–2264.

[37] A.A. Shklyaev, K.E. Ponomarev, Strain-induced Ge segregation on Si at high temperatures, J. Cryst. Growth 413 (2015) 94–99.

[38] J. Fuhr, P. Müller, Strain distribution due to surface domains: a self-consistent approach with respect to surface elasticity, Beilstein J. Nanotechnol. 6 (2015) 321–326.

[39] A.A. Shklyaev, A.E. Budazhapova, Ge deposition on Si(100) in the conditions close to dynamic equilibrium between islands growth and their decay, Appl. Surf. Sci. 360 (2016) 1023–1029.

[40] Z. Gai, R.G. Zhao, X. Li, W.S. Yang, Faceting and nanoscale faceting of Ge(hhl) surfaces around (113), Phys. Rev. B 58 (1998) 4572.

[41] A.A. Shklyaev, K.N. Romanyuk, A.V. Latyshev, Epitaxial Ge growth on Si(111) covered with ultrathin SiO_2 films, JSEMAT 3 (2013) 195–204.

[42] A. Shklyaev, L. Bolotov, V. Poborchii, T. Tada, Properties of three-dimensional structures prepared by Ge dewetting from Si(111) at high temperatures, J. Appl. Phys. 117 (2015) 205303.

[43] J.M. MacLeod, J.A. Lipton-Duffin, U. Lanke, S.G. Urquhart, F. Rosei, Shape transition in very large germanium islands on Si(111), Appl. Phys. Lett. 94 (2009) 103109.

[44] A.A Shklyaev, AV. Latyshev, Formation of lateral nanowires by Ge deposition on Si(111) at high temperatures, J. Cryst. Growth 441 (2016) 84–88.

[45] J. Tersoff, R.M. Tromp, Shape transition in growth of strained islands: spontaneous formation of quantum wires, Phys. Rev. Lett. 70 (1993) 2782.

[46] Y. Nakamura, A. Murayama, M. Ichikawa, Epitaxial growth of high quality Ge films on Si(001) substrates by nanocontact epitaxy, Cryst. Growth Des. 11 (2011) 3301–3305.

[47] D. Leonhardt, S.M. Han, Dislocation reduction in heteroepitaxial Ge on Si using SiO_2 lined etch pits and epitaxial lateral overgrowth, Appl. Phys. Lett. 99 (2011) 111911.

[48] W. Ikeda, Y. Nakamura, S. Okamoto, S. Takeuchi, J. Kikkawa, M. Ichikawa, et al., Characterization of Ge films on Si(001) substrates grown by nanocontact epitaxy, Jpn. J. Appl. Phys. 52 (2013) 095503.

[49] M. Henzler, Correlation between surface structure and surface states at the clean germanium (111) surface, J. Appl. Phys. 40 (1969) 3758–3765.

[50] B.Z. Olshanetsky, S.M. Repinsky, A.A. Shklyaev, LEED investigation of germanium surfaces cleaned by sublimation of sulphide films; structural transitions on clean Ge(110) surface, Surf. Sci. 64 (1977) 224.

[51] M. Kuzmin, M.J.P. Punkkinen, P. Laukkanen, J.J.K. Lang, J. Dahl, V. Tuominen, et al., Surface core-level shifts on Ge(111) c (2×8): experiment and theory, Phys. Rev. B 83 (2011) 245319.

[52] M. Henzler, The roughness of cleaved semiconductor surfaces, Surf. Sci. 36 (1973) 109–122.

[53] B.Z. Olshanetsky, S.M. Repinsky, A.A. Shklyaev, LEED studies of vicinal surfaces of germanium, Surf. Sci. 69 (1977) 205–217.

[54] A.A. Shklyaev, F.N. Dultsev, K.P. Mogilnikov, A.V. Latyshev, M. Ichikawa, Electroluminescence of dislocation-rich Si layers grown using oxidized Si surfaces, J. Phys. D Appl. Phys. 44 (2011) 025402.

[55] I.I. Kurkina, I.V. Antonova, A.A. Shklyaev, S.A. Smagulova, M. Ichikawa, Luminescence and deep-level transient spectroscopy of grown dislocation-rich Si layers, AIP Adv. 2 (2012) 032152.

[56] A.I. Yakimov, A.V. Dvurechenskii, A.I. Nikiforov, S.V. Chaikovskii, S.A. Tiis, Ge/Si photodiodes with embedded arrays of Ge quantum dots for the near infrared (1.3–1.5 μm) region, Semiconductors 37 (2003) 1345–1349.

[57] V. Kuryliuk, O. Korotchenkov, A. Cantarero, Carrier confinement in Ge/Si quantum dots grown with an intermediate ultrathin oxide layer, Phys. Rev. B 85 (2012) 075406.

[58] V.V. Kuryliuk, O.A. Korotchenkov, Features of the stress-strain state of Si/SiO_2/Ge heterostructures with germanium nanoislands of a limited density, Semiconductors 47 (2013) 1031–1036.

[59] A.O. Podolian, V.V. Kuryliuk, A.B. Nadtochiy, S.V. Kondratenko, O.A. Korotchenkov, Yu.N. Kozyrev, et al., Photovoltage performance of Ge/Si nanostructures grown on intermediate ultrathin SiO_x layers, Adv. Mater. Res. 276 (2011) 159–166.

[60] A.A. Shklyaev, O.A. Shegai, Y. Nakamura, M. Ichikawa, Impact ionization of excitons in Ge/Si structures with Ge quantum dots grown on the oxidized Si(100) surfaces, J. Appl. Phys. 115 (2014) 203702.

14

Monte Carlo Simulation of Semiconductor Nanostructure Growth

I.G. Neizvestny[1,2], N.L. Shwartz[1,2]

[1]RZHANOV INSTITUTE OF SEMICONDUCTOR PHYSICS OF THE SIBERIAN BRANCH OF THE RUSSIAN ACADEMY OF SCIENCES, NOVOSIBIRSK, RUSSIA [2]NOVOSIBIRSK STATE TECHNICAL UNIVERSITY, NOVOSIBIRSK, RUSSIA

14.1 Introduction

Modern objects of nanoelectronics are multilayer and porous structures, nanoclusters, and nanowires that can be obtained by different techniques: molecular beam epitaxy (MBE), chemical vapor deposition (CVD), and its variant, atomic layer deposition (ALD). Simulation can be considered as an instrument for the analysis of nanostructure formation mechanisms on the atomic scale. The nanometer size of electron device elements requires elucidation of atom migration details during formation of these structures. Using the Monte Carlo (MC) technique, one can study growth kinetics in the system with sizes comparable to the sizes of real objects, and the duration of model processes corresponds to real times. The use of lattice MC models appreciably increases processing speed. MC methods are widely used for modeling processes in crystals and on their surfaces [1–3]. It is worth noting that MC simulation takes into account spatio-temporal fluctuations that may be important in growth processes.

The first MC simulations of atomic processes on a semiconductor surface during deposition were performed on the Kossel crystal in the standard *solid-on-solid* model [1]. It was believed in that model that the substrate and the growing layer had a simple cubic lattice. Even this simplified model has proved very useful for the analysis of morphology transformation during MBE growth [3]. Comparison of simulated step-density oscillations with experimental RHEED oscillations during MBE helped clarify the role of steps in oscillation damping, and demonstrated a self-organizing stationary asynchronous structure formation (structure with islands nucleated at different times), continuously generated during growth [4]. It has been shown in Ref. [5] that an unusual step-density (RHEED) oscillation shape in the form of a beat can be due not only to the instability of the deposited atomic flux, but also to the substrate surface relief (the presence of opposite steps and step echelons). The introduction of additional energy barriers for atoms crossing the monoatomic step—Schwoebel step-edge barriers—allowed realization of the formation of a 3D compact island; and the Stransky–Krastanov growth mode showed when, at the initial growth stage, the

Advances in Semiconductor Nanostructures. DOI: http://dx.doi.org/10.1016/B978-0-12-810512-2.00014-7

wetting layer was created and then 3D islands grew [6]. The influence of Schwoebel barriers and diffusion anisotropy on growing layer morphology was analyzed in Ref. [7]. The increase in atomic surface anisotropy was shown to be responsible for slowing down step-density oscillation damping, and an oscillation phase shift in anisotropy similar to the anisotropy on (001) reconstructed surfaces. The existence of the Schwoebel barrier increases step-density oscillation damping. An estimation of the Schwoebel barrier, which is necessary for the absence of oscillations, was done.

The substrate crystal structure turned out to be important for nanostructure simulation, and the complex structure shape encouraged the transition to three-dimensional (3D) models. Previously, we developed a program package, SilSim3D, for simulation of epitaxial growth on a diamond-like crystal 3D lattice [8]. Morphology transformations of porous substrates during deposition and annealing were studied using SilSim3D. Modeling of pore overgrowth on substrates with (001) and (111) orientation was accomplished. The differences in solid layer morphology grown on porous Si(111) and Si(100) substrates were detected. Morphology transformations during epitaxial growth on these two surfaces differ greatly, due to the diverse in diffusion characteristics of these surfaces [8–9]. The sintering process of spongy films with randomly distributed pores was analyzed in Ref. [10]. The sintering rate was demonstrated to be nonmonotonous in time for any film porosity and a different lattice coordination number. The metastable states of the system, dependent on time and temperature of the annealing process, were revealed. Investigation of step behavior on the vicinal Si (111) surface with nonequidistant steps revealed the step width of periodic variations due to sublimation during annealing [11]. The diamond-like crystal structure results in asymmetry in atom fluxes to the upper or lower terraces, with the sign of asymmetry depending on the step structure. The essential effect of vacancies, created during sublimation on (111) terraces, for atom migration length was revealed [12]. Simulation of layer-by-layer growth of multicomponent films by ALD was carried out. The procedures that can help to solve the main ALD problem—slow and uncontrollable new layer nucleation at the initial stage of the process—were proposed on the basis of simulation [13].

The SilSim3D software, designed to study the atomic processes that take place during nanostructure formation in the 3D subsurface layer of crystal substrates, allows simulation of systems containing up to seven atom species [14]. The mathematical model takes into account random walks of particles on lattice sites, chemical transformations, adsorption, and desorption. The activation energies of all elementary events are the model system parameters, and they determine its physico-chemical nature. Below the simulated results of silicon nanocluster (Si-nc) formation in silicon dioxide layers, and Si and GaAs nanowhisker growth are presented.

14.2 Silicon Nanoclusters in Silicon Dioxide

Photoluminescence in visible and infrared spectral regions (700–900 nm) [15] from Si-ncs embedded in a SiO_2 matrix is promising for optoelectronic devices based on silicon [16]. Different techniques are used to obtain such structures [17–18], but the final stage of Si-nc formation is high-temperature annealing. The annealing step is necessary since it removes structural defects inevitably created during low temperature deposition or ion implantation. High-temperature annealing results in phase separation of SiO_x layers into Si and SiO_2, with Si cluster sizes being determined by temperature, annealing time, and the degree of nonstoichiometry (x value) [19]. At high temperatures, the important role of silicon monoxide in silicon transfer in a Si-SiO_2 system was revealed [17,20]. During high-temperature annealing of silicon suboxide layers, along with oxygen and silicon atom diffusion, silicon monoxide is created [17]. At high temperatures, processes of SiO generation, dissociation, and diffusion can play an important role in silicon nanoclusters formation during phase separation.

A kinetic MC model of Si-nc formation during high-temperature annealing of Si suboxide layers was suggested [21–24]. The model takes into account, along with silicon and oxygen atom diffusion, the processes of silicon monoxide creation and dissociation. Using this model, examination of SiO_x layer annealing in the temperature range $1173K \leq T \leq 1573K$ for $1 \leq x \leq 2$, and analysis of the influence of silicon monoxide on silicon cluster aggregation were carried out. To describe the Si-SiO_2 system, we took into account that a tetrahedron is the main element of silicon and silica crystal structure. Since our model is a lattice with a diamond-like crystal structure, we consider only crystalline Si clusters. The dependence of Si nanocluster structure on annealing temperature was beyond the scope of this work. The use of a partially filled diamond-like crystal lattice for amorphous SiO_2 layer modeling enabled arrangement of the correct short-range order corresponding to the SiO_2 matrix [25]. In silicon dioxide, the Si atom is in the center and four oxygen atoms are at the corners of the tetrahedron. The neighboring SiO_4 tetrahedra share one oxygen atom; therefore, silicon is quadrivalent and oxygen is bivalent. To simulate SiO_x layer annealing, a three-component system was considered: (1) Si, (2) O, (3) SiO. The possible elementary events in the model system were adsorption, desorption, diffusion, and chemical reactions. The probability of the event was calculated according to the expression: $P_{event} = \upsilon_0 \exp(-E_{event}/k_BT)$, where k_B is the Boltzmann constant; T is temperature (K); υ_0 is the attempt frequency (for diffusion hop $\upsilon_0 = 10^{13}$ s^{-1}: Debye frequency of atom oscillation in a potential well); and E_{event} is the energy barrier for the given event. The input parameters of the model were binding energies between atoms E_{i-j}, sublimation energies for each material $E_{sub,i}$ and activation energies of the chemical reactions $E_{r,i}$. According to literary data [26–27], the following relationship between covalent bond energies was fulfilled: $E_{Si-O} > E_{Si-Si} > E_{O-O}$. Simulation was carried out using the following energies of single covalent bonds: $E_{Si\text{-}O} = 2.5$ eV, $E_{Si-Si} = 1.1$ eV, and $E_{O\text{-}O} = 0.6$ eV. Considering the silicon monoxide molecule, mobile and volatile, the binding energy of SiO for all configurations was chosen as 1.4 eV, and the SiO desorption energy was equal to 1.6 eV. To achieve silicon agglomeration into compact 3D clusters, the increase in the energy barrier for a diffusion hop of the silicon atom belonging to silicon clusters or their interface, was taken into consideration. This shift imitated the increase in a potential barrier for the Si diffusion hop at Si-ncs due to a strain at the Si–SiO_2 interface. Chemical reactions for the creation of silicon monoxide ($Si + O \rightarrow SiO + vac$) and its dissociation ($SiO + vac \rightarrow Si + O$) were included in the model, where vac is the empty lattice site (vacancy). The relation between the activation energies of SiO formation and dissociation were chosen based on the temperature dependence of pressure of gaseous silicon monoxide generated during high-temperature annealing of solid phase silicon and silicon dioxide mixtures at the ratio of 1:1 and 1:2 [28].

The initial state of the model layer was a random distribution of silicon and oxygen atoms with the Si:O ratio of 1:x ($1 \leq x \leq 2$) over 60% of lattice sites (40% of lattice sites were vacant). A semi-infinite crystal with cyclic boundary conditions in the lateral direction was under consideration. Two annealing alternatives were analyzed: annealing in an open system, or in a closed system. The nanocluster growth rate dependence on the initial SiO_x layer composition was analyzed. The Si-nc size and growth rate increased with Si supersaturation (x decreasing). The difference in nanocluster sizes after the same duration of annealing of the mixture with a different composition is clearly seen in Fig. 14.1. The simulation demonstrated that silicon started to gather into clusters only when an excessive silicon value in a SiO_x layer exceeded some definite value: $x < 1.8$. At high temperatures, silicon monoxide continuously generated and decayed at the Si-nc/SiO_2 interface. SiO can play the role of an additional silicon carrier between clusters. The influence of monoxide on silicon cluster aggregation depends on the stage of Si-nc formation. At the initial stage of SiO layer annealing, the silicon monoxide concentration noticeably exceeded the equilibrium value; so, the SiO contribution to the Si gathering process is high. The creation of silicon monoxide per se did not accelerate Si-nc growth, as SiO generated at the Si-nc/SiO_2 interface, and due to that the diffusion movement could decay at an arbitrary lattice site. However, with an increasing diffusion barrier for SiO at a Si-nc, the probability of SiO decay at the interface rose, since SiO molecules stayed near the cluster longer. One can suggest that the strain induced at the Si–SiO_2 interface, due to the volume mismatch, may create a local minimum for the SiO molecule: an increase in a potential barrier for SiO diffusion hops similar to the energy barrier increase in the diffusion hop for Si atoms belonging to the interface. In this case, the presence of mobile SiO molecules can accelerate the silicon aggregation process. We analyzed how the variation of these barriers influenced the kinetics of Si-nc formation. At higher Si binding energies, small clusters were more stable, and their enlargement started later. The creation of SiO at the nanocluster interface, on the contrary, resulted in low crystal

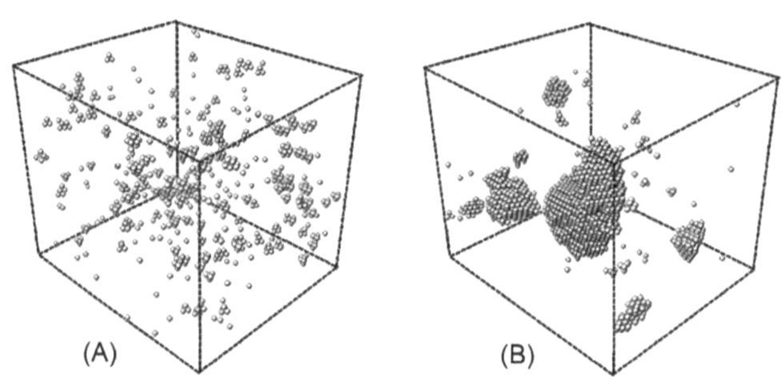

FIGURE 14.1 Three-dimensional views of Si-nc formed after SiO_x layer annealing at $T = 1373K$ during 1 s: (A) $x = 1.7$, (B) $x = 1.5$. Only silicon atoms belonging to clusters are shown. *From E.A. Mikhantiev, I.G. Neizvestny, S.V. Usenkov, N.L. Shwartz, Silicon monoxide role in silicon nanocluster formation during Si-rich oxide layer annealing–Monte Carlo simulation, Comput. Mater. Sci. 90 (2014) 99 with permission.*

dissolution. The increase in the diffusion hop barrier for SiO resulted in an acceleration of the Si-nc aggregation rate because the increase in critical nuclei size, as silicon monoxide, continuously generated and decayed at the Si-nc/SiO_2 interface, easily destroyed Si nuclear by moving silicon from one cluster to the other. This case corresponds to a decrease in the number of small clusters, and an increase in average cluster size.

SiO affected not only the Si-nc nucleation stage, but also the latest stage of slow cluster enlargement, when silicon monoxides with a relatively low diffusion activation energy acted as an additional silicon carrier between clusters. Three stages of Si-nc growth were observed: nucleation, coalescence of neighboring nanoclusters, and slow enlargement due to the diffusion exchange between long distant clusters.

A self-assembled Si cluster layer can be created during thin SiO_x film annealing on the Si substrate. The absorbing role of the substrate in Si-nc formation was illustrated in a computing experiment; annealing in the closed system of a uniform SiO_x film ($x = 1$) on the Si substrate. The zone denuded of Si-ncs over the substrate is clearly seen in Fig. 14.2A, since a portion of excessive silicon atoms is incorporated into the substrate. The plots in Fig. 14.2B demonstrate Si redistribution in the SiO layer during annealing. At the initial stage, silicon in clusters was nearly uniformly distributed across the layer. Later, self-assembled Si clusters and the denuded zone near the substrate were detected. The nanocluster-free oxide layer formation was due to a transfer of redundant silicon to the substrate. The absorbing Si–SiO_2 interface provoked the creation of a zone denuded of Si-ncs. The denuded zone expanded

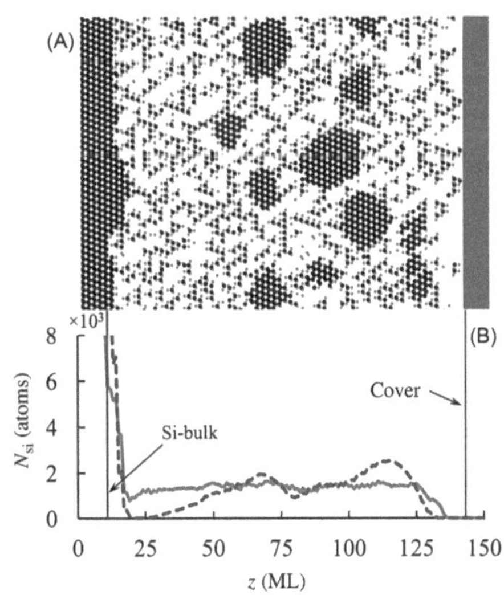

FIGURE 14.2 (A) Cross-section of a SiO_x ($x = 1.0$) layer on the Si substrate after 20 s of annealing at $T = 1225K$; silicon atoms are marked in *black* color, and oxygen in *gray* color; *gray rectangle* is reflecting cover. (B) Profile of a number of silicon atoms gathered in clusters in the direction perpendicular to the substrate after 0.1 s (*solid line*) and 20 s (*dotted line*). *From E.A. Mikhantiev, I.G. Neizvestny, S.V. Usenkov, N.L. Shwartz, Silicon monoxide role in silicon nanocluster formation during Si-rich oxide layer annealing—Monte Carlo simulation, Comput. Mater. Sci. 90 (2014) 99 with permission.*

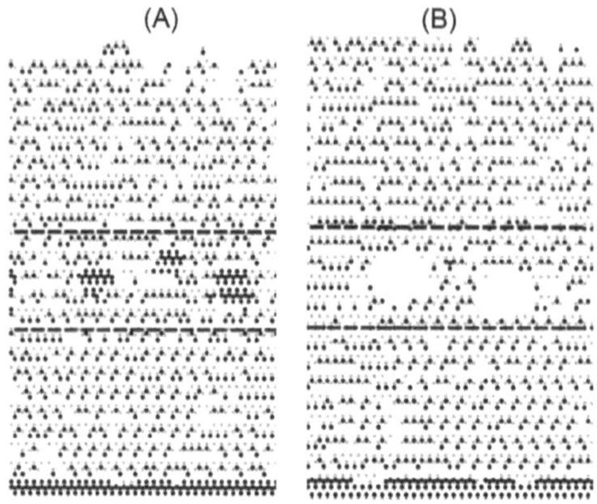

FIGURE 14.3 Cross-sections of the thin SiO_2 film containing a suboxide silicon layer on the Si substrate after annealing at 1323K (A), and 1423K (B). Silicon atoms are marked in *black* and oxygen is in *gray*; *dashed lines* indicate location of the substoichiometric layer. *From E.A. Mikhantiev, I.G. Neizvestny, S.V. Usenkov, N.L. Shwartz, Silicon monoxide role in silicon nanocluster formation during Si-rich oxide layer annealing–Monte Carlo simulation, Comput. Mater. Sci. 90 (2014) 99 with permission.*

and Si-ncs enlarged with annealing time. By varying annealing time and temperature, it is possible to obtain Si-ncs located at a certain distance from the $Si-SiO_2$ interface.

A simulation, imitating annealing of the SiO_2 film containing a layer with excess silicon, created by ion implantation, was carried out. The initial model system was SiO_2 film on a silicon substrate, with a 4.7 nm layer of nonstoichiometric composition SiO_x ($x = 1.5$). Annealing was fulfilled in the open system. In the SiO_x-layer region after annealing at a temperature of $T = 1323K$, Si-ncs were formed, and at $T = 1423K$ cavities were created (Fig. 14.3). Observations of film composition and morphology during annealing revealed that, in the region corresponding to the SiO_x layer, many SiO molecules were generated. These SiO molecules could either dissociate and take part in Si-nc formation, or be removed from the system due to diffusion through vacant sites in the silicon dioxide matrix with a following desorption. The character of the process depends on the temperature regime. The temperature variation from 1323K to 1424K in the experimental annealing process of silicon-implanted SiO_2 films also dramatically changed the result: instead of excessive silicon aggregation into nanoclusters, cavity formation was observed [17]. Valid relations between the activation energies of elementary processes made it possible to achieve agreement between simulation and experiment.

14.3 Nanowhisker Growth

Nanowhiskers (NWs), filamentary crystals, are nanocrystals with a high (length/diameter) aspect ratio possessing unique electronic and optical properties and exceptional mechanical strength, and are promising nanoelectronic elements. Generally, nanowhiskers are grown from catalytic seed drops according to the vapor–liquid–solid (VLS) mechanism [29]. The semiconductor material is deposited on a substrate activated by catalyst drops.

Supersaturation of the drops with the material from a flux with a following crystallization at the drop–substrate interface results in NW growth. Semiconductor NWs can be grown from various materials using different catalysts, but the main catalyst is gold [30–32]. Gold is attractive due to its ability to form low temperature eutectic binary alloys.

14.4 Si Nanowhiskers

Nowadays, Si NWs are used in field effect transistors, diodes, emitters, chemical, and biological sensors [33–36]. Single vertical Si whiskers are used as tips for atomic-force microscopy, ordered arrays of Si NWs may serve as electron and ion emitters in field emission displays [37], and as a coating with regulated (controllable) wettability [38].

In this section, the lattice MC model of Si NW formation according to the VLS growth mechanism is briefly described. NW growth rate dependences on temperature, flux intensity, and catalyst drop size, are presented, as well as NW morphology dependence on deposition conditions. The effect of silicon wetting with a catalyst matter on the character of NW growth is examined. The process of Si-Ge heterostructure formation during nanowire growth is analyzed.

The basic stages of VLS growth are included in the model: silicon dissolution into a catalyst (Au) drop resulting in eutectic alloy formation, silicon diffusion through the drop, and crystallization at the drop–crystal interface. A three-component system was considered in the model: silicon in solid Si(s) and liquid states Si(liq), and Au (Fig. 14.4). A substrate with a diamond crystal structure and surface orientation (111) activated with a catalyst drop represents the initial model layer. The simulated layer is assumed to overlie a semi-infinite crystal wafer. In the lateral directions, cyclic boundary conditions are applied. Particles (atoms or molecules) from the flux could be deposited perpendicularly to the substrate, or at arbitrary angles. The semiconductor material arrives at the drop directly from the flux (adsorption

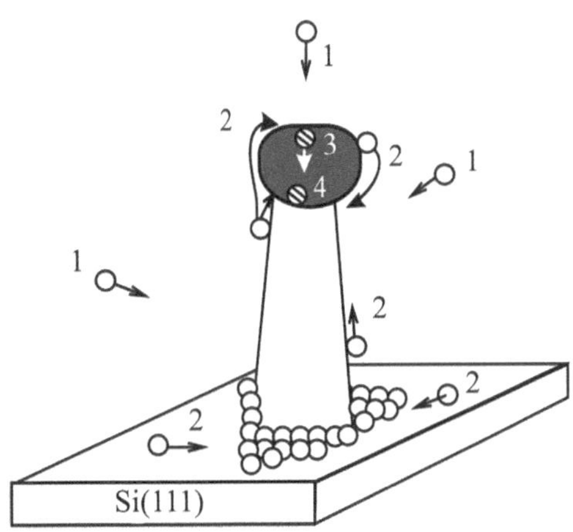

FIGURE 14.4 Scheme of the model system: (1) Si adsorption; (2) Si surface diffusion; (3) Si(s) dissolution into the Au drop, and Si(liq) diffusion through the drop; (4) Si (liq) crystallization at the interface. Si(s) is marked in *white*, and gold is in dark *gray*.

growth mode) and by diffusion via the substrate surface (diffusion growth mode). NWs grew due to: (1) dissolution of semiconductor material in the drop with a following crystallization at the drop–whisker interface (the classical VLS mechanism); (2) atom incorporation along the triple VLS line at sites provided with incomplete wetting (incorporation mechanism).

The input parameters of the model growth process were: type of deposition process (MBE or CVD), flux intensity F (ML/s), and substrate temperature T (K). In the case of the MBE process, the whisker material is deposited in the atomic form on the surface at a certain angle. When we consider CVD growth, the semiconductor matter arrives at the surface at arbitrary angles imitating the vapor phase. The energy parameters of the model are energy of covalent bonds E_{i-j} (i, j, matter sorts), activation energies of chemical reactions ($E_{r,i}$), and desorption ($E_{\mathrm{sub},i}$). No more than two reactants and no more than two reaction products are involved in chemical reactions. Chemical reactions in the gas phase were not considered.

The scheme of the model process is presented in the following way. Silicon atoms Si(s) were dissolved into the gold drop according to the reaction: Si(s) + Au→Si(liq) + Au. In the lattice model, transport of semiconductor atoms within the liquid drop occurred by the exchange reaction. Silicon diffusion through the drop was simulated by exchange diffusion Si (liq) + Au→Si(liq) + Au. The reaction activation energy depends on the reactant neighborhood. Liquid silicon crystallized only when contacting solid phase silicon: Si(liq) + Si(s)→ Si(s) + Si(s).

Using this model, NW growth on the Si(111) surface activated by gold was realized [39–45]. NW growth rate dependencies on temperature and seed drop diameter at different growth regimes were obtained [42]. The ranges of model growth conditions suitable for the growth of nanowhiskers and nanotubes were determined [43]. The formation of axial Si-Ge heterojunctions (HJs) in NW was examined [44–45]. A number of effects, observed experimentally, were found in the model system: NW growth, faceting the NW side wall surface, Au droplet slipping from the crystal top, and NW branching [41]. The branching effect in the model system is illustrated in Fig. 14.5. The drop-seed migration from the top to the lateral facet can cause branching of the growing crystal. Nanowhisker branching may also be due to drop splitting. Each droplet can initiate the growth of a separate NW branch.

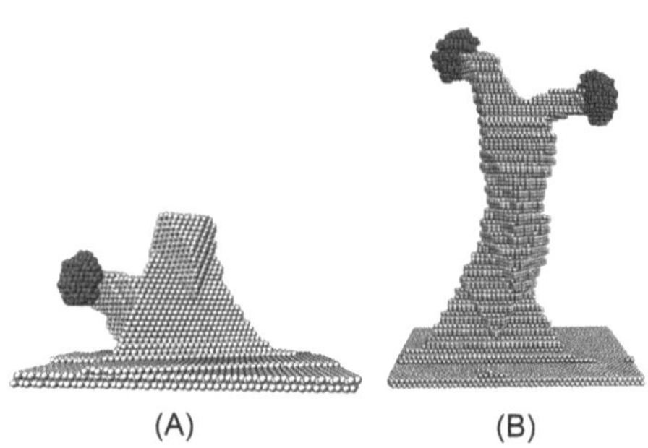

FIGURE 14.5 Three-dimensional images of NWs demonstrating the branching effect due to drop migration (A), and drop splitting (B).

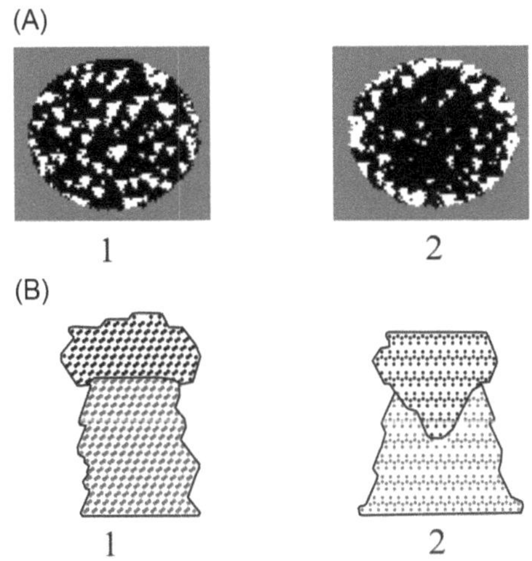

FIGURE 14.6 (A) Plan view cross-sections of the drop–whisker interface for low (1) and high (2) diffusion growth components; (B) cross-sections of the NW fragment with a catalyst drop after CVD (1) and MBE (2) growth.

The character of NW new layer nucleation determines the quality of the drop whisker interface and the further morphology of the growing whisker. Fig. 14.6A illustrates nucleus distribution at the drop–NW interface depending on the diffusion component contribution to the whisker growth process. When the diffusion component is low, nuclei are randomly distributed at the interface. An increase in diffusion components results in a higher nuclei concentration along the triple line. A similar situation is expected in the case of MBE growth. The interface shape depends on the type of growth process (Fig. 14.6B). The cup-shaped interface typical of MBE growth is determined by a high diffusion growth component, and the flat interface corresponds to the adsorption-induced growth mode typical of CVD growth.

The wetting between NW and catalyst materials significantly affects NW morphology. For a certain choice of catalyst and substrate materials, the formation of hollow whiskers is possible. The conditions for tube-like nanowhisker formation are weak wetting between a substrate and catalyst materials, and a considerable number of atoms being incorporated along the drop–whisker triple line. In this case, the atoms, incorporated under the drop, form a ring along the triple line which prevents matter incorporation under the drop and promotes cavity formation. It should be noted that the higher nuclear concentration along the triple line may be the reason for nanotube-like growth. Since growth conditions influence the incorporation regime, the dependence of NW morphology on temperature and flux intensity was analyzed. The regions of growth conditions corresponding to a different NW morphology are demonstrated in Fig. 14.7. It was revealed that, under certain temperatures and fluxes, hollow whisker formation is possible.

The dependences of nanowhisker growth rate on diameter were obtained [42]. These dependences are different for various growth modes. In the diffusion mode, the growth rate

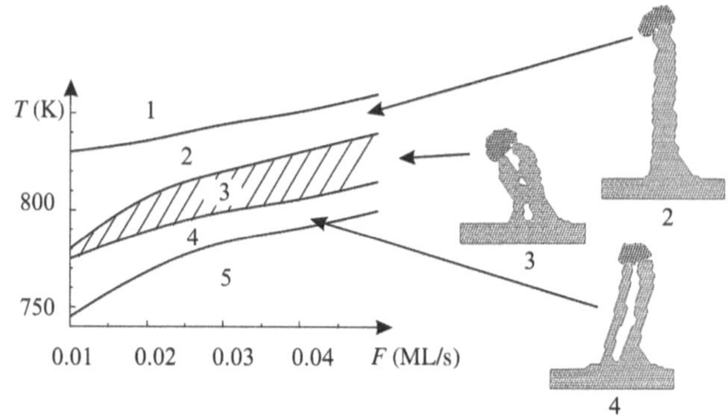

FIGURE 14.7 Growth conditions of NWs of various morphologies in the *T*–*F* coordinates ($d = 4.8$ nm): (1) no growth; (2) solid NWs; (3) NWs with small cavities; (4) nanotubes; (5) no growth.

is proportional to $1/d^2$ if the growth occurs by the VLS mechanism, or the growth rate is $\sim 1/d$, if the growth is dominated by incorporation of the material along the triple line. In the adsorption mode, the growth rate does not depend on the diameter, and is only determined by the external flux. Thus, the growth rate dependences on diameter allow one to identify the NW growth mode and mechanism.

At the end of this section, the simulation results of Si-Ge HJ formation in NWs are presented. The dependences of axial HJ abruptness on NW diameter, flux intensity, solubility of the semiconductor material in a catalyst drop, and duration of Ge deposition were obtained [45]. To imitate MBE growth of axial Ge-Si HJs, Si and Ge atomic fluxes (F_{Si} and F_{Ge}, correspondingly) were switched alternately to grow silicon and germanium parts of the whisker. The growth conditions were chosen in accordance with the experimental work [46]. A model whisker with axial heterojunctions is shown in Fig. 14.8. Thin Ge layers were embedded into a silicon NW during growth. The germanium concentration profile along the NW axis is shown in Fig. 14.8. One can see that the interface between Si and Ge segments is not abrupt, with the embedded layer composition being Ge_xSi_{1-x}. The composition of the Ge_xSi_{1-x} axial heterojunction was found to be dependent on the flux ratio, duration of Si and Ge deposition, and catalyst drop diameter. The kinetics of drop composition variation after flux switching were analyzed. The dependence of germanium percentage in a Ge_xSi_{1-x} layer on Ge deposition time is the result of a gradual change of liquid drop composition after switching the fluxes (the phenomenon of residual concentration).

The effects of different factors for heterojunction abruptness were investigated. In the Ge concentration profiles in Fig. 14.8, one can see leading and trailing edges. The thickness of these regions corresponds to the Si-Ge HJ width. The HJ width can be expressed as: $\alpha = Vt_{st}$, where V is NW growth rate, and t_{st} is time required for complete Ge substitution for Si in the catalyst drop (time required for the saturation of Ge_xSi_{1-x} composition). In the general case, both V and t_{st} depend on the catalyst drop diameter. The simulation demonstrated t_{st} to be in direct proportion to the diameter. Taking into account the $V(d)$ dependence in different

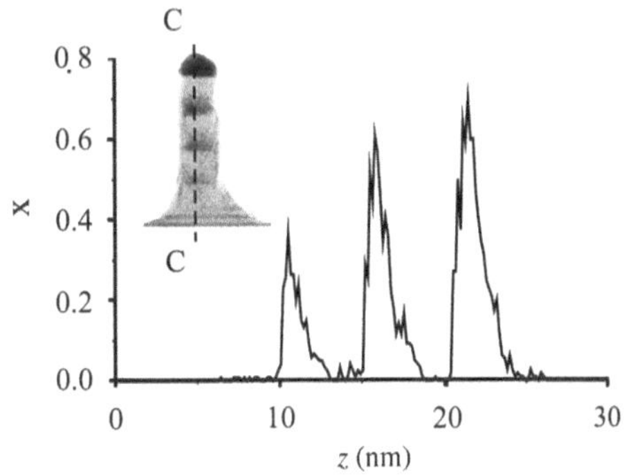

FIGURE 14.8 Si NW with three axial Ge layers: $F_{Si} = 0.05$ ML/s, $F_{Ge} = 0.01$ ML/s, $d = 6$ nm, duration of Ge layers deposition $t_{Ge1} = 20$ s, $t_{Ge2} = 40$ s, $t_{Ge3} = 60$ s. Ge concentration profile along the C–C line, x is the Ge fraction in the Ge_xSi_{1-x} layer. Si(s) is marked in *light gray*, Ge(s) is in *dark gray*.

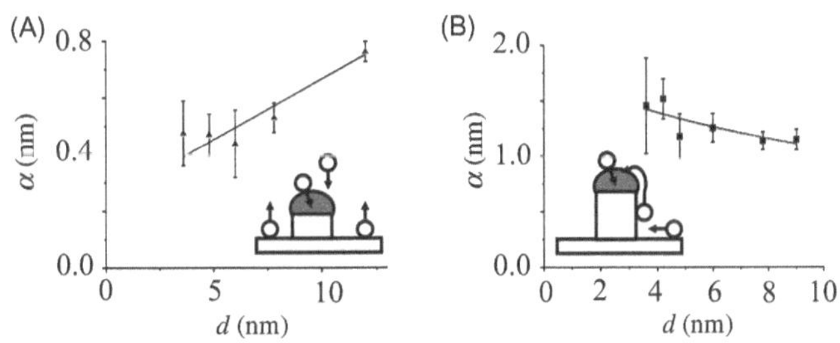

FIGURE 14.9 Dependence of HJ abruptness α on the NW diameter in adsorption (A), and diffusion (B) growth modes. The schemes of adsorption and diffusion regimes are illustrated in the insert.

(growth regimes, the dependence of HJ abruptness α on diameter was obtained: α is $\sim d$ for the adsorption growth mode and is independent from d or $\sim 1/d$ for the diffusion growth mode) (Fig. 14.9). The increase of dependence $\alpha(d)$ was shown for CVD-grown NWs with axial Si-Ge heterojunctions experimentally [47]. Relation $\alpha \sim 1/d$ is the theoretical prediction for NW growth by MBE because, in MBE, the diffusion component predominates. So, the junction abruptness was found to be dependent on the nanowhisker diameter: in the adsorption-induced growth mode, abruptness decreases with diameter, and it increases in the diffusion-induced mode. It means that the abrupt Si-Ge heterojunction in nanowires with a small diameter can be obtained only in the CVD process with a negligible diffusion growth component.

The process of radial Si-Ge heterostructure formation during nanowire growth was studied [45]. It was demonstrated that radial growth can be stimulated by adding chemical species that decrease semiconductor material solubility in a catalyst drop and/or reduce the silicon adatom diffusion length along NW walls. Any of these factors leads to Si shell formation around a Ge core.

14.5 Au-Catalyzed and Self-Catalyzed GaAs Nanowhisker Growth

The nanowhiskers based on III–V compounds are of interest due to their promising application in modern nano- and optoelectronic devices [48–50]. Gold remains a popular catalyst for GaAs NW growth, providing the maximum growth rate for long vertical nanocrystals [51–54]. Au-assisted growth, being a more controllable and stable technique, is a prospect for growing NWs in nanoelectronic applications. However, Au creates traps in a semiconductor, which lead to degradation of the optical characteristics of structures based on NWs. Self-catalyzed NW growth with one of the nanocrystal components, Ga, serving as a catalyst is promising [55–58]. Usually, vertical NWs are grown on the GaAs(111)B surface. However, GaAs NW can be grown both on GaAs substrates of different surface orientations [55–57,59], and on the Si(111) substrate [60–62]. With any substrate, a necessary growth condition is the presence of a thin silicon dioxide layer on the surface. The oxide film can be considered as a mask for the subsequent growth of NWs, with the growth direction set by the crystalline substrate orientation. Self-catalyzed growth is known to be more sensitive to the arsenic flux than growth from gold seed droplets [63]. Nanowhisker growth characteristics become markedly different when the catalyst is replaced (gold with gallium).

Theoretical studies of NW growth mechanisms [63–66] helped to understand a number of experimental results in self-catalyzed growth. The effect of arsenic on growth characteristics was carefully studied in Ref. [66]. However, the self-catalyzed growth mechanism includes many complex processes, and an exact analytical solution cannot be obtained without these being taken into account [66]. MC simulation is a suitable method for elucidating the kinetics and growth details of GaAs NW growth.

All the processes included in the MC model of GaAs NW growth are shown in Fig. 14.10. The liquid phase in the lattice growth model of GaAs nanowhiskers was imitated similar to Ref. [43], where single-component NW growth was described. When simulating the growth of NWs based on III–V compounds, it is necessary to take into account the different

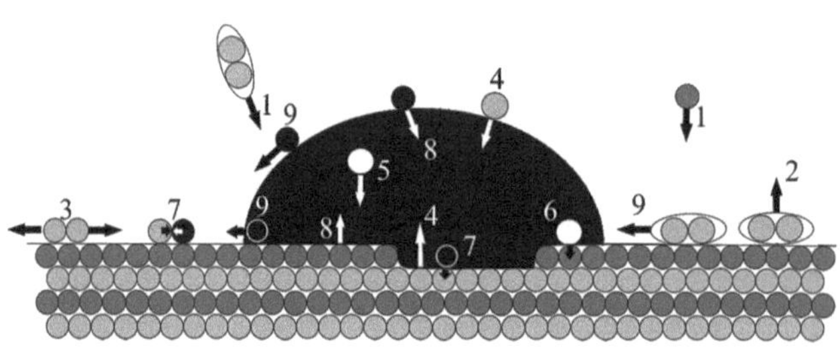

FIGURE 14.10 Model system and possible processes in this system: (1) As_2 molecule and Ga atom deposition; (2) As_2 desorption; (3) As_2 dissociation; (4) As dissolution into a drop; (5) diffusion of atoms within a drop; (6) As crystallization at the drop–crystal interface; (7) Ga crystallization; (8) Ga dissolution into a drop; (9) Ga, As, and As_2 diffusion over the surface. *Black* color denotes the catalyst material (Au/Ga(liq)), *dark gray* color corresponds to Ga (s), *black* shows Ga(liq), *light gray* denotes As(s), and *white* corresponds to As(liq). *From M.V. Knyzeva, A.G. Nastovjak, I.G. Neizvestny, N.L. Shwartz, Simulated growth of GaAs nanowires: catalytic and self-catalyzed growth, Semiconductors 49 (2015) 60–68 with permission [67].*

solubility of the components in a catalyst drop, and the different volatilities of Group-III and Group-V components, as well as the variation in the surface tension of a drop with its composition.

To simulate GaAs NW growth, we considered the six-component system: As_2, Ga(s), Ga(liq), As(s), As(liq), and Au (for gold-assisted growth), or M_f (for self-catalyzed growth), i.e., molecular arsenic, gallium and arsenic in both solid and liquid phases, and gold or the material of the masking film. The following elementary events were considered: Ga and As_2 adsorption, molecular arsenic dissociation and formation on the surface, As_2, Ga(s), Ga(liq) desorption, atomic arsenic and gallium dissolution into a catalytic drop, diffusion through a drop, Ga(liq) and As(liq) crystallization at the drop–GaAs interface, and diffusion hops of all components (Fig. 14.10). When describing the interaction between Ga and As atoms, on the basis of the published data [68,69], the following relationships between the bonding energies were chosen: $E_{Ga-Ga} < E_{As-As} < E_{Ga-As}$. The atomic processes occurring in the liquid drop and at the drop–crystal interface were simulated in the lattice model via several reactions.

Upon Au-assisted growth, flows of atomic Ga and molecular As_2 were deposited onto the surface of GaAs (111)B with drops of gold. Molecular arsenic could be desorbed from the surface or dissociated. Arsenic atoms were either incorporated into the GaAs surface or they were dissolved in the gold drop, and diffused toward the drop–crystal interface. It was assumed that gold has a catalytic effect on arsenic, i.e., the energy barrier for the dissociation of molecular arsenic into atoms becomes noticeably lower on a gold drop surface. In addition, the decrease in the diffusion coefficient of As_2 along the surface of gold was taken into account.

Gallium dissolution in a gold drop was described by the reaction Ga(s) + Au→Ga(liq) + Au, and arsenic dissolution, by the reaction As(s) + Au→As(liq) + Au. The diffusion of gallium and arsenic in the liquid phase was described by exchange reactions Ga(liq) + Au→Au + Ga(liq), and As(liq) + Au→Au + As(liq). Gallium crystallization at the drop–crystal interface occurred only upon contact with the solid phase of arsenic: Ga(liq) + As(s)→Ga(s) + As(s), and arsenic crystallization occurred only upon contact with the solid phase of gallium: As(liq) + Ga(s)→As(s) + Ga(s). The parameters determining the drop composition are the activation energies of semiconductor material dissolution in the gold drop, the activation energies of gallium/arsenic diffusion in the drop, and the activation energy of crystallization at the drop–crystal interface.

Gold-assisted growth was simulated at temperatures of $T = 750–950$K, gallium deposition rates of $F_{Ga} = 0.3–3.5$ ML/s, and various As_2 flux intensities. The initial catalyst drop model diameters were in the range from 4 to 8.4 nm. The dependences of the NW growth rate on temperature, catalyst drop size, and intensity of Ga and As_2 fluxes were obtained. NW growth was only observed in a certain range of temperatures and gallium to arsenic flux ratios. The temperature dependence of the NW growth rate is shown in Fig. 14.11. At temperatures below 700K, gold drops were overgrown with a gallium arsenide layer, and no NW growth was observed. At temperatures above 900K, NWs did not grow either. The termination of NW growth at high temperatures is due to a decrease in the concentration of dissolved arsenic in the drop because of higher As_2 desorption. The optimal growth temperature,

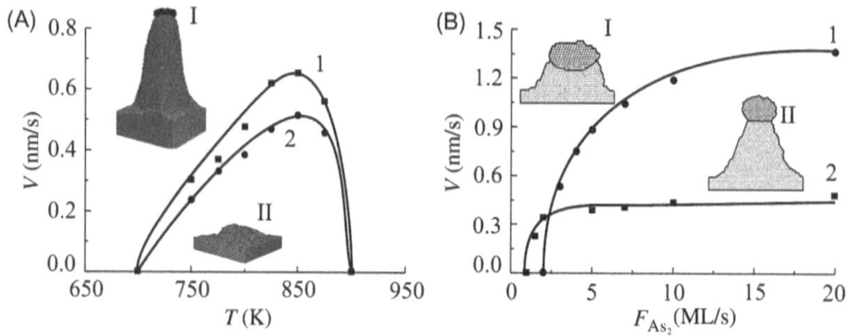

FIGURE 14.11 (A) NW growth rate versus temperature: 1−*d* = 5.3 nm, 2−*d* = 7.4 nm; the insert shows a 3D image of the model surface at the simultaneous deposition of Ga and As_2 at T = (I) 850K and (II) 700K; (B) NW growth rate versus arsenic flux at T = 850K for two gallium flow rates: (1)F_{Ga} = 1.5 ML/s; (2) F_{Ga} = 0.5 ML/s; in the insert−NW cross-sections at F_{Ga} = 0.5 ML/s for: F_{As_2} = 2 ML/s (*I*), and F_{As_2} = 5 ML/s (II). *From M.V. Knyzeva, A.G. Nastovjak, I.G. Neizvestny, N.L. Shwartz, Simulated growth of GaAs nanowires: catalytic and self-catalyzed growth, Semiconductors 49 (2015) 60–68 with permission [67].*

corresponding to the maximum growth rate is 850K. It should be noted that the standard GaAs nanowire growth temperature in the experiments is 850K. At a given Ga flux rate, the NW growth rate nonlinearly depends on the arsenic flux (Fig. 14.11B). At a low As_2 flux intensity, most of the arsenic does not reach the drop due to desorption, and no growth occurs. The situation is similar to that upon annealing. With the arsenic flux increase, the growth rate increases, becomes saturated, and ceases to depend on the arsenic flux. The quality of the drop–NW interface heavily depends on the flux ratio F_{As_2}/F_{Ga}. At a small flux ratio, the interface is cup-like. The interface becomes smoother with a considerable excess of arsenic. Under conditions of arsenic excess, the NW growth rate increased with the gallium flow rate. When choosing optimal growth conditions, it is necessary to control not only the growth rate, but also the interface quality. The interface quality is determined both by the ratio between the gallium and arsenic flux intensities, and by the gallium flux value.

The growth rate dependences on the catalyst drop diameter and the density of NW arrangement were analyzed. The decreasing growth rate dependences on the drop diameter and on NW density were obtained, which is in agreement with the experimental data [70], and is evidence of the diffusion growth mode. All the model experiments corresponded to the case in which the distances between NWs are smaller than the Ga diffusion length over the substrate. With a decreasing NW density, the amount of gallium i.e., gathered from the substrate and is involved in the growth process, increases. This is, in fact, equivalent to an increase in the gallium flux, which leads to a higher NW growth rate. The growth rate values of model NW (0.2–1.4 nm/s) were found to be close to the experimental values (0.8–1.2 nm/s).

In the case of self-catalyzed growth, gallium drops served as a catalyst, and the GaAs surface was covered with a film acting as a mask. The self-catalyzed NW growth was simulated at temperatures of T = 750–950K, gallium deposition rates of F_{Ga} = 1–6 ML/s, and a flux intensity ratio of $F_{As_2}/F_{Ga} = 0.25 - 4$. The range of growth conditions was chosen according to the experimental data on self-catalyzed GaAs NW growth [55,56,71]. A semi-infinite GaAs (111)B crystal covered with a film (imitating the mask) with a round through-hole, served as the initial substrate. At first, only Ga atoms were deposited onto the surface. Being weakly bound to the film material, gallium atoms gathered into a Ga cluster only on the GaAs surface at the bottom of the opening. Upon reaching a certain size, the solid cluster melted. The deposited dose of gallium depended on the through-hole diameter. Then, gallium atoms and arsenic molecules were simultaneously deposited onto this surface. Depending on the ratio between the arsenic to gallium fluxes, nanostructures with various morphologies were created. In the conditions of an excessive arsenic flux, a GaAs nanocrystal (quantum dot), instead of a nanowhisker, was formed from the Ga drop. At a certain Ga-to-arsenic flux ratio and a temperature range 750−950K, NW growth was observed.

The growth rate dependences on the arsenic and gallium fluxes were obtained. In contrast to catalytic growth, the NW growth rate linearly depended on the As_2 flux intensity over a wide range of arsenic flux (Fig. 14.12). The *dotted line* corresponds to a completely consumed Ga drop by arsenic flux (growth rate in Fig. 14.12 was determined at the same time). A dependence of this kind is characteristic for self-catalyzed growth [60]. At a fixed arsenic flux, the growth rate nonlinearly decreased with increasing gallium flux. The decrease in the growth rate is due to the increasing gallium drop size. With the Ga flux increase, starting from a certain F_{As_2}/F_{Ga} ratio corresponding to the NW growth mode, the gallium drop size increased linearly. At a certain flux ratio, the drop size remained constant for a long time. This mode corresponded to stable NW growth.

The influence of flux ratio on the morphology and self-catalyzed growth rate is illustrated in Fig. 14.13. We analyzed how NW length L varies with time at F_{As_2}/F_{Ga} (Fig. 14.13A). After a short nonlinear section, the linear dependence $L(t)$, which corresponded to a constant growth rate, was observed. Stable growth was commonly observed under these conditions, with a practically constant diameter along the NW growth axis. At a gallium excess, the Ga drop increases during deposition, NW becomes thicker and, as a consequence, the growth

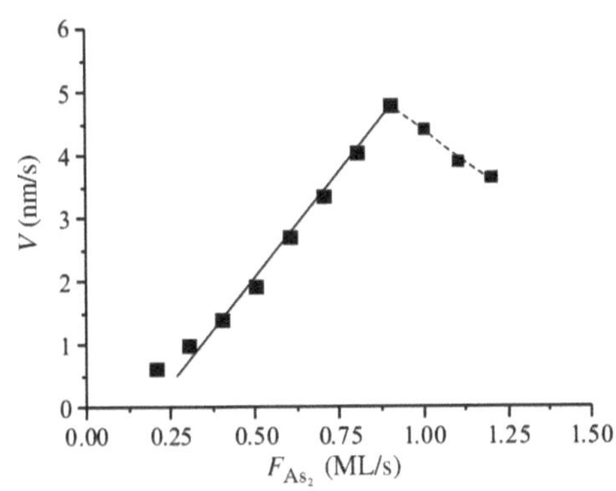

FIGURE 14.12 Dependence of the NW growth rate on arsenic flux at $T = 900K$, $F_{Ga} = 0.6$ ML/s.

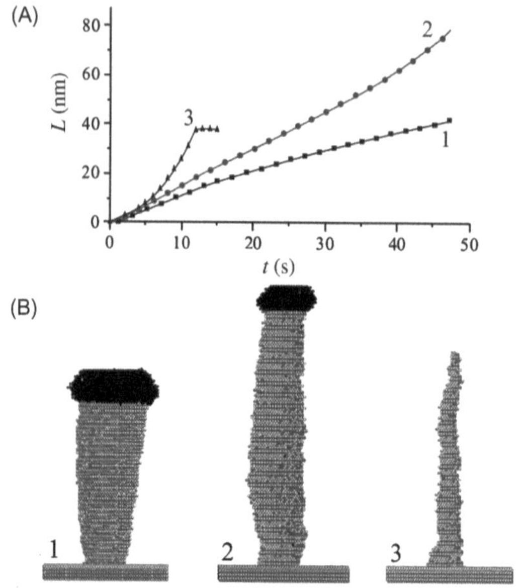

FIGURE 14.13 (A) Dependence of the NW length on time at various F_{As_2}/F_{Ga}; (B) NW view after 30 s deposition (1, 2) and at the moment drop consumption (3); *black* color is for Ga(liq), *dark gray* is Ga(s), *light gray* denotes As(s), and gray is film-mask material. $T = 900K$, $F(As_2) = 0.4$ ML/s; (1) $F_{As2}/F_{Ga} = 0.4$, (2) $F_{As_2}/F_{Ga} = 0.7$, (3) $F_{As2}/F_{Ga} = 2$.

rate decreases. With increasing F_{As_2}/F_{Ga}, the growth rate became higher. However, starting from certain F_{As_2}/F_{Ga} values, NW growth ceased due to gallium drop consumption.

Fig. 14.14 demonstrates how temperature affects NW growth. At low temperatures, the whisker grows slowly due to Ga drop size enlargement. With a temperature rise, the growth rate increases, reaching its maximum at some critical temperature T_m. This temperature value depends on the F_{As_2}/F_{Ga} value. The growth rate temperature dependence determined at the moment of gallium drop consumption demonstrates a pronounced maximum at $T = T_m$. Temperature T_m is not optimal, since the whisker does not reach its maximum possible length at the moment of drop consumption. There are optimal conditions corresponding to long NW growth with uniform thickness. With a temperature rise, the nuclear size increases at the drop–crystal interface and decreases drop supersaturation with arsenic. At a temperature higher than 960K, NW growth stopped due to gallium drop evaporation and GaAs crystal destruction.

Simulated characteristics of Au-assisted and self-catalyzed growth were compared to the experimental data. The optimal growth temperature, in the case of self-catalyzed GaAs NW growth, was higher than that for catalytic growth both in the simulation and the experiment. Self-catalyzed growth was shown to be more sensitive to the gallium-to-arsenic flux ratio than catalytic growth, since this parameter directly affects the gallium drop size. At high arsenic fluxes, the gallium drop was consumed completely, and NW growth stopped. The effects

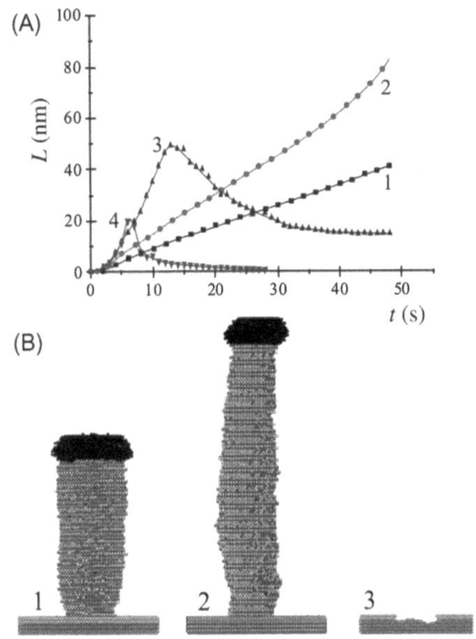

FIGURE 14.14 (A) NW length versus deposition time at $F_{As_2}/F_{Ga} = 0.7$ and various temperatures; (B) The model surface view after 30 s deposition. (1) T – 850K, (2) 900K, (3) 950K, (4) 970K.

and dependences observed in the model system are in qualitative agreement with the experimental data. The suggested MC model may be useful in an analysis of the growth of NWs based on other III–V semiconductors.

References

[1] G.H. Gilmer, P. Bennema, Simulation of crystal growth with surface diffusion, J. Appl. Phys. 43 (1972) 1347–1360.

[2] A. Sadiq, K. Binder, Diffusion of absorbed atoms in ordered and disordered monolayers at surfaces, Surf. Sci. 128 (1983) 350–382.

[3] S. Clarke, D.D. Vvedensky, Origin of reflection high-energy electron-diffraction intensity oscillations during molecular-beam epitaxy: a computational modeling approach, Phys. Rev. Lett. 58 (1987) 2235.

[4] Z. Sh Yanovitskaya, I.G. Neizvestnyi, N.L. Shwartz, M.I. Katkov, I.P. Rhyzhenkov, Desynchronization mode of 2D-island creation on the vicinal surface during MBE growth, Appl. Surf. Sci. 123/124 (1998) 729–733.

[5] M.I. Katkov, I.G. Neizvestny, I.P. Ryzhenkov, N.L. Shwartz, Z. Sh Yanovitskaya, Smoothing of the surface relief during MBE process as a cause of RHEED oscillation distortions, Phys. Low-Dimens. Struct. 5/6 (1997) 13–21.

[6] D.V. Brunev, I.G. Neizvestny, N.L. Shwartz, Z. Sh Yanovitskaya, Schwoebel barriers and quantum dot lateral size equalization during epitaxial growth, Nanotechnology N12 (2001) 413–416.

[7] I.G. Neizvestny, N.L. Shwartz, Z.S. Yanovitskaya, Influence of Schwoebel barrier and diffusion anisotropy on step density oscillation amplitude during epitaxial growth, Comput. Mater. Sci. 36 (2006) 180–183.

[8] I.G. Neizvestny, N.L. Shwartz, Z. Sh Yanovitskaya, A.V. Zverev, 3D-model of epitaxial growth on porous {111} and {100} Si surfaces, Comput. Phys. Commun. 147 (2002) 255.

[9] A.V. Chemakin, I.G. Neizvestny, N.L. Shwartz, Z.S. Yanovitskaja, A.V. Zverev, Evolution of porous Si (111) and Si (001) surfaces during epitaxy: simulation, Phys. Low-Dimens. Struct. 9/10 (2002) 7.

[10] I.G. Neizvestny, N.L. Shwartz, Z. Sh Yanovitskaya, A.V. Zverev, Monte Carlo simulation of porous layers sintering, Key Eng. Mater. 352 (2007) 5.

[11] I.G. Neizvestnyi, N.L. Shwartz, Z. Sh Yanovitskaya, A.V. Zverev, Monte Carlo simulation of steps behavior on Si(111) surface during sublimation, Thin Solid Films 380 (2000) 61–63.

[12] A.V. Zverev, I.G. Neizvestny, N.L. Shvartz, Z. Sh Yanovitskaja, The effects of monovacancies on the terrace width during sublimation from the (111) surface of a diamond-like crystal, Semiconductors 37 (2003) 649–655.

[13] I.G. Neizvestny, N.L. Shwartz, Z.S. Yanovitskaya, A.V. Zverev, Simulation of surface relief effect on ALD process, Comput. Mater. Sci. 36 (2006) 36.

[14] A.V. Zverev, K. Yu Zinchenko, N.L. Shwartz, Z. Sh Yanovitskaja, A Monte Carlo simulation of the processes of nanostructure growth: the time-scale event-scheduling algorithm, Nanotechnol. Russ. 4 (2009) 215.

[15] P. Mutti, G. Ghislotti, S. Bertoni, L. Bolondi, G.F. Cerofolini, L. Meda, et al., Room-temperature visible luminescence from silicon nanocrystals in silicon implanted SiO_2 layers, Appl. Phys. Lett. 66 (1995) 851.

[16] L. Pavesi, L.D. Negro, C. Mazzoleni, G. Franzo, F. Priolo, Optical gain in silicon nanocrystals, Nature 408 (2000) 440.

[17] V. Beyer, J. von Borany и, K.-H. Heinig, Dissociation of Si^+ ion implanted and as-grown thin SiO_2 layers during annealing in ultra-pure neutral ambient by emanation of SiO, J. Appl. Phys. 101 (2007) 053516.

[18] A.N. Karpov, D.V. Marin, V.A. Volodin, J. Jedrzejewski, G.A. Kachurin, E. Savir, et al., SiO_x layer formation during plasma sputtering of Si and SiO_2 targets, Semiconductors 42 (2008) 731–736.

[19] D. Comedi, O.H.Y. Zalloum, E.A. Irving, J. Wojcik, T. Roschuk, M.J. Flynn, et al., X-ray-diffraction study of crystalline Si nanocluster formation in annealed silicon-rich silicon oxides, J. Appl. Phys. 99 (2006) 023518.

[20] D. Tsoukalas, C. Tsamis и, P. Normand, Diffusivity measurements of silicon in silicon dioxide layers using isotopically pure material, J. Appl. Phys. 89 (2001) 7809.

[21] E.A. Mikhant'ev, I.G. Neizvestnyi, S.V. Usenkov, N.L. Shvarts, Modeling the formation of silicon nanoclusters during annealing SiO_x layers, Optoelectronics Instrum. Data Process. 47 (5) (2011) 490–497.

[22] A.N. Karpov, E.A. Mikhantiev, S.V. Usenkov, N.L. Shwartz, Izvestiya Vysshikh Uchebnykh Zavedenii, Mater. Elektron. Tekh. 1 (2012) 33–39 (in Russian).

[23] E.A. Mikhantiev, I.G. Neizvestny, S.V. Usenkov, N.L. Shwartz, Silicon monoxide role in silicon nanocluster formation during Si-rich oxide layer annealing—Monte Carlo simulation, Comput. Mater. Sci. 90 (2014) 99.

[24] E.A. Mikhant'ev, I.G. Neizvestny, S.V. Usenkov, N.L. Shwartz, Monte Carlo simulation of the effect of silicon monoxide on silicon-nanocluster formation, Semiconductors 48 (7) (2014) 891–898.

[25] A.V. Zverev, I.G. Neizvestnyi, N.L. Shwartz, Z. Sh Yanovitskaya, Lattice Monte Carlo model of SiO_x layers, Nanotechnol. Russ. 3 (2008) 367–376.

[26] V.M. Babich, N.I. Blezkan, E.F. Venger, Oxygen in Silicon Monocrystals, Interpress Ltd, Kiev, 1997, p. 240 (in Russian).

[27] G.Ya Krasnikov, N.A. Zaitsev, I.V. Matyushkin, Mathematical simulation of the kinetics of high-temperature silicon oxidation and the structure of the boundary layer in the Si-SiO_2 system, Semiconductors 37 (2003) 44.

[28] D. Bahloul-Hourlier, P. Perrot, Thermodynamics of the Au–Si–O system: application to the synthesis and growth of silicon–silicon dioxide nanowires, J. Phase Equilib. Diffus. 28 (2007) 150.

[29] R.S. Wagner, W.C. Ellis, Vapor–liquid–solid mechanism of single crystal growth, Appl. Phys. Lett. 4 (1964) 89.

[30] E.I. Givargizov, Fundamental aspects of VLS growth, J. Cryst. Growth 31 (1975) 20–30.

[31] J. Kikkawa, Y. Ohno, S. Takeda, Growth rate of silicon nanowires, Appl. Phys. Lett. 86 (2005) 123109.

[32] N. Zakharov, P. Werner, L. Sokolov, U. Gosele, Growth of Si whiskers by MBE: mechanism and peculiarities, Phys. E 37 (2007) 148.

[33] J. Goldberger, A.I. Hochbaum, R. Fan, P. Yang, Silicon vertically integrated nanowire field effect transistors, Nano Lett. 6 (2006) 973.

[34] C. Thelander, P. Agarwal, S. Brongersma, J. Eymery, L.F. Feiner, A. Forchet, et al., Nanowire-based one-dimensional electronics, Mater. Today 9 (10) (2006) 28.

[35] Y.C. Cui, Q. Wei, H. Park, C.M. Lieber, Nanowire nanosensors for highly sensitive and selective detection of biological and chemical species, Science 293 (2001) 1289.

[36] F. Patolsky, B.P. Timko, G. Zheng, Ch. M. Lieber, Nanowire-based nanoelectronic devices in the life sciences, MRS Bull. 32 (2007) 142.

[37] E.I. Givargizov, Controlled growth of filamentary crystals and fabrication of single-crystal whisker probes, Crystallogr. Rep. 51 (5) (2006) 888.

[38] J.-G. Fan, D. Dyer, G. Zhang, Y.-P. Zhao, Nanocarpet effect: pattern formation during the wetting of vertically aligned nanorod arrays, Nanoletters 4 (11) (2004) 2133.

[39] A.G. Nastovjak, I.G. Neizvestny, S.V. Usenkov, N.L. Shwartz, Special features of the crystallization process in Monte Carlo simulation of nanowhisker growth, Russ. Phys. J. 52 (2009) 1170–1176.

[40] A.G. Nastovjak, I.G. Neizvestny, N.L. Shwartz, Effect of growth conditions and catalyst material on nanowhisker morphology: Monte Carlo simulation, Solid State Phenom. 156–158 (2010) 235.

[41] A.G. Nastovjak, I.G. Neizvestny, N.L. Shwartz, Z. Sh Yanovitskaya, Monte Carlo simulation of growth of nanowhiskers, Semiconductors 44 (1) (2010) 127.

[42] A.G. Nastovjak, I.G. Neizvestny, N.L. Shwartz, E.S. Sheremet, Mechanisms of nanowhisker formation: Monte Carlo simulation, Optoelectron. Instrum. Data Process. 45 (2009) 342–347.

[43] A.G. Nastovjak, I.G. Neizvestny, N.L. Shwartz, Possibilities of Monte Carlo simulation for examination of nanowhisker growth, Pure Appl. Chem. 82 (2010) 2017.

[44] A.G. Nastovjak, I.G. Neizvestny, N.L. Shwartz, Simulation of growth of silicon nanowhiskers with Ge-Si heterojunctions, J. Surf. Investig. X-ray Synchrotron Neutron Tech. 5 (5) (2011) 870.

[45] A.G. Nastovjak, I.G. Neizvestny, N.L. Shwartz, Peculiarities of axial and radial Ge–Si heterojunction formation in nanowires: Monte Carlo simulation, Pure Appl. Chem. 84 (2012) 2619.

[46] N.D. Zakharov, P. Werner, G. Gerth, L. Schubert, L. Sokolov, U. Gosele, Growth phenomena of Si and Si/Ge nanowires on Si (111) by molecular beam epitaxy, J. Cryst. Growth 290 (2006) 6.

[47] T.E. Clark, P. Nimmatoori, K.-K. Lew, L. Pan, J.M. Redwing, E.C. Dickey, Diameter dependent growth rate and interfacial abruptness in vapor–liquid–solid $Si/Si_{1-x}Ge_x$ heterostructure nanowires, Nano Lett. 8 (2008) 1246–1252.

[48] C. Colombo, M. Heis, M. Gratzel, A.F. Morral, Gallium arsenide p-i-n radial structures for photovoltaic applications, Appl. Phys. Lett. 94 (2009) 173108.

[49] M. Heiss, Y. Fontana, A. Gustafsson, G. Wüst, C. Magen, D.D. O'Regan, et al., Self-assembled quantum dots in a nanowire system for quantum photonics, Nat. Mater. 12 (2013) 439.

[50] P. Krogstrup, H.I. Jorgensen, M. Heiss, O. Demichel, J.V. Holm, M. Aagesen, et al., Single-nanowire solar cells beyond the Shockley-Queisser limit, Nat. Photonics 7 (2013) 306.

[51] B. Wacaser, K. Dick, J. Johansson, M. Borgstrom, K. Deppert, L. Samuelson, Preferential interface nucleation: an expansion of the VLS growth mechanism for nanowires, Adv. Mater. 21 (2009) 153.

[52] G.E. Cirlin, A.A. Tonkikh, Yu.B. Samsonenko, I.P. Soshnikov, N.K. Polyakov, V.G. Dubrovskii, et al., A3B5 nanowhiskers: MBE growth and properties, Czech. J. Phys. 56 (2006) 13.

[53] J.C. Harmand, M. Tchernycheva, G. Patriarche, L. Travers, F. Glas, G. Cirlin, GaAs nanowires formed by Au-assisted molecular beam epitaxy: effect of growth temperature, J. Cryst. Growth 301–302 (2007) 853.

[54] M.C. Plante, R.R. LaPierre, Growth mechanisms of GaAs nanowires by gas source molecular beam epitaxy, J. Cryst. Growth 286 (2006) 394.

[55] C. Colombo, D. Spirkoska, M. Frimmer, G. Abstreiter, A. Fontcuberta i Morral, Ga-assisted catalyst-free growth mechanism of GaAs nanowires by molecular beam epitaxy, Phys. Rev. B 77 (2008) 155326.

[56] A.F. Morral, C. Colombo, G. Abstreiter, J. Arbiol, J.R. Morante, Nucleation mechanism of gallium-assisted molecular beam epitaxy growth of gallium arsenide nanowires, Appl. Phys. Lett. 92 (2008) 063112.

[57] A.F. Morral, D. Spirkoska, J. Arbiol, M. Heigoldt, J.R. Morante, G. Abstreiter, Prismatic quantum heterostructures synthesized on molecular beam epitaxy GaAs nanowires, Small 4 (2008) 899.

[58] D. Spirkoska, J. Arbiol, A. Gustafsson, S. Conesa-Boj, F. Glas, I. Zardo, et al., Structural and optical properties of high quality zinc-blende/wurtzite GaAs nanowire heterostructures, Phys. Rev. B 80 (2009) 245325.

[59] S. Ambrosini, M. Fanetti, V. Grillo, A. Franciosi, S. Rubini, Self-catalyzed GaAs nanowire growth on Si-treated GaAs (100) substrates, J. Appl. Phys. 109 (2011) 094306.

[60] M.R. Ramdani, J.C. Harmand, F. Glas, G. Patriarche, L. Travers, Arsenic pathways in self-catalyzed growth of GaAs nanowires, Cryst. Growth Design 13 (2013) 91.

[61] D. Rudolph, S. Hertenberger, S. Bolte, W. Paosangthong, D. Spirkoska, M. Döblinger, et al., Direct observation of a noncatalytic growth regime for GaAs nanowires, Nano Lett. 11 (2011) 3848.

[62] G.E. Cirlin, V.G. Dubrovskii, Y.B. Samsonenko, A.D. Bouravleuv, K. Durose, Y.Y. Proskuryakov, et al., Self-catalyzed, pure zincblende GaAs nanowires grown on Si (111) by molecular beam epitaxy, Phys. Rev. B 82 (2010) 035302.

[63] F. Glas, M.R. Ramdani, G. Patriarche, J.-C. Harmand, Predictive modeling of self-catalyzed III–V nanowire growth, Phys. Rev. B 88 (2013) 195304.

[64] N.V. Sibirev, M. Tchernycheva, M.A. Timofeeva, J.-C. Harmand, G.E. Cirlin, V.G. Dubrovskii, Influence of shadow effect on the growth and shape of InAs nanowires, J. Appl. Phys 111 (2012) 104317.

[65] V.G. Dubrovskii, M.A. Timofeeva, M. Tchernycheva, A.D. Bolshakov, Lateral growth and shape of semiconductor nanowires, Semiconductors 47 (2013) 50.

[66] P. Krogstrup, H.I. Jorgensen, E. Johnson, M.H. Madsen, C.B. Sorensen, A.F. Morral, et al., Advances in the theory of III–V nanowire growth dynamics, J. Phys. D Appl. Phys. 46 (2013) 313001.

[67] M.V. Knyzeva, A.G. Nastovjak, I.G. Neizvestny, N.L. Shwartz, Simulated growth of GaAs nanowires: catalytic and self-catalyzed growth, Semiconductors 49 (2015) 60–68.

[68] Z.Q. Wang, D. Stroud, Monte Carlo study of liquid GaAs: bulk and surface properties, Phys. Rev. B 42 (1990) 5353.

[69] D.A. Murdick, X.W. Zhou, H.N.G. Wadley, D. Nguyen-Manh, R. Drautz, D.G. Pettifor, Analytic bond-order potential for the gallium arsenide system, Phys. Rev. B 73 (2006) 045206.

[70] V.G. Dubrovskii, G.E. Cirlin, I.P. Soshnikov, A.A. Tonkikh, N.V. Sibirev, Yu. B. Samsonenko, et al., Diffusion-induced growth of GaAs nanowhiskers during molecular beam epitaxy: theory and experiment, Phys. Rev. B 71 (2005) 205325.

[71] L.T.T. Giang, C. Bougerol, H. Mariette, R. Songmuang, Intrinsic limits governing MBE growth of Ga-assisted GaAs nanowires on Si(111), J. Cryst. Growth 364 (2013) 118.

PART

III

Radiation Effects on Semiconductor Structures

15

The Energy Pulse-Oriented Crystallization Phenomenon in Solids (Laser Annealing)

A.V. Dvurechenskii[1,2]

[1]RZHANOV INSTITUTE OF SEMICONDUCTOR PHYSICS OF THE SIBERIAN BRANCH OF THE RUSSIAN ACADEMY OF SCIENCES, NOVOSIBIRSK, RUSSIA [2]NOVOSIBIRSK STATE UNIVERSITY, NOVOSIBIRSK, RUSSIA

15.1 Introduction

The history of the discovery of energy pulse-oriented crystallization of solids is based on studies of ion implantation in semiconductors, i.e., injection of dopant atoms into a solid by their ionization and subsequent acceleration to the energies required to deliver atoms to the required depth of the material used. This approach was proposed by M.M. Bredov, A.R. Regel, and R.F. Komarova (Konoplyova), based on experiments at the Ioffe Laboratory. The results of those experiments were published in 1954 in the *Proceedings of the USSR Academy of Sciences* on the recommendation of A.F. Ioffe. There were many potential advantages of the ion-implantation technique compared with the material doping methods available at that time: the low temperature of the ion implantation process; implantation of any of the elements on the periodic table (including insoluble and poorly soluble elements); high accuracy of the doping process in concentration levels, as well as in projected range, by changing the ion current and the energy of the beam particles; uniformity of doping by means of beam/substrate scanning; precise tuning of the depth distribution of elements (multistage implantation); simplicity of masking by introduction of masks and, at the same time, the possibility of implantation through multilayer metal and dielectric coatings.

However, implementation of the proposed technique faced multiple difficulties and problems so that, sometimes, it seemed that such an intricate situation would force the approach to be abandoned. The main issue was the vast number of defects generated in the material during forced implantation of even a single element by means of acceleration devices. The number of displacement atoms during implantation is proportional to the energy transferred to the target atoms in a cascade of elastic interactions with nuclei, and is inversely proportional to the threshold energy of atom displacement from the lattice node. The defects introduced strongly change the properties of the materials, especially semiconductors, since they are most sensitive to external influences even of weak particle beams.

Advances in Semiconductor Nanostructures. DOI: http://dx.doi.org/10.1016/B978-0-12-810512-2.00015-9

The defects were, in fact, disguising the manifestation of material doping changes of properties due to the introduction of an element on the periodic table.

All attempts to resolve this problem using the usual method of high temperature treatment failed. Heating resulted in defect transformation: point defects in irradiated materials transformed into extended ones (twins, dislocations), most of them persisted up to annealing at the highest temperatures. Heating to those high temperatures initiated irreversible changes in the properties of the original material. Thus, the idea of ion beam doping itself was received with skepticism. It was suggested that ion beam doping would create a pure source of necessary elements only in the near-surface area of a solid, which then would diffuse the implanted element from the damaged layer into the matrix material. This assumption significantly reduced the potential for ion beam doping, especially with regard to obtaining thin (nanometer) doping layers and high gradients in depth distribution of the implanted elements. These parameters are often required for the development of many electronic devices and circuits.

The research into defect formation and modification, and crystal-to-amorphous transition under ion irradiation [1–3], led to the first successes in solving the problem of material doping and obtaining the expected ion-implantation technique advantages. The recrystallization temperature of layers amorphized by ion implantation turned out to be noticeably lower than the annealing temperature of many point and extended defects of crystal structures. The crystalline quality of the recrystallized amorphous layer appeared to be sufficiently high. The necessary doping level does not always fit the crystal - to- amorphous transition after ion implantation. In such cases, the material was exposed to additional irradiation by ions of the target material, or by ions of inert gases, in order to transform the doping layer into an amorphous state. Subsequent recrystallization of the amorphous layer provided activation of doping elements and recovery of a rather perfect crystalline material. Nevertheless, the structure remained imperfect because of tails in defect depth distribution at ion implantation.

The discovery of the *Energy Pulse-Oriented Crystallization Phenomenon in Solids* (*Laser Annealing*) became a breakthrough event in removing defects. This was also the title of a series of studies of the energy pulse–solids interaction. The scientists involved in these studies—I.B. Khaibullin, E.I. Shtyrkov, M.F. Galyautdinov, R.M. Bayazitov, L.S. Smirnov, L.N. Aleksandrov, G.A. Kachurin, A.V. Dvurechenskii, Yu.V. Kovalchuk, Yu.V. Pogorelskii, and Yu. V. Kopaev, were awarded the USSR State Prize in 1988. By experiments on new approaches to removing defects, it was discovered that the nanosecond pulsed laser action at $\sim 10^7$ W/cm^2 radiation power density on ion-implanted semiconductor wafers with an amorphous layer results in the removal of (recrystallization) damage and amorphization in ion-implanted silicon [4–10]. This phenomenon was named "energy pulsed-oriented crystallization." The term "laser annealing," which is widely used in publications, refers broadly to changes in structure and composition, as well as to the electric and optical characteristics of solids under pulsed laser action with different pulse durations. Very short pulsed treatment of ion-irradiated semiconductor layers with electromagnetic radiation resulted in recovery of the crystalline structure and undisturbed original depth distribution of implanted ions into a solid.

The rate of transformation of an amorphous layer into a monocrystalline state was an intriguing detail that interested researchers working in various fields of laser annealing. This transformation is only possible due to epitaxy—the oriented growth of the surface crystalline layer during which the crystallographic orientation of the growing layer fits the substrate crystallographic orientation. The rate of transformation of an amorphous layer into a monocrystalline area can be estimated using the following relationship: $v \sim d/\tau_p$, where d is the amorphous layer thickness, and τ_p is the energy pulse duration. At $d \approx 0.1\ \mu m$ and $\tau_p \approx 10^{-7}$ s, the speed of crystallization front movement is $V \approx 1$ m/s. This value significantly exceeds typical values of crystal growth speed. Expansion of the variation in physical process parameters always suggests there will be new system properties, existing theories will be verified, and there will be new ideas to evaluate. With respect to laser annealing, there was huge interest in understanding annealing mechanisms in different time regimes, from very short pulse (ps) lasers, through to continuous wave lasers, and rapid thermal annealing. After publication of the first experiments on laser annealing, studies of the processes were also carried out as part of joint activities between the USSR Academy of Sciences and the Academy of Sciences of the German Democratic Republic. The scientists L.S. Smirnov, A.V. Dvurechenskii, I.B. Khaibullin, Ya.V. Fattakhov, E. Wieser, R. Groetzschel, D. Panknin, and M. Voelskow were awarded the International Prize of the USSR and GDR Academies of Sciences in 1988 for their joint work, entitled *Development of the Physical Fundamentals of Ion Beam and Energy Pulse Modification of Microelectronics Materials.*

15.2 Heating and Cooling During Laser Annealing

When optical radiation passes through condensed matter, light energy transforms into another kind of energy: excitation of electrons transit to higher energy levels, and there is excitation of lattice vibrations (absorption of infrared radiation by optical phonons). The electron–hole pair generation (when photon energy exceeds the width of the band gap) is the dominant light absorption mechanism in semiconductors and insulators. In metals, conduction electrons are excited. The subsequent relaxation of excited electrons is accompanied by the transfer of energy to other electrons (electron–electron interaction), as well as to lattice vibration modes, which, in turn, transfer their energy to portions corresponding to the energy of acoustic phonons in a solid. The typical time required for the excited electrons to achieve equilibrium with a lattice (electron–phonon and phonon–phonon interaction time) for widely used materials lies between 10^{-13} and 10^{-12} s at temperatures and excitation densities generally used for laser annealing. Therefore, when the pulsed duration of laser radiation exceeds these values, light absorption will lead to a transfer of energy to the lattice, and the material will heat during the impulse action. This phenomenon is described here.

Let us consider an idealized rectangular laser pulse. Let τ_p denote the pulse duration, and J_0 the radiation intensity. The light beam falls perpendicular to the solid surface with a surface reflection coefficient of R, and an absorption coefficient of α. The fundamental law of light absorption connects the intensity J of a light beam passed through a layer of absorbing medium with thickness z, and the intensity of a falling radiation beam J_0: $J = J_0\ (1-R)\ \exp(-\alpha z)$.

Thermal material properties are described by specific heat capacity C_v, density ρ, and thermal conductivity κ. The thermal diffusivity $D = \kappa/C_v\rho$ characterizes the length $(2D\tau_p)^{1/2}$ of heat diffusion during the pulse action. In order to evaluate the heating and cooling rates by order of magnitude, let us assume that these parameters do not depend on temperature, and ignore the latent heat during phase transitions. There are two limiting cases for solving the one-dimensional problem of heat propagation into a semi-infinite body, if the light beam is considerably greater than the thickness of the heated layer [11].

Case A: Light absorption length α^{-1} is small in comparison to heat diffusion length: $\alpha(2D\tau_p)^{1/2} \gg 1$. In this case, the absorbed energy of pulse radiation $(1 - R)J_0\tau_p$ heats a layer with a thickness of $(2D\tau_p)^{1/2}$. The average increase in temperature in a layer is:

$$\Delta T \approx (1 - R)J_0\tau_p/C_v\rho(2D\tau_p)^{1/2} \tag{15.1}$$

After the pulse action is completed, heat diffuses deeper into the substrate. The order of magnitude of the cooling time is equal to τ_p. Thus, the heating and cooling rates can be estimated using Eq. (15.1) as $\Delta T/\tau_p$.

Case B: Light absorption length α^{-1} is large in comparison to heat diffusion length $\alpha(2D\tau_p)^{1/2} \ll 1$. This results in an exponentially decreasing temperature profile with a typical length of α^{-1}. The heating rate is $\Delta T(z)/\tau_p$. The cooling rate is more important, because it determines the structure and composition of a surface layer after pulsed laser action. In order to cool that layer, the length of diffusion of heat into the substrate must exceed α^{-1}. The cooling time τ_c can be roughly estimated using the following correlation: $\alpha^{-1} = (2D\tau_c)^{1/2}$. The order of magnitude of cooling rate is:

$$\Delta T(z) = \{(1 - R)\alpha J_0\tau_p \exp(-\alpha z)\}/C_v\rho \tag{15.2}$$

The results of the calculation of the energy required for heating silicon up to 1000°C by pulsed laser radiation of varying pulse duration at a wavelength of $\lambda = 0.5\ \mu m$ are presented in Table 15.1 as an example. The reflection coefficient and the value inverse to the absorption coefficient are for Si $R = 0.35$, and $\alpha^{-1} = 10^{-4}$ cm. These calculations lead to the following conclusion. The length of heat diffusion in metal conductors exceeds light penetration

Table 15.1 The Intensity of a Radiation Beam J_0 and Energy E Required for Heating Silicon up to 1000°C by a Pulsed Laser Radiation of Varying Pulse Duration at Wavelength $\lambda = 0.5\ \mu m$ [11]

Material $\Delta T = 10^3$°C	τ_p (s)	$(2D\tau_p)^{1/2}$ (cm)	J_0 (W/cm^2)	E (J/cm^2)
Si	3×10^{-6}	2×10^{-3}	1.1×10^6	3.3
	3×10^{-8}	2×10^{-4}	1.1×10^7	0.33
	3×10^{-11}	0.6×10^{-5}	5×10^9	0.15

thickness at pulse durations of an order of tens of picoseconds (Case A). In silicon, Case A is reached at $\tau_p \geq 30$ ns. Heating and cooling rates in these examples exceed 10^8 °C/s.

15.3 Solid-Phase Crystallization

The process of solid-phase crystallization will be considered using the example of tetrahedrally coordinated semiconductors (Si, Ge, GaAs, and others) after ion beam bombardment resulting in crystal-to-amorphous state transition in the near-surface layer (e.g., an a-Si-c-Si structure is amorphous Si on a crystal substrate). The amorphous materials have no long-range order in the network of atom location, and lack of long-range order is random by nature. A short-range order in the network of atom location can exist. Tetrahedrally coordinated amorphous semiconductors have the same short-range order as the crystal. This is what distinguishes them from glass-like systems where short-range order is essentially a frozen liquid.

Sufficiently slow heating of semiconductor structures with an amorphous layer leads to solid-phase epitaxial crystallization, which is the ordering of the atom's location at the crystal–amorphous interface, and movement of this boundary to the surface. The solid-phase crystallization front movement rate is described by the activation law [12]:

$$V = V_0 \exp(-E_0/kT) \tag{15.3}$$

For example, for Si $V_0 = 1.55 \times 10^6$ cm/s, and $E_0 = 2.3$ eV, Eq. (15.3) allows estimation of the maximum rate of solid-phase epitaxial crystallization at temperatures close to the melting temperature. In silicon, this velocity equals ≈ 0.1 cm/s. With a typical thickness of an amorphous layer of ≈ 0.1 μm, the time required for solid-phase crystallization will be $\approx 10^{-4}$ s. Hence, the solid-phase epitaxial crystallization of an amorphous layer can take place at a pulsed laser duration of $\geq 10^{-4}$ s in an a-Si-c-Si structure. At an opposite inequality, the time required for solid-phase crystallization is insufficient. Therefore, laser annealing of an a-Si-c-Si structure with shorter pulses, and turning the amorphous layer to a crystalline state is possible only through the formation of a liquid phase, crystallization of which can occur at higher rates. The higher values of the diffusion coefficients of atoms in a liquid phase allow promotion of the process of their ordering on the amorphous–crystalline interface. For silicon, the limiting value of the liquid-phase crystallization front movement rate reaches 15 m/s. Exceeding this value leads to solidification of the melt with amorphous state formation, because the atoms in the melt do not have enough time to form an ordered structure.

15.4 Melting and Liquid-Phase Crystallization

High heating rates lead to high spatial gradients of temperature, and provide a possibility to melt the area of $(2D\tau_p)^{1/2}$, or an area of radiation absorption in accordance with cases A and B. The process of amorphous layer melting and subsequent crystallization provides a way to trace the structural transformations occurring.

The melting of amorphous tetrahedrally coordinated semiconductors should result in changes of local symmetry around every atom, as well as coordination (the coordination number equals four in an amorphous state and six in a liquid state). This provides the basis for an assumption on the first order phase transition in a system "amorphous state with a covalent bonding to liquid state with metal bonding." Each phase is characterized by its own dependence for Gibbs energy on temperature. The assumption on the first order phase transition means that there is a transition temperature at which the Gibbs energy of a covalent amorphous semiconductor is equal to the Gibbs energy of liquid metal. The temperature of such a transition is the melting temperature T_a of an amorphous semiconductor (Fig. 15.1). For elementary semiconductors, the parameters known from experiments are sufficient to calculate the temperature of such a transition. The calculations of temperature dependence of Gibbs energy for Ge and Si in amorphous, crystalline, and liquid states yielded the following result: $T_a/T_m = 0.8$, where T_m is the crystal melting temperature [13,14].

Experiments confirmed that T_a is less than T_m. In Refs. [15,16] the following conclusion was stated: the melting temperature for an a-Si obtained by ion irradiation equals 1100 ± 100 K. The T_a quantity characterizes the metastable state (amorphous state). Therefore, it cannot be expected to keep T_a constant under changing techniques and conditions of amorphous semiconductor preparation. For an a-Si, the received T_a values lie between 1100K and 1450K ($T_m = 1700$K). Such a scatter apparently reflects the change in T_a, depending on the conditions of amorphous silicon preparation, as well as on the parameters of pulse laser action. In cases of slow heating (long pulses), structural rearrangement during pulse action can cause changes in T_a in noncrystalline material.

If the energy of pulse action is sufficient only to melt the amorphous layer, then the liquid formed will be in a supercooled state. There are two paths of further melt crystallization: (1) homogeneous nucleation; and (2) growth of formed nuclei and epitaxy on the melt–crystalline substrate interface. Latent heat of L_m is released during crystallization.

The formation of crystallites in a homogeneous melt has a latent period (waiting time). This time strongly depends on supercooling ($T_m - T$) which, under nonstationary conditions

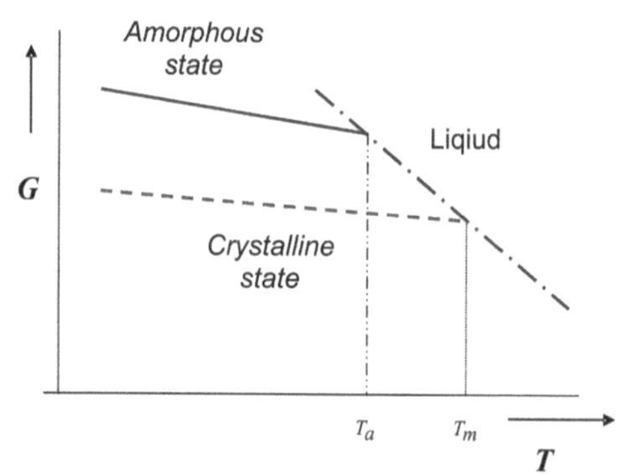

FIGURE 15.1 Schematic temperature dependence of isobaric Gibbs free energy relationships of four coordinated amorphous, crystalline covalent solids and six coordinated metallic liquid phases [13,14].

of pulse action, depends on time. The selected T time for the appearance of a nucleus of a new phase is described by the formula [17]:

$$\tau(T) = A\{T^{1/2}/|T_m - T|\}\exp[(0.5W_{cr} + U)/kT] \tag{15.4}$$

where U is the activation energy of transition of an atom from melt into crystal, and W_{cr} is the energy formation of a critical nucleus. The A constant is determined experimentally from the condition of equality of $\tau(T)$ and the minimum waiting time for nucleation to start. At lower values of melt cooling time, solidification will occur with formation of an amorphous phase.

The melt–crystal substrate interface movement speed depending on temperature is described by the law of normal growth:

$$V = C\exp(-U/kT)\{1 - \exp[-L_m(T_m - T)/kTT_m]\} \tag{15.5}$$

The C factor is determined by comparison of experiments. The intensity of latent heat release is proportional to the front movement speed.

Fig. 15.2 shows the melt front movement during laser pulse action with different energies E ($\tau_p = 300$ nanoseconds) on an a-Si-c-Si structure with subsequent cooling. At an energy density sufficient only for melting an amorphous Si layer (but insufficient for melting crystalline Si), the forming melt is supercooled ($T_m - T$ reaches 500–600K). Melting begins at the surface, and the "solid–melt" front moves to the bulk of the a-Si-c-Si structure. Some time after melting starts, formation of crystallites in the melt near the surface occurs, because the melt existence time is longer just near the surface. Therefore, condition (15.4) of crystal phase nucleation is met earlier than in other areas of the melt.

Heat released by crystallization increases the melting front movement speed, and also partially reduces supercooling. Nucleation rate, therefore, decreases, and the formed nuclei begin to grow.

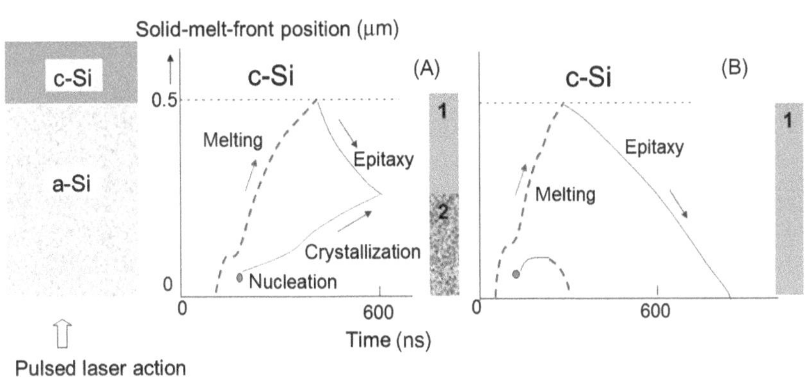

FIGURE 15.2 Transient melting and crystallization at pulsed laser annealing of an amorphous Si layer on a crystalline Si substrate with energy density E_1 (A) and E_2, (B) ($E_1 < E_2$).

At the same time, the melting process continues in the matrix up to the amorphous layer–crystal interface. Upon reaching this border, epitaxial crystallization of the supercooled melt begins. Thus, after some time, the crystallization fronts at the melt move towards each other. The melt becomes warmer with the release of latent heat, and the crystallization rate decreases as the fronts converge. The resulting structure, after solidification, includes a fine-grained (as a result of homogeneous nucleation) and large-block polycrystal, as well as a partially recrystallized layer.

Obtaining a completely monocrystal layer after pulsed laser annealing is of interest for practical applications. To do this, it is necessary to increase the energy density of the laser pulse. It is worth noting that with an increase of E, there are conditions under which an intermediate crystallization phenomenon is observed. Some time after the melting starts, formation of crystallites near to the surface in accordance with Eq. (15.4) takes place. However, the subsequent absorption of radiation by crystals results in their melting (Fig. 15.2B). Epitaxial crystallization, as in the previous case, starts at the melt–crystal substrate interface and ends at the structure surface. The crystal–melt front movement rate is determined by the melt supercooling (Eq. 15.5). In turn, the supercooling quantity after the completion of the laser pulse is controlled by heat diffusion to the substrate. The epitaxial growth rate value can be obtained by differentiating a function describing the dependence of the crystal–melt front position on time.

15.5 Self-Sustained (Explosive) Crystallization

There is one more interesting phenomenon that occurs during recrystallization of layers of amorphous semiconductors. This phenomenon is crystallization of an amorphous film outside the energy pulse action area. Typically, a recrystallized film is represented by a periodic sequence of fine- and coarse-particle polycrystal areas. The physical mechanism leading to the formation of such a structure is based on side propagation of the melting front from the energy pulse action area, due to the heat released by crystallization.

This happens because high rates of homogeneous nucleation lead to intense latent heat release. As a result, the melt is heated (supercooling reduces), and the conditions necessary for crystallite growth are created. During this growth, the intensity of latent heat release decreases, and the natural energy dissipation into the substrate once again leads to the formation of a strongly supercooled melt and, as a result, to an increase in homogeneous nucleation rate. This way, the process can repeat itself several times and result in periodic structures, an example of which is shown in Fig. 15.3.

In fact, the main approach to describe the processes of the phenomenon of energy pulse-oriented crystallization of solids is based on a thermal model including the processes of melting, nucleation, and growth of crystallites from a supercooled melt. Structural transformations during melting occur under the generation dense electron–hole plasma formed during absorption of powerful light radiation. The electron–hole pair's lifetime at a high excitation level is determined by the Auger recombination process. Taking into account the

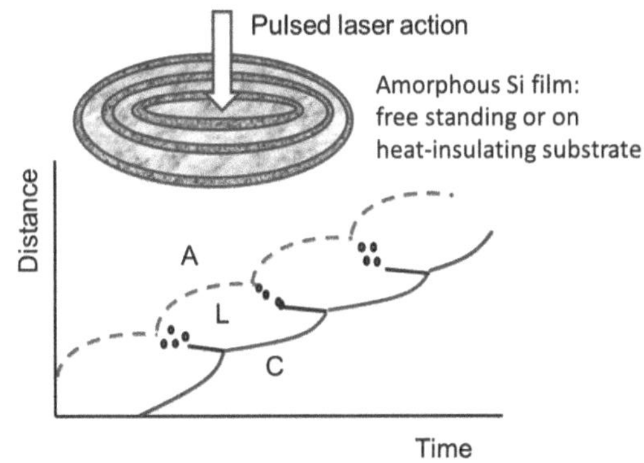

FIGURE 15.3 Amorphous–liquid (A-L) and crystallization (L-C) front position at a distance from the central part of a free-standing a-Si film (or a-Si film on a heat-insulating substrate in insert) versus the time after pulsed laser action in the inner circle. Points show nucleation of the crystals in supercooled molten Si.

electron–hole plasma contribution to phase transitions occurring at pulse heating of semiconductor layers was proposed in Ref. [18]. It was assumed that plasma influences only the melting process via reduction of temperature and enthalpy of transition of a solid into a liquid state associated with the plasma contribution to the free energy of a system. The plasma mechanism should apparently manifest itself in areas of picosecond and femtosecond pulse durations.

15.6 Dopant Element Solubility and Spatial Distribution

With solid-phase laser annealing it was possible to keep doping elements at the depth of spatial distribution in the same depth position as before the pulse action. This appeared to be one of the important advantages of laser annealing of layers after ion implantation. Diffusion lengths under such regimes can be easily estimated. For example, for As in Si, the maximum diffusion coefficient in a solid phase (close to the melting point) is $\sim 10^{-11}$ cm^2/s. At a pulse duration of $\sim 10^{-3}$ s, sufficient for solid-phase crystallization of the amorphous layer, the diffusion length is $\sim 10^{-7}$ cm. Obviously, such a redistribution can be neglected in many cases.

As for solubility, during solid-phase crystallization of an amorphous layer, an increase in dopant concentration in lattice nodes by several times in comparison with the equilibrium limit of solubility was observed. The increase in solubility was due to high concentrations of nonequilibrium defects during amorphous layer recrystallization, which provided embedding of doping elements in lattice nodes.

Even stronger effects of exceeding the equilibrium solubility limit were observed during liquid-phase crystallization, when laser radiation provided high crystallization rates. The high speed of the crystal–melt front movement leads to doping elements being captured in lattice nodes of a growing crystal; this explains the solubility increase.

Table 15.2 Equilibrium Solubility C_{so} and Distribution Coefficient k_0 in Si as Compared With Same Nonequilibrium Values C_{mo} and k_m at Laser Pulsed Action [10]

Element	C_{so} (cm^{-3})	C_{mo} (cm^{-3})	k_0	k_m
As	1.5×10^{21}	6×10^{21}	0.3	1
Sb	7×10^{19}	1.3×10^{21}	0.023	0.7
Bi	8×10^{17}	4×10^{20}	0.0007	0.4
Ga	4.5×10^{19}	5×10^{20}	0.008	0.2
In	8×10^{17}	1.5×10^{20}	0.0004	0.15

The magnitude of limiting solubility at pulse laser action is of specific interest. Let us consider covalent Si and Ge semiconductors characterized by the conservative nature of their solubility: as the temperature decreases from T_m to a certain T_n, the solubility increases, and then drops. This solubility curve is a result of overlapping temperature and concentration dependences of the distribution coefficient. The distribution coefficient (or capture coefficient) is the ratio of concentration of doping elements in the growing crystal and melt.

During energy pulse-oriented crystallization of near-surface Si layers with rates up to several m/s, the concentration of As, Sb, Bi, Ga, and In doping elements in crystal exceed the equilibrium value by 3 to 500 times (see Table 15.2). The largest increase in solubility occurs for elements with a small equilibrium value of distribution coefficient. The nonequilibrium distribution coefficient tends to one as the crystallization rate increases.

15.7 Melting of Nanocrystals Embedded in a Crystal Matrix

Heterosystems which are crystal matrices with coherently embedded nanocrystals (quantum dot heterostructures) are usually formed on the basis of epitaxial technologies (molecular beam or vapor-phase epitaxy). In such structures, nanocrystal growth occurs in the Stranski–Krastanov regime, which provides for ordering of nanocrystals by size (a self-organization process). Nevertheless, dispersion in the size of nanocrystals exists. In Ge–Si heterostructures with Ge quantum dots, dispersion is about 17%. The size dispersion inevitably leads to the scatter of energy levels of charge carriers in an array of quantum dots. As a result, the discrete electron spectrum necessary for device applications of nanostructures with quantum dots can transform into a quasi-continuous one, losing the main advantages provided by the phenomenon of three-dimensional quantization in zero-dimensional electronic systems.

An approach to applying the method of pulse laser modification of Ge nanocrystals (quantum dots) in Si for reduction of their dispersion in size was suggested in Ref. [19]. The structures were grown using the molecular beam epitaxy technique, and included one layer of Ge nanocrystals embedded in crystalline Si.

The energy spectrum of holes in Ge–Si heterostructures with self-organized Ge quantum dots before and after laser pulse action was analyzed on the basis of measurements of temperature and frequency dependences of high frequency conduction, as well as volt-farad characteristics of Schottky diodes. To control the element composition of Ge nanoclusters and mechanical stresses in structures, combinational light scattering spectroscopy was used. The energy density of monopulse emission was 1 J/cm^2, which corresponded to the melting threshold of the Si crystal surface with 80 ns pulse duration of a ruby laser ($\lambda = 694$ nm). The melting threshold was determined with in situ methods based on the dynamics of Si reflective ability and peak surface temperature. Treatments with 1 and 10 ruby laser pulses were used. The research resulted in the following conclusion: pulse laser action leads to deepening of the energy levels of holes in Ge nanoclusters relative to the threshold of the Si valence zone, and to reduction in their energy spread (Fig. 15.4A). Dispersion of energy levels associated with quantum dot size fluctuation in an ensemble decreases by two times after treating with 10 pulses. In addition, the width of the peak of Raman scattering of light at Ge–Ge bond oscillations decreases by 1.5 (Fig. 15.4B). The available results indicate an increase in homogeneity in size in an array of quantum dots after nanosecond pulsed laser irradiation.

A model of melting and crystallization of Ge nanocrystals in a crystal silicon matrix under nonstationary conditions of heating at pulse laser action was developed in Ref. [20]. The model describes the kinetics of melting and crystallization of nanocrystals with consideration for their sizes, and shapes, as well as for elastic deformations caused by the difference in lattice constants of Ge nanocrystals and the Si matrix. It is shown that the melting temperature of nanocrystals increases as their size decreases. Taking into account the contribution of elastic deformations, the temperature of the Ge melting point can be above the melting point

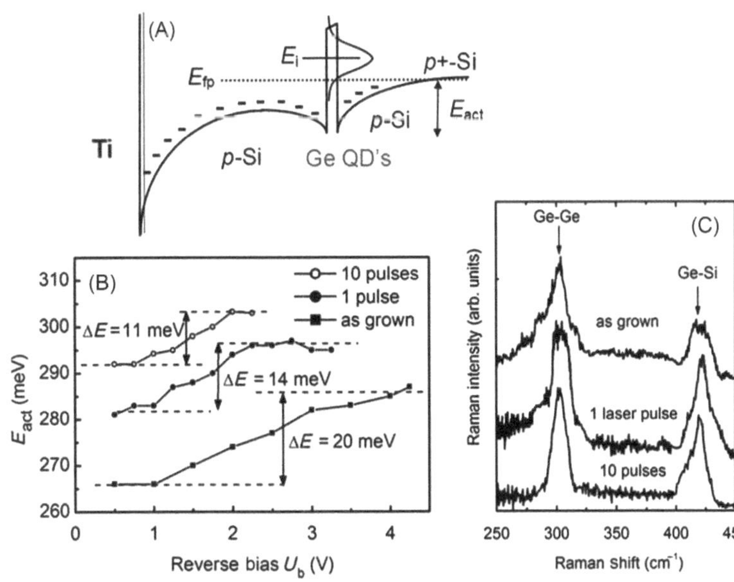

FIGURE 15.4 (A) The valence band profile of a Si Schottky diode with a layer of Ge quantum dots (QDs); (B) Bias dependence activation energies of hole emission rate before and after pulsed laser annealing; (C) Raman spectra of a Si–Ge quantum dot heterostructure before and after pulsed laser annealing [19].

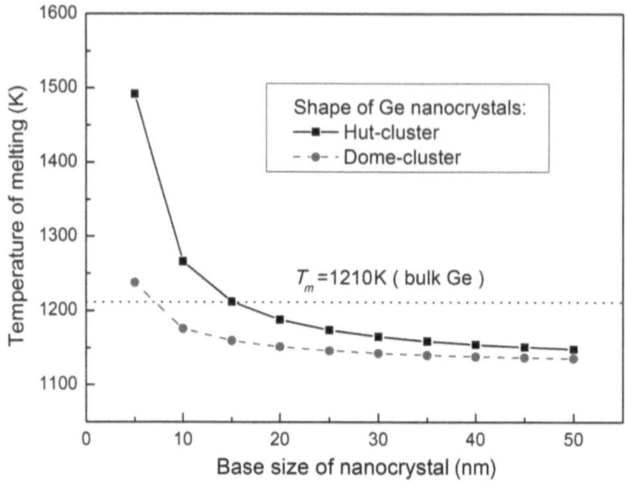

FIGURE 15.5 Calculated melting temperature of Ge nanocrystals embedded in Si versus size, shape, and strain due to the lattice mismatch of Ge and Si [20].

(size less than 15 nm), as well as beyond the melting point (size more than 15 nm), of bulk Ge (Fig. 15.5). The following result was obtained: at pulsed laser action close to experimental values (pulse duration of 80 ns, energy density of 0.85 J/cm^2), nanocrystals with >15 nm begin to melt, whereas smaller-sized nanocrystals remain crystalline. Melting of "large" nanocrystals is accompanied by a decrease in their sizes, because of partial dissolution of Ge in the Si matrix during the time of melt crystallization after laser action. As a result, the initial dispersion of nanocrystals by size, determined by regimes of their growth, should be decreased. At the same time, their layer density should not change significantly because complete dissolution of large clusters is unlikely. We should note that, according to model data, the regime of melting selective nanocrystal sizes occurs over a quite narrow range of energy densities of 0.8–0.9 J/cm^2. Higher energy densities of pulsed laser radiation correspond to the regime of complete nanocrystal melting, regardless of their size. As a nanocrystal's size decreases, its surface-to-volume ratio increases. Thus, at a constant time of melt existence small clusters are able to dissolve in a matrix surrounding Si more easily. This can lead to complete dissociation of nanocrystals smaller than a certain typical size determined by Si diffusion length in the Ge melt. Thus, for both the considered regimes of pulsed laser action, we should expect reduction in nanocrystal size dispersion. But, in the latter case, a significant decrease in the layer density of nanocrystals should be observed because of their complete dissolution.

The kinetics of phase transformation at pulse laser action developed in Ref. [20] demonstrate the possibility of reducing the size dispersion in heterostructures with quantum dot bulk material which has a melting temperature below the melting temperature of the surrounding matrix.

15.8 Conclusions

The pulse-oriented crystallization phenomena is fundamental for a new direction in condensed matter physics. This direction includes investigation of fast structural transitions in crystals, nonequilibrium crystal growth and dopant capture processes, formation of new phases, as well as precise tuning of doping element depth distribution. In relation to the technological applications, ion-implantation technology is currently the main technology for semiconductor doping in electronics manufacturing. Pulsed (laser) annealing is the fundamental technology used by the leading global manufacturers of various circuits and devices, in particular, large-scale matrices of field-effect transistors used for LCD displays. The advantages of laser treatment in modifying near-surface layers and films without modification of the bulk properties are highly attractive. The short duration of effect enables local modification of properties of a selected surface, keeping the properties of the other areas of a system used unchanged. Significant progress has been achieved in solving important microelectronic problems: production of thin doped semiconductor layers (for electric contacts, source, and drain areas in field-effect transistors), crystallization of semiconductor films on amorphous dielectric substrates, and strengthening of phase synthesis of surface layers in metal alloys.

Many scientists from various countries contributed to research on the laser annealing effect. Initially, the annealing of ion-alloyed layers with laser radiation pulses was proposed by scientists from the Physical-Technical Institute of the Kazan Scientific Center of the USSR Academy of Sciences (KPhTI) [4,6]. Similar work was independently started in the Institute of Semiconductor Physics of the Siberian Branch of the USSR Academy of Sciences (ISP, Novosibirsk) [5,7–9]. In 1977, the results of Russian research were presented at the Soviet–American seminar on ion implantation in New York State University at Albany, USA. The Russian delegation included the following scientists: I.B. Khaibullin (KPhTI, Kazan), L.S. Smirnov, N.N. Gerasimenko, G.A. Kachurin, A.V. Dvurechenskii (ISP, Novosibirsk), V.S. Vavilov, V.V. Krasnopevtsev, I.P. Akimchenko, and A.N. Georgobiani (Lebedev Institute of the Russian Academy of Sciences, Moscow). It was, in fact, the first international event presenting results on pulsed laser annealing in reports by I.B. Khaibullin, A.V. Dvurechenskii, and G.A. Kachurin. American colleagues immediately evaluated the potential of the phenomenon discovered for both fundamental and applied studies. After this Russian–American seminar, the worldwide scientific community was involved in large-scale experiments with physical processes in solids under laser and light (noncoherent flash lamp) radiation, as well as electron and ion beams. The first topical symposium on *Laser-Solid Interaction and Laser Processing* was held in Boston, USA, in 1978. Attendance at the symposium numbered in excess of 300 scientists and technologists from 10 countries, representing over 60 research institutions [21]. The European topical international conference *Energy Pulse and Particle Beam Modification of Materials* started in 1984 and was then held in 1987 and 1989 in Dresden [22,23]. As for the practical results, rapid thermal annealing systems based on second and millisecond light pulses have gained wide recognition within research and development organizations, and the electronics industry.

References

[1] J.W. Mayer, L. Eriksson, J.A. Davis, Ion Implantation in Semiconductors (Silicon and Germanium), Academic Press, New York, NY, 1970.

[2] V.V. Bolotov, A.V. Vasiljev, A.V. Dvurechensky, G.A. Kachurin, N.B. Pridachin, L.S. Smirnov, et al., A Survey of Semiconductors Radiation Techniques, Mir, Moscow, 1983.

[3] M. Nastasi, J.W. Mayer, Ion Implantation and Synthesis of Materials, Springer-Verlag, Berlin, 2006.

[4] E.I. Shtyrkov, I.B. Khaibullin, M.M. Zaripov, M.F. Galiautdinov, R.M. Bayazitov, Local laser annealing of ion-implanted semiconductor layers, Semiconductors 9 (1976) 1309–1311 (Original in Russian. Fiz. Tekhn. Poluprovodn. 8 (1975) 2000–2002)

[5] G.A Kachurin, N.B. Pridachin, L.S. Smirnov, Annealing of radiation defects with pulsed laser irradiation, Semiconductors 9 (1975) 946–947 (Original in Russian Fizi. Tekhn. Poluprovodn. 8 (1975) 1428–1429).

[6] I.B. Khaibullin, E.I. Shtyrkov, MM Zaripov, M.F. Galiautdinov, G.G. Zakirov, On the factor of implanted-impurity utilization under laser annealing of ion-doped layers on silicon, Semiconductors 11 (1977) 190–194.

[7] G.A Kachurin, E.V. Nidaev, A.V. Khodyachikh, L.A. Kovaleva, Annealing of implanted layers with a scanning laser beam, Semiconductors 10 (1976) 1128–1129.

[8] A. Kh Antonenko, N.N. Gerasimenko, A.V. Dvurechenskii, L.S. Smirnov, G.M Tseitlin, Distribution of an implanted impurity in silicon after laser annealing, Semiconductors 10 (1976) 265–266.

[9] A.V. Dvurechensky, G.A. Kachurin, A. Kh Antonenko, The mechanism of impurity redistribution at laser-annealing of ion-implanted semiconductors, Rad. Eff. 37 (1978) 179–181.

[10] A.V. Dvurechenskii, G.A. Kachurin, E.V. Nidaev, L.S. Smirnov, Pulsed Annealing of Semiconductor Materials, Nauka, Moscow, 1982 (in Russian)

[11] N. Blombergen, Fundamentals of laser-solid interaction, in: H.J. Leamy, J.M. Poate (Eds.), Laser-Solid Interactions and Laser Processing-1978, American Institute of Physics, New York, NY, 1979, pp. 1–10.

[12] L. Csepregi, J.W. Mayer, T.W. Sigmon, Regrowth behavior of ion implanted amorphous layers on (111) silicon, Appl. Phys. Lett. 29 (1976) 92–93.

[13] F. Spaepen, D. Turnbull, Kinetics of motion of crystal-melt interfaces, in: H.J. Leamy, J.M. Poate (Eds.), Laser-Solid Interactions and Laser Processing-1978, American Institute of Physics, New York, NY, 1979, pp. 73–84.

[14] B.G. Bagley, H.S. Chen, A calculation of the thermodynamic first order amorphous semiconductor to metallic liquid transition temperature, in: H.J. Leamy, J.M. Poate (Eds.), Laser-Solid Interactions and Laser Processing-1978, American Institute of Physics, New York, NY, 1979, pp. 97–101.

[15] A.V. Dvurechenskii, T.N. Mustafin, L.S. Smirnov, H.-D. Geiler, G. Goetz, U. Juhn, Influence of the thickness of damage layers on the migration of dopands during laser annealing in implanted silicon, Phys. Stat. Sol. (a) 63 (1981) 203–206.

[16] A.V. Dvurechenskii, R. Groetzschel, N.M. Igonina, B.A. Koval, N.I. Lebedeva, Diffusion of impurities in undercooled melt of pulse heated ion-implanted Si, Phys. Stst. Sol. (a) 84 (1984) 171–177.

[17] V. Yu. Balandin, L.N. Aleksandrov, A.V. Dvurechenskii, O.A. Kulyasova, Simulation of structural transformations in multilayers systems on silicon by pulsed nanosecond annealing, Rev. Solid State Sci. 5 (1991) 1–14.

[18] V. Yu. Balandin, A.V. Dvurechenskii, L.N. Aleksandrov, Plasma effect at nanosecond pulsed annealing of amorphous layers of silicon, Poverhnost 7 (1988) 79–86 (in Russian)

[19] A.I. Yakimov, A.V. Dvurechenskii, V.A. Volodin, M.D. Efremov, A.I. Nikiforov, G. Yu. Mikhalyov, et al., Effect of pulsed laser action on hole energy spectrum of Ge/Si self-assembled quantum dots, Phys. Rev. B 72 (2005)115318-7

[20] V.A. Zinoviev, A.V. Dvurechenskii, Zh.V. Smagina, G.D. Ivlev, E.I. Gatskevich, V.L. Malevich, Nanocrystals embedded in a crystal matrix heated by nanosecond laser pulses, JETP 72 (2012) 436–444.

[21] A.V. Dvurechensky, G.A. Kachurin, T.N. Mustafin, L.S. Smirnov, Laser annealing of ion-implanted semiconductors, in: H.J. Leamy, J.M. Poate (Eds.), Laser-Solid Interactions and Laser Processing-1978, American Institute of Physics, New York, NY, 1979, pp. 245–258.

[22] A.V. Dvurechenskii, R. Groetzschel, V.P. Popov, Impurity effects at ion beam induced disordering of highly doped silicon, in: K. Hennig (Ed.), Energy Pulse and Particle Beam Modification of materials, Akademie-Verlag, Berlin, 1988, pp. 44–49.

[23] V.A. Geisler, R. Groetzschel, A.V. Dvurechenskii, T.V. Kurochkina, Yu.A. Manzhosov, Raman scattering in heavily doped silicon layers prepared with laser anneling, in: K. Hennig (Ed.), Energy Pulse and Particle Beam Modification of Materials, Akademie-Verlag, Berlin, 1990, pp. 149–151.

[24] O. Bostanjoglo, Time-resolved TEM of pulsed crystallization of amorphous Si and Ge films, Phys. Stat. Sol. (a) 70 (1982) 473–481.

16

Universality of the {113} Habit Plane in Si for Mixed Aggregation of Vacancies and Self-Interstitial Atoms Provided by Topological Bond Defect Formation

L.I. Fedina[1], A.K. Gutakovskii[1], A.V. Latyshev[1,2], A.L. Aseev[1]

[1]RZHANOV INSTITUTE OF SEMICONDUCTOR PHYSICS OF THE SIBERIAN BRANCH OF THE RUSSIAN ACADEMY OF SCIENCES, NOVOSIBIRSK, RUSSIA

[2]NOVOSIBIRSK STATE UNIVERSITY, NOVOSIBIRSK, RUSSIA

16.1 Introduction

Interest in topological defects related to the disturbance of atomic bond geometry while retaining the coordination of atoms is conditioned by their significant impact on the electronic, optical, magnetic, and other properties of the crystal. In two-dimensional materials, i.e., graphene, topological-bond defects (*TBD*) play a key role in the nanoengineering of material properties [1–4]. In 3D crystals, *TBD* are an integral part of various dislocation core structures and grain boundaries. They constitute the essence of the amorphous material structure. Although the role of *TBD* in the change of optical properties is of special importance, for solar cell technology and silicon-based light-emitting devices that role is not quite clear. To a great extent, this is due to the complexity of *TBD* research in 3D materials, which are based primarily on first-principle calculations [5,6].

From the viewpoint of possible *TBD*, for visualization under a HREM with direct electron irradiation (in situ HREM), 2D graphene is an ideal material. All that is needed is an appropriate aberration-corrected microscope with sub-ångström resolution (0.08 nm) allowing visualization of single carbon atoms at a small distance (0.142 nm). However, not only does in situ HREM for *TBD* visualization during irradiation of 3D crystals (e.g., Si) require labor-consuming simulation of HREM images associated with calculation of extended defect structures in large cells (at least 5000 atoms), but it also requires extensive knowledge of the processes running in a thin crystal under irradiation. Knowledge accumulated by our institute over the past years through use of the high-voltage electron microscope (HVEM) irradiation

Advances in Semiconductor Nanostructures. DOI: http://dx.doi.org/10.1016/B978-0-12-810512-2.00016-0

techniques (in situ HVEM)[1] [7] supports in many ways the subsequent successful use of the in situ HREM technique to understand the mechanism of extended {113} defect formation in silicon. This important interstitial-type of defect occurs in silicon under any nonequilibrium conditions (up to $T = 0.5T_{\text{melting}}$), relating to the generation of point defects. The reasons for its prevailing formation have not yet been explained, and recent attempts to prove it directly in the course of large-scale parallel molecular dynamics (*MD*) calculations have failed [8]. Instead of the {113} defect, quite a different defect, {111}, was found to exist. We previously used in situ HREM irradiation to demonstrate that the {111} defect occurs in the cores of both extrinsic and intrinsic Frank partial dislocations, and also as an isolated interstitial-type of defect with a displacement vector of $(a/5)\langle 111 \rangle$ due to hampered self-interstitial atoms (*Is*) embedding into lattice points at $T < 0.5T_{\text{melting}}$ [9]. In this case, one-dimensional $\langle 110 \rangle$ chains of *Is* are formed in the defect plane and bonded with the Si matrix by means of doubled fivefold and single eightfold atomic rings that can be visualized in a (110) HREM cross-sectional view. In that defect, the energy accumulated in *Is* chains is extraordinarily low and does not exceed 0.3–0.4 eV per atom, but the probability of its formation both upon exposure to electron irradiation and after implantation is only 10% compared to {113} defects [9].

The equilibrium structure of the interstitial-type *Is*-{113} defect is well established for electron irradiation at $T = 450°C$ [10], although by default it is considered to be common for ion implantation. *Is* chains in the plane of that defect are bonded with the Si matrix by means of *TBD* consisting of fivefold, sixfold (hexagonal), sevenfold, and eightfold rings (5-6_h-7-8 structure). However, within the interstitial concept, as discussed in a review paper [11] devoted to the contemporary problems of ion implantation of Si, neither ab initio calculations predicting solely compact *Is* configurations nor experimental techniques (DLTS, EPR, PAS) detecting a great variety of clusters of vacancies (Vs) are able to explain the evolution mechanism of the extended {113} defect. Irrespective of implantation, the problem of small cluster evolution is now becoming a driver for further development of semiconductor technologies.

The application of in situ HVEM and HREM techniques enabled us to formulate the idea of mixed *Vs* and *Is* aggregation in the {113} plane as far back as 1996 [12,13]. At a later stage it was confirmed by direct observation of ordered *I-V* pairs in the nearest neighboring atomic chains constituting the {113} plane during in situ HREM irradiation of very thin crystals at room temperature [14,15]. However, in thicker crystals, an amorphous phase emerged in the {113} planes [16], indicating the formation of more complex *Vs-Is* clusters, which is more characteristic of ion implantation than of electron irradiation. Indeed, quite recently, we managed to demonstrate that during hot (600°C) implantation of Er^+ ions,[2] an ordered aggregation of V_2-$2I$ clusters takes place in the {113} plane [17].

[1] In various years, investigations were conducted using JEOL-1250 high-voltage electron microscopes of the Institute of Microstructure Physics, Halle (Saale), Germany; Oxford University; Baikov Institute of Metallurgy and Materials Science, Moscow; and Electron Microscopy Centre for Materials Science, Belgium (University of Antwerp RUCA-EMAT).

[2] Implantation of Er^+ ions in silicon was conducted by N.A. Sobolev (Ioffe Institute, Saint Petersburg), while ex situ HREM research was carried out by N.D. Zakharov (Institute of Microstructure Physics, Halle (Saale), Germany), and HREM image analysis, image simulation, and defect structure calculations were performed by L.I. Fedina.

In this study, we used in situ and ex situ HREM, molecular mechanics calculations of defect structures (HyperChem 7.0) in large cells (5000 atoms), followed by HREM image simulations (MUSLI software [18]) to demonstrate the universality of the {113} plane for mixed *Vs* and *Is* aggregation in the course of electron irradiation and implantation of heavy erbium ions. The key aspect of this novel research is [110]-HREM visualization of nearest neighboring atomic chains (atomic columns) constituting the {113} plane which accumulate multiple vacancies, and ⟨110⟩-split interstitials as second nearest neighbors before these should transform into *TBD*.

16.2 Basic Results of Point Defect Aggregation in Silicon Obtained by In Situ HVEM

When applying the in situ electron irradiation technique in the JEOL-1250 HVEM, where the electron energy is equal to 1 MeV, and heating of the sample within a temperature range of 20–1000°C is possible, we have achieved remarkable progress in understanding the common pattern of *Is* aggregation in Si and Ge [7]. Over a long period, this problem has been the subject of violent disputes due to the multitude of extended interstitial-type defects (perfect and partial dislocation loops, oxidation-induced stacking faults, and {113} defects) occurring in silicon under various process exposures (implantation, oxidation, and prolonged annealing), and the nontrivial atomic structure of these defects. The fundamental difference between extrinsic dislocation-type defects and {113} defects, as demonstrated by us in Refs. [7,19], is the structural positions of self-interstitial atoms in the defect plane: lattice points for all *Is* in the dislocation loop (except the dislocation core), and pure interstitial positions for them in the {113} defect. We have shown that under high-energy electron irradiation when both vacancy and self-interstitial atoms are extremely mobile, the transition of an interstitial from the dislocation core to a lattice point in the loop plane involves overcoming an energy barrier of 1.3 + 0.2 eV, and takes place at a temperature exceeding $0.5T_{melting}$ [19]. Therefore, the primary form of an extended *Is*-defect existing at lower temperatures should be a metastable {113} defect. Their formation in silicon during electron irradiation is mainly due to the inequality of coefficients of *Vs* and *Is* interaction with a real crystal surface $k_{is} \ll k_{vs}$ [7], which is based on a difference in relaxation volumes of the vacancy and self-interstitial atom [20–22]. This results in the formation of dominating *Is* excess in Si during electron irradiation. There is only a limited zone for *Is* drain close to the surface where excessive *Vs* may exist. Its depth strongly depends on the oxide – nitride thickness on the silicon surface. For example, at an oxide thickness of ~1 nm, the *Is* drain zone is equal to ~300 nm, while at an oxide thickness of 15–30 nm it narrows down to 10 nm. However, the *Vs* drain zone, where the {113} defects form, amounts to ~500 nm [7].

Thus, the prevailing *Vs* drain to the surface during in situ electron irradiation enabled us to study the kinetics of {113} defect growth thoroughly, depending on the type and thickness of coating films on the Si surface, as well as on irradiation intensity and temperature [7,13,21,22]. High intensity HVEM irradiation enabled us to achieve a point defect generation rate comparable to that of light-ion implantation. Of special importance is the fact that, with an elevation of the irradiation temperature up to $T = 450°C$, we managed to distinguish long-lasting (up to 45 min) stationary stages of {113} defect growth in ⟨332⟩ directions, which were governed either by *Vs* surface drain, or by the reaction of point defect annihilation ($V+I=0$). This enabled us to apply the theory of quasi-chemical reactions, and to show that energy barriers for recombination of charged point defects prevailing in *n*- and *p*-type Si under electron irradiation should be different [21,22].

At reduced irradiation temperatures ($T<100°C$), stationary stages of {113} defect growth were never observed [7]. As *Is* drain to the silicon surface occurs within a rather narrow near-surface region only, observation of *Vs* clustering under in situ HVEM irradiation is extremely hampered, although its indications manifest themselves in the kinetics of nucleation and the disappearance of {113} defects [12,13]. It was discovered that for *n*- and *p*-type Si with various concentrations of dopant atoms not exceeding 10^{19} cm^{-3}, a ring distribution of small {113} defects is formed within a temperature range between 20°C and 300°C, centered along the electron beam axis (Fig. 16.1). That distribution appears due to the disappearance of primary {113} defects in the center of the irradiated area, owing to a nonuniform intensity across the electron beam. In the center, the intensity is equal to 6×10^{19} electrons/cm^2 s, and on the periphery it is 10 times lower. Unlike extra-pure FZ-Si, where a ring pattern appears almost immediately (Fig. 16.1A and B), in Cz-Si with high phosphorus concentration (1×10^{19} cm^{-3}) such a distribution is never observed (Fig. 16.1F and G). Consecutive stages of ring pattern formation are clearly visible in low-doped *n*-type Cz-Si (Fig. 16.1C–E). One can see that as irradiation proceeds, the primary {113} defects disappear and secondary clusters of extremely high density emerge in the center. This effect had been observed earlier and was attributed to ingress of impurities from the surface of the sample [23]. However, possible ingress of impurities during our experiments was excluded, as all thinned experimental samples were coated with SiO_2 or Si_3N_4 layers with a thickness of 15–30 nm on both sides [21,22]. It later turned out that this trivial and frequently observed ring pattern of {113} defect formation during HVEM irradiation contained nontrivial physics related to a mixed aggregation of point defects in the {113} planes.

The fact that the ring pattern formation is suppressed by an increase in the concentration of dopant atoms capable of interacting with vacancies (oxygen, phosphorus) testifies to the possibility of *Vs* accumulation and clustering in the central zone when dopants are absent. The study of nucleation kinetics in FZ-Si reveals that the incubation time τ_d only slightly depends on irradiation intensity for primary {113} defects, and decreases abruptly with the intensity increase for secondary defects (Fig. 16.2A). At the same time, the rate of {113} defect nucleation C_d changes in reverse order; it decreases for primary defects with the intensity increase, and increases for the secondary defects (Fig. 16.2B). Such behavior strongly suggests the varying nature of primary and secondary defects. However, to prove this, full-scale atomistic research on the point defect clustering process is needed, and this can be provided only by the in situ HREM irradiation technique.

Indeed, the very first experiments have demonstrated that thin silicon crystals (up to 10 nm) coated with a layer of Si_3N_4 (3–5 nm) feature *Vs* clustering in the {113} planes, but predominantly interstitial-type {113} defects are observed with an increase in sample thickness. This enabled us to formulate the idea of a mixed point defect aggregation inevitability in the {113} planes [12].

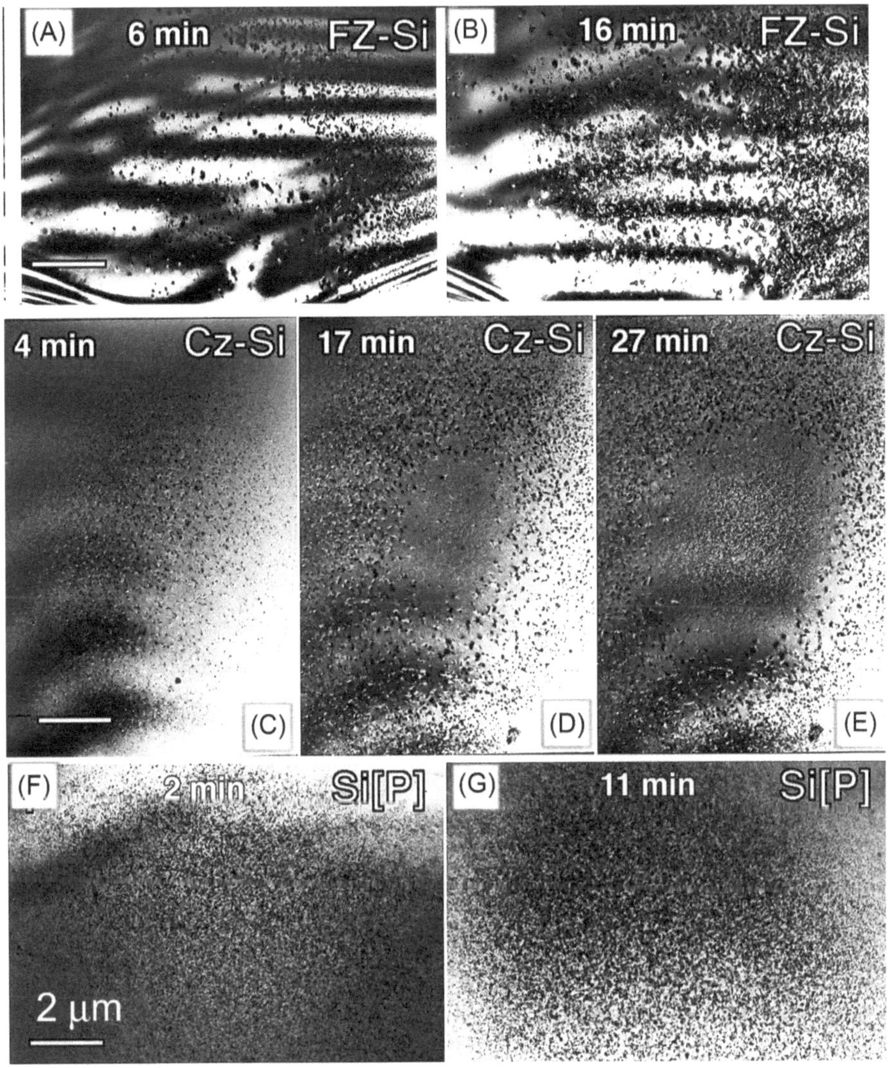

FIGURE 16.1 Formation of {113} defects in FZ-Si (A and B); in Cz-Si with a phosphorus concentration of 1×10^{15}cm^{-3} (C–E); and 1×10^{19} cm^{-3} (G and H) during HVEM irradiation at $T = 25°C$, and $I = 6 \times 10^{19}$ electrons/cm^2 s^1. The irradiation time is shown in the photographs. The alternating *light and dark strips* are thickness contours.

16.3 Ordering of Close Correlated *I-V* Pairs in Si in {113} Planes

The equilibrium structure of the extended interstitial-type {113} defects is well established, both experimentally [10] and theoretically [24,25]. It is reliably detected under electron irradiation at $T = 450°C$ [10], while for ion implantation it remains uninterpreted in most cases

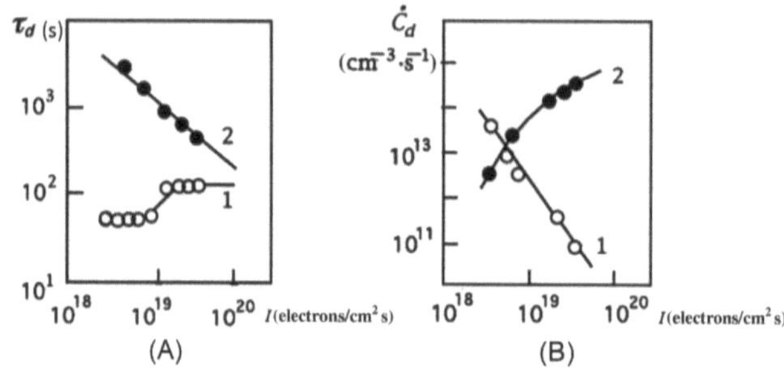

FIGURE 16.2 Incubation time τ_d (A), and nucleation rate C_d (B), for primary (1) and secondary (2) {113} defects versus HVEM irradiation intensity (*I*).

due to the complexity of the HREM images that fail to correspond to the equilibrium structure [26–28]. Nevertheless, this structure is believed to emerge every time, although the first-principle calculations are not able to reproduce it based on small *Is* cluster evolution [29,30]. A keen interest in this defect is explained by the fact that in the presence of *Is* clusters or {113} defects, the diffusion of dopant atoms in Si is five to six orders faster [11,26,31]. An understanding of small *Is* cluster evolution should provide a solution for a number of problems related to the formation of small *p–n* junctions in electronic components measuring less than 20 nm [11].

It was established from ab initio calculations that the ⟨110⟩-split (dumbbell) configuration is the lower energy structure for the neutral self-interstitial ($I_{\langle 110 \rangle}$) compared to other *I* configurations in the tetrahedral or hexagonal position [32,33]. However, experimentally the primary states of *I* were not detectable during irradiation due to their extremely high mobility. In 1997, after we had stated the idea of mixed *Vs* and *Is* aggregation [12], the first theoretical studies [34,35] demonstrated that due to the low symmetry of $I_{\langle 110 \rangle}$ and *V*, a stable close correlated *I-V* pair could emerge in the course of their interaction. The next stage of transformation, which was called incomplete recombination of the *I-V* pair, involved the emergence of *TBD* consisting of crossed double fivefold and sevenfold rings (*TBD* 5-7). The calculated value of the energy barrier (E_r) for a close correlated *I-V* pair recombination amounted to 1.2 eV at a lifetime τ of about 1 h. According to other calculations, those values were significantly lower: an E_r of about 0.4 eV, and τ of several microseconds [36]. Nevertheless, this *TBD* 5-7 complex was demonstrated to have the lowest formation energy of all known point defects in silicon [32–35], but the issue of close correlated *I-V* pair detectability remained. This resulted in numerous attempts [38–42] to construct the {113} defect through *TBD* 5-7 formation, based on unlikely (about 5 eV) 90-degree bond switching in the Si lattice [37]. The current concept of the mechanism of {113} defect formation includes the idea of spontaneous *TBD* 5-7 formation around the *Is* chain that is somehow introduced in the {113} plane [41,42]. The reason why the physically consistent idea of *TBD* formation from the close correlated *I-V* pair has not seen widespread development is solely due to the difficulties of its experimental detection in 3D material.

In this section, we will demonstrate that during in situ HREM irradiation, owing to the ordering of multiple *I-V* pairs, when *V* and *I* turn out to be localized as second-nearest neighbors in nearest neighboring atomic chains along the electron beam incidence direction [110], these can be distinctly visualized by amplifying the signal from the stack of *I-V* pairs. However, the *I-V* pair's visualization can be implemented in thin crystals only where there are tensile strains introduced by dislocation cores or heterointerfaces.

Experiments were carried out using a JEM-4000EX microscope operated at 400 keV, which significantly exceeds the threshold voltage of 200 keV for the formation of {113} defects in the transmission electron microscope [43]. The electron flux density in the center of the irradiation area was equal to about 10^{20} electrons/cm^2 s^1. All in situ HREM irradiation experiments were conducted at room temperature. Transparent (110)-Cz-Si specimens with a high concentration of dislocations introduced through high-dose implantation of boron and phosphorus, followed by annealing at $T = 1150°C$, were prepared by chemical etching. The dislocation density in the samples amounted to about 10^{12} cm^{-2} which was necessary for observation of the point defect clustering in very thin crystals. The samples had no special coating on the surface. Most dislocations observed were of an extrinsic type with a displacement vector of $a/2\langle 110\rangle$ corresponding to perfect 60-degree dislocations.

Fig. 16.3A and B show two types of [110]-HREM images of the {113} defect discovered in Cz-Si during in situ electron irradiation close to two dislocation cores. One {113} defect (Fig. 16.3A) emerged at a distance of about 3 nm from the dislocation core (outside the figure), another one emerged close to the other dislocation core and has an interstitial-type structure (*TBD* 5-6_h-7-8), (Fig. 16.3B). Other variations of HREM images featured no {113} defects, although defects in the {111} and {100} planes were observed close to other dislocation cores (Fig. 16.3C and D).

We may conclude that local strains induced by dislocation cores serve as point defect traps, thus, without dislocations their aggregation cannot be observed in uncoated and very thin samples. It is also obvious that the core of 60-degree dislocation does not determine the type of extended defect emerging under irradiation. Therefore, we believe that the two different HREM images of the {113} defect represent various stages of its formation. As the second {113} defect (Fig. 16.3B) has an equilibrium structure, it is logical to assume that the first image is its initial stage.

According to HREM image simulations, the sample thickness in this case did not exceed around 5 nm. Since the HREM images were obtained at the Scherzer defocus condition, their analysis allows straightforward interpretation. White spots correspond to channels in the Si lattice viewed in the [110] direction, and the dark columns correspond to the projections of [110] atomic chains on the (110) image plane. They are generally known as atomic columns. The upper left corner of Fig. 16.3A is a schematic representation of atom positions inside the atomic columns. In Fig. 16.3A, the image of the {113} defect is formed by paired alternation of atomic columns, one of them featuring 40% elongation in the [001] direction, and the other 30% shortening in the same direction. Such variation of the atomic column length suggests that their elongation and shortening is caused by the emergence of opposite-type defects in the nearest neighboring [110] chains. Elongation of the columns is due to $I_{[110]}$

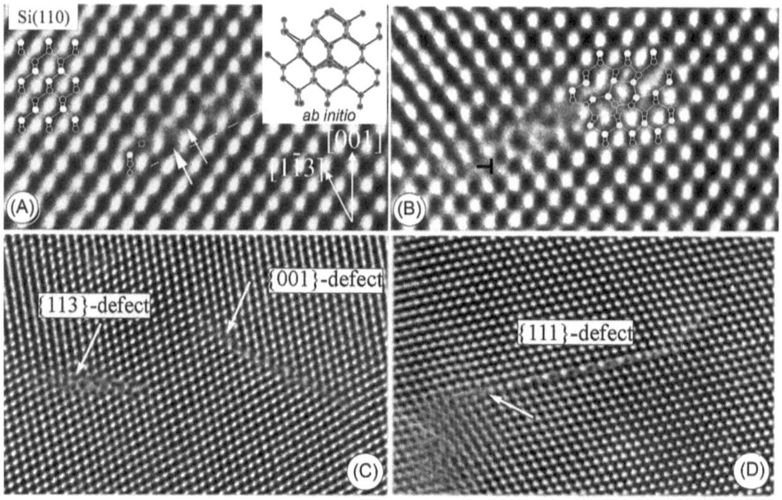

FIGURE 16.3 (A and B) HREM images of {113} defects observed during in situ HREM irradiation close to two cores of 60-degree dislocations. The pair of *arrows* in (A) show nearest neighboring [110] atomic chains constituting the (1$\bar{1}$3) plane and accumulating multiple (*I*-*V*) pairs. (C and D) HREM images of various extended defects emerging at the cores of 60-degree dislocations. The *white arrows* indicate the dislocation core position.

accumulation along the [110] atomic chains, where the split axis of $I_{[110]}$ coincides with the axis of the chains accumulating the defects. According to first-principle calculations [32,44], the center of dumbbell-$I_{\langle 110\rangle}$ displaces by 0.76 Å in the ⟨001⟩ direction, and hence should elongate the atomic column by 50%. In contrast, accumulation of vacancies in the chain should lead to a shortening of the atomic column length. The insert in Fig. 16.3A, right corner, shows ab initio calculation[3] of a single *I*-*V* pair according to its location in Si as per the [110]-HREM data.

HREM image calculations enabled us to prove that single $I_{[110]}$ and *V* cannot be visualized at a crystal thickness of 5 nm. This meant that the observed image in Fig. 16.3A is attributed to multiple *I* and *V* localization in nearest neighboring atomic chains that provide their exact position as second nearest neighbors to form stable close correlated *V*-*I* pairs, as is proposed by first-principle calculations. In order to determine possible multiple $I_{[110]}$ and *V* locations along the [110] chains, we used HyperChem 7.0 software to develop various *I*-*V* pair models within a cell size of 42 × 42 × 42 Å (about 5000 atoms), which were used for HREM image simulations after molecular mechanics calculations (Mm + optimization). The model could be considered reliable, if good agreement of experimental and theoretical images is achieved. In our models presented in Fig. 16.4, the length of dumbbell ($I_{[110]}$) −2.43 Å corresponded to the data of ab initio calculations [32], with displacement of its center in the [001] direction amounting to 0.52 Å, which was less than 0.76 Å for the straightforward calculations. Correct

[3] Ab initio calculations were performed by I. Supryadkina under the supervision of K.K. Abgaryan and D.I. Bazhanov at the Dorodnicyn Computing Centre of the Russian Academy of Sciences.

V relaxation at Mm + optimization (for the case of classic potential) cannot be reproduced, but since it is low [32], all *V* in our models remained without relaxation. However, these restrictions did not prevent us from obtaining satisfactory agreement of experimental and simulated HREM images.

Extensive HREM image simulations enabled us to establish that at the initial stage (Fig. 16.4A and B) paired chains in the (1$\bar{1}$3) plane contain eight *V* and $I_{[110]}$, where every *V* and $I_{[110]}$ occur at a distance of the second nearest neighbor, i.e., they constitute close *V-I* pairs. As the axis of all dumbbells should be directed along the axis of the chain accumulating the defects, the *V-I* pairs cannot be located closer than 7.68 Å without significant deviation of dumbbells from the [110] direction. This resulted in disagreement of the calculated and experimental HREM image. According to ab initio calculations [34,35] for a single *V-I*

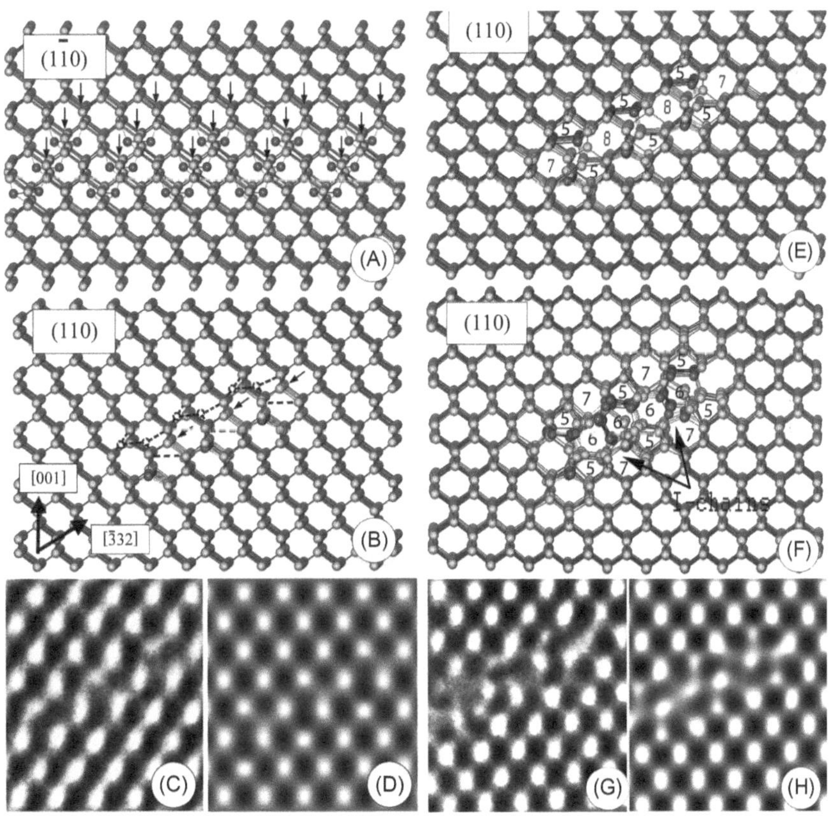

FIGURE 16.4 (A and B) A model of ordered *I-V* pairs in (1$\bar{1}$0) (A), and (110) (B) sections. The *red* (dark gray in print versions) doubled spheres are $I_{[110]}$. The *small arrows* indicate *V* positions relative to $I_{[110]}$. The *dotted lines* in (B) indicate the TBD 5-8 formation path. (C and D) Experimental (C) and simulated (D) HREM images for model (B). (E and F) Models of primary TBD in the (110) section together with additional *V* and $I_{[110]}$ (E) and *Is* chains incorporated into eightfold rings (F). (G and H) Experimental HREM image (G) versus simulated one (H) for model (F). Defocus value is −360 Å, and crystal thickness is 42 Å for both simulated HREM images presented in (D) and (H).

pair, the *V* and $I_{[110]}$ position at a distance of the second nearest neighbors is stable and not accompanied by spontaneous recombination.

HREM image simulation (Fig. 16.4D) for this model in the (110) section (Fig. 16.4B) has shown that it reproduces the experimental image (Fig. 16.4C) very well. In order to attain the next stage, which is essentially a recombination stage of the *I-V* pairs and involves the formation of an ordered *TBD* array, all $I_{[110]}$ need to be reoriented relative to the respective *V* in each pair. The $I_{[110]}$ reorientation path and bond switching are shown by *dotted lines* in Fig. 16.4B. During this stage, the defect structure features an equal number of *V* and $I_{[110]}$, so a zero-type extended defect is formed. After relaxation of that model due to the tight location of *I-V* pairs, an ordered *TBD* 5-8 array with a strict arrangement of eightfold rings in the $[\bar{3}32]$ direction appears in the $(1\bar{1}3)$ plane (Fig. 16.4E). Sevenfold rings corresponding to *TBD* 5-7 from a single *I-V* pair [34] remain only at the extended defect's edges.

After $I_{[110]}$ reorientation, a space for incorporation of new *V* and $I_{[110]}$ vacates in the initial chains (Fig. 16.4E). Additional point defects are required to form eightfold channels wherein continuous *Is* chains can be built to realize the interstitial-type $(1\bar{1}3)$ defect with the (5-6_h-7-8) topological structure (Fig. 16.4F). In the model shown in Fig. 16.4F, the number of *Is* does not exceed 55% in the left *Is* chain and 73% in the right *Is* chain of the required quantity of atoms in a perfect atomic chain corresponding to the crystal thickness. As can be seen from a comparison of the simulated and experimental HREM images, the best agreement is achieved for the right *Is* chain (Fig 16.4G and H).

It should be stressed that multiple *V* and $I_{[110]}$ ordering observed experimentally is implemented along two directions of [110] and $[\bar{3}32]$ that constitute the $(1\bar{1}3)$ plane of the extended defect. This strongly suggests that elastic interaction of a large number of *V* and $I_{[110]}$ acts as a driving force for their ordering in the $(1\bar{1}3)$ plane preceding the complete recombination of defects. In other words, the native step-chain structure of the {113} plane in silicon possessing a great number of bonds per unit length predetermines close spacing of multiple *V* and $I_{[110]}$ in the paired neighboring chains necessary for creation of the ordered *TBD* 5-8 array that should restore the perfect structure of Si through bond switching, or accommodate *Is* excess.

Thus, at the final stage of evolution, the ordered array of *I-V* pairs transforms into the (5-6_h-7-8) topological structure known as the interstitial {113} defect. But before its emergence, the array of *I-V* pairs should transform into the *TBD* 5-8 array, which implies a collective (step-by-step) $I_{[110]}$ reorientation accompanied by incorporation of additional *V* and $I_{[110]}$ into the basic initial atomic chains, and excessive *Is* into the eightfold channels. So, $I_{[110]}$ reorientation is the limiting stage of growth of *Is* chains along the ⟨110⟩ direction, ultimately determining the rod-like shape of the {113} defect (Fig. 16.5A). In situ HVEM investigations imply [7,13] that growth of the {113} defect along the defect axis in the ⟨110⟩ direction at an initial stage is of a profound activation nature, as against the growth rate in the ⟨332⟩ direction, does not depend on the temperature, and reflects high *I* mobility under electron irradiation (Fig. 16.5B). The activation energy of the {113} defect growth in the ⟨110⟩ direction, and hence the barrier for $I_{[110]}$ reorientation, is equal to about 0.4 eV. This value agrees with the energy barrier of 0.4 eV for recombination of close correlated *I-V* pairs from the *MD* calculations [13], and disagrees with calculations (about 1.2 eV) using the tight-binding potential for neutral defects [34]. Since electron irradiation under heavy crystal ionization is dominated by charged point defects, the 0.4 eV barrier appears to be the most realistic one [21].

The above case of observed $V\text{-}I_{[110]}$ ordering in the {113} plane close to the core of 60-degree dislocation is not unique. Similar HREM images (Fig. 16.5C and D) consisting of pairs of elongated and shortened atomic columns were observed earlier during irradiation of the strained $Si/Si_{0.87}Ge_{0.13}/(001)Si$ heterostructure (in the Si layer) [20], as well as close to the core of extrinsic Frank dislocation. In both cases, the regions of observed pairs feature tensile strains. In the first case, the defect is located in the silicon part at a distance of 35 nm from the interface, and in the second case at 2 nm from the dislocation core. According to HREM image simulations, the thickness of irradiated crystal for these cases amounted to 5 nm (Fig. 16.5C) and 7 nm (Fig. 16.5D), respectively. As one can see, only elongated columns are well visualized with the crystal thickness increase, while shortening of columns implying accumulation of vacancies is poorly detected. Besides, a large number of closely spaced V and $I_{[110]}$ cause strain build-up in the crystal, thus strongly hindering interpretation of {113} defect HREM images. This is supported by recent *MD* calculations of the amorphization process in Si [45], demonstrating that the strains resulting from accumulation of closely correlated (second nearest neighbors) $V\text{-}I_{[110]}$ pairs are significantly higher than those resulting from accumulation of their *TBD* 5-7, and hence the amorphization threshold for $V\text{-}I_{[110]}$ pairs decreases. These data and the fact that HREM visualization of the ordered $V\text{-}I_{[110]}$ pairs is easier to detect in the regions with tensile strains testify to their importance in $V\text{-}I_{[110]}$ pair stabilization. Besides, *MD* calculations demonstrate the preferred formation of interstitial {113} defects, when tensile strains are available in the crystal [8]. As for the electronic

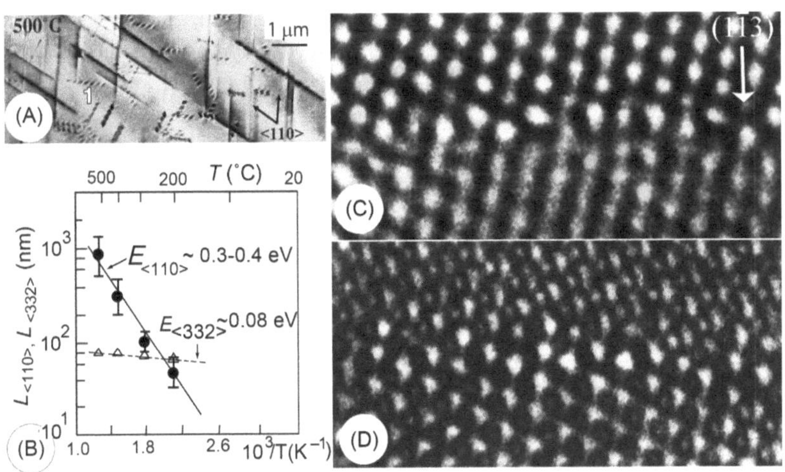

FIGURE 16.5 (A and B) Rod-like {113} defects in Si observed during in situ HVEM irradiation at $T = 500°C$ for 8 min (A), and temperature dependencies of their growth rates along the ⟨110⟩ and ⟨332⟩ directions. (C and D) Variations of HREM images of ordered $V\text{-}I$ pairs in the $(1\bar{1}3)$ plane observed by in situ electron irradiation of strained $Si/Si_{0.83}Ge_{0.13}/(001)Si$ heterostructure (in the Si layer) (C), and a Si crystal containing extrinsic Frank partial dislocations (D). Both the hetero-interface and dislocation core are out of images (see the text).

structure of these defects, it is not quite understood at present since it depends largely on the local atomic structure of the defect, the density of eightfold rings, and their alternation with *Is* chains [46]. An isolated ⟨110⟩-*Is* chain surrounded by two eightfold rings creates states in the band gap, which are induced by strains around one, but at the same time the structure consisting of three successive *Is* chains does not display any deep states. However, the general trend is a narrowing of the band gap for the {113} defect in the central point of the Brillouin zone capable of causing localization of carriers and their recombination, similar to quantum dot behavior.

16.4 *Vs* and *Is* Aggregation in the {113} Plane in Si Foils Covered With Si_3N_4 Films

In order to investigate *Vs* and *Is* aggregation in more detail, thinned (110)-FZ-Si foils used for in situ HREM experiments were covered with thin amorphous layers of Si_3N_4 with a thickness of about 3–5 nm on both sides. As was established earlier from in situ HVEM experiments, the availability of coatings results in a significant growth of point defect concentrations in the irradiated samples, due to their surface drain restriction [7]. In this case, a narrow Si layer with *Vs* excess is formed close to the surface, allowing observation of the vacancy clustering [12]. However, for a good HREM image, the crystal thickness during in situ experiments shifts toward higher values (8–10 nm) due to the availability of the coating. The experiments were carried out in JEM-4000EX with an acceleration voltage of 400 keV at room temperature, and irradiation intensity of about 10^{20} electrons/cm^2 s^1.

Fig. 16.6A–D show a series of HREM images for the 8 nm FZ-Si sample using an exposure time varying between 1 and 15 min. This thickness region still allows straightforward interpretation of HREM images where the dark columns represent the projections of atomic chains, and the bright spots are channels in the Si lattice along the [110] direction. As one can see, an increase of the channel brightness in the $(1\bar{1}3)$ planes is observed as early as during the first minute. After 2 min, a sequence of doubled white spots is formed in these planes due to weakening of the contrast of atomic columns. This is a direct indication of vacancy accumulation in the [110] atomic chains localized in the $(1\bar{1}3)$ planes. An increase in exposure time leads to the emergence of a nonordered dark contrast in the defect planes and its complex evolution (Fig. 16.6C). Finally, disappearance of the defects and restoration of the crystal structure takes place after 15 min (Fig. 16.6D). Contrast alteration and subsequent defect disappearance obviously imply interaction of aggregated vacancies with *Is*, which is accompanied by a recombination of defects. However, to prove this obvious evolution of extended defects and understand which type of vacancy clusters emerge at the initial stage, we should create a model of the defective structure and calculate its HREM image.

Using image calculations, we searched through a number of models with vacancy chain structures before we decided in favor of the model consisting of divacancies (V_2). This model turned out to be the only adequate one to reproduce the experimental HREM images by simulation. Fig. 16.7 shows the models of V_2-{113} defects at the initial stage (Fig. 16.7A), and at

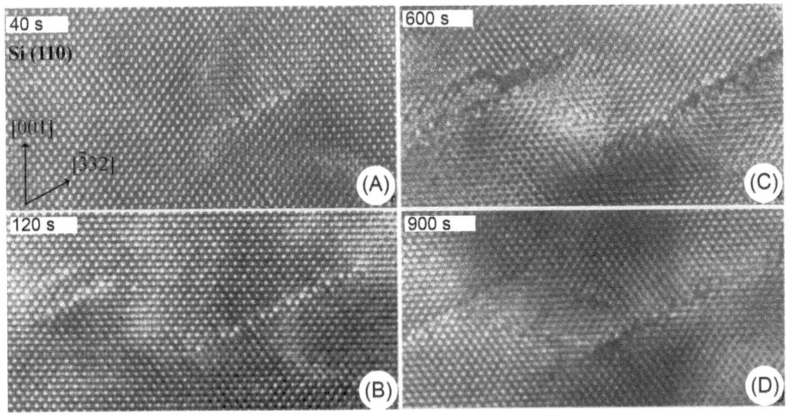

FIGURE 16.6 Consecutive stages of V_2-{113} defects formation in an 8 nm FZ-Si crystal coated with thin Si_3N_4 film during in situ electron irradiation in JEM400EX at $T = 25°C$. The irradiation time is shown in the photos.

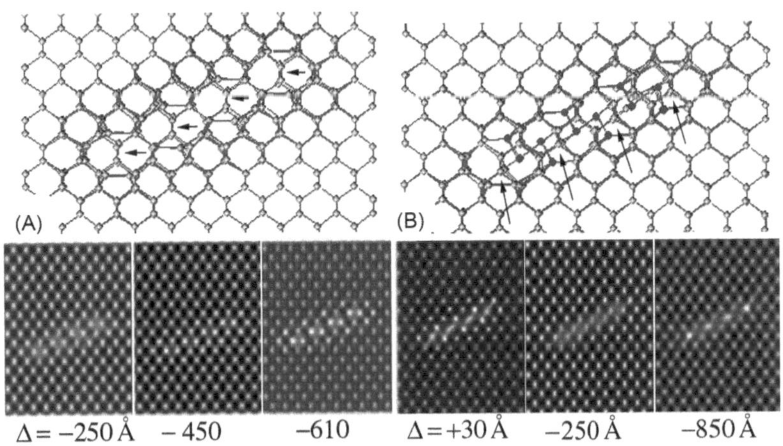

FIGURE 16.7 Models of a V_2-{113} defect at the initial stage (A), and at subsequent incorporation of an equal number of *Is* shown by *red* (dark gray in print versions) balls (B). The lower panels are maps of the simulated HREM images at a crystal thickness of 76.9 Å, depending on image defocus value Δ, (Å). The *arrows* indicate V_2 positions.

subsequent incorporation of an equal number of *Is* (Fig. 16.7A). There are maps below each model of the respective simulated HREM images. As determined from image simulations, the distance between V_2 in the chains is equal to 7.68 Å. During Mm + optimization of the V_2-{113} defect model, we used the predicted by ab initio calculation V_2-mode with pairing bonds [47], although the latest *first-principle* calculations of V_2 in a cell with 1000 atoms showed that V_2 with resonant bonds display higher stability [48]. The same study demonstrated that upon application of compression along the atomic chain axis, the V_2 mode with pairing bonds becomes stable. Strains introduced into the thin crystal by means of the Si_3N_4 film are possibly rather important for observation of this V_2 mode. The model obtained of ordered V_2 array in the $(1\bar{1}3)$ plane consists of fivefold and eightfold rings (*TBD* 5-8) provided by V_2 mode with pairing bonds (Fig. 16.7A). Based on the above, the same *TBD* 5-8 structure is also formed in the {113} plane during transformation of the ordered array of *I-V* pairs. However, the type of extended defects is different in these cases: for V_2-clustering, it is a vacancy-type defect, and for *I-V* pair accumulation, it is a zero-type defect. In both cases, during in situ electron irradiation *TBD* 5-8 structures are metastable and provide for point defect recombination or {113} defect formation with the (5-6_h-7-8) structure, depending on the local concentration of *Is*. Note, that the 5-8 structure related to a single V_2 is also observed during electron irradiation of graphene [2,3]. However, as distinct from 3D Si, many other *TBD* structures in 2D graphene are formed following 90-degree rotation of C–C bonds. In 3D Si, such rotation involves a high-energy barrier (about 5 eV [37]), and therefore is very difficult to implement.

During our in situ experiments, spontaneous *TBD* formation was never observed without participation of *Vs* and *Is*. Comparison of simulated and experimental HREM images for the V_2-{113} defect presented in Fig. 16.7A shows their good agreement over a wide range of defocus values, and hence testifies to the adequacy of the model developed. The fact that the V_2-{113} defects slowly disappear during electron irradiation suggests a rather low concentration of *Is* in the 8-nm Si crystal where their surface drain is realized. Due to a continuous change in the defect structure, the stable V_2-*Is* cluster configuration cannot be determined through HREM image calculations. Yet, we managed to achieve certain agreement of experimental and simulated HREM images for the recombination stage (compare Figs. 16.6D and 16.7B). The emergence of X-shaped configurations alternating with broadened white spots in the {113} defect plane is distinctly reproduced on simulated HREM images. This just confirms the mixed nature of V_2- and *Is*-interaction at the stage of recombination. However, to provide a more detailed analysis of the V_2-*Is* recombination stage, first-principle calculations are required. There are no data of this sort in current publications.

As the thickness of the irradiated area increases, the nature of point defect clustering in FZ-Si becomes much more complex. Fig. 16.8A and B present experimental HREM images of two {113} defects (a and b) after 25- and 45-min irradiation, along with microdiffraction patterns obtained during in situ HVEM irradiation in similar conditions, as well as defect models and their theoretical HREM images. The presence of amorphous rings on the microdiffraction pattern (an insert in Fig. 16.8A) corresponds to the initial amorphous structure of defects a and b on the HREM image. Disappearance of the amorphous rings under prolonged irradiation (Fig. 16.8B), and ordering of the structure of defects a and b on the HREM image suggest ordering of an amorphous phase. As can be seen from the models developed for the simulated HREM images (Fig. 16.8B), defect a transforms into an interstitial {113} defect with a (5-6_h-7-8) topological structure [10], while defect b transforms into a structure consisting of fivefold and eightfold channels with two incorporated *Is* chains. This model is based on continuous vacancy chains (V_n), and is also consistent with the structure of defect a in the local area with the incorporated *Is* chains. However, they are fundamentally of different types: a is an interstitial-type defect, and b, if we exclude *Is* chains, is a vacancy-type defect and a zero-type defect in the *Is* chain region. The comparison of experimental

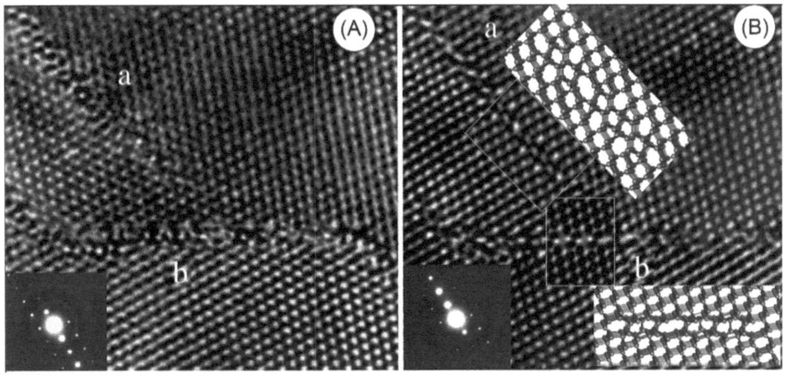

FIGURE 16.8 Transformation of the atomic structure of the {113} defects during prolonged in situ HREM irradiation of FZ-Si foil covered with 5 nm of Si_3N_4 films: (A) is 25 min, and (B) is 45 min. Inserts in the lower left corner of (A) and (B) are microdiffraction patterns taken from {113} defects during in situ HVEM irradiation under similar conditions. Models and theoretical HREM images in (B) for defects a and b are shown by inserts. The calculated thickness of the FZ-Si crystal is 192 Å and defocus value Δ is −350 Å.

and theoretical HREM images inserted into the planes of defects a and b (Fig. 16.8B, inserts in *white frames*) shows good agreement and hence confirms the adequacy of the models. At first glance, the 5-8 structure observed for the b defect is similar to that of a divacancy. Yet, they are different in the number of dangling bonds—two per each V_2 in V_2 chains, and only two at the edges of continuous V_n chains. We assume that the V_2-{113} defect is a metastable stage of *Vs* clustering during electron irradiation. However, the V_n-{113} defect consists completely of a *TBD* 5-8 array, which after *nI* incorporation as we will show below transforms into a zero-V_n-nI-{113} defect (*TBD* 5-6_h-7-8), indistinguishable from the interstitial-type {113} defect having the same 5-6_h-7-8 structure.

The zero-V_n-nI-{113} defect structure was thoroughly studied through first-principle calculations by a team of theorists (R. Jones, University of Exeter, Exeter, UK) [49], in pursuance of our idea of mixed *Vs* and *Is* aggregation in the {113} plane [16,20,50]. They demonstrated that initial stabilization of the V_n-{113} defect structure requires an additional shift in the ⟨332⟩ direction and 0.05 nm compression of eightfold rings in the direction perpendicular to the defect plane. The presence of such a small compression (0.03 nm in our experiment), which is characteristic of a vacancy platelet in the {113} plane, enabled us to detect this defect in a strained SiGe/Si heterostructure (in the SiGe layer) with a 0.5% misfit by in situ HREM irradiation [20]. According to ab initio calculations [49], all bond lengths in the V_n-{113} defect fall within 99−115%, while angles fall within 85−134% relative to values for a perfect Si lattice. The shift entails a drop in the full energy of the V_n-{113} defect up to 1 eV/vacancy, which is comparable to hexavacancy energy in Si. Moreover, this energy is just one-third of that of an isolated vacancy [51], and only slightly lower than that of the formation of a large cluster of 35 vacancies [52]. Therefore, this defect provides an efficient sink for mobile vacancies. While studying the electronic states' density for a fully relaxed V_n-{113} defect [49], it was discovered that strained Si−Si bonds form unfilled states in the upper half of the band gap. The electronic structure along the vacancy chain is such that the band of empty states crosses the whole band gap and creates deep states, which can be detected by deep level spectroscopy. These calculations also showed that the zero-V_n-nI-{113} defect is surprisingly stable. The per-atom energy of its formation amounts to 0.7 eV (about 58 MeV/Å^2), which is only slightly less than the energy of the V_n-{113} defect of about 1 eV/atom (about 80 MeV/Å^2). It should be noted that the energy of the interstitial-type *Is*-{113} defect also falls within the interval of 0.5–0.7 eV/atom [53], depending on the number of *Is* chains per unit length in the defect plane. So, irrespective of the type of point defects aggregated in the {113} habit plane, their excessive energy, owing to the emerging *TBD*, appears to be low, and hence provides for the formation of various types of extended defects in this plane. These calculations fully confirm our previously stated idea that *Vs* and *Is* in Si undergo mixed aggregation in the {113} plane [12].

It should be noted that during HREM analysis *TBD* are always visualized in the (110) plane perpendicular to the [110] atomic chains accumulating point defects. This is because the HREM signal from multiple localized point defects is accumulated only in the chains running parallel to the electron beam incidence. The signal from *TBD* in any other ⟨110⟩ chain is suppressed by the superimposed HREM image from the perfect crystalline structure, which is similar to the impossibility of detecting a single *V or I* in the [110] chain.

16.5 V_2-2*I* Cluster Aggregation in the {113} Plane in Si Under Hot Implantation of Erbium Ions

In order to prove that the phenomenon of mixed *Is* and *Vs* aggregation in the {113} plane in silicon is of a fundamental nature, we have analyzed the structure of defects in Si implanted with erbium ions. Er^+ ions[2] were selected, because this heavy ion creates a dense cascade of displacements and does not substitute the silicon atom during subsequent annealing, i.e., it does not create excessive *Is*. This suggests the possibility of studying *Is* and *Vs* aggregation under given conditions of approximate equality, which cannot be achieved during implantation of dopant atoms, which inevitably results in excessive *Is*. In the course of our previous studies of hot (900°C) implantation of Xe^+ ions, we observed the formation of narrow dislocation dipoles representing the final stage of {113} defect evolution at a temperature exceeding their stability limit of $T \sim 700$°C [7].

Erbium ions at a dose of 10^{14} cm^{-2} and energy of 2 MeV were implanted in Si(100) wafers at $T = 600$°C. The ex situ HREM investigations were conducted using the JEOL-4000EX microscope[2] at an acceleration voltage of 400 keV. Thin specimens in [110] cross-sections were prepared by Ar^+ ion milling. HREM image simulations were performed using MUSLI software [18]. Various structures of extended defects were obtained within a cell size of 43 Å × 42 Å × 42 Å by molecular mechanic calculations. Differences in the lengths of bonds for defective configurations obtained after Mm + optimization, and from the data of ab initio calculations [32], amounted to 0.2 Å, which allowed good agreement to be achieved between the calculated and experimental HREM images.

The HREM analysis demonstrated that the density of extended defects in the sample was $(3-5) \times 10^{11}$ cm^{-2}, and their size varied over a wide range between 0.8 and 15 nm. Fig. 16.9 shows the experimental HREM images of small clusters and {113} defects in a Si sample in the (110) cross-section. These images were obtained from an area with a thickness of less than 10 nm at Scherzer defocus, allowing straightforward interpretation. The small clusters in Fig. 16.9A appear as doubled bright spots corresponding to the disappearance of the atomic chain contrast, while the extended defects in Fig. 16.9B–D are characterized by a sequence of doubled bright spots in the {113} plane. At first sight, these HREM images appear to be quite similar to the V_2-{113} images described in Section 16.4. However, V_2-array is not stable under electron irradiation and recombine rather quickly by means of interaction with *Is*. In this case, the observed defect structures were very stable. This meant that V_2 clustering is indeed observed in the {113} planes during implantation of heavy erbium ions. During closer analysis of HREM images in Fig. 16.9B–D, it was recognized that the four atomic columns surrounding each V_2 are not of equal length. Two of them are 20–40% longer than the others, and the column elongation coincides with the [001] direction. Since the thickness of the crystal where the HREM image was obtained did not exceed 10 nm, such elongation, as in the case of observation of ordered *I-V* pairs, could be ascribed to the emergence of two $I_{[110]}$ around each V_2 which provided for the observed structure stability. In this thickness region, all HREM images of the {113} defects are practically identical, irrespective of their size.

For the development of a structural model of this new type of {113} defect, we used a divacancy model supplemented with an equal number of $I_{[110]}$. This model of a V_2-2*I*-{113}

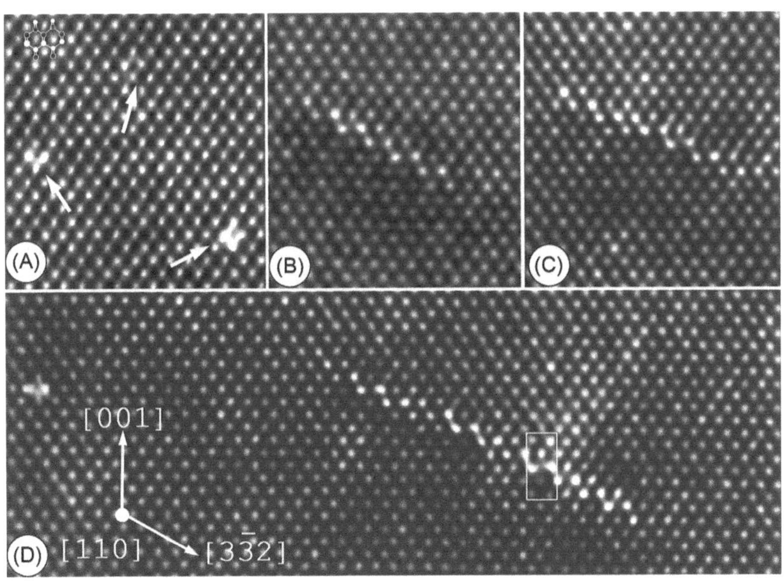

FIGURE 16.9 [110]-HREM images of structural defects in Si implanted with Er$^+$ ions at $T = 600°$C. The *arrows* indicate small clusters. The image in the *white frame* represents a translational cell within the structure of an extended {113} defect.

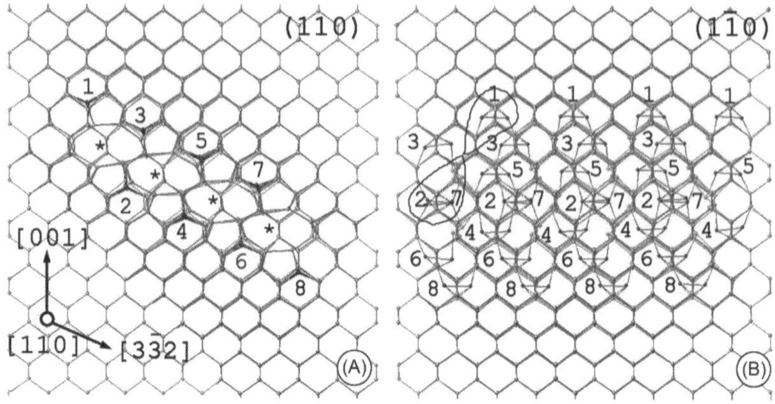

FIGURE 16.10 A model of a V_2-2I-{113} defect shown in two sections: (110) in cross-section of chains accumulating V_2 and $I_{[110]}$ (A), and (1$\bar{1}$0) along the ones. (B). In (A) numbers 1–8 indicate the chains containing $I_{[110]}$. Asterisks mark the chains where V_2 is located and *red* (gray in print versions) balls show $I_{[110]}$. In (B), the bold eight-shaped loop indicates the positions of two $I_{[110]}$ relative to V_2.

defect is shown in Fig. 16.10A and B in two sections: (110), perpendicularly to the atomic chains accumulating V_2 and I; and (1$\bar{1}$0) along these chains, respectively. The first one is used to calculate the [110]-HREM image as it is exactly displayed in the microscope, and the second one is required to demonstrate which of two $I_{[110]}$ along the chains form the V_2-2I cluster (shown as an eight-shaped loop in Fig. 16.10B). As is seen in the last figure, all dumbbells are oriented strictly along the [110] atomic chains and occur at the second nearest neighbor distance relative to each V_2 on both sides. The length of each dumbbell after molecular mechanics relaxation is equal to 2.43 Å, which is close to 2.45 Å, in accordance with first-principle calculations [32]. At the same time, displacement of the dumbbell center in the [001] direction is equal to only 0.5 Å, which is less than 0.7 Å, obtained by ab initio calculations [32]. Such a clear V_2-2I ordering had never before been observed in implanted layers in Si.

The V_2-2I-{113} defect model is characterized by the presence of two dangling bonds per each V_2. Owing to lattice relaxation around V_2 with the pairing mode, the *TBD* 5-8 already emerges in the (110) plane at the stage of cluster formation. Ordering of multiple V_2 and 2$I_{[110]}$ within the nearest neighboring atomic chains constituting the {113} plane, as in the case of electron irradiation, provides for a decrease in the elastic energy of the crystal due to their close locations, and a possibility of subsequent recombination. G.D. Watkins, based on observations on nonreorientable V_2 during EPR studies [54], earlier suggested the hypothetical possibility of a single V_2 formation pinned with 2I. This idea was recently taken up once again in relation to the problem of DLTS identification of small clusters in Si irradiated with fast neutrons [55], but no direct evidence has been obtained so far.

The fact that the V_2-2I-{113} defect introduced by hot erbium ion implantation is stable at $T = 600°C$ suggests some gain in V_2-2I cluster configuration energy compared to a single V_2. There are currently no data regarding the formation energy of V_2-2I clusters in publications.

Preliminary ab initio calculations[4] for a single V_2-$2I$ cluster in a cell consisting of 64 atoms demonstrated that this cluster indeed has a slight gain in formation energy. Even during the interaction of the first I and V_2, the cluster formation energy decreases from 5.3 to 4.9 eV. However, first-principle calculations for the extended V_2-$2I$ structure are very complex, require multiscale parallel calculations, and have not yet been implemented. Therefore, the model of the extended V_2-$2I$-{113} defect obtained by us using molecular mechanics calculations is an important step on the way toward understanding the process of defect clustering in implanted layers. A comparison of experimental and simulated HREM images, as well as models of the extended V_2-$2I$-{113} defects, and single V_2-$2I$ clusters obtained by molecular mechanics and ab initio calculations, respectively, is presented in Fig. 16.11.

It can be seen that $I_{[110]}$ positions relative to V_2 in the extended defect and single cluster are similar; hence, the extended model provides good agreement of the simulated and experimental HREM images. A slight difference is observed in the elongation of atomic columns. Nevertheless, the map of simulated HREM images for the V_2-$2I$-{113} defect presented in Fig. 16.12 shows that they fit well with the experimental images of defects in Figs. 16.9–16.10 at defocus values from (−250) to (−550) Å.

Additionally, using the V_2-$2I$-{113} model for image simulations, we have analyzed HREM images of {113} defects observed at a thick part of a Si specimen implanted with erbium. One of these typical experimental images, in comparison with a simulated one, is presented in Fig. 16.13. As can be seen, the experimental image of a {113} defect in this case strongly differs from that obtained in a thin part of the Si specimen (see Figs. 16.9–16.10), but it is reproduced very well by simulation. Furthermore, this simulated HREM image fits the experimental image well of the {113} defect after Si^+ ion implantation presented in Ref. [56], which was ascribed an equilibrium interstitial-type {113} defect structure.

A good agreement of calculated and experimental HREM images in a wide range of defocus values and crystal thicknesses indicates that the V_2 $2I$ {113} structure is indeed stable under ion implantation up to $T = 600°C$, unlike electron irradiation at room temperature, when this defect reflects point defect recombination in the {113} plane under a small amount of excess *Is*.

We have also studied the subsequent transformation of the V_2-$2I$-{113} defect into the topological structure, and its HREM detectability through image simulation of a corresponding model (Fig. 16.14A–C). Fig. 16.14 also shows other types of *TBD* for comparison: one formed from continuous V_n-chains surrounded on two sides by self-interstitials (V_n-nI-{113} defect—Fig. 16.14D–F); and a classic interstitial-type {113} defect (I_s-{113} defect; Fig. 16.14G and H). Simulated HREM images of all types of *TBD* are shown for two crystal thicknesses: 77 and 192 Å, corresponding to the direct lattice image and its inversion. As it appears from Fig. 16.14A–C and 16.14D–F, both structures (V_2-$2I$-{113} and V_n-nI-{113})transform into *TBD* (5-6_h-7-8) after reorientation of all $I_{[110]}$ around the nearest neighboring *V*, but the

[4] Ab initio calculations were performed by I. Supryadkina under the supervision of K.K. Abgaryan and D.I. Bazhanov. This work was done within the framework of the joint project (OFI-m)-RFFI No.13-01-12110 between the Dorodnicyn Computing Centre of the Russian Academy of Sciences and the Rzhanov Institute of Semiconductor Physics (Siberian Branch of the Russian Academy of Sciences).

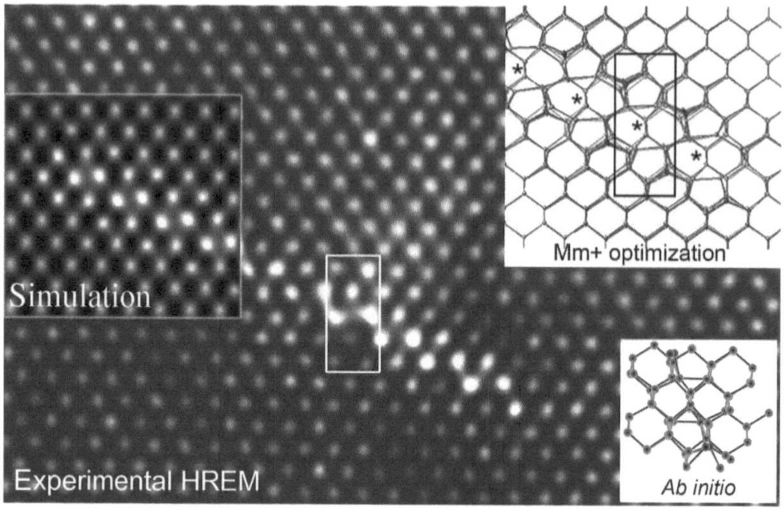

FIGURE 16.11 Experimental and simulated HREM (left insert) images of a V_2-2*I*-{113} defect. The rectangular frames show a translational cell of the {113} defect on the image and Mm + optimized model used for image simulation. The asterisks indicate V_2 positions and *red* (gray in print versions) balls are $I_{[110]}$. Lower right: model of a single V_2-2*I* cluster from first-principle calculations.

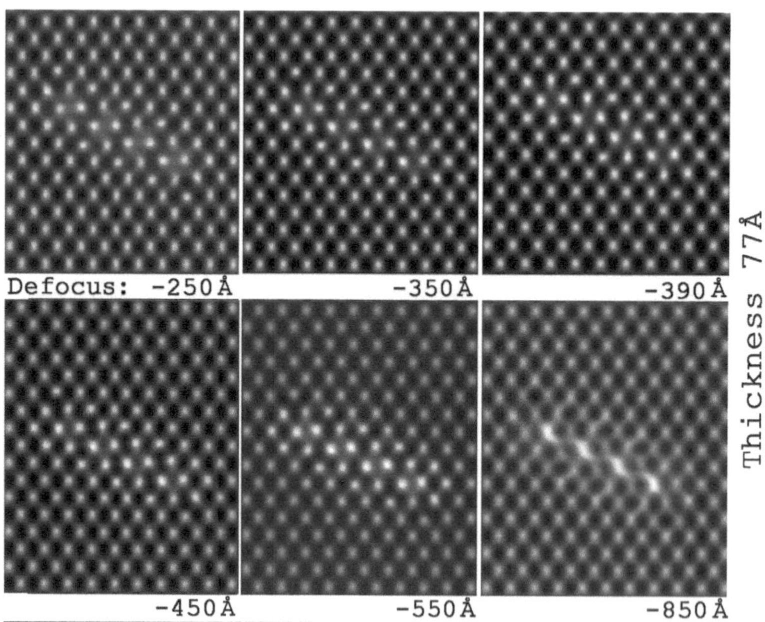

FIGURE 16.12 The map of calculated HREM images of the V_2-2*I*-{113} defect for the crystal thickness of 77 Å at various defocus values.

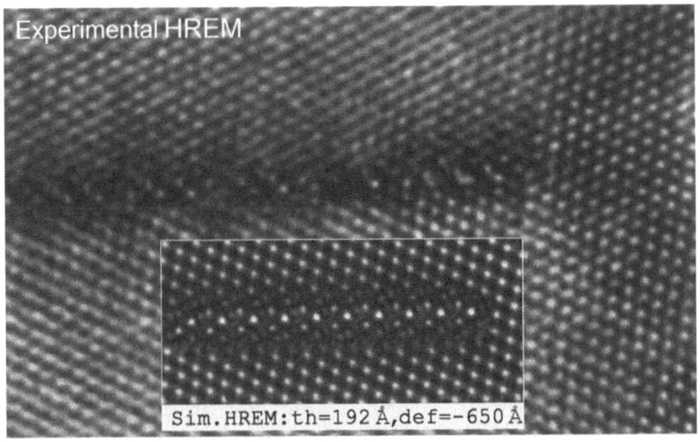

FIGURE 16.13 Experimental and simulated (insert) HREM images of the V_2-2*I*-{113} defect observed in a thick part of a specimen implanted with Er^+ ions. The thickness of the crystal is 192 Å and the defocus value is −650 Å.

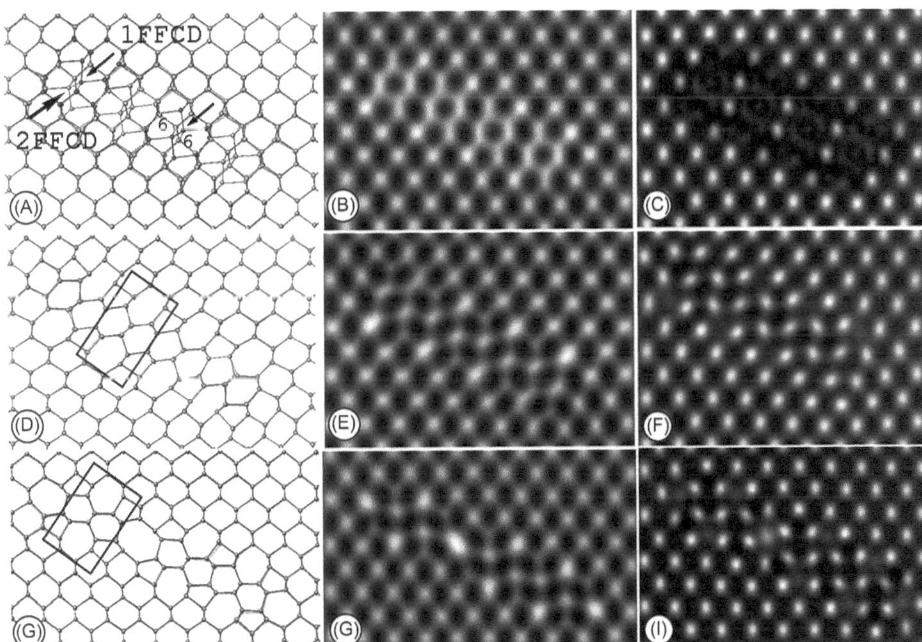

FIGURE 16.14 Models and simulated HREM images of TBD corresponding to the transformation of V_2-2*I*-{113} (A–C), V_n-*nI*-{113}, (D–F) structures, and of classical interstitial-type I_s-{113}-defects (G–I) for crystal thicknesses of 77 and 192 Å at a defocus value of −350 Å.

former is best visible in HREM only in a thick crystal. It can be seen from Fig. 16.14 that the simulated image of the I_s-{113} defect also possessing a (5-6_h-7-8) structure is practically indistinguishable from the image of a V_n-*nI*-{113} defect, although the former is an interstitial defect, and the latter is a zero-type corresponding to a recombination stage. As discussed earlier, the energy of both V_n-nI-{113} and I_s-{113} defects fall within the interval of 0.5−0.7 eV/atom [49,53]. This indicates that the type of implanted ion and local concentrations of *Is* and *Vs* play important roles in the formation of the {113} defect structure in implanted layers.

16.6 Conclusions

The data obtained demonstrate the universality of the {113} habit plane for mixed *Vs* and *Is* aggregation, irrespective of the method of their generation in Si. Since the ⟨110⟩ atomic chains represent the basic structural element of the Si lattice, a nearest neighbor pair of them becomes the place for multiple *V* and *I* aggregation to decrease crystal energy. Not only do the paired atomic chains ensure the second nearest neighbor distance between *V* and *I* necessary for close correlated *I-V* pair formation, but also the cross-linking of all *TBD* emerging by transformation of tightly ordered (*I-V*) pairs in the ⟨110⟩ and ⟨332⟩ directions constituting the {113} plane. Thus, the {113} plane structure in Si, and its step-chain nature with a great number of bonds per unit length, predetermines the possibility of close V (V_2) and $I_{[110]}$ location in nearest neighboring atomic chains that leads to the formation of a universal set of *TBD* 5-8, wherein all atoms retain their coordination and are characterized by low (0.7−1.0 eV) energy. The 5-8 array provides for subsequent crystal structure restoration or excessive *Is* incorporation to build up the 5-6_h-7-8 topological structure known as the {113} interstitial-type of defect. However, we have shown that the same 5-6_h-7-8 structure presenting a zero-type V_n-nI-{113} defect is formed from continuous V_n chains surrounded on two sides by self-interstitials. This indicates that the local concentration of *Vs* and *Is*, as well as irradiation conditions, play important roles for {113} defect structure formation. Such a complex mechanism of point defect aggregation in silicon is conditioned by a low symmetry of the primary point defects, resulting in decreased crystal energy for any mixed clustering of such defects in the {113} plane within a temperature range of $T < 0.5T_{\text{melting}}$. The results provide a solution for a long-standing problem in Si related to the prevailing formation of {113} defects during various process exposures, and are a basis for reconsidering some existing concepts in implantation physics.

Acknowledgment

We gratefully acknowledge partial financial support from RSF (project No 14-22-00143).

References

[1] H. Terrones, R. Lv, M. Terrones, M.S. Dresselhaus, The role of defects and doping in 2D graphene sheets and 1D nanoribbons, Rep. Prog. Phys. 75 (2012) 062501−062530.

[2] A.V. Krasheninnikov, K. Nordlund, Ion and electron irradiation-induced effects in nanostructured materials, J. Appl. Phys. 107 (2010) 071301−071370.

[3] J.H. Warner, Dislocation-driven deformations in graphene, Science 337 (2012) 209−212.

[4] O. Ori, F. Cataldo, M.V. Putz, Topological anisotropy of Stone-Wales waves in graphenic fragments, Int. J. Mol. Sci. 12 (2011) 7934–7949.

[5] R.M. Nieminen, Supercell methods for defect calculations, in: D.A. Drabold, S.K. Estreicher (Eds.), Theory of Defects in Semiconductors, Topics in Applied Physics, 104, Springer-Verlag, Berlin Heidelberg, 2007, pp. 29–68.

[6] S.A. Ghasemi, M. Amsler, R.G. Hennig, S. Roy, S. Goedecker, T.J. Lenosky, et al., Energy landscape of silicon systems and its description by force fields, tight binding schemes, density functional methods, and quantum Monte Carlo methods, Phys. Rev. B 81 (2010) 214107.

[7] A. Aseev, L. Fedina, D. Hoehl, H. Barsch, Clusters of Interstitial Atoms in Silicon and Germanium, Academy Verlag, Berlin, 1994.

[8] S.S. Kapur, T. Sinno, Detailed microscopic analysis of self-interstitial aggregation in silicon. I. Direct molecular dynamics simulations of aggregation, Phys. Rev. B 82 (2010) 045205.

[9] L. Fedina, A. Gutakovskii, A. Aseev, J. Van Landuyt, J. Vanhellemont, On the mechanism of {111}-defect formation in silicon studied by in situ electron irradiation in a high resolution electron microscope, Philos. Mag. A 77 (1998) 423–435.

[10] S. Takeda, M. Kohyama, K. Ibe, Interstitial defects on (113) in Si and Ge line defect configuration incorporated with a self-interstitial chain, Philos. Mag. A 70 (1994) 287–312.

[11] S. Libertino, A. La Magna, Damage formation and evolution in ion-implanted crystalline Si, in: H. Bernas (Ed.), Materials Science With Ion Beams. Topics Appl. Physics, vol. 116, 2010, pp. 147–212.

[12] L. Fedina, J. Van Landuyt, J. Vanhellemont, A. Aseev, Observation of vacancy clustering in Si crystals during in situ electron irradiation in a high voltage electron microscope, in: Materials Research Society Symposium Proceedings. Wuhan, 1996, pp. 189–194.L. Fedina, J. Van Landuyt, et al., Observation of vacancy clustering in FZ-Si during in situ electron irradiation in a high voltage electron microscope, Nucl. Instrum. Methods Phys. Res. B 112 (1996) 133–138.

[13] L. Fedina, A. Gutakovskii, A. Aseev, J. Van Landuyt, J. Vanhellemont, Intrinsic point defect clustering in Si: a study by HVEM and HREM in situ electron irradiation, in: P.L. Gi (Ed.), In-Situ Microscopy in Materials Research: Leading International Research in Electron and Scanning Probe Microscopies, Springer Science & Business Media, New York, NY, 1997, pp. 63–92.

[14] I. Fedina, S.A. Song, A.L. Chuvilin, A.K. Gutakovskii, A.V. Latyshev, On the mechanism of {113}-defect formation in Si, in: A.G. Cullis, J.L. Hutchison (Eds.), Microscopy of Semiconducting Materials, Proceedings of the 14th Conference, April 11-14 2005, vol. 107, A.G. Springer Proceedings in Physics, Oxford, 2005, pp. 359–362.

[15] L.I. Fedina, S.A. Song, A.L. Chuvilin, A.K. Gutakovskii, A.V. Latyshev, The Mechanism of {113} defect formation in silicon: clustering of interstitial–vacancy pairs studied by in situ high-resolution electron microscope irradiation, Microsc. Microanal. 19 (2013) 1–5.

[16] L. Fedina, F. Gutakovskii, A. Aseev, J. Van Landuyt, J. Vanhellemont, Extended defects formation in Si crystals by clustering of intrinsic point defects studied by in-situ electron irradiation in an HREM, Phys. Stat. Sol. (a) 171 (1999) 147–157.

[17] L. Fedina, N. Zakharov, I. Supryadkina, N. Sobolev, A.V. Latyshev, Ordered aggregation of (V_2-2I) clusters in the {113} plane in Si implanted with 2 MeV of hot Er^+ ions, prepared for submission to Appl. Phys. Lett.

[18] A. Chuvilin, U. Kaiser, On the peculiarities of CBED pattern formation revealed by multislice simulation, Ultramicroscopy 104 (2005) 73–82.

[19] L. Fedina, A. Aseev, Study of point defects interaction with dislocations in silicon by means of irradiation in an electron microscope, Phys. Stat. Sol. (a) 95 (1986) 517–529.

[20] L. Fedina, O. Lebedev, G. Van Tendeloo, J. Van Landuyt, O. Mironov, E. Parker, In situ HREM irradiation study of point defect clustering in in MBE-grown strained Ge_xSi_{1-x}/(001)Si heterostructure, Phys. Rev. B 61 (2000) 10336–10345.

[21] L. Fedina, Recombination of point defects and their interaction with the surface in the course of clusterization of these defects in Si, Semiconductors 35 (2001) 1072–1080.

[22] L.I. Fedina, On the roles of bulk recombination of intrinsic point defects in Si and their interaction with the surface in the course of {113} defect growth: a kinetic study by in situ HVEM irradiation, in: D.J. Fisher (Ed.), Defects and Diffusion in Semiconductors, vol. 210–212, Trans Tech Publications, Zurich, 2002, pp. 21–36.

[23] R. Oshima, G.C. Hua, Characterization of Czochralski Si crystals by HVEM, Ultramicroscopy 39 (1991) 160–170.

[24] A. Parisini, A. Bourret, Diamond hexagonal silicon and {113} defects. Energy calculation and new defect models, Philos. Mag. A 67 (1993) 605–630.

[25] M. Kohyama, S. Takeda, Tight-binding study of the {113} planar interstitial defects in Si, Phys. Rev. B 51 (1995) 13111.

[26] D.J. Eaglesham, P.A. Stolk, H.-J. Gossmann, J.M. Poate, Implantation and transient B diffusion in Si: the source of the interstitials, Appl. Phys. Lett. 65 (1994) 2305.

[27] A. Agarwal, T.E. Haynes, D.J. Eaglesham, H.-J. Gossmann, D.C. Jacobson, J.M. Poate, et al., Interstitial defects in silicon from 1–5 keV Si^+ ion implantation, Appl. Phys. Lett. 70 (1997) 3332–3339.

[28] G.Z. Pan, R.P. Ostroumov, Y.G. Lian, K.N. Tu, K.L. Wang, {113} Defect-engineered silicon light-emitting diodes, Electron Devices Meeting, 2004, IEDM Technical Digest. IEEE International, 2004, pp. 343–346.

[29] S. Lee, G.S. Hwang, Structure and stability of small compact self-interstitial clusters in crystalline silicon, Phys. Rev. B 77 (2008) 085210.

[30] F. Schiettekatte, S. Roorda, R. Poirier, M.O. Fortin, S. Chazal, R. Héliou, Direct evidence for 8-interstitial-controlled nucleation of extended defects in c-Si, Appl. Phys. Lett. 77 (2000) 4322–4324.

[31] M. Jaraiz, G.H. Gilmer, J.M. Poate, T.D. de la Rubia, Atomistic calculations of ion implantation in Si: point defect and transient enhanced diffusion phenomena, Appl. Phys. Lett. 68 (1996) 409–411.

[32] O.K. Al-Mushadani, R.J. Needs, Free-energy calculations of intrinsic point defects in silicon, Phys. Rev. B 68 (2003) 235205.

[33] S. Goedecker, T. Deutsch, L. Billard, A fourfold coordinated point defect in silicon, Phys. Rev. Lett. 88 (2002) 235501–235505.

[34] M. Tang, L. Colombo, J. Zhu, T. Diaz de la Rubia, Intrinsic point defects in crystalline silicon: tight-binding molecular dynamics studies of self-diffusion, interstitial-vacancy recombination, and formation volumes, Phys. Rev. B 55 (1997) 14279.

[35] F. Gargoni, C. Gatti, L. Colombo, Formation and annihilation of a bond defect in silicon: an ab initio quantum-mechanical characterization, Phys. Rev. B 57 (1998) 170.

[36] L.A. Marqués, L. Pelaz, J. Hernández, J. Barbolla, G.H. Gilmer, Stability of defects in crystalline silicon and their role in amorphization, Phys. Rev. B 64 (2001) 045214.

[37] M.I. Heggie, Self-diffusion, deformation and melting in silicon—microscopic link, Philos. Mag. Lett. 58 (1988) 75–80.

[38] D.M. Stock, B. Weber, K. Gärtner, Role of the bond defect for structural transformations between crystalline and amorphous silicon: a molecular-dynamics study, Phys. Rev. B 61 (2000) 8150.

[39] Y.A. Du, R.G. Hennig, T.J. Lenosky, J.W. Wilkins, From compact point defects to extended structures in silicon, Eur. Phys. J. B 57 (2007) 229–234.

[40] L.A. Marqués, L. Pelaz, I. Santos, P. López, M. Aboy, Structural transformations from point to extended defects in silicon: a molecular dynamics study, Phys. Rev. B 78 (2008) 193201.

[41] H. Park, J.W. Wilkins, A topological point defect regulates the evolution of extended defects in irradiated silicon, Appl. Phys. Lett. 98 (2011) 171915–171919.

[42] L.A. Marqués, M. Aboy, K.J. Dudeck, G.A. Botton, A.P. Knights, R.M. Gwilliam, Modeling and experimental characterization of stepped and v-shaped {311} defects in silicon, J. Appl. Phys. 115 (2014) 143514–143533.

[43] L. Fedina, A.L. Aseev, S.G. Denisenko, L.S. Smirnov, Electron microscopy data for threshold energy of point defect creation in silicon, in: H.J. Bardeleben (Ed.), Defects in Semiconductors, Trans Tech Publications Ltd, Aedermannsdorf, 1986, pp. 1123–1128.

[44] M. Gharaibeh, S.K. Estreicher, P.A. Fedders, P. Ordejón, Self-interstitial–hydrogen complexes in Si, Phys. Rev. B 64 (2001) 235211.

[45] G. Lulli, E. Albertazzi, S. Balboni, L. Colombo, Defect-induced homogeneous amorphization of silicon: the role of defect structure and population, J. Phys. Condens. Matter 18 (2006) 2077–2088.

[46] J.P. Goss, T.A.G. Eberlein, R. Jones, N. Pinho, A.T. Blumenau, T. Frauenheim, et al., Planar interstitial aggregates in Si, J. Phys. Condens. Matter 14 (2002) 12843–12853.

[47] G.D. Watkins, Vacancies and interstitials and their interactions with impurities in c-Si, Phys. B 376–377 (2006) 50–53.

[48] J. Iwata, K. Shiraishi, A. Oshiyama, Large-scale density-functional calculations on silicon divacancies, Phys. Rev. B 77 (2008) 115208.

[49] J.P. Goss, P.R. Briddon, R. Jones, Calculated properties of a {113} planar vacancy aggregate in Si, J. Phys. Condens. Matter 16 (2004) 3311–3318.

[50] L. Fedina, A. Gutakovskii, A. Aseev, In situ HREM irradiation study of an intrinsic point defects clustering in FZ-Si, Cryst. Res. Technol. 35 (2000) 775–786.

[51] M.I.J. Probert, M.C. Payne, Improving the convergence of defect calculations in supercells: an ab initio study of the neutral silicon vacancy, Phys. Rev. B 67 (2003) 075204.

[52] A. Bongiorno, L. Colombo, T. Diaz de la Rubia, Structural and binding properties of vacancy clusters in silicon, Europhys. Lett. 43 (1998) 695–700.

[53] J.P. Goss, P.R. Briddon, R. Jones, Calculated properties of a {113} planar vacancy aggregate in Si, J. Phys. Condens. Matter 14 (2002) 12843–12853.

[54] G.D. Watkins, in: F.L. Vook (Ed.), Radiation Effects in Semiconductors, Plenum Press, Inc., New York, NY, 1968, p. 67 cited by I. Arimura, R.R. Freeman, Rapid annealing in *n*-type silicon following pulsed 10 MeV electron irradiation, J. Appl. Phys. 40 (1969) 2570.

[55] R.M. Fleming, C.H. Seager, E. Bielejec, G. Vizkelethy, D.V. Lang, J.M. Campbell, Defect annealing in neutron and ion damaged silicon: influence of defect clusters and doping, J. Appl. Phys. 107 (2010) 053712.

[56] P.A. Stolk, H.-J. Gossmann, D.J. Eaglesham, D.C. Jacobson, C.S. Rafferty, G.H. Gilmer, et al., Physical mechanisms of transient enhanced dopant diffusion in ion-implanted silicon, J. Appl. Phys. 81 (1997) 6031.

17

Silicon-on-Insulator Structures Produced by Ion-Beam Synthesis and Hydrogen Transfer

I.E. Tyschenko, V.P. Popov

RZHANOV INSTITUTE OF SEMICONDUCTOR PHYSICS OF THE SIBERIAN BRANCH OF THE RUSSIAN ACADEMY OF SCIENCES, NOVOSIBIRSK, RUSSIA

17.1 Introduction

Silicon-on-insulator (SOI) structures are an alternative to silicon substrates, and consist of a thin monocrystalline silicon layer separated from a silicon substrate by a dielectric layer, which is usually silicon dioxide. The advantages of SOI structures, compared with the bulk silicon wafer primarily due to the isolation from a silicon substrate, include complete isolation of individual integrated circuit elements, resistance to latch-up, etc. In the first place, the transition to SOI structures was considered as a way of increasing radiation resistance of devices made for military and aerospace applications. However, in recent decades the commercial use of SOI structures has grown exponentially. They have become the basis for the creation of super-large integrated circuits.

In the early stages of manufacturing of SOI structures, the main problem was associated with the formation of a top silicon layer. Its quality would not yield the quality of bulk silicon. It had to have a single crystal structure and a dislocation density not higher than $10^4\ cm^{-2}$. The second problem was to achieve a high-quality interface "silicon-insulator." From a number of proposed methods for creating SOI structures, only two have reached industrial application: ion-beam synthesis (industrial technology SIMOX) and hydrogen transfer of silicon layers (industrial technology by Smart-Cut).

17.2 Ion-Beam Synthesis of SOI Structures

Ion-beam synthesis is a method in which the formation of new phases occurs in layers irradiated by ions. In 1961, Guseva and Alexandria observed for the first time the formation of new phases in ion-implanted layers at ion doses that provide the stoichiometric composition of the synthesized compounds [1]. Then, a number of investigators demonstrated the possibility of ion-beam synthesis of different materials. In the 1980s, Reeson and Hemment

Advances in Semiconductor Nanostructures. DOI: http://dx.doi.org/10.1016/B978-0-12-810512-2.00017-2

prepared buried dielectric layers in silicon using this method [2]. Under the conditions of ion-beam synthesis of SOI structures, formation of the dielectric phase occurs as a result of implantation of high doses of chemically active ions (O^+ and/or N^+) in the monocrystalline Si matrix. The ion dose is about 10^{18} cm^{-2}, that is several orders of magnitude higher than doses typically used in silicon technology for local doping. The concentration of chemically active atoms in the ion distribution corresponding to the stoichiometric composition of SiO_2 or Si_3N_4 was the criterion for the ion dose option. Since the early 1980s, ion-beam synthesis of buried dielectric layers has been actively studied. At this time, the developmental success of the method was generally predetermined by the progressing accelerator technique for industrial applications. The creation of high-current sources providing implantation of chemically active ions was required. The NV-200 accelerator was produced by the Eaton Company (USA) at the request of the Japanese company NTT. It was intended for the production of SOI wafers by the SIMOX method (Separation by Implanted Oxygen). The ion current density during implantation reached 100 $\mu A/cm^2$, and the target temperature reached $\geq 600°C$ in several seconds.

In the mid-1980s, scientific research in the field of ion-beam synthesis was determined by available technical ability. High-dose implantation with chemically active ions accompanied by strong heating of the target stimulates the formation of a new phase during irradiation. This leads to a change in the stopping power of the ions during implantation, because of the modification of target composition. At that, substrate swelling may also occur. Long-time ion bombardment was also followed by silicon target sputtering. Therefore, solving the problems connected with the laws of atom penetration and synthesized phase distribution during high-temperature high-dose ion implantation, were the key tasks in ion-beam synthesis of SOI structures. The formation of continuous dielectric layers was considered to be a result of the implantation of stoichiometric and over-stoichiometric N^+ or O^+ ion doses. On the other hand, the initial stage of dielectric phase nucleation and its dependence on competing sinks in the silicon substrate could only be investigated by implantation of chemically active ions at a substoichiometric dose. The solution to these problems was very important from a practical point of view, namely, to reduce radiation-thermal stress on silicon during ion-beam synthesis of dielectric layers, as well as to produce thin dielectric layers.

In the mid-1980s, the vast majority of experiments on ion-beam synthesis were associated with SIMOX technology. Only a small number of studies were devoted to the synthesis of silicon nitride. On the other hand, silicon nitride films are now playing an increasingly important role in the transition to nanometer dimensions in electronics. In the Institute of Semiconductor Physics, Siberian Branch of the Russian Academy of Science (ISP SB RAS), studies of the initial stage of Si_3N_4 phase nucleation during ion-beam synthesis of buried dielectric layers were started by Kachurin et al. [3–5]. The technical resource for these studies was provided by ion accelerator "Vesuvius-5," developed at the S.A. Vekshinsky Research Institute for Vacuum Techniques (Moscow). The accelerator design allowed placing a furnace into the take-up chamber to heat specimens to high temperatures during implantation, irrespective of the ion-beam power. This allowed them to determine the mechanism of phase nucleation during ion-beam synthesis of silicon nitride, and to define the role of nonequilibrium point defects in this process. The data obtained allowed the formation of continuous buried Si_3N_4 layers in silicon at N^+ ion doses below those that corresponded to the stoichiometric nitrogen concentration. High top silicon film perfection was achieved, too.

The features of nitrogen profiles in silicon as a function of the implantation temperature were investigated in early experiments [3–5]. Nitrogen ions to an energy of 130 or 140 keV with doses of 5×10^{16}–5×10^{17} cm^{-2} were implanted in a single-crystal silicon target at implantation temperatures of $T_i = 700$–$1100°C$. The ion current density was about 1–5 mA/cm^2. The use of independent heating of samples during implantation, and small ion current densities allowed the suppression of nonstationary silicon target heating due to ion-beam power. It was found that there was a critical implantation temperature (about 900°C) above which nitrogen drains from thc buricd layer to the silicon surface (Fig. 17.1A).

The implantation temperature affects nitrogen penetration in silicon during subsequent high-temperature annealing, too. Annealing of the samples implanted at relatively low temperatures ($T_i < 900°C$) at 1000–1100°C accompanied by flow of nitrogen in the implantation area formed a Gaussian-like profile that had narrower straggling and greater impurity concentration at maximum distribution compared to those from the original profile (Fig. 17.1B). After subsequent annealing, the lower the implantation temperature is, the higher the nitrogen peak concentration at a depth of the projective ion range (R_p) is. As the implantation temperature increases to 900°C, the nitrogen flow to thc silicon surface becomes dominant. The implantation temperature of 900°C, at which nitrogen redistributes

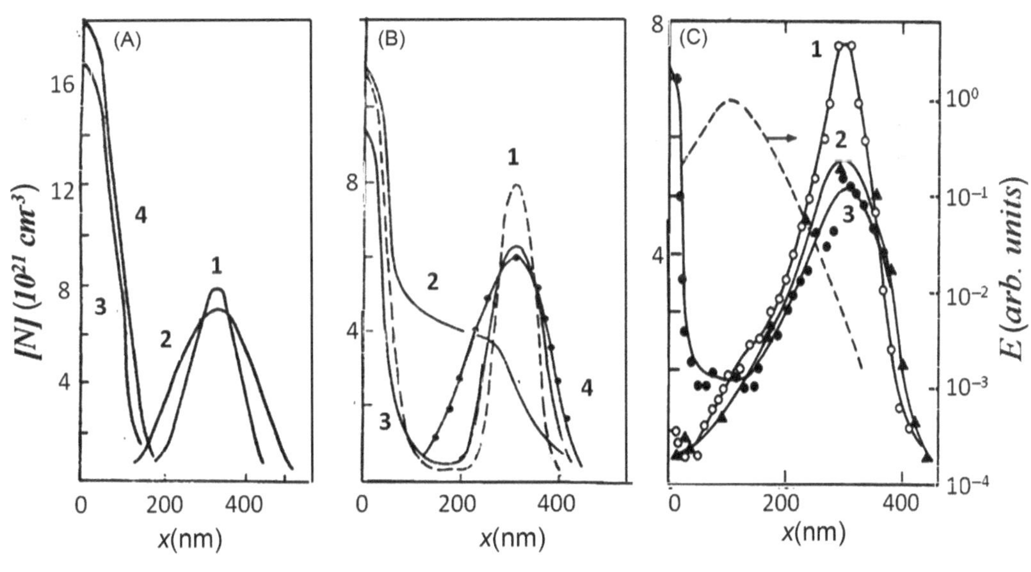

FIGURE 17.1 Nitrogen depth distribution in silicon after N^+ implantation to dose ~1.5×10^{17}cm^{-2} at (A) T_i: (1) 700°C; (2) 900°C; (3) 1000°C; (4) 1100°C; (B) T_i: (1) <300°C; (2) 700°C; (3) 900°C after subsequent annealing at $T_a = 1150°C$ for 3 h in vacuum; (4) <300°C without subsequent annealing. (C) Nitrogen depth distribution in silicon preimplanted with Ar^+ at an energy of 145 keV to dose 10^{16} cm^{-2} at $T_i = 800°C$. N^+ ion implantation temperature T_i: (1) 900°C, (2) 1000°C, (3) 1100°C. The *dashed line* shows the distribution of elastic losses for Ar^+.

from a depth of R_p to the surface, agrees with a critical implantation temperature for introduction of residual structural damage (full dislocation loops and dipoles) by light ions [6]. The observed effect suggests that secondary radiation defects in silicon are sinks for nitrogen in the buried layer.

To confirm the role of secondary defects as a sink for nitrogen, profiles of nitrogen atoms in silicon preimplanted with Ar^+ ions were investigated as a function of the implantation temperature [5]. First, the samples were implanted with Ar^+ ions at an energy of 145 keV to a dose of $10^{16}\ cm^{-2}$ and at implantation temperature $T_i = 800°C$. Under these conditions, nucleation of Frank dislocation loops in the displacement spikes of Ar^+ ions takes place. Their depth-distribution is similar to that of the elastic energy losses of Ar^+ ions (*dashed curve* in Fig. 17.1C). The implantation of N^+ ions was carried out using the modes described earlier. Fig. 17.1C shows the profiles of nitrogen subsequently implanted at temperatures of 900°C, 1000°C, and 1100°C obtained by Auger spectroscopy. The profiles of nitrogen ion-implanted at $T_i = 900°C$ have a Gauss-like shape. No diffusive spreading of profiles is observed. It suggests the presence of nitrogen capture centers in the ion-implanted region. At that, the distance between the centers should be less than the diffusion length of nitrogen atoms. If competing sinks are not present in the silicon bulk, the redistribution of nitrogen to the silicon surface becomes preferential (curve 3, Fig. 17.1C).

In order to discover whether the above-mentioned competition between sinks can result in the experimentally observed nitrogen profiles during high-temperature ion implantation, a numerical simulation was made [7]. The computation was designed to display the following: generation of point defects and the appearance of dislocation loops under the influence of incident ions; the transfer of nitrogen from the incident ion beam to the silicon crystal; the diffusion of point defects and nitrogen atoms in silicon; the growth of the dislocation loop size as a result of trapping excess mobile point defects; shrinking of the loops by the evaporation of atoms from their edges; and condensation of nitrogen on the loops whose size and concentration were functions of time and depth during ion bombardment. The main parameters of nitrogen atoms and point defects were taken from famous literary sources.

The calculations showed three principal places in the silicon bulk for nitrogen precipitation: (1) the region of maximal supersaturation with nitrogen atoms corresponding to the N-ion projective range (R_p^N); (2) the elastic energy loss peak of the nitrogen ions (R_d^N); and (3) the region of the dislocation loops location at the same depth position as the elastic energy loss peak of the argon ions (R_d^{Ar}). These three factors form the multimodal distribution of the nitrogen atoms. It was shown that the increasing radius of the dislocation loops with subsequent implantation of high dose of nitrogen ions is not at the depth of the maximum defect concentration R_d^{Ar}, but at the depth of the nitrogen projective range R_p^N (Fig. 17.2).

Further investigation showed that the defects are not only sinks for nitrogen atoms, but also the centers of Si_3N_4 phase nucleation. This means that the achievement of stoichiometric nitrogen concentration in the implanted layer is not a prerequisite for the formation of stoichiometric Si_3N_4. This was confirmed by the spin-dependent recombination (SDR) measurements [8]. For lower doses ($<10^{16}\ cm^{-2}$) or lower implantation temperatures ($<750°C$), a single isotropic line ($g = 2.0035 \pm 0.0005$) was observed. Just such a SDR signal

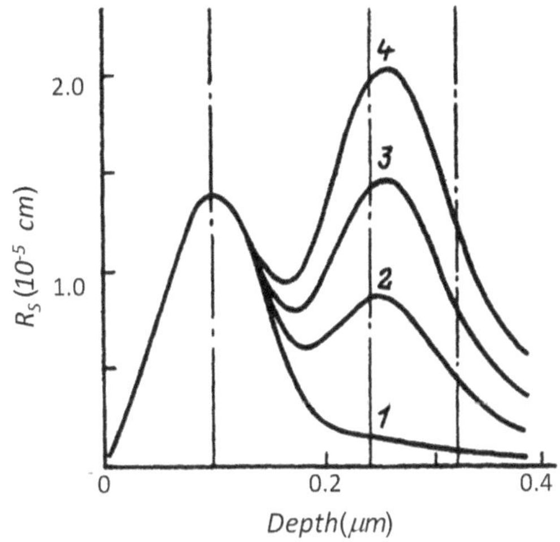

FIGURE 17.2 The radius of dislocation loops formed in silicon by Ar^+ ion implantation at an energy of 135 keV to dose 10^{14} cm^{-2} at 800°C as a depth function after N^+ ion implantation at energy 135 keV to dose: (1) 10^{16}, (2) 10^{17}, (3) 2×10^{17}, and (4) 3×10^{17} cm^2 at 1000°C. The dotted-dashed lines represent R_d^{Ar}, R_d^N, and R_p^N.

is usually detected in silicon with dislocations. As both nitrogen ion doses increased to $\geq 5 \times 10^{16}$ cm^{-2}, and the implantation temperature rose above 800°C, an intense anisotropic spectrum was observed. It corresponds to a defect with electronic spin $S = 1/2$ and the ⟨111⟩ axially symmetric g-tensor with the principal values of $g_{||} = 2.0010 \pm 0.0005$ and $g_{\perp} = 2.0086 \pm 0.0005$. These g-factor parameters coincide with those of the defects at the Si/SiO_2 (P_{bO} center), or Si/Si_3N_4 (P_{bN} center) interfaces. Observation of P_{bN} centers in the layers implanted with high-dose nitrogen ions at high temperatures indicates the formation of stoichiometric Si_3N_4 inclusions within the amorphous silicon nitride phase. Note, the indicated dose of 5×10^{16} cm^{-2} is two orders of magnitude lower than that corresponding to the stoichiometric composition of Si_3N_4.

On the base of the results above, the ~0.2 μm thick continuous buried Si_3N_4 layers were synthesized by implantation of the substoichiometric nitrogen ion dose. First, in order to form the sinks for subsequently implanted nitrogen, preimplantation of N^+ ions at an energy of 135 keV to a dose of 1×10^{17} cm^{-2} was carried out in a silicon target heated to a temperature of 600°C. The next portion of nitrogen ions (about 7×10^{17} N^+/cm^2) was implanted at a temperature of 1000°C. The implantation conditions used allowed both the accumulation of nitrogen inside the silicon substrate at the depth of the ion range, and the prevention of dislocation generation in the top silicon on the insulator. Fig. 17.3 shows this structure, with its comparison with the structure performed by the implantation of full ion dose at a temperature of 850°C.

Thus, the displayed heterogeneous mechanism of phase nucleation during ion-beam synthesis of silicon nitride in silicon was of great practical importance. First, it provided the synthesis of buried dielectric layers at ion doses several times lower than those usually used to achieve stoichiometric concentration of nitrogen in silicon nitride. Second, reducing the

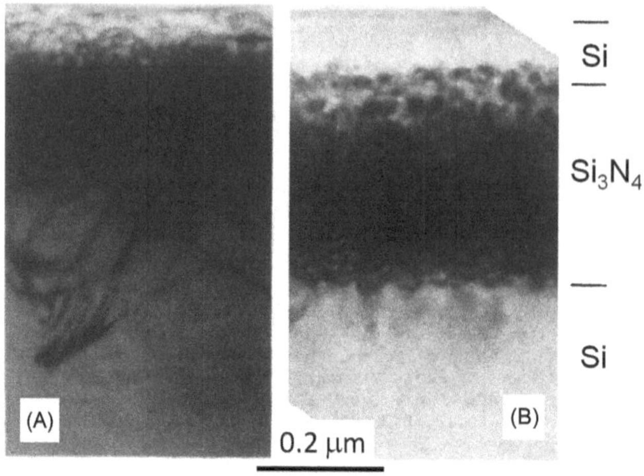

FIGURE 17.3 A cross-section electron microscope image of the SOI structures formed by N^+ ion implantation in silicon at a full dose of about 8×10^{17} cm^{-2}: (A) $T_i = 850°C$, (B) $1 \times 10^{17} N^+/cm^2$ at $T_i = 600°C$, and then 7×10^{17} N^+/cm^2 at $T_i = 1000°C$.

dislocation density in the top silicon layers can be achieved at a 200–300°C lower temperature in comparison with that traditionally used. Third, there is a possibility of forming thin dielectric layers at the required depth.

The experimentally observed disagreements between the structural properties of ion-beam synthesized silicon nitride and those from Si_3N_4 prepared by gaseous-phase methods can also be explained in terms of the interaction of nitrogen with generated defects. The structure transformations in amorphous silicon nitride during its crystallization were studied elsewhere by Edelman et al. [9–11]. It was found that β-Si_3N_4 is more energetically favorable compared to α-Si_3N_4. β-Si_3N_4 was formed as a result of heating powders of amorphous silicon nitride at a temperature of ≥1450°C. α-Si_3N_4 is usually observed as a result of epitaxial crystallization of amorphous silicon nitride films deposited on a monocrystalline silicon substrate at a temperature of about 1200°C. However, in the case of ion-beam synthesis, both the α- and β-Si_3N_4 phases are observed simultaneously just after high-temperature ion implantation of N^+ ions. At that, there was enhanced crystallization of nitride at low temperatures. The β-nitride phase was detected during implantation of N^+ ions at a dose rate of 50–200 $\mu A/cm^2$ that provided local heating of the target to a temperature of 650–800°C. Subsequent annealing of these layers was accompanied by the formation of a mixture of the two phases. α-Si_3N_4 was also obtained after high-temperature annealing of silicon samples implanted with a high dose of N^+ ions at a low current density.

There has been a lot of speculation about the role of dose rate in ion-beam synthesis of β-Si_3N_4. For example, it was suggested that some profound dynamic effect favors the nucleation of the β-phase. The thermal peaks of N^+ ions that caused local melting were also proposed as a possible reason for the stimulated nucleation of β-Si_3N_4. Ionization resulting in enhanced diffusion was also assumed to explain the effect of dose rate on nitride crystallization. However, these suggestions did not explain all results. It should be noted that in the literature, β-Si_3N_4 was usually observed after implantation of N^+ ions under conditions of nonstationary heating of the silicon target by ion-beam power. In our works [12,13], in order to detect the crystallization mechanism of silicon nitride as a function of the current density, heating the samples was carried out independently from the ion-beam power. In the experiments, the N^+ current beam density was varied from 3 to 100 $\mu A/cm^2$, and the target temperature was changed over a range of 600−1100°C. Under these conditions, the formation of amorphous silicon nitride occurred at an implantation temperature of $\leq$700°C, independent from the current density. Crystallization of silicon nitride took place under implantation temperatures above 700°C. At that temperature, the structure of crystalline Si_3N_4 was dependent on the ion dose rate. The low current density (very few $\mu A/cm^2$) caused the formation of α-Si_3N_4 precipitates, and provided growth of single-crystal matrix-oriented buried α-Si_3N_4 layers during subsequent annealing. As the implantation temperature rose to 900°C, full implanted nitrogen was included in the crystalline α-Si_3N_4.

High dose rate implantation led, in the early stage, to the formation of β-Si_3N_4 precipitates which enlarged with increasing dose and combined with the β-Si_3N_4. The formation of a continuous β-Si_3N_4 layer with a thickness of about 0.25 μm was observed after implantation of a nitrogen dose of 10^{18} cm^{-2}. It formed a block structure composed of random crystallites with a mean size of about 2−3 μm. It was assumed that the role of implantation temperature consists in providing sufficient mobility to the nitrogen atoms and suppression of defect accumulation in silicon. An increase in the dose rate favors the accumulation of defects which serve as nucleation sites for nitride, and enhances the supply rate of nitrogen atoms to the growing precipitates. The rise in current density with the implantation temperature remaining constant means an increase in the defect generation rate, resulting in a more damaged Si lattice with many more sites for nitride nucleation. On the other hand, the high current density provides a diffusion flux of nitrogen atoms to the nuclei in excess of the Si lattice's ability to govern Si_3N_4 growth. As a result, the energetically more favorable β-Si_3N_4 forms.

The results obtained allow one to suppress silicon nitride polycrystallinity during high-temperature treatments, and to synthesize single-crystal silicon nitride with an electrical strength of about 10^6 V/cm.

17.3 Wafer Bonding and Hydrogen Transfer

In spite of the great achievement in SIMOX development, a part of the structures produced with this technology took up about 10% of the world market for SOI. Long-time heating at the limit temperatures and low recovery because of wafer warping are the basic reasons for very expensive SIMOX substrates. Other reasons for the limited application of SIMOX in microelectronics are the high degree of roughness of the Si/SiO_2 interface (>1 nm), high dislocation density in the top silicon layer (about 10^5 cm^{-2}), and high nonuniformity of the buried dielectric layer. Wafer bonding and hydrogen transfer of the layers are methods free of the shortcomings inherent in SIMOX technology.

Wafer bonding is more suitable to grow together not only silicon wafers, but also those materials with unfavorable compositions and topography [14]. In this method, one of two bonded wafers is thinned to several microns by mechanical polishing. This method was named BESOI (Bonded and back side silicon-on-insulator). The thick limit of the bonded wafers and the high roughness of the work layer are the main disadvantages of these structures. Such wafers with a micrometer-thick silicon layer on the insulator are commonly used in micromechanical applications.

Another method of producing SOI wafers is using a porous Si material. First, the epitaxial silicon layer is grown on porous silicon and this is followed by oxidation. The oxidized wafer is bonded to the silicon substrate, and then splitting the layers is carried out within porous silicon by means of a water jet. At the final stage, in order to remove surface roughness, the prepared structures are heated in ambient hydrogen. This technology was named *ELTRAN* (Epitaxial Layer Transfer). This allows the creation of SOI structures with a 10-nm thick top silicon layer, and with an atomically smooth silicon surface. *ELTRAN* technology was successfully used by Cannon Inc., Japan. However, since 2007, the production of the wafers has been ended due to their high price.

The silicon layer hydrogen transfer method was proposed in the early 1990s by Bruel from the Scientific Centre LETI, Grenoble, and was widely recognized [15]. The technology developed using this method was named Smart-Cut. A phenomenon of radiation blistering on surfaces implanted with high doses of hydrogen, deuterium, or helium ions discovered by Primak and Kaminsky [16,17] is the basis of this method. The first investigation of the blistering effect in early 1970 were provided for the creation of thermonuclear reactors. Blistering was shown as one of the first destructive effects to the wall of fusion reactors. The problem consists in the exfoliation of the surface layer of the target under increasing ion dose; this effect was used by Bruel.

Hydrogen bubble formation in silicon is connected with special features occurring in the interaction of hydrogen with radiation defects. The hydrogen atoms can form Si–H bonds in silicon as a result of both their interaction with the dangling silicon bonds, and their incorporation in relaxed Si–Si bonds. These Si–H bonds are stable up to 350°C. In this case, the formation of H_2^* or H_n^* complexes takes place. It is energetically favorable because the full energy of the H_2^* complex is 1.2 eV lower than that of two separated neutral hydrogen atoms. H_n^* complexes are observed as an extended planar defect in silicon (the so-called platelet defect). They are formed as a result of the aggregation of H_2^* complexes. Annealing the element defects results both in the reconstruction of dangling bonds, and in hydrogen leaving the bound state. In a perfect silicon lattice, the H_2^* configuration corresponding to the tetrahedral interstitial position is more stable than the separated monatomic hydrogen radicals. As the hydrogen concentration in a platelet increases, H_2 molecules are formed. Thereby, hydrogen accumulation and the force of H_2 molecules favor platelet defect transformation in hydrogen gas bubbles.

An increase in gas pressure within the platelet results in its expansion. Initially, the platelets grow in the lateral direction. However, this process is limited by van der Waals forces which act between two hydrogen-covered surfaces of a platelet defect. Lateral growth of the platelets occurs until the defect radius reaches its critical value r_{crit}. At that, the gas pressure is higher than the surface energy γ_p. It is the starting point of anisotropic growth of a hydrogen bubble.

From an experimental point of view, it is very important to determinate the condition where the height h of a hydrogen bubble reaches its maximum. As the bubble height becomes equal to its radius r_0, the silicon layer above the bubble cracks. It happens because of a warp in the silicon layer due to an increase in bending energy, compared to the elastic silicon energy. The bubble radius, as a function of annealing temperature T, annealing time t, and hydrogen ion dose ϕ_0 can be expressed as follows [18]:

$$r(t, T, \phi_0) = 3\pi kTE\phi_0[1 - \exp(-\vartheta t)]/16P^2(1 - \nu^2) \qquad (17.1)$$

where E is Young's modulus, $\vartheta = \tau_r^{-1}\exp(-(\mathrm{E}_a)/kT)$ is the generation rate of H_2 molecules, $\tau_r \sim 10^{-12}$ is the frequency ratio, ν is the Poisson ratio, P is the gas pressure inside the platelets, and E_a is the activation energy of expected time for the bubble's appearance. It can be seen from Eq. (17.1) that there is a hydrogen ion dose region in which the bubble radius rapidly increases as the temperature reaches 500°C.

The growth of microspores and micro-cracks occurs during annealing, and results in the formation of a weakened zone in the direction parallel to the sample surface. It happens because of the anisotropic distribution of Si–H bonds. Moderate additional heating or mechanical effects lead to the layer splitting precisely inside the vacancy generation area. This effect was used by Bruel for the development of the Smart-Cut method.

The idea of using hydrogen splitting for the creation of SOI structures was very efficient compared to the ion-beam method of synthesis. It allowed one to eliminate the basic short coming of SOI technology used at the time, namely, to reduce radiation thermal load on silicon. For most semiconductor materials, the critical dose for the formation of hydrogen bubbles is about 10^{16}–10^{17} cm^{-2}. It is 20–100 times lower than the oxygen dose used in SIMOX technology to produce a continuous silicon dioxide layer. Smart-Cut did not require the development of new ion accelerators for industrial applications. The required hydrogen doses could easily be obtained with existing implanters. The only problem with the realization of this technology was suppression of the blistering effect. A method to eliminate this effect by increasing the thickness of the transferred layer, increasing the energy of hydrogen ions to a few MeV, was suggested by Reutov and Ibragimov in 1983 [19]. Instead, Bruel proposed covering the silicon surface with a thick SiO_2 layer. This prevented blistering by distancing the region of micro-cracks from the surface. At that, the silicon substrate was transferred together with a SiO_2 layer. Fig. 17.4 shows a schematic representation of the Smart-Cut method.

Despite all its advantages, the Smart-Cut method has a number of weak points that limit its application. First, implantation of hydrogen occurs through the SiO_2 layer which is then used as a built-in insulator for the SOI structure. In order to remove structural defects in the hydrogen-ion implanted Si and SiO_2 layers, it is necessary to anneal the SOI wafers at a temperature of 1100–1200°C. Second, the necessity for distancing the region of micro-crack

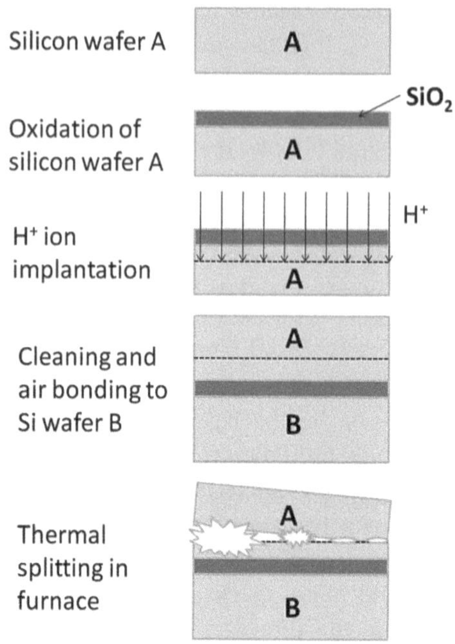

FIGURE 17.4 Schematic of the SOI structure preparation by the Smart-Cut method.

generation from SiO_2 constrains the ion energy. Increasing ion energy is accompanied by an increase in hydrogen straggling. This requires a corresponding increase in the dose of hydrogen ions. As a result, an increase in transferred layer roughness occurs. Fabrication of SOI structures with a nanometer-thick top silicon layer by Smart-Cut technology is also problematic. This is mainly due to the imperfections of the interface between Si and thermal SiO_2. Under Smart-Cut conditions, the Si/SiO_2 interface thickness can reach several nanometers.

The shortcomings of Smart-Cut were taken into account in the DeleCut technology developed in ISP SB RAS. A schematic of this method is presented in Fig. 17.5. The difference in surface energies of a pair of hydrophilic surfaces (Si/SiO_2) and a pair of hydrophobic surfaces (Si/Si) as a function of the temperature was used as the base of the DeleCut method. It was shown elsewhere [19] that the magnitude of the surface energies of hydrophilic Si/SiO_2 surfaces is significantly higher than that of hydrophobic Si–Si surfaces within the temperature range of 20–500°C. In the case of clean surfaces, this difference reaches its maximum at temperatures of 150–300°C. These temperatures are substantially below those used in the Smart-Cut technology. From the point of view of the DeleCut technology, the creation of SOI structures can be considered as a hydrophilic surface splitting process and the bonding of hydrophobic surfaces in a specified temperature range. To realize these processes, it was necessary to solve two problems: (1) to fabricate external high-quality hydrophilic surfaces; (2) to create two internal hydrophobic surfaces in the neighboring atomic planes of the silicon wafer. The temperature and high structural quality of the surfaces are the parameters which determine the magnitude of the surface energy in both cases. Maximum possible

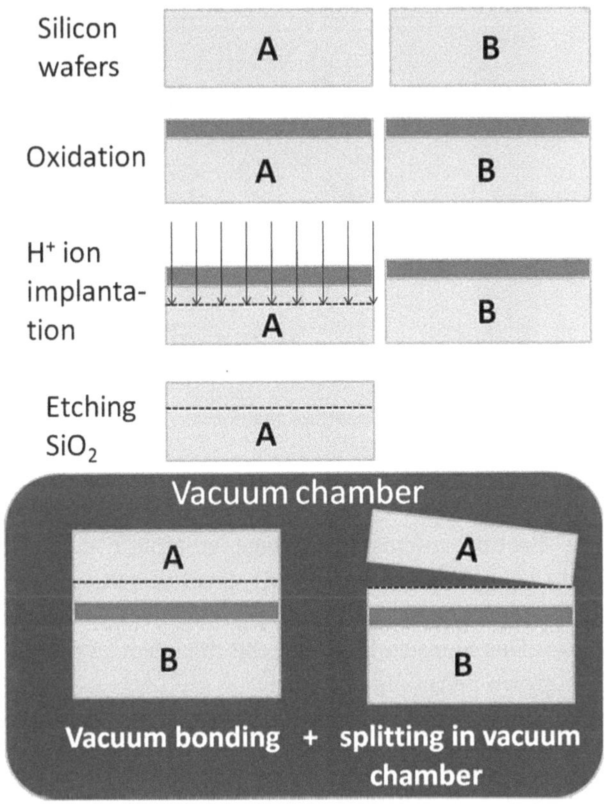

FIGURE 17.5 Schematic of the SOI structure preparation by the DeleCut method.

purity of matched surfaces has been achieved through the use of a number of special cleaning, hydrophilic treatments, and bonding in a vacuum. Internal hydrophobic surfaces in the neighboring atomic planes parallel to the silicon wafer surface were formed by hydrogen ion implantation. The formation of such surfaces occurred through the formation of Si–H–H–Si bonds in the implanted layer due to the capture of hydrogen on the stretched and weakened Si–Si bonds perpendicular to the surface.

Compared to Smart-Cut technology, the DeleCut method developed at ISP SB RAS (Novosibirsk) has the following advantages and additional features: (1) it allows a reduction in hydrogen ion energy and, as a result, a decrease in the thickness of the transferred silicon layer; (2) it avoids the silicon dioxide layer radiation; (3) it uses a smaller dose of hydrogen ions and, consequently, decreases the exposure time; (4) lower roughness of the SOI layer is achieved, and that is accompanied by reducing a radiation-thermal load on the structure during its fabrication; (5) the defect concentration at the Si/SiO_2 interface decreases; (6) it is possible to completely suppress the formation of bubbles on the SOI surface; (7) the yield of the structure is increased; and (8) bonding and exfoliation can be carried out for one stage in a vacuum.

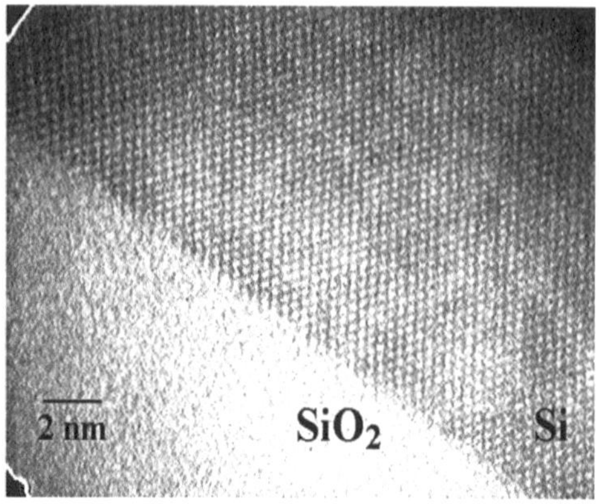

FIGURE 17.6 An electron microscope image of the interface between the transferred Si layer and the buried SiO_2 of the SOI structure produced by the DeleCut method.

Fig. 17.6 shows an electron microscope image at atomic resolution of the interface between the transferred silicon film and buried SiO_2 layer of a SOI structure produced by the DeleCut method. It is clearly seen that the interlayer thickness at the bonding Si/SiO_2 interface is about 0.2–0.3 nm. This is much less than the thickness of the interlayer between the substrate and thermally grown SiO_2 (about 3 nm).

The DeleCut method allows creation of both SOI structures with nanometer-thick layers, and combination of the capabilities of hydrogen transfer and ion-beam synthesis methods to produce SOI structures with a composite buried dielectric.

17.4 Nanometer-Thick SOI Structures

Silicon-based devices developed in recent years and utilizing quantum effects require the creation of SOI structures with silicon and dielectric layers of 3–5 nm thickness. For a long time it has been believed that the creation of single crystal films of nanometer thickness on disordered (amorphous) substrates is impossible from the point of view of thermodynamics. The first breakthrough in this direction was made by ion-beam synthesis of the dielectric embedded by step-by-step implantation of oxygen ions into monocrystalline silicon and subsequent oxidation [20]. It allowed one to reduce the top silicon layer thickness on the ion-beam- synthesized silicon dioxide to 30 nm. However, the roughness of the Si/SiO_2 interface did not allow creation of the top silicon layers below 10 nm.

The DeleCut method, combined with subsequent low-temperature oxidation of silicon layers in an ambient of dry oxygen, allowed creation of SOI structures with a silicon thickness of 40, 30, 10, and 3 nm [21]. The layers produced possessed atomically flat heterophase boundaries and were elastic stressed. Fig. 17.7 shows an electron microscope image of a cross-sectional SOI structure with a 4 nm thick silicon layer. It can be seen that the Si/SiO_2 interlayer thickness is not more than 1–2 monolayers.

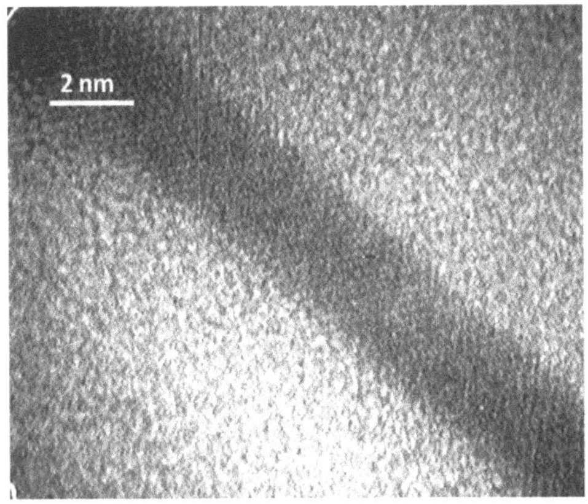

FIGURE 17.7 An electron microscope image of cross-sectional SOI structure with the 4 nm thick top silicon film produced by the DeleCut method.

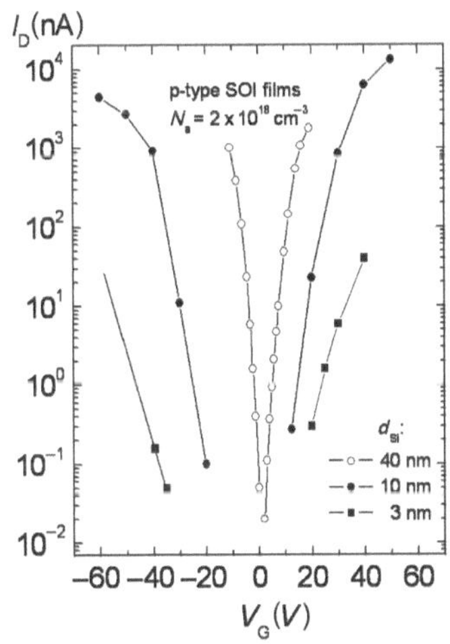

FIGURE 17.8 The channel current as a gate voltage function of a pseudo-MOS transistor formed on SOI structures with the Si layer thickness of 40, 10, and 3 nm.

To investigate the electrical properties of prepared SOI structures, the source-gate current–voltage characteristics of pseudo-MOS transistors were measured. These characteristics are shown in Fig. 17.8. It can be seen from the figure that, as the top silicon layer thickness decreases, the channel current reduces. Current reduction can be provided both by a decrease in the carrier mobility, and by a quantization effect. In the case of the 7 nm thick Si layer buried within wide-band gap material, quantum correction of the charge-carrier concentration and conductance can be above 20%. The dimensional quantization effect results in an increase in the energy gap of the local spectrum for electron and hole states, and a change of the state density. This effect can be taken into account by correct selection of the substrate material, and orientation to obtain the carriers' minimal effective mass along their transport direction.

17.5 SOI Structures With the Nitrogenated Buried SiO_2 Layer

Nitrogenation of thermal oxide as a result of N atom interaction with excess silicon atoms leads to the elimination of silicon–silicon bonds. This effect is accompanied by a decrease in the hole trap concentration in SiO_2. A number of experiments show that nitrogenation of a SiO_2 layer allows one to increase its breakdown voltage, to reduce the surface state density on the Si/SiO_2 interface, to improve $I-V$ and $C-V$ characteristics, to create a diffusion barrier for boron from the polycrystalline gate to the gate dielectric, and to raise the radiation resistance of MOS devices. Hemment was one of the first to investigate SOI structures with a buried dielectric layer composed of silicon dioxide and silicon nitride layers [22]. Next investigations of nitrogenation of the buried silicon dioxide layers were also connected with ion-beam synthesized SOI structures. As a rule, nitrogenation was carried out by nitrogen ion implantation through the top silicon layer. It was accompanied both by radiation defect generation in silicon, and Si/SiO_2 interface degradation.

To produce SOI structures with a nitrogenated buried SiO_2 layer, a new method including ion-beam synthesis and DeleCut technology was developed at ISP SB RAS, Novosibirsk [23]. This method allows us to maintain a high-quality top silicon layer. The basis of this method is the possibility to segregate nitrogen from silicon and silicon dioxide to the Si/SiO_2 interface.

Fig. 17.9 shows a schematic representation of the formation of SOI structures with a buried SiO_2 layer nitrogenated by nitrogen ion implantation. As an initial material, silicon wafers with n-type conductivity, resistivity of 5–10 Ω cm, and orientation (100) were used. Initially, a SiO_2 film with a thickness of 300 nm was thermally grown on the first wafer. Then, N^+ ions were

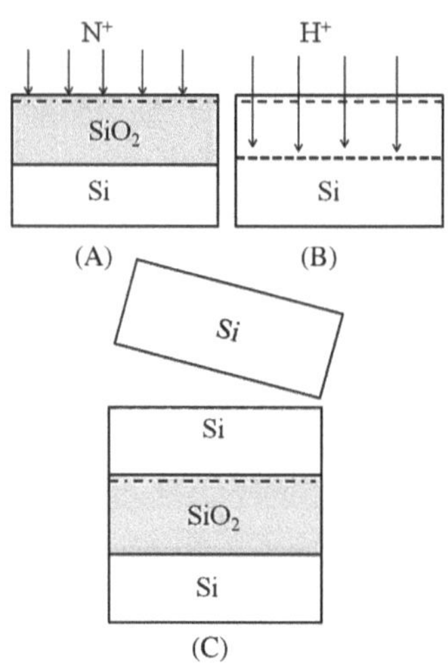

FIGURE 17.9 Schematic image of SOI structure formation with a nitrogenated insulator: (A) N^+ ion implantation into the SiO_2 layer grown thermally on the first wafer; (B) H^+ ion implantation into the second wafer; and (C) Bonding of the wafers implanted sides and the Si layer transfer.

implanted into the silicon dioxide layer (Fig. 17.9A). N^+ ion parameters were chosen such that they could provide, during subsequent annealing, the formation of a silicon nitride or silicon oxinitride layer with a thickness of at least one monolayer at the bonding interface. For this, a maximum N^+ ion distribution should be at a distance comparable to the length of nitrogen diffusion in SiO_2. The temperature range most favorable for complete nitrogen segregation from silicon dioxide to the Si/SiO_2 interface is about 1000–1200°C. These heating conditions were determined from the results of nitrogen diffusion in silicon dioxide [24]. It is also necessary to take into account that silicon nitride can crystallize at high temperatures [25]. These conditions specify the epitaxial growth of an intermediate dielectric layer from the top silicon formation. In the experiments [26], the ion energy was varied from 20 to 40 keV and the ion dose was varied in thc rangc of $(1-6) \times 10^{15}\ cm^{-2}$. The energies of N^+ ions used ensured the formation of Gaussian-like profiles with a maximum at a depth from 70 to 130 nm, respectively; the content of nitrogen atoms at the maximum of distribution was 0.15–1.15 at %.

Hydrogen ions with an energy of 140 keV and a dose of $2 \times 10^{16}\ cm^{-2}$ were implanted into the second wafer (Fig. 17.9B). Then, in a vacuum chamber, the implanted sides of the wafers were joined with simultaneous exfoliation of the silicon layer over the internal hydrophobic surface formed by implantation of H^+ ions; this layer was then transferred from the second wafer to the first (Fig. 17.9C). The processes of wafer bonding, exfoliation, and transfer of a silicon layer were accomplished within a single stage at a temperature of 200°C in a vacuum chamber with the pressure at 10^2 Pa. The final annealing was performed for 1 h at a temperature of 1100°C. The transferred Si layer thickness was ~600 nm. Then, as a result of subsequent oxidation, the transferred silicon layer thickness was made to be 350–370 nm.

Fig. 17.10 shows an electron microscope cross-section image of the SOI structure formed with the SiO_2 layer nitrogenated by N^+ ion implantation at an energy of 40 keV to a dose of $6 \times 10^{15}\ cm^{-2}$ after subsequent annealing at 1100°C for 30 min in N_2 gas ambient. It is clearly seen that the bonding interface is not atomically flat as it was in the case of the nitrogen-free SOI structure (see Fig. 17.6). The intermediate layer growth between the top silicon layer, buried SiO_2, and its lattice is coherent with that of the top silicon film.

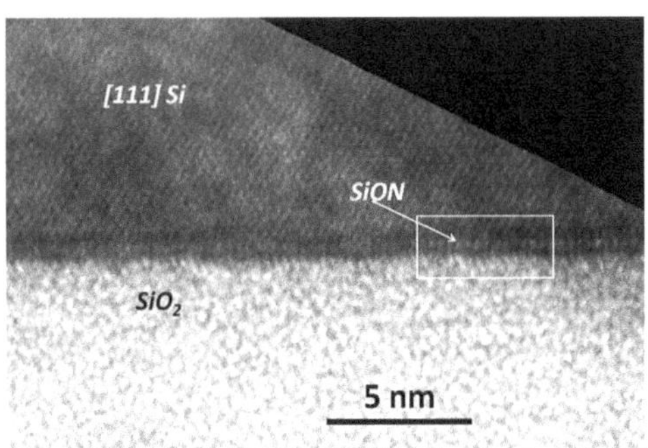

FIGURE 17.10 A high-resolution electron microscope image of the cross-section of the SOI structure with the buried dielectric implanted with N^+ ions at energy 40 keV to a dose of $6 \times 10^{15}\ cm^{-2}$ after subsequent annealing at 1000°C for 30 min at N_2 ambient.

Nitrogen diffusion toward the interface occurs because of the weak chemical interaction of nitrogen atoms inside the silicon oxide matrix. Silicon dangling bonds on the Si/SiO_2 interface make this attractive to the formation of Si–N bonds. A possible structure of nitrogen atom arrangement at the Si/SiO_2 interface may correspond to the calculated atom structure of the Si/Si_3N_4 interface [27].

In order to study the charge accumulation in the buried silicon dioxide layer and at the Si/SiO_2 interfaces, the capacitance–voltage (C–V) characteristics of SOI-based MOS capacitors were measured as a function of the probing signal frequency in the range of 10^2–10^6 Hz at room temperature.

In Fig. 17.11, we show the high-frequency (1 MHz) C–V characteristics of SOI capacitors, which do not contain nitrogen in the buried insulator, and also the structures with the SiO_2 layer implanted with N^+ ions at an energy of 40 keV to a dose of $(1-6)\times 10^{15}$ cm^{-2} after annealing for 1 h at 1100°C. The characteristics obtained are typical of semiconductor –insulator–semiconductor structures, which are represented by two MOS capacitors connected "back-to-back." It can be seen from Fig. 17.11 that nitrogen implantation into the SiO_2 layer with doses of $(1-2)\times 10^{15}$ cm^{-2} bring about additional shifts in the flat band voltage (V_{FB}) for the upper (V_{FB1}) and lower (V_{FB2}) MOS capacitors. These shifts indicate an increase in the positive fixed charge in the insulator. A further increase in the nitrogen content in SiO_2 corresponding to an ion dose of 3×10^{15} cm^{-2} leads again to a negative shift of V_{FB} at positive values of the voltage applied to the gate. At a dose of N^+ ions 6×10^{15} cm^{-2}, the V_{FB} reaches a value corresponding to the flat band voltage for nonnitrogenated structures. In the region of negative voltages at the gate, V_{FB2} shifts to positive values. This is a consequence of a decrease in the positive charge density in the insulator near the interface with the silicon substrate.

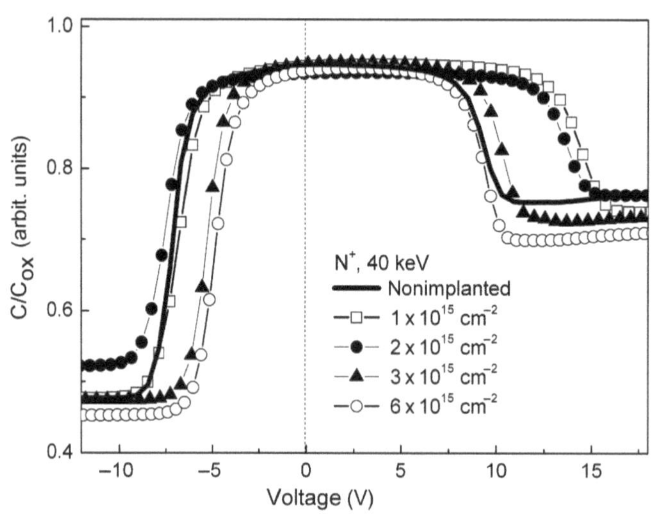

FIGURE 17.11 High-frequency (10^6 Hz) C–V characteristics of the Al/n-Si/SiO_2/n-Si/Al structures with a nonimplanted SiO_2 layer and a SiO_2 layer implanted with N^+ at an energy of 40 keV to a dose of $(1-6)\times 10^{15}$.

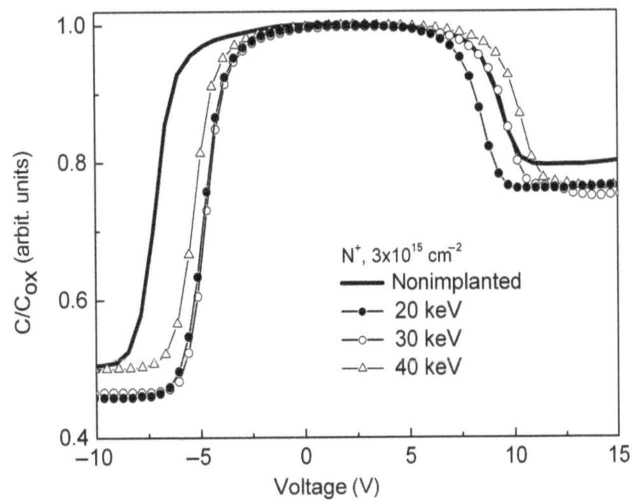

FIGURE 17.12 High-frequency (10^6 Hz) $C-V$ characteristics of the Al/n-Si/SiO_2/n-Si/Al structures with a nonimplanted SiO_2 layer and a SiO_2 layer implanted with N^+ ions to a dose of 3×10^{15} cm^{-2} and energy of 20, 30, and 40 keV.

Fig. 17.12 shows the high-frequency (1 MHz) $C-V$ characteristics of SOI capacitors, which do not contain nitrogen in the insulator layer, and also the MOS structures with the SiO_2 layer implanted with N^+ ions at energies of 20, 30, and 40 keV to a dose of 3×10^{15} cm^{-2} and then annealed for 1 h at a temperature of 1100°C. It can be seen that a decrease in the nitrogen ion energy is accompanied by a positive shift in the flat band voltage V_{FB2} in the region of negative biases, and by a negative shift of V_{FB1} in the region of positive biases. This denotes a decrease in the positive charge density in the silicon dioxide layer near the respective Si/SiO_2 interface. The fixed charge density in the oxide corresponding to each interface was estimated by the shift in the respective flat band voltage. Its value was about $(3-6) \times 10^{11}$ cm^{-2}. The results obtained were explained on the basis of the experimental data obtained by the photoelectron spectroscopy method [28]. According to these studies, excess over-stoichiometric silicon exists in silicon oxide near the Si/SiO_2 interface. Excitation of excess Si–Si bonds can bring about the capture of holes and a fixed positive charge in the oxide. Thermal oxide nitrogenation leads to the elimination of Si–Si bonds as a result of N atoms interaction with excess silicon atoms. This effect is accompanied by a decrease in the hole concentration of traps in SiO_2. In the case under consideration, implantation of low nitrogen doses is not sufficient to compensate for the existing Si–Si bonds, and leads only to generation of additional defects, including those near the interface. Apparently, it is this effect that is the cause of an increase in the positive fixed charge in structures implanted with low N^+ ion doses. An increase in the ion dose to $(4-6) \times 10^{15}$ cm^{-2} is found to be sufficient for compensation of Si–Si bonds near the interfaces by the formation of Si–N–Si and Si–N–O bonds. At that, a decrease in the positive fixed charge in the insulator near the Si/SiO_2 interface is observed.

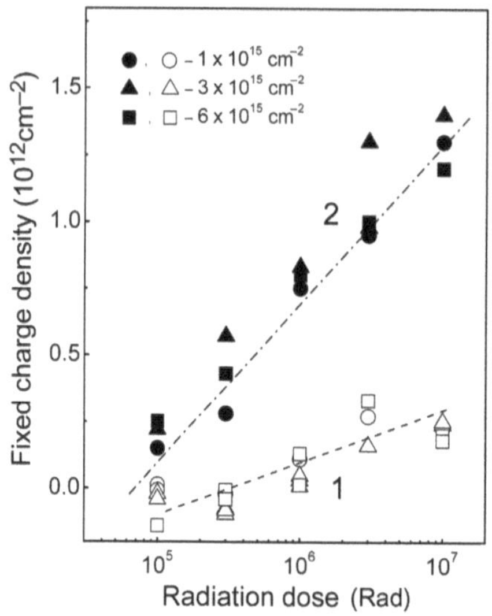

FIGURE 17.13 An effective fixed charge density in nitrogenated buried silicon dioxide near the top (1), and bottom (2) Si/SiO_2 interfaces of a DeleCut SOI structure as a function of radiation dose.

Analysis of the dependence of capacitance and capacitive conductance as a function of voltage for different probing signal frequencies allows one to determinate the surface states density, their energy distribution in the upper half of the conduction band, the characteristic times of recombination–generation processes, and cross-sections σ_n for electron captures by surface states [26]. It was shown that, as the concentration of nitrogen atoms within the buried oxide increases, the cross-section for capture of the majority of charge carriers at the bonding boundary decreases by more than two orders of magnitude. If the N^+ ion energy is lowered to 20 keV (and the nitrogen-containing layer approaches the Si/SiO_2 interface), the decrease in σ_n (compared to the case of nonnitrogenated bonding boundary) can be as large as three orders of magnitude.

DeleCut SOI structures exhibit improved resistance to ionizing radiation up to a dose of 10^6–10^7 rad. Fig. 17.13 shows the effective charge density in the N^+ ion implanted silicon dioxide reduced to the Si/SiO_2 interface as a function of irradiation dose. It is clearly seen that the buried silicon dioxide layer nitrogenation is accompanied by a decrease in the generation rate of fixed charges under irradiation at an order of magnitude.

17.6 SOI Structures With a Ge Layer Embedded on the Si/SiO_2 Interface

The use of germanium for the purpose of modifying thin silicon film electrical properties is due to a number of advantages and, first of all, to high hole mobility which is higher by a factor of four than that of silicon. In addition, germanium in silicon creates a potential well for holes. The hole mobility in the potential well is much higher than that in the silicon matrix. Traditionally, molecular beam epitaxy (MBE) is used for the preparation of SiGe heterostructures. However, MBE is a low-temperature method (400–500°C) that is of little use in current silicon technology utilizing temperature treatments up to 1100°C.

Enhanced diffusion of germanium from silicon dioxide to the Si/SiO_2 interface was observed elsewhere [29]. At that, the accumulation of Ge atoms on the interface in the sites, coherent with these in the top silicon layer, takes place. It allowed us to create nanometer-thick intermediate germanium layers on the bonding interface of SOI structures [30,31]. The method developed is like that presented in Fig. 17.9.

In our experiments, silicon–germanium structures on insulator (SGOI) were prepared in the following way. First, the 220-nm thick thermally grown SiO_2 film was implanted with Ge^+ ions at an energy of 40 keV to a dose about 8×10^{15} cm^{-2}. The resulting Ge distribution peaked at about 50 nm below the top SiO_2 surface. Afterwards, under conditions of wafer bonding processing, the 500-nm thick silicon film was transferred to this substrate. As a result, a SOI structure with a buried silicon dioxide layer containing germanium near the bonding interface was prepared. Subsequent anneals were performed in the temperature range from 500°C to 1100°C for 30 min in N_2 ambient. It was obtained from the RBS spectra, and as the annealing temperature was increased to 800°C, the ion implanted germanium redistribution toward the bonding Si/SiO_2 interface took place. A comparison of the random and aligned spectra shows that part of the Ge atoms were in coherent silicon lattice positions (see Fig. 17.14). As the annealing temperature reached 1100°C, this part reached ~90%. The

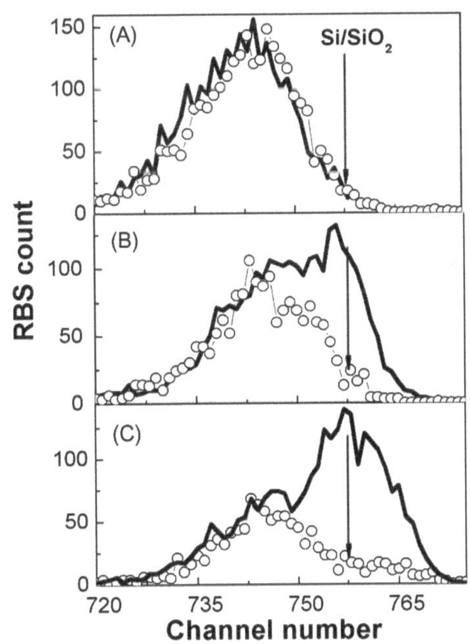

FIGURE 17.14 The random (*solid line*) and aligned (*circles*) (with respect to the ⟨100⟩-direction in Si) spectra of He^+ ions backscattered from Ge atoms implanted into the SiO_2 layer of the SOI structure, on an enlarged scale. The annealing temperatures were T_a = (A) 700°C, (B) 1000°C, and (C) 1100°C. The position of the Si/SiO_2 interface is indicated.

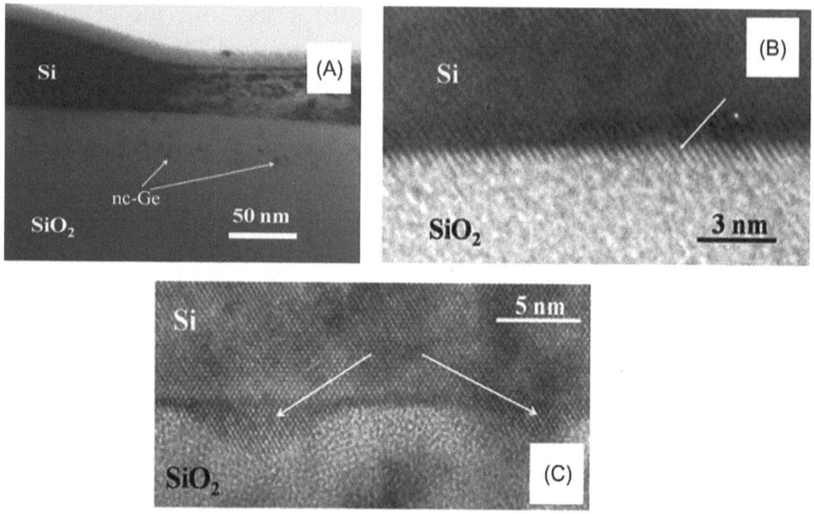

FIGURE 17.15 An electron microscope image of a cross-section of the formed SOI structure with the SiO_2 layer implanted with Ge^+ ions after annealing at temperatures of: (A) 1000°C, and (B and C) 1100°C for (A and B) 30 and (C) 120 min.

Ge atom concentration located at regular lattice positions as a function of the annealing temperature was well fitted by a linear function with a slope of 0.76 ± 0.01 eV. The activation energy obtained was much lower than that of germanium diffusion in the SiO_2 matrix.

XTEM analysis showed that Ge nanocrystal formation is localized below the bonding interface in a 40-nm thick layer after annealing at a temperature of 1000°C over 30 min (Fig. 17.15A). Their mean radius was about 2.5 nm. As the annealing temperature rose to 1100°C, no nanocrystals were detected in the ion-implanted layer, but a clearly marked intermediate layer was seen at the bonding interface. Its thickness was about 1.3 nm. The lattice orientation of this layer was coherent to the top silicon film orientation (Fig. 17.15B). An increase in the annealing time from 30 to 120 min at a temperature of 1100°C resulted both in the intermediate layer thinning, and in the formation of dome-shaped epitaxially grown nanocrystals at the bonding interface (Fig. 17.15C). Their lattice orientation was also coherent with the silicon lattice orientation.

The step-by-step Raman measurements were also used to study the Ge atom position near the Si/SiO_2 interface. For this purpose, the top silicon layer was thinned to 60 and 3 nm. The results obtained were in good agreement with the RBS and XTEM data. Fig. 17.16A shows the Raman spectra at the frequency of optical phonon mode from the structure with a top silicon layer of 60 nm thickness (curve 1). The figure also shows the Raman spectrum from the respective structure after removing the top silicon layer (curve 2). Two peaks at the frequencies of 300 and 420 cm^{-1} corresponding to Raman scattering on the optical phonon modes of the Ge–Ge and Ge–Si bonds, respectively, are clearly seen in the spectra along with the peak at 520 cm^{-1} corresponding to the optical phonon mode in monocrystalline silicon. Two low-intensity peaks at frequencies of 280 and 400 cm^{-1} are seen, too. They correspond to the

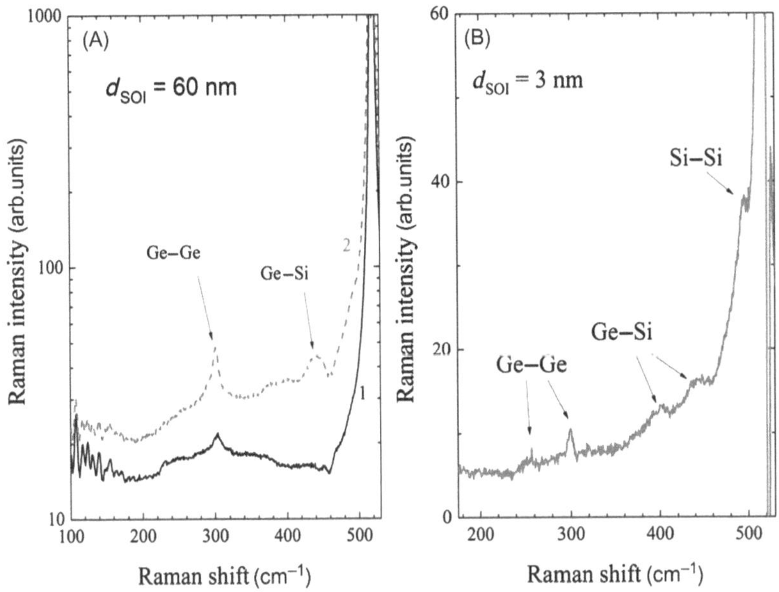

FIGURE 17.16 Raman spectra from the SOI structure with the SiO_2 layer implanted with Ge^+ ions first at the energy of 40 keV to a dose of 8×10^{15} cm^{-2}, and then at an energy of 60 keV to a dose of 1.6×10^{16} cm^{-2} after subsequent annealing at a temperature of 1100°C for 30 min. The thickness of the silicon layer on insulator was: (A) 60 nm (curve 2), 0 nm (curve 1); (B) 3 nm.

scattering on the optical phonon modes of the Ge–Ge and Ge–Si bonds in an amorphous matrix. The relation between the intensities of the respective peaks in the monocrystalline and amorphous phases means that the majority of germanium atoms stay in the regular positions. As the top silicon layer was removed, the Raman peak at a frequency of 420 cm^{-1} disappeared. At that, the intensity of the peak at 300 cm^{-1} was visible (see curves 1 and 2). It suggests germanium growth from the interface deep into the SiO_2 layer.

A decrease in Si layer thickness to 3 nm (Fig. 17.16B) results in a reduction of intensity of the 280, 300, 400, and 420 cm^{-1} peaks. At that, the formation of a weak-intensity shoulder is seen in the low-frequency part (about 500 cm^{-1}) of the main Raman peak. The origin of this shoulder can be provided by optical phonon localization in the nanometer-thick silicon layer.

Thermodynamic analysis of the $Si/Ge/SiO_2$ system shows that the solid-phase epitaxial growth of the Ge layer on the Si/SiO_2 interface is not energy-favorable. Indeed, taking into account the Frank–van der Merwe law, the next relation between the surface energies has to be realized: $\gamma_{BC} > \gamma_{AB} + \gamma_{AC}$, where $\gamma_{BC} = 1.44$ J/m^2 is the interface energy between SiO_2 and Si, $\gamma_{AB} = 0.88$ J/m^2 is the interface energy between Ge and SiO_2, and $\gamma_{AC} = 1.1$ J/m^2 is the interface energy between Ge and Si. The Ge interlayer growth can be created as a result of the formation of a SiGe solid solution at the Si/SiO_2 interface due to Ge accumulation and, next, melting of this solution. The molten Ge layer formation on the Si/SiO_2 interface satisfies the Bauer–van der Merwe condition. Under these conditions, the interlayer formation is a result of liquid Ge crystallization. The excessive increase of annealing duration at the fixed temperature

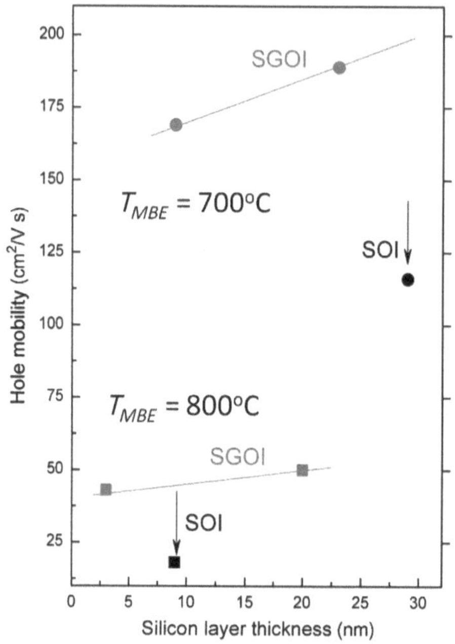

FIGURE 17.17 Hall hole mobility as a function of silicon–germanium (SGOI) layer thickness at MBE processing temperatures 700°C and 800°C. The respective Hall mobility obtained from the germanium-free layers (SOI) is shown with the *arrows*.

$T = 1100°C$ is accompanied by a loss of germanium on the interface, because of its diffusion in silicon. At that, wetting layer instability occurs. This is accompanied by solid film dissolution in separate islands, and by the formation of domes at the border splicing (see Fig. 17.15C).

As noted above, the SiGe layers have a higher carrier mobility compared to that in the silicon layers. In order to measure the electrical properties of nanometer-thick silicon films, structures with modulation doping are often used [32]. In our studies [30], the structures with modulation doping were grown by the MBE method. A wetting layer of undoped silicon with a thickness of 5 nm was grown first; then, a 5-nm thick silicon layer doped with boron at a concentration of $[B] = 5 \times 10^{18}$ cm^{-3} or 1×10^{19} cm^{-3} was grown; and, in the final stage, an encapsulating layer of undoped silicon with a thickness of 5 nm was again grown. The epitaxy was performed at temperatures of $T_{MBE} = 700°C$ or 800°C. Finally, Hall bridges were formed on the structures obtained. The Hall mobility, as a function of silicon thickness, is shown in Fig. 17.17. A comparative analysis of the data obtained indicates that there is an appreciable increase in the hole mobility of the SGOI structures. It can be seen that, even at higher MBE temperatures, T_{MBE} is detrimental to the charge carriers mobility as a whole; the hole mobility values μ_p exceed the values of μ_p by a factor of 2–3 in conventional SOI structures, even with larger top-layer thickness values. Thus, e.g., at $T_{MBE} = 700°C$ and the bulk level of modulation doping 5×10^{18} cm^{-3}, the hole mobility values were $\mu_p = 170–190$ cm^2/(V s) in SGOI structures with the initial top silicon layer thickness $d_{SOI} = 15$ nm. In the corresponding SOI structures, even with the thickness of the top Si layer being 30 nm, the hole mobility values did not exceed 115 $cm^2/(V\ s)$. A decrease in d_{SOI} from 15 to 10 nm is accompanied by a decrease in μ_p by 50%. A detrimental factor for charge-carrier mobility is the MBE temperature. These data are consistent with the available published data [33], and are caused by the quality of the modulation-doping layers formed in the course of MBE, rather than by the quality of the SOI structures. However, in spite of this, even at $T_{MBE} = 800°C$ and a bulk level of modulation doping of $1 \times 10^{19}\ cm^{-3}$ in SGOI structures with $d_{SOI} = 3$ nm, the hole mobility exceeded μ_p by a factor of ~2.4 in the corresponding SOI films with a thickness of 10 nm. It is worth noting that a typical T_{MBE} does not exceed 600°C owing to diffusion-related broadening of the boundaries for the growing layers. For example, it was shown elsewhere [33] that an increase in T_{MBE} to 700°C leads to a decrease in the hole mobility in the SiGe-on-insulator layers with a thickness of 16 nm from 200 to 90 $cm^2/(V\ s)$. The corresponding comparative analysis shows that the SGOI structures obtained by the ion implantation method with subsequent hydrogen-assisted transfer feature a higher thermal stability and retain hole mobility at a level of 190 $cm^2/(V\ s)$, even at $T_{MBE} = 700°C$.

The capacitance–voltage analysis of SOI structures with the buried silicon dioxide layer implanted with G^+ ions shows a decrease in the positive charge density in the insulator. At the same time, an increase in the surface state density at the bonding Si/SiO_2 interface was observed. In MOS capacitors based on germanium-containing SOI structures, ionizing irradiation leads to a shift in the flat band voltage two times lower than that corresponding to Ge-free structures. Thus, SOI structures containing germanium in the buried SiO_2 layer have a high radiation resistance. This effect can be provided by the formation of small Ge nanocrystals within the SiO_2 matrix which are electron traps. The presence of negative charge traps compensates positive charges generated in SiO_2 under radiation.

Thus, the development of SOI technology has opened up new opportunities for the creation of a new class of silicon-based devices. The study of charge carriers transport properties in nanometer-thick Si films is of great importance from the point of view of creating nanotransistors with complete depletion. The development of SOI methods also allows creation of radiation-resistant electronic devices.

17.7 Conclusion

Ion-beam synthesis and hydrogen transfer methods are considered as the most favorable techniques for creating SOI structures. The specificity of DeleCut technology allows a combination of the capabilities of these two methods. It opens up new possibilities for the formation of SOI structures possessing unique properties. SOI structures, produced by N^+ ion implantation of SiO_2 film followed by subsequent hydrogen transfer of the top Si layer, demonstrate both the low fixed charge density in buried silicon dioxide, and improved resistance to ionizing radiation according to these in nitrogen-free SOI structures. Ge^+ ion implantation of SiO_2 layers and then hydrogen transfer of Si films followed by high-temperature treatments allowed realization of the intermediate nanometer-thick Ge layer at the bonding interface of the SOI structure. Such types of nanometer-thick silicon–germanium-on-insulator layers have a hole mobility higher than that in respective silicon layers by a factor of 2–3. The method developed provides an atomically flat Si/SiO_2 bonding interface, and is suitable for nanoelectronic applications.

References

[1] M.I. Guseva, B.G. Alexandria, The ion-beam density effect on the concentration of isotopic targets prepared with electromagnetic separator, Techn. Phys. 31 (1961) 867–875 (in Russian)

[2] K.J. Reeson, P.L.F. Hemment, Buried layers of silicon oxy-nitride fabricated using ion beam synthesis, Nucl. Instrum. Meth. Phys. Res. B 32 (1988) 427–432.

[3] G.A Kachurin, I.E. Tyschenko, V.P. Popov, S.A Tijs, AE Plotnikov, Implantation of nitrogen in silicon at 700-1100°C, Semiconductors 23 (1989) 271–273.

[4] G.A. Kachurin, I.E. Tyschenko, V.P. Popov, S.A. Tijs, Interaction of nitrogen implanted into silicon with defects in the buried layer, Phys. Stat. Sol. (A) 113 (1989) K165–K169.

[5] G.A. Kachurin, I.E. Tyschenko, S.A. Tijs, AE Plotnikov, Effect of competing precipitation centers on the distribution of nitrogen implanted in Si during the formation of buried layers, Semiconductors 29 (1995) 256–258.

[6] G.A. Kachurin, I.E. Tyschenko, L.I. Fedina, High-temperature ion implantation in silicon, Nucl. Instrum. Meth. Phys. Res. B 68 (1992) 323–330.

[7] G.V. Gadiyak, G.A. Kachurin, I.E. Tyschenko, Effect of competing trapping centers (sinks) on the evolution of implanted-nitrogen distribution in silicon—a numerical simulation, Semiconductors 30 (1996) 1019–1027.

[8] A.A. Karanovich, A.V. Dvurechenskii, I.E. Tyschenko, G.A. Kachurin, Centers of spin-dependent recombination in structures formed by N^+ ion implantation into Si, Nucl. Instrum. Meth. Phys. Res. B 55 (1991) 630–632.

[9] F.L. Edelman, V.Z. Latuta, B.N. Zaitsev, A.A. Khoromenko, The structure of silicon nitride films. I. Stoichiometric silicon nitride, Phys. Stat. Sol. (A) 50 (1978) 573–578.

[10] F.L. Edelman, B.N. Zaitsev, V.Z. Latuta, A.A. Khoromenko, The structure of silicon nitride films. II. Non-stoichiometric silicon nitride, Phys. Stat. Sol. (A) 51 (1979) 49–56.

[11] F.L. Edelman, The structure of silicon nitride films. III. Oxidation of silicon nitride—defects, Phys. Stat. Sol. (A) 51 (1979) 375–381.

[12] G.A. Kachurin, I.E. Tyschenko, AE Plotnikov, V.P. Popov, Growth of single-crystal α-Si_3N_4 in buried layers formed by low-rate implantation of N^+ ions in heated silicon, Semiconductors 26 (1992) 779–781.

[13] G.A. Kachurin, V.D. Akhmetov, I.E. Tyschenko, A.E. Plotnikov, Roles of implantation temperature and ion dose rate in ion-beam synthesis of buried Si_3N_4 layers, Nucl. Instrum. Meth. Phys. Res. B 74 (1993) 399–404.

[14] Q.-Y. Tong, U. Goesele, 10158-0012. Semiconductor Wafer Bonding. Science and Technology, John Wiley & Sons, Inc, New York, NY, 1999, p. 297.

[15] M. Bruel, Silicon on insulator material technology, Electron. Lett. 31 (1995) 1201–1202.

[16] W. Primak, Radiation-induced cavities and exfoliation, J. Appl. Phys. 34 (1963) 3630–3631.

[17] M. Kaminsky, Atomic-collision sequences revealed in the sputtering of metal monocrystals by ion bombardment in the Rutherford collision region, Bull. Am. Phys. Soc. 8 (1963) 428.

[18] K. Mitani, U.M. Gösele, Formation of interface bubbles in bonded silicon wafers—a thermodynamic model, Appl. Phys. A 54 (1992) 543–552.

[19] V.F. Reutov, Sh.Sh. Ibragimov, Method of Manufacturing of Thin Silicon Films, Patent of Soviet Union № SU 1282757 (30.12.1983).

[20] P.L.F. Hemment, K.J. Reeson, J.A. Kilner, R.J. Chater, C. March, G.R. Booker, et al., Novel dielectric/silicon planar structures formed by ion synthesis, Nucl. Instrum. Meth. Phys. Res. B 21 (1987) 129–133.

[21] V.P. Popov, A.L. Aseev, I.V. Antonova, Yu.V. Nastaushev, T.A. Gavrilova, O.V. Naumova, Low dimension properties of nanostructures on ultrathin layers of silicon formed by oxidation of ion cut SOI wafers and electron lithography, in: F. Balestra, et al. (Eds.), Progress in SOI Structures and Devices, Operating at Extreme Conditions, Kluwer Academic Publishers, Norwell, MA, 2002, pp. 87–91.

[22] P.L.F. Hemment, Silicon on insulator by high dose implantation, Mater. Res. Soc. Symp. Proc. 33 (1984) 41–51.

[23] V.P. Popov, I.E. Tyschenko, N.V. Dudchenko, Method of manufacturing of silicon-on-insulator structure, Bull. Inventions Useful Model 5 (2010) 812–813 (in Russian)

[24] W.J.M.J. Josquin, The application of nitrogen ion implantation in silicon technology, Nucl. Instrum. Meth. 209/210 (2003) 581–587.

[25] G.A Kachurin, I.E. Tyschenko, Ion synthesis of buried dielectric layers for silicon-on-insulator structures, Russ. Microelectron. 23 (1994) 3–12 (in Russian)

[26] I.E. Tyschenko, V.P. Popov, Silicon-on-insulator structures with a nitrogenated buried SiO_2 layer: preparation and properties, Semiconductors 45 (2011) 335–342.

[27] K. Kato, D. Matsushita, K. Muraoka, Y. Nakasaki, Agglomeration and uniform growth of Si_3N_4 on Si (100): experiments and first-principles calculations, Phys. Rev. B 78 (2008) 085321.

[28] V.A. Gritsenko, Atomic structure of the amorphous nonstoichiometric silicon oxides and nitrides, Phys. Usp. 51 (2008) 699–708.

[29] L. Rebohle, I.E. Tyschenko, J. von Borany, B. Schmidt, R. Grötzschel, A. Markwitz, et al., Strong blue- and violet light emission from silicon- and germanium-implanted silicon-dioxide films, Mater. Res. Soc. Symp. Proc. 486 (1998) 175–181.

[30] I.E. Tyschenko, M Voelskow, AG. Cherkov, V.P. Popov, Growth and electrical properties of the (Si/Ge)-on-insulator structures formed by ion implantation and subsequent hydrogen-assisted transfer, Semiconductors 43 (2009) 58–63.

[31] I.E. Tyschenko, M. Voelskow, A.G. Cherkov, V.P. Popov, Nanometer-thick SGOI structures produced by Ge^+ ion implantation of SiO_2 film and subsequent hydrogen transfer of Si layer, Nucl. Instrum. Meth. Phys. Res. B 267 (2009) 1277–1280.

[32] M. Myronov, T. Irisawa, S. Koh, O.A. Mironov, T.E. Whall, E.H.C. Parker, et al., Temperature dependence of transport properties of high mobility holes in Ge quantum wells, J. Appl. Phys. 97 (2005) 083701.

[33] T. Uehara, H. Matsubara, R. Nakane, S. Sugahara, S.-I. Takagi, Ultrathin Ge-on-insulator metal source/drain p-channel metal–oxide–semiconductor field-effect transistors fabricated by low-temperature molecular-beam epitaxy, Jpn. J. Appl. Phys. 46 (2007) 2117–2121.

PART IV

Electronic Advanced Materials

18

Superminiature Radiation Sources Based on Semiconductor Nanostructures

V.A. Haisler[1], A.V. Haisler[1], I.A. Derebezov[1], A.S. Yaroshevich[1], A.K. Bakarov[1], D.V. Dmitriev[1], A.K. Kalagin[1], A.I. Toropov[1], M.M. Kachanova[1], Yu.A. Zhivodkov[1], T.A. Gavrilova[1], A.S. Medvedev[1], L.A. Nenasheva[1], O.I. Semenova[1], K.V. Grachev[1], V.K. Sandyrev[1], A.S. Kozhukhov[1], D.V. Sheglov[1], D.B. Tretyakov[1], I.I. Beterov[1], V.M. Entin[1], I.I. Ryabtsev[1], A.V. Latyshev[1,2], A.L. Aseev[1]

[1]RZHANOV INSTITUTE OF SEMICONDUCTOR PHYSICS OF THE SIBERAIN BRANCH OF THE RUSSIAN ACADEMY OF SCIENCES, NOVOSIBIRSK, RUSSIA [2]NOVOSIBIRSK STATE UNIVERSITY, NOVOSIBIRSK, RUSSIA

The last two decades have been characterized by the extremely "aggressive" behavior of semiconductor technologies in developing the market for emitters with a wide range of applications and a wide spectral range.

The spectral range from ultraviolet to the submillimeter (THz) region has already been mastered. The steadily expanding list of applications for semiconductor emitters apparently contains hundreds of items. Some of the most promising are high-power LED emitters that are used for lighting and backlighting monitors and TV screens, and powerful matrix laser modules used in processing plants and for pumping solid state lasers. Less bright emitters with a power of $\sim 1-10^2$ mW visibly or invisibly accompany us in everyday life, providing the work of laser printers, DVD systems, mobile phones, and fiber optic communication lines.

Subminiature semiconductor emitters are a new development in optoelectronics, opening prospects for creating high-performance optical information systems, quantum cryptography systems, miniaturized optical power, and frequency standards. Vertical cavity surface-emitting lasers (VCSEL), which have been commercial products for some years and whose annual production is now about 10^8 articles, should be referred to as such subminiature semiconductor emitters [1]. Approximately half of these lasers are used in local area network optical fiber systems that solve the "last mile problem" at a rate of about 1 Gbit/s; the other half are used in high-precision hand-held computer manipulators (mice). This is of course only an initial, pilot phase of VCSELs commercialization. Lasers of this type are of great interest for creating high-speed data transmission systems in the supercomputers that are under development at present, and in other information systems, where a one channel transmission rate that can reach tens of Gbit/s, and even Tbit/s, is an achievable milestone if matrix-type VCSELs are used. The number

Advances in Semiconductor Nanostructures. DOI: http://dx.doi.org/10.1016/B978-0-12-810512-2.00018-4

of other developments where VCSELs form the basis of sensor devices, laser printers, and miniature atomic frequency standards can be interesting as well. Commercialization of these developments may be expected in the near future.

The absolute limit of emitter miniaturization is creating single-photon emitters (SPE) and emitters of entangled photon pairs (EEPPs) [2–5]. Regarding SPE, we are certainly not speaking about weakening of the luminous flux to a level when one photon is emitted on average within a specified time period. Such a problem is solved over a few minutes by using filters which attenuate radiation. However, the solution will be a light flux with classical Poisson statistics, when one photon is averagely emitted over the time interval, and the probability of this event is $1/e$. The probability that there will be no photon over the interval is exactly the same, $1/e$, the probability of emission of two photons over the interval is $1/2e$, and of three photons $1/6e$, etc. The aim is to create a nonclassical emitter generating Fock photon states with a strictly specified number of photons, when one, and only one, photon is generated over a specified time interval. Efficient SPEs can find a variety of applications in quantum cryptography systems, quantum computing, creating optical power standards, and conducting precision spectral experiments, including experiments in quantum electrodynamics. The EEPP can also be an important resource for a number of protocols in quantum cryptography and quantum computing. The results from recent years are convincing in that effective SPE and EEPP can be implemented within semiconductor technology, but practical implementation will require years of intensive research. We should not forget that it took 17 years to develop from the first ideas of VCSEL to their introduction to the market. The new task will take up more time, because the problem is much more serious.

This chapter discusses the principles of subminiature semiconductor emitters, and offers research results of the performance for those emitters that have been developed and manufactured at the SB RAS Institute of Semiconductor Physics for the last 5 years.

18.1 Vertical Cavity Surface-Emitting Lasers

Even now, lasers of this type hold a number of world records not only in the semiconductor field, but also in all laser technologies. Those records include the minimum threshold current generated by laser: unit μA. Vertical cavity surface-emitting lasers are the smallest commercial lasers developed at the present time. The record of laser high frequency current modulation is tens of GHz, which is of paramount importance in the development of high-speed optical information systems, and is also among the world's best achievements.

The operating principle of semiconductor VCSELs is the same as that of conventional stripe semiconductor lasers: a Fabry–Perot cavity is used in both types of lasers, and the gain of quantum active region is achieved by electron and hole injection, and recombination in that area. The principal difference in the lasers is the method of forming a laser cavity. The semiconductor stripe laser includes a Fabry–Perot cavity formed by two mirrors, obtained by cleaving the wafer along the crystallographic directions belonging to the type [110]. Thus, the cavity axis lies in the plane of the semiconductor wafer, and laser radiation

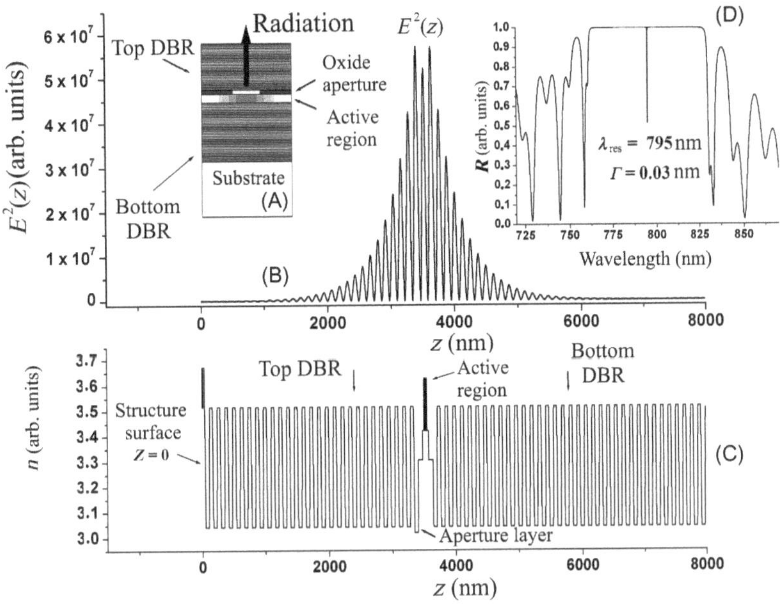

FIGURE 18.1 The optical characteristics of the laser structure, predicted by the one-dimensional model: (A) laser structure; (B) distribution of the squared amplitude of the electric field of the light wave $E^2(z)$ in the laser cavity; (C) reflectivity spectrum of the full VCSEL structure; (D) refractive index $n(z)$ in the laser cavity.

is parallel to the substrate plane. The Fabry–Perot cavity in VCSELs is made from two Bragg mirrors, which are formed in a unified process of laser structure growth or postgrowth processes. The layers of the Bragg mirrors are arranged parallel to the original substrate, and the resonator axis and direction of radiation are perpendicular (vertical) with respect to the plane of the wafer; this defines the lasers name: vertical cavity surface-emitting lasers.

The VCSEL structure is illustrated in Fig. 18.1A. Two semiconductor Bragg reflectors form the laser cavity. Between the Bragg mirrors there are semiconductor layers with a total thickness of λ, which form the active area of the laser. The active VCSEL area contains a few (usually two or three) nondoped quantum wells, which are located near the maximum of the standing electromagnetic wave of the laser microcavity (Fig. 18.1B and C). The charge carriers are injected into the active area through the p-doped upper output semiconductor mirror and the n-doped bottom mirror. An oxide aperture $AlGaO_x$ is formed in the course of selective oxidation of high Al content in the $Al_xGa_{1-x}As$ layer. The ring of the $AlGaO_x$ is located directly above the active area, and ensures effective current and optical confinement in the VCSEL structure.

The VCSEL structure may provide the following number of advantages [1,6,7]:

1. Ultralow threshold operation is expected from its small cavity volume, reaching microampere levels;
2. $(I-I_{th})/I_{th} > 100$ is possible, where I is the driving current, and I_{th} is the threshold current;

3. Wavelength and thresholds are relatively insensitive to temperature variation;
4. Dynamic single-mode operation is possible;
5. High relaxation frequency provides high-speed modulation capability;
6. Long device lifetime is possible due to the completely embedded active region and passivated surfaces;
7. High-power conversion efficiency is possible, i.e., >50%;
8. Vertical emission from a substrate can be achieved;
9. Easy coupling to optical fibers due to good mode matching from the single-mode through thick multi-mode fibers;
10. A number of laser devices can be fabricated by fully monolithic processes yielding very low cost chip production;
11. The initial probe test can be carried out before separating devices into discrete chips;
12. Easy bonding and mounting is possible;
13. Cheap module and package cost;
14. Densely packed and precisely arranged two-dimensional (2D) laser arrays can be formed;
15. Vertical stack integration of multithin film functional optical devices can be made to the VCSEL resonator, taking advantage of micro-machining technology and providing polarization-independent characteristics;
16. Compatible integration together with LSIs is achievable.

It should be noted that, along with the above-mentioned advantages, this type of laser is characterized by one substantial "drawback." This "flaw" is that very sophisticated multilayer and precision heterostructures must be used to create VCSELs. The growth of such structures is a very difficult task, even for modern technology of molecular beam epitaxy and epitaxy from metal–organic compounds. The number of different layers in the laser structure exceeds 10^3, and the required accuracy of the layer setting thickness is 1% or even 0.1%. While growing the VCSEL structures, it is necessary to use a set of calibrations for the growth process equipment, to allow selection of the optimal growth modes. Particular attention is paid to the control of semiconductor alloy composition and layer thickness. Several test structures are grown immediately prior to the growth of the complete structure, in order to determine the accuracy of the choice of various laser element growth modes, such as active region and Bragg mirrors, and to determine the accuracy of the setting position of the Fabry–Perot laser resonance. When the analysis of test structures and the correction of growth modes are made, the complete structure can be grown.

As noted, one of the most promising options for applying VCSEL is its use as a source of radiation in a new generation chip scale atomic clock. Previously mentioned unique characteristics of VCSEL such as small size, cost, and performance perfectly meet the requirements for emitters in miniature atomic frequency standards [8–12].

Available atomic clocks are based on resonant lamps emitting at the frequency of optical transitions in Rb and Cs atoms, and on absorbing cells in the microwave resonator tuned to the frequency of the transition between sublevels of the superfine structure of these atoms ground state. Such devices are rather heavy (several kilograms), large (several cubic decimeters), have inefficient energy use (with energy consumption of tens of watts), and are expensive; therefore, they do not find many applications in standard devices and instruments.

In chip scale atomic clocks, resonant lamps are replaced by low-power high-stability VCSELs. The reference signal is obtained with the use of a small-scale absorbing cell in the magnetic screen, while the microwave resonator is not used at all. The injection current of the laser diode emitting at an optical frequency ω_L is modulated by a microwave generator at a frequency f; as a result, lateral frequencies $\omega_L - f$ and $\omega_L + f$ arise in the laser spectrum. These frequencies initiate optical transitions from two superfine sublevels of the ground state of atoms to the global excited state. During reconstruction of the generator frequency, a narrow interference resonance of coherent population trapping appears, and it can be used to stabilize the frequency of the microwave generator with relative accuracy from 10^{-10} to 10^{-12}. This offers prospects for creating a new class of atomic clocks with a low weight (tens of grams), small size (several cubic centimeters), and low energy consumption (tens of milliwatts). Such devices can substantially improve the performance of a large number of telecommunication and navigation systems.

Here we present a numerical simulation and investigation of the generation characteristics of single-mode VCSEL experimental samples with a wavelength of 795 nm, which seem to be promising for miniature atomic clocks operating on $5S_{1/2} \rightarrow 5P_{1/2}$ transitions of Rb^{87} atoms.

The VCSEL structure was developed based on the following requirements for the emitter of the miniature atomic clock:

1. Rigorously specified wavelength (795 nm) corresponding to the $5S_{1/2} \rightarrow 5P_{1/2}$ transition of Rb^{87} atoms, and the possibility of exact tuning of the laser wavelength to this resonance;
2. Linearly polarized radiation; stable single-mode regime of generation on the TEM_{00} fundamental longitudinal mode;
3. Minimum threshold and operation currents (≈ 1 mA) with output power in the single-mode regime approximately equal to 0.1 mW.

The optical characteristics of the VCSEL developed were calculated on the base of the one-dimensional model of characteristic matrices, and the CAMFR rigorous three-dimensional model of eigenmodes. The laser structure is illustrated in Fig. 18.1A. The laser is a multilayer semiconductor structure based on the $Al_xGa_{1-x}As$ alloy, which contains a total of 1156 $Al_xGa_{1-x}As$ layers with different compositions. Between the Bragg mirrors, there are semiconductor layers with a total thickness of λ, which form the active laser area. The active VCSEL area contains three nondoped $Al_{0.07}Ga_{0.93}As$ quantum wells 8-nm thick, which are located near the maximum of the standing electromagnetic wave of the laser microcavity (Fig. 18.1B and C). An oxide $AlGaO_x$ layer forms the laser aperture. The aperture size A, defined by the inner diameter of the oxide ring, was varied from 1 to 10 μm, with a step of 1 μm in computations and experiments. The period of the Bragg mirrors embraces $\sim\lambda/4$ layers of $Al_{0.94}Ga_{0.06}As$ and $Al_{0.23}Ga_{0.77}As$ with interface inserts on the boundaries between

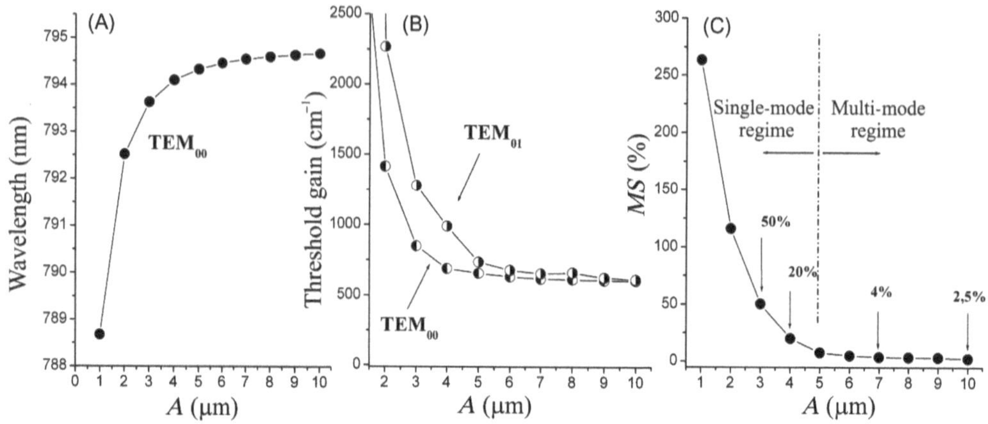

FIGURE 18.2 The characteristics of the laser structure predicted by the three-dimensional model: (A) resonant wavelength of the VCSEL versus the oxide aperture size *A*; (B) threshold gain for the fundamental mode TEM_{00} (G_{thF}) and the first transverse mode TEM_{01} (G_{thT}) versus the oxide aperture size *A*; (C) modal stability (*MS*) parameter versus the aperture size *A*.

the layers; these interface inserts are formed by thin (2 nm) $Al_xGa_{1-x}As$ layers with a variable composition. The interface inserts are used to reduce the ohmic resistance of the Bragg mirrors. The top Bragg mirror contains 28 periods, and the bottom Bragg mirror has 35.5 periods, this ensures a high level of reflection factors of the mirrors (99.79% and 99.93%, respectively, as is predicted by the one-dimensional model with an allowance for absorption on free charge carriers). This factor ensures a high figure of merit of the microcavity, a low level of threshold amplification, and a low level of VCSEL threshold and operation currents. The reflection spectrum of the full laser structure, demonstrating a resonant wavelength equal to 795 nm, is shown in Fig. 18.1D.

The laser structure characteristics calculated by the three-dimensional model are plotted in Fig. 18.2. As the oxide aperture size *A* is reduced, the lateral size of the mode decreases; the reduction of the mode volume leads to a short-wave shift of the resonant wavelength of the laser. This effect is demonstrated in Fig. 18.2A. As can be seen from the figure, the effect of the short-wave shift is fairly significant for lasers with a small oxide aperture; its value reaches several nanometers, which should be taken into account in developing a VCSEL with a rigorously specified wavelength of 795 nm. Fig. 18.2B shows the calculated threshold gain for the fundamental mode TEM_{00} (G_{thF}) and the first transverse mode TEM_{01} (G_{thT}) as a function of oxide aperture size *A*. It can be seen that the level of the threshold gain for these modes almost coincides to $A > 5\,\mu m$. As *A* is reduced, the loss of amplification is more intense for the transverse mode, which results in a more drastic increase in the threshold gain for the TEM_{01} mode than for the TEM_{00} mode. The relative difference between the threshold gain values for these modes is determined by the modal stability (*MS*) parameter:

$$MS = 100(G_{thT} - G_{thF})/G_{thF} \tag{18.1}$$

The *MS* parameter is plotted in Fig. 18.2C as a function of *A*. If the *MS* parameter is small (G_{thT} and G_{thF} differ by several percent), the VCSEL operates in the multi-mode regime. Single-mode operation is only possible if G_{thT} is much higher than G_{thF}, i.e., in the case of high values of the *MS* parameter. For the VCSEL under study, the boundary separating the single-mode and multi-mode regimes of laser operation is indicated in Fig. 18.2C by the dotted-and-dashed vertical line. Thus, a stable single-mode regime of VCSEL operation can be expected at $A \leq 5$ μm.

Various fabrication stages of the experimental VCSEL samples are shown in Fig. 18.3. The initial laser structure was grown on the Riber C21 molecular beam epitaxial (MBE) facility on *n*-GaAs substrates with orientation (001); the substrate diameter was 76 mm. Before growing the full laser structure, we performed careful calibrations of the growth rates of the $Al_xGa_{1-x}As$ layers, which provided the accuracy for specifying the layer thickness to within 1%. The full laser structure was grown with substrate rotation, which ensured the best homogeneity of layer thicknesses over the structure area. Fig. 18.3A shows a cross-section of the grown full laser structure, which was obtained by scanning electron microscopy (SEM). An analysis of the laser cavity resonant wavelength distribution over the structure area shows that the wavelength is ≈798 nm at the center of the structure, and decreases to ≈792 nm

FIGURE 18.3 Stages of VCSELs fabrication: (A) MBE-grown initial structure (SEM image); (B) laser mesa structure (SEM image); (C) ready VCSELs; (D) wafer containing ≈2800 VCSELs; (E) VCSEL near the generation threshold ($I \approx I_{th}$); (F) VCSEL in the generation regime ($I \approx 3I_{th}$).

toward the structure's edge. Thus, the structure is highly homogeneous in terms of the grown layer thicknesses ($\approx 0.7\%$) and contains regions with the necessary wavelength $\lambda_{res} \approx 795$ nm.

For fabricating laser emitters, we formed mesas of height $\approx 6\ \mu m$ (Fig. 18.3B) on the grown structure; the mesa diameter was varied at a step of 1 μm for the purpose of obtaining VCSELs with different oxide aperture sizes. After that, the VCSEL oxide aperture was formed by means of selective $Al_{0.98}Ga_{0.02}As$ layer oxidation in the atmosphere of water and nitrogen vapors at $T = 420°C$ over 30 min. The oxidation depth was $\sim 20\ \mu m$, which made it possible to form lasers with apertures from 1 to 10 μm with a step of 1 μm. Further technological steps included the formation of dielectric layers and metallic contact areas. The ready-made laser structures are illustrated in Fig. 18.3C–F.

The results of studying the characteristics of the ready-made VCSELs (at $T = 295$K) are summarized in Fig. 18.4.

The lasers are characterized by a low level of threshold current I_{th}, which varies from 150 to 800 μA, depending on the laser aperture diameter (Fig. 18.4A). The quantum efficiency of the fabricated VCSELs is 0.3–0.4 μW/μA. Inspection of the spectral characteristics (Fig. 18.4B and C) shows that lasers with $A \leq 5\ \mu m$ demonstrate a stable single-mode generation regime

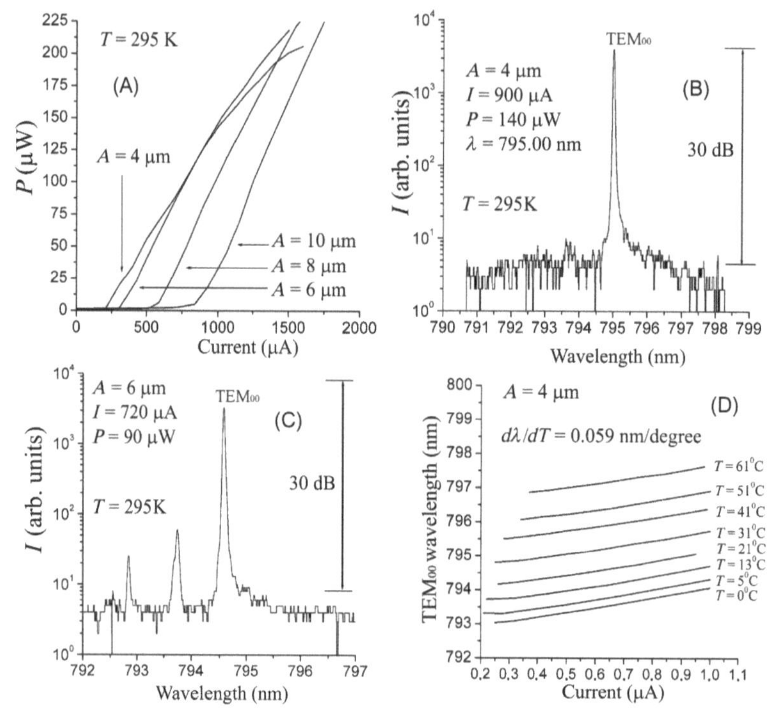

FIGURE 18.4 Characteristics of the laser structures examined: (A) family of watt–ampere characteristics for VCSELs with different values of the oxide aperture A; (B) emission spectrum of the VCSEL with $A = 4\ \mu m$ for $I = 900\ \mu A$ and $P = 219\ \mu W$; (C) emission spectrum of the VCSEL with $A = 6\ \mu m$ for $I = 720\ \mu A$ and $P = 124\ \mu W$; (D) VCSEL wavelength versus injection current and temperature.

in the entire range of pumping current examined (0–1.5 mA), while lasers with $A = 6\ \mu m$ exhibit single-mode operation only when the threshold current is slightly exceeded. Finally, lasers with large apertures operate in the multi-mode regime at all levels of pumping. The experimental data obtained are in good agreement with the calculated results described above (Fig. 18.2C). Fig. 18.4B shows the emission spectrum of a VCSEL with $A = 4\ \mu m$, operating in the single-mode regime at a wavelength of 795 nm with output power of 200 μW. The maximum power of this VCSEL emission is 350 μW at a current of 1.5 mA. The VCSEL wavelength can be accurately tuned to the $5S_{1/2} \rightarrow 5P_{1/2}$ transition of Rb^{87} atoms by changing the external temperature, or by changing the injection level, which also alters the temperature regime of laser operation. Fig. 18.4D shows the results of studying the laser wavelength reconstruction due to changes in the injection current and temperature for a VCSEL with aperture $A = 4\ \mu m$. The temperature factor of laser wavelength variation is $d\lambda/dT = 0.059$ nm/deg, which is in good agreement with the data obtained in Refs. [1,6].

Thus, the set of experimental data presented (stable single-mode operation regime with an output power up to 350 μW at low operation currents of ≈1.5 mA, a wavelength of 795 nm, and the possibility of accurate tuning of the latter by means of changing the temperature and injection current) demonstrates numerous prospects for using the lasers developed in miniature atomic clocks.

18.2 Single-Photon Emitters

The absolute limit in light emitter miniaturization is the development and design of SPEs. Effective SPEs are most important for systems of quantum cryptography and quantum calculations; they are also of great interest for producing precise spectral equipment and optical power standards. An ideal SPE is defined as a photon source in which only one photon is emitted under the action of a control signal. This emitter is a source of single photon Fock states, and can be implemented only on the base of isolated quantum systems, such as a single atom, a molecule, a color center, or an "artificial" atom, a single semiconductor quantum dot (QD).

The implementation of effective SPE devices is one of the most difficult problems in nanotechnology, because it requires the following operations:

1. Localization of a quantum system;
2. Its efficient pumping;
3. Collection of radiation.

All of the above-mentioned problems can, in principle, be solved by means of modern semiconductor technology, which makes it possible to integrate a single semiconductor QD into a semiconductor microcavity. An obvious advantage of semiconductor SPEs is the possibility of pumping QDs by means of an electric current. This provides a means for the development of entirely solid state emitters in the form of miniature light emitting diodes. Recent work in this field has been focused on the search for optimal growth technologies and the positioning of QDs, and on the development of a microcavity design that ensures the highest external quantum yield and the smallest emitted radiation divergence [13–19].

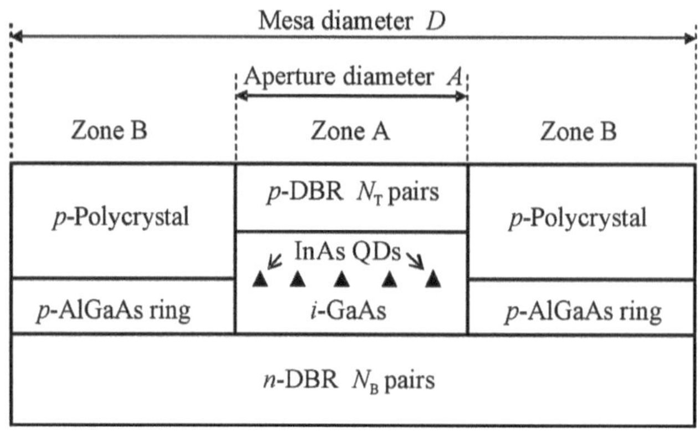

FIGURE 18.5 Layout of the quantum dot-based single-photon emitter with a Bragg microcavity.

Here, we report on the design and implementation of an entirely semiconductor Bragg microcavity for a current-pumped SPE, in which an $Al_xGa_{1-x}As$ ring provides a high level of electrical and optical confinement, which can also be used for selective positioning of QDs within the current and optical aperture. In contrast to previous designs [13–15], the microcavity does not contain oxide aperture layers which cause high levels of mechanical strain. As a result, this structure is able to withstand thermal cycling, and can operate at cryogenic temperatures.

The layout of the Bragg microcavity structure for QD-based SPEs is shown in Fig. 18.5.

The structure contains two semiconductor distributed Bragg reflectors (DBRs) with *p*-type and *n*-type doping. The top output mirror consists of N_T pairs of alternating GaAs and $Al_{0.9}Ga_{0.1}As$ quarter-wave layers, and the bottom mirror consists of N_B pairs of alternating GaAs and AlAs quarter-wave layers. Between the mirrors there is a λ thick undoped GaAs layer. In the middle of this layer, at the antinode of the cavity standing wave, there is a layer of InAs QDs, which are excited on passing an electric current through the $p-i-n$ structure. The QDs are shown as triangles in Fig. 18.5. The *i*-GaAs layer with QDs is shaped as a disk with diameter *A*. This disk is surrounded by a ring of *p*-type $Al_xGa_{1-x}As$ with outer diameter *D* and thickness $\lambda/4$. This ring defines the current and optical aperture of the emitter (zone A). Above this ring (zone B), there is a polycrystalline ring with *p*-type conductivity, from which holes are injected into the aperture region. Owing to the presence of the wide-gap $Al_xGa_{1-x}As$ ring ($x \approx 0.6$), the current flows predominantly through the aperture region formed by the GaAs layer containing InAs QDs. Calculations of the current–voltage characteristics for heterostructures formed in zones A and B suggest that the efficiency of the current confinement, as given by the ratio $K_{CC} = j_A/j_B$ of the current densities in these zones, is at least 10^3.

An extremely important parameter of a SPE is the external quantum efficiency. In the case of emitters based on single semiconductor QDs, it is determined as the ratio of the number of photons in the output signal to the number of electron–hole pairs injected into the QDs. The external quantum efficiency $\eta_{external}$ of an emitter is determined by the product of two factors:

$$\eta_{external} = \eta_{internal} \times \eta_{extraction} \tag{18.2}$$

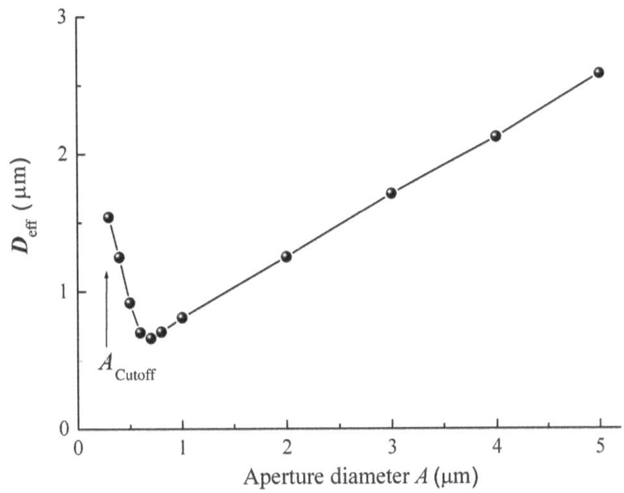

FIGURE 18.6 Dependence of the effective mode diameter D_{eff} on the aperture diameter A for a microcavity with $N_B = 15$ and $N_T = 2$ periods in the bottom and top distributed Bragg reflectors, respectively.

In high-perfection defectless QDs, the first factor, which is the internal quantum efficiency, may be as high as a unit [17–19]. Thus, the external quantum efficiency of a SPE is determined by the radiation extraction coefficient $\eta_{extraction}$. The microcavity design was optimized in order to obtain the highest value of $\eta_{external} \approx \eta_{extraction}$ for the fundamental mode. The main characteristics of the fundamental mode, such as mode frequency ω_{cavity}, the distribution of electric and magnetic fields, and the electromagnetic energy flux (the Poynting vector), the mode Q factor, mode volume V and effective diameter D_{eff}, and the Purcell factor F_P were calculated. The calculations were performed using the one-dimensional characteristic matrix model [20], and the CAMFR rigorous three-dimensional model of eigenmodes [21]. The external quantum efficiency of the emitter was calculated as [5,19]:

$$\eta_{external} \approx \eta_{extraction} = \beta_{couple} \times \eta_{conf} \tag{18.3}$$

Here, β_{couple} is the efficiency of photon coupling to the fundamental mode, which is determined by the Purcell factor [19], and η_{conf} is the configurational parameter of the fundamental mode, which determines the ratio of the emitter output flux power to the net energy loss rate in the microcavity. To obtain high values of β_{couple}, one needs a high Q and a small mode volume V simultaneously. Mode volume V is determined by the efficiency of axial and lateral localization of the electromagnetic field energy in the microcavity. GaAs/AlGaAs Bragg mirrors provide efficient energy localization in the axial direction. Efficient localization of the electromagnetic field energy in the lateral direction is provided by the $Al_{0.6}Ga_{0.4}As$ ring, which has a lower refractive index compared to GaAs and InAs, forming the aperture region. The efficiency of optical localization in the lateral direction for this type of microcavity is illustrated in Fig. 18.6, which shows effective mode diameter D_{eff} as a function of aperture diameter A. Effective diameter D_{eff} is determined according to the following expression [19]:

$$D_{eff} = \frac{\int \varepsilon(r)|E(r)|^2 dr}{\max\left(\varepsilon(r)|E(r)|^2\right)} \tag{18.4}$$

where $\varepsilon(r)$ is the dielectric constant, and $E(r)$ is the electric field of the mode. Integration is carried out along the lateral direction r within the microcavity cross-section corresponding to the antinode of the standing wave.

Dependence $D_{eff}(A)$ is nonmonotonic. As aperture diameter A decreases, the mode diameter D_{eff} decreases as well. However, for very small diameters $A < \lambda$, the aperture no longer provides efficient localization of the mode energy in the lateral direction; in this region, the mode diameter increases with decreasing A. The minimum value of D_{eff} is attained for $A \approx 0.8$ μm. For $A_{cutoff} \approx 0.35$ μm, the cutoff of the fundamental mode occurs, and for $A \rightarrow 0$ it transforms into a plane wave. Up to the values $A \approx 0.7$ μm, the optical confinement efficiency $K_{OC} = D_{eff}/A < 1$.

Mode volume V was calculated as:

$$V = \frac{\iiint_V \varepsilon(\bar{r})|E(\bar{r})|^2 d^3\bar{r}}{\max(\varepsilon(\bar{r})|E(\bar{r})|^2)} \tag{18.5}$$

where integration is carried out over the microcavity volume.

The Q factor was calculated as follows:

$$Q = \frac{\iiint \frac{D \cdot E + B \cdot H}{2} \cdot dV}{\sum \iint S \cdot dS} \cdot \omega_{cavity} \tag{18.6}$$

Here, D and B are the inductions and E, and H the electric and magnetic field strengths, respectively; ω_{cavity} is the fundamental mode frequency; and S is the Poynting vector. Integration is carried out over the microcavity volume and over the surfaces bounding the microcavity in calculating the mode energy (numerator) and the energy loss rate (denominator).

The Purcell factor F_P and parameters β_{couple} and β_{conf} were calculated according to the following equations:

$$F_P = \frac{3Q(\lambda/n)^3}{4\pi^2 V} \tag{18.7}$$

$$\beta_{couple} = \frac{F_P}{F_P + 1} \tag{18.8}$$

$$\eta_{\text{conf}} = \frac{\iint S_{\text{out}} \cdot dS}{\Sigma \iint S \cdot dS} \tag{18.9}$$

Integration in the numerator and denominator is carried out over the top surface, and over all surfaces bounding the microcavity, respectively. Fig. 18.7 shows the calculated external quantum efficiency η_{external} for a microcavity containing $N_B = 15$ pairs of layers in the bottom mirror (the mirror reflectivity equals 97.8%) for different aperture diameters A, and numbers of layer pairs N_T in the top mirror.

The highest value of $\eta_{\text{external}} = 59\%$ is attained for $A = 0.8\ \mu\text{m}$ and $N_T = 8$. For $A < 0.8\ \mu\text{m}$; there is an increase in D_{eff} (Fig. 18.6), which causes an increase in modal volume V and a reduction in F_P, β_{couple}, and η_{external}. A decrease in η_{external} for $A > 0.8\ \mu\text{m}$ is also related to an increase in D_{eff} and V, and a reduction in β_{couple}. For constant A, an increase in $N_T > 8$ causes a reduction in the external quantum efficiency η_{external} due to a decrease in η_{conf}, which cannot exceed 0.5 for a symmetrical cavity ($N_B = N_T$). A decrease in $N_T < 8$ leads to a decrease in η_{external} due to a considerable drop in the Q factor. Thus, for $A = 0.8\ \mu\text{m}$ and $N_T = 8$, $Q \approx 500$, while for $A = 0.8\ \mu\text{m}$ and $N_T = 2$, $Q \approx 80$. This decrease in Q entails a decrease in F_P, β_{couple}, and η_{external}.

A further increase in the external quantum efficiency can be attained in microcavities with extremely high Q factors in the order of 10^4. In this case, F_P may exceed 10, β_{couple} may be higher than 90%, and η_{conf} also may exceed 90% in asymmetrical microcavities. Calculations for microcavities with $N_B = 35$ (reflection coefficient 99.85%) and N_T in the

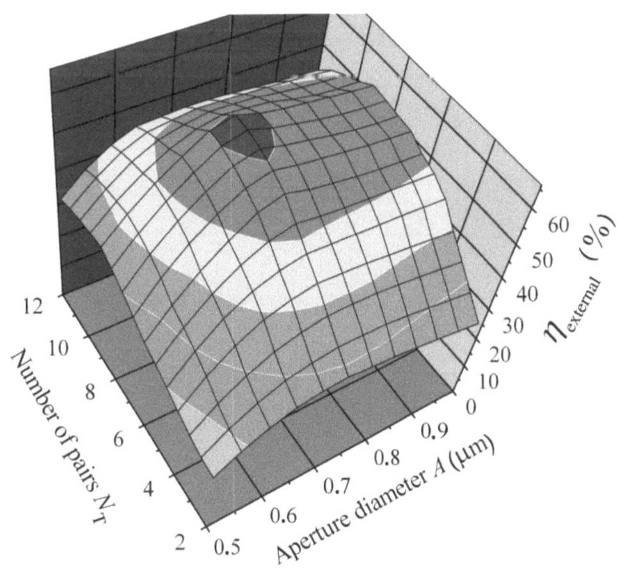

FIGURE 18.7 External quantum efficiency for a microcavity with $N_B = 15$ pairs of layers in the bottom mirror as a function of the aperture diameter A, and the number of pairs of layers N_T in the top mirror.

range of 15–25 yield values of $\eta_{\mathrm{external}} \geq 80\%$. The highest calculated value of $\eta_{\mathrm{external}} = 88.4\%$ was found for a microcavity with $N_B = 35$ and $N_T = 19$.

The radiation divergence for the fundamental mode of the microcavity was also calculated. Divergence angle α_{div} was defined with respect to the $1/e^2$ level of the power density of emitted radiation. The calculation results for a microcavity with $N_B = 15$ and $N_T = 2$ are shown in Fig. 18.8, with the dependence of the numerical aperture $NA = \sin(\alpha_{\mathrm{div}})$ on microcavity aperture diameter A. The largest divergence angle ($NA \approx 0.21$) is attained for $A = 0.8$ μm, when effective modal diameter D_{eff} is minimal (see Fig. 18.6). According to Fig. 18.8, the designed microcavity features low divergence of emitted radiation ($NA \leq 0.2$), which ensures that the radiation can be efficiently coupled into a standard optical fiber.

Prototype emitters were made on the basis of the microcavity design described above. The emitters are fabricated in several steps. First, an n-type doped bottom DBR and then a p-type doped $Al_xGa_{1-x}As$ $\lambda/4$ layer are grown by molecular beam epitaxy on a (001)-oriented n-GaAs substrate. The growth was performed using a Riber C21 system.

Next, a p-$Al_xGa_{1-x}As$ ring with an outer diameter of D and an inner diameter of A (Fig. 18.5) was formed by wet etching. The mesa diameter D was 18 μm, and the aperture diameter A varied between 1 and 10 μm. The etching depth was set somewhat larger than the depth of the p-$Al_xGa_{1-x}As$ layer. The actual etching depth was determined using an atomic-force microscope, and was taken into account when performing the regrowth. This is the next step, in which an undoped GaAs layer containing InAs QDs is produced by MBE. The QDs layer was grown by the Stranski–Krastanov mechanism [17–19]. For this purpose, undoped InAs is deposited on the GaAs surface. Upon reaching the critical InAs layer thickness (~1.8 monolayers (MLs)), an array of self-assembled InAs QDs is formed. The QD array is overgrown with an undoped $\lambda/2$ thick layer of GaAs. On the surface of this layer, a p-type

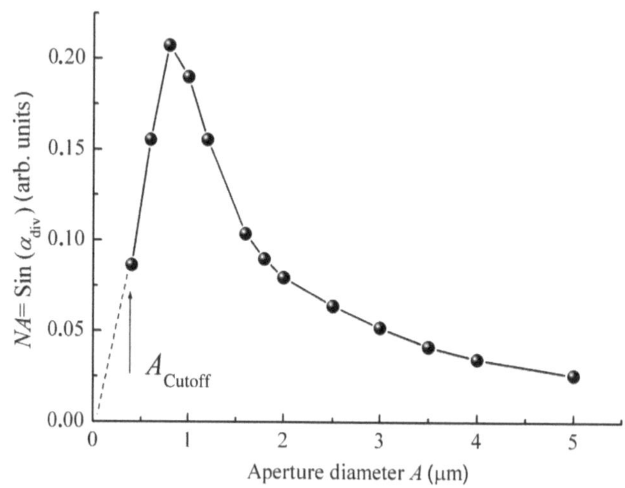

FIGURE 18.8 Dependence of the numerical aperture $NA = \sin(\alpha_{\mathrm{div}})$ on the aperture diameter A for a microcavity with $N_B = 15$ and $N_T = 2$.

doped top DBR was grown. In order to attain the required optical characteristics of microcavities and the necessary QDs density, the growth conditions were carefully chosen and corrected, taking into account the photoluminescence and reflectivity data obtained for the test structures.

A series of experiments, in which the QDs density, determined by the InAs layer thickness and the duration of growth interruption between the end of InAs growth and the beginning of GaAs growth, was varied over a broad range from about 10^7–10^{10} cm^{-2}, were carried out. On the basis of these results, structures with optimum QDs density, in which peaks corresponding to the emission of single QDs appear within the resonance band in the electroluminescence spectra, were selected. The QDs density in these structures was ~10^8–10^9 cm^{-2}. The deviation of the experimentally determined resonance wavelength of the microcavities from the calculated value did not exceed 1–2%. At the final stage of the emitter fabrication process, dielectric layers were deposited and contact pads were formed (Fig. 18.9). QDs within inner diameter A (zone A, Fig. 18.5) are located in the region of a p–n junction and, thus, can be efficiently excited upon the application of positive bias to the structure. As was noted above, the current density in zone B is much lower and, furthermore, QDs in this zone cannot be excited upon the passage of an electrical current, because they are located in the p-region of the structure. The regrowth conditions in zones A and B differ considerably from each other. In zone A, regrowth begins with a GaAs layer and proceeds in the normal mode. In this case, high-quality QDs and perfect DBRs are formed, which is confirmed by the data on photoluminescence and reflectivity. In zone B, regrowth begins with a layer of $Al_xGa_{1-x}As$ ($x = 0.6$), which contains nearly unavoidable aluminum oxides. This causes a significant deterioration in the quality of the grown layers, and leads to the

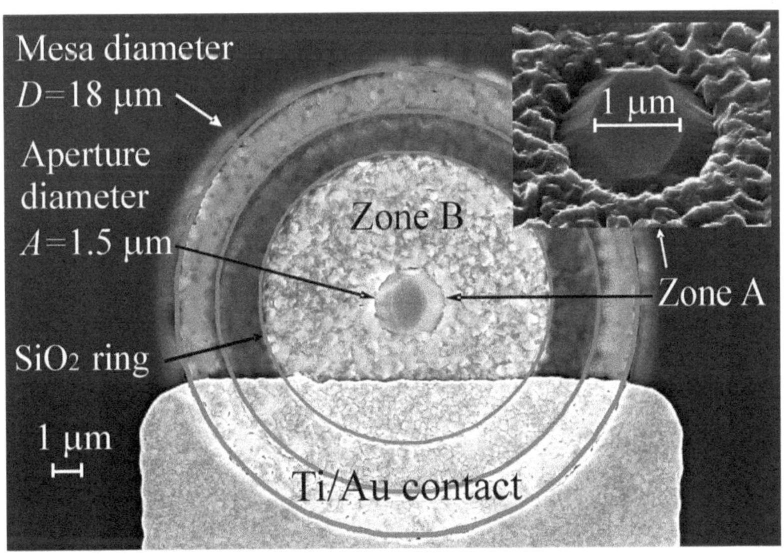

FIGURE 18.9 Scanning electron microscope image of an emitter with $D = 18$ μm, $A \approx 1.5$ μm, $N_B = 15$, and $N_T = 2$.

formation of polycrystalline regions (Figs. 18.5 and 18.9). Under these conditions, either the formation of QDs does not occur, or defective QDs with zero internal quantum efficiency are formed. This is confirmed by the photoluminescence spectra. The two above-mentioned factors ensure efficient positioning of the QDs within aperture A.

The image of an emitter obtained using a scanning electron microscope is shown in Fig. 18.9. Here, the single crystalline aperture region (zone A) and the polycrystalline region (zone B) are clearly seen. The aperture diameter A is $\approx 1.5\ \mu m$. In the small InAs QDs investigated here (emitting in the spectral range of 0.9–1.0 μm), there are only ground electron and hole levels. The electron level can be occupied by two electrons with opposite spins (the electron spin component along the growth axis z may be $\pm 1/2$). In the valence band, only heavy holes (with the momentum component along the growth axis equal to $\pm 3/2$) are localized [19]. Thus, the emission spectrum of a single InAs QD can consist of a number of peaks corresponding to different charge states of the QD and corresponding to the recombination of an exciton X, a biexciton XX, or a trion X^+ (one electron and two holes in the QD), or X^- (two electrons and one hole) [17–19]. For the lowest excitation level, peaks corresponding to excitons (X) appear first. With increasing excitation power, the peaks of biexcitons XX, trions X^+, and X^- appear as well; all peaks can be present in the spectra simultaneously. Recombination in any of the above-mentioned charge states of a QD (X, XX, X^+, or X^-) can be used to implement a SPE. However, in order to obtain an emitter with high efficiency, it is necessary for the spectrum to contain only one peak. The spectral filtering of these peaks is difficult, because all of them occur within a spectral range just a few nanometers wide. Thus, the optimum conditions correspond to low excitation levels, when the emission spectrum features only one peak, corresponding to the exciton recombination (X). The prototype emitters investigated here contained $N_B = 15$ periods in the bottom DBR, and $N_T = 2$ periods in the top DBR, which provides a value of $Q \approx 80$ and a resonance linewidth of $\Gamma \approx 12$ nm. Structures were obtained where, at cryogenic temperatures and low excitation levels, narrow peaks corresponding to the emission of single InAs QDs appear in the electroluminescence spectrum within the resonance band.

The electroluminescence spectra of an emitter recorded for $T = 10$K and different levels of electric current excitation are shown in Fig. 18.10. At the lowest excitation level, the spectrum contains a single line X, corresponding to exciton recombination (Fig. 18.10, the bottom curve). As the pump-level increases, additional peaks XX, X^+, and X^- appear in the spectrum (Fig. 18.10, the top three curves). These results experimentally support the feasibility of the creation of a SPE based on the microcavity design developed.

18.3 Emitters of Entangled Photon Pairs

Quantum entanglement [22–25] is a quantum mechanical phenomenon in which the quantum states of two or more objects are interdependent. This interdependence is retained even if these objects are separated in space beyond the limits of any known interactions. For example, one can get a pair of photons in an entangled state and then, if when

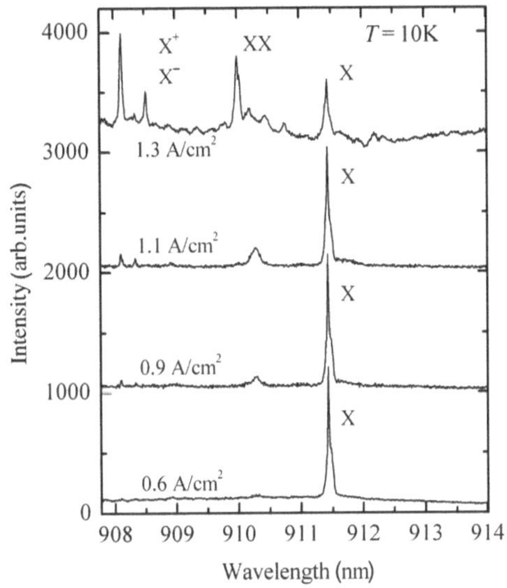

FIGURE 18.10 Electroluminescence spectra of an emitter recorded at $T = 10K$ for pump-current densities of 0.6, 0.9, 1.1, and 1.3 A/cm^2.

measuring the spin of the first particle helicity is positive, the second helicity is always negative, and vice versa. EEPPs can be used in both quantum computing systems, and in the implementation of a number of quantum cryptography protocols. Generation of entangled photon pairs can be, possibly, the result of various physical processes, starting with "space" processes, such as annihilation of positronium (parapositronium consisting of an electron and a positron with opposite spins), and the Higgs boson decay. In the first case, the entangled photon pair and an energy of about 0.51 MeV are reached during $\sim 10^{-10}$ s. The entangled photon pair and an energy of about 63 GeV appear in the second case during $\sim 10^{-22}$ s, which corresponds to a frequency of $\nu \approx 1.5 \cdot 10^{25}$ Hz and a wavelength of $\lambda \approx 1.96 \cdot 10^{-17}$ m (about 100 times smaller than the classical electron radius $r_o \approx 2.81 \cdot 10^{-15}$ m). In the visible and near-IR range, entangled photon pairs can occur as a result of parametric down-conversion in nonlinear optical crystals (β-BaB_2O_4, LiB_3O_5, $KTiOPO_4$, and others), and also as a result of two-photon cascade emission in atoms (e.g., of Ca). Note that, as a result of these processes, the number of photon pairs described by the classical Poisson distribution are generated.

Semiconductor QDs are also promising for the fabrication of EEPP [5,24–30]. Such pairs can be emitted by semiconductor QDs in the biexciton–exciton recombination cascade if the exciton states are energy degenerate, or if their splitting ΔE_{FS} do not exceed the natural width of exciton levels $\Gamma_X = \hbar/\tau_X$ (where τ_X is the exciton lifetime). In this case, a pair of polarization-entangled photons is emitted [5]. In actual QDs, exciton-state splitting ΔE_{FS} are usually much larger than the natural width of exciton levels Γ_X. This is caused by the deviation of the QD shape from the ideal, the presence of a piezopotential induced by built-in mechanical strains, and the formation of solid solutions in the interface regions [5,29,30]. The investigation of the factors that determine ΔE_{FS} and the conditions required for the suppression of the exciton-state splitting to the level of Γ_X are important for the development of EEPP based on semiconductor QDs.

Here, we present the results of a study of the fine structure of exciton states in InAs QDs grown using molecular beam epitaxy by the Stranski–Krastanov method with short growth interruptions.

The samples under study were grown in a Riber C21 MBE machine on (001)-oriented GaAs substrates. A 0.3 μm-thick GaAs buffer layer was grown on the substrate surface. The structure contained a 160-nm thick GaAs layer clad between two 50-nm thick $Al_{0.6}Ga_{0.4}As$ layers to prevent the diffusion of photoexcited charge carriers. At the center of the GaAs layer, where the major fraction of the excitation laser power was absorbed, a layer of InAs QDs was placed. The top $Al_{0.6}Ga_{0.4}As$ layer was capped by a 10-nm thick GaAs protective layer. InAs QDs were grown by the Stranski–Krastanov mechanism [5,17–19] at a temperature of T = 430°C. A critical-thickness layer of InAs (~1.8 MLs) was grown on the GaAs surface. Upon attaining this thickness, the formation of an array of self-assembled InAs QDs commenced. The growth rate of the InAs layer was 0.02 ML/s. The As/In element flux ratio was kept to a level of 100. The transition from the two-dimensional to the three-dimensional growth mode was monitored by high-energy electron diffraction. Upon attaining critical thickness, the deposition of the InAs layer was halted and growth was interrupted for a period τ_{GI}. During this time, an array of InAs QDs formed via the Ostwald mechanism [31–34]. Next, the layer of InAs QDs was overgrown by GaAs. The growth interruption time τ_{GI} was varied from several seconds to several hundred seconds. The value of τ_{GI} determines, to a large extent, the QD array parameters. An increase in τ_{GI} to 10^1–10^2 s leads to high-density QD array formation (up to 10^{10} cm^{-2}), and both the average QD size and size dispersion increase. Meanwhile, using short growth interruptions (a few seconds) is an efficient method of obtaining QD arrays of ultralow densities (to 10^6 cm^{-2}). In this case, QDs of minimal size and small size dispersion are formed. The InAs QD structures were studied by macro- and microphotoluminescence. The area of the excitation laser spot on the surface of the sample was 3000 and 3 μm^2, respectively. The luminescence was excited by 532 nm radiation from a continuous-wave Nd:YAG laser. The radiation power density on the sample surface was varied between 0.01 and 100 W/cm^2. The samples were placed in He cryostats where a temperature of $T = 10K$ was maintained. The luminescence was recorded using a single monochromator equipped with a cooled p–i–n Ge photodiode (macroluminescence instrument), or a triple TriVista-555 monochromator with a cooled Si matrix photodetector (microluminescence instrument).

Fig. 18.11 shows the spectra of macroluminescence and microluminescence (insets) for two structures with InAs QDs. In the fabrication of the first structure (Fig. 18.11A), the duration of the growth interruption was $\tau_{GI} = 120$ s. In this case, a high-density QD array (~10^{10} cm^{-2}) was formed. The second structure (Fig. 18.11B) was fabricated with a growth interruption of $\tau_{GI} = 3$ s, which provided a very low density of QDs (at a level of ~10^6 cm^{-2}).

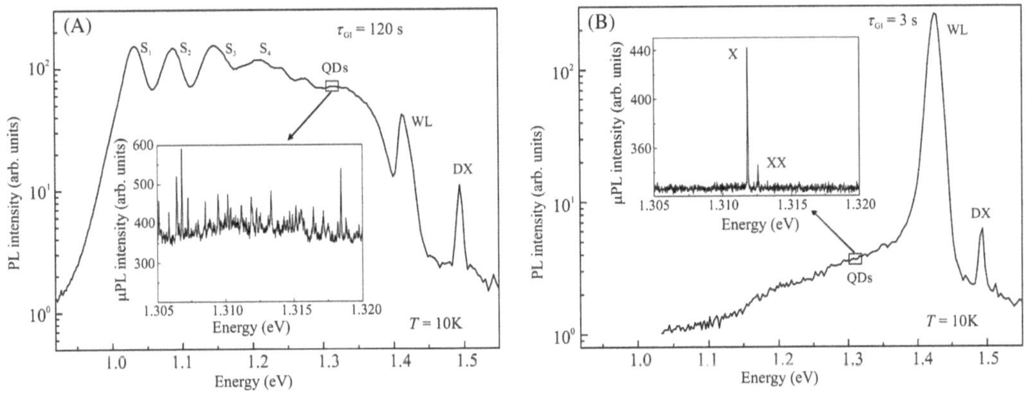

FIGURE 18.11 Spectra of macroluminescence and microluminescence (insets) for two structures with InAs QDs at $T = 10K$. The growth interruption time τ_{GI} was 120 s (A), and 3 s (B).

The macroluminescence spectra contain the lines corresponding to bound excitons in GaAs (DX), luminescence from the ~1.8 ML-thick InAs wetting layer, which forms a quantum well (WL), and the broad bands corresponding to the QD luminescence. For the first sample, the intense QD luminescence band appears in the energy range of 0.9–1.4 eV. This band is formed by the emission from $\sim 10^6$ QDs; for this reason, the spectrum is continuous. The luminescence peaks designated as S1, S2, etc., in Fig. 18.11A correspond to a QD multimodal size distribution. The sizes of QDs forming subassemblies S1, S2, etc., differ by one monolayer [17]. For the second sample, the QD luminescence is present as a weak low energy shoulder of the wetting-layer line in the energy range of 1.1–1.4 eV. The reduction of intensity and narrowing of the energy range is a manifestation of a considerable drop in QD density, and of a decrease in the dispersion of their sizes. The insets in Fig. 18.11 show the microluminescence spectra of the two samples under study in the energy ranges indicated by rectangular boxes in the macroluminescence spectra. The narrow lines corresponding to the luminescence of individual QDs are clearly seen. The luminescence spectrum of the first sample contains dozens of lines from different QDs forming the continuum. This hinders the interpretation of these lines and analysis of their characteristics. For the second sample, where the density of QDs is low ($\sim 10^6\ cm^{-2}$), the average spacing between the dots is about 10 μm; meanwhile, the diameter of the laser spot on the sample surface is ~2 μm. Thus, the luminescence of single InAs QDs can be excited and analyzed reliably without the need for additional nanotechnological processing steps, such as the formation of submicrometer mesas or submicrometer apertures in metal layers deposited on the surface of the structure [29,30,35]. The microluminescence spectrum of the second sample (Fig. 18.11B, inset) contains two narrow lines corresponding to the exciton (X) and biexciton (XX) luminescence in a single InAs QD. The interpretation of the luminescence lines of single QDs was based on the dependences of their intensities on the excitation power density. At the lowest laser power densities, exciton line X appears first, and its intensity varies linearly with the power

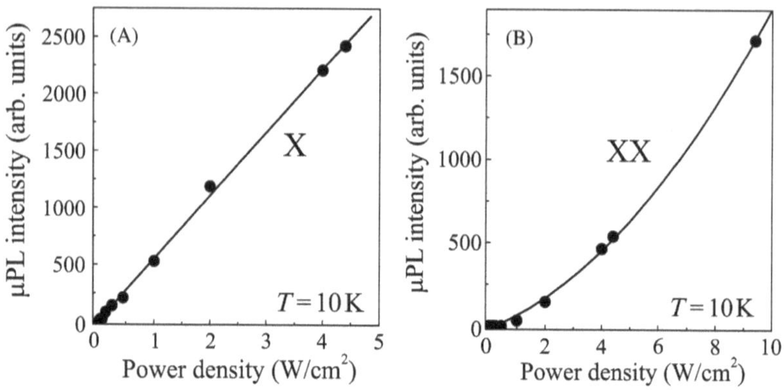

FIGURE 18.12 Intensities of the: (A) exciton (X); and (B) biexciton (XX) luminescence lines for a single InAs QD versus the excitation power density at $T = 10\text{K}$. *Circles* show experimental points, and *solid lines* correspond to the linear and quadratic approximations.

density. At higher power densities, the biexciton line XX appears in the spectra, its intensity increasing quadratically with the power density [36]. The dependences of X and XX intensities on the excitation power density are shown in Fig. 18.12.

Small InAs QDs under study have only ground-state electron and hole levels. The electron spin component along the growth axis (z axis) equals $\pm 1/2$. In the valence band, only heavy holes are confined, the component of their angular momentum along the growth axis is $\pm 3/2$ [5]. Therefore, four different energy-degenerate exciton states exist, with a total angular momentum of $M = 2$, 1, -1, and -2. The fourfold degeneracy is lifted owing to the electron–hole exchange interaction, which leads to the formation of two pairs of exciton states with a total angular momentum of $M = 2$, -2 (optically inactive excitons), and $M = 1$, -1 (optically active excitons). The latter pair of exciton states is twofold energy-degenerate. In real QDs, the twofold degeneracy of the optically active excitons is also lifted as a result of the QD shape deviation from the ideal, as well as the influence of the piezopotential induced by built-in mechanical strains [29,30]. The magnitude ΔE_{FS} of the exciton-level splitting is determined by a number of factors, such as the extent to which the QD shape deviates from the ideal, and the role of the piezopotential, whose importance grows with increasing QD size [5,29,30]. The value of ΔE_{FS} is affected by electric and magnetic fields, as well as by post-growth treatments like high-temperature annealing [5,37]. When ΔE_{FS} considerably exceeds Γ_X, the radiation of exciton and biexciton is linearly polarized along the [110] and $[1\bar{1}0]$ crystallographic directions [5,29,30]. Fig. 18.13 shows the energy diagram of the exciton states in InAs QDs, and describes exciton and biexciton radiative recombination processes for the case of $\Delta E_{FS} \gg \Gamma_X$.

In a QD containing two electrons and two holes (the biexciton state), the first electron–hole pair recombines within the characteristic time τ_{XX} with the emission of a photon polarized along the [110] or $[1\bar{1}0]$ crystallographic directions (XX[110] and XX$[1\bar{1}0]$ in

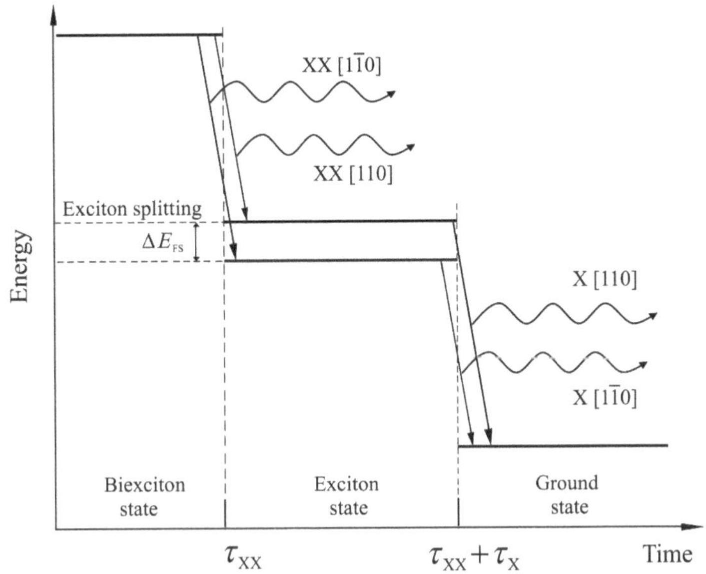

FIGURE 18.13 Energy diagram of the exciton states in an InAs QD, and the scheme of exciton and biexciton radiative recombination processes for the case $\Delta E_{FS} > \Gamma_X$.

Fig. 18.13). In the process, the QD makes a transition to the exciton state and now contains one electron–hole pair. This pair recombines within the characteristic time τ_X, emitting a photon with the same polarization along the [110] or [1$\bar{1}$0] directions (in Fig. 18.13, these photons are designated as X[110] and X[1$\bar{1}$0]). According to the energy diagram in Fig. 18.13, the energies of XX[110] and XX[1$\bar{1}$0] photons, as well as those of X[110] and X[1$\bar{1}$0] photons, differ by ΔE_{FS}. Correspondingly, one expects that the luminescence spectra recorded for polarizations oriented along the [110] and [1$\bar{1}$0] directions should contain X[110] and X[1$\bar{1}$0] exciton lines, and XX[110] and XX[1$\bar{1}$0] biexciton lines, which are shifted by ΔE_{FS} with respect to each other. Thus, it is possible to determine the exciton-state splitting parameter ΔE_{FS} directly from the emission spectra (see Fig. 18.14). In this study, the value of ΔE_{FS} was determined as the difference between the energies of X[110] and X[1$\bar{1}$0] photons emitted upon the exciton recombination in QDs. As was mentioned previously, ideal QDs with energy-degenerate exciton states ($\Delta E_{FS} = 0$) and QDs with a small exciton-state splitting ($\Delta E_{FS} \approx \Gamma_X$) may serve as emitters of entangled photon pairs [5].

When these conditions are satisfied, pairs of polarization-entangled photons are emitted in the process of biexciton–exciton cascade recombination. According to the published data, the exciton lifetime τ_X in InAs QDs at cryogenic temperatures is about 10^{-9} s [18,36,38]. This sets the width of exciton line X at the level of 10^{-6} eV. Thus, the fabrication of EEPPs requires InAs QDs in which ΔE_{FS} does not exceed a few microelectronvolts. For the characterization of QDs satisfying this requirement, the value of ΔE_{FS} also needs to be determined with an error of a few microelectronvolts.

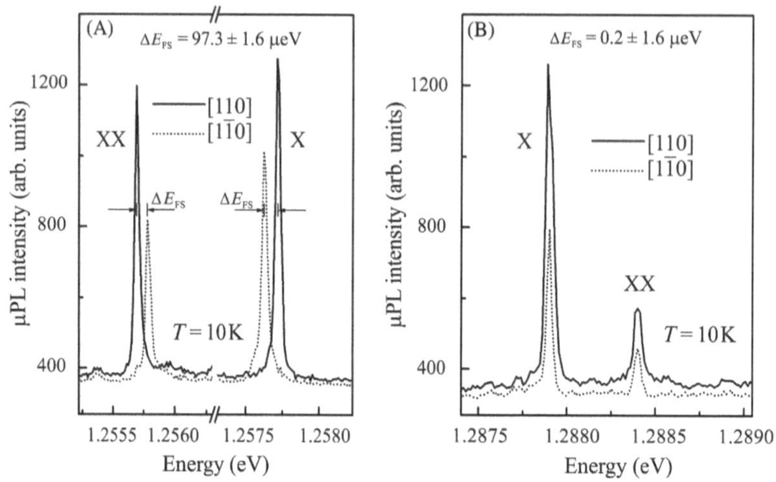

FIGURE 18.14 Microluminescence spectra of two InAs QDs at $T = 10$K: (A) QD with $\Delta E_{FS} = 97$ μeV; and (B) QD where the exciton-state splitting does not exceed the measurement error (± 1.6 μeV).

In order to improve the accuracy with which ΔE_{FS} is determined, we approximated the experimentally recorded contours of the luminescence lines by Lorentzian contours, and the peak of the Lorentzian contour was taken to be the peak of the line. Using this procedure, we determined the values of ΔE_{FS} with the error $\sigma_{FS} = \pm 1.6$ μeV, which is comparable to the natural width Γ_X of the exciton lines. Parameter ΔE_{FS} was determined for 135 individual QDs (see Figs. 18.14 and 18.15).

The experimental results for ΔE_{FS} plotted against the QD exciton energy are summarized in Fig. 18.15. Let us discuss the main features of the dependence obtained. With a decrease in the exciton energy (i.e., with an increase in the QD size), ΔE_{FS} increases steadily from small values to those as high as ~ 100 μeV for exciton energies around 1.22 eV. A large scattering in the values of ΔE_{FS} for a given exciton energy can be seen. These features were already reported previously [5,29,30]. The increase in ΔE_{FS} is related to an increase in piezopotential with a growing QD size. Large fluctuations in ΔE_{FS} are related to the deviation of QDs shape from the ideal, i.e., a reduction of QD symmetry from C_{4v} to C_{2v}, and fluctuations in the aspect ratio of the vertical and lateral QDs sizes. The most important feature of the dependence obtained is the existence of a group of QDs with exciton energies in the range of 1.3–1.4 eV, where the magnitude of ΔE_{FS} is close to zero within the measurement accuracy (± 1.6 μeV). This group consists of the smallest QDs, where the piezopotential does not cause a noticeable splitting of the exciton states. This group is indicated by a rectangular box in Fig. 18.15. The inset in Fig. 18.15 shows the histogram describing the distribution of the QDs with respect to the exciton-state splitting in the range of ΔE_{FS} from -10 to 40 μeV. The column width is twice the measurement error (3.2 μeV), and is comparable to the natural width Γ_X of the exciton levels. One can see that the maximum of this distribution

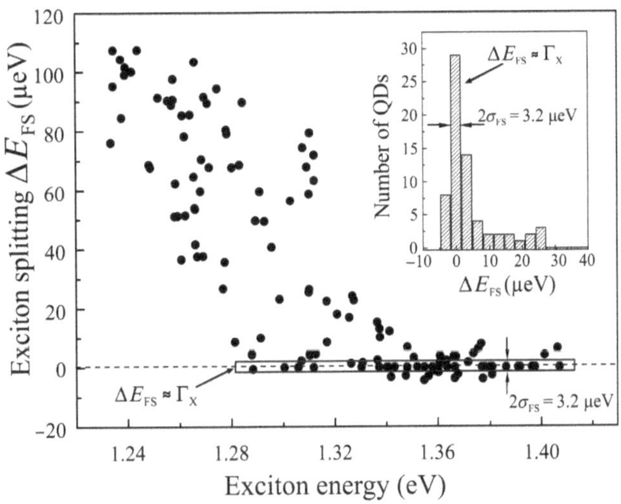

FIGURE 18.15 Data on the exciton-state splitting ΔE_{FS} for 135 InAs QDs. The *rectangular box* indicates experimental points for which the value of ΔE_{FS} is comparable to the natural width of the exciton states. Inset: Histogram for the distribution of QDs with respect to the exciton-state splitting in the range of ΔE_{FS} from −10 to 40 μeV. The column width is twice the measurement error (3.2 μeV), comparable to the natural width Γ_X of the exciton levels.

corresponds to the QDs with $\Delta E_{FS} = 0 \pm 1.6$ μeV. Evidently, this group of QDs is promising for the fabrication of EEPPs. Thus, we have demonstrated that structures with InAs QDs grown via the Stranski–Krastanov mechanism with short growth interruptions feature subassemblies of QDs with $\Delta E_{FS} \approx \Gamma_X$, which are suitable for the fabrication of EEPPs.

18.4 Conclusion

The cumulative experimental results presented demonstrate great potential for semiconductor technology in the development of effective subminiature emitters. Due to semiconductor emitters, our future is reliably confirmed by light. Some very important and highly organized, though not the biggest part of this light, will be emitted by VCSELs, SPEs and, of course, EEPPs.

Acknowledgment

The work was financially supported by the RAS Presidium (Programm No. 24 "Basic research in nanotechnologies and nanomaterials"), by the Ministry of Science and Education of the Russian Federation (agreement No. 14.621.21.0004 and by Russian Science Foundation (projects No. 14-22-00143 and No. Sh-2138.2014.8). Some of the experiments were performed using the equipment of the Nanostructures Collective Center.

References

[1] R. Michalzik (Ed.), VCSELs: Fundamentals, Technology and Applications of Vertical-Cavity Surface-Emitting Lasers, Springer-Verlag, Berlin Heidelberg, 2013.

[2] D. Bouwmeester, A.K. Ekert, A. Zeilinger, The Physics of Quantum Information, Springer-Verlag, Berlin Heidelberg, 2010.

[3] N. Gisin, G. Ribordy, W. Tittel, H. Zbinden, Quantum cryptography, Rev. Modern Phys. 74 (2002) 145–195.

[4] D.F. Walls, G.J. Milburn, Quantum Optics, Springer-Verlag, Berlin, 2008.

[5] P. Michler (Ed.), Single Semiconductor Quantum Dots, Springer-Verlag, Berlin, 2009.

[6] K. Iga, Vertical–cavity surface–emitting laser: its conception and evolution, Jpn. J. Appl. Phys. 47 (2008) 1–10.

[7] S.A. Blokhin, J.A. Lott, A. Mutig, G. Fiol, N.N. Ledentsov, M.V. Maximov, et al., Oxide-confined 850 nm VCSELs operating at bit rates up to 40 Gbit/s, Electron. Lett. 45 (2009) 389–390.

[8] F. Gruet, A. Al-Samaneh, E. Kroemer, L. Bimboes, D. Miletic, C. Affolderbach, et al., Metrological characterization of custom-designed 894.6 nm VCSELs for miniature atomic clocks, Opt. Express 21 (2013) 5781–5792.

[9] B. Tan, Y. Tian, H. Lin, J. Chen, S. Gu, Noise suppression in coherent population trapping atomic clock by differential magneto-optic rotation detection, Opt. Lett. 40 (2015) 3703–3706.

[10] S. Knappe, V. Shah, P.D.D. Schwindt, L. Hollberg, J. Kitching, L.-A. Liew, et al., A microfabricated atomic clock, Appl. Phys. Lett. 85 (2004) 1460–1462.

[11] L.-A. Liew, S. Knappe, J. Moreland, H. Robinson, L. Hollberg, J. Kitching, Microfabricated alkali atom vapor cells, Appl. Phys. Lett. 84 (2004) 2694–2695.

[12] S. Knappe, P.D.D. Schwindt, V. Shah, L. Hollberg, J. Kitching, A chip-scale atomic clock based on ^{87}Rb with improved frequency stability, Opt. Express 13 (2005) 1249–1253.

[13] A. Lochmann, E. Stock, O. Schulz, F. Hopfer, D. Bimberg, V.A. Haisler, et al., Electrically driven single quantum dot polarised single photon emitter, Electron. Lett. 42 (2006) 774–775.

[14] A. Lochmann, E. Stock, J.A. Tofflinger, W. Unrau, A. Toropov, A. Bakarov, et al., Electrically pumped, micro-cavity based single photon source driven at 1 GHz, Electron. Lett. 45 (2009) 566–567.

[15] M.C. Münnix, A. Lochmann, D. Bimberg, V.A. Haisler, Quantum-dot-based single photon emitters, IEEE J. Quant. Electron. 45 (2009) 1084–1088.

[16] A. Strittmatter, A. Schliwa, J.-H. Schulze, T.D. Germann, A. Dreismann, O. Hitzemann, et al., Lateral positioning of InGaAs quantum dots using a buried stressor, Appl. Phys. Lett. 100 (2012) 093111.

[17] D. Bimberg (Ed.), Semiconductor Nanostructures, Springer-Verlag, Berlin, 2008.

[18] Z.M. Wang (Ed.), Self-Assembled Quantum Dots, Springer Science + Business Media, LLC, New York, NY, 2008.

[19] P. Michler (Ed.), Single Quantum Dots: Fundamentals, Applications and Concepts, Springer-Verlag, Berlin, 2003.

[20] P. Yeh, Optical Waves in Layered Media, John Wiley & Sons, Singapore, 1991.

[21] P. Bienstman, R. Baets, J. Vukusic, A. Larsson, M.J. Noble, M. Brunner, et al., Comparison of optical VCSEL models on the simulation of oxide-confined devices, IEEE J. Quant. Electron. 37 (2001) 1618–1631.

[22] I. Bengtsson, K. Zyczkowski, Geometry of Quantum States: An Introduction to Quantum Entanglement, Cambridge University Press, Cambridge, 2008.

[23] M.A. Nielsen, I.L. Chuang, Quantum Computation and Quantum Information, Cambridge University Press, Cambridge, 2010.

[24] D. Bimberg, E. Stock, A. Lochmann, A. Schliwa, J.A. Töfflinger, W. Unrau, et al., Quantum dots for single- and entangled-photon emitters, IEEE Photonics J. 1 (2009) 58–68.

[25] O. Benson, C. Santori, M. Pelton, Y. Yamamoto, Regulated and entangled photons from a single quantum dot, Phys. Rev. Lett. 84 (2000) 2513–2516.

[26] R.M. Stevenson, R.J. Young, P. Atkinson, K. Cooper, D.A. Ritchie, A.J. Shields, A semiconductor source of triggered entangled photon pairs, Nature 439 (2006) 179–182.

[27] A. Mohan, M. Felici, P. Gallo, B. Dwir, A. Rudra, J. Faist, et al., Polarization-entangled photons produced with high-symmetry site-controlled quantum dots, Nat. Photonics 4 (2010) 302–306.

[28] R.M. Stevenson, C.L. Salter, J. Nilsson, A.J. Bennet, M.B. Ward, I. Farrer, et al., Indistinguishable entangled photons generated by a light-emitting diode, Phys. Rev. Lett. 108 (2012) 040503.

[29] R. Seguin, A. Schliwa, T.D. Germann, S. Rodt, K. Potschke, A. Strittmatter, et al., Control of fine-structure splitting and excitonic binding energies in selected individual InAs/ GaAs quantum dots, Appl. Phys. Lett. 89 (2006) 263109.

[30] R. Seguin, A. Schliwa, S. Rodt, K. Potschke, U.W. Pohl, D. Bimberg, Quantum dot size dependence of exciton fine—structure splitting, Phys. E 32 (2006) 101–103.

[31] L.H. Li, N. Chauvin, G. Patriarche, B. Alloing, A. Fiore, Growth–interruption–induced low–density InAs quantum dots on GaAs, J. Appl. Phys. 104 (2008) 083508.

[32] T.J. Krzyzewski, T.S. Jones, Ripening and annealing effects in InAs/GaAs (001) quantum dot formation, J. Appl. Phys. 96 (2004) 668–674.

[33] L. Muller-Kirsch, R. Heitz, U.W. Pohl, D. Bimberg, Temporal evolution of GaSb/GaAs quantum dot formation, Appl. Phys. Lett. 79 (2001) 1027–1029.

[34] U.W. Pohl, K. Potschke, A. Schliwa, F. Guffarth, D. Bimberg, N.D. Zakharov, et al., Evolution of a multimodal distribution of self-organized InAs/GaAs quantum dots, Phys. Rev. B 72 (2005) 245332.

[35] E. Moreau, I. Robert, J.M. Gérard, I. Abram, L. Manin, V. Thierry-Mieg, Single-mode solid-state single photon source based on isolated quantum dots in pillar microcavities, Appl. Phys. Lett. 79 (2001) 2865–2867.

[36] R.M. Thompson, R.M. Stevenson, A.J. Shields, I. Farrer, C.J. Lobo, D.A. Ritchie, et al., Single-photon emission from exciton complexes in individual quantum dots, Phys. Rev. B 64 (2001) 201302.

[37] T. Mano, M. Abbarchi, T. Kuroda, B. McSkimming, A. Ohtake, K. Mitsuishi, et al., Self-assembly of symmetric GaAs quantum dots on (111)A substrates: suppression of fine-structure splitting, Appl. Phys. Express 3 (2010) 065203.

[38] J.W. Tomm, T. Elsaesser, Yu. I. Mazur, H. Kissel, G.G. Tarasov, Z. Ya Zhuchenko, et al., Transient luminescence of dense InAs/GaAs quantum dot arrays, Phys. Rev. B 67 (2003) 045326.

19

Three-Dimensional Systems and Nanostructures: Technology, Physics and Applications

V.Ya. Prinz

RZHANOV INSTITUTE OF SEMICONDUCTOR PHYSICS OF THE SIBERIAN BRANCH OF THE RUSSIAN ACADEMY OF SCIENCE, NOVOSIBIRSK, RUSSIA

This chapter reviews the main results achieved in over the last decade. Original methods and technologies for the fabrication of 3D semiconductor, graphene, and polymer nanostructures are described. Quantum properties of formed nanostructures are analyzed. Attention is focused on nanomaterials and devices intended for applications such as: electromagnetic metamaterials, graphene field emitters, high-power electrostatic nanoactuators, fast-response anemometric sensors and smart systems based on such sensors, plasmon nanostructures, nanotools and nanomaterials for biology and medicine, and special polymer nanostructures formed by imprint lithography and 3D printing.

19.1 The Technology of 3D Nanostructures

Progress in new technologies, materials, and devices has always determined economic development and the possibilities for mankind at all times since the Stone Age. Nanotechnology is the absolute technology to ensure progress in all life spheres. Presently, progress in electronics, medicine, material science, mechanics, mechanical engineering, and space exploration largely depends on nanotechnology. The passage from centimeter and millimeter to micrometer device sizes has enabled microelectronics that have dramatically changed our lives. High reproducibility and simultaneous parallel production of some billion transistors has enabled mass fabrication of products with reduced costs. Today, for every inhabitant on Earth, there are almost one billion transistors that cost US$10, the sum the first transistors cost. That is, over half a century the cost of one transistor has reduced by a billion times!

The passage from micrometer to nanometer sizes has allowed the formation of new materials and devices with substantially improved properties. Nanotechnology is subject to the same general requirements which are imposed on microtechnology. These are the possibility of mass fabrication of products, and high precision and reproducibility of nanostructures (practically with atomic accuracy). Our technology of three-dimensional (3D) nanostructures [1], which has allowed the passage from planar to functionally more rich 3D micro- and

Advances in Semiconductor Nanostructures. DOI: http://dx.doi.org/10.1016/B978-0-12-810512-2.00019-6

nanostructures, meet all those requirements. The basics of the technology and 3D nanostructures formed using it were described in Ref. [2].

In the present chapter, I give an brief overview of the results that have received wide recognition during the last years.

We were first to free films of nanometer and angstrom thicknesses from semiconductor substrate bonding [1,2] (previously, only films several hundred nanometers thick could be freed from the substrate). From such films, we have suggested the formation of 3D nanostructures driven by lattice-mismatch induced elastic strains in films made up of layers of dissimilar materials [1,2]. The technology developed has allowed us to fabricate precise nanoshells and systems [3–5]. For instance, from a freed film about 5 Å thick formed by one GaAs monolayer and by one InAs monolayer, using an elastic strain-driven process, we have obtained nanotubes with diameters down to 2 nm [2]. The technology is a molecular one, with molecular monolayers of semiconductors, metals, and dielectrics playing a key role. Planar monolayers are transformed into 3D monolayer configurations (bent, rolled, corrugated, buckled, etc.). The technology is based on well-established methods allowing precise control over the structures formed.

Our priority in the development of this technology is acknowledged in many tutorials published in Russia and abroad, and in numerous reviews (the citation index of our first publication [1] now exceeds 480 references). We have demonstrated the possibility of creating a new class of semiconductor nanostructures: nanotubes, helices, rings, corrugations, etc., with characteristic sizes ranging from 100 μm to 2 nm [3–5]. Figs. 19.1–19.3 illustrate the transformation of planar heterofilms into the simplest 3D structures, and the scalability of such structures. The driving force for the transformation is the elastic strain due to lattice mismatch between the films component layers. The diameter, D, of rolled heterostructures is defined by the thickness, d, of the heterofilm, and by the mismatch between the lattice constants of the layers, $\Delta a/a$. For a bilayer film, $D \approx d/(\Delta a/a)$. The parameters of the film are controlled with high accuracy by epitaxial growth conditions; this fact enables fabrication of tubes with precisely controlled diameters ranging from several hundred micrometers to

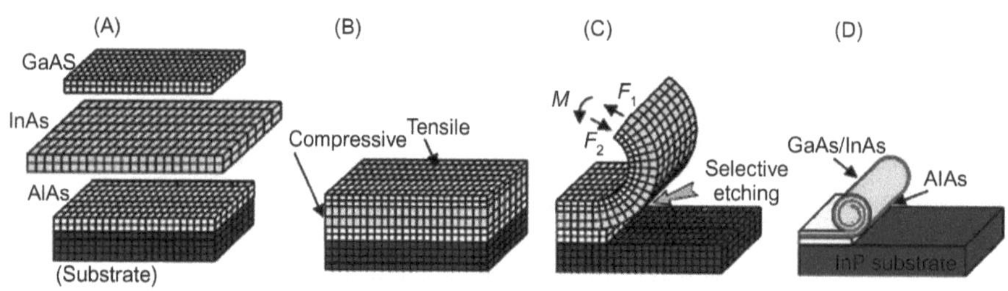

FIGURE 19.1 Schematic illustrating the formation process of simplest nanostructures—nanotubes. (A) Layers of GaAs and InAs with different lattice constants in free state (each of the layers contains two molecular monolayers); (B) layers of GaAs and InAs grown on an InP substrate. The crystal lattices of thin layers adapt themselves to be matched to the substrate lattice in the course of epitaxial growth (pseudomorphic growth condition); (C) bending of GaAs/InAs film on its freeing from substrate after selective removal of AlAs sacrificial layer; (D) self-rolling of GaAs/InAs film into a tube-scroll (Figure D is given not to scale).

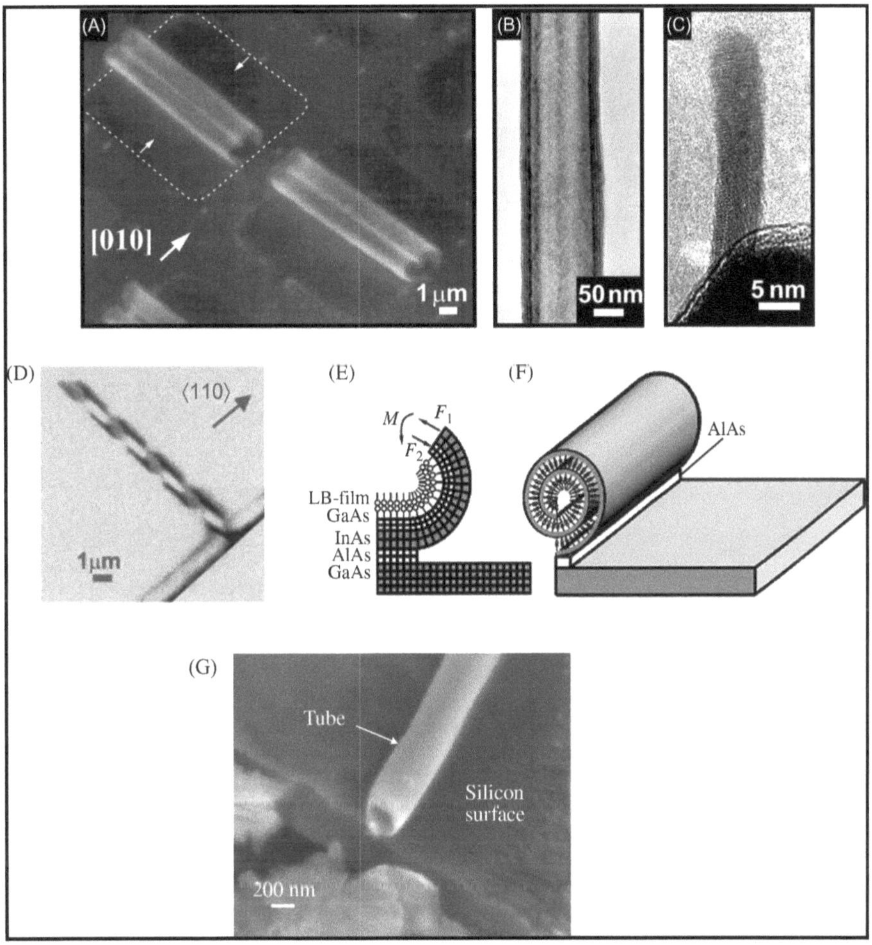

FIGURE 19.2 Photographs of micro- and nanotubes taken on a scanning electron microscope: (A) double-barrel; (B) six-coil nanotube rolled from a 4ML GaAs+4ML $In_xGa_{1-x}As$ ($x = 0.6$) film; (C) nanotube rolled from a 2ML GaAs+1ML InAs film; (D) InGaAs/GaAs microhelices; (E,F) schematic illustrating the formation process of hybrid nanotubes from a hybrid InAs/GaAs/Langmur-Blodgett film. The separation between the turns of the coil is precisely defined by the thickness of the Langmur-Blodgett film; (G) TEM image of a metal Au/Ti nanotube, tube length 10 cm, tube diameter 200 nm.

several nanometers. The high quality of grown heterostructures has enabled rolled heterostructures several centimeters long with smooth, uniform thickness walls to be produced. Nanocorrugated semiconductor structures were fabricated as shown in Ref. [4]. Fig. 19.3 illustrates a nanocorrugated (nanobuckled) film prepared from a compressed, initially flat film locally detached from the substrate. Such corrugations can be formed because locally released compressed planar films are unstable in their free state, and readily undergo buckling. The shape results from simple elongation of the release films. If the ripples can be controlled, they can be used to advantage in nanoscale devices and electronics, opening up a new opportunity in graphene engineering: strain-based devices or "straintronics."

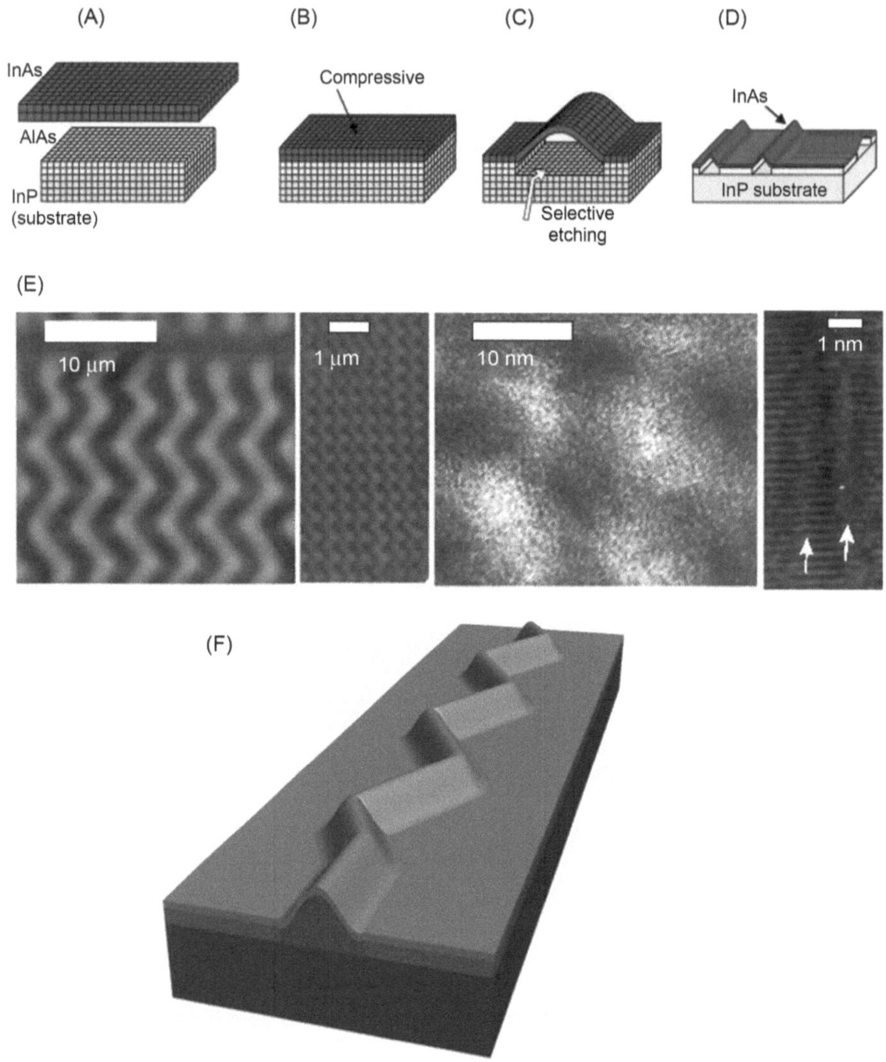

FIGURE 19.3 (A–D) Schematic illustrating the formation of corrugations proceeding on detachment of compressed layers from substrates. (E) Demonstration of scalability of the formation process of corrugations made from InGaAs/AlAs/GaAs structures. Elastic interaction between corrugations and their mutual correlation is evident. (F) In the last TEM image, the *arrows* point to corrugations. The transverse lines are atomic planes.

In Ref. [6], we numerically examined shells formed as a result of buckling of compressed hybrid graphene/semiconductor films. Fig. 19.4 shows three stable shapes of buckled graphene/InGaAs films freed from the GaAs substrate in the central circular region, and fixed to a substrate around the contour. The first buckling mode in Fig. 19.4A is interesting due to the fact that, at its vertex, the graphene layer is in tension; yet, even more important here is the circumstance that such a structure provides a basis for creating quantum dots. The simplest method for forming such quantum dots is based on using the field effect. Providing a gate located over the vertex,

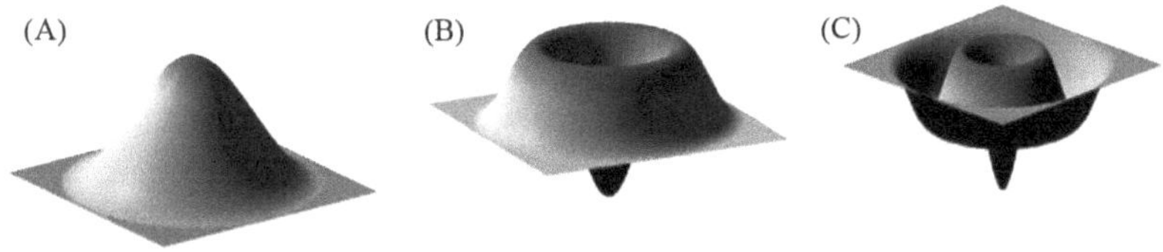

FIGURE 19.4 Tunable devices. The first mode (A), the second mode (B) the third mode (C) monolayer graphene over the surface of initially strained InGaAs monolayers.

one can control the position of Fermi level in the graphene layer, thus generating circular *p*- and *n*-regions. Even more interesting is the second buckling mode illustrated by Fig. 19.4B. For an initial value of elastic strain $\varepsilon_0 = -3\%$ in the 2-nm thick InGaAs plate of radius 60 nm, the strain in the graphene monolayer of this buckling shape amounts to 2%. The strain is distributed non-uniformly across the graphene monolayer. Such structures offer promise in forming ideal graphene quantum rings and other functional elements [6]. It is important to note that in this case the edge states will not appear, because of the absence of abrupt ends on the graphene sheet. In such structures, giant gradient-induced phenomena in strong magnetic fields and 3D quantum phenomena will emerge [7]. (In a uniform magnetic field a gradient of the normal-to-surface field component up to 10^9 T/cm can be obtained. The magnitude of the pseudomagnetic field can reach a value of about 100 T [6,8].) It should be noted here that the technology is based on tens of original methods, from directional film rolling to ultra-selective etching of sacrificial layers, and methods of deformation-free supercritical drying of structures [1–5].

Based on the technology described above, we have fabricated many new materials, devices and sensors, including two-dimensional (2D) metamaterials, polarizers for THz radiation, new metamaterials, nanotools for biology and medicine (nanoneedles, probes, syringes, neuroprobes, chips, scalpels), ultra-fast response anemometric nanosensors for studies in fluid dynamics, smart systems for quick registration and suppression of turbulence, supersensitive magnetic nanosensors, nanoprinters, field emitters, and electrostatic nanoactuators [9]. Ten RU patents were received; 2 of those patents were included in the list of the 100 best RU patents. Developments made abroad on the basis of nanostructures produced include a laser and an optical resonator based on a multicoil tube, a plasmon-based cylindrical metamaterial for observing objects smaller in size than wavelength, a nanotransformer, a nanoactuator based on helices and magnetic particles, a catalyst-based nanoactuator, nanochanneled fluidic chips, force sensors, nanorobots, etc. [10–14].

Apart from device structures, we have for the first time fabricated structures with quantum wells, and structures with 2D electron gas freed from structure intended for fundamental research [1,15].

Obviously, the future belongs to advanced technologies permitting precise control of nanostructures in terms of their sizes and location with atomic and molecular accuracy.

19.2 Nanoimprint Lithography

Advanced technology can be made cheap with mass parallel production of nanostructures over large areas. A most appropriate large-area technology is nanoimprint lithography. In the world, imprint lithography and continuous stamp forming technology on flexible polymer films with resolutions from several micrometers to tens of micrometers and several nanometers are under development. Today, they are used to fabricate electronic paper, flexible electronic circuits, flexible large-area displays, smart glasses with controlled transparency and reflection coefficients, flexible solar cell panels, photodetectors, and self-cleaning superhydrophobic micro- and nanostructured coatings and surfaces (see, e.g., Ref. [16]). In many applications, the best characteristics were achieved using nanoimprint lithography.

At Rzhanov Institute of Semiconductor Physics, Russian Academy of Science, many technological processes implemented on a Nano Imprinter Eitre 6 nanoimprint lithography facility (Sweden) were mastered and improved (Fig. 19.5); moreover, new processes were proposed. The facility allows imprints to be made on the surface of polymer films as large as 150 mm in diameter, with implement heating up to 200°C and pressures up to 80 bar during presswork. The minimum possible element size is down to 10 nm.

(A)

(B)

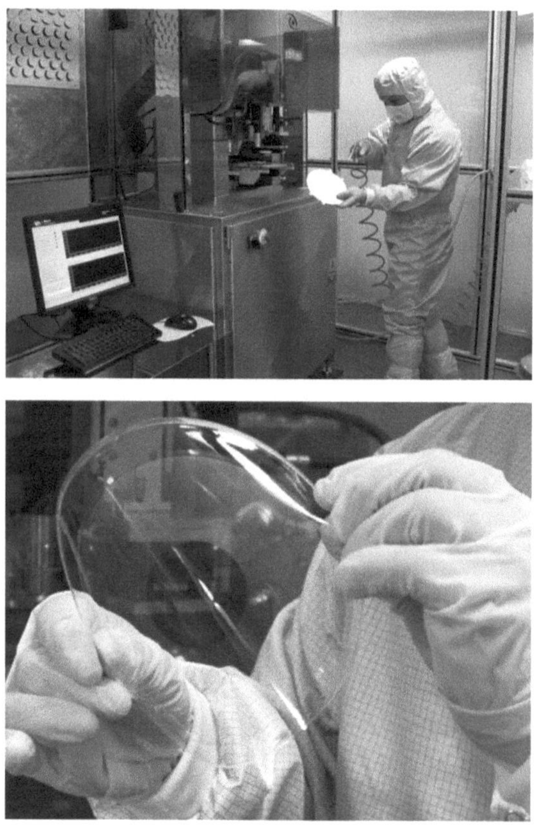

FIGURE 19.5 (A) Photograph of the Eitre 6 facility installed in a class-10 clean room. (B) Example of a stamped polymer film.

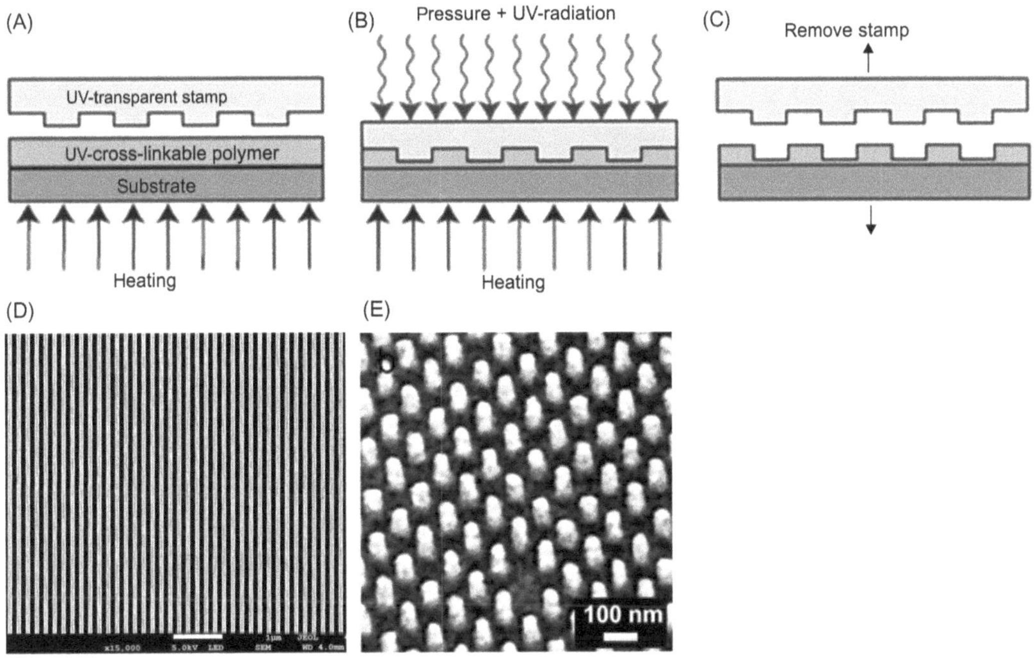

FIGURE 19.6 (A–C) Schematic illustrating the formation of a lithographic pattern with the help of a stamp. SEM image of a formed array of (D) 70-nm wide nanostrips and (E) 50-nm diameter polymer nanopillars.

A useful possibility here is a simple, cheap technology for forming high-quality metal elements on the large surface area of semiconductors and polymer films (up to 160 cm^2). We found that at certain cooling and pressure-unloading conditions, the polymer experiences a nonuniform shear deformation, and stamp edges lift up the imprint edges, thus forming polymer tube shells which present 3D objects that cannot be produced by conventional imprint lithography. Using already known and specially developed polymer thermocontracting materials, we have developed methods enabling not only a size reduction of nanostamp elements, but also a reconfiguration of those elements. New formation methods for 3D hybrid metal–polymer elements have been proposed. Cooling and pressure-unloading conditions of thermoplastic nanoimprint lithography have been optimized; that optimization has allowed the formation of a high-quality grating replica for 180 nm pitch subwavelength gratings, and a large wavelength-to-strip width ratio (in excess of 10^6) (Fig. 19.6). High-quality two-level subwavelength gratings, which can be used as extreme broadband polarizers and plasmon nanostructures, were fabricated (Figs. 19.6 and 19.7) [17]. A technology for dry resistless formation of single- and multilevel nanostructures (metal–semiconductor, Au/Ti-polymer) and systems with typical feature sizes from 70 down to 10 nm has been proposed.

Nanoimprint lithography can be perfectly matched to the technology of 3D nanostructures. Moreover, using nanoimprint lithography, we have transferred the technology of 3D nanostructures onto polymer and hybrid films.

Below, we give a description of various applications for the technology of 3D nanostructures.

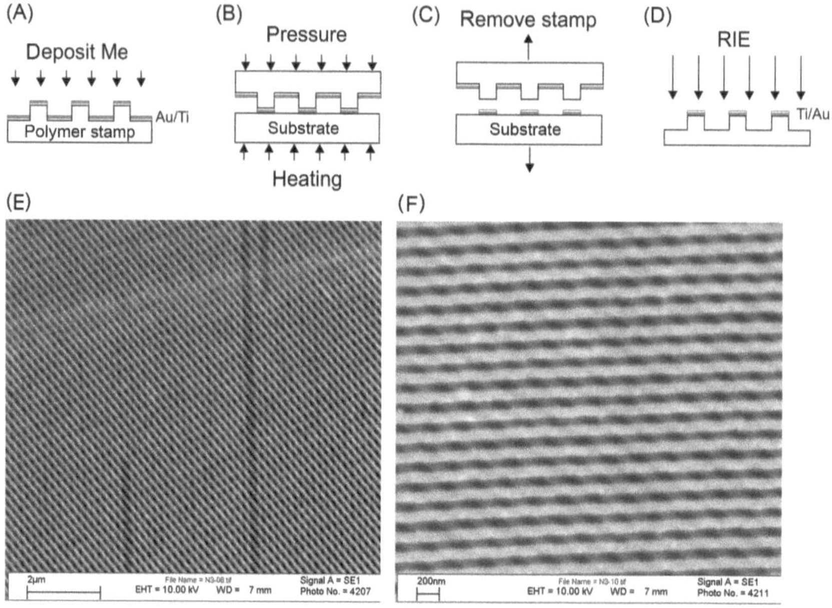

FIGURE 19.7 (A–F) Schematic illustrating the transfer of Au/Ti metal films from stamp to a polymer substrate, SEM images of the formed multilayer nanostructure. The array of upper-level strips is oriented to the array of lower-level strips at an angle of 45 degrees.

19.3 Electromagnetic Nanomaterials

Electromagnetic properties of ordinary materials are defined by the atoms and molecules of the materials or, more precisely, by the manner in which their electrons interact with electromagnetic waves. Unlike in the case of ordinary materials, building blocks of metamaterials (supermaterials) are larger objects (electromagnetic resonators, normally prepared in the form of metal strips, split rings, and helices with sizes substantially smaller than radiation wavelength). If the frequency of an incident electromagnetic wave is close to resonance frequency, rather intense currents arise in the resonators; those induced currents generate secondary electromagnetic fields that amplify or weaken the initial wave. As a result, in metamaterials we have quite a different situation compared with traditional materials. The coefficient of refraction may assume negative and almost zero values, and the Doppler and Vavilov–Cherenkov effects become reversed. Cardinal changes are expected to occur when activities aimed at the development of superlenses with less-than-wavelength resolution, polarizers, and invisibility cloaks in which light rounds metamaterial-coated objects to render them invisible are achieved. The main problems in the field of metamaterials are: (1) passage from 2D to 3D resonance elements; (2) passage from small to large areas; (3) formation of quick-response dynamically controlled electromagnetic metamaterials; (4) passage to optical metamaterials whose unit cell has nanometer sizes; and (5) passage from metamaterials to devices. The need to solve these problems has become especially important in the optical

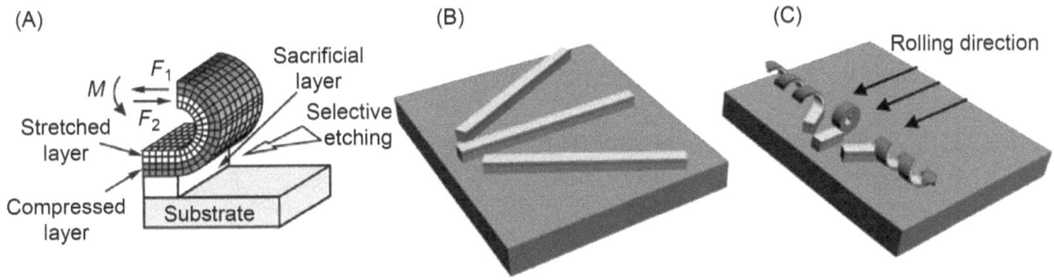

FIGURE 19.8 (A) Bending of a strained bifilm detached from the substrate. The forces *F1* and *F2* are induced by the stretched and compressed layers, respectively; these forces give rise to the bending moment *M* that rolls the bifilm. Strains in the pseudomorphous epitaxial layers are due to the lattice mismatch. The curvature radius *R* of the rolled bifilm is inversely proportional to the relative lattice mismatch a/a, where a is the lattice constant and a is the lattice mismatch between layers (for instance, for InAs and GaAs $a/a = 7.2\%$, for Ge and Si $a/a = 3.7\%$). (B) Strained strips of different orientations on the substrate. (C) Rolling of strips into rings and helices.

region in connection with advances in nanophotonics. In fact, to date little has been done to solve these problems.

For the first time, we have passed from 2D resonance elements to more efficient 3D resonance elements, and to the formation, on the basis of the latter elements, of 2D and 3D metamaterials for the THz range, which presently almost completely lacks a means of control for radiation. Moreover, from the very beginning of our activities in this field [18,19], we have put our hopes on the helix as a unit resonator for metamaterials (Fig. 19.8).

The helix is a unique element. First, it efficiently interacts both with the electric and magnetic fields in electromagnetic waves. Second, a conducting helix simultaneously generates an electric and magnetic dipole, thus making special combining unnecessary, in a single structure, planar elements are capable of generating electric or magnetic dipoles separately (Fig. 19.9).

Such combining is difficult to perform in the submicrometer region, and it is impossible in the nanoregion. Yet, negative index metamaterials cannot be obtained without such combining. The passage to 3D elements has allowed us to organize an efficient interaction of the elements with electric and magnetic fields in electromagnetic waves. Moreover, it has enabled the creation of an original system based on nanomaterials in which dynamic control over metamaterial electromagnetic properties is based on resonant photon "tunneling." The latter research field is extremely important because superfast control of radiation over small sizes is critically required for progress in nanophotonics.

We have created seven new electromagnetic metamaterials with 3D resonators, including chiral isotropic and anisotropic resonators, high-frequency magnetic resonators, tubular resonators with helical conductivity, and resonators with polymer-embedded elements (Fig. 19.10). These materials are intended for operation in the THz radiation range, where a means of control for radiation is almost completely lacking. The chiral materials created possess giant optical activity: they rotate the plate of polarization through 36 degrees as the radiation propagates a material thickness 10 times smaller than radiation wavelength; they have

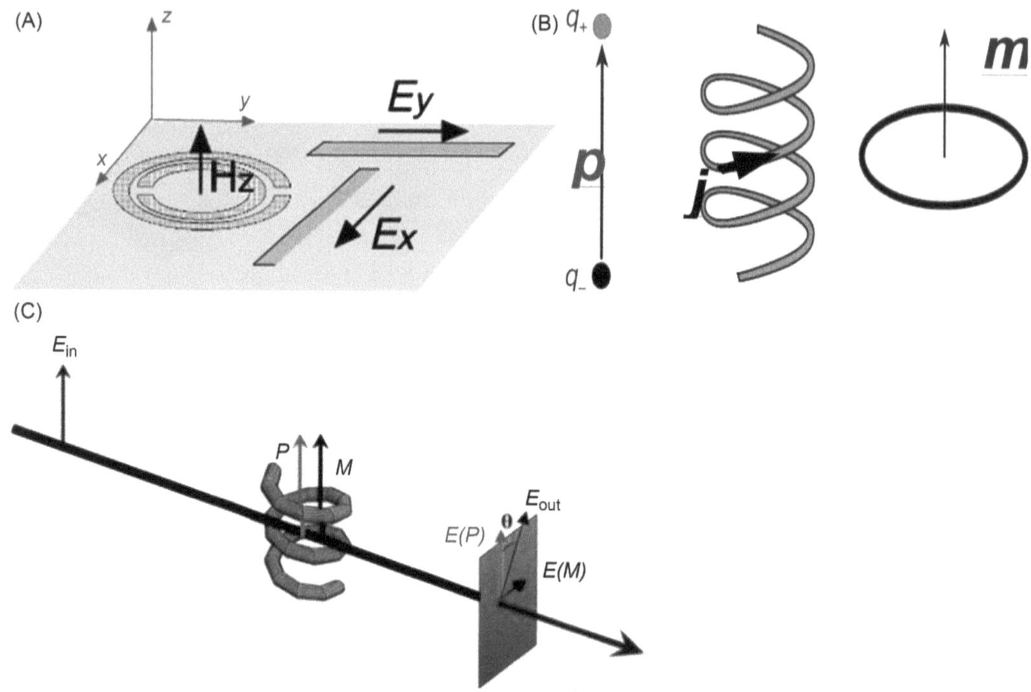

FIGURE 19.9 (A) Planar electric dipoles—strips and magnetic dipoles—split rings (here, the electromagnetic waves, directed normally to the surface, excites just an electric dipole, and it does not excite a magnetic dipole), (B) combined electric (*p*) and magnetic (*m*) dipoles in helices, (C) electric field of electromagnetic waves falling onto a helix simultaneously excites an electric dipole and a magnetic dipole. The polarization direction of secondary radiation is formed by electric and magnetic dipoles, which circumstance provides for polarization angle rotation (optical activity).

a negative refraction index; and they are functionally richer than metamaterials that contain planar elements. A monolayer of such a metamaterial rotates the polarization plane by more than 45 degrees [20]. It can be shown that at any conditions a change of radiation frequency as small as 0.5% causes a change of polarization rotation angle by 19 degrees. Metamaterials based on 3D precise electromagnetic resonators possess unique properties; for instance, the spectral dependence of polarization in such materials exhibits sharp resonance peaks. The occurrence of such peaks simplifies the control of polarization and its superfast modulation, and open fresh possibilities in data communication and the formation of "smart" metamaterials. A possibility arises for continuous superfast control of transmitted radiation polarization. Note that creation of such metamaterials is only possible with the help of our technology. High accuracy in the formation of elements presents a key requirement for the creation of negative index metamaterials and, most of all, invisibility cloaks.

The technology admits sequential applications of many layers, or creation of 3D metamaterials [20].

Two-dimensional THz isotropic materials with compensated chirality and identical values of resonant dielectric and magnetic permittivities were developed. The sharp reduction of

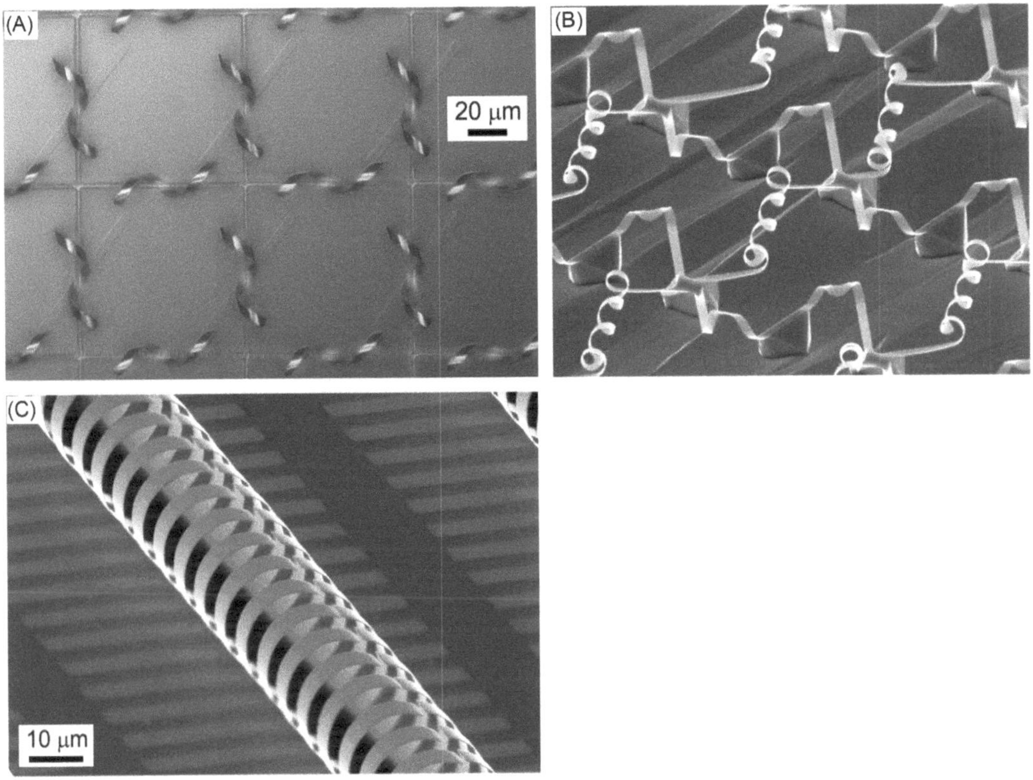

FIGURE 19.10 (A–C) Formation examples of THz metamaterials.

reflection at resonance frequency achieved is a most significant property of helix arrays with optimized parameters [21].

For the first time, for an original system based on a 2D chiral THz metamaterial (metamaterial-transparent GaAs substrate-thin GaAs wafer), it was established that the addition of a GaAs plate disposed from the substrate at a tunnel distance for photons leads to the emergence of giant peaks in polarization spectra. It was shown that, in metamaterials based on screw-tubes, the radiation polarization shows changes as large as 160 degrees as the radiation frequency deviates by 0.5%. Undoubtedly, this strong effect will find useful applications. The investigation and deciphering of the discovered effect opens previously unknown possibilities in superfast continuous control of polarization. Here, we put considerable hopes on an original system involving an array of suspended metal–semiconductor helices (Fig. 19.10).

The experimental activities aimed at the formation and study of metamaterial properties were accompanied by theoretical and simulation studies of those properties [22,23]. We would like to emphasize here that not only in the THz spectral region, but also in the IR range, metamaterials still remain the only precise metamaterials created to be obtained by mass fabrication methods.

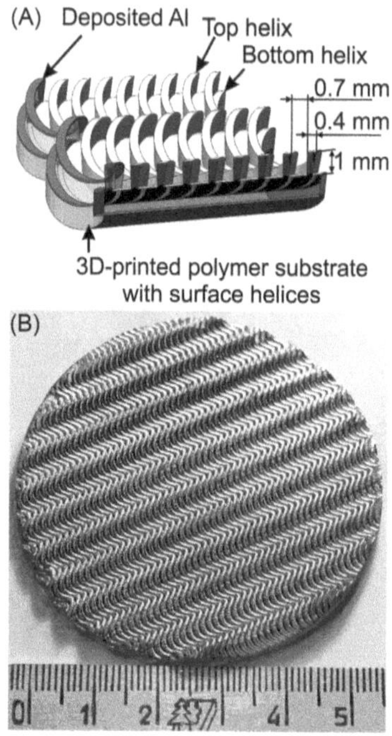

FIGURE 19.11 Scheme (A) and photography (B) of the formed structure. The production time of the structure is less than 1 h.

With the help of nanoimprint lithography and methods allowing transfer of metal structures from a stamp onto a polymer surface, we have created two-level plasmon metamaterials formed by an array of 70 nm wide polymer-embedded Au/Ti nanostrips. It can be shown that, both in absorption and ellipsometric spectra, localized Au—air and Au—polymer plasmon resonances are manifested. Undoubtedly, the occurrence of sharp peaks in the transmission spectrum is a circumstance significant for the creation of selective filters and control systems.

Additive technology or 3D printing is rapidly developing now. We developed a technological method [24] for the formation of microwave and terahertz chiral metamaterials based on helical polymer and metal—polymer resonant elements, and showed giant rotation of the polarization plane of microwave radiation by the structures formed. This method can be used for cheap mass fabrication of metamaterials of large area (Fig. 19.11). The technology developed includes: (1) formation of a stamp with periodical relief structure by 3D printing; (2) imprinting of polymer film; and (3) shadow deposition of metal. For simplicity, the imprinting of polymer film can be omitted, and in this case the metal is deposited directly on the 3D printed relief structure. Fig. 19.11A shows a 3D model of the relief structure formed. A single helix can be imagined as a metal ribbon "sinusoid" lying on the wavy surface. The radius and shape of the cross-section of the helix are determined by the amplitude of this sinusoid and the wavy profile depth.

Additive technologies have opened up unprecedented possibilities in the formation of 3D materials and devices. Examples are given by results on the formation of new

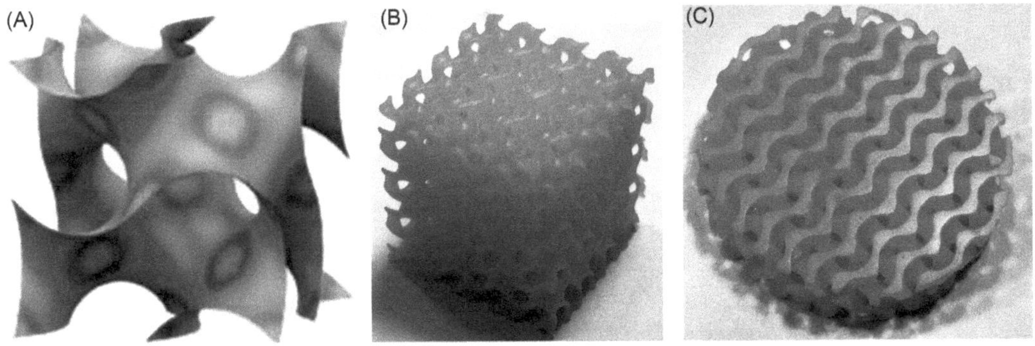

FIGURE 19.12 The model gyroid (A) and a photo of the printed structures: (B) gyroid crystals and (C) complementary gyroid.

electromagnetic materials with properties that have never been found in natural materials, or even have not been theoretically known so far. In recent years, with the advent of additive technologies, realization of some ideas in topological photonics has become possible. Gyroid photonic crystals and metamaterials, which have earlier been treated just theoretically, have been implemented using 3D printing technology. The gyroid is a 3D triply periodic minimal surface that divides space in labyrinths of opposite chiralities (Fig. 19.12A). We have developed a model gyroid material, which was then formed using 3D printing (Fig. 19.12B) and examined using microwave techniques.

In the present study, for developing the 3D model gyroid, we used software for constructing 3D models. However, that software was not intended for constructing analytically defined surfaces. Therefore, we were forced to solve several relevant problems: (1) partition of the unit cell of a cube-inscribed gyroid-type surface in eight simpler (primitive) identical surfaces connected with each other via displacements and rotations; (2) construction of the primitive surface in terms of curves being arcs located in the cube faces; (3) attachment of thickness to the surface, yielding a 3D body; (4) matching the primitive 3D bodies together, yielding a 3D model gyroid.

In matching the primitive 3D bodies together, we encountered the problem of mismatched interfaces, which was solved by introducing additional curves whose tangent lines were normal at the interfaces to the faces of the primitive cubes. It is possible to vary the wall thickness and, in this way, to alter the fill factor of the material structure.

The gyroid and the gyroid mesh were printed out using the parallel optical digital printing technique implemented with the digital light processor of a Carima DP110 3D printer. An advantageous feature of the printer is the possibility of simultaneous formation of each layer with surface area up to 70 cm^2. Reflection, absorption, and polarization spectra of formed materials were measured in the frequency range of 120–150 GHz.

19.4 Polymer Nanomaterials

With the help of imprint lithography, we have fabricated nanostructured polymer surfaces presenting arrays of 50 nm diameter *nanopillars* with a 100-nm pitch (Fig. 19.13). These

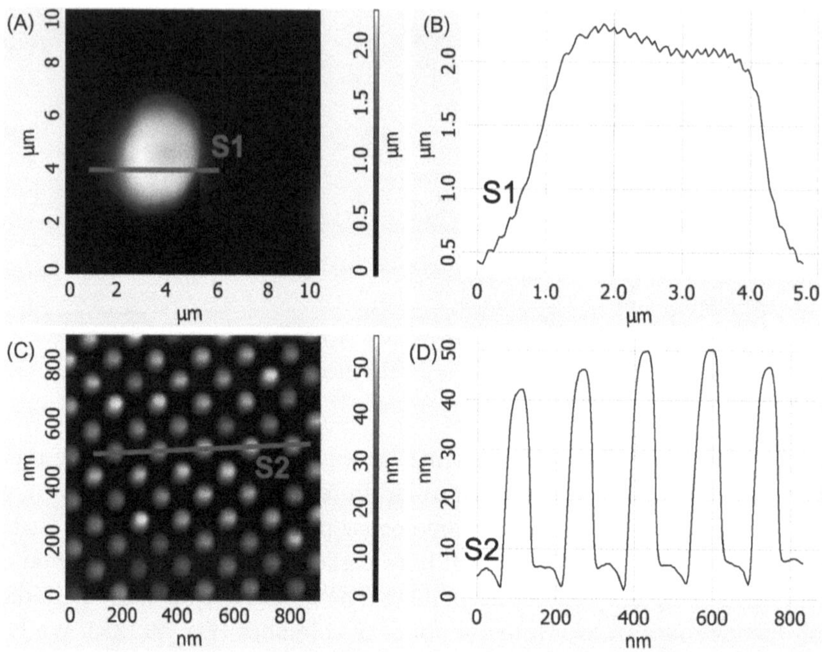

FIGURE 19.13 Example of a hierarchic multilevel gecko-adhesive structure. The structure is formed by an array of 4-µm diameter micropillars, with the nanostructured top of each micropillars containing an array of 50-nm diameter nanopillars whose pitch is 100 nm.

surfaces have formed the basis for the creation of a gesco-adhesive structure whose micro- and nanostructured surface resembles the structure of gesco legs, and provides adhesion to surfaces due to Van der Waals forces.

The dry adhesives created with an area up to 150 cm^2 provide adhesion to various solid surfaces with forces of 14–20 N/cm^2, they operate over a broad range of pressures and temperatures, and preserve their adhesive properties after many (several hundreds) attachment–detachment cycles (Fig. 19.13) [25]. The regime of multiple attachment–detachment cycles to various surfaces (glass, plastic, metal) is important for use of the adhesives in severe service conditions. The fields of application of such adhesives touch upon almost all areas of human life, from household appliances and medicine, to the creation of robots and special appliances for astronaut and military outfits. We have also identified other possible applications for adhesives in technological operations.

Also, nanostructured polymer surfaces were used by us in fabricating 3D matrices for regenerative medicine. The fabricated nanostructured surfaces (a nanostrip array with 180-nm pitch and an array of 50-nm diameter pillars with 100-nm pitch) proved capable of stimulating the differentiation of stem cells into bone cells (Fig. 19.14).

A technology for rapid formation of bone tissue was proposed [26]. This technology includes the formation, from bioresorbable polymer, of planar films with a microstructure that imitates Havers channels, and of related lacunas arranged in a nanogroove array; this

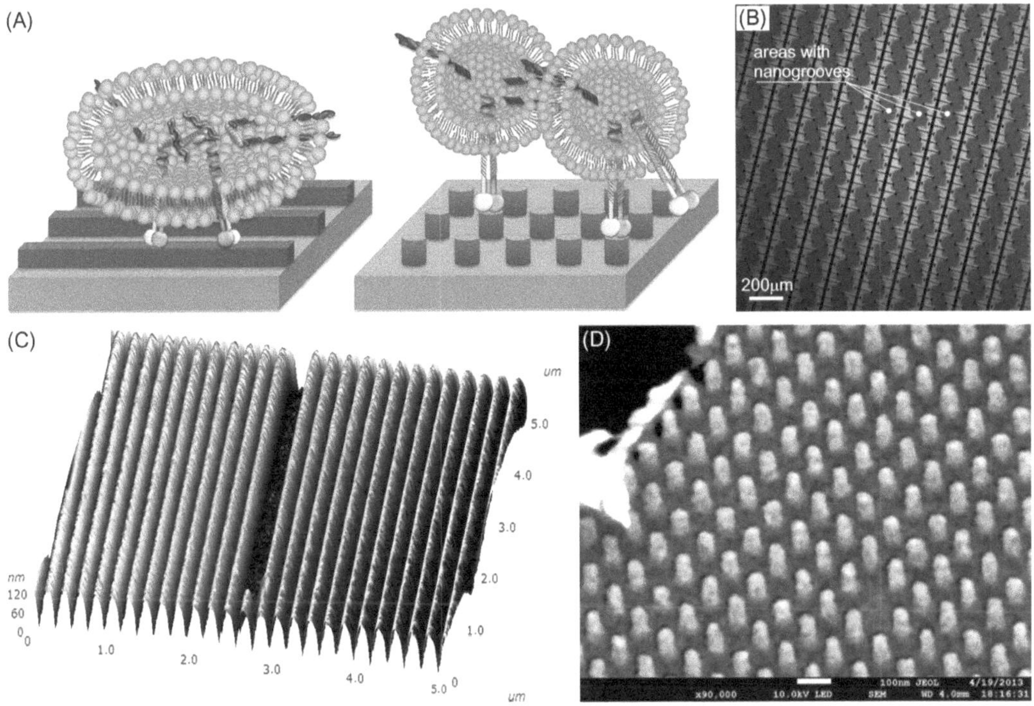

FIGURE 19.14 (A) Schematic illustrating the attachment of a cell to a nanostructured matrix, (B) photograph of a multilevel micro- and nanostructured polylactide film. (C) The nanostructured region presents an array of nanogrooves oriented to the channels at an angle of 45 degrees, (D) AFM image of a polylactide matrix obtained on an NT-MDT atomic force microscope and on an electron microscope. The pitch of the groove array is 180 nm, and the spike diameter varies from 50 to 30 nm.

array is intended to stimulate the differentiation of stem cells into bone cells (Fig. 19.14). A new feature here is that simultaneous growth of bone cells can be held in nutrient solutions on large film areas. Films with a grown cell layer are then cut into parts to be put in piles. The accomplishment of the bone formation process proceeds via the interaction and union of cell layers.

19.5 Sensors and Actuators

Our original results on the formation of new micro- (nano-) electromechanical systems (MEMS/NEMS) are outlined in Ref. [9]. By using the capabilities offered by the film rolling technology developed, we have fabricated tubular anemometric sensors and heaters (Fig. 19.15) [27–29] with a time constant almost 1000 times smaller than that of traditional hot-wire sensors and heaters. Also, a smart system for superfast registration and suppression of turbulence was proposed (the time constant of traditional hot-wire sensors is 1 ms, whereas that of tubular amemometric sensors is about 1 μs).

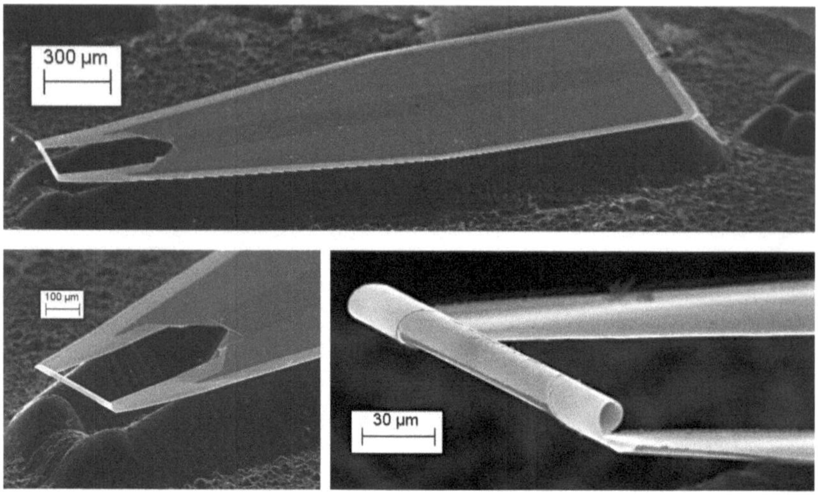

FIGURE 19.15 Suspended anemometric sensors with metal contacts.

Tubular structures may also find other applications. Such applications include, e.g., medical–biological applications, in which tubular needles and syringes can be used for penetration into cells, for delivering chemicals and DNA solutions into cells in a precise noninvasive manner, for making intra-cell diagnostics, and for registering neuron-generated signals.

In our experiments, we used a new class of atomically sharp tubular needles (see Fig. 19.16) intended for low-invasive piercing of cells, with the aim of introducing biological materials into the cells [30]. For the first time, an array of vertical tube-needles connected to micropumps was fabricated [31]. Stability calculations for tubular (multiwall) needles were performed. Those calculations were made for needles whose walls were prepared from graphene–semiconductor hybrid structures and graphene sheets, with a spike in the form of an open trapezoidal cylindrical shell. The possibility of mass fabrication of such needles and chips substantially widens the spectrum of accessible operations, both on the surface of living cells and inside them, and it opens prospects for precise efficient manipulation of individual cells.

The nanotubes developed were tested at the Institute of Cytology and Genetics of the Siberian Branch of the Russian Academy of Sciences. A chip with microtubes was fastened to a manipulator to pierce mice oocytes. Microinjections were made into a monolayer of HEP-2 cell lines using a prepared atomic force microscopy chip with a microtube fastened to a cantilever (Fig. 19.17).

A new universal approach to building high-power electrostatic micro- and nanoactuators was proposed. This approach is based on the possibility of the formation of micro- and nanocorrugated systems (Figs. 19.3 and 19.18). Here, pilot samples of original nanoactuators were fabricated, including a new class of variable-rigidity electrostatic actuators. The problem of creating high-powered nanoactuators suitable for the formation of microrobotic, intelligent, and nanoelectromechanical systems was solved. The actuators ensure displacements ranging from several hundred microns to several nanometers. In contrast to piezoelectric

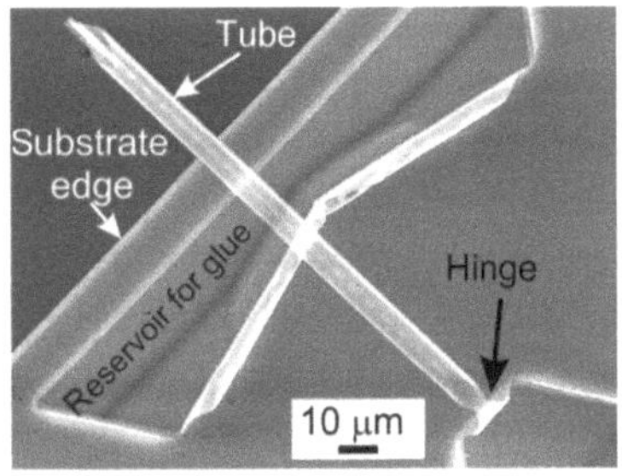

FIGURE 19.16 Si/SiGe tubes-needles for piercing cells.

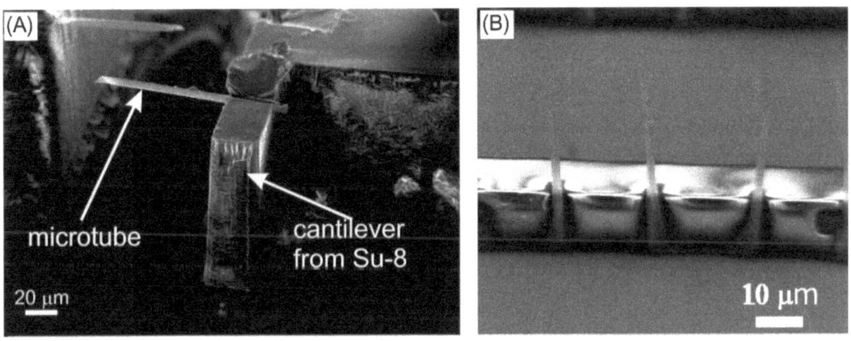

FIGURE 19.17 (A) Cantilever for an atomic force microscope with a SiGe/Ge microtube; (B) array of vertical tubes with nanowalls (in the course of TEM measurements, the tubes take on charge and experience bending).

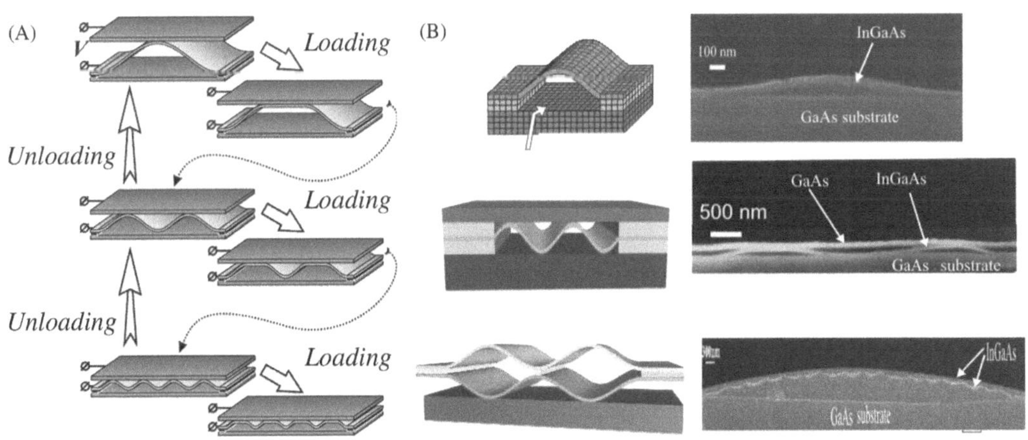

FIGURE 19.18 (A, B) Schematic illustrating the operation of the electrostatic nanoactuator (motor).

actuators operating in this displacement range, here the total amount of actuators developed is a billion times smaller, whereas the control voltage is 100 times lower. An elastically corrugated spring introduced into the gap solves the instability problem of electrostatic actuators with narrow gaps. The possibility of forming nanogaps between electrodes substantially increases the magnitude of the attractive force acting between the electrodes, thus increasing the forces developed by nanoactuators.

Laboratory samples of micro- and nanoactuators were formed from open corrugated multilayer heterostructures based on A3B5 and SiGe/Si compounds, and hybrid metal–semiconductor heterostructures. Characteristics of the nanoactuators proved to be in good agreement with the results of numerical calculations (Fig. 19.18) [9,32].

Electrostatic micro- and nanoactuators and arrays of such actuators form the basis for the creation of various MEMS/NEMS, and they can serve as an alternative to piezoelectrically driven actuators in various systems, such as tunnel and atomic force microscopes, and ultra-small-sized manipulators and robots.

Numerical simulations and experiments have shown that the variable-rigidity electrostatic nanoactuator (motor) is capable of producing substantial displacements and generating forces up to 10^5 N/cm^{-2}.

The nanogap between the electrodes allows one to exercise control over the actuator via the application of rather small bias voltages (5 V for 100 nm wide gap and 6 nm thickness of corrugated film), which makes such actuators highly promising for applications.

19.6 Growth and Functionalization of Graphene Structures

Graphene is a new modern material relevant to the field of nanotechnology. For applications of this material, one has to be capable of growing single-crystal graphene sheets on large substrate areas (more than 1 cm^2), and forming conducting graphene films with high charge-carrier mobility values on insulating substrates. Also, the possibility of functionalization of graphene sheets during their exposure to chemical agents, the possibility of the creation of heterostructures involving graphene sheets, and the formation of metamaterials and devices on the basis of such heterostructures are required.

Fig. 19.19 shows a photo of an advanced growth facility developed for growing high-quality graphene sheets from a low-pressure argon–methane–hydrogen mixture on copper or semiconductor substrates.

The activities in this field are aimed at the formation of single-crystal graphene sheets. Studies of graphene growth processes were performed, from the formation of rhomb-like islands to that of continuous monolayer graphene films. Unexpected regularities in the formation and growth of graphene islands at an early stage of growth were identified. Processes following a sharp reduction in temperature were analyzed. Quality diagnostics of grown graphene layers were performed by means of Raman spectroscopy (Fig. 19.20). For the first time, with the help of grown graphene structures, the uniformity problem for field emission from carbon materials was solved (because of the scatter of tube sizes and properties, field emitters prepared from arrays of carbon tubes fail to provide the required uniform emission). For forming precise graphene field emitters, we used grown layers and our technology of 3D structures [33]. An optimized array of precise SiO_2/Ni/graphene field emitters is shown in Fig. 19.21. The sizes and properties of 3D graphene–semiconductor nanoshells were described elsewhere [6].

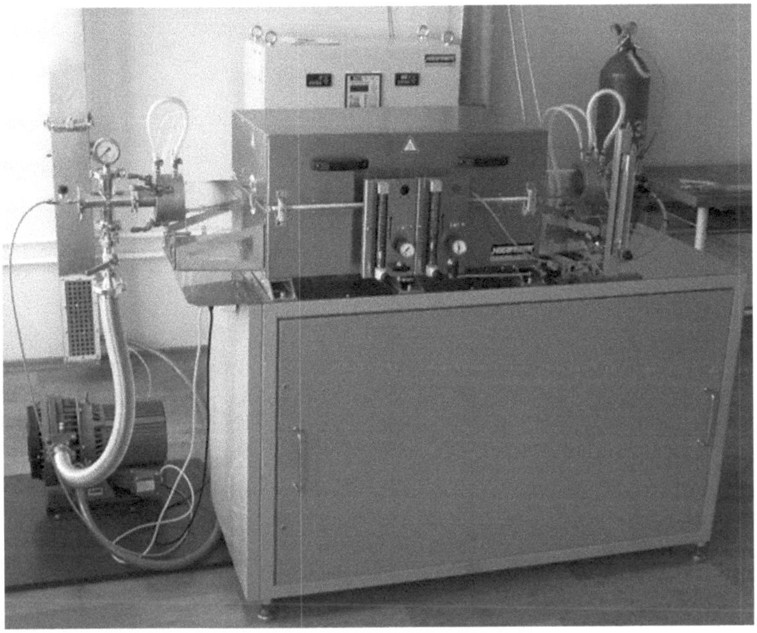

FIGURE 19.19 An advanced facility for growing graphene at low pressures of CH_4.

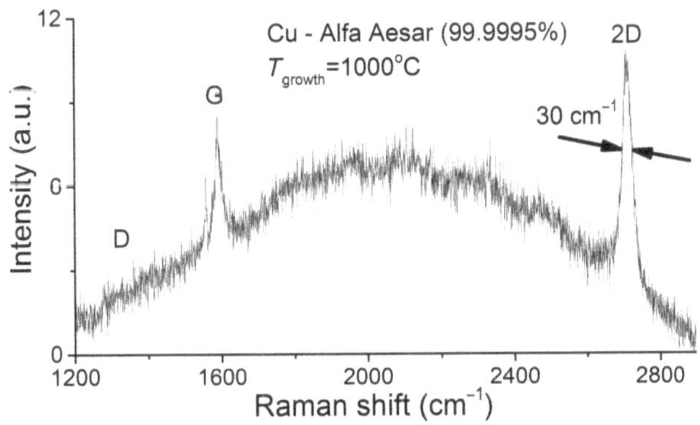

FIGURE 19.20 Raman spectra of a graphene film grown on a copper substrate.

The optimization of conditions for graphene growth on Cu substrates was undertaken [34]. Substrates capable of preserving high mobility in graphene sheets were identified. On a preliminary annealed Alfa Aesar smooth copper foil (purity 99.999%), at a temperature of 1000°C, and in time intervals of half an hour, high-quality, large-area (20 cm^2) graphene sheets were grown in low-pressure (7 torr) atmospheres of Ar, H_2, and CH_4. The Raman spectrum measured of a graphene sheet on copper substrate is shown in Fig. 19.20. Here, a narrow 2D peak of less than 35 cm^{-1} bandwidth, typical of a monolayer graphene, is distinctly seen. By using a modified method for freeing graphene sheets from the copper

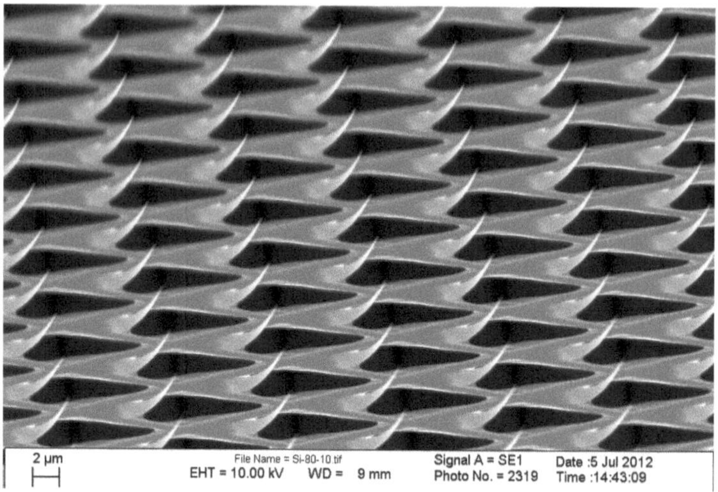

FIGURE 19.21 SEM image of an array of precise graphene field emitters.

substrate (with the use of polycarbonate film and etching of copper in special solutions), the whole grown graphene monolayer could be detached from the copper and transferred onto a SiO_2–Si substrate for subsequent measurement of charge-carrier mobility.

Two-dimensional materials such as graphene, functionalized graphene, transition-metal dichalcogenides, etc., and also, structures based on these materials, demonstrate many new unique properties offering promise for a broad range of applications. Today, all over the world, technologies for obtaining such materials are evolving rapidly. The properties of such materials are under study. The fabrication methods of various structures (heterostructures) based on such materials are being developed. New applications of such materials and structures continue to emerge.

In this area, we have adhered to our own line of research. Technologies for the creation of new 2D graphene- and multigraphene-based materials and structures using intercalation and chemical functionalization have been developed [35–42]. These include, first of all, the creation and study of fluorographene-based self-formed quantum structures; creation and study of graphene and multigraphene quantum dot arrays in fluorographene matrix; arrays of quantum antidots formed from hydrated graphene; hybrid superlattices; and other structures with controlled electronic properties. Modification of the degree of functionalization of the materials allows variation in the density of the arrays of quantum dots and antidots, and their sizes.

An original technology for creation of graphene quantum dots in a fluorographene matrix was developed. That technology is based on incomplete (partial) fluorination of graphene or multigraphene sheets (thickness up to 10 nm) in an aqueous solution of hydrofluoric acid [35–38]. A study of nonstationary electronic processes related to the charge capture on quantum dots has already demonstrated new and interesting properties of such systems. First of all, it was shown that the time constant for emission of charge carriers from quantum dots depends on the thickness of the functionalized graphene layer structure [42].

Another line of research is related to the creation and study of various functional units, vertical heterostructures, and test device structures using new materials and films that were obtained in our laboratory. We have developed an original technology for the fabrication of high-resistive substrates with a graphene sheet on the surface, in which charge carriers have a very high mobility at room temperature, up to 40000 $cm^2/V\,s$ [39–42], this value being indicative of the atomic smoothness of the substrate. The substrates were formed using intercalation of *N*-methylpyrrolidone (organic polar solution) into multigraphene films followed by subsequent annealing. As a result, films as thick as 10 nm, with a resistivity value six or seven orders of magnitude higher than graphene resistivity, were obtained. Then, the electric conduction in the upper monolayer (graphene) was restored by treatment of the structures in hydrofluoric acid vapor for a few minutes. It should be noted here that additional treatment of the surface in hydrofluoric acid opens new interesting possibilities in the formation of 3D nanostructures and devices with conducting nanostructured layers on highly resistive substrates. Other original developments include arrays of quantum antidots prepared from hydrated graphene or multigraphene, also created using intercalation [41], and the fluorination of graphene suspension, yielding thin uniform fluorographene films. The latter technology has potential in a broad range of applications.

An approach based on chemical modification of the graphene which opens up new possibilities for fabrication of electronic and optoelectronic devices was also implemented [43]. This technologically simple approach to the synthesis of lateral graphene-based heterostructures with thin potential barriers aimed to control the current in the graphene channel. Thin potential barriers emerged at the boundaries of domains of the original graphene at the initial stage of its chemical modification. Fluorination and hydrogenation of graphene and its interaction with intercalated organic molecules were used to form the barriers. The channels of transistors fabricated from such heterostructures exhibited efficient current control (four to five orders of magnitude Fig. 19.22) under a varying gate voltage.

It was found that films fabricated from fluorinated graphene suspension are very promising for dielectric layers for graphene, flexible, and printed electronics [44,45]. The electrical and structural properties of the films suggest their use as insulating elements in thin-film nano- and microelectronic device structures. In particular, it was found that films prepared from the fluorinated suspension display rather high breakdown voltages (field strength of $(1-3)\times 10^6$ V/cm), and ultralow densities of charges in the film and at the interface with the silicon substrate in metal–insulator–semiconductor structures ($(1-5)\times 10^{10}$ cm^{-2}). Such excellent characteristics of dielectric film can be compared only to well-developed SiO_2 layers. The films from the fluorinated suspension are cheap, practically feasible, and easy to produce. Suspensions of fluorinated graphene with nanometer size flakes are of interest for the development of 2D ink-jet printing technologies, and production of thermally and chemically stable dielectric films for nanoelectronics. The printing of fluorinated graphene films on flexible substrates is demonstrated in Fig. 19.23.

One more attractive property of fluorinated graphene films for the first time revealed and analyzed in Ref. [46], is a bipolar reversible resistive effect detected in films created from a partially fluorinated graphene suspension, which makes them promising for resistive

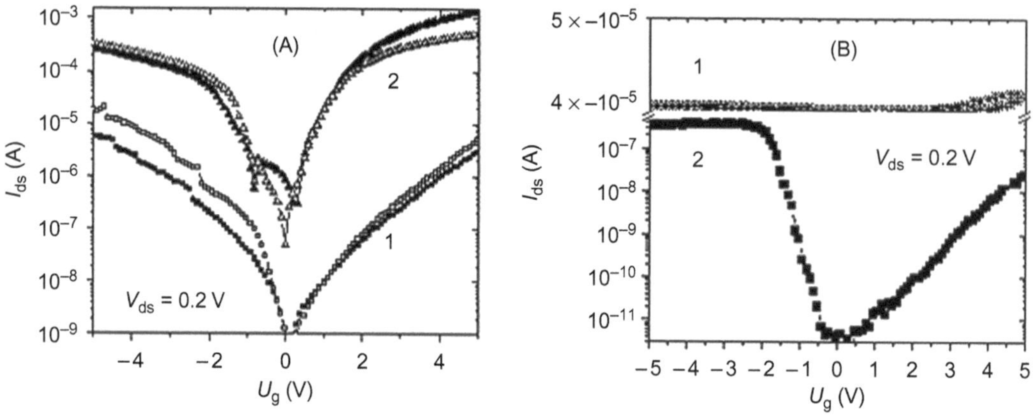

FIGURE 19.22 Current I_{ds} versus gate voltage V_g for different transistor structures: (A) Superlattice graphene *N*-methylpyrrolidone with thickness of 4 nm, annealed at 160°C before (1) and after (2) repairing of the top graphene layer conductivity. (B) Few-layer graphene with thickness of 4 nm before (1) and after (2) fluorination in an aqueous solution of hydrofluoric acid with fluorination degree about 20%.

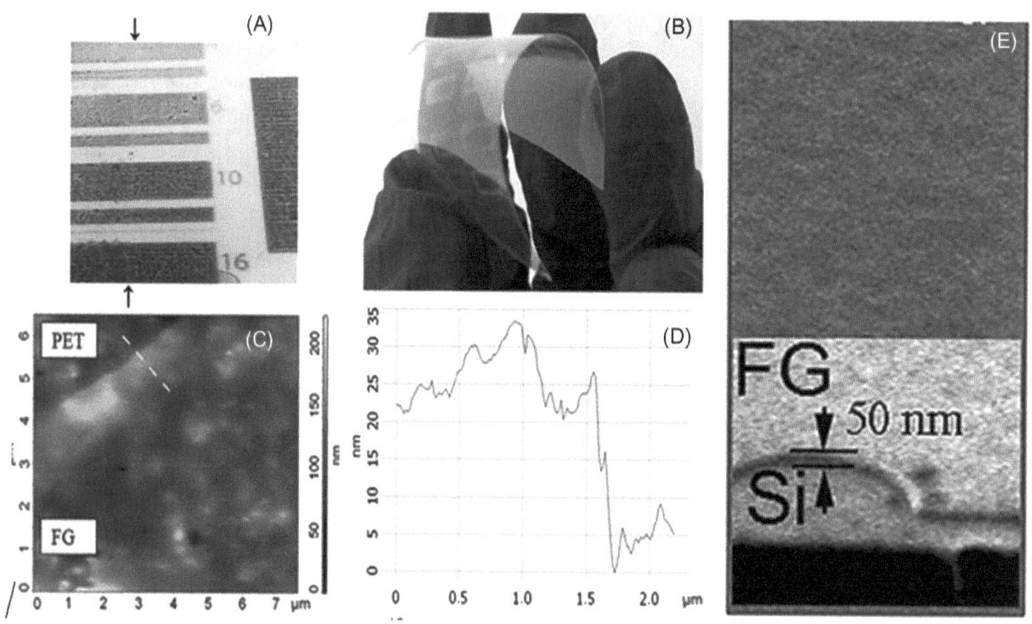

FIGURE 19.23 (A) Images of graphene oxide printed layers (4, 6, 10, 16 printed cycles) with an additional FG layer (8 cycles) printed on top of the left half of the wafer; the boundary of the FG layer is slightly seen in the image and for better visualization it is marked with *arrows*. (B) FG printed layers (10 cycles) on the PET substrate. An AFM image (C) and height profile (D) near the edge of the FG film printed with the use of inkjet printer FUJIFILM Gimatrix DMP-2831 on the PET substrate (10 printed cycles). SEM images of films created from fluorinated graphene suspension by drops on Si substrate. The inset in (E) shows an image of the film edge taken at the 45-degree angle to the surface; the film thickness indicated in the figure was equal to 20 nm.

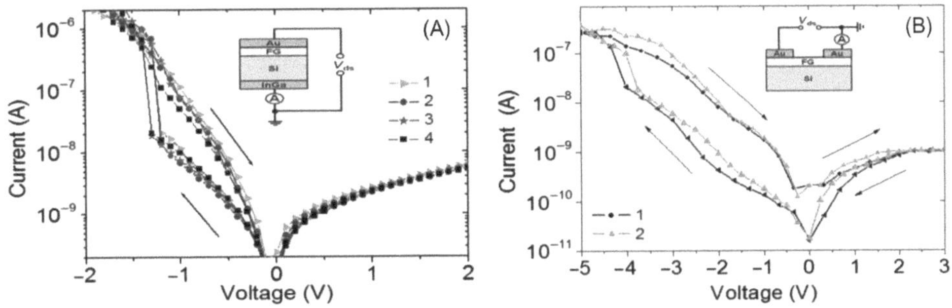

FIGURE 19.24 Current–voltage characteristics measured at daylight conditions of (A) vertical Au/FG/Si/InGa structures and (B) lateral Au/FG/Au structures measured several times at two voltage sweeps at the temperature of 300 K.

memory applications. The fluorination degree for the films demonstrating the resistive effect was estimated as ~50%. The relation of the currents in low- and high-resistance states was one to two orders of magnitude. The resistive switching effect was observed in a lateral and vertical configuration of current measurements (Fig. 19.24). The switching effect was observed without changes for at least 160 cycles of measurements.

19.7 Quantum Properties of 3D Semiconductor Nanostructures

In two-dimensional electron gas (2DEG) nanoshells, bending deformation induces considerable unusual changes in the electronic spectrum of the structure, thus offering fresh possibilities in designing highly efficient quantum systems [47]. In Ref. [48], simulated data for quantum properties of InAs/GaAs nanotubes were reported, and a considerable change of band gap energy and spatial separation of charge carriers (electrons and holes) in tube walls were demonstrated. It was found that, as a result of bending, monolayer-thick films fundamentally acquire new properties. For instance, a corrugated bilayer nanofilm, containing a sequence of strongly compressed and stretched local regions, forms a system of deep potential wells (Figs. 19.25 and 19.26) [47]. Fig. 19.26 shows results obtained for an InAs/GaAs corrugated bilayer film about 3 nm thick [47]. Such films can be made conductive even without performing monolayer electronic passivation of the surface, so that the effective mass approximation still can be used in calculating the quantum properties of the films. Unlike the bulk material, free-standing superthin InAs and GaAs films are not direct-band ones because, as a result of the small effective electron mass in the Γ-minimum, the main quantum level of Γ-minimum electrons ascends over the quantum level of X-minimum electrons. On the other hand, the bending deformation of a corrugated film causes a shift and splitting of the X-minimum, which leads to the formation of a giant potential relief and causes electron localization (see Fig. 19.26B) [47].

Micro- and nanosized conducting shells prepared from various strained semiconductor heterosystems are promising objects for revealing new phenomena. Specific features inherent

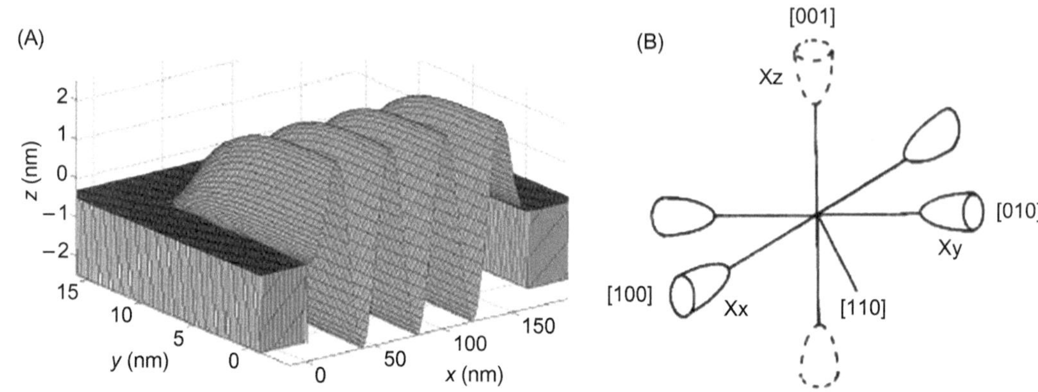

FIGURE 19.25 (A) Schematic of a nanocorrugated structure, (B) Arrangement of X values with respect to crystallographic axes. The direction [001] is normal to the plane of film. Bending-induced shifts of band edges in the X_z and X_x, X_y valleys differ substantially.

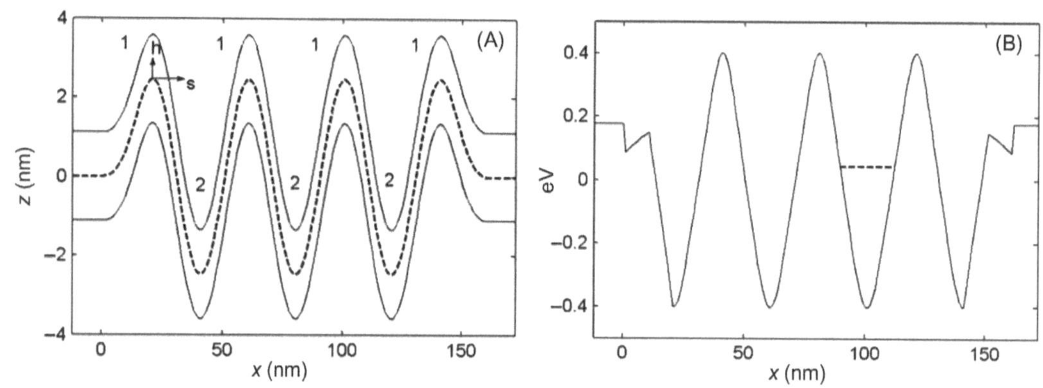

FIGURE 19.26 (A) Cross-sectional diagram of a corrugated bilayer film; in the upper film: 1—stretched, 2—compressed regions; (B) Potential relief and quantum level in the upper film. Film thickness—3 nm.

in semiconductor shells and radically affecting transport of charge carriers in such shells are the following: (1) strong gradient of normal magnetic field component (about 10^6 T/m), which defines the Lorentz force acting on charges and, as a consequence, the occurrence of magneto-transport effects caused by this force [49]; (2) huge crystal-lattice strains of the order of several percent emerge at intricate profiles of mechanical stresses, including periodic and sign-alternating profiles [47]; and (3) geometric potential due to shell curvature. At small curvature radii comparable with the Fermi length of charge carriers, the geometric potential induces substantial modifications into the band spectrum. To date, microshells with 2DEG on a cylindrical surface in a GaAs quantum well with electron mobility of ~100000 cm^2/V s have been fabricated (Fig. 19.27) [7]. For the first time, a ballistic electron transport regime in cylindrical

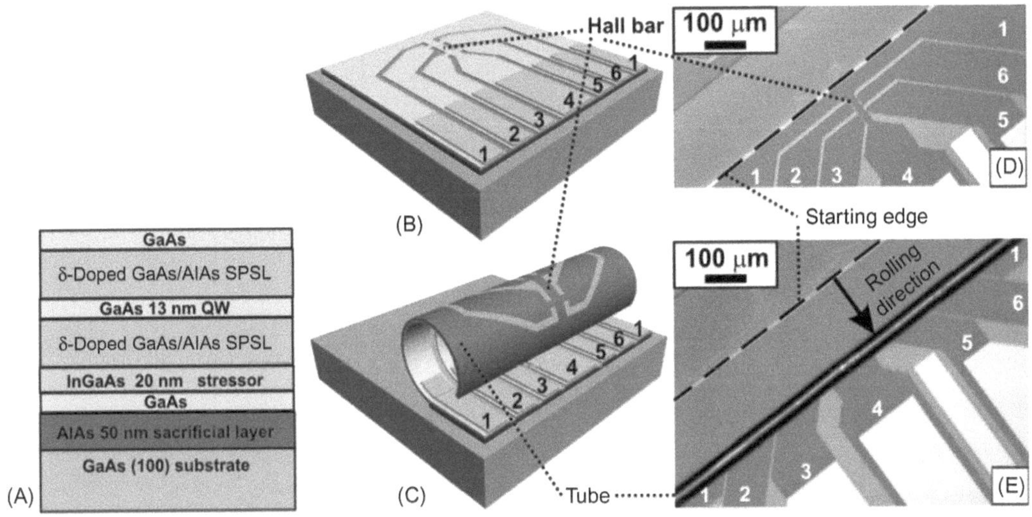

FIGURE 19.27 (A) Sequence of layers in the initial multilayer epitaxial heterostructure with a GaAs quantum well; (B–E) arrangement of the contacts to the rolled Hall bar; samples: left—schematic images, right—photos, top—initial planar structure, bottom—the same structure in rolled state.

2DEG was demonstrated [50]. It was experimentally shown that, as an electric current flows in a 2DEG on a cylindrical surface in the direction of a magnetic field vector, the magneto resistance in diffusion mode exhibits a strong asymmetry, the resistance R_{xx} along one edge of the sample becomes zero, whereas the resistance along the opposite edge retains its finite value comparable with the value of Hall resistance R_{xy}. The latter effect is manifested already in classical magnetic fields, and it remains preserved in quantizing magnetic fields (Fig. 19.28). In classical fields, the magneto resistance asymmetry phenomenon can be understood as a manifestation of the so-called static skin effect, an exponential dependence of current density on the transverse coordinate in a 2D system placed in a nonuniform magnetic field. The explanation for quantizing fields is based on the fact that, due to the nonuniformity of the surface–normal magnetic field component, Landau levels experience bending. This bending leads to the appearance of additional lines at the intersection of Landau levels with the Fermi level and, hence, to the formation of conducting channels analogous to edge states yet going in the "bulk." Together with the edge states, these channels form closed contours whose shape depends on tube radius and magnetic field strength. The magnitude of measured resistance is defined by the relative location of these contours and potential contacts [49].

Helical nanomembranes containing high-mobility 2DEG manifest features in longitudinal resistance corresponding to integer and fractional quantum Hall effects (experiments were carried out in fields up to 28 T). Interpretation of the Hall measurements in a 2DEG on a helical nanomembrane is nontrivial. For this new geometry, an edge state scheme was discussed and calculations in terms of the Landauer–Buttiker formalism were performed (Fig. 19.29). The calculated results were in good agreement with the experimental results [51].

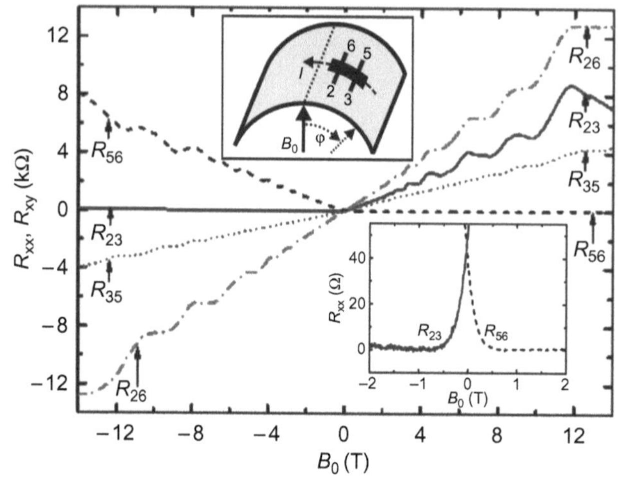

FIGURE 19.28 Magnetoresistance of cylindrical 2DEG. The upper inset explains the experiment geometry, and the inset at the bottom shows magnetoresistance data (R_{xx}) on an enlarged scale. The subscripts in the labels to the experimental curves indicate the number of the contacts used to perform the magnetoresistance measurements.

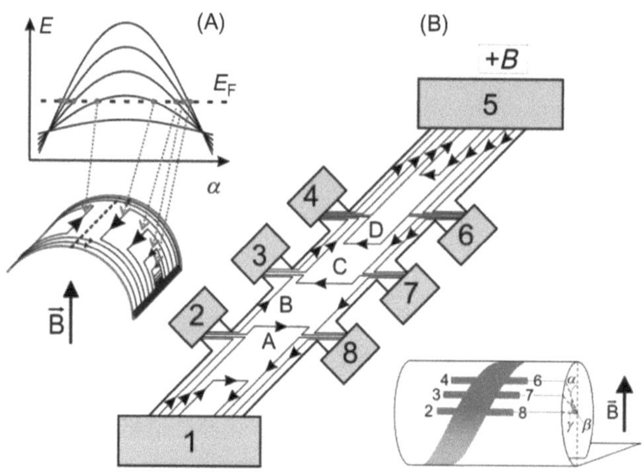

FIGURE 19.29 Schematic illustration of the edge state formation on a cylindrical surface and distribution of the edge states on a helicalHall bar.

Recently, the technique of semiconductor shell fabrication was brought to a level permitting design and fabrication of device prototypes. A schematic illustrating the fabrication process of a three-axis Hall sensor is shown in Fig. 19.30. In laboratory samples of such sensors, the Hall elements in two microtubes are located so that, in an external magnetic field, the voltages applied to those elements are chosen to be in proportion to the three orthogonal components of the magnetic vector. The sensitivity of the sensors to the magnetic

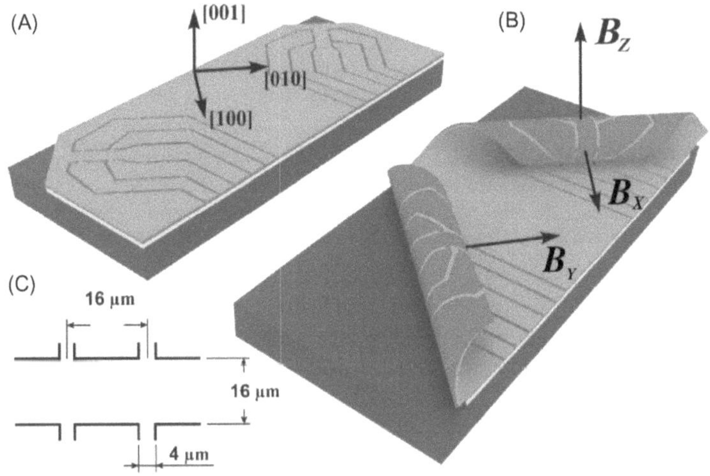

FIGURE 19.30 Schematic illustrating the formation of a three-axis Hall sensor: lithographically patterned film (A) before and (B) after rolling; (C) nominal Hall-bar sizes. The *arrows* in (A) indicate crystallographic directions, (B) magnetic vector components.

field reaches 1000 Ω/T, which is comparable to the sensitivity value of the best commercially available Hall sensors [51].

In conclusion, we would like to note that the foreground areas for our laboratory in this field include the following points:

- Original nanotechnologies, nanomaterials, and nanodevices whose creation occurred in our laboratory;
- Novel approaches to exerting superfast control over electromagnetic radiation;
- Original metamaterials, and nanophotonic and nanoplasmonic elements;
- Quantum processes in 3D nanostructures with 2DEG;
- Nanoimprint lithography and 3D printing methods;
- Growth and functionalization of graphene structures, fluorographene, and graphene-based hybrid structures;
- Original characterization techniques and instruments, including microwave diagnostics of multilayer semiconductor structures.

Acknowledgments

We would like to emphasize here that the majority of activities in our laboratory have been possible only due to close collaboration with the laboratories of molecular beam epitaxy, headed by Dr. A.I. Toropov and Dr. V.V. Preobrazhenskii, and with the laboratory of nanodiagnostics and nanolithography, which in different years was headed by Acad. A.L. Aseev and Prof. Dr. Sc. A.V. Latyshev.

Works on the formation of 3D structures by 3D printing were funded by Russian Science Foundation (project No. 15-12-00050).

References

[1] V.Ya Prinz, V.A. Seleznev, A.K. Gutakovsky, A.V. Chehovskiy, V.V. Preobrazenskii, M.A. Putyato, et al., Free-standing and overgrown InGaAs/GaAs nanotubes, nanohelices and their arrays, Phys. E 6 (1–4) (2000) 828.

[2] V.Ya Prinz, Self-formed precise 3D nanostructures for future nanoelectronic and nanomechanic devices, in: A.L. Aseev (Ed.), Nanotechnologies in Semiconductor Electronics, Russian Academy of Sciences, Siberian Branch, 2004, p. 85. <http://www.sibai.ru/camoformiruyushhiesya-preczizionnyie-3d-nanostrukturyi-dlya-budushhix-priborov-nanoelektroniki-i-nanomexaniki.html/>.

[3] V.Ya Prinz, D. Grützmacher, A. Beyer, C. David, B. Ketterer, E. Deccard, A new technique for fabricating three-dimensional micro- and nanostructures of various shapes superlattices, Nanotechnology 12 (2001) 399.

[4] V.Ya Prinz, A new concept in fabricating building blocks for nanoelectronic and nanomechanic devices, Microelectron. Eng. 69 (2–4) (2003) 466.

[5] V.Ya Prinz, Precise, molecularly thin semiconductor shells: from nanotubes to nanocorrugated quantum systems, Phys. Stat. Sol. (B) 243 (13) (2006) 3333.

[6] R.M. Taziev, V.Ya Prinz, Buckling of a single-layered graphene sheet on an initially strained InGaAs thin plate, Nanotechnology 22 (2011) 305705.

[7] B. Vorob'ev, K.-J. Friedland, H. Kostial, R. Hey, U. Jahn, E. Wiebicke, et al., Giant asymmetry of the longitudinal magnetoresistance in high-mobility two-dimensional electron gas on a cylindrical surface, Phys. Rev. B 75 (20) (2007) 205309.

[8] N. Levy, S.A. Burke, K.L. Meaker, M. Panlasigui, A. Zettl, F. Guinea, et al., Strain-induced pseudo-magnetic fields greater than 300 tesla in graphene nanobubbles, Science 329 (2010) 544.

[9] V.Y. Prinz, V.A. Seleznev, A.V. Prinz, A.V. Kopylov, Topical review: 3D heterostructures and systems for novel MEMS/NEMS, Sci. Technol. Adv. Mater. 10 (2009) 034502.

[10] X. Li, Strain induced semiconductor nanotubes: from formation process to device applications, J. Phys. D Appl. Phys. 41 (2008) 193001.

[11] C. Vervacke, C.C. Bof Bufon, D.J. Thurmer, O.G. Schmidt, Three-dimensional chemical sensors based on rolled-up hybrid nanomembranes, RSC Adv. 4 (2014) 9723.

[12] D. Karnaushenko, N. Münzenrieder, D.D. Karnaushenko, B. Koch, A.K. Meyer, S. Baunack, et al., Biomimetic microelectronics for regenerative neuronal cuff implants, Adv. Mater. 27 (2015) 6797.

[13] R. Arayanarakool, A.K. Meyer, L. Helbig, S. Sanchez, O.G. Schmidt, Tailoring three-dimensional architectures by rolled-up nanotechnology for mimicking microvasculatures, Lab Chip 15 (2015) 2981.

[14] S. Schwaiger, A. Rottler, S. Mendach, Rolled-up metamaterials, Adv. Optoelectron. Article ID 782864 (2012) 10 pp.

[15] V.Ya. Prinz, V.A. Seleznev, A.K. Gutakovsky, A.I. Toropov, V.S. Aliev, Rolled InAs/GaAs/Ta_2O_5 hetero-films containing InAs quantum dots, in: 11th International Symposium "Nanostructures: Physics and Technology", St. Petersburg, Russia, 2003.

[16] H.J. Lim, et al., Roller nano-imprint lithography for flexible electronic devices of a sub-micronscale, Microelectron. Eng. 88 (2011) 2017.

[17] A. Gayduk, V. Prinz, V. Seleznev, S. Rechkunov, Large-area multilayer infrared nano-wire grid polarizers, Infrared Phys. Technol. 75 (2016) 77–81.

[18] E.V. Naumova, V.Ya. Prinz, S.V. Golod, V.S. Seleznev, Polarization rotation of THz radiation by an array of helices, in: Abstract Book of the Joint 31st International Conference on Infrared Millimeter Waves and 14th International Conference on Terahertz Electronics, Shanghai, China, 2006, p. 351.

[19] E.V. Naumova, V.Ya. Prinz, V.A. Seleznev, S.V. Golod, R.A. Soots, V.V. Kubarev, et al., Fabrication of metamaterials on the basis of precise micro- and nanoshells, in: First International Congress on Advanced Electromagnetic Materials in Microwaves and Optics, Rome, Italy, 2007, p. 74.

[20] E.V. Naumova, V.Ya Prinz, S.V. Golod, V.A. Seleznev, R.A. Soots, V.V. Kubarev, Manufacturing chiral electromagnetic metamaterials by directional rolling of strained heterofilms, J. Opt. A Pure Appl. Opt. 11 (2009) 074010.

[21] I.V. Semchenko, S.A. Khakhomov, V.A. Asadchii, E.V. Naumova, V.Ya Prinz, S.V. Golod, et al., Study of the properties of weakly reflecting metamaterials with compensated chirality, Crystallogr. Rep. 56 (3) (2011) 366–373.

[22] V.A. Seifi, V.Ya Prinz, Spectral characteristics of microwave and infrared metamaterials with 3D resonators, Optoelectronics, Instrumentation and Data Processing 50 (3) (2014) 283–286.

[23] A.A.Bocharov, V.Ya. Prinz, Chiral metamaterial system with giant optical activity, in: Nanostructures: Physics and Technology, St. Petersburg, 2014.

[24] V.Ya. Prinz, S.V. Golod, Provisional Application for a Patent of the Russian Federation, №2015103139 Filed on January 30, 2015 (Favorable decision on February 18, 2016).

[25] V.A. Seleznev, V.Ya. Prinz, I.A. Korneev Fabrication of gecko-inspired reversible adhesives by nanoimprint, in: Lithography 22nd International symposium Nanostructures: Physics and Technology, St. Petersburg, June 23–27, 2014, p. 254.

[26] V.Ya. Prinz, P.M. Larionov, M.A. Sadovoi, E.V. Mamonova, Bioresorbable Polymer Cell Matrix, Patent RU 2563621, April 27, 2015.

[27] V.A. Seleznev, V.Ya Prinz, V.M. Aniskin, A.A. Maslov, Generation and registration of disturbances in a gas flow. 1. Formation of arrays of tubular microheaters and microsensors, J. Appl. Mech. Tech. Phys. 50 (2009) 291.

[28] A.N. Shiplyuk, V.M. Aniskin, V.A. Seleznev, V.Ya Prinz, A.A. Maslov, R.S. Matvienko, Generation and registration of disturbances in a gas flow. 2. Experiments with arrays of tubular microheaters and microsensors, J. Appl. Mech. Tech. Phys. 50 (3) (2009) 454.

[29] V.A. Seleznev, V.Ya. Prinz, A Method for Fabrication of Gas or Liquid Flow Velocity Sensor, RU Patent No. 2353998, 2009.

[30] V.Ya. Prinz, S.V. Golod, A.V. Prinz, Integrated Hollow Needle and a Method to Fabricate It, RU Patent No. 2341299, 2008.

[31] A.V. Prinz, A method for fabricating arrays of long vertical semiconductor microtubes, Nano Mikrosist. Tekh. 10 (2013) 5.

[32] V.Ya. Prinz, A.V. Prinz, A.V. Kopylov, Electrostatic Micro-Nano-Actuator, RU Patent No. 2374746, 2009.

[33] V.Ya. Prinz, S.V. Mutilin, S.V. Golod, A Method for Fabrication of Graphene Field Emitters, RU Patent No. 2400858, 2010.

[34] A.A. Pachnevich, V.Ya Prinz, S.V. Golod, Surface melting of copper during graphene growth by chemical vapour deposition, J. Phys. D Appl. Phys. 48 (43) (2015) 435303.

[35] I. Antonova, S. Mutilin, V. Seleznev, R. Soots, V. Volodin, V. Prinz, Extremely high response of electrostatically exfoliated few-layered graphene flakes to ammonia adsorption, Nanotechnology 22 (28) (2011) 285502.

[36] I.V. Antonova, V.Ya Prinz, Benefits of few-layer graphene for applications (Chapter 16). ISBN-13: 978-1466591189, 2768 pp. Handbook of Graphene Science: Vol. 2, Nanostructure and Atomic Arrangement, Taylor and Francis Books-CRC Press, Boca Raton, FL, 2016.

[37] N.A. Nebogatikova, I.V. Antonova, V.A. Volodin, V.Ya Prinz, Functionalization of graphene and few-layer graphene with aqueous solution of hydrofluoric acid, Phys. E 52 (2013) 106.

[38] N.A. Nebogatikova, I.V. Antonova, V.Ya Prinz, V.B. Timofeev, S.A. Smagulova, Graphene quantum dots in fluorographene matrix formed by means of chemical functionalization, Carbon 77 (2014) 1095–1103.

[39] I.A. Kotin, I.V. Antonova, A.I. Komonov, V.A. Seleznev, R.A. Soots, V.Ya. Prinz, High carrier mobility in graphene on atomically flat high-resistivity layer, J. Phys. D Appl. Phys. 46 (2013) 285303.

[40] I.V. Antonova, I.A. Kotin, R.A. Soots, V.A. Volodin, V.Ya Prinz, Tunable properties of few layer graphene-*N*-methylpyrrolidone hybrid structures, Nanotechnology 23 (2012) 315601.

[41] I.A. Kotin, I.V. Antonova, R.A. Soots, V.A. Volodin, V.Ya Prinz, Layered structures based on hydrated graphene sheets with high charge-carrier mobility, Russ. Nanotechnol. 8 (9–10) (2013) 26.

[42] I.V. Antonova, Vertical heterostructures based on graphene and other single-layer materials, Semiconductor 50 (2016) 66–82.

[43] I.V. Antonova, I.A. Kotin, N.A. Nebogatikova, V. Ya Prinz, Modulation of current in self-forming lateral graphene-based heterostructures, Tech. Phys. Lett. 41 (2015) 950–953.

[44] N.A. Nebogatikova, I.V. Antonova, V.Ya Prinz, I.I. Kurkina, G.N. Aleksandrov, V.B. Timofeev, et al., Fluorinated graphene dielectric films obtained from functionalized graphene suspension: preparation and properties, Phys. Chem. Chem. Phys. 17 (2015) 13257–13266.

[45] N.A. Nebogatikova, I.V. Antonova, I.I. Kurkina, R.A. Soots, V.I. Vdovin, V.B. Timofeev, et al., Graphene flakes fragmentation in suspension in the course of fluorination, Nanotechnology 27 (2016) 205601.

[46] I.I. Kurkina, I.V. Antonova, N.A. Nebogatikova, A.N. Kapitonov, S.A. Smagulova, Resistive switching effect and traps in partially fluorinated graphene films, J. Phys. D Appl. Phys. 49 (2016) 095303.

[47] V.M. Osadchii, V.Ya Prinz, Quantum confinement in nanocorrugated semiconductor films, Phys. Rev. B 72 (2005) 033313.

[48] V.M. Osadchii, V.Ya Prinz, Scrolled heterostructures, JETP Lett. 72 (2000) 312.

[49] L.I. Magarill, D.A. Romanov, A.V. Chaplik, Ballistic transport and spin-orbit interaction of 2-dimensional electrons on a cylindrical surface, JETP 86 (1998) 771.

[50] K.-J. Friedland, R. Hey, H. Kostial, U. Jahn, E. Wiebicke, K.H. Ploog, et al., Magnetotransport in two-dimensional electron gases on cylindrical surfaces, in: H. Takayanagi, J. Nitta, H. Nakano (Eds.), Proceedings of the International Symposium on Mesoscopic Superconductivity and Spintronics, Controllable Quantum States, Atsugi, Japan, 2006, World Scientific, 2008, pp. 288–294, <arXiv:cond-mat/0703665>.

[51] A. Vorob'ev, A. Chesnitskiy, A. Toropov, V. Prinz, Three-axis Hall transducer based on semiconductor microtubes, Appl. Phys. Lett. 103 (17) (2013) 173513.

[52] I.V. Antonova, N.A. Nebogatikova, V.Ya Prinz, Self-organized arrays of graphene and few-layer graphene quantum dots in fluorographene matrix: formation of quantum dots and charge spectroscopy, Appl. Phys. Lett. 104 (19) (2014) 193108.

[53] J.S. Vorobyova, A.B. Vorob'ev, V.Y. Prinz, A.I. Toropov, D.K. Maude, Magnetotransport in two--dimensional electron gas in helical nanomembranes, Nano Lett. 15 (3) (2015) 1673–1678.

20

The Nature of Defects Responsible for Transport in a Hafnia-Based Resistive Random Access Memory Element

D.R. Islamov[1,2], T.V. Perevalov[1,2], V.A. Gritsenko[1,2], V.Sh. Aliev[1], A.A. Saraev[3], V.V. Kaichev[3], E.V. Ivanova[4], M.V. Zamoryanskaya[4], A. Chin[5]

[1]RZHANOV INSTITUTE OF SEMICONDUCTOR PHYSICS OF THE SIBERIAN BRANCH OF THE RUSSIAN ACADEMY OF SCIENCES, NOVOSIBIRSK, RUSSIA [2]NOVOSIBIRSK STATE UNIVERSITY, NOVOSIBIRSK, RUSSIA [3]BORESKOV INSTITUTE OF CATALYSIS, SIBERIAN BRANCH OF THE RUSSIAN ACADEMY OF SCIENCES, NOVOSIBIRSK, RUSSIA [4]IOFFE PHYSICAL-TECHNICAL INSTITUTE, RUSSIAN ACADEMY OF SCIENCES, ST. PETERSBURG, RUSSIA [5]NATIONAL CHIAO TUNG UNIVERSITY, HSINCHU, TAIWAN

20.1 ReRAM: the Next Generation of Non-Volatile Memory

The major driving force in the current development of microelectronics is the search for a universal memory, which would combine the most favorable properties of both high-speed dynamic random access memory and nonvolatile flash memory. The candidates for universal memory are ferroelectric memory, phase-change memory, magneto-resistive memory, and resistive memory. Each option has advantages and disadvantages. One of the most promising options for universal memory is resistive random access memory (ReRAM) [1–5], which has many advantages: a simple metal–insulator–metal (MIM) structure, a small memory cell, the potential for 3D integration, high read-and-write operation speeds, low power consumption, and the ability to save information for a long time. A ReRAM operation is principally based on switching back and forth from the insulating medium's high resistance state (HRS) to a low resistance state (LRS) when current flows. Hafnia HfO_2 and hafnium suboxides $HfO_{x<2}$ are the most promising materials for use as the active medium of resistive memory [6,7]. Hafnia conductivity is limited by trap ionization [6] when ReRAM is switched to the HRS state. However, the atomic structure of the defect that affects localization and charge transport still remains unclear. Currently, the accepted hypothesis is that oxygen vacancies are responsible for charge transport in dielectric and ReRAM switching [5]. Although the theory of the atomic and electronic structure of oxygen vacancies in hafnia was investigated in many studies [8–15], direct experimental data regarding the presence of oxygen vacancies in hafnia have not been reported.

Advances in Semiconductor Nanostructures. DOI: http://dx.doi.org/10.1016/B978-0-12-810512-2.00020-2

Here, we clearly show both experimentally and theoretically, that oxygen vacancies are responsible for HRS charge transport in resistive memory elements based on HfO_2. This study also shows that the charge transport mechanism is described in terms of percolation through hafnium suboxides when hafnia-based ReRAM is switched to the LRS state. This study describes electronic transport in the ReRAM insulator medium completely.

20.2 ReRAM Element Fabrication and Characterization

Our devices were fabricated as follows. $HfO_{x<2}$ films were produced using ion beam sputtering deposition. For substrates, we used silicon plates, Si(100), with a resistivity of 4.5 $\Omega \times$ cm, deep-cleaned using a technique provided by the RCA Company. The target was sputtered by a beam of Ar^+ ions, while highly pure oxygen ($O_2 > 99.999\%$) was simultaneously delivered into the area near the target and substrate. The composition (x-parameter) of the HfO_x film was defined using a gas flow controller to regulate the partial oxygen pressure. In our experiments, we grew two sets of HfO_x samples at partial oxygen pressures of 9×10^{-3} Pa and 2×10^{-3} Pa. Under such conditions, we produced samples of near-stoichiometric ($x \approx 2$) and nonstoichiometric compositions ($x < 2$).

The valence band spectra of grown hafnia films were obtained using a SPECS X-ray photoelectron spectral machine with monochromatic AlK_α radiation ($h\nu = 1486.74$ eV). The defects in HfO_2/Si films were generated by Ar^+ ion bombardment with an energy of 2.4 keV and an ion current density of approximately 10 $\mu A/cm^2$. It is known that sputtering by Ar^+ preferentially removes oxygen from hafnium oxides, leading to the formation of hafnium suboxides ($HfO_{x<2}$) [16]. The [O]/[Hf] atomic ratio was estimated according to XPS peak areas combined with Scofield cross-sections. The electron energy loss spectra (EELS) were measured as described in Ref. [17]. The [O]/[Hf] atomic ratio for nonstoichiometric and Ar^+-bombarded hafnia film is approximately 1.9.

The cathodoluminescence (CL) spectra of the HfO_x samples were measured using a Camebax electron probe microanalyzer. The CL spectra were recorded in the 1.5–3.8 eV range at 300K. The sample was irradiated by a beam of 1000 eV electrons, penetrating to a depth of 0.2 μm with a current density of 1.2 A/cm^2.

Transport measurements were recorded for metal-insulator semiconductor (MIS) and MIM structures of Si/HfO_x/Ni and Si/TaN/HfO_x/Ni, respectively. To fabricate Si/HfO_x/Ni structures, the 20-nm thick amorphous hafnia ($HfO_{x<2}$) was deposited on an n-type Si wafer using the atomic layer deposition system. Tetrakis dimethyl amino hafnium (TDMAHf) and water vapor were used as precursors at a chamber temperature of 250°C for HfO_x film deposition. For the ReRAM measurements, Si/TaN/HfO_x/Ni structures were used. To fabricate these structures, we deposited 8-nm thick amorphous hafnia on 100-nm thick TaN films, and these films onto Si wafers using physical vapor deposition. A pure HfO_2 target was bombarded by an electron beam, and HfO_2 molecules were deposited on the wafer. We did not apply any postdeposition annealing to produce the most nonstoichiometric films. Structural analysis showed that the resulting HfO_x films were amorphous. All samples for transport measurements were equipped with round 50-nm thick Ni gates with a radius of 70 μm.

Transport measurements were performed using semiconductor parameter analyzers and precision LCR meters. The current compliance of 1 μA was set for ReRAM measurements.

The *ab initio* simulations of electronic and optical properties were performed based on density functional theory using Quantum-ESPRESSO software [18]. We adopted a hybrid functional B3LYP to reproduce accurate band gap values. We used 81-atom cubic hafnia (c-HfO_2) supercells to simulate the oxygen vacancy. The choice of cubic phase is justified according to its simple structure.

20.3 Transport Properties in High Resistance State

Fig. 20.1A shows a typical hysteresis of experimental current–voltage characteristics of TaN/HfO_x/Ni MIM structures at room temperature. The memory window (i.e., current difference) is more than three orders at a bias voltage of 0.5 V. To study the electronic transport mechanism in hafnia, we performed transport measurements in simpler metal–insulator–semiconductor structures, specifically, *n*-Si/HfO_x/Ni [19]. The experimental current–voltage characteristics in *n*-Si/HfO_x/Ni structures, measured at different temperatures with a positive bias applied to the Ni contact, are shown in Fig. 20.1B, graphed by characters in different colors. We attempted to qualitatively explain the experimental data by using different models, such as the Frenkel trap ionization model [20], the Hill trap ionization model [21], and the model of multiphonon

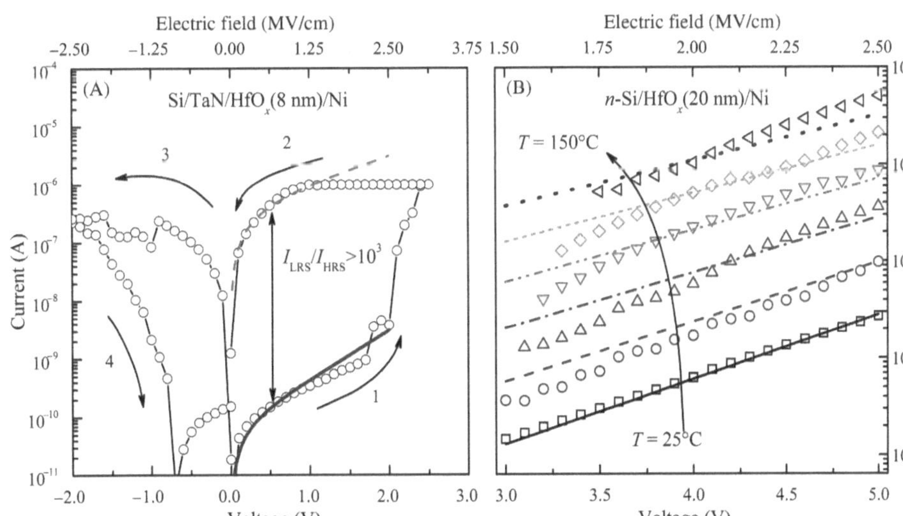

FIGURE 20.1 Experimental current–voltage characteristics (characters) and simulations (lines). (A) ReRAM current–voltage characteristic hysteresis in Si/TaN/HfO_x/Ni MIM structures at room temperature. The *blue solid line* (dark gray in print versions) represents HRS simulation by phonon-assisted tunneling between traps at room temperature. The *green dashed line* (light gray in print versions) shows LRS simulation by phonon-assisted tunneling between traps at room temperature. The simulation procedure gives the following parameters of HfO_x: $W_{opt} = 2.5$ eV, $W_t = 1.25$ eV, $N_{HRS} = 1.9 \times 10^{21}cm^{-3}$, $N_{LRS} = 1.1 \times 10^{22}cm^{-3}$. (B) Experimental current–voltage characteristics of an *n*-Si/HfO_x/Ni structure, and simulation by phonon-assisted tunneling between traps (Eq. 20.1) at different temperatures $T = 25–150°C$, $\Delta T = 25°C$.

ionization of an isolated trap [22]. However, the fitting procedures involved in using these models returned nonphysical fitting parameter values.

To describe the experiments quantitatively and qualitatively, we performed simulations based on the model of phonon-assisted tunneling between traps [23]:

$$J = \frac{e}{s^2}\frac{n_t}{N_t}\left(1 - \frac{n_t}{N_t}\right)P,$$
$$P = \frac{\sqrt{2\pi}hW_t}{m * s^2\sqrt{kT(W_{opt} - W_t)}}\exp\left(-\frac{W_{opt} - W_t}{2kT}\right)\exp\left(-\frac{2s\sqrt{2m * W_t}}{\mathrm{h}}\right)\sinh\left(\frac{eFs}{2kT}\right), \tag{20.1}$$

where J is the current density, e is the electron charge, n_t is the density of filled traps, P is the probability of electron tunneling between traps per second, $\hbar$ is the Planck constant, W_t is the trap thermal excitation energy, W_{opt} is the trap optical excitation energy, $s = N_t^{-1/3}$ is the distance between traps, k is the Boltzmann constant, and F is the electric field. Analysis shows that taking $n_t/N_t = 1/2$ allows us to use uniform electric field approximation $F = V/d$, where V is voltage bias applied to MIM structure, and d is the dielectric layer thickness. The results of this multiparameter fitting procedure are shown in Fig. 20.1B, graphed in *solid lines*. This procedure yielded the values of different transport parameters, such as $N_t = 2.5 \times 10^{20}$ cm^{-3}, $W_t = 1.25$ eV, $W_{opt} = 2W_t = 2.5$ eV, and $m^* = 0.8m_e$. Quantitatively, there is full agreement between the model of phonon-assisted tunneling between traps and the experimental data. The electron trap energy in hafnium oxide was previously determined in Ref. [24] to be 1.2 eV. The trap thermal energy value of 1.25 eV that we obtained is close to that of $W_t = 1.36$ eV observed in the experiment in [25]. Furthermore, the trap optical energy value of $W_{opt} = 2.5$ eV is close to the calculated value of 2.35 eV for the negatively charged oxygen vacancy in hafnia reported in Ref. [14].

The configuration diagram of a negatively charged oxygen vacancy (electron trap) in hafnia is shown in Fig. 20.2. The vertical transition with the value of 2.5 eV corresponds to the optical trap excitation. The thermal trap energy value was 1.25 eV.

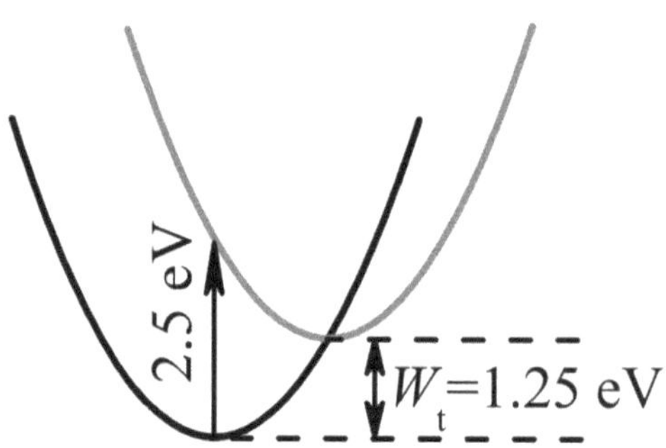

FIGURE 20.2 Configuration coordination energy diagram of the trap ionization process on a negatively charged oxygen vacancy in hafnia. Lower (*black*) term represents the filled ground state, upper (*red* (gray in print versions)) term denotes the empty excited state.

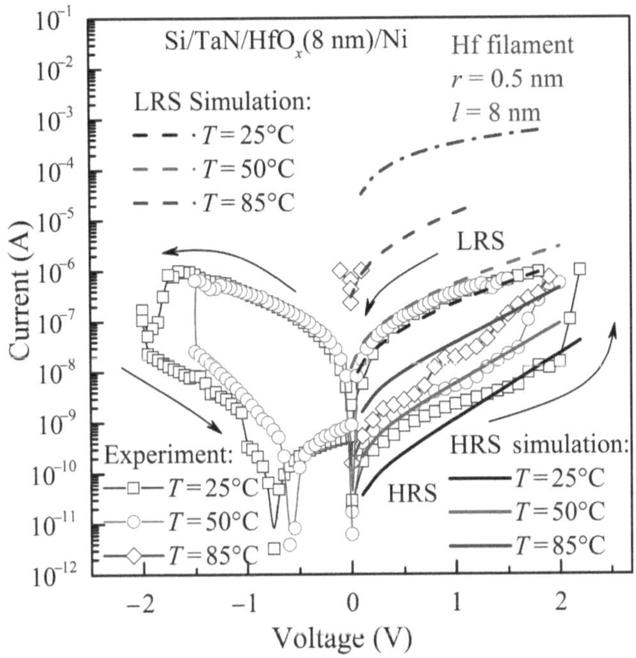

FIGURE 20.3 Experimental ReRAM current–voltage characteristic hysteresis (characters) in Si/TaN/HfO_x/Ni structures at different temperatures. *Solid lines* represent HRS simulation by phonon-assisted tunneling between traps at different temperatures with the following parameter values: $W_{opt} = 2.5$ eV, $W_t = 1.25$ eV, $N_t = 5.5 \times 10^{20} cm^{-3}$, and $m^* = 0.9m_e$. *Dashed lines* represent LRS simulations in terms of a percolation model. *Green dashed-dotted line* (dark gray in print versions) models current–voltage characteristics of a metal Hf filament of diameter of 1 nm and length 8 nm.

We applied the same fitting procedures to the data obtained in our experiments of current–voltage characteristics in MIM structures. The experimental data (graphed in color) of current–voltage measurements at different temperatures are shown in Fig. 20.3. The *solid-colored lines* indicate the results of HRS simulations of phonon-assisted tunneling between traps (Eq. 20.1). MIM structures have the following transport parameter values: $N_t = 5.5 \times 10^{20}$ cm^{-3}, $W_t = 1.25$ eV, $W_{opt} = 2.5$ eV $= 2W_t$, and $m^* = 0.9m_e$. The difference between MIM and MIS structures in effective mass is due to the bulk charge which is adequately addressed in Ref. [26]. However, it is important to notice that the trap energy parameters are invariants of the grown structures. Consequently, we found that the nature of charge-carrier transport in the HRS of ReRAM structures is phonon-assisted tunneling between traps.

20.4 Electronic Structure of Defects in the Active Layers of ReRAM Element

To examine the defects in atomic and electron structures, we measured the X-ray photoelectron spectra of the valence band, cathodoluminescence, and EELS in nonstoichiometric

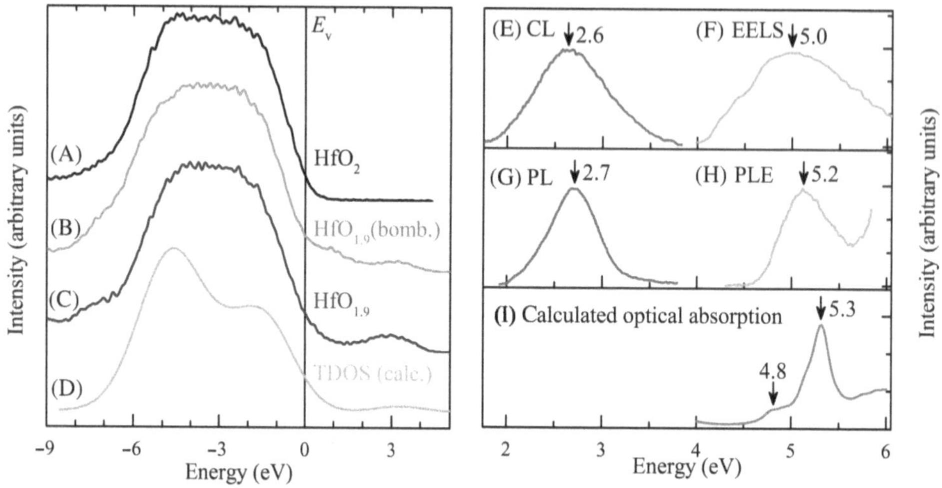

FIGURE 20.4 Experimental and calculated spectra in hafnia. (A) Experimental valence band X-ray photoelectron spectra for amorphous stoichiometric HfO_2; (B) bombed hafnia ($HfO_{1.9}$); (C) nonstoichiometric hafnia ($HfO_{1.9}$); and (D) calculated TDOS for c-HfO_2 with an oxygen vacancy. (E) Cathodoluminescence spectra of $HfO_{1.9}$. (F) Electron energy loss spectra of $HfO_{1.9}$. (G) Photoluminescence spectra. (H) Photoluminescence excitation spectra of $HfO_{x<2}$ from Ref. [28]. (I) Calculated optical absorption spectra for hafnia with oxygen vacancy.

hafnia. The experimental XPS spectra of the valence band obtained before and after Ar^+ ion bombarding of the stoichiometric amorphous Si/HfO_2 film, as well as the spectra of nonstoichiometric Si/$HfO_{x<2}$, are shown in Fig. 20.4A–C. The valence band XPS spectra for both HfO_x films have a rather intense peak, nearly 3 eV above the top of the hafnia valence band. The peaks in this region might have been caused by electron photoemission from the *5d* level of metallic hafnium. The Ar^+ ions irradiating Si/HfO_x have an additional XPS peak, 1 eV higher than the top of the valence band. According to our calculations, the peak is related neither to neutral oxygen polyvacancies, nor to charged monovacancy. The calculated total density of states (TDOS) of the valence band for c-HfO_2, in considering the oxygen monovacancy, agrees with the valence band width in a defect level position (a feature of TDOS at 3.0 eV in Fig. 20.4D). These results prove the presence of oxygen vacancies in amorphous nonstoichiometric hafnia and hafnia films following Ar^+ ion bombardment.

Experimental CL and EELS in nonstoichiometric $HfO_{1.9}$ are shown in Fig. 20.4E–H. A peak with a value of 2.6 eV was observed in the CL HfO_x spectrum (Fig. 20.4E). Previously, a cathodoluminescence band with a value of 2.7 eV was observed in Ref. [27]. Excitation in the EELS (Fig. 20.4F) is observed in $HfO_{1.9}$ at 5.0 eV, although the margin of error for determining EELS peak energy values is 0.3 eV. The photoluminescence spectrum and excitation photoluminescence band spectrum 2.7 eV (PLE) in HfO_x, obtained from the scientific literature [28], are also presented in Fig. 20.4G and H, respectively. A peak of 2.7 eV is also observed in the PL spectrum of HfO_x. Luminescence band excitation with a maximum value of 2.7 eV is located at 5.2 eV.

To identify the defect responsible for excitation at 5.2 eV, we examined the optical absorption of an oxygen vacancy in hafnium oxide. The calculated hafnium oxide optical

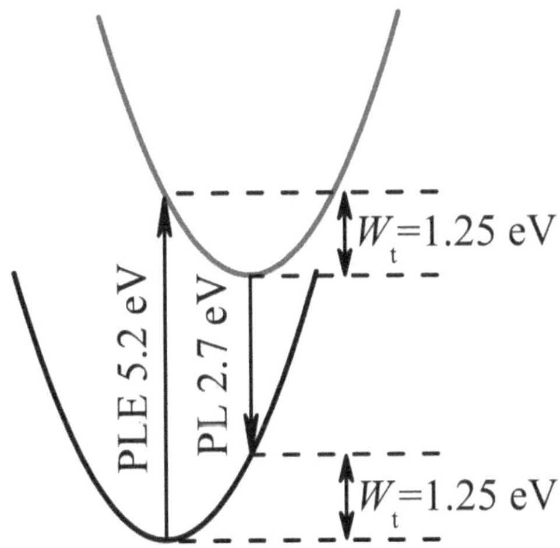

FIGURE 20.5 Configuration coordination energy diagram of optical transition on neutral oxygen vacancy in hafnia. The lower (*black*) term is the filled ground state, the upper (*red* (gray in print versions)) term represents the filled excited state.

absorption spectrum is shown in Fig. 20.4I. The dominant peak has a value of 5.3 eV. The oxygen vacancy excitation corresponds to the optical transition of 5.3 eV. The calculation shows a wide absorption spectrum with energy values between 4.7 and 4.8 eV. The optical absorption in hafnium oxide with similar energy values was experimentally observed in Ref. [24]. A previous quantum mechanical simulation established that the absorption in hafnium oxide within 4.7–4.8 eV is connected to the transition of electrons from the top of the valence band to the excited state of oxygen vacancy [10].

A diagram of the neutral oxygen vacancy configuration is shown in Fig. 20.5. A vertical transition with an energy value of 5.2 eV corresponds to optical electron excitation, and the transition with an energy value of 2.7 eV corresponds to the radiating transition into equilibrium. The energy distributed to phonons corresponds, in our model, to the thermal trap energy. These data indicate that the luminescence band of 2.7 eV and the excitation band of 5.2 eV are conditioned by intrinsic defects, namely, oxygen vacancies in $HfO_{1.9}$.

20.5 Transport Properties in Low Resistance State

The most commonly used LRS model in ReRAM structures consists of a metal filament of approximately 1–10 nm [29,30]. This metal filament must have linear current–voltage characteristics. This approach exhibits two "bottlenecks": (1) it cannot describe ReRAM in GeO_x and SiO_x-based structures [31,32], and (2) the experimentally measured current is much lower than that predicted by Ohm's law. To illustrate the second case, we calculated the current through a pure metal Hf wire, 1 nm in diameter and 8 nm long. The results are shown in Fig. 20.3 by a *dashed-dotted green line* (dark gray in print versions). The calculations are approximately 10^3 times higher than those obtained in the experiments. Therefore, we suppose that LRS conductivity is conditioned by the presence of the nonstoichiometric HfO_x

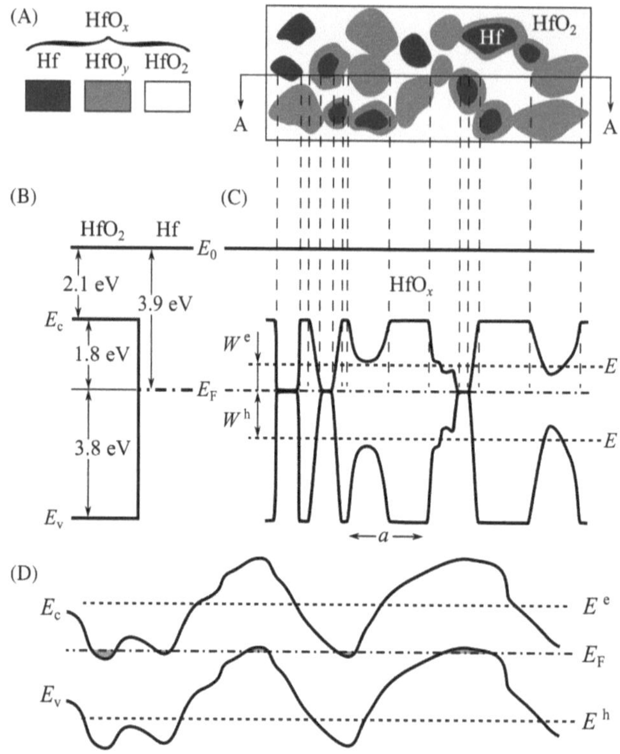

FIGURE 20.6 Percolation model in electron systems with large-scale potential fluctuations. (A) Schematic planar illustration of Hf/HfO$_x$ ($x<2$) space-modulated by chemical composition structure. (B) Flat band energy diagram of HfO$_2$/Hf structure. (C) Energy diagram of the structure with large-scale potential fluctuations. (D) Energy diagram of large-scale potential fluctuations in a semiconductor layer [37]. $E^{e,h}$ and $W^{e,h}$ are percolation levels and percolation threshold of the electrons and holes respectively.

phase [33]. The outlook on the structure of nonstoichiometric oxides and nitrides was developed in Refs. [34–37] for SiO$_x$, SiN$_x$, and SiO$_x$N$_y$. In the random mixture model, SiO$_x$ consists of a mixture of two phases, Si and SiO$_2$. In the random bonding (RB) model, SiO$_x$ consists of a mixture of five types of tetrahedron (Si-Si$_4$, Si-SiO$_3$, Si-Si$_2$O$_2$, Si-Si$_3$O, and Si-O$_4$), the statistics of which are described using binominal distribution. An intermediate SiO$_x$ structural model is proposed in [35]. According to the intermediate model (IM), SiO$_x$ consists, as in the RB model, of five tetrahedron types; however, tetrahedron distribution is not described according to random binominal distribution.

A 2D structural image of nonstoichiometric HfO$_x$, regarding the IM, is presented in Fig. 20.6A. According to this model, the filament in hafnium oxide consists of a mixture of metallic hafnium and hafnium suboxide HfO$_y$ clusters sprinkled into HfO$_2$. Fig. 20.6C is an energy diagram of HfO$_x$ in the IM model. According to this diagram, spatial fluctuations of HfO$_x$ chemical composition lead to spatial fluctuations in local band gap width. The maximal fluctuation of the energy scale is equal to the HfO$_2$ band gap width of $E_g = 5.6$ eV [38]. The work function of metallic hafnium is 3.9 eV. The maximal fluctuation scale of the HfO_x conduction band is 1.8 eV, which equals the electron barrier height of Hf/HfO_2. The maximal fluctuation scale of the HfO_x valence band is 3.8 eV, the hole barrier height of Hf/HfO_2 (Fig. 20.6B).

The large-scale fluctuations of the bottom of conduction band E_c and the top of valence band E_v are close to the model developed by Éfros and Shklovskii [39,40], as shown in Fig. 20.6D. The charge transport in such electron systems can be described according to percolation. This model assumes that excited electrons are thrown to flow level E^e and, rounding a random potential, they transfer the charge. The hole conductivity is realized through the excitation of electrons with energy E^h to the Fermi level, with the formation of a hole-type quasi-particle with a further charge transfer according to the quasi-particle obtained. In other words, to be involved in the transport processes, electrons and holes must overcome energy thresholds ($W^{e,h}$ here and $W^e \neq W^h$ in general).

The current–voltage characteristics are exponentials [39]:

$$I = I_0(T)\exp\left(\frac{\left(CeFaV_0^{\nu}\right)^{\frac{1}{1+\nu}}}{kT}\right), \tag{20.2}$$

$$I_0(T) \sim \exp\left(-\frac{W}{kT}\right), \tag{20.3}$$

$$\begin{gathered} kT\left(\frac{kT}{V_0}\right)^{\nu} \ll eFa \ll V_0, \\ kT \ll V_0, \end{gathered} \tag{20.4}$$

where a is the space scale of fluctuations, V_0 is the energy fluctuation amplitude, C is the numeric constant, and ν is the critical index. The constants' values were derived from Monte Carlo simulations and evaluated at $C \sim 0.25$ [40], and $\nu = 0.9$ [39]. The fitting procedure for LRS simulations by the percolation model in Eq. (20.2) return the combination value as $CaV_0^{\nu} = 0.45$ nm eV^{ν}, which corresponds to $V_0 = 1.9$ eV when $a = 1$ nm and $C = 0.25$. Simulations in preexponent $I_0(T)$ (Eq. 20.2) correspond to the percolation threshold of $W \approx 1$ eV. Assuming that $W \leq V_0 \leq 1.8$ eV, one can estimate the space size of nanoscale fluctuations as $a \approx 1–2$ nm. The found values satisfy the model limits in Eq. (20.4).

We also attempted to simulate the LRS regarding phonon-assisted tunneling between traps, as we did for HRS. The results are shown in Fig. 20.1A by a *dashed green line* (light gray in print versions). Using the same values for energy and effective mass obtained from HRS simulations, we found that the results from the LRS simulations and the experimental data were in very good agreement using the value of $N_{LRS} = 1.1 \times 10^{22}\ \text{cm}^{-3}$. We understand that the trap concentration is close to the density of atoms in metals $\sim 10^{22}\ \text{cm}^{-3}$, and that considering vacancies with such a high concentration is inappropriate; hence, the N_{LRS} value obtained is outside the model limits. However, surprisingly, phonon-assisted tunneling between traps formally describes LRS conduction.

The previous experiments in charge transfer have demonstrated that hafnium oxide conductivity is bipolar (or two-band) [41–44]; electrons are injected from a negatively shifted contact in the dielectric, and holes are injected from a positively shifted electrode in the dielectric. HfO_2 conductivity in an HRS was examined in these previous experiments. In our model, the filament conductivity in LRS can also be studied using electrons and holes. For reasons of simplicity, the current study was limited to considering monopolar electron conductivity both in the HRS and LRS.

20.6 Conclusion

To sum up, we examined the transport mechanisms of hafnia, demonstrating that transport in an HRS is described according to phonon-assisted tunneling between traps. Simulating the current–voltage characteristics of this model, and comparing experimental data with calculations, revealed the energy parameters of the traps in hafnia: the thermal excitation energy of 1.25 eV and the optical excitation energy of 2.5 eV. Oxygen vacancies and the vacancy-conditioned luminescence band of 2.7 eV were identified in the current study by using hafnium oxide photoelectron spectroscopy and quantum–chemical simulations. The photoelectron valence band spectra in stoichiometric HfO_2 and slightly nonstoichiometric $HfO_{x<2}$ were investigated. The photoelectron spectroscopy data revealed electron states in the HfO_x band gap. Quantum mechanical simulation indicated that these states are conditioned by oxygen vacancies. These simulations established that the cathodoluminescence band of 2.7 eV in HfO_x and the excitation peak of 5.2 eV are also conditioned by oxygen vacancies. These results facilitated the determination that oxygen vacancies act as charge-carrier traps. We also demonstrated that charge transport in LRS is described according to the percolation model in electron systems exhibiting potential large-scale fluctuations. Our results shed light on the processes in ReRAM, and are useful for producing higher-quality ReRAM devices in the future.

Recently, it has been demonstrated that HfO_x-based ReRAM elements exhibit intermediate or multilevel states between HRS and RLS. The study of the transport mechanisms in these states is beyond the scope of the current work, and is a subject of further research.

Acknowledgment

This work was supported by the Russian Science Foundation (Grant No. 16-19-00002) and the Ministry of Science and Technology, Taiwan. The modeling was performed on a supercomputing cluster of the Institute of Semiconductor Physics, Siberian Branch of the Russian Academy of Sciences.

References

[1] D.B. Strukov, G.S. Snider, D.R. Stewart, R.S. Williams, The missing memristor found, Nature 453 (2008) 80–83.

[2] J.J. Yang, et al., Memristive switching mechanism for metal/oxide/metal nanodevices, Nat. Nanotechnol. 3 (2008) 429–433.

[3] M.-J. Lee, et al., Electrical manipulation of nanofilaments in transition-metal-oxides for resistance-based memory, Nano Lett. 9 (2009) 1476–1481.

[4] J. Borghetti, et al., 'Memristive' switches enable 'stateful' logic operations via material implication, Nature 464 (2010) 873–876.

[5] M.-J. Lee, et al., A fast, high-endurance and scalable non-volatile memory device made from asymmetricTa_2O_{5-x}/TaO_{2-x} bilayer structures, Nat. Mater. 10 (2011) 625–630.

[6] Z. Wang, et al., Transport properties of HfO_{2-x} based resistive-switching memories, Phys. Rev. B 85 (2012) 1953229.

[7] L. Goux, et al., Evidences of oxygen-mediated resistive-switching mechanism in TiN\HfO_2\Pt cells, Appl. Phys. Lett. 97 (2010) 243509.

[8] A.S. Foster, F.L. Gejo, A.L. Shluger, R.M. Nieminen, Vacancy and interstitial defects in hafnia, Phys. Rev. B 65 (2002) 174117.

[9] K. Xiong, J. Robertson, M.C. Gibson, S.J. Clark, Defect energy levels in HfO_2 high-dielectric-constant gate oxide, Appl. Phys. Lett. 87 (2005) 183505.

[10] P. Broqvist, A. Pasquarello, Oxygen vacancy in monoclinic HfO_2: a consistent interpretation of trap assisted conduction, direct electron injection, and optical absorption experiments, Appl. Phys. Lett. 89 (2006) 262904.

[11] D.M. Ramo, A.L. Shluger, J.L. Gavartin, G. Bersuker, Theoretical prediction of intrinsic self-trapping of electrons and holes in monoclinic HfO_2, Phys. Rev. Lett. 99 (2007) 155504.

[12] S. Guha, V. Narayan, Oxygen vacancies in high dielectric constant oxide-semiconductor films, Phys. Rev. Lett. 98 (2007) 196101.

[13] J.X. Zheng, et al., Fist-principles study of native point defects in hafnia and zirconia, Phys. Rev. B 75 (2007) 104112.

[14] D.M. Ramo, J.L. Gavartin, A.L. Shluger, Spectroscopic properties of oxygen vacancies in monoclinic HfO_2 with periodic and embedded cluster density functional theory, Phys. Rev. B 75 (2007) 205336.

[15] H. Jiang, R.I. Gomez-Abal, P. Rinke, M. Scheffler, Electronic band structure of zirconia and hafnia polymorphs from the GW perspective, Phys. Rev. B 81 (2010) 085119.

[16] T.P. Smirnova, et al., Composition and structure of hafnia films on silicon, Inorg. Mater. 44 (2008) 965–970.

[17] V.A. Gritsenko, I.P. Petrenko, S.N. Svitasheva, H. Wong, Excess silicon at the Si_3N_4/SiO_2 interface, Appl. Phys. Lett. 72 (1997) 462–464.

[18] P. Giannozzi, et al., Quantum espresso: a modular and open-source software project for quantum simulations of materials, J. Phys. Condens. Matter 21 (2009) 395502.

[19] D.R. Islamov, V.A. Gritsenko, C.H. Cheng, A. Chin, Origin of traps and charge transport mechanism in hafnia, Appl. Phys. Lett. 105 (2014) 222901.

[20] J. Frenkel, On pre-breakdown phenomena in insulators and electronic semi-conductors, Phys. Rev. 54 (1938) 647–648.

[21] R.M. Hill, Poole-Frenkel conduction in amorphous solids, Philos. Mag. 23 (1970) 59–86.

[22] S. Makram-Ebeid, M. Lannoo, Quantum model for phonon-assisted tunnel ionization of deep levels in a semiconductor, Phys. Rev. B 28 (1982) 6406–6424.

[23] K.A. Nasyrov, V.A. Gritsenko, Charge transport in dielectrics via tunneling between traps, J. Appl. Phys. 109 (2011) 093705.

[24] H. Takeuchi, D. Ha, T.-J. King, Observation of bulk HfO_2 defects by spectroscopic ellipsometry, J. Vac. Sci. Technol. A 22 (2004) 1337–1341.

[25] D.S. Jeong, C.S. Hwang, Tunneling current from a metal electrode to many traps in an insulator, Phys. Rev. B 71 (2005) 165327.

[26] K.A. Nasyrov, S.S. Shaimeev, V.A. Gritsenko, J.H. Han, Phonon-coupled trap-assisted charge injection in metal-nitride-oxide-silicon/silicon-oxide-nitride-oxide-silicon structures, J. Appl. Phys. 105 (2009) 123709.

[27] A.A. Rastorguev, et al., Luminescence of intrinsic and extrinsic defects in hafnium oxide films, Phys. Rev. B 76 (2007) 235315.

[28] T. Ito, et al., Similarities in photoluminescence in hafnia and zirconia induced by ultraviolet photons, Jpn. J. Appl. Phys. 97 (2005) 054104.

[29] G. Bersuker, et al. Metal oxide RRAM switching mechanism based on conductive filament microscopic properties, in: IEDM10, 2010, pp. 456–459 (19.6.1–19.6.4).

[30] T.-H. Hou, et al., Evolution of RESET current and filament morphology in low-power HfO_2 unipolar resistive switching memory, Appl. Phys. Lett. 98 (2011) 103511.

[31] A.V. Shaposhnikov, et al., Mechanism of GeO_2 resistive switching based on the multi-phonon assisted tunneling between traps, Appl. Phys. Lett. 100 (2012) 243506.

[32] A. Mehonic, et al., Resistive switching in silicon suboxide films, J. Appl. Phys. 111 (2012) 074507.

[33] D.R. Islamov, V.A. Gritsenko, C.H. Cheng, A. Chin, Percolation conductivity in hafnium sub-oxides, Appl. Phys. Lett. 105 (2014) 262903.

[34] K. Hubner, Variation of semiconductor properties through the Si-SiO_x region of Si-SiO_2 interface, J. Non-Crystalline Solids 35–36 (1980) 1011–1014.

[35] Yu.N. Novikov, V.A. Gritsenko, Short-range order in amorphous SiO_x by x ray photoelectron spectroscopy, J. Appl. Phys. 110 (2011) 014107.

[36] V.A. Gritsenko, R.W.M. Kwok, H. Wong, J.B. Xu, Short-range order in non-stoichiometric amorphous silicon oxynitride and silicon-rich nitride, J. Non-Crystalline Solids 297 (2002) 96–101.

[37] V.A. Gritsenko, et al., Short range order and nature of defects and traps in amorphous silicon oxynitride, Phys. Rev. Lett. 81 (1998) 1054–1057.

[38] V.V. Afanas'ev, Internal Photoemission Spectroscopy: Principles and Applications, Elsevier, Amsterdam, 2008.

[39] B.I. Shklovskii, A.L. Éfros, Percolation theory and conductivity of strongly inhomogeneous media, Sov. Phys. Usp. 18 (1975) 845–862.

[40] B.I. Shklovskii, Percolation conductivity in strong electric fields, Sov. Phys. Semicond. 13 (1979) 53.

[41] D.R. Islamov, V.A. Gritsenko, C.H. Cheng, A. Chin, Bipolar conductivity in amorphous HfO_2, Appl. Phys. Lett. 99 (2011) 072109.

[42] V.A. Gritsenko, T.V. Perevalov, D.R. Islamov, Electronic properties of hafnium oxide: a contribution from defects and traps, Phys. Rep. 613 (2016) 1–20.

[43] T. Ando, et al., On the electron and hole tunneling in a HfO_2 gate stack with extreme interfacial-layer scaling, IEEE Electron Device Lett. 32 (2011) 865–867.

[44] L. Vandelli, et al., A physical model of the temperature dependence of the current through SiO_2/HfO_2 stacks, IEEE Trans. Electron Devices 58 (2011) 2878–2887.

21

The Optical Multiplexor Based on Multiple Coupled Waveguides in Silicon-on-Insulator Structures

A.V. Tsarev[1,2]

[1]RZHANOV INSTITUTE OF SEMICONDUCTOR PHYSICS OF THE SIBERIAN BRANCH OF THE RUSSIAN ACADEMY OF SCIENCES, NOVOSIBIRSK, RUSSIA [2]NOVOSIBIRSK STATE UNIVERSITY, NOVOSIBIRSK, RUSSIA

Silicon photonics [1] create a large number of optical devices intended for different applications from telecommunications to sensors. None of them are ideal, but each has some advantages related to the others. Thus, among the urgent tasks for modern research, one can find both improvements in technology and novel structure design. The current chapter provides a brief description of new photonic structures in silicon-on-insulator (SOI) technology working by multisplitting (MS) of optical beams which can be used as a tunable filter or a reconfigurable optical add-drop multiplexer.

We focus on filtering devices based on the interference of multiple optical signals (with the constant optical path difference) which are split and combined in the optical pass (see Table 21.1). The structure parameters used for the simulations are listed below.

h, W, and N_{Si} are the height, width, and refractive index of silicon 3D wire;
H_f, W_f, and N_f are the height, width, and refractive index of polymer 3D waveguide;
H_g and N_c are the height and refractive index of the silica buffer between the silicon wire and polymer waveguide;
dd, L_G, and L_{gap} are the taper tip, taper length, and spacing between taper tips;
N_p and N_g are phase and group indexes of the fundamental guided mode in the silicon wire;
w and N_d are the width and refractive index of the central part of a combined 2D waveguide profile for modified effective index method (MEIM);
N_b is a refractive index of the basic part (with width W) of a combined 2D waveguide profile for MEIM;
N_s is a refractive index of a 2D waveguide (with width W) for effective index method (EIM);
d, L_g, and R are the gap, coupling length, and radius of curvature of a directional coupler for the multisplitting array;
M is the number of waveguides in the array (the number of power splitters in the MS filter);

Advances in Semiconductor Nanostructures. DOI: http://dx.doi.org/10.1016/B978-0-12-810512-2.00021-4

Table 21.1 Comparison of Tunable Devices by Different Technology

Type of Device Technology	Tuning Range Estimation	Power Consumption[a]	IL[b] (dB)
AWG	$\lambda_m \cdot \delta Np/Ng$	$M(M-1) \cdot P_{00}$	−1.0
Ring resonator	$\lambda_m \cdot \delta Np/Ng$	$4 \cdot P_{00}$	−0.5
Ladder filter*	$\lambda_m \cdot \delta N_{pa}/Ng \cdot (L_a/\Delta S) - \lambda_m \cdot \delta N_{pb}/Ng \cdot (L_b/\Delta S)$	$2 \cdot M \cdot P_{00}$	−2.7
MR filter	$\lambda_m \cdot \delta Np/Ng \cdot (L_a + L_b)/\Delta S + \lambda_m \cdot LL/\Delta S \cdot \Delta N/Ng$	$4 \cdot M \cdot P_{00}$	−0.5
MS filter	$\lambda_m \cdot \delta Np/Ng \cdot (L_a + L_b)/\Delta S + \lambda_m \cdot LL/\Delta S \cdot \Delta N/Ng$	$4 \cdot M \cdot P_{00}$	−0.7

[a]Estimation of maximum possible power to switch the device across FSR, which corresponds to the 2π phase shift.
[b]Theoretical drop loss.

ΔT and P_{00} are the temperature change in phase shifters for fine tuning and heating power to produce a $\pi/2$ phase shift;
$\Delta T_0 - \Delta T_3$ and $\Delta T_0' - \Delta T_3'$ are the temperature change in the phase shifters for wide tuning of the Drop and Through signals;
ΔS, L_a, L_b, and LL are the pass-length difference, the period of the splitter positions along Input and Drop waveguides, and the length of phase shifters along the crossed waveguide array;
δN_p and ΔN are refractive index increments for fine tuning (in the bus waveguide), and for wide tuning (in the crossed waveguides).

An arrayed waveguide grating (AWG) utilizes a "free-space" focusing medium and could be manufactured on SOI [1,2]. AWG accomplishes wavelength tuning by controlling the pass difference, say, by the phase shifting triangular array with a step increase of heater electrode length [3,4] which make it possible to tune in both wavelength directions (up and down). A triangular type of heater limits the tuning range, and number of tunable channels, and results in high power consumption which is proportional to the square number of waveguides constituting the array grating (see Table 21.1). The internal loss depends on the structure design, and the actual loss on SOI structures can be around 0.6 dB [2]. AWG has a unique multichannel feature and offers the possibility of a flattened spectral response (IL = −2 dB) [1].

Optical filters based on ring resonators are compact, and they provide a flattened spectral response using multiple rings. A single ring requires minimum power for wavelength tuning [5,6], but it only provides a small tuning range of 8–10 nm at the telecom wavelength which is restricted by the available change of refractive index under external signals (see Table 21.1). This limitation could be overcome only by extremely high heating, or by coupled ring resonators having different free spectral ranges FSR1 and FSR2 which are tuned by the Vernier effect [7]. The last option provides a wide tuning range [7] at low power consumption, but it disturbs signals at intermediate drop wavelengths of ring resonators (divisible by FSR1 and FSR2).

On the contrary, ladder-type [8–10] and multireflector (MR) [11–14] filtering devices provide wide and efficient optical wavelength tuning by optical phase shifters, without applying the Vernier effect. These devices accomplish filtering by splitting and adding multiple optical bunches in the output optical waveguide.

The ladder filter is constructed by multiple multimode interference couplers, and has limitations in the internal loss (−2.7 dB) and in the number of crossing waveguides (<10) [10] which limit the filter linewidth. Using a multiplication factor $L_{a,b}/\Delta S > 1$ the filter can be tuned in both wavelength directions within the total FSR by applying a small index increment (by the multiplication factor $(L_a + L_b)/\Delta S$) in the phase shifters in two interferometer arms (see Table 21.1).

The multireflector filter carries out power splitting using partial reflectors built by deep trenches [12,13] or hole rows [15,16]. In comparison with the ladder filter [10], it provides a larger number of wavelength channels, and smaller internal loss [13,14]. It provides wide wavelength switching by phase shifters for wide and fine tuning [11−14]. The multireflector elements work only with wide optical waveguides, which is rather complicated to accomplish on SOI structures characterized by high index contrast. We have resolved this problem using wide quasi-monomode optical waveguides which have a small optical loss for the fundamental mode, and artificial loss for all high order modes. Depending on the technological possibilities, it may be heterogeneous optical waveguides with high concentrations of free charges on the waveguide sides [14] or strip and grating loaded waveguides [17,18] that are most useful. The multireflector filtering elements have a moderate tuning power which is proportional to the number of power splitters (see Table 21.1). They require high-quality multiple partial reflectors with a small parasitic scattering which have not yet been experimentally demonstrated.

Multiple power splitting by directional couplers is currently utilized in the lattice Mach−Zehnder filter (MZF), which consists of a series of Mach−Zehnder interferometers with a constant path-length difference and varying coupling ratios [19] for filter apodization. They are used as fixed filtering devices with a small wavelength channel number.

Recently, a multisplitting filter design, which integrates the advantages of multireflector and lattice MZF technologies, has been proposed [20,21]. The MS filter utilizes directional couplers for power splitting and MR-like structure design for optical filtering. A short description of new multisplitting filtering elements based on multiple directional couplers is the aim of this chapter.

21.1 Multisplitting Filter-Multiplexer

It is evident that the MS approach is also suitable for the construction of an optical multiplexer. Unfortunately, such a MS design [20] naturally contains multiple silicon wire crossings which may strongly disturb device performance. In order to eliminate this effect, and to provide negligible scattering, the waveguide crossings could be developed by means of a vertical up and down coupling through the silica buffer of tapered Si wires with the upper optical waveguide [22−26].

A general view of a novel multisplitting multiplexer is shown in Fig. 21.1. The input signal (In) comes into the bottom Si wire waveguide and is split into subbeams by multiple adiabatic directional couplers with a varying coupling ratio (different gap width) which is

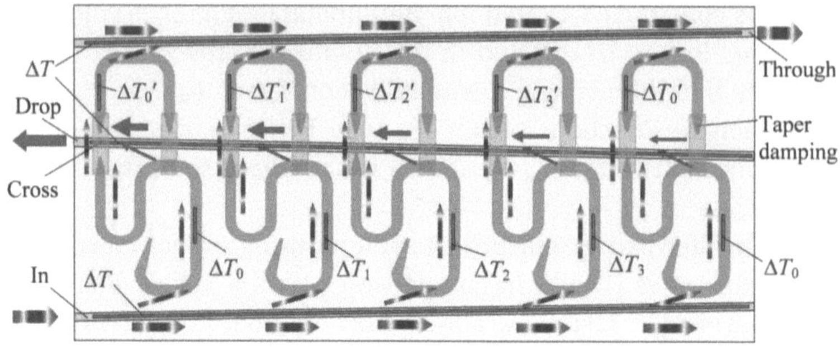

FIGURE 21.1 A principal view of a multisplitting tunable multiplexer.

responsible for filter apodization. Each of the subbeams goes further along the crossed waveguides (vertical in structure design), and is combined in the output wire waveguide (Drop) by an appropriate set of similar couplers.

The structure has a constant optical path difference of $\Delta S = L_a + L_b$. Thus, at drop optical wavelengths, all the subbeams add in the phase along the output wire waveguide (Drop) and produce efficient filtering. All other wavelengths pass further along the coupler, cross the Drop waveguide, and further combine in the output wire waveguide (Through) by an appropriate set of similar directional couplers. These couplers also have a varying coupling ratio (generally inverse to the In and Drop bus waveguides) in order to increase the Through signal efficiency. The structure design provides that most of the split subbeams have the same pass difference on the way from input (In) to output (Through). Thus, all the optical wavelengths which are not filtered by the Drop channel will be transmitted to the Through output. As a result, the structure presented in Fig. 21.1 could be used as the optical multiplexer, as well as the optical filter.

Device examination is very complicated, and the study has been divided into several separate parts, each of which are focused on one key point. Part A is devoted to the approach of multilayer waveguide crossing [22–26], which is also suitable for construction of a damper. Part B discusses the MEIM [27], which is optimal for direct 2D modeling of the device. The spectrum properties of the MS filter and its thermo-optic tuning are presented in Parts C and D, respectively. The main results are summarized in Section 21.2.

21.1.1 Part A: Multilayer Waveguide Crossing

It is known that a good solution for the problem of cross-scattering is arranging the waveguide crossing in different spatial layers. For example, an efficient crossing has been produced in high-contrast silicon wires on the basis of tunnel coupling of two vertically overlapped inverse tapers [22,23]. Practical development of this requires polysilicon or silicon nitride deposition technology and precision manufacturing of two oppositely directed tapers located in different layers, one under the other, in strict position along the optical pass.

Recently, an alternative and more technological wire crossing has been made available (see Fig. 21.2) [24–26]. It uses the known technology of a fiber-to-wire waveguide coupler

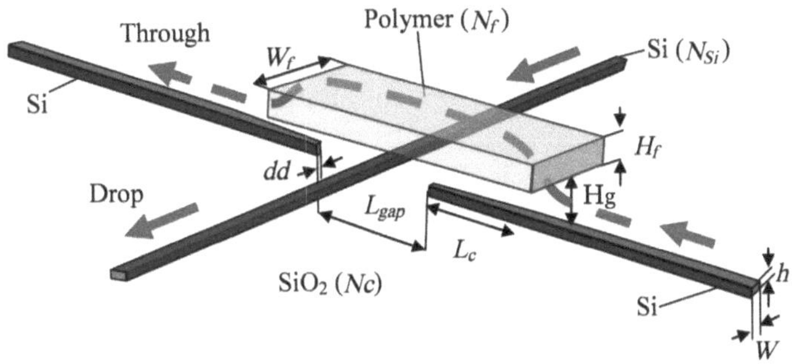

FIGURE 21.2 The basic scheme of multilayer effective crossing of two silicon wires.

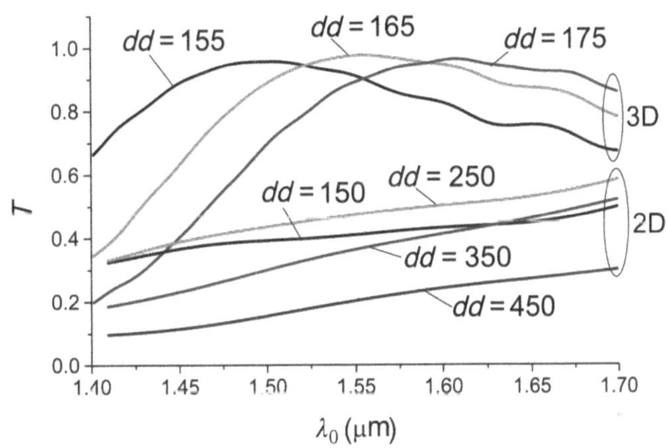

FIGURE 21.3 Wavelength dependence of silicon wire crossing efficiency on the 3D and 2D cases for different taper width *dd* (in nm). FDTD simulations: $W_f = 1.5$ μm, $H_f = 1.7$ μm, $H_g = 340$ nm, $L_{gap} = 3$ μm, $L_c = 45$ μm, $N_f = 1.56$, $N_{Si} = 3.478$, $N_c = 1.4$, $h = 220$ nm, $W = 450$ nm.

[28] based on the vertical coupling of a Si wire taper with the upper polymer waveguide through the silica buffer. In our device, two such structures are stacked together and the crossing silicon wire is placed between them (see Fig. 21.2). This optical element (with optimal design parameters) provides effective energy transmission between the basic waveguide (a thin tapered photonic wire with a high refractive index) to the additional top waveguide (a two-mode thick polymer SU-8 waveguide with a low refractive index), and back to the basic tapered wire (see "Through" direction in Fig. 21.2). The power transmittance is inherently wavelength-dependent. Thus, the structure parameters have to be adjusted to the particular wavelength band to be utilized by the device. Direct modeling by the 3D finite difference time domain (FDTD) method [24–26] shows that at telecom wavelengths the transmitting efficiency of the Through signal (see Fig. 21.2) can reach 98% [24] for an optimum taper tip of *dd* = 165 nm (see Fig. 21.3).

Note that, just in the middle of the structure, about 99% of the total optical energy is concentrated in the upper polymer waveguide. Thus, a crosswire waveguide (Drop in Fig. 21.1) has negligible scattering at the crossing area and, besides, one can use half of the structure (with the single Si taper) as an efficient way of escaping the power from the Si layer, thus making a damper for the optical signals (Damping in Fig. 21.1).

It is most important that the Drop Si wire is separated by the silica buffer from the other crossing Si wire, and from the additional top polymer waveguide (see Fig. 21.2). Thus, the proposed structure has a small loss (<0.1 dB) for a Through direction, and an almost lossless silicon wire crossing in the Drop direction, which provides a negligible total loss (<0.1 dB) [25] even for a high crossing number (>1000).

21.1.2 Part B: Modified Effective Index Method

The structure design and analysis of the proposed multiplexer can be made using various optical methods, among which the FDTD [29] is the most popular, as it provides the prospect of a numerical experiment. Unfortunately, large 3D structures are very complicated for numerical simulation as they have a very high demand for computer resources. Thus, for the purpose of modeling, real 3D waveguides are often substituted by 2D analogs under the approximation of the EIM [30], which makes it possible to simulate waveguide structures within a reasonable time.

The principal EIM limitation comes from the fact that the real 3D waveguide (see the insert in Fig. 21.4A), with two different cross-dimensions (height h and width W), is examined by the 2D waveguide (see the insert in Fig. 21.4B) with only one size parameter (width W and index N_s). Thus, the wavelength dependence of the mode effective index in 3D and 2D waveguides has to be principally different (see Fig. 21.4). This results in a major error in studying the device impulse excitation by the 2D FDTD method, due to the wrong group

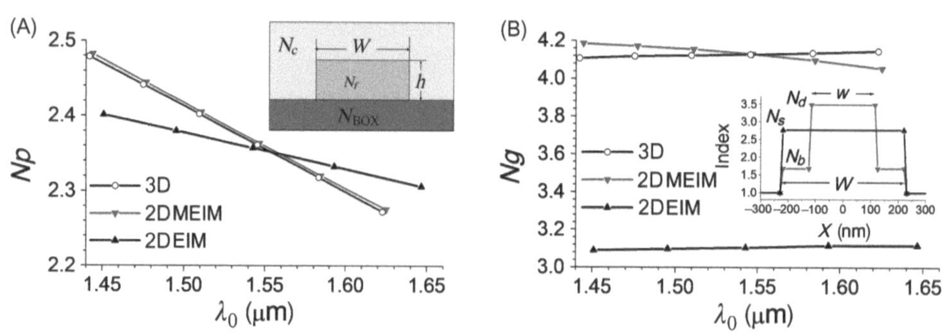

FIGURE 21.4 Comparison of EIM and MEIM. (A) Effective mode indexes as a function of optical wavelength λ_0 for different FDTD simulation cases. In the insert, the initial 3D waveguide structure used in 3D simulations is presented; (B) group mode indexes in a ring as a function of optical wavelength λ_0 for different FDTD simulation cases. In the insert, the refractive index profile used in 2D simulations is presented.

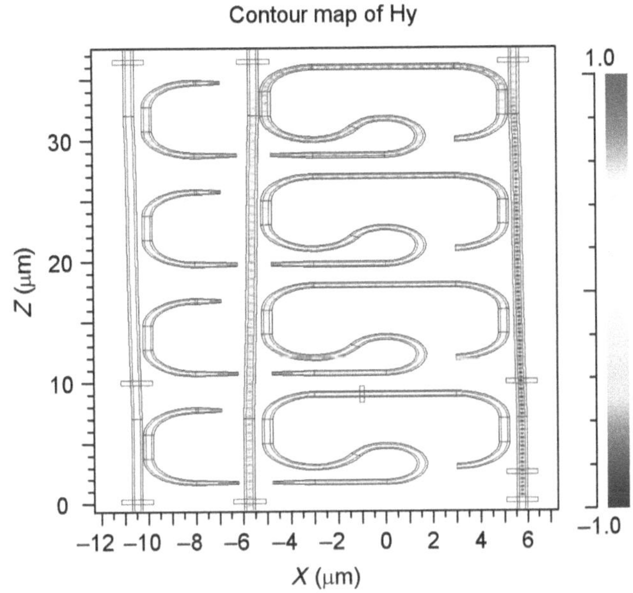

FIGURE 21.5 Optical wave propagation through the MS multiplexer with four directional couplers at every waveguide bus. Two-dimensional FDTD simulation.

index N_g being determined by EIM (see Fig. 21.5). This creates a restriction in the implementation of EIM for studying optical filtering devices where the group index plays an important role, as it determines the FSR for the given path difference [31]:

$$FSR = \lambda^2/(N_g \Delta S). \tag{21.1}$$

We overcome this limitation in our study of the MS multiplexer by implementing the above-mentioned MEIM which correctly describes, in a 2D case, both the phase and the group indexes in the 3D wire waveguide built in thin SOI structures.

In order to prove the advantages of this method, we applied EIM and MEIM to the classical example of a ring resonator built using a silicon wire on SOI with refractive index $N_{Si} = 3.478$, height $h = 220$ nm, width $W = 450$ nm, and radius 2.5 μm surrounded by silica with a refractive index of 1.4. This waveguide supports a single TE polarization mode. We studied the pulse excitation of the structure by FDTD, and found the set of the Drop wavelength. By using the known ring circumference, we determined the phase and group indexes (see Fig. 21.4), as well as FSR for three important cases: namely, for the direct 3D FDTD initial structure simulation; for the 2D FDTD simulations under classical EIM; and under novel MEIM. Besides, we disregarded material dispersion as it was much smaller (14 times) than the structure index dispersion which was strongly dependent on waveguide dimensions h and W. It is known that 3D and 2D waveguides (under EIM approximation for the fixed value of 2D waveguide index N_s) have a strongly different slope of phase index versus optical wavelength (see Fig. 21.4A) and, thus, EIM gives the wrong group index N_g value [31] (see Fig. 21.4B):

$$N_g = N_p - \lambda \partial N_p / \partial \lambda. \tag{21.2}$$

To provide better agreement between the phase and group indexes in 2D and 3D waveguides, MEIM utilizes a combined index profile [27] (see the insert in Fig. 21.4B) containing two spatial parameters as in the actual 3D waveguide (see the insert in Fig. 21.4A). Namely, the central part with refractive index $N_d = N_{Si}$ has a width w (which is around the 3D waveguide height h), and is mainly responsible for group index N_g. The basic part has the same width, W, as in the 3D waveguide, and a refractive index N_b which is mainly responsible for the phase index N_p. By variation of w and N_b, we can fit, in the wide wavelength range, the effective index of the combined 2D waveguide to the 3D silicon wire waveguide mode index (see Fig. 21.4).

As a result, this combined structure (with $w = 250$ nm and $N_b = 1.537$) provides much better agreement than EIM ($N_s = 2.780$) of the group and phase indexes in 2D and 3D waveguides (see Fig. 21.4) and thus, it inherently provides better accuracy than EIM in studying optical devices in the frequency domain. As a result, 2D FDTD simulation plus MEIM fits (with error 1–2%) the FSR parameters measured by 3D FDTD well. Note that the traditional EIM provides an error in FSR of about 30–35%. These results prove that MEIM is an excellent numerical tool to examine complicated 3D structures under 2D approximation.

21.1.3 Part C: Studying the MS Multiplexer by MEIM and by 2D FTDT

Now we apply the MEIM to study the multi-splitter multiplexer by 2D FDTD which has 4, 8, 16, or 32 directional couplers in each bus. It is limited by our computer resources. Although the actual devices may have a larger number of couplers ($M > 32$), their main features can be correctly revealed by a smaller number of M (by scaling the results). A typical structure design for $M = 4$ is presented in Fig. 21.5. Every directional coupler contains a straight part ($L_g = 2$ μm) and two curved small radius ($R = 2$ μm) sections. In order to decrease parasitic signals for the Drop channel (the central vertical waveguide in Fig. 21.5), the waveguide crossing is accomplished by tapered waveguides with a tip of $dd = 250$ nm. It corresponds to the maximum transmittance in the cross-direction (see Fig. 21.3). The damping regions (see Fig. 21.1) are simulated as taper sections with a large imaginary part (equal to 0.4) of the refractive index.

The filter apodization by varying gap width d in different couplers (see Fig. 21.6) is arranged using a predetermined small bus waveguide curvature, as described in Ref. [20]. Namely, for the given Gaussian apodization function (to provide high sidelobe suppression), we find the dependence of the splitting ratio on d for deflected (R_c) and transmitted (T_c) waves (see Fig. 21.6A). Then, we determine the required dependence of the splitting ratio and gap width d on the coupler number n (see Fig. 21.6B), and simulate the filter response

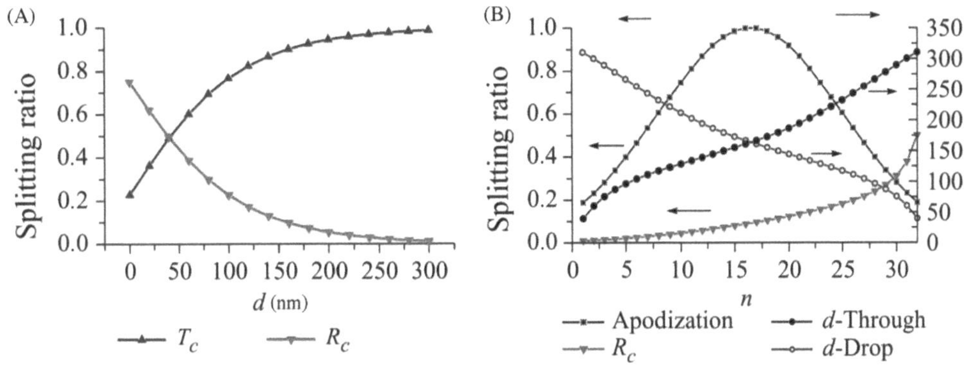

FIGURE 21.6 The data for providing the desired apodization function of MS filter with 32 directional couplers based on SOI wire waveguides. (A) Splitting ratio of reflected (R_c) and transmitted (T_c) power as a function of coupling gap d; (B) Main design parameters as a function of splitter number. $L_g = 2$ μm, $R = 2$ μm. Two-dimensional FDTD simulation.

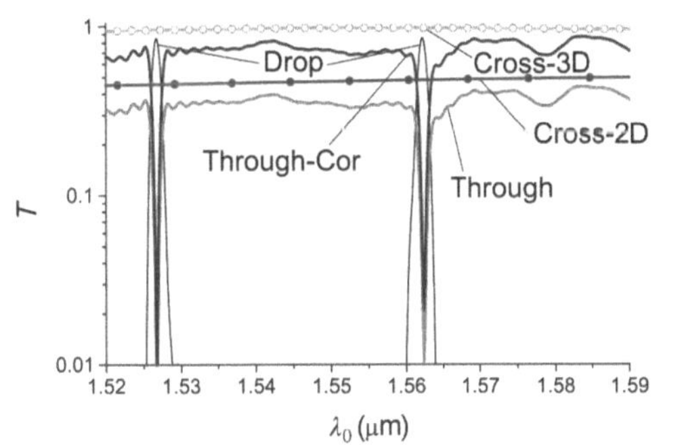

FIGURE 21.7 Wavelength dependence of transmitted power for Drop and Through channels of the optical multiplexer with 32 directional couplers. Additional curves indicated by Cross-2D and Cross-3D presents waveguide crossing transmittance efficiency for 2D and 3D simulations, respectively.

by a standard FDTD procedure [32] launching a short pulse and performing Furies analysis. Note that the joint use of MEIM and 2D FDTD provides the correct mode dispersion contribution, as well as the wavelength dependence of coupling ratio and phase delay in all power splitters.

Typical wavelength dependences of transmitting power efficiency are presented in Fig. 21.7. One can see that the MS filter provides about 85% power transmission at the Drop wavelength, corresponding to the condition at which all the subbeams coming from different couplers are constrictively interfered within the Output waveguide. The transmitted Through channel power is only about 35%. This moderate efficiency is a result of the additional loss for all optical beams which have to cross the Drop waveguide by means of two 2D taper couplers. Nevertheless, one has to take into account that this additional loss (see Cross-2D in Fig. 21.7) is a result of 2D wire-crossing modeling (realized in a single layer).

The multilayer waveguide crossing, which results in the much smaller transmitting loss (see Cross-3D in Fig. 21.7), will be used in the actual device. Thus, in order to predict the wavelength response of an actual multiplexer, it is possible to correct the results of 2D FDTD simulations of the MS multiplexer taking into account the data of an actual crossing loss in the 3D design. The corrected results are presented in Fig. 21.7 as the curve Through-Cor. One can see that the proposed multiplexer provides high efficiency for both the Drop and Through signals, as well as good sidelobe suppression.

21.1.4 Part D: Wavelength Tuning by TO Effect

One has to point out that the proposed multiplexer has a common tuning feature with the previously studied multireflector filtering devices [11–14]. Thus, the Drop wavelength can be tuned by the optical phase shifters which control the refractive indexes in the bus waveguides (δNp) for fine tuning, and in the crossed waveguides (ΔN) for wide tuning [20].

$$\delta\lambda = \lambda_m \cdot \delta Np/Ng \cdot (L_a + L_b)/\Delta S + \lambda_m \cdot LL/\Delta S \cdot \Delta N/Ng, \tag{21.3}$$

where LL is the phase shifter length along the crossed waveguides. The phase shifters for fine tuning can control a small part (about 1/4) of the FSR due to a small available change in the refractive index (δNp). Phase shifters for wide tuning have the multiplication coefficient $LL/(2Ls)$ which provide wide tuning even for a small increment index ΔN.

The aim of these phase shifters is to produce a controllable linear phase delay of optical subbeams coming from "Input" to "Output" of the device [20]. In the multiplexer filter, we can accomplish filtering by the thermo-optical effect, namely, by changing the temperature in the appropriate waveguide set shown in Fig. 21.1. Fine tuning of the device is realized by increasing the temperature (ΔT) in "In," "Drop," and "Through" waveguides. The results are presented in Fig. 21.8, and they show that this filter has a moderate tuning rate $\Delta\lambda/\Delta T = 0.065$ nm/°C, which is typical of optical filters on silicon, say, of ring resonators.

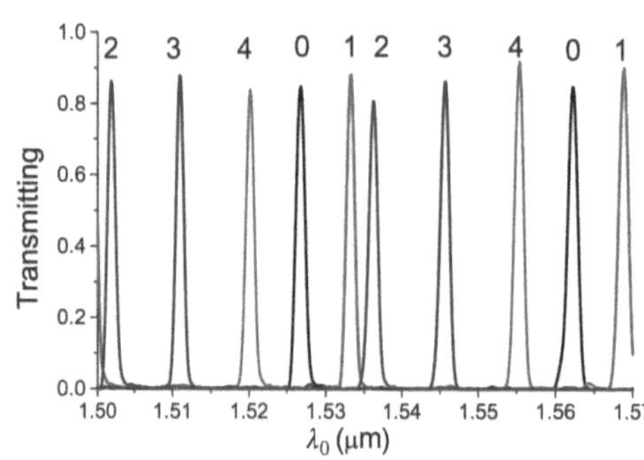

FIGURE 21.8 Wideband Drop wavelength tuning by different temperature increments for fine tuning and digital wide tuning (with step FSR/4). (0) $\Delta T = 0°C$; (1) $\Delta T = 100°C$; (2) $\Delta T = 0°C°$, $\Delta T_1 = 1 \times \Delta T_{00}$, $\Delta T_2 = 2 \times \Delta T_{00}$, $\Delta T_3 = 3 \times \Delta T_{00}$; (3) $\Delta T = 0°C$, $\Delta T_1 = 2 \times \Delta T_{00}$, $\Delta T_2 = 0 \times \Delta T_{00}$, $\Delta T_3 = 2 \times \Delta T_{00}$; (4) $\Delta T = 0°C$, $\Delta T_1 = 1 \times \Delta T_{00}$, $\Delta T_2 = 2 \times \Delta T_{00}$, $\Delta T_3 = 3 \times \Delta T_{00}$; $LL = 2$ μm; $M = 32$; Total length $L = 258$ μm; $L_g = 2$ μm; $R = 2$ μm; $\delta\lambda \sim 1.4$ nm; FSR ~ 35 nm. Two-dimensional FDTD simulation.

Table 21.2 Phase Shifting Increments ($\Delta\phi/2\pi$) of Heaters for Digital Tuning According to a Manifold Module (2π) Scheme

Tuning Step ($n/\Delta\varphi$)/Number of the Heater	1	2	3	4	5	6	7	8	9
0 FSR	0	0	0	0	0	0	0	0	0
1/4 FSR	0	1/4	2/4	3/4	0	1/4	2/4	3/4	0
1/2 FSR	0	2	0	2	0	2	0	2	0
3/4 FSR	0	3/4	2/4	1/4	0	3/4	2/4	1/4	0

For reasonable temperature changes (<140°C), this fine tuning (see curves 0 and 1 for $\Delta T = 0$°C and 100°C, respectively) could provide only 25% of the total FSR which, in our case, is about 35 nm.

Wide tuning is provided by digital wavelength switching according to a manifold module (2π) scheme, where all the phase shifts in subbeams vary linearly within the 2π range [12,13]. An example of such a scheme that provides wide tuning with the step about FSR/4 is presented in Table 21.2 [20]. The desirable Drop wavelength tuning is presented in the left column of Table 21.2. Other elements of Table 21.2 describe the value of phase shift changes (divided by 2π) for different heaters (n) numbered from 1 to 9 (optionally). One can see that the phase shift increments are a periodic function versus the heater number, and heaters with the same phase shifts can be connected in groups. Thus, the optical wavelength could be digitally tuned with an FSR/4 step applying three appropriate signals to the different heater groups, shown in Fig. 21.1 as ΔT_0–ΔT_3. Note that, in order to provide a constant pass difference for all subbeams for the Through output, one has to adjust the temperature increments for Drop and Through tuning ($\Delta T_j + \Delta T_j$ = constant, where $j = 0, 1, 2, 3$).

We examined this scheme with an FSR/4 wavelength step by digital phase shifting of the crossed waveguide array (see "vertical" heaters on the waveguide in Fig. 21.1). We arranged the phase shifters in four periodic groups with different temperature increments (ΔT_0, ΔT_1, ΔT_2, ΔT_3). We determined them in units of ΔT_{00}, i.e., the temperature increment corresponding to the $\pi/2$ phase shift and power P_{00} at the central optical wavelength. For all these simulations, we have $\Delta T_0 = 0$. Further monotonic tuning within every one-quarter of FSR is accomplished by controlling the temperature (ΔT) in the fine tuning phase shifters.

For FDTD simulations, we use compact structures with a small phase shifter length ($LL = 3$ μm). For practical use, a larger heater length, which provides a small handling temperature, is preferable. For example, $\Delta T_{00} \sim 21.5$ °C for $LL = 100$ μm.

Typical multiplexer Drop responses are presented in Fig. 21.8. The filter is digitally tuned with step FSR/4 using the appropriate temperature increments ΔT_1, ΔT_2, and ΔT_3 in the vertical waveguides shown in Fig. 21.1. The results in Figs. 21.7 and 21.8 demonstrate that this MS device could be used as a widely tunable optical filter or multiplexer with a small internal insertion loss (−0.7 dB) and high sidelobe suppression (below −20 dB). The Drop optical wavelength could be tuned within the total FSR applying four controlling signals: namely, three for digital wide tuning (ΔT_1, ΔT_2, ΔT_3), and one for monotonic fine tuning (ΔT). FSR depends on the optical path difference and, for a compact device ($L_g = 2$ μm and

$R = 2\ \mu m$), it is about 35 nm. At the central telecom optical wavelength ($\sim 1.55\ \mu m$), the filter tuning rate is about 0.065 nm/°C, and the full width half maximum (FWHM) $\delta\lambda$ is about $1.2 \cdot FSR/M$. It is fundamental that FSR can be increased by reducing L_g and R, and/or using the slanted waveguide arrangement (similar to the multireflector device [12–14]), which can multiply the increase by a factor $(L_a + L_b)/\Delta S$, increasing the FSR and tuning rate at the expense of increasing the filter size proportional to L_a and L_b.

The heating power P_{00} required to produce a $\pi/2$ phase shift depends on the structure and the heater parameters. Taking into account the known results ([33], Fig. 2), one can obtain $P_{00} = 6.9$ mW and switching time 8 μs for a 2 μm thickness of buried oxide (BOX), and a 0.5 μm heater width. The total power consumption of the MS filter depends on the power P_{00} and power splitter number M. Similar to multireflector devices [11], it consists of two parts P_f and P_w, where $P_f = M \cdot P_{00}$ is the maximum power for fine tuning within one-quarter of FSR, and $Pw = 3/2\, M\, P_{00}$ is the maximum power for wide tuning with a quarter FSR step. For our device ($M = 32$), thus, the maximum total power $(P_f + P_w)$ is about 550 mW. The switching power can be reduced by optimization of the waveguide-heater arrangement [33,34]. For the particular case of a large number of couplers ($M \sim 900$), which are needed to provide a small linewidth (~ 0.05 nm), the maximum operation power will be high enough (~ 16 W). Nevertheless, this device can be used as an interrogator from fiber brag grating (FBG) sensors [35]. Namely, using the special heater signal form [36], one can provide a fast linear rise of ΔT in the time, thus giving about 1 pm spectral resolution with about 1 ms readout time of total FSR. It provides a one-per-second interrogation of the total 35 nm spectrum range by an average heating power of about 16 mW. These results prove that MS filtering devices could be tuned within the total FSR by a small modulation magnitude, using moderate power consumption.

21.2 Conclusions

We have discussed a general concept of a novel tunable optical filter and multiplexer with multisplitting elements based on directional couplers in a typical silicon wire waveguide on a thin SOI as is used in silicon photonics. In order to test this device by 2D FDTD correctly, we have to introduce the MEIM. It utilized the combined index profile of a 2D waveguide with two spatial parameters. It correctly describes, in a 3D case (with about a 1% error), both the phase and the group index of the 3D silicon waveguide, and thus makes it possible to simulate complicated SOI waveguide structures by 2D FDTD. The proposed multiplexer contains multiple crossings of silicon wires. In order to decrease parasitic scattering such crossings are arranged by vertical coupling through a silica buffer of tapered silicon wire with a thick upper polymer waveguide.

Numerical modeling by the FDTD method proves the general concept of a new type of optical filter and multiplexer, and has demonstrated that a compact (0.26 mm) structure with 32 directional couplers with a varying gap (from 40 to 310 nm to provide filter apodization) will provide filter resolution of about 1.4 nm, internal insertion loss of −0.7 dB, and sidelobe suppression below −20 dB. The filter can be tuned within FSR of about 35 nm at an optical wavelength around 1550 nm with a moderate temperature change (<140°C) in four optical phase shifter sets. Scaling these data for larger structures, we find that the filter FWHM can be about 0.05 nm for a 0.7 cm device.

The first experimental investigation of the proposed device was undertaken jointly with the University of St Andrews, Scotland, UK [39]. Our results prove that our numerical approach (MEIM plus 2D FDTD) well describes the spectrum properties of optical filters on the basis of multiply-coupled optical waveguides (silicon wires) in SOI structures. For example, the discrepancy in the measured and modeling results (by 2D FDTD and MEIM) of the FSR was around 3% (in comparison with 43% error for the case of a traditional EIM approach).

The proposed optical filter can be a competitor for AWG from the point of view of the tuning range and operational power, a competitor for the ladder filter from the point of view of linewidth and device size, and a competitor for the ring resonator from the point of view of monotonic wavelength tuning without the Vernier effect [19]. It could be constructed by modern CMOS-compatible technology [1,5,19,34,37,38]. It can find practical applications in photonics, including fiber optical networks, and fiber brag grating sensor interrogation [36].

Acknowledgments

The author is very thankful to RSoft Design Group, Inc. for providing the user license for FDTD simulations [32]. The author would like to express his gratitude to Alexander V. Zuravlev, ISP SB RAS Executive interpreter for his necessary textual corrections to the English version of this chapter.

References

[1] W. Bogaerts, S. Selvaraja, P. Dumon, J. Brouckaert, K. De Vos, D. Van Thourhout, et al., Silicon-on-insulator spectral filters fabricated with CMOS technology, J. Sel. Top. Quant. Electron. 16 (2010) 33–44.

[2] S. Pathak, D. Van Thourhout, W. Bogaerts, Design trade-offs for silicon-on-insulator-based AWGs for (de) multiplexer applications, Opt. Lett. 38 (2013) 2961–2964.

[3] T. Watanabe, Y. Inoue, K. Kaneko, N. Ooba, T. Kurihara, Polymeric arrayed-waveguide grating multiplexer with wide tuning range, Electron. Lett. 33 (1997) 1547–1548.

[4] S. Toyoda, N. Ooba, K. Kitoh, T. Kurihara, T. Maruno, Wide tuning range and low operating power AWG-based thermo-optic wavelength tunable filter using polymer waveguide, Electron. Lett. 37 (2001) 1130–1132.

[5] S. Li Qing, Y. Mohammad, Siva, A. Ali, Design and demonstration of compact, wide bandwidth coupled-resonator filters on a silicon-on-insulator platform, Opt. Express 17 (2009) 2247–2254.

[6] A. Melloni, M. Martinelli, Synthesis of direct-coupled-resonators bandpass filters for WDM systems, J. Lightw. Technol. 20 (2002) 296–303.

[7] P. Prabhathan, J. Zhang, V.M. Murukeshan, H. Zhang, Sh Cheni, Discrete and fine wavelength tunable thermo-optic WSS for low power consumption C+ L band tunability, IEEE Photonics Technol. Lett. 24 (2012) 152–154.

[8] S. Matsuo, Y. Yoshikuni, T. Segawa, Y. Ohiso, H. Okamoto, A widely tunable optical filter using ladder-type structure, IEEE Photonics Technol. Lett. 15 (2003) 1114–1116.

[9] S. Matsuo, S.-H. Jeong, T. Segawa, H. Okamoto, Y. Kawaguchi, Y. Kondo, et al., Digitally tunable ring laser using ladder filter and ring resonator, IEEE J. Sel. Top. Quant. Electron. 11 (2005) 924–930.

[10] S.-H. Jeong, Sh Matsuo, Y. Yoshikuni, T. Segawa, Y. Ohiso, H. Suzuki, Analysis and demonstration of single-passband response and tuning characteristics in a chirped ladder interferometric filter for a widely tunable laser diode, Appl. Opt. 44 (2005) 6007–6016.

[11] V.M.N. Passaro, F. Magno, A.V. Tsarev, Investigation of thermo-optic effect and multi-reflector tunable filter/multiplexer in SOI waveguides, Opt. Express 13 (2005) 3429–3437.

[12] A.V. Tsarev, V.M.N. Passaro, F. Magno, Widely tunable reconfigurable optical add/drop multiplexers in silicon-on-insulator technology: a new approach, in: M. Vittorio, N. Passaro (Eds.), Silicon Photonics, Research Signpost, Trivandrum, Kerala, 2006, pp. 47–77 (Chapter 3).

[13] A.V. Tsarev, Multiplexers for WDM with nanophotonic reflectors—a new way for control of multi-hundred optical spectral channels, J. Nano and Microsyst. Tech. (Ru) 4 (2007) 51–55.

[14] A.V. Tsarev, Thin heterogeneous optical silicon-on-insulator waveguides and their application in reconfigurable optical multiplexers, Quant. Electron. 38 (2008) 445–451.

[15] P. Pottier, S. Mastroiacovo, R.M. De La Rue, Power and polarization beam-splitters, mirrors, and integrated interferometers based on air-hole photonic crystals and lateral large index-contrast waveguides, Opt. Express 14 (2006) 5617–5633.

[16] A.V. Tsarev, Power beam splitter based on photonic crystal row of holes and Brewster effect in SOI waveguides, Opt. Lett. 35 (2010) 968–970.

[17] A.V. Tsarev, New wide strip and grating loaded quasi-single-mode waveguide on SOI, Opt. Express 17 (2009) 13095–13101.

[18] A.V. Tsarev, Optical properties of wide strip and grating loaded single mode channel waveguides, Quant. Electron. 39 (2009) 1169–1174.

[19] K. Yamada, T. Shoji, T. Tsuchizawa, T. Watanabe, J. Takahashi, S. Itabashi, Silicon-wire-based ultrasmall lattice filters with wide free spectral ranges, Opt. Lett. 28 (2003) 1663–1664.

[20] A. Tsarev, Numerical modeling of the optical multiplexer on SOI constructed by multiple coupled waveguides, IEEE J. Sel. Top. Quant. Electron. 20 (2014) 77–84.

[21] A. Tsarev, Numerical simulation of multi-splitting widely tunable filter on SOI, in: Proc. SPIE, vol. 8781, Integrated Optics: Physics and Simulations, 2013, 878106–878106-13.

[22] R. Sun, M. Beals, A. Pomerene, J. Cheng, C. Hong, L. Kimerling, et al., Impedance matching vertical optical waveguide couplers for dense high index contrast circuits, Opt. Express 16 (2008) 11682–11690.

[23] A.M. Jones, C.T. DeRose, A.L. Lentine, D.C. Trotter, A.L. Starbuck, R.A. Norwood, Ultra-low crosstalk, CMOS compatible waveguide crossings for densely integrated photonic interconnection networks, Opt. Express 21 (2013) 12002–12013.

[24] A.V. Tsarev, Efficient silicon wire waveguide crossing with negligible loss and crosstalk, Opt. Express 19 (2011) 13732–13737.

[25] A. Tsarev, E. Kolosovsky, Almost lossless multiple crossing of silicon wires by means of vertical coupling with a polymer strip waveguide, in: Proc. SPIE, vol. 8781, Integrated Optics: Physics and Simulations, 2013, 878112–878112-12.

[26] A.V. Tsarev, E.A. Kolosovskii, Analysis of light propagation for a crossing of thin silicon wires using vertical tunnelling coupling with a thick optical channel waveguide, Quant. Electron. 43 (2013) 744–750.

[27] A. Tsarev, Modified effective index method to fit the phase and group index of 3D photonic wire waveguide, Opt. Lett. 38 (2013) 293–295.

[28] G. Roelkens, P. Dumon, W. Bogaerts, D. Van Thourhout, R. Baets, Efficient silicon-on-insulator fiber coupler fabricate during 248-nm-deep UV lithography, IEEE Photonics Technol. Lett. 17 (2005) 2613–2615.

[29] K.S. Yee, Numerical solution of initial boundary value problems involving Maxwell's equations in isotropic media, IEEE Trans. Antennas Propagat. AP-14 (1966) 302–307.

[30] K.S. Chiang, Dual effective-index method for the analysis of rectangular dielectric waveguides, Appl. Opt. 25 (1986) 2169–2174.

[31] W. Bogaerts, P. De Heyn, T. Van Vaerenbergh, K. De Vos, S. Kumar Selvaraja, T. Claes, et al., Silicon microring resonators, Laser Photon. Rev. 6 (2012) 47–73.

[32] Rsoft Photonic CAD Suite, version 8.0, 2007, <https://optics.synopsys.com/rsoft/>.

[33] A.H. Atabaki, E. Shah Hosseini, A.A. Eftekhar, S. Yegnanarayanan, A. Adibi, Optimization of metallic microheaters for high-speed reconfigurable silicon photonics, Opt. Express 18 (2010) 18312–18323.

[34] D. Dai, L. Yang, S. He, Ultrasmall thermally tunable microring resonator with a submicrometer heater on Si nanowires, J. Lightw. Technol. 26 (2008) 704–709.

[35] A.D. Kersey, M.A. Davis, H.J. Patrick, M. LeBlanc, K.P. Koo, C.G. Askins, et al., Fiber grating sensors, J. Lightw. Technol. 15 (1997) 1442–1463.

[36] V.M.N. Passaro, A.V. Tsarev, F. De Leonardis, Wavelength interrogator for optical sensors based on a novel thermo-optic tunable filter in SOI, J. Lightw. Technol. 13 (2012) 2143–2150.

[37] W. Bogaerts, R. Baets, P. Dumon, V. Wiaux, S. Beckx, D. Taillaert, et al., Nanophotonic waveguides in silicon-on-insulator fabricated with CMOS technology, J. Lightw. Technol. 23 (2005) 401–412.

[38] T. Tsuchizawa, K. Yamada, H. Fukuda, T. Watanabe, J.-I. Takahashi, M. Takahashi, et al., Microphotonics devices based on silicon microfabrication technology, IEEE J. Sel. Top. Quant. Electron. 11 (2005) 232–240.

[39] L. O'Faolain, A. Tsarev, Experimental demonstration of original optical filter based on multiply coupled waveguides, Opt. Lett. 39 (2014) 3627–3629.

Index

Note: Page numbers followed by "*f*" and "*t*" refer to figures and tables, respectively.

Kolibri 360 GmbH
Ferdinand-Jühlke-Straße 7
99095 Erfurt, Deutschland
produktsicherheit@kolibri360.de